Geophysical Monograph Series

Including
IUGG Volumes
Maurice Ewing Volumes
Mineral Physics Volumes

Geophysical Monograph 177

Ocean Modeling in an Eddying Regime

Matthew W. Hecht

Hiroyasu Hasumi

Editors

American Geophysical Union
Washington, DC

Library of Congress Cataloging-in-Publication Data

Ocean modeling in an eddying regime / Matthew W. Hecht, Hiroyasu Hasumi, editors.
     p. cm. — (Geophysical monograph; 177)
   Includes bibliographical references and index.
   ISBN 978-0-87590-442-9
  1. Oceanography—Mathematical models. 2. Ocean circulation—Mathematical models.
   GC10.4.M36026 2008
   551.4601'5118—dc22

ISBN: 978-0-87590-442-9
ISSN: 0065-8448

Cover image: Sea surface temperature from a 1/10° global ocean simulation using the LANL Parallel Ocean Program (POP) (courtesy of Mathew E. Maltrud).

CONTENTS

Section 3: Modeling at the Mesoscale
State of the Art and Future Directions

PREFACE

Only a few years after the advent of primitive equation ocean modeling, a new understanding of the importance of oceanic variability began to emerge from observational analysis. The concept of ocean circulation dominated by steady mean currents gave way to a more complex picture in which the prominence and influence of temporal variability were recognized. Accordingly, even in its first decade, ocean modeling began to address the role of eddy variability. Our physical understanding of the influence of variability on the large-scale mean circulation has been built on observations, theory, and modeling.

Earlier ocean modeling of eddying regimes was necessarily performed within idealized models. With advances in computing power and technique, realistic simulation with primitive equation ocean models became possible. Regional simulations with ocean models generating a level of mesoscale variability comparable to that seen from satellite-based altimetry were first completed about a decade ago.

When we refer to a modeled ocean circulation as "strongly eddying," we mean that the spectrum of variability is comparable to that observed at the so-called mesoscale, as defined by the first internal Rossby radius of deformation. In the language of the trade, this is sometimes referred to as eddy-resolving ocean modeling, even if it is only "resolved" in this limited sense.

This monograph is the first to survey progress during this period of realistic simulation in a strongly eddying regime.

The volume opens with an introduction by Frank Bryan, that touches on the content of each of the 20 papers that follow. The rest of the volume is organized into three sections. The first section contains papers that attempt to explain the physical processes involving eddies. These papers can perhaps be thought of as the most recent efforts in a fruitful and long-running pursuit that can be traced back to the early intersection of ocean modeling and the observation of oceanic variability.

The second section addresses realistic modeling of ocean basins. Comparisons with observations are particularly critical here, so the section opens with a paper addressing the subject of appropriate metrics. The section closes with some of the first results from a coupled climate model with an eddying ocean.

The third and final section of the monograph identifies the fronts upon which we anticipate important progress in our field, as seen from our present vantage point. Future advances will bring existing assumptions into question, and so the section opens with a reconsideration of the equations of ocean models.

The editors would like to take this opportunity to acknowledge a number of people whose contributions are greatly appreciated. The staff at AGU Books, including in particular Dawn Seigler, Maxine Aldred, and Carole Saylor, has been most helpful. We would not have attempted this project without the encouragement of Allan Graubard, who was Acquisitions Editor when this monograph was first proposed, before Jeffrey Robbins took over from Allan. Chunzai Wang served as our Oversight Editor. We thank Morrison Bennett, as well as the staff of AGU Books, for the copy editing.

All the papers in this monograph received full peer review, and we thank our many reviewers for their essential contributions.

M.H. gratefully acknowledges the support of the Climate Change Prediction Program within the U.S. Department of Energy's Office of Science. I also acknowledge Bill Holland and Bob Malone, who not only assembled fine programs in ocean modeling, but figure among my most valued mentors.

H.H. acknowledges the late Nobuo Suginohara not only for his personal encouragement to me, but also for his invaluable contribution to Japan's ocean sciences.

Finally, we wish to acknowledge the debt we owe to our late colleague, Peter Killworth, who died on January 28, 2008, of motor neuron disease. Peter influenced ocean modeling and physical oceanography most deeply, and that influence extended to our topic of ocean modeling (or modelling, as Peter would quite sensibly insist) in an eddying regime. From the success of the early Fine Resolution Antarctic Model to the ongoing and ambitious effort to develop a nonhydrostatic finite element based ocean model, Peter made essential contributions to our field. The editors join with all of the authors in dedicating this work to the memory of Peter Killworth.

Ocean Modeling in an Eddying Regime
Geophysical Monograph Series 177
Copyright 2008 by the American Geophysical Union.
10.1029/177GM01

Matthew W. Hecht
Hiroyasu Hasumi

Introduction: Ocean Modeling—Eddy or Not

Frank O. Bryan

National Center for Atmospheric Research, Boulder, Colorado, USA

Mesoscale eddies are a ubiquitous feature of the ocean circulation, yet are absent from the ocean components of nearly every contemporary climate system model. Progress in ocean modeling over the last decade and advances in high performance computing will soon remedy this dichotomy. Eddy-resolving ocean models are bringing new insights into the physical processes operating in the ocean and will soon find applications in Earth system modeling and weather forecasting. The chapters in this volume provide surveys of the formulation of these models, their ability to simulate the present state of the ocean, and the challenges that remain in refining their accuracy and precision.

1. INTRODUCTION

The atmospheric components of Earth System models simulate climate by explicitly predicting, then averaging over, the day-to-day weather for periods of years to centuries. The weather systems, arising spontaneously as instabilities of the flow established by the equator to pole gradients in solar heating, have spatial scales of O(1,000 km) and time scales of a few days to a week. The weather systems are more than simply noise on the climate state: Their transports of energy, water, and angular momentum are fundamental in establishing the character of the atmospheric circulation and Earth's climate [*Peixoto and Oort*, 1992]. The ocean has weather systems, i.e., mesoscale eddies, dynamically analogous to those in the atmosphere. The majority of the kinetic energy in the ocean is contained in mesoscale eddies and, as in the atmosphere, they are a central part of the dynamical balances of major current systems and in the transport of energy and material through the ocean [*McWilliams*, this volume]. Yet, out of 24 coupled climate models presented in the recent Fourth Assessment Report of the IPCC [*Solomon et al.*, 2007], only one [*K-1 Model Developers*, 2004] attempts to explicitly simulate them. Instead, the flow in nearly every contemporary ocean climate model is represented as essentially laminar, and the effects of ocean mesoscale eddies are entirely parameterized.

Why does this fundamental difference in the approach to modeling atmospheric and oceanic climate dynamics exist? The answer is primarily one of computational economics. Ocean mesoscale eddies have characteristic length scales of O(100 km) or less. With a factor of ten additional grid points required in each horizontal direction and a corresponding decrease in the time step, approximately 1,000 times more computational degrees of freedom are required for an eddy-resolving ocean model compared to an atmosphere model to simulate the global domain for the same period. This computational cost has been impractically high for climate length integrations. With the impending availability of petascale class supercomputers, we will witness a transition in this state of affairs. Fully coupled climate simulations with ocean component models having a resolution of 10 km or better can be expected in the next couple of years. The papers in this volume document the formulation of ocean models of this class, their ability to simulate the various current systems and processes operating in the world ocean, and the challenges that remain in the representation of processes at yet smaller ranges of scales.

2. CURRENT CAPABILITIES

The recognition that ocean mesoscale eddies are a ubiquitous feature of the ocean general circulation led to several

Ocean Modeling in an Eddying Regime
Geophysical Monograph Series 177
Copyright 2008 by the American Geophysical Union.
10.1029/177GM02

international research programs during the 1970s and early 1980s to investigate their structure and dynamics [*Robinson*, 1983]. The eddy mixing parameterizations used in ocean climate models today are largely based on the theory and understanding developed from those observational programs and the accompanying process modeling studies. These early eddy-resolving ocean model experiments were generally carried out in highly idealized basin configurations and most often using the quasi-geostrophic equations. Under this set of approximations, the basic ocean stratification was prescribed rather than predicted [see *Holland*, [1985] for a comprehensive review].

During the era of the World Ocean Circulation Experiment of the late 1980s and 1990s, basin- to global-scale "eddy-permitting" ocean simulations based on the hydrostatic primitive equations [see *Griffies and Adcroft*, this volume] became computationally practical [*Böning and Semtner*, 2001]. In these models, mesoscale eddies were generally not well resolved spatially, but the explicit dissipation used in the models was reduced to a level where instabilities could at least grow and the flow transitioned from laminar to weakly turbulent. In contrast to the first generation of eddy-resolving models, these simulations included full thermodynamic forcing, and the role of eddies in heat and property transports and water mass formation could begin to be investigated.

Since the turn of the century, we have seen the emergence of basin- to global-scale ocean simulations in which the mesoscale appears to be resolved well enough that global metrics of the simulated mean flow and its variability compare well with available observations [*McClean*, this volume]. The improvements in the simulated circulation have resulted not only from the more accurate representation of mesoscale eddies but also from improvements in the representation of features of the mean flow (such as western boundary currents and their extensions) that have similarly small scales. Perhaps even more fundamentally, the geometry of the ocean domain imposes a tremendous range of scales that can influence the general circulation. Indeed, the first example of a fractal curve described by *Mandlebrot* [1982] is the boundary of the ocean domain—the coastline! Increases in resolution have allowed a more accurate treatment of the basic geometry of the ocean domain, especially narrow straits and passages between basins.

Improvements in the fidelity of the simulated circulation in each of the major ocean basins as model resolution is increased are described in the chapters of this volume. The Southern Ocean is the region where the role of eddies in the fundamental balances of the circulation has been most clearly demonstrated and where the open channel geometry allows the most direct application of theories of eddy–mean flow interaction developed in the atmospheric context [*Ivchenko*

et al., this volume; *Tanaka and Hasumi*, this volume]. While state-of-the-art parameterizations of ocean eddies can be tuned to allow coarse resolution models to obtain realistic transports and water mass structures in the Southern Ocean, *Hallberg and Gnanadesikan* [2006] have shown that the transient response to changing forcing (as might occur with a shift in the westerly winds accompanying global warming) is fundamentally different in models with explicitly resolved versus parameterized mesoscale eddies. *Sakamoto and Hasumi* [this volume] show an analogous difference in the response to global warming of the upper ocean circulation in the tropical Pacific for eddy-resolving versus eddy-parameterized models.

One of the most notable systematic biases in coarse-resolution and eddy-permitting ocean models has been a poor simulation of the path of western boundary currents, especially the position of their separation from the coast [*Chassignet and Marshall*, this volume]. As described in the chapters by *Hecht and Smith* [this volume] for the North Atlantic and *Suzuki et al.* [this volume] for the North Pacific, a dramatic improvement in the fidelity of ocean simulations is seen when resolution is increased to around 10 km. The simulations remain quite sensitive to various modeling choices, however, and there remain significant differences between the simulations realized by different models. While eastern boundary regions are less energetic, the important processes setting the basic structure of the flow occur on similarly small scales [*Capet et al.*, this volume].

As mentioned above, the increases in resolution have facilitated not only a better representation of mesoscale eddies but also a more accurate representation of the basic geometry of the ocean domain. *Masumoto et al.* [this volume] provide a prominent example of this for the Indian Ocean where the water mass properties are strongly influenced by flow through a series of narrow passages in the Indonesian Throughflow. The simulation of exchanges between the subpolar regions and the Arctic has also been shown to be significantly improved with increases in resolution to below 10 km [*Maslowski et al.*, this volume], with important implications for the heat budget and ice distributions in that basin.

3. FUTURE DIRECTIONS

The improvements in the fidelity of ocean models achieved over the last several years represent a significant milestone in ocean research but should, by no means, be considered an end point. Global coupled climate simulations have just recently entered the "eddy-permitting" regime, and eddy-resolving stand-alone ocean simulations have been integrated for less than a century [*Masumoto et al.*, 2004]. The next several years should see a broader application of eddy-resolving models not just in studies of ocean dynamics but also in the broader context of Earth system modeling including biogeochemical processes [*Oschlies*, this volume] and operational environmental forecasting systems [*Hurlburt et al.*, this volume].

There is no indication that the "eddy-resolving" solutions obtained to date have converged even for quasi-geostrophic dynamics [*Siegel et al.*, 2001], and the current generation of models remain quite sensitive to any number of choices in model configuration and sub-grid-scale closure [*Hecht et al.*, this volume]. That is, we are still far from "large eddy simulation" for global ocean models. It is inevitable that models with increasingly finer resolution will be constructed, although potentially more resource-efficient methods such as nested grids [*Capet et al.*, this volume] or adaptive grid methods [*Piggott et al.*, this volume] may see increased application.

Moving into the eddying regime presents new challenges and opportunities in developing parameterizations of processes occurring at yet smaller scales. *Fox-Kemper and Menemenlis* [this volume] and *Hecht et al.* [this volume] describe several approaches for subgrid-scale closures more appropriate for the situation where the turbulent cascade is partially resolved. The expanded range of explicitly resolved scales opens the opportunity for putting parameterizations of processes such as inertial wave excitation and propagation [*Klein*, this volume], gravity currents [*Legg et al.*, this volume], and mixed-layer restratification [*Thomas et al.*, this volume] on a firmer physical foundation.

Acknowledgments. NCAR is supported by the National Science Foundation.

REFERENCES

Böning, C. W., and A. J. Semtner (2001), High-resolution modeling of the thermohaline and wind-driven circulation, in *Ocean Circulation and Climate*, edited by G. Siedler, J. Church, and J. Gould, pp. 59–71, Academic, New York.

Hallberg, R., and A. Gnanadesikan (2006), The role of eddies in determining the structure and response of the wind-driven Southern Hemisphere overturning: Results from the Modeling Eddies in the Southern Ocean (MESO) Project, *J. Phys. Oceanogr.*, 36, 2232–2252.

Holland, W. R. (1985), Simulation of mesoscale variability in mid-latitude gyres, in: *Issues in Atmospheric and Oceanic Modeling. Part A: Climate Dynamics*, edited by S. Manabe, pp. 479–523, Academic, New York.

K-1 Model Developers (2004), K-1 Coupled GCM (MIROC) Description. Center for Climate System Research, University of Tokyo. Available at: www.ccsr.u-tokyo.ac.jp/kyosei/hasumi/MIROC/tech-repo.pdf.

Mandelbrot, B. (1982), *The Fractal Geometry of Nature*, 420 pp., W.H. Freeman, New York.

Masumoto, Y., H. Sasaki, T. Kagimoto, N. Komori, A. Ishida, Y. Sasai, T. Miyama, T. Motoi, H. Mitsudera, K. Takhashi, H. Sakuma, and T. Yamagata (2004), A fifty-year simulation of the world ocean—Preliminary outcomes of OFES (OGCM for the Earth Simulator), *J. Earth Simulator*, 1, 35–56.

Peixoto, J. P., and A. H. Oort (1992), *Physics of Climate*, 520 pp., American Institute of Physics, New York.

Robinson, A. R. (1983), *Eddies in Marine Science*, 609 pp., Springer, Berlin.

Siegel, A., J. B. Weiss, J. Toomre, J. C. McWilliams, P. S. Berloff, and I. Yavneh (2001), Eddies and vortices in ocean basin dynamics, *Geophys. Res. Lett.*, 28, 3183–3186.

Solomon, S., D. Qin, M. Manning, Z. Chen, M. Marquis, K. B. Averyt, M. Tignor, and H. L. Miller (2007), *Climate Change 2007: The Physical Science Basis. Contribution of Working Group I to the Fourth Assessment Report of the Intergovernmental Panel on Climate Change*, 996 pp., Cambridge University Press, Cambridge.

Frank Bryan, Climate and Global Dynamics Division, National Center for Atmospheric Research, Boulder, CO 80307, USA.

The Nature and Consequences of Oceanic Eddies

James C. McWilliams

Department of Atmospheric and Oceanic Sciences, University of California, Los Angeles, California, USA

Mesoscale eddies are everywhere in the ocean. They provide important material and dynamical fluxes for the equilibrium balances of the general circulation and climate. Eddy effects are necessary ingredients for oceanic general circulation models, whether parameterized or, as is increasingly done, explicitly resolved in the simulations. The primary characteristics of eddies are summarized in this paper, and the principal roles of eddies in the dynamics of large-scale circulation are described. Those eddy roles are the following: maintenance of boundary and equatorial currents; lateral and vertical buoyancy fluxes, isopycnal form stress, and available potential energy conversion; Reynolds stress, kinetic energy conversion, and current rectification; material dispersion and mixing; energy cascades and dissipation routes; Lagrangian mean advection; surface-layer density restratification; ventilation and subduction; stratification control; frontogenesis; topographic form stress; biological pumping and quenching; and generation of intrinsic climate variability.

1. INTRODUCTION

Mesoscale eddies pervade the ocean (e.g., Plates 1 and 2 and *Stammer* [1997]), and they usually account for the peak in the kinetic energy spectrum [*Wunsch and Stammer*, 1995]. For the most part, they arise from instabilities of more directly forced, persistent currents. Consequently, they play a central role in limiting the strength of the persistent currents. How eddies play this role can be expressed qualitatively as an effective eddy viscosity that spreads and dissipates the parent currents (although the language of eddy diffusion is notoriously subtle in fluid dynamics). Similarly, eddies disperse material concentrations, largely along isentropic/isopycnal surfaces in the stratified interior, and this role can be expressed as an anisotropic eddy diffusion[1] as well as, somewhat more subtly, an eddy-induced advection. Of course, in the nonstationary, inhomogeneous four-dimensional structure of the oceanic general circulation and climate, such simple

Ocean Modeling in an Eddying Regime
Geophysical Monograph Series 177
Copyright 2008 by the American Geophysical Union.
10.1029/177GM03

characterizations of the effects of eddies—i.e., the eddy-averaged fluxes due to eddy-fluctuation covariances—are far from the whole story. This essay surveys the conceptual landscape of eddy effects at a qualitative level. It does not delve deeply into how these effects are, or should be, represented in models either by parameterizations or direct simulation. These subjects are large ones, while the essay is relatively short and much less than a comprehensive review.

Oceanic spectra are broad band. The mesoscale range is centered around horizontal scales L of tens and hundreds of kilometers. L is close to the first baroclinic deformation radius, $R_d \approx NH/f$ outside the tropics or $R_d \approx \sqrt{NH/\beta}$ near the equator [*Stammer*, 1997]. N is the buoyancy frequency in the main pycnocline, H its depth, f the Coriolis frequency, and β the meridional gradient of f. R_d is much smaller than most basin widths. The most energetic vertical scales are relatively large, comparable to H or the oceanic depth D. The evolutionary timescales range from $(\beta L)^{-1}$ ~ days—for larger scale barotopic Rossby waves with $L > R_d$—to $R_d V^{-1}$ (V is the horizontal velocity) or $(\beta R_d)^{-1}$ ~ weeks and months—for baroclinic eddies [*Chelton et al.*, 2007]—and even to years—for some coherent vortices like Meddies [*McWilliams*, 1985] that survive until some destructive encounter with another strong current or steep topography. The spatial distribution of eddy energy is quite inhomogeneous with the highest energy near the major persistent currents, e.g., the Gulf Stream and Antarctic Circumpolar Current [*Ducet et al.*, 2000].

For present purposes, it is useful to decompose oceanic currents and material concentration variations into four components: large-scale, mesoscale, submesoscale, and microscale. Climate and general circulation models simulate the large-scale distributions, with mesoscale eddy effects either parameterized or at least partially resolved (following the strategy of large eddy simulation; *Pope* [2000]). We must acknowledge, however, that we are still several computer generations away from being able to fully and routinely resolve the mesoscale in global climate models (e.g., with a horizontal grid scale of 5 km and a model integration time of centuries).

Submesoscale eddies are defined as smaller than the mesoscale while still significantly influenced by Earth's rotation and density stratification. They are not resolved in modern oceanic general circulation models (OGCMs). Determining which submesoscale effects need to be parameterized in OGCMs is still very much a future frontier. Among the most plausibly important effects are density restratification, especially in the surface layer [*Rudnick and Ferrari*, 1999], and a route to energy dissipation for the mesoscale [*Muller et al.*, 2005]. Parameterization of microscale effects are important for the turbulent boundary layers at the top, bottom, and sides (if any) and for the interior diapycnal mixing of materials across stably stratified isopycnal surfaces.

These are interesting times in the science of eddies and the general circulation. The theoretical framework for eddy dynamics is well established, as are the observational and modeling bases for eddy structure and phenomenology. But the observational basis for eddy fluxes is still quite sparse, and no measurement techniques in prospect are likely to change this greatly. There is a modeling path for establishing "eddy-flux truth," but this path probably requires much higher resolutions than are commonly used as yet, and probably further refinements in submesoscale and microscale parameterizations are necessary as well to become demonstrably robust and reproducible. The papers in this book comprise a status report on eddy-resolving ocean modeling.

2. EDDY THEORY AND MODELING

Mesoscale eddy flows are understood by oceanographers to satisfy approximately the diagnostic force balances, viz., geostrophic in the horizontal and hydrostatic in the vertical. This is a consequence of their dynamical parameter regime with small Rossby number, Froude number, and aspect ratio:

$$Ro = \frac{V}{fL}, Fr = \frac{V}{NH}, \lambda = \frac{H}{L} \ll 1. \tag{1}$$

In this regime, the asymptotic dynamical model is quasi-geostrophy:

$$[\partial_t + \mathbf{v}_g \cdot \nabla_h] q_{qg} = NCE, \tag{2}$$

where

$$q_{qg} = f_0 + \beta(y - y_0) + \frac{1}{f_0}\nabla_h^2\phi + f_0\partial_z\left(\frac{b}{N^2(z)}\right)$$
$$- \frac{f_0(D - D_0)}{H_b}\delta[z, -D_0] + \frac{f_0 b}{H_s N^2}\delta[z, 0],$$

$$\mathbf{v}_h \approx \mathbf{v}_g = f_0^{-1}\hat{\mathbf{z}} \times \nabla_h\phi,$$

$$b = \partial_z\phi \approx g\left(\tilde{\alpha}_0(T - T_0) - \tilde{\beta}_0(S - S_0)\right). \tag{3}$$

The subscript 0 denotes a constant reference value. The subscript h denotes the horizontal component of a vector, and x and y are the zonal (eastward) and meridional (northward) horizontal coordinates. The upward vertical coordinate is z; $\hat{\mathbf{z}}$ is its unit vector; $z = 0$ is mean sea level; and $z = -D(\mathbf{x}_h)$ is the bottom. $\mathbf{v}_g$ is geostrophic velocity; ϕ is geopotential function (dynamic pressure); b is buoyancy variation around the mean stratification, $\int N^2(z)dz$; T is temperature; and S is salinity. g is gravitational acceleration; $\tilde{\alpha}$ and $\tilde{\beta}$ are thermal-expansion and haline-contraction coefficients; and q_{qg} is quasi-geostrophic potential vorticity. NCE indicates nonconservative effects associated with submesoscale and microscale stirring and mixing parameterizations, which usually have small rates except near the surface and bottom where they convey boundary stress and material fluxes into the ocean. H_s and H_b are constant vertical layer thicknesses adjacent to the surface and bottom, and $\delta[a,b]$ is a discrete delta function equal to one when $a = b$ and zero otherwise.

Almost all our understanding of eddy dynamics and phenomena has its roots in quasi-geostrophic theory and its extensive body of analytical and computational solutions [*Pedlosky*, 1987; *McWilliams*, 2006; *Vallis*, 2006]. This provides paradigms for

— *wave and vortex propagation* due to β and due to horizontal gradients of topography $\nabla_h D$ and other persistent components of potential vorticity $\nabla_h\langle q\rangle$ (brackets denote an average over eddies);
— long-lived *coherent vortices* at both the mesoscale and smaller;

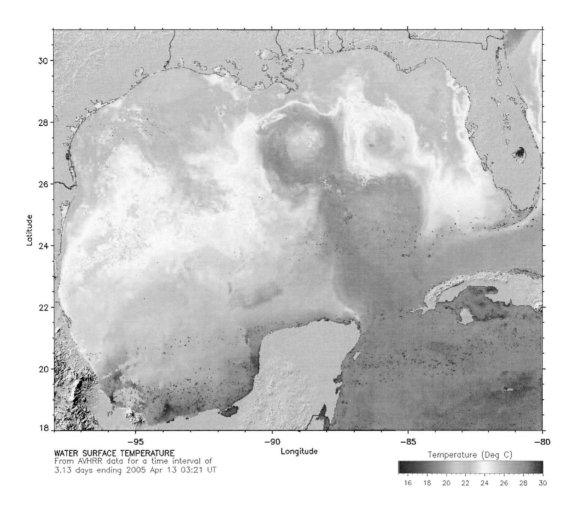

Plate 1. Anticyclonic eddies in the Gulf of Mexico detaching from the Loop Current sector of the Gulf Stream between the Yucatan and Florida Straits in April 2005. Eddy generation occurs quasi-annually and typically is followed by westward propagation across the basin [*Hurlburt and Thomson*, 1980; *Sturgis and Leben*, 2000]. Their diameters are approximately 250 km (courtesy of the Ocean Remote Sensing Group, Johns Hopkins University Applied Physics Laboratory).

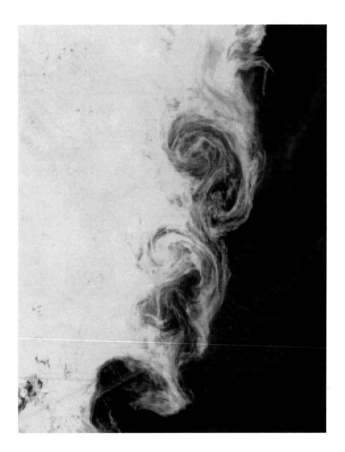

Plate 2. Oceanic eddies in Davis Strait (north of the Labrador Sea, west of Greenland) during June 2002. The eddies are made visible by sea-ice fragments. The eddies arise as an instability of the boundary current along the ice margin. Their diameters are approximately 25 km (courtesy of Jacques Descloirest, NASA Goddard Space Flight Center).

the arrest scale is $L_\beta = (V'/\beta)'^{/1/2}$ [*Rhines*, 1975], with typical values of hundreds of kilometers in the ocean (i.e., larger than L_d). Barotropic currents, in particular, can be dissipated through bottom boundary layer drag, which provides one important route to dissipation for eddy energy. There may also be an important forward eddy energy cascade into the submesoscale in the interior associated with a breakdown of the geostrophic and even more general diagnostic force balances [*Muller et al.*, 2005; *Capet et al.*, 2008a; Molemaker et al., Balanced and unbalanced routes to dissipation in an equilibrated Eady flow, submitted to *Journal of Fluid Mechanics*, 2007, hereinafter referred to as Molemaker et al., submitted manuscript, 2007], followed by further forward cascade to the microscale and dissipation scale. *Scott and Wang* [2005] show observational evidence for such a forward cascade at least with respect to eddy kinetic energy. Eddy routes to dissipation are a necessary accompaniment for eddy depletion of the general circulation through mean-current instabilities (sections 3.2 and 3.6).

3.5. Eddy-Induced Lagrangian Transport

Any vector can be decomposed with respect to any other vector. For eddy transport, it is relevant to decompose eddy buoyancy flux $\langle v'b' \rangle$ relative to the mean buoyancy gradient $\nabla \langle b \rangle$:

$$\langle v'b' \rangle = -\frac{\langle v'b' \rangle \times \nabla \langle b \rangle}{|\nabla \langle b \rangle|^2} \times \nabla \langle b \rangle + \frac{\langle v'b' \rangle \cdot \nabla \langle b \rangle}{|\nabla \langle b \rangle|^2} \nabla \langle b \rangle$$

$$\equiv \Psi^* \times \nabla \langle b \rangle + \mathbf{R}^*[\langle b \rangle]. \qquad (5)$$

The separate terms represent eddy fluxes along and across mean buoyancy surfaces (i.e., isopycnal and diapycnal, with the former component also referred to as the skew flux, perpendicular to the mean buoyancy gradient; *Griffies* [1998]). The final relation in (5) implicitly defines the eddy-induced streamfunction Ψ^* and residual buoyancy flux $\mathbf{R}^*[\langle b \rangle]$. We further define an eddy-induced velocity,

$$\mathbf{v}^* = \nabla \times \Psi^*. \qquad (6)$$

We can readily show that $\mathbf{v}^*$ is nondivergent, $\nabla \cdot \mathbf{v}^* = 0$ (as is the Eulerian velocity in an incompressible fluid) and that it provides an eddy-induced buoyancy advection in the mean buoyancy equation,[3]

$$\nabla \cdot \langle v'b' \rangle = \mathbf{v}^* \cdot \nabla \langle b \rangle + \nabla \cdot \mathbf{R}^*[\langle b \rangle]. \qquad (7)$$

This formal decomposition of the eddy buoyancy flux allows a statement of the physical hypothesis that the di-

apycnal flux $\mathbf{R}^*[\langle b \rangle]$ is very small in the stably stratified, nearly adiabatic interior of the ocean (i.e., outside the turbulent boundary layers) where b is almost conserved along trajectories over several eddy fluctuation cycles [*Gent and McWilliams*, 1990; *Gent et al.*, 1995]. This implies that the only important eddy effect on the baroclinic dynamics of the circulation is an additional buoyancy advection by $\mathbf{v}^*$, which can thus be considered part of the Lagrangian mean flow[4] when it is added to $\langle v \rangle$. When eddies act to extract mean available potential energy (section 3.2), $\mathbf{v}^*$ is an overturning circulation acting to flatten mean isopycnal surfaces. A conspicuous example of this is the meridional overturning cell in the Antarctic Circumpolar Current that is opposite to the Eulerian-mean Deacon Cell (i.e., $\hat{\mathbf{x}} \cdot \Psi^* > 0$ in the center of the current).

The interpretation of eddy-induced advection extends to the eddy flux of any material concentration c with the decomposition,

$$\langle v'c' \rangle \equiv \Psi^* \times \nabla \langle c \rangle + \mathbf{R}^*[\langle c \rangle]. \qquad (8)$$

Again, the eddy-induced Lagrangian velocity $\mathbf{v}^*$, defined in terms of the eddy buoyancy flux, contributes an advective tendency for $\langle c \rangle$, but now, the residual material flux $\mathbf{R}^*[\langle c \rangle]$ includes fluxes both along the mean buoyancy surfaces as well as across them [$\mathbf{R}^*$ here is defined by making a decomposition of $\langle v'c' \rangle$ perpendicular and parallel to $\langle \nabla \langle c \rangle \rangle$, analogous to (5) relative to $\langle \nabla \langle b \rangle \rangle$, and then by subtracting the first term in (8)]. The isopycnal component is not expected to be small and is often interpreted and modeled as isopycnal eddy diffusion of $\langle c \rangle$ (section 3.3; *Redi* [1982]). The diapycnal component of $\mathbf{R}^*[\langle c \rangle]$ is still expected to be small if eddy trajectories are approximately confined to isopycnal surfaces.

This quasi-adiabatic characterization of eddy fluxes does not hold within the turbulent boundary layers [*Tandon and Garett*, 1996], and the eddy material fluxes in (5) and (8) both develop stronger diapycnal components and rotate toward a boundary-parallel orientation to preclude any boundary-normal flux [*Ferrari et al.*, 2008].

3.6. Eddy Reynolds Stress, Kinetic Energy Conversion, and Current Rectification

Barotropic instability allows eddies to grow at the expense of the mean current (section 3.1) when the horizontal Reynolds stress $\langle v'_h v'_h \rangle$ has a negative tensor correlation with the mean horizontal shear $\nabla \langle v_h \rangle$, i.e., it acts in the sense of an eddy flux down the mean gradient, consistent with a positive horizontal eddy viscosity, $v_h > 0$, and it effects a conversion from mean to eddy kinetic energy. Vertical Reynolds stress is much less likely to be significant for eddies because their vertical velocity w' is so weak (section 2). The Loop Current in the Gulf of Mexico exhibits this behavior, with finite fluctuation amplitudes leading to the detachment of large, anticyclonic eddies (Figure 1; *Hurlburt and Thompson* [1980]). There is similar, if limited, evidence for down-gradient momentum flux in the Gulf Stream along the western boundary [*Dewar and Bane*, 1985] in the location where such an effect is expected from gyre theory (section 3.1). However, there is a competing paradigm for $\langle \mathbf{v}'_h \mathbf{v}'_h \rangle$, viz., it can accelerate a mean current by β-induced Rossby wave radiation away from a source region of eddy fluctuations [*Haidvogel and Rhines*, 1983; *Berloff*, 2005]. There is observational evidence for this effect, e.g., near the offshore Kuroshio Current [*Tai and White*, 1990]. In particular, this type of "negative viscosity" ($v_h < 0$), eddy-rectification behavior is common for baroclinically unstable (section 3.2), eastward jets [*Marshall*, 1981; *McWilliams and Chow*, 1981]. It is an example of spatially nonlocal eddy-mean interaction where wave propagation establishes a connection between the mean flows in the separated locations of wave generation and dissipation.

3.7. Eddy Restratification

The eddy-induced conversion of mean available potential energy and associated Lagrangian mean circulation (sections 3.2 and 3.5) have the effect of restratifying the ocean by flattening the tilted isopycnal surfaces. This effect is widespread for the tilted pycnocline surfaces associated with baroclinic, geostrophic mean currents. It can have an especially dramatic effect after episodes of deep convective boundary-layer mixing that usually occur with significant lateral inhomogeneity (e.g., in the Labrador and Mediterranean Seas). In this configuration, the isopycnal surfaces surrounding the convection zone are very steeply tilted, and they geostrophically support a strong, vertically sheared rim current. The ensuing baroclinic instability of the rim current leads to efficient lateral and vertical eddy buoyancy fluxes $\langle v'b' \rangle$ and restratifies the convectively mixed water mass [*Joes and Marshall*, 1997; *Katsman et al.*, 2004], typically on timescales of weeks and months. Another situation where this effect is important is in the surface boundary layer when there are horizontal buoyancy gradients $\nabla\langle b \rangle$, weak geostrophic vertical shear, and weak stratification. The resulting baroclinic instability occurs on rather small scales (often more submesoscale than mesoscale), and it acts to restratify the "mixed layer" and thus competes against the microscale turbulent mixing generated by surface wind and buoyancy fluxes [*Boccaletti et al.*, 2007; *Capet et al.*, 2008a]. This eddy restratification effect may be particularly evident after strong surface fluxes abate.

3.8. Eddy Ventilation and Subduction

Where isopycnal surfaces intersect either the surface or bottom boundaries, a quasi-adiabatic pathway opens into the oceanic interior. This can provide a ventilation of the interior region to anomalous material concentrations generated through microscale boundary-layer mixing [*Armi*, 1978; *Garrett*, 1979] and surface fluxes. While boundary-normal eddy fluxes are zero at the boundaries (section 3.5), they increase into the interior on the scale of the boundary-layer thickness [*Ferrari et al.*, 2008]. The situation where surface-layer materials are carried down into the interior is referred to as eddy subduction, and it may be accomplished either by eddy mixing or by eddy-induced advection (sections 3.3 and 3.5) [*Marshall*, 1997; *Hazeleger and Drijfhout*, 2000].

3.9. Eddy-Controlled Stratification

Implicit in the eddy roles in material transport and buoyancy flux (sections 3.2, 3.3, and 3.5) are eddy influences on the 3D distribution of $\langle b \rangle$, including its vertical profile, i.e., the stratification. The idea of eddy control of stratification is explicitly addressed in several idealized models where eddy fluxes establish the pycnocline slope beneath the diabatic surface boundary layer. With the surface horizontal gradient $\nabla_h\langle b \rangle$ strongly constrained by large-scale air-sea interaction (i.e., by atmospheric climatological distributions), the vertical gradient $\partial_z\langle b \rangle$ is determined from $\nabla_h\langle b \rangle$ divided by the pycnocline slope. This characterization is most fully developed for flows without zonal boundaries, hence relevant to the Antarctic Circumpolar Current [*Marshall et al.*, 2002; *Marshall and Radko*, 2003; *Cessi and Fantini*, 2004; *Olbers and Visbeck*, 2005] as well as to the tropospheric Westerly Winds.

3.10. Eddy-Induced Frontogenesis

As analyzed by *Hoskins and Bretherton* [1972], a horizontal strain flow induces frontogenesis by sharpening horizontal buoyancy gradients, especially near the surface boundary. In the ocean, this process occurs both due to the large-scale circulation (e.g., the subtropical fronts near the equatorward edge of subtropical wind gyres) and a fortiori due to mesoscale eddies because of their larger strain rates [*Capet et al.*, 2008a]. Frontogenesis involves a conversion of available potential energy into kinetic energy (section 3.2) on increasingly smaller horizontal scales reaching into the submesoscale, and the ensuing frontal instabilities can further feed into a forward eddy energy cascade (section 3.4; Molemaker et al., submitted manuscript, 2007). Frontogenesis is an effective means of upper-ocean restratification (section 3.7), and it contributes to eddy ventilation and sub-duction partly through secondary circulation in the cross-frontal plane (section 3.8; *Thomas et al.* [2007]). The surface quasi-geostrophic model—i.e., (2) and (3) with q_{qg} including the final term with surface buoyancy variations but having a uniform interior value—depicts the mesoscale frontogenesis process fairly well but excludes some important frontal instability behaviors [*Held et al.*, 1995]; it also captures the forward cascade of available potential energy and conversion into kinetic energy that occurs in a turbulent field of surface fronts *Capet et al.* [2008a, 2008b].

3.11. Topographic Form Stress and Rectification

Eddies play at least two important roles in combination with topography in shaping the large-scale circulation. The most direct influence is by providing a drag force at the bottom, viz., the integrated horizontal pressure force against the bottom, the form stress $\langle \phi' \nabla_h D' \rangle$. The form stress be rewritten as $f_0 \hat{z} \times \langle v'_g D' \rangle$ with the geostrophic approximation (cf., isopycnal form stress in section 3.2). As D' is steady in time in this context, the relevant eddies are also the steady deviations from the large-scale flow, i.e., the standing eddies. Form stress provides the principal balancing force to the wind stress in the Antarctic Circumpolar Current [*Treguir and McWilliams*, 1990; *Wolff et al*, 1991], and it probably is significant in other places where mean currents extend to the bottom and cross topographic contours. A different type of influence is the effect of transient eddies over a topographic slope to generate rectified currents aligned with the topographic contours (sometimes referred to as the Neptune effect) [*Holloway*, 1992; *Adcock and Marshall*, 1999], somewhat analogous to β-induced rectification (section 3.6). It seems likely that the Neptune effect will generate primarily abyssal currents in the interior and on continental slopes, as they arise from bottom eddies in contact with the topography, but the theoretical basis for predicting their vertical structure is still rudimentary [*Merryfield and Holloway*, 1999].

3.12. Eddy Pumping and Quenching

In many places phytoplankton growth is limited by the supply of nutrients from the oceanic interior into the euphotic zone, and the cycle of plankton consumption and senescence is a biological pump exporting biogeochemical materials into the oceanic interior. Eddy fluxes can contribute to these exchanges through large-scale subduction and ventilation (section 3.8) and mesoscale and submesoscale secondary circulation near fronts (section 3.10). Eddy pumping may also contribute [*Falkowski et al.*, 1991; *McGillicuddy and Robinson*, 1997; *Benitez-Nelson et al.*, 2007] when the pycnocline

is geostrophically elevated inside cyclonic eddies and brings its resident nutrients closer to the surface so growth and cycling can occur. After growth depletes the local nutrients, they can be replenished by isopycnal eddy material fluxes from adjacent regions. Another process is eddy quenching in the biologically active, subtropical eastern boundary upwelling currents where offshore eddy subduction along descending isopycnals removes nutrients from the euphotic zone before plankton growth fully depletes them (Gruber et al., Eddy-induced reduction of biological productivity in upwelling systems, submitted to *Nature*, 2007).

3.13. Eddy-Induced Climate Variability

As eddy effects are an essential part of the dynamics of the large-scale circulation, it can be expected that they will modulate the oceanic variability induced by climate variability in the surface forcing. However, eddies may also be a source of intrinsic climate variability by modifying the large-scale circulation, hence the surface temperature field, hence air-sea fluxes, hence climate. The first stage of this sequence is demonstrated in idealized wind-gyre models with conspicuously high grid resolution and large Reynolds number [*Berloff and McWilliams*, 1999; *Berloff et al.*, 2007]: Even with a steady wind stress forcing, the basin-scale circulation changes significantly on decadal time scales, modulated by the eddy fluxes in the separated boundary-current extension and recirculation regions. An analogous spontaneous decadal variability also occurs in the Antarctic Circumpolar Current, albeit by a different mechanism involving eddies and topography [*Hogg and Blundell*, 2006]. As yet, OGCMs and global climate models have not been configured to examine how important this effect might be. Nor have some of the other major current systems—equatorial currents, thermohaline circulation, etc.—yet been investigated for this behavior.

4. CONCLUSIONS

Eddy processes and eddy fluxes have many potential consequences for the oceanic general circulation and climate. The evolution in oceanic modeling toward routinely including eddies through finer scale grid resolution is therefore a significant advance. At present, there are many theoretical ideas and idealized computational demonstrations of eddy effects on the large-scale circulation and material distributions and on climate, but as yet few certainties about how they occur in the real ocean and how they will manifest in more realistic OGCM simulations. Nevertheless, it is now technically feasible to systematically carry out the research that should resolve these uncertainties within the coming years.

Acknowledgments. I am grateful to Pavel Berloff, Bill Dewar, and John Marshall for helpful suggestions about this essay.

Notes

1. More precisely, advection can stir material fields, irreversibly entangle their iso-surfaces, and transfer fluctuation variance to ever-finer scales, but only molecular diffusion can complete the mixing process that ultimately removes the variance of material gradients.
2. A common view is that diapycnal material mixing rates in the pycnocline are mostly due to breaking internal gravity waves [*Gregg*, 1989]. The associated κ value is $\sim 10^{-5}$ m^2 s^{-1}, which is 8 orders of magnitude smaller than the mesoscale isopycnal value above. The extreme anisotropy and disparity of mesoscale and microscale mixing efficiencies are the bases for the quasi-adiabatic hypothesis for eddy material fluxes [*Redi*, 1982; *Gent and McWilliams*, 1990].
3. There is an alternative view that the most useful definition of the eddy-induced transport velocity should be based on the material conservation of potential vorticity and its eddy flux, $\langle v'q' \rangle$, rather than buoyancy as in (5). The differences between these velocities are often small in large-scale flows, as b variations are often the dominant influence in q variations, and the present evidence is mixed about which perspective is preferable [*Marshall et al.*, 1999; *Drijfhout and Hazeleger*, 2001].
4. This is also sometimes called the residual mean flow, following *Andrews and McIntyre* [1976], although there are subtle distinctions between these quantities based on the averaging operators.

REFERENCES

Adcock, S. T., and D. P. Marshall (2000), Interactions between geostrophic eddies and the mean circulation over large-scale bottom topography, *J. Phys. Oceanogr.*, *30*, 3223–3238.

Andrews, D. G., and M. McIntyre (1976), Planetary waves in horizontal and vertical shear: The generalized Eliassen-Palm relation and the mean zonal acceleration, *J. Atmos. Sci.*, *33*, 2041–2048.

Armi, L. (1978), Some evidence for boundary mixing in the deep ocean, *J. Geophys. Res.*, *83*, 1971–1979.

Benitz-Nelson, C. R. et al. (2007), Mesoscale eddies drive increased silica export in the Subtropical Pacific Ocean, *Science*, *316*, 1017–1021.

Berloff, P. S., and J. C. McWilliams (1999), Large-scale, low-frequency variability in wind-driven ocean gyres, *J. Phys. Oceanogr.*, *29*, 1925–1949.

Berloff, P. S. (2005), On rectification of randomly forced flows, *J. Mar. Res.*, *63*, 497–527.

Berloff, P., A. Hogg, and W. Dewar (2007), The turbulent oscillator: A mechanism of low-frequency variability of the wind-driven ocean gyres, *J. Phys. Oceanogr.*, *37*, 2363–2386.

Boccaletti, G., R. Ferrari, and B. Fox-Kemper (2007), Mixed layer instabilities and restratification, *J. Phys. Oceanogr.*, *37*, 2228–2250.

Capet, X., J. C. McWilliams, M. J. Molemaker, and A. Shchepetkin (2008a), Mesoscale to submesoscale transition in the California Current System: (I) Flow structure, eddy flux, and observational tests. (II) Frontal processes. (III) Energy balance and flux, *J. Phys. Oceanogr.*, in press.

Capet, X., P. Klein, B. L. Hua, G. Lapeyre, and J. C. McWilliams (2008b), Surface kinetic energy transfer in SQG flows, *J. Fluid Mech.*, in press.

Cessi, P., and M. Fantini (2004), The eddy-driven thermocline, *J. Phys. Oceanogr.*, *34*, 2642–2658.

Charney, J. G. (1971), Geostrophic turbulence, *J. Atmos. Sci.*, *28*, 1087–1095.

Chelton, D. B., M. G. Schlax, R. M. Samelson, and R. A. deSzoeke (2007), Global observations of large oceanic eddies, *Geophys. Res. Lett.*, *34*, L15606, doi:10.1029/2007GL030812.

Dewar, W. K., and J. M. Bane (1985), The subsurface energetics of the Gulf Stream near the Charleston bump, *J. Phys. Oceanogr.*, *15*, 1771–1789.

Drijfhout, S. S., and W. Hazeleger (2001), Eddy mixing of potential vorticity versus thickness in an isopycnic ocean model, *J. Phys. Oceanogr.*, *31*, 481–505.

Ducet, N., P. Y. Le Traon, and G. Reverdin (2000), Global high-resolution mapping of ocean circulation from TOPEX/Poseidon and ERS-1 and -2, *J. Geophys. Res.*, *105*, 19,477–19,498.

Falkowski, P. G., D. Ziemann, Z. Kolber, and P. K. Bienfang (1991), Role of eddy pumping in enhancing primary production in the ocean, *Nature*, *352*, 55–58.

Ferrari, R., J. C. McWilliams, V. Canuto, and M. Dubovikov (2008), Parameterization of eddy fluxes near oceanic boundaries, *J. Clim.*, in press.

Garrett, C. (1979), Comments on "Some evidence for boundary mixing in the deep ocean" by L. Armi, *J. Geophys. Res.*, *84*, 5095–5096.

Gent, P. R., and J. C. McWilliams (1990), Isopycnal mixing in ocean circulation models, *J. Phys. Oceanogr.*, *20*, 150–155.

Gent, P. R., J. Willebrand, T. J. McDougall, and J. C. McWilliams (1995), Parameterizing eddy-induced tracer transports in ocean circulation models, *J. Phys. Oceanogr.*, *25*, 463–474.

Gill, A. E., J. S. A. Green, and A. J. Simmons (1974), Energy partition in the large-scale ocean circulation and the production of mid-ocean eddies, *Deep-Sea Res.*, *21*, 499–528.

Gregg, M. C. (1989), Scaling turbulent dissipation in the thermocline, *J. Geophys. Res.*, *94*, 9686–9698.

Griffies, S. (1998), The Gent-McWilliams skew flux, *J. Phys. Oceanogr.*, *28*, 831–841.

Haidvogel, D. B., and P. B. Rhines (1983), Waves and circulation driven by oscillatory winds in an idealized ocean basin, *Geophys. Astrophys. Fluid Dyn.*, *25*, 1–63.

Hazeleger, W., and S. S. Drijfhout (2000), Eddy subduction in a model of the subtropical gyre, *J. Phys. Oceanogr.*, *30*, 677–695.

Held, I., R. Pierrehumbert, S. Garner, and K. Swanson (1995), Surface quasigeostrophic dynamics, *J. Fluid Mech.*, *282*, 1–20.

Hogg, A. M., and J. R. Blundell (2006), Interdecadal variability of the Southern Ocean, *J. Phys. Oceanogr.*, *36*, 1626–1645.

Holloway, G. (1992), Representing topographic stress for large-scale ocean models, *J. Phys. Oceanogr.*, *22*, 1033–1046.

Hoskins, B. J., and F. P. Bretherton (1972), Atmospheric frontogenesis models: Mathematical formulation and solution, *J. Atmos. Sci.*, *29*, 11–37.

Hurlburt, H. E., and J. D. Thompson (1980), A numerical study of Loop Current intrusions and eddy shedding, *J. Phys. Oceanogr.*, *10*, 1611–1651.

Jones, H., and J. Marshall (1997), Restratification after deep convection, *J. Phys. Oceanogr.*, *27*, 2276–2287.

Katsman, C. A., M. A. Spall, and R. S. Pickart (2004), Boundary current eddies and their role in the restratification of the Labrador Sea, *J. Phys. Oceanogr.*, *34*, 1967–1983.

Krauss, W., and C. Böning (1987), Lagrangian properties of eddy fields in the northern North Atlantic as deduced from satellite-tracked buoys. *J. Mar. Res.*, *45*, 259–291.

Larichev, V. D., and I. M. Held (1995), Eddy amplitudes and fluxes in a homogeneous model of fully developed baroclinic instability, *J. Phys. Oceanogr.*, *25*, 2285–2297.

Marshall, D. P. (1997), Subduction of water masses in an eddying ocean, *J. Mar. Res.*, *55*, 201–222.

Marshall, D. P., R. G. Williams, and M. M. Lee (1999), The relation between eddy-induced transport and isopycnic gradients of potential vorticity, *J. Phys. Oceanogr.*, *29*, 1571–1578.

Marshall, J. C. (1981), On the parameterization of geostrophic eddies in the ocean, *J. Phys. Oceanogr.*, *11*, 257–271.

Marshall, J. C., H. Jones, R. Karsten, and R. Wardle (2002), Can eddies set ocean stratification?, *J. Phys. Oceanogr.*, *32*, 26–38.

Marshall, J. C., and T. Radko (2003), Residual mean solutions for the Antarctic Circumpolar Current and its associated overturning circulation, *J. Phys. Ocean.*, *33*, 2341–2354.

McGillicuddy, D. J., and A. R. Robinson (1997), Eddy-induced nutrient supply and new production in the Sargasso Sea, *Deep-Sea Res. I*, *44*, 1427–1449.

McWilliams, J. C., and J. H. S. Chow (1981), Equilibrium geostrophic turbulence: I. A reference solution in a beta-plane channel, *J. Phys. Oceanogr.*, *11*, 921–949.

McWilliams, J. C. (1985), Submesoscale, coherent vortices in the ocean, *Rev. Geophys.*, *23*, 165–182.

McWilliams, J. C. (1996), Modeling the oceanic general circulation. *Annu. Rev. Fluid Mech.*, *28*, 1–34.

McWilliams, J. C. (2006), *Fundamentals of Geophysical Fluid Dynamics*, Cambridge University Press, Cambridge, 249 pp.

Merryfield, W. J., and G. Holloway, (1999), Eddy fluxes and topography in stratified quasigeostrophic models, *J. Fluid Mech.*, *380*, 59–80.

Muller, P., J. C. McWilliams, and M. J. Molemaker (2005), Routes to dissipation in the ocean: The 2D/3D turbulence conundrum, in *Marine Turbulence: Theories, Observations and Models*, edited by H. Baumert, J. Simpson, and J. Sundermann, pp. 397–405, Cambridge University Press, Cambridge.

Munk, W. H. (1950), On the wind-driven ocean circulation, *J. Meteorol.*, *7*, 79–93.

Olbers, D., and M. Visbeck (2005), A model of the zonally averaged stratification and overturning in the Southern Ocean, *J. Phys. Oceanogr.*, *35*, 1190–1205.

Pedlosky, J. (1987), *Geophysical Fluid Dynamics*, Springer, New York, 710 pp.

Pope, S. B. (2000), *Turbulent Flows*, Cambridge University Press, Cambridge, 771 pp.

Redi, M. H. (1982), Oceanic isopycnal mixing by coordinate rotation, *J. Phys. Oceanogr.*, *12*, 1154–1158.

Rhines, P. B. (1975), Waves and turbulence on a beta-plane, *J. Fluid Mech.*, *69*, 417–443.

Rhines, P. B., and W. R. Young (1982), Homogenization of potential vorticity in planetary gyres, *J. Fluid Mech.*, *122*, 347–367.

Rudnick, D., and R. Ferrari (1999), Compensation of horizontal temperature and salinity gradients in the ocean mixed layer, *Science*, *283*, 526–529.

Salmon, R. (1982), Geostrophic turbulence, *Topics in Ocean Physics*, Proc. Int. Sch. Phys. 'Enrico Fermi,' Varenna, Italy, pp. 30–78.

Scott, R. B., and F. Wang (2005), Direct evidence of an oceanic inverse kinetic energy cascade from satellite altimetry, *J. Phys. Oceanogr.*, *35*, 1650–1666.

Stammer, D. (1997), Global characteristics of ocean variability estimated from regional TOPEX/POSEIDON altimeter measurements, *J. Phys. Oceanogr.*, *27*, 1743–1769.

Sturges, W., and R. Leben (2000), Frequency of ring separations from the loop current in the Gulf of Mexico: A revised estimate, *J. Phys. Oceanogr.*, *30*, 1814–1819.

Sundermeyer, M., and J. Price (1998), Lateral mixing and the North Atlantic Tracer Release Experiment: Observations and numerical simulations of Lagrangian particles and a passive tracer, *J. Geophys. Res.*, *103*, 21481–21497.

Tai, C. K., and W. B. White (1990), Eddy variability in the Kuroshio Extension as revealed by Geosat Altimetry: Energy propagation away from the jet, Reynolds stress, and seasonal cycle, *J. Phys. Oceanogr.*, *20*, 1761–1777.

Tandon, A., and C. Garrett (1996), On a recent parameterization of mesoscale eddies, *J. Phys. Oceanogr.*, *26*, 406–411.

Thomas, L. N., A. Tandon, and A. Mahadevan (2007), Submesoscale processes and dynamics, this volume.

Treguier, A. M., and J. C. McWilliams (1990), Topographic influences on wind-driven, stratified flow in a β-plane channel: An idealized model of the Antarctic Circumpolar Current, *J. Phys. Oceanogr.*, *20*, 324–343.

Vallis, G. K. (2006), *Atmospheric and Oceanic Fluid Dynamics: Fundamentals and Large-Scale Circulation*, Cambridge University Press, Cambridge, 745 pp.

Wolff, J. O., E. Maier-Reimer, and D. J. Olbers (1991), Wind-driven flow over topography in a zonal β-plane channel: A quasigeostrophic model of the Antarctic Circumpolar Current, *J. Phys. Oceanogr.*, *21*, 236–264.

Wunsch, C., and D. Stammer (1995), The global frequency-wavenumber spectrum of oceanic variability estimated from TOPEX/POSEIDON altimetric measurements, *J. Geophys. Res.*, *100*, 24895–24910.

Wunsch, C., and R. Ferrari (2004), Vertical mixing, energy, and the general circulation of the oceans, *Annu. Rev. Fluid Mech.*, *36*, 281–314.

J. C. McWilliams, Department of Atmospheric and Oceanic Sciences, University of California, Los Angeles, CA, 90095-1565, USA. (jcm@atmos.ucla.edu)

Submesoscale Processes and Dynamics

Leif N. Thomas

Woods Hole Oceanographic Institution, Woods Hole, Massachusetts, USA

Amit Tandon

Physics Department and Department of Estuarine and Ocean Sciences, University of Massachusetts, Dartmouth, North Dartmouth, Massachusetts, USA

Amala Mahadevan

Department of Earth Sciences, Boston University, Boston, Massachusetts, USA

Increased spatial resolution in recent observations and modeling has revealed a richness of structure and processes on lateral scales of a kilometer in the upper ocean. Processes at this scale, termed submesoscale, are distinguished by order-one ($O(1)$) Rossby and Richardson numbers; their dynamics are distinct from those of the largely quasi-geostrophic mesoscale as well as fully three-dimensional, small-scale processes. Submesoscale processes make an important contribution to the vertical flux of mass, buoyancy, and tracers in the upper ocean. They flux potential vorticity through the mixed layer, enhance communication between the pycnocline and surface, and play a crucial role in changing the upper-ocean stratification and mixed-layer structure on a timescale of days. In this review, we present a synthesis of upper-ocean submesoscale processes that arise in the presence of lateral buoyancy gradients. We describe their generation through frontogenesis, unforced instabilities, and forced motions due to buoyancy loss or down-front winds. Using the semi-geostrophic (SG) framework, we present physical arguments to help interpret several key aspects of submesoscale flows. These include the development of narrow elongated regions with $O(1)$ Rossby and Richardson numbers through frontogenesis, intense vertical velocities with a downward bias at these sites, and secondary circulations that redistribute buoyancy to stratify the mixed layer. We review some of the first parameterizations for submesoscale processes that attempt to capture their contribution to, first, vertical buoyancy fluxes and restratification by mixed-layer instabilities and, second, the exchange of potential vorticity between the wind- and buoyancy-forced surface mixed layer, and pycnocline. Submesoscale processes are emerging as vital for the transport of biogeochemical properties, for

Ocean Modeling in an Eddying Regime
Geophysical Monograph Series 177
Copyright 2008 by the American Geophysical Union.
10.1029/177GM04

generating spatial heterogeneity that is critical for biogeochemical processes and mixing, and for the transfer of energy from meso- to small scales. Several studies are in progress to model, measure, analyze, understand, and parameterize these motions.

1. INTRODUCTION

The oceanic mesoscale flow field, characterized by a horizontal length scale of 10 to 100 km, has been studied extensively for its dynamics and its contribution to the lateral transport of heat, momentum, and tracers by means of eddies. Similarly, three-dimensional processes at small lengthscales less than a kilometer (0.1–100 m) have been investigated for their contribution to mixing and energy dissipation. However, submesoscales (~1 km) that lie intermediate to meso- and small-scale three-dimensional motions are less understood and have only more recently been brought to light through observational, modeling, and analytical studies. The submesoscale, characterized by $\mathcal{O}(1)$ Rossby number dynamics, is not described appropriately by the traditional quasi-geostrophic theory that applies to mesoscales. It is not fully three-dimensional and nonhydrostatic, either, but is inevitably crucial to bridging the meso- and smaller scales through processes and dynamics that we are just beginning to understand. The objective of this article was to review and synthesize the understanding of submesoscales put forth through recent diverse studies.

Our discussion will focus on the upper ocean where submesoscale processes are particularly dominant due to the presence of lateral density gradients, vertical shear, weak stratification, a surface boundary that is conducive to frontogenesis, and a relatively small Rossby radius based on the mixed-layer depth. This is not to say that submesoscale phenomenon occurs solely in the upper ocean. In the ocean interior and abyss, there are submesoscale coherent vortices [*McWilliams*, 1985] and balanced flows associated with the oceanic vortical mode, which are thought to play an important role in the isopycnal stirring of tracers [*Kunze and Stanford*, 1993; *Kunze*, 2001; *Polzin and Ferrari*, 2003; *Sundermeyer and Lelong*, 2005]. Furthermore, internal gravity waves can vary on the submesoscale, but will not be described in this review.

The motivation to study submesoscale processes comes from several factors. As the geometrical aspect (depth to length) ratio and Rossby number Ro, associated with meso- and larger scale flow are $\ll 1$, and the Richardson number $Ri \gg 1$, the associated vertical velocities are 10^{-3} to 10^{-4} times smaller than the horizontal velocities, which are typically 0.1 m s^{-1}. However, localized submesoscale regions de-

velop in which Ro and Ri are $\mathcal{O}(1)$. At these sites, submesoscale dynamics generate vertical velocities of $\mathcal{O}(10^{-3})$ m s^{-1} or ~ 100 m day^{-1} that are typically an order of magnitude larger than those associated with the mesoscale. In addition, at the submesoscale, there is a marked asymmetry in the strength of upwelling versus downwelling and anticyclonic versus cyclonic vorticity, with an enhancement of downward velocity and cyclonic vorticity.

Owing to their large vertical velocities, submesoscale processes can be instrumental in transferring properties and tracers, vertically, between the surface ocean and the interior. This vertical transport plays an important role in supplying nutrients to the euphotic zone for phytoplankton production and exchanging gases between the atmosphere and the ocean. Understanding how vertical exchange is achieved between the biologically active, but nutrient-depleted, surface euphotic layer and the nutrient-replete thermocline has been a long-standing question for the carbon cycle and biogeochemistry of the upper ocean. For example, estimates of new production (phytoplankton production relying on a fresh rather than recycled supply of nutrients) based on oxygen utilization and cycling rates [*Platt and Harrison*, 1985; *Jenkins and Goldman*, 1985; *Emerson et al.*, 1997] and helium fluxes [*Jenkins*, 1988] are much higher in the subtropical gyres than can be accounted for through the physical circulation in global carbon cycle models [*Najjar et al.*, 1992; *Maier-Reimer*, 1993]. Several studies, such as those of *McGillicuddy and Robinson* [1997] and *McGillicuddy et al.* [1998], suggest that mesoscale eddies act to pump nutrients to the euphotic zone. However, a basinwide estimate for the eddy-pumping fluxes [*Oschlies*, 2002a; *Martin and Pondaven*, 2003; *Oschlies*, 2007] turns out to be inadequate in supplying the nutrient flux required to sustain the observed levels of productivity in the subtropical gyres. As described above, vertical velocities associated with submesoscale features are much stronger than their mesoscale counterparts, suggesting that submesoscale vertical fluxes of nutrients may play a critical role in enhancing productivity not only in the subtropical gyres but in the world ocean as a whole.

A large part of the ocean's kinetic energy resides at meso- and larger scales. At these scales, oceanic flow is largely two-dimensional and in a state of hydrostatic and geostrophic balance from which it is difficult to extract energy.

A major conundrum [*McWilliams et al.*, 2001; *McWilliams*, 2003], therefore, is how energy is transferred from the mesoscale to the small scale at which it can be dissipated through three-dimensional processes. The strong ageostrophic flow at submesoscales can extract energy from the balanced state and transfer it to smaller scales. *Charney* [1971] argued that large-scale stirring would induce a forward enstrophy cascade consistent with a kinetic energy spectrum of slope −3. For the oceanic context, numerical simulations have shown that the quasi two-dimensional mesoscale flow field is characterized by kinetic energy spectra with a slope of −3 [*Capet et al.*, 2008a; *Klein et al.*, 2007]. Three-dimensional numerical simulations at progressively finer resolutions show that resolving submesoscale processes leads to flattening the kinetic energy spectra slope to −2 [*Capet et al.*, 2008a] and a transfer of energy to larger as well as smaller scales [*Boccaletti et al.*, 2007].

Yet, another factor associated with submesoscale instabilities is the flux of potential vorticity to and from the surface to the interior ocean and the change in stratification of the mixed layer. Submesoscale instabilities in the mixed layer are shown to hasten restratification and buoyancy transport several-fold as compared to what can be achieved through mesoscale baroclinic instability [*Fox-Kemper et al.*, 2007]. Hence, their contribution to eddy transport can be significant. Present-day global circulation models do not resolve submesocales; conceivably, this is the reason for the dearth of restratifying processes and mixed layers that are far too deep in the models [*Oschlies*, 2002b; *Hallberg*, 2003; *Fox-Kemper et al.*, 2007]. Hence, parameterizing these processes is of interest to climate modeling. Similarly, the cumulative vertical flux of potential vorticity through submesoscale processes can alter the potential vorticity budget of the thermocline and mixed layer [*Thomas*, 2005, 2007]. Submesoscale dynamics provide a pathway between the surface boundary layer, where properties are changed by friction and diabatic processes, and the interior, which is largely adiabatic and conserves properties.

Resolving submesoscales within the mesoscale field has been a challenge for models and observations, but one that is being currently met through improvements in technology. Hydrographic surveys using towed vehicles (such as a SeaSoar) equipped with conductivity-temperature-depth sensors have revealed submesoscale features in the upper ocean associated with compensated and uncompensated ocean fronts [e.g., *Pollard and Regier*, 1992; *Rudnick and Luyten*, 1996; *Rudnick and Ferrari*, 1999; *Lee et al.*, 2006b]. Shipboard acoustic Doppler current profiler velocity measurements show that the distribution of the relative vorticity in the upper ocean is skewed to positive values, hinting at the presence of submesoscale flows with stronger cyclonic versus anticyclonic vorticity [*Rudnick*, 2001]. Recent observations centered around a drifter show the role played by the baroclinicity in setting the stratification within the mixed layer (Hosegood et al., Restratification of the surface mixed layer with submesoscale lateral gradients: Diagnosing the importance of the horizontal dimension, submitted to *Journal of Physical Oceanography*, 2007). Similar measurements made following mixed-layer Lagrangian floats have captured rapid (occurring over a day) changes in the mixed-layer stratification that cannot be ascribed to heating or cooling, and hence, are thought to result from submesoscale processes [*Lee et al.*, 2006a]. Further examples of submesoscale variability are seen in high-resolution velocity fields from radar [*Shay et al.*, 2003], sea-surface temperature fields from satellites [*Flament et al.*, 1985; *Capet et al.*, 2008b], a proliferation of cyclonic vortices revealed by sunglitter on the sea surface [*Munk et al.*, 2000], and biogeochemical sampling along ship transects. We are at an exciting juncture because we are now able to achieve the required resolution in models and observations to capture this scale. The results from high-resolution numerical modeling and analytical studies, several of which are discussed in this review, suggest that both forced and unforced instabilities drive submesoscale processes.

We begin section 2 by defining the term submesoscale and describing phenomena with which it is associated. Furthermore, we examine mechanisms that generate submesoscales in the upper ocean. In section 3, we present a mathematical framework for understanding the secondary circulation associated with fronts where submesoscale processes are found to be active. This framework is used to provide a dynamical explanation for several key features of submesoscale phenomena. In section 4, we discuss the implications of submesoscale phenomena, which include mixed-layer restratification, vertical transport and biogeochemical fluxes, and potential vorticity fluxes. Finally, we provide a discussion of outstanding questions and possible connections to other areas.

2. PHENOMENOLOGY

2.1. What are Submesoscales?

An active flow field in the upper ocean generates localized regions, typically along filaments or outcropping isopycnals, within which the relative vertical vorticity $\zeta = v_x - u_y$ equals or exceeds the planetary vorticity f, and the vertical shear can be quite strong. The dynamics within these regions differs from mesoscale dynamics characterized by small Rossby numbers ($Ro \ll 1$) and large Richardson numbers ($Ri \gg 1$). We thus define submesoscale flows based on dynamics, as those where the gradient Rossby number, $Ro = |\zeta|/f$, and the gradient Richardson number, $Ri = N^2/|\partial_z \mathbf{u}_h|^2$, are both $\mathcal{O}(1)$, where $\mathbf{u}_h$ is the horizontal velocity, $N^2 = b_z$ is the square of the buoyancy frequency, $b = -g\rho/\rho_o$ is the buoyancy, ρ is the density, g is the acceleration due to gravity, and ρ_o is a reference density. If we introduce bulk Richardson and Rossby numbers: $Ri_b = N^2H^2/U^2$ and $Ro_b = U/fL$ (U, H, and L are the characteristic speed, vertical lengthscale, and horizontal lengthscale of the velocity field, respectively), it then follows that submesoscale flows with $Ro_b = Ri_b = \mathcal{O}(1)$ are also characterized by an order-one Burger number $Bu = N^2H^2/f^2L^2$. It is worth noting that the bulk Richardson number can be related to the Froude number $Fr = U/NH = 1/\sqrt{Ri_b}$. Submesoscale flows, as we define them here, are hence characterized by $Fr \geq 1$. This parameter range with $Ro_b = \mathcal{O}(1)$ is also relevant to the transition from geostrophic to stratified turbulence [e.g., *Waite and Bartello*, 2006].

The characteristic vertical extent H of upper-ocean submesoscale phenomena typically scales with the mixed-layer depth h_{ml}. The condition that $Bu \sim 1$ implies that the horizontal lengthscale associated with submesoscale processes scales with the mixed-layer Rossby radius of deformation

$$L \sim L_{ml} = \frac{N_{ml}h_{ml}}{f}, \qquad (1)$$

where N_{ml} is the buoyancy frequency in the mixed layer. The weak stratification and limited vertical extent of mixed layers makes the characteristic length of submesoscale flows small relative to the first baroclinic Rossby radius of deformation that defines the mesoscale. For example, a mixed layer of depth $h_{ml} = 100$ m, with $N_{ml} = 10^{-3}$ s^{-1} in the mid-latitudes ($f = 1 \times 10^{-4}$ s^{-1}), yields $L \sim L_{ml} = 1$ km. Given this small horizontal scale, it only takes a relatively weak velocity of $U = 0.1$ m s^{-1} to yield a Rossby number of order-one, indicating that such modest submesoscale flows can be susceptible to nonlinear dynamics and ageostrophic effects.

Another way to state the $Bu = 1$ condition is that the aspect ratio of submesoscale flows, $\Gamma = H/L$, scales as f/N. For typical oceanic conditions, $f/N \ll 1$, so that $\Gamma \ll 1$. Scaling the vertical momentum equation shows that the hydrostatic balance is accurate to $O(Ro^2\Gamma)$. Hence, although $Ro = \mathcal{O}(1)$ for submesoscale flows, $\Gamma \ll 1$ in the upper ocean, and these processes can, to a good approximation, be considered hydrostatic. Indeed, non-hydrostatic effects are difficult to detect in submesoscale model simulations at horizontal grid resolutions of 500 m [*Mahadevan*, 2006].

Another perspective used to understand submesocale processes is in terms of Ertel's potential vorticity (PV)

$$q = (f + \zeta)N^2 + \omega_h \cdot \nabla_h b, \qquad (2)$$

where $\omega_h = (w_y - v_z, u_z - w_x)$ is the horizontal vorticity and $\nabla_h b$ is the horizontal buoyancy gradient. At large scales and in the interior, q is dominated by the planetary PV: fN^2. At mesoscales, the contribution from the vertical vorticity, ζN^2, becomes important. In the upper ocean, in the presence of density gradients arising from frontal filaments or outcropping isopycnals, the contribution to q from the horizontal buoyancy gradient and the thermal wind shear $(u_z^{tw}, v_z^{tw}) = (-b_y/f, b_x/f)$ becomes important and always tends to lower the PV. More specifically, the contribution to the PV from horizontal buoyancy gradients is as large as the planetary PV and significantly reduces the magnitude of the total PV when the Richardson number is $\mathcal{O}(1)$, that is, when the flow is submesoscale according to our dynamical definition [*Tandon and Garrett*, 1994; *Thomas*, 2007]. Submesoscale dynamics are thus intimately linked with processes that modify the PV, such as forcing by wind stress and buoyancy fluxes and advection of PV by eddies. This chapter therefore often uses PV as a means to help interpret submesoscale physics.

Next, we describe mechanisms that are known to be active in generating submesoscales including (1) frontogenesis; (2) unforced instabilities, such as the ageostrophic baroclinic instability [*Molemaker et al.*, 2005; *Boccaletti et al.*, 2007]; and (3) forced motion, such as flows affected by buoyancy fluxes or friction at boundaries. We will use the results from a numerical model to individually demonstrate the above submesoscale mechanisms. Although the submesoscale conditions are localized in space and time, the mesoscale flow field is crucial in generating them. In the ocean, it is likely that more than one submesoscale mechanism acts in tandem with mesoscale dynamics to produce a complex submesoscale structure within the fabric of the mesoscale flow field.

2.2. Frontogenesis

Consider the flow field generated by a geostrophically balanced front in the upper mixed layer of the ocean overlying a pycnocline. As the front becomes unstable and meanders, the nonlinear interaction of the lateral velocity shear and buoyancy gradient locally intensify the across-front buoyancy gradient. Strong frontogenetic action pinches outcropping isopycnals together, generating narrow regions in which the lateral shear and relative vorticity become very large, and the Ro and Ri become $\mathcal{O}(1)$. At these sites, the lateral strain rate $S \equiv ((u_x - v_y)^2 + (v_x - u_y)^2)^{1/2}$ is also large, and strong ageostrophic overturning circulation generates intense vertical velocities. In Figure 1, we plot the density, horizontal and vertical velocities, strain rate, Ro, and Ri from a frontal region in a model simulation. The model was initialized with an across-front density variation of 0.27 kg m^{-3}

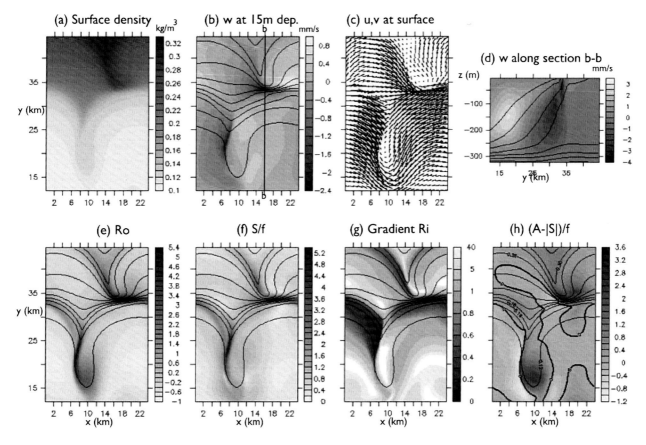

Figure 1. A region in the model domain where spontaneous frontogenesis has set up large shear and relative vorticity ζ, high lateral strain rate S, and a strong ageostrophic secondary circulation. (a) Surface density, (b) vertical velocity at 15 m depth, (c) surface u, v velocities, (d) vertical section through the front at $x = 16$ km showing vertical velocity (light tones indicate upward, dark tones downward) and isopycnals (black contours), (e) $Ro = \zeta/f$, (f) S/f, (g) gradient Ri, and (h) $(A - |S|)/f$ with the zero contour shown as a dark black line. Light black contours indicate surface density in all cases.

across 20 km (i.e., $|b_y| \approx 10^{-7}$ s^{-2}), over a deep mixed layer extending to 250 m and allowed to evolve in an east–west periodic channel with solid southern and northern boundaries. Submesoscale frontogenesis is more easily seen when the mixed layer is deep because the horizontal scale, which is dependent on h_{ml}, is larger and more readily resolved in the numerical model. Here, the mixed layer is taken to be 250 m deep so as to exaggerate frontogenesis. As the baroclinically unstable front meanders, the lateral buoyancy gradient is spontaneously locally intensified in certain regions, as in Figure 1, generating submesoscale conditions at sites approximately 5 km in width. This mechanism is ubiquitous to the upper ocean due to the presence of lateral buoyancy gradients and generates submesoscale phenomena when intensification can proceed without excessive frictional damping, or in a model with sufficient numerical resolution and minimal viscosity.

2.3. Unforced Instabilities

The instabilities in the mixed-layer regime where $Ro = \mathcal{O}(1)$ and $Ri = Ro^{-1/2} = \mathcal{O}(1)$ are different from the geostrophic baroclinic mode in several respects. The ageostrophic baroclinic instability problem of a sheared rotating stratified flow in thermal wind balance with a constant horizontal buoyancy gradient was investigated using hydrostatic [*Stone*, 1966, 1970] and non-hydrostatic equations [*Stone*, 1971] for finite values of Ro. *Molemaker et al.* [2005] extend these analyses by examining them in the context of loss of balance that leads to a forward energy cascade. Their instability analysis is applicable to the mixed-layer regime and shows two distinct instabilities. The largest growth rates arise for a geostrophic mode (large Ro Eady mode), which is mostly balanced and well captured by hydrostatic equations [*Stone*, 1966, 1970]. *Boccaletti et al.* [2007]

investigate an instability problem similar to *Molemaker et al.* [2005] and *Stone* [2007], although with a reduced gravity bottom boundary condition. They show that baroclinic instability arises at the mixed-layer Rossby radius of deformation, that is, (1), resulting in eddies of size $L = 1 - 10$ km in the mixed layer. While *Boccaletti et al.* [2007] call this instability ageostrophic (as the instability arises in a high Ro and small Ri regime), this mixed-layer instability (MLI) is the geostrophic mode of *Molemaker et al.* [2005] and is akin to the Eady mode for large Ro, small Ri. MLI, or the geostrophic mode, is mostly in balance (meaning that it can be diagnosed from PV), and its timescale, given by $Ri^{1/2}/f$, is $O(1/f)$, as $Ri \sim 1$. This timescale is much shorter than that of mesoscale baroclinic instability, and MLI hastens the slumping of fronts and hence leads to rapid restratification of the mixed layer.

In Figure 2, we show the evolution of the mixed-layer (balanced) mode or MLI in a model simulation as it rapidly restratifies the mixed layer. The model is initialized with a north-south buoyancy gradient in the mixed layer, which decays with depth in the upper thermocline. The east-west oriented front is seen to develop submesoscale meanders that merge and expand in lengthscale to the internal Rossby radius. As this occurs, lighter water overrides the denser side of the front, and as the frontal outcrop is pushed back, the front slumps and the mixed layer restratifies. The simulation replicates the phenomena in the work of *Boccaletti et al.* [2007]. In this unforced case, the submesoscale wiggles at the front grow at $O(f^{-1})$ timescales between days 5 and 10 of the simulation, and transcend to mesoscale meanders within a couple of days by the upscale transfer of energy. The evolution of meanders then progresses at mesoscale timescales, but submesoscale processes continue to dominate the secondary circulation and the vertical fluxes at the front, leading to restratification. The values of the local Ro and gradient Ri during the evolution of the MLI suggest that the instability is largely driven by the ageostrophic component of the flow, although it is largely in gradient wind balance.

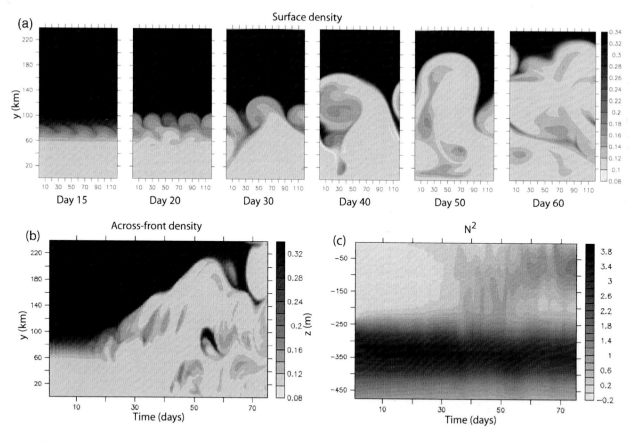

Figure 2. (a) Sequential figures of the surface density showing the evolution of the mixed layer instability. (b) Hovmöller plot of the across-front surface density at $x = 60$ km. (c) Hovmöller plot showing the evolution of the square of the buoyancy frequency N^2 over time averaged across the section at $x = 60$ km. The density is expressed in units of kg m^{-3}, while N^2 is multiplied by 10^5 and is expressed in s^{-2}.

A departure from balanced dynamics, in which submesoscale dynamics most likely play an important role, facilitates a forward cascade of energy [McWilliams et al., 2001]. The nongeostrophic, unbalanced mode of Molemaker et al. [2005] requires that the difference between the absolute vorticity, $A = f + v_x - u_y$, and the magnitude of the strain rate, that is, $A - |S|$, changes sign. The frontal simulation in Figure 1 shows that this criteria is well satisfied in regions where active frontogenesis and MLI occur. However, diagnosis of the unbalanced mode remains a challenge because a method is required to decompose the flow into its balanced and unbalanced components. The imbalance itself has to be characterized as either a departure from a hydrostatic and cyclostrophically adjusted state (pressure imbalance) or as a departure from a state that is fully diagnosable from the Ertel potential vorticity (PV imbalance). These different imbalance characterizations lead to different decompositions. PV-based decompositions for nonhydrostatic equations have been suggested by McKiver and Dritschel [2008] and Viúdez and Dritschel [2004]. In submesoscale-resolving numerical simulations, Mahadevan and Tandon [2006] used the quasi-geostrophic omega equation to diagnose the balanced vertical velocities and compared these to the vertical velocities from the three-dimensional model. Likewise, one may diagnose the three-dimensional ageostrophic stream function (described in section 3.1.3) to represent the balanced flow and compare it with the ageostrophic secondary circulation (ASC) from the model. Capet et al. [2008a] calculate normalized departures from gradient wind balance in simulations of the California Current system. By successively increasing the horizontal model resolution, they see a greater departure from gradient wind balance, indicative of unbalanced instabilities.

2.4. Forced Motions

Applying destabilizing atmospheric forcing to baroclinic mesoscale features can lead to the generation of submesoscale motions. Destabilizing atmospheric forcing is defined as that which tends to reduce the stratification, Richardson number, and, more generally, the PV of the upper ocean. Such forcing can arise from heat or salt fluxes that extract buoyancy from the ocean or by down-front wind stress.

2.4.1. Buoyancy loss. Uniformly cooling a mesoscale flow or differentially cooling a laterally homogeneous layer can trigger submesoscale instabilities. The case of differential cooling was investigated by Haine and Marshall [1998] using non-hydrostatic numerical simulations run in two- and three-dimensional configurations. In both configurations, the differential cooling creates a lateral buoyancy gradient that forms a baroclinic jet through geostrophic adjustment. Due to the destabilizing forcing, the PV of the baroclinic jet is drawn to negative values, which, in the two-dimensional simulations, triggers symmetric instability of submesoscale ($\mathcal{O}(1 \text{ km})$) width that mixes the PV to create a mixed layer with nearly zero PV, nonzero stratification, and an order-one Richardson number. In the three-dimensional configuration, the jet goes unstable to baroclinic waves that have similar properties to the geostrophic mode described by Molemaker et al. [2005] and the MLI of Boccaletti et al. [2007], that is, with horizontal scales ~5 km and timescales of development of $\mathcal{O}(1 \text{ day})$. The baroclinic instability of Haine and Marshall [1998] rapidly restratifies the mixed layer and grows in horizontal scale, as seen in the numerical experiments of Boccaletti et al. [2007] and the solution presented in Figure 2.

The case of uniform cooling of a mesoscale flow was investigated using nonhydrostatic numerical simulations by Legg et al. [1998] and Legg and McWilliams [2001] for a flow consisting of an initial mesoscale eddy field and by Yoshikawa et al. [2001] for an ocean front. When a single cyclonic eddy is cooled by Legg et al. [1998], erosion of stratification reduces the Rossby radius of deformation to a value smaller than the radius of the eddy and allows for the formation of submesoscale baroclinic waves on the edges of the eddy. Numerical experiments in which a mixture of baroclinic cyclonic and anticyclonic eddies are cooled were analyzed by Legg and McWilliams [2001]. In these experiments, the barotropic eddy kinetic energy is observed to increase at a faster rate as compared to the case without cooling, suggesting that cooling, by triggering submesoscale baroclinic instability, provides an additional means of converting available potential energy at mesoscales to barotropic eddy kinetic energy.

Yoshikawa et al. [2001] detailed the way in which destabilizing buoyancy fluxes at an ocean front can enhance the frontal vertical circulation. Comparing forced and unforced model runs, they showed that lateral strain associated with baroclinic instabilities was stronger for fronts forced by cooling and resulted in more intense frontogenesis. Using the semi-geostrophic omega equation, Yoshikawa et al. [2001] demonstrated that the amplified frontogenetic strain, combined with the cooling-induced low PV in the mixed layer, causes an enhancement in the strength of the frontal vertical circulation. We will elaborate on the dynamics of this important result in section 3.1.1.

2.4.2. Down-front wind stress. Down-front winds, that is, winds blowing in the direction of the surface frontal jet, reduce the stratification and Richardson number in the surface boundary layer and thus provide favorable conditions for submesoscale phenomena [Thomas and Lee, 2005; Thomas,

2005]. Non-hydrostatic simulations show that for this wind orientation, Ekman flow advects denser water over light, destabilizing the water column and triggering convective mixing [*Thomas and Lee*, 2005]. The strength of this mixing is set by the magnitude of the surface lateral buoyancy gradient and the Ekman transport, the latter being a function of the vorticity of the flow. This is because flows with $\mathcal{O}(1)$ Rossby number exhibit nonlinear Ekman dynamics in which the advection of momentum plays a role in the Ekman balance. This causes the Ekman transport to vary inversely with the absolute, rather than planetary, vorticity [*Stern*, 1965; *Niiler*, 1969; *Thomas and Rhines*, 2002]. One of the more striking results of nonlinear Ekman dynamics is that a spatially uniform wind stress will induce Ekman pumping or suction if it is blowing over a current with lateral variations in its vorticity field. For example, a down-front wind blowing over a frontal jet will drive surface Ekman transport from the dense to light side of the front. The transport is weaker (stronger) on the cyclonic (anticyclonic) side of the jet and hence induces a secondary circulation with convergence and downwelling on the dense side of the front and divergence and upwelling at the center of the jet. Localized mixing caused by Ekman-driven convection at the front drives an additional secondary circulation. For down-front winds, frontal intensification can occur when the two secondary circulations modify the vertical vorticity and lateral buoyancy gradient in such a way as to enhance the nonlinear Ekman pumping or suction and Ekman-driven convection, thus reinforcing the secondary circulations themselves. This positive feedback mechanism can lead to the growth of multiple fronts in a mixed layer with an initially spatially uniform horizontal buoyancy gradient forced by down-front winds [*Thomas and Lee*, 2005].

The submesoscale-resolving, wind-forced model simulations of *Mahadevan and Tandon* [2006] and *Capet et al.* [2008a] show a preponderance of positive vorticity at submesoscales. Downward velocities are more intense and narrowly confined in elongated regions relative to upward velocities. The simulations of *Mahadevan and Tandon* [2006] demonstrate how even a weak, sustained downfront wind stress (of magnitude 0.025 N m^{-2}) generates a profusion of frontal structures that exhibit very large strain rates, relative vorticity, and vertical velocities in the surface mixed layer. These characteristics can be explained by SG dynamics, as shown in section 3.1.1. In contrast to the upper ocean, the velocity structure in the stratified pycnocline remains largely mesoscale in character.

Convective mixing induced by down-front winds leads to a reduction in the PV. This modification of the PV is due to an upward frictional PV flux that extracts PV from the ocean. The numerical simulations of *Thomas* [2005] demonstrate how this frictional PV flux is transmitted through the oceanic surface layer by submesoscale secondary circulations that downwell low PV from the surface while upwelling high PV from the pycnocline, yielding a vertical eddy PV flux that scales with the surface frictional PV flux. Although these numerical experiments were two-dimensional, fully three-dimensional simulations of wind-forced fronts have also shown such a vertical exchange of PV by submesoscale frontal circulations [*Capet et al.*, 2008b; *Mahadevan and Tandon*, 2006; *Thomas*, 2007]. In the work of *Thomas* [2007], the subduction of the low PV surface water results in the formation of a submesoscale coherent vortex or intrathermocline eddy. The vortex was shown to exert an along-isopycal eddy PV flux that scales with the wind-driven frictional PV flux at the surface. The eddy PV flux drove an eddy-induced transport or bolus velocity down the outcropping isopycal, which had an effect on the large-scale mean flow. The correspondence of the eddy and frictional PV fluxes in both the two- and three-dimensional simulations suggests that a possible approach for parameterizing wind-forced submesoscale phenomena is to use a parameterization scheme based on PV fluxes, a subject that will be discussed further in section 4.2.

3. SUBMESOSCALE DYNAMICS

The timescale of variability for submesoscale flows is often not very distinct from the period of near-inertial internal gravity waves, but unlike internal gravity waves, submesoscale flows are to a large extent balanced. This implies that submesoscale dynamics are determined by a single scalar field from which all other variables (density, horizontal and vertical velocity, etc.) can be determined using an invertibility principle [*McKiver and Dritschel*, 2008]. In the most familiar approximate forms of the equations governing balanced flows, there are two choices for the controlling scalar field: the horizontal streamfunction (e.g., the balanced equations of *Gent and McWilliams* [1983]) and the potential vorticity (e.g., the quasi- and semigeostrophic models of *Charney* [1948] and *Eliassen* [1948], respectively). Of the models listed above, the quasi-geostrophic (QG) model places the most severe restriction on the Rossby and Richardson numbers. QG theory assumes that $Ro \ll 1$ and $1/Ri \ll 1$; consequently, it is not designed to accurately describe the dynamics of submesoscale phenomena. A thorough study of the advantages and disadvantages of each model listed above can be found in *McWilliams and Gent* [1980]. In general, for flows with significant curvature, the balance equations are the most accurate [*Gent et al.*, 1994]. However, for relatively straight flows, the SG equations are accurate and provide insights into the dynamics of the intense fronts and vertical circulations typical of submesoscale features in a relatively simple manner. In this section, we will describe some of the key features of submesoscale phenomena, namely frontogenesis, strong vertical circulation, restratification, forward cascade through frontogenesis, and nonlinear Ekman effects using the dynamical framework of the semigeostrophic equations.

3.1. Semigeostrophic Dynamics

In SG theory, the flow is decomposed into geostrophic and ageostrophic components $\mathbf{u} = \mathbf{u}_g + \mathbf{u}_{ag}$, where the geostrophic velocity $\mathbf{u}_g \equiv \hat{\mathbf{k}} \times \nabla_h p / f$ and p is the pressure. The SG equations are

$$\frac{D\mathbf{u}_g}{Dt} = -f\hat{\mathbf{k}} \times \mathbf{u}_{ag} \qquad (3)$$

$$0 = -\frac{1}{\rho_o}\frac{\partial p}{\partial z} + b \qquad (4)$$

$$\frac{Db}{Dt} = 0 \qquad (5)$$

$$\nabla \cdot \mathbf{u}_{ag} = 0, \qquad (6)$$

with $D/Dt = \partial/\partial t + (\mathbf{u}_g + \mathbf{u}_{ag}) \cdot \nabla$. These are valid if $D^2 u/Dt^2 \ll f^2 u$ and $D^2 v/Dt^2 \ll f^2 v$, that is, if the Lagrangian timescale of variability of the flow is much longer than an inertial period [Hoskins, 1975].

3.1.1. Two-dimensional vertical circulation.
Consider a front in the $y - z$ plane, that is, $b_y \neq 0$ and $b_x = 0$ where the along-front velocity u is purely geostrophic, that is, $u = u_g$, and in thermal wind balance, $fu_{gz} = -b_y$. The two-dimensional ageostrophic circulation can be described by an across-front overturning streamfunction ψ where $(v_{ag}, w) = (\psi_z, -\psi_y)$. As first derived by Eliassen [1948] and Sawyer [1956], a single equation for ψ can be constructed by combining the y derivative of the buoyancy equation (5), with the z derivative of the zonal component of (3), yielding

$$F_2^2\frac{\partial^2\psi}{\partial z^2} + 2S_2^2\frac{\partial^2\psi}{\partial z\partial y} + N^2\frac{\partial^2\psi}{\partial y^2} = -2Q_2^g, \qquad (7)$$

where $N^2 = b_z$, $S_2^2 = -b_y = fu_{gz}$, $F_2^2 = f(f - u_{gy})$, and Q_2^g is the y component of the Q vector

$$\mathbf{Q}^g = (Q_1^g, Q_2^g) = \left(-\frac{\partial\mathbf{u}_g}{\partial x}\cdot\nabla b, -\frac{\partial\mathbf{u}_g}{\partial y}\cdot\nabla b\right) \qquad (8)$$

introduced by Hoskins et al. [1978]. A geostrophic flow with a nonzero Q vector will modify the magnitude of the horizontal buoyancy gradient following the equation

$$\frac{D}{Dt}|\nabla_h b|^2 = \mathbf{Q}^g \cdot \nabla_h b \qquad (9)$$

and will consequently disrupt the thermal wind balance of the flow. To restore geostrophy, an ASC is required, and its solution is governed by (7). The sign of the potential vorticity of the two-dimensional geostrophic flow

$$q_{2D} = \frac{1}{f}\left(F_2^2 N^2 - S_2^4\right) = fN^2\left[1 + Ro_{2D} - \frac{1}{Ri_{2D}}\right], \qquad (10)$$

where $Ro_{2D} = -u_{gy}/f$ and $Ri_{2D} = N^2/(u_{gz})^2$, determines the canonical form of (7). When $fq_{2D} > 0$, (7) is elliptic, but when $fq_{2D} < 0$, (7) is hyperbolic and has an entirely different character. Although (7) is formally valid only for two-dimensional ageostrophic motions, it is useful for diagnosing the importance of various mechanisms in generating an ASC and large vertical velocities in the presence of lateral buoyancy gradients. A three-dimensional version of this equation with diabatic and frictional effects is thus presented and discussed in section 3.1.3.

The solution to (7) can be found using the method of Green's functions. The Green's function for ψ satisfies the following equation:

$$F_2^2\frac{\partial^2 G}{\partial z^2} + 2S_2^2\frac{\partial^2 G}{\partial z\partial y} + N^2\frac{\partial^2 G}{\partial y^2} = \delta(y - \mathcal{Y}, z - \mathcal{Z}), \qquad (11)$$

which, for any Q-vector distribution yields the ageostrophic circulation: $\psi = -2\iint G(y - \mathcal{Y}, z - \mathcal{Z})Q_2^g(\mathcal{Y}, \mathcal{Z})d\mathcal{Y}d\mathcal{Z} + \psi_h$, where ψ_h is a homogeneous solution to (7) satisfying the boundary conditions. The solution to (11) for constant coefficients is

$$G = \frac{1}{4\pi\sqrt{fq_{2D}}}\log|\text{Arg}|, \qquad (12)$$

where

$$\text{Arg} = \frac{[(y - \mathcal{Y}) - (z - \mathcal{Z})S_2^2/F_2^2]^2}{L_{SG}^2} + \frac{(z - \mathcal{Z})^2}{H^2},$$

and

$$L_{SG} = H\frac{\sqrt{fq_{2D}}}{F_2^2} \qquad (13)$$

is the semigeostrophic Rossby radius of deformation and H is a characteristic vertical lengthscale of the flow [*Eliassen*, 1951; *Hakim and Keyser*, 2001]. G is plotted in Figure 3 for a buoyancy field that decreases in the y direction ($S_2^2 > 0$). Streamlines take the shape of tilted ellipses oriented at an angle $\theta = 0.5 \tan^{-1}[2S_2^2/(N^2 - F_2^2)]$. For typical conditions, $F_2^2 \ll |S_2^2| \ll N^2$, the ellipses are oriented parallel to isopycnals, $\tan\theta \approx S_2^2/N^2$, and the secondary circulations form slantwise motions along isopycnals. The sense of the secondary circulation of G is thermally direct, tending to flatten isopycnals. The Green's function is driven by a right-hand-side forcing of (7) corresponding to a negative point-source Q vector so that $\mathbf{Q}^g \cdot \nabla_h b > 0$, indicating that this secondary circulation is associated with frontogenetic forcing, that is, $D|\nabla_h b|^2/Dt > 0$. If either the buoyancy gradient or the direction of the Q vector were reversed, a thermally indirect circulation would be induced that tends to steepen isopycnals.

The vertical velocity associated with (12) is $-\partial G/\partial y$, or

$$w = -\frac{F_2^4}{2\pi(fq_{2D})^{3/2}H^2}\frac{(y - \mathcal{Y})}{\text{Arg}} \qquad (14)$$

and has a magnitude that varies greatly with the PV and vertical vorticity of the geostrophic flow. This solution suggests that for a given geostrophic forcing, in regions of low PV (where the Richardson number, stratification, or absolute vorticity of the fluid is small), the vertical circulation is strong. This amplification of the vertical circulation in regions of low PV was also noted by *Yoshikawa et al.* [2001], as noted in section 2.4.2. The solution given by (14) also suggests that w will be stronger in regions of cyclonic versus anticyclonic vorticity when both regions have the same PV. The enhancement of the vertical velocity in areas of low PV and high vertical vorticity is a consequence of the reduction in the characteristic horizontal lengthscale of the circulation in accordance with (13). This reduction in lengthscale is more pronounced for flows with high Rossby number and low Richardson number.

3.1.2. Horizontal deformation and frontal collapse. A simple yet dynamically insightful model for the formation of intense fronts in the submesoscale regime ($Ro \sim Ri \sim 1$) is the horizontal deformation model of *Hoskins and Bretherton* [1972]. In this model, the geostrophic flow is decomposed into two parts,

$$(u_g, v_g) = (\alpha x + u'_g(y, z, t), -\alpha y). \qquad (15)$$

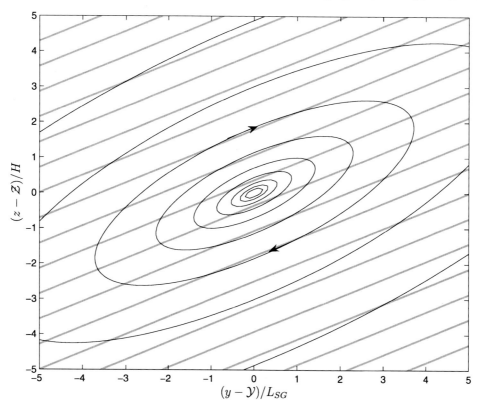

Figure 3. Ageostrophic secondary circulation G driven by a negative point-source Q vector, $Q_2^g < 0$, at $y = Y, z = Z$. Isopycnals (gray contours) slant upward to the north due to a southward buoyancy gradient. For this frontogenetic forcing, $\mathbf{Q}^g \cdot \nabla_h b > 0$, the circulation is thermally direct and tends to restratify the fluid.

The parts are associated with a deformation field with confluence $\alpha = $ constant > 0, and a time-evolving, two-dimensional zonal flow u'_g, respectively. In the context of submesoscale phenomena, the deformation field could be associated with a mesoscale eddy field, while u'_g could be interpreted as an evolving submesoscale frontal feature. The equations are solved by introducing a geostrophic coordinate following the method of *Hoskins and Bretherton* [1972].

The solution for a front with an initial laterally varying buoyancy field $B = \nabla b/2 \tanh(y/L)$, with $\Delta b = 0.0016$ m s^{-2} and $L = 4$ km, in a background stratification of $N_o^2 = 4.2 \times 10^{-6}$ s^{-2} and forced by a confluence $\alpha = 0.1 f$ ($f = 1 \times 10^{-4}$ s^{-1}) is shown in Figure 4. As the geostrophic forcing is frontogenetic, a thermally direct secondary circulation with an upward vertical buoyancy flux $\overline{wb} > 0$ (the overline denotes a lateral average) is induced. The ageostrophic flow is itself frontogenetic and leads to the formation of a frontal discon-

tinuity in the zonal velocity and buoyancy field that first appears at the horizontal boundaries in a finite amount of time. During frontogenesis, asymmetric vertical vorticity and vertical velocity distributions are generated. Near the upper (lower) boundaries, regions of intense cyclonic vorticity and downward (upward) vertical velocities coincide, consistent with the argument presented in section 3.1.3 that the characteristic lateral lengthscale of the secondary circulation (13) is compressed in areas of high absolute vorticity. The formation of the discontinuity in the zonal velocity field causes the horizontal wavenumber spectrum of u to flatten in time, asymptoting to a $l^{-8/3}$ spectrum as the time of frontal collapse approaches [*Andrews and Hoskins*, 1978]. As the frontal discontinuity forms, the maximum in the Rossby and inverse Richardson numbers grows rapidly, taking on values greater than one. The evolution of Ro and Ri^{-1} closely track one another. This is a consequence of conservation of PV. For large

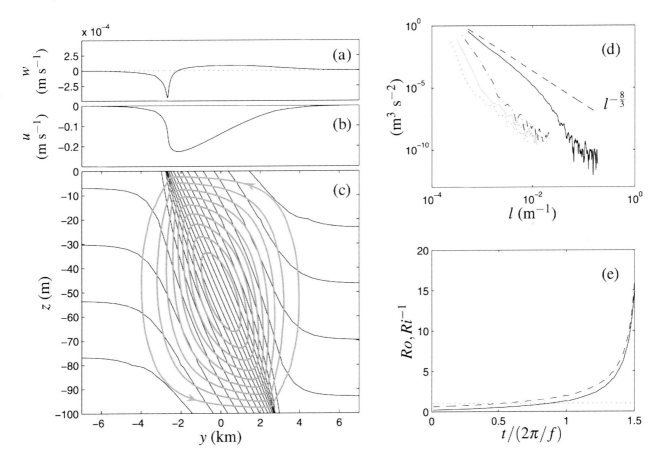

Figure 4. Solution of the horizontal deformation model for an initial buoyancy field $B = \Delta b/2 \tanh(y/L)$: (a) w and (b) u at $z = -2.5$ m, and (c) the overturning streamfunction ψ (gray contours) and buoyancy b (black contours) at $t = 1.5$ inertial periods when frontal collapse at the boundaries is about to occur. (d) The horizontal spectrum of u at $z = 0$ for $t = 0.26$ (dotted), 0.67 (gray), 1.09 (dot–dash), and 1.50 (black) inertial periods. The dashed line denotes a $-8/3$ spectral slope. (e) Evolution of the maximum in the Rossby (solid) and inverse Richardson (dashed) number of the flow at $z = 0$. A dotted line at $Ro = Ri^{-1} = 1$ is plotted for reference.

Rossby numbers, the PV given by (10) is approximately $q_{2D} \approx fN^2(Ro_{2D} - Ri_{2D}^{-1})$ so that for the PV to be conserved, $Ro_{2D} \propto Ri_{2D}^{-1}$. Many of these features of the deformation model (asymmetric vorticity and w distributions, restratifying secondary circulations, flatter kinetic energy spectra, and coincident regions of high Rossby and low Richardson numbers) are typical of submesoscale flows and suggests that the simple physics encompassed in this model is quite relevant to the dynamics of submesoscale phenomena. It should be noted, however, that the semi-geostrophic equations can only be integrated until the first frontal discontinuity forms. Therefore, while semigeostrophic theory provides an accurate description of submesoscale frontal dynamics before frontal collapse, it cannot capture the continuing steepening of fronts after frontal collapse [*Boyd*, 1992].

3.1.3. Three-dimensional overturning circulation with frictional and diabatic effects. The semi-geostrophic equations can be used to study fully three-dimensional flows as well, provided that the cross-stream lengthscale is much less than the radius of curvature of the flow [*Gent et al.*, 1994]. In particular, semigeostrophic theory is useful for understanding the three-dimensional ageostrophic secondary circulation, which can concisely be described in terms of a vector streamfunction (ϕ, ψ) such that

$$u_{ag} = \frac{\partial \phi}{\partial z}, \quad v_{ag} = \frac{\partial \psi}{\partial z}, \quad w = -\left(\frac{\partial \phi}{\partial x} + \frac{\partial \psi}{\partial y}\right) \quad (16)$$

[*Hoskins and Draghici*, 1977]. If friction or diabatic effects are included, then the governing equation for the ageostrophic circulation is

$$\begin{pmatrix} \mathcal{L}_{11} & \mathcal{L}_{12} \\ \mathcal{L}_{21} & \mathcal{L}_{22} \end{pmatrix} \begin{pmatrix} \phi \\ \psi \end{pmatrix} = \underbrace{-2 \begin{pmatrix} Q_1^g \\ Q_2^g \end{pmatrix}}_{I} + \underbrace{f \frac{\partial}{\partial z} \begin{pmatrix} Y \\ -X \end{pmatrix}}_{II} - \underbrace{\begin{pmatrix} \frac{\partial \mathcal{D}}{\partial x} \\ \frac{\partial \mathcal{D}}{\partial y} \end{pmatrix}}_{III},$$

$$(17)$$

where (X, Y) are the components of the horizontal frictional force and $\mathcal{D}$ represents diabatic processes, that is,

$$\mathbf{F} = (X, Y) \quad \text{and} \quad \mathcal{D} \equiv \frac{Db}{Dt}. \quad (18)$$

For example, if lateral mixing of momentum and buoyancy are negligible, $(X, Y) = \partial_z(\tau^x, \tau^y)/\rho_o$ and $\mathcal{D} = -\partial_z F^B$, where (τ^x, τ^y) and F^B are the turbulent stress and vertical

buoyancy flux, respectively. The differential operator on the left hand side of (17) is

$$\mathcal{L}_{11} = F_1^2 \partial_{zz} - 2S_1^2 \partial_{xz} + N^2 \partial_{xx}$$
$$\mathcal{L}_{12} = -C^2 \partial_{zz} + 2S_2^2 \partial_{xz} + N^2 \partial_{xy}$$
$$\mathcal{L}_{21} = -C^2 \partial_{zz} - 2S_1^2 \partial_{yz} + N^2 \partial_{xy}$$
$$\mathcal{L}_{22} = F_2^2 \partial_{zz} + 2S_2^2 \partial_{yz} + N^2 \partial_{yy},$$

where, N^2, F_2^2, and S_2^2 are the same as in (7), and the coefficients F_1^2, S_1^2, and C^2 are functions of the stratification, shear, and confluence or difluence of the geostrophic flow as follows:

$$F_1^2 = f\left(f + \frac{\partial v_g}{\partial x}\right), \quad S_1^2 = f\frac{\partial v_g}{\partial z},$$
$$C^2 = -f\frac{\partial v_g}{\partial y} = f\frac{\partial u_g}{\partial x}. \quad (19)$$

As shown by *Hoskins and Draghici* [1977], equation (17) forced solely by the "geostrophic forcing" term I can be converted to the semigeostrophic omega equation.

Similar to the "geostrophic forcing", vertically varying frictional forces or laterally varying buoyancy sources or sinks (terms II and III respectively) will disrupt the thermal wind balance and hence drive ageostrophic circulations [*Eliassen*, 1951]. In addition, it is worth noticing that the equations governing nonlinear Ekman dynamics are encompassed in (17) when term II is balanced on the left hand side by terms with coefficients F_1^2 and F_2^2. For example, if the geostrophic flow were purely zonal and forced by a zonal wind stress τ_w^x, then the meridional Ekman transport $M_e^y = \int_{-\infty}^{0} v\,dz$ can be found by vertically integrating term II twice and dividing by F_2^2 (assuming that u_g is approximately constant through the thickness of the Ekman layer) yielding

$$M_e^y = -\frac{\tau_w^x}{\rho_o(f - \partial u_g/\partial y|_{z=0})}, \quad (20)$$

which is the solution of *Stern* [1965] and *Niiler* [1969] alluded to in section 2.4.2. Term III is important at wind-forced fronts where Ekman flow can advect dense water over light, generating convection and turbulent mixing of buoyancy at the front. The strength of this mixing is set by a wind-driven buoyancy flux

$$F_{wind}^B = \mathbf{M}_e \cdot \nabla_h b|_{z=0}, \quad (21)$$

which is a function of both the lateral buoyancy gradient and, as a result of (20), the vertical vorticity of the front. This mixing drives (through term III) frontogenetic ageostrophic secondary circulations [*Thomas and Lee*, 2005].

For submesoscale flows in the upper ocean, all three forcing terms of (17) are likely to play a role in driving an overturning circulation. In inferring the vertical velocities from high-resolution hydrographic and velocity observations in the upper ocean, it is often assumed that term I dominates the dynamics (i.e., the vertical velocity follows the omega equation) and that the ASC is well described by the quasi-geostrophic version of (17), that is with $F_1^2 \to f^2$, $F_2^2 \to f^2$, $C^2 \to 0$, $N^2 \to N^2$, $S_1^2 \to 0$, and $S_2^2 \to 0$ [e.g., *Pollard and Regier*, 1992; *Viúdez et al.*, 1996; *Rudnick*, 1996; *Allen and Smeed*, 1996; *Pinot et al.*, 1996; *Shearman et al.*, 1999; *Vélez-Belchí et al.*, 2005]. The validity of these assumptions comes into question for submesoscale flows, especially if the flows are exposed to atmospheric forcing. The QG omega equation does not work well for diagnosing vertical velocities when tested with a high-resolution numerical simulation that generates $\mathcal{O}(1)Ro$, as it misdiagnoses the position and sign of the most intense vertical velocities near the surface [*Mahadevan and Tandon*, 2006]. To account for departures from QG dynamics for high Rossby number flows, methods for diagnosing the vertical velocity from observations have been employed that use the inviscid, adiabatic semigeostrophic omega equation [*Naveira Garabato et al.*, 2001] or the generalized omega equation [*Pallàs Sanz and Viúdez*, 2005]. For forced upper ocean flows where friction and diabatic processes can be important, methods for inferring the overturning circulation using variants of (17) have been utilized [e.g., *Nagai et al.*, 2006; L. N. Thomas, C. M. Lee, and Y. Yoshikawa, The subpolar front of the Japan/East Sea II: Inverse method for determinig the frontal vertical circulation, submitted to *Journal of Physical Oceanography*, 2008].

To yield a unique solution for the ASC for given boundary conditions, (17) must be elliptic. Equation (17) switches from being elliptic to hyperbolic, and the semigeostrophic limit of balance is crossed when the three-dimensional semigeostrophic PV

$$q_{3D} = \left[\left(F_1^2 F_2^2 - C^4 \right) N^2 + 2C^2 S_1^2 S_2^2 - F_2^2 S_1^4 - F_1^2 S_2^4 \right] / f^3 \tag{22}$$

multiplied by the Coriolis parameter becomes negative, that is $fq_{3D} < 0$. *Hoskins* [1975] demonstrated that the semigeostrophic equations without friction and diabatic terms, that is (3)–(6), can be manipulated to form a conservation law for q_{3D}, $Dq_{3D}/Dt = 0$. Although q_{3D} is materially conserved in an inviscid adiabatic fluid similar to the Ertel PV given by (2), it differs conceptually from q in that its conservation depends on the approximations used in semigeostrophic theory [*Viúdez*, 2005]. Having said this, the material invariance of (22) is consequential, as it im-

plies that in an adiabatic inviscid geostrophic flow that initially has $fq_{3D} > 0$ everywhere, the semigeostrophic equations can be integrated forward for all times, as the semigeostrophic limit of balance can never be crossed. When friction or diabatic effects are present, this is not always the case, as both q_{3D} and the Ertel PV (2) can be driven to negative values under certain conditions, as described in the next section. In three-dimensional numerical simulations [e.g., *Mahadevan and Tandon*, 2006], the regions where the PV changes sign tend to coincide with the sites of most intense submesoscale activity and downwelling.

3.2. Frictional or Diabatic Modification of the Potential Vorticity

Changes in the PV arise from convergences or divergences of PV fluxes

$$\frac{\partial q}{\partial t} = -\nabla \cdot \left(\mathbf{u}q + \mathbf{J}^{na} \right), \tag{23}$$

where q is the Ertel PV given by (2) and

$$\mathbf{J}^{na} = \nabla b \times \mathbf{F} - \mathcal{D}(f\hat{k} + \nabla \times \mathbf{u}) \tag{24}$$

is the nonadvective PV flux [*Marshall and Nurser*, 1992]. *Thomas* [2005] shows that friction or diabatic processes, that is (18), acting at the sea surface will result in a reduction of the PV when

$$fJ_z^{na} = f\left[\nabla_h b \times \mathbf{F} \cdot \hat{k} - \mathcal{D}(f + \zeta) \right]\Big|_{z=0} > 0. \tag{25}$$

Destabilizing atmospheric buoyancy fluxes reduce the buoyancy in the upper ocean $\mathcal{D} = Db/Dt < 0$, which, for inertially stable flows, $f(f + \zeta) > 0$, results in a diabatic PV flux that satisfies condition specified by (25) and reduces the PV. Friction can either input or extract PV from the fluid depending on the orientation of the frictional force and the lateral buoyancy gradient. Down-front winds drive PV fluxes that meet the condition specified by (25) and, as illustrated in nonhydrostatic simulations, Ekman-driven convection ensues to mix the stratification and reduce the PV [*Thomas and Lee*, 2005]. Friction injects PV into the fluid when a baroclinic current is forced by upfront winds or during frictional spindown by vertical mixing of momentum [*Boccaletti et al.*, 2007; (L. N. Thomas and R. Ferrari, Friction, frontogenesis and the stratification of the surface mixed layer, submitted to *Journal of Physical Oceanography*, 2007)]. Regardless of whether friction increases or decreases the PV, frictional modification of PV at the sea surface is largest in regions with strong lateral buoyancy gradients. Consequently, submesoscale phenomena with their enhanced baroclinicity are especially prone to frictional PV change.

4. IMPLICATIONS

As highlighted in the previous section, in the submesoscale regime $Ro \sim Ri \sim 1$, vertical motions are enhanced. In this section, we discuss implications of the submesoscale vertical fluxes on budgets of the buoyancy, PV, and biogeochemical properties along with possible approaches to parameterization.

4.1. Effect on Mixed-Layer Stratification

The MLI described in section 2.3 release available potential energy from upper ocean fronts by inducing an upward buoyancy flux $\overline{w'b'}$ that tends to restratify the mixed layer, for example, Figure 2c. Submesoscale vertical buoyancy fluxes play an important role in the buoyancy budget of the mixed layer by competing with, or augmenting, buoyancy fluxes associated with small-scale turbulent motions. This may be expressed as

$$\partial_t \overline{b} + \nabla_h \cdot \left(\overline{\mathbf{u}'_h b'} + \overline{\mathbf{u}}_h \overline{b} \right) + \partial_z \left(\overline{w'b'} + \overline{w}\,\overline{b} \right) = \overline{\mathcal{D}}. \quad (26)$$

where the overline denotes a lateral average, primes denote the deviation from that average, and the turbulent effects are encompassed in the diabatic term $\overline{\mathcal{D}}$. The horizontal and vertical eddy fluxes of buoyancy on the left-hand side of (26) have contributions from both mesoscale and submesoscale flows. Submesoscale-resolving numerical experiments show that the vertical buoyancy flux $\overline{w'b'}$ is dominated by submesoscale eddies, while the lateral eddy buoyancy flux $\overline{\mathbf{u}'_h b'}$ is largely mesoscale [Capet et al., 2008a; Fox-Kemper et al., 2007]. While $\overline{w'b'}$ acts to redistribute buoyancy rather than generate a net input of buoyancy (as would be induced by an actual surface heat flux), expressing the vertical buoyancy flux in units of a heat flux gives the reader an appreciation for the strong restratifying capacity of submesoscale flows and motivates a parameterization. The submesoscale vertical buoyancy fluxes simulated in the numerical experiments of Capet et al. [2008a] had values equivalent to heat fluxes of $\mathcal{O}$ (100 W m^{-2}). However, Capet et al. [2008a] find that in their forced model simulations, the sea surface temperature is not very sensitive to increasing grid resolution because the increase in restratifying submesoscale vertical buoyancy fluxes, although significant, is offset by an equivalent change in $\overline{\mathcal{D}}$ that tends to destratify the mixed layer.

Fox-Kemper et al. [2007] propose parameterizing the submesoscale vertical buoyancy flux using an overturning streamfunction, that is $\overline{w'b'} = \Psi \overline{b}_y$ (y indicates the cross-front direction). Using both dynamical and scaling arguments, they suggest that Ψ takes the form

$$\Psi = C_e \mu(z) \frac{H^2 \overline{b}_y^{\overline{z}}}{|f|}, \quad (27)$$

where H is the mixed layer depth and $\overline{b}_y^{\overline{z}}$ is the highest resolution cross-front buoyancy gradient in a non-submesoscale permitting simulation vertically averaged over the mixed layer. In performing a suite of numerical experiments with varying H, f, and lateral buoyancy gradients, some forced by a diurnally varying heat flux, but none by wind stress, Fox-Kemper et al. [2007] showed that the buoyancy fluxes predicted by (27) compare well with those of the numerical solutions when the constant C_e is in the range of 0.04 to 0.06 and the vertical structure function is $\mu(z) = -4z(z + H)/H^2$. Owing to its strong dependence on H, implementation of (27) in a global circulation model has the greatest impact in regions with deep mixed layers and is found to alleviate the problem of overly deep mixed layers at high latitudes (B. Fox-Kemper, personal communication, 2007).

4.2. Frictionally Driven Eddy Potential Vorticity Fluxes

As described in section 2.4.2, down-front, wind-forced baroclinic flows form submesoscale circulations that advect PV and induce eddy PV fluxes. In two-dimensional simulations, submesoscale ASCs drive vertical PV fluxes that scale with the surface frictional PV flux, i.e.,

$$\overline{w'q'} \propto \overline{\nabla_h b \times \mathbf{F}} \cdot \hat{k}|_{z=0}, \quad (28)$$

where q is the full Ertel PV, the overline denotes a lateral average, and primes denote a deviation from that average [Thomas, 2005]. At a fully three-dimensional front, an analogous result holds: along-isopycnal eddy PV fluxes on isopycnal surfaces that outcrop at a front that is forced with down-front winds scale with the frictional PV flux averaged over the outcrop window of the isopycnal (see Figure 5). Thus,

$$\int \int_{z=z_b}^{z=z_t} q''v'' \, dz \, dx \sim \int \int_A \nabla_h b \times \mathbf{F} \cdot \hat{k}|_{z=0} \, dy \, dx \quad (29)$$

where z_t, z_b are the top and bottom depths of the isopycnal layer, A is the area of the outcrop, x is the along-front direction, and the double primes denote the deviation from the thickness-weighted isopycnal mean, i.e., $v'' = v - \overline{vh}^x/\overline{h}^x$, $h = z_t - z_b$ (the overline denotes a x average) [Thomas, 2007]. Results from a high-resolution numerical simulation demonstrating this are shown in Figure 5 to illustrate the correspondence between eddy and frictional PV fluxes. The details of the initial configuration and numerics of the simulation can be found in the work of Thomas [2007]. The key element of the experiment is that the front is initially aligned in the x di-

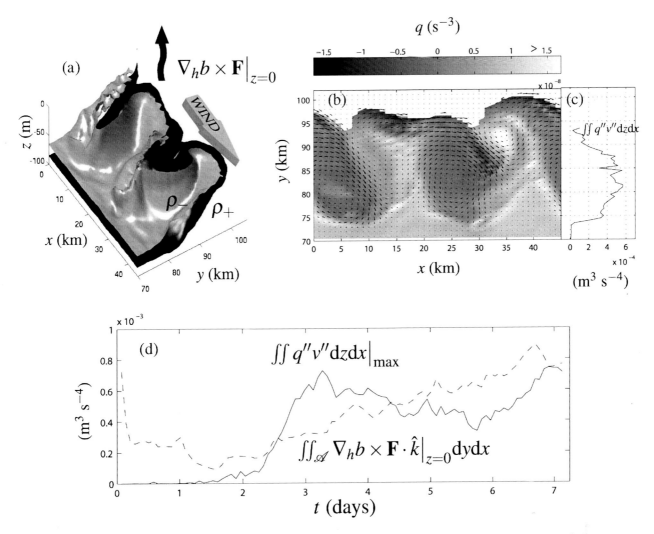

Figure 5. An example of submesoscale eddy PV fluxes driven by winds. (a) Down-front winds of strength 0.2 N m^{-2} forcing a front induce an upward frictional PV flux, triggering frontal instabilities that distort the bounding frontal isopycnal surfaces ρ_- (gray) and ρ_+ (black) ($\Delta\rho = \rho_+ - \rho_- = 0.2$ kg m^{-3}), shown here $t = 4.1$ days after the onset of the winds. (b) Isopycnal map of the PV (shades) and velocity (vectors) averaged in the vertical over the isopycnal layer shown in (a) illustrates the manner in which the instabilities subduct low PV from the surface while upwelling high PV from the pycnocline. (c) The correlation of the velocity and PV fields results in a net positive meridional eddy PV flux along the isopycnal layer $\iint q''v''dzdx > 0$. (d) A time series of the maximum value of the eddy PV flux with respect to y (solid) and the frictional PV flux integrated over the outcrop area (dashed) reveal that the two fluxes scale with one another after the initial growth of the instabilities, i.e. $t > 3$ days.

rection and is forced by a spatially uniform down-front wind that persists through the 7.3-day-long experiment (this differs from the experiment described by *Thomas* [2007] where the wind stress is shut off after 3.6 days). As can be seen in Figure 5b, the destruction of PV by the winds results in the formation of low, even negative, PV fluid where the frontal isopycnals outcrop into the surface Ekman layer ($z > -40$ m and 100 km $> y > 90$ km). Streamers of low PV surface fluid are subducted into the interior by the submesoscale instabilities, while high PV interior fluid is drawn to the surface so

that a positive eddy PV flux is induced at the y locations of the eddies. Once the eddies grow to finite amplitude, the relation shown in (29) holds, Figure 5d. In an experiment where the winds are turned off, (29) also holds, but the eddy PV flux, which persists after the winds cease, scales with the frictional PV flux at the time of the wind forcing, suggesting that there is a temporally nonlocal relation between the two fluxes [*Thomas*, 2007].

Isopycnal eddy PV fluxes are important because they drive an eddy-induced transport velocity (also known as a bolus velocity, which can be related to the residual circulation of the transformed Eulerian mean equations; *Plumb and Ferrari* [2005]) that can play an important role in the advection of tracers [*Greatbatch*, 1998]. When the tracer is the buoyancy, advection by the eddy transport velocity can result in restratification. Indeed, the process of mixed layer restratification by MLI described by Fox-Kemper et al. [2007] and summarized in section 4.1 is expressed in terms of the flattening of isopycnals by an overturning stream function associated with the residual circulation. In the wind-forced problem, the eddy transport velocity $v^* = \overline{vh}^x / \overline{h} - \overline{v}^x$ was found to scale with the eddy PV flux, i.e., $v^* \approx -\iint \overline{q''v''} dz dx / (\Delta b L_x f)$, where Δb is the buoyancy difference across the isopycnal layer and L_x is the zonal width of the domain [*Thomas*, 2007]. As the eddy PV flux also scaled with the surface frictional PV flux, i.e., (29), and the magnitude of frictional force is set by the strength of the wind stress τ and Ekman depth δ_e, that is $|\mathbf{F}| \sim \tau / (\rho_o \delta_e)$, it was shown that the eddy transport velocity scaled with the Ekman flow $v_e \sim \tau / (\rho f \delta_e)$ [*Thomas*, 2007]. This result suggests that a closure scheme for submesoscale eddy PV fluxes and residual circulation at wind-forced baroclinic currents would have a

dependence on the wind stress, which would be in contrast to parameterizations based on down-gradient PV fluxes that make no explicit reference to atmospheric forcing.

4.3. Effect on Biogeochemistry

Submesoscales processes affect ocean biogeochemistry in multiple ways. First, as submesoscale vertical velocities are substantially higher than their mesoscale counterparts, the vertical fluxes they support enhance phytoplankton productivity and air-sea gas exchange. Phytoplankton productivity, and the strength of the biological pump, is typically limited by the availability of nutrients that are plentiful at depth, and light, which is at the surface. Modeling studies of fronts show that the upward flux of nutrients into the euphotic layer and consequently, new production, intensify with progressively higher horizontal model resolution (increased from 40 to 10 km by *Mahadevan and Archer* [2000] and from 6 to 2 km by *Lévy et al.* [2001]), signifying that submesoscale processes are at play. Second, as vertical motion is strongly linked to horizontal strain at submesoscales, regions of high strain in which tracer filaments are stretched and stirred by lateral motions at the surface

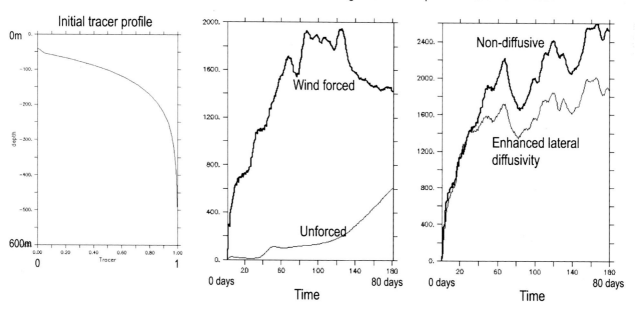

Figure 6. The domain-wide convergence of the vertical flux, or $\frac{1}{\tau}\overline{c'}|_{c'>0}$, which denotes the rate of nutrient uptake, is plotted as a function of time for two numerical experiments. In the middle panel, we compare the nutrient uptake rate for a model run that develops submesoscale structure due to a constant down-front wind, with one that is unforced and lacks submesoscale structure. In the right panel, we show another experiment where we compare the uptake rate between two cases: one in which the lateral diffusivity of the tracer is zero in the numerical model and a second in which it is enhanced to 100 m^2 s^{-1}. The lateral variability in the distribution of the tracer is shown in Figure 7. The left panel shows the initial tracer profile $c_0(z)$. In both experiments, the tracer concentrations are normalized between 0 and 1.

are also regions of intense up or downwelling. The vertical velocities, in turn, can affect the lateral stirring of a phytoplankton bloom and enhance phytoplankton productivity through nutrient supply in filaments as well as carbon export through the subduction of organic matter and the air-sea transfer of carbon dioxide. A further effect of submesoscale processes on ocean biogeochemistry is that they generate lateral gradients at scales of ~1 km and create spatial heterogeneity in property

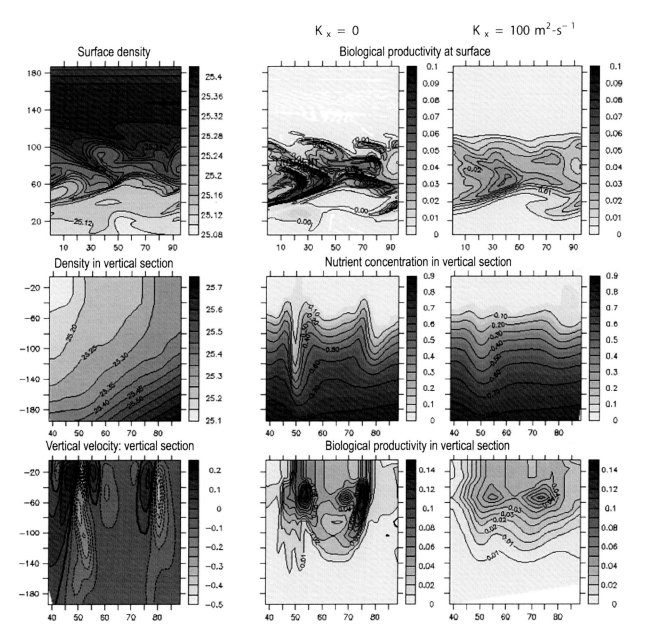

Figure 7. Modeling the nutrient and biological productivity due to submesoscale vertical fluxes shows a reduction in productivity when the lateral diffusivity of the biological tracers is increased to $100 \, m^2 s^{-1}$. The leftmost column shows the physical variables, density anomaly (in $kg \, m^{-3}$) and vertical velocity (in $mm \, s^{-1}$). These are the same for the two simulations compared. The middle column shows how the vertical submesoscale velocities supply nutrient and fuel biological productivity in the euphotic zone. The nutrient concentration is normalized between 0 and 1, and biological productivity is the rate of uptake of nutrient. The right column shows simulations with increased lateral diffusivity of the biogeochemical tracers. In this case, vertical nutrient transport is suppressed as pathways for transport are obliterated by diffusivity.

distributions. A number of biological processes are affected by distribution patchiness, as they are limited by access to nutrients, oxygen, or properties that are involved in a reaction.

Biogeochemical properties typically exhibit a large vertical concentration gradient because they are altered by processes that are depth-dependent but limited by the rate of vertical exchange. The ratio of the timescale of vertical advection to that of the processes altering the tracers (such as biological production) is an important parameter (termed the Damköhler number), which affects not only the biogeochemical tracer flux and the time-averaged vertical tracer profile but also the spatial heterogeneity of the tracer distribution in the upper ocean. Submesoscale processes increase lateral spatial heterogeneity of biogeochemical properties either by introducing submesoscale concentration anomalies due to vertical advection or by the lateral straining and drawing out of filaments. In the former case, an increase in the altering process (or reaction) timescale shifts the spatial variance towards larger scales [*Mahadevan and Campbell*, 2002]. In the latter, slower reaction timescales offer greater scope for filamentation and increased submsoscale spatial variance [*Abraham*, 1998].

We quantify the rate of vertical exchange in the upper ocean in numerical experiments by tracking the cumulative divergence of the vertical flux of a tracer c, with an initial vertical concentration profile $c_0(z)$. The profile $c_0(z)$ is chosen to represent the mean profile of a phytoplankton nutrient like nitrate; it is abundant at depth, but depleted from the surface, with a strong vertical gradient between 50 and 250 m (Figure 6). The rate of supply of the tracer for phytoplankton production, $-w\frac{\partial c}{\partial z}$, is the convergence of the vertical flux. Our numerical experiments are set within an east-west periodic domain, with a north-south density gradient representative of an upper ocean front. As the tracer is advected in the frontal flow field, we restore its concentration to the initial profile $c_0(z)$ with an e-folding timescale τ, such that

$$\frac{\partial c}{\partial t} + u\frac{\partial c}{\partial x} + v\frac{\partial c}{\partial y} + w\frac{\partial c}{\partial z} = -\frac{1}{\tau}(c - c_0(z)). \quad (30)$$

Faster restoring, that is smaller τ, maintains smaller values of $c - c_0(z)$, sharper vertical gradients, and thus higher vertical fluxes. When $c > c_0(z)$, the negative right-hand side term denotes the uptake of nutrient by phytoplankton. We quantify the convergence of vertical tracer flux over the entire domain by integrating the uptake of the nutrient-like tracer by phytoplankton. By writing $c = c_0(z) + c'$ and noting that the domain-wide average of c' (denoted by an overbar) is zero, that is $\overline{c'} = 0$, we get, for $\frac{\partial c}{\partial z} < 0$,

$$\overline{w^{(+)}\partial c/\partial z} = -\frac{1}{\tau}\overline{(c - c_0(z))}|_{c>c_0}, \quad (31)$$

where $w^{(+)}$ refers to positive (upward) w. Thus, the convergence of vertical flux, which is the rate of supply of nutrient, is calculated by evaluating the magnitude of the right hand side $\frac{1}{\tau}\overline{c'}|_{c'>0}$ of (31), specifically when $c' > 0$.

An enhancement of the rate of uptake of a nutrient-like tracer by wind-forced submesoscale instabilities can be seen in Figure 6 where we compare the vertical divergence of tracer flux from a numerical experiment in which submesoscale structure is generated by surface wind to a case without wind forcing and submesoscale structure. In these simulations, the mixed layer is 50 m deep. The tracer is restored to its initial profile $c_0(z)$ on a timescale $\tau = 3$ days. The convergence of vertical flux of tracer is the amount of tracer that is removed per time step to perform the described restoration of the tracer profile. By treating the tracers identically in both the wind-forced and unforced cases, the difference in the nutrient uptake can be attributed entirely to the differences in the vertical velocities between the model runs, which turns out to be substantial.

In another experiment (Figure 7), the lateral variability in tracer distribution introduced by submesoscale processes is suppressed by increasing the lateral diffusivity of the tracer. The result is that narrow pathways for vertical transport are obliterated, and vertical tracer flux is substantially diminished (Figure 6, right panel). When the tracer is considered to be nitrate and its vertical transport results in phytoplankton productivity, the lateral diffusion of tracer diminishes phytoplankton productivity in the model.

5. DISCUSSION

As described in section 4.1, restratification of the mixed layer by submesoscale flows is accomplished at the expense of the available potential energy (APE) stored in the larger scale baroclinic currents in which they form. This characteristic of submesoscale flows, which is a consequence of their strong secondary circulations, hints at their potentially important role in the energy budget of the ocean, a subject that is explored in this section.

The power spectrum of kinetic energy is often used to characterize the energy distribution in terms of lengthscales. While three-dimensional turbulence spectra show a characteristic −5/3 power law (or −5/3 slope in log-log space) and the cascade of energy to smaller scales, stratified quasi-geostrophic mesoscale flow (geostrophic turbulence) spectra have a −3 slope [e.g., *Vallis*, 2006] and a reverse energy cascade. With increasing grid resolution in a primitive equation model, *Capet et al.* [2008a] find the spectral slope transitions from −3 at mesoscale resolutions to −2 at submesoscales in the upper ocean (also seen in *Legg and McWilliams* [2001]). *Boccaletti et al.* [2007] and *Fox-Kemper et al.* [2007] show that the energy peak triggered by mixed layer instability is transferred to both larger and smaller scales. The reverse cascade is physically manifest through the evolution of mixed layer eddies into larger ones, while the forward cascade is intimately tied to frontogenesis and frontal instabilities.

Finite depth surface quasi-geostrophic (SQG) models have been used to explain the transition from −3 to −5/3 slope in atmospheric spectra derived from aircraft observations [*Tulloch and Smith*, 2006]. SQG has also been advanced by *Lapeyre et al.* [2006] as appropriate for $O(1)$ Rossby number flows with scales smaller than mesoscales. For the oceanic case, simulations by *Klein et al.* [2007] show a transition from −3 to −5/3 slope in the EKE spectrum from the interior to the upper ocean, with a reverse cascade at larger scales and a forward energy cascade at $O(1)$ km. As shown in section 3.1.2, the frontogenetic motion described by semi-geostrophic equations also produces a forward energy cascade, as smaller lengthscale structures arise as frontogenesis occurs. In contrast to the QG equations, the SG equations do not neglect advection by ageostrophic flow. (Capet et al., Mesoscale to submesoscale transition in the California Current System: Energy balance and flux, submitted to *Journal of Physical Oceanography*, 2007) show that although their contribution to the eddy kinetic energy (EKE) is small, the forward kinetic energy (KE) flux is entirely associated with advection by the horizontally divergent, ageostrophic component of the flow. Although the ageostrophic flow is weaker than the geostrophic flow, it plays a critical role in the forward cascade at small scales and in frontogenesis (Molemaker et al., Balanced and unbalanced routes to dissipation in an equilibrated Eady flow, submitted to *Journal of Fluid Mechanics*, 2007). It also follows that the vertical velocity distribution in the SQG and SG models is different and will likely result in very different vertical tracer fluxes. The SQG^{+1} extension model of *Hakim et al.* [2002] allows for cyclonic/anticyclonic asymmetry and an up/down asymmetry in the vertical velocity similar to the SG model because both include higher order contributions to the potential vorticity.

As discussed in section 3.1.2 and by *Andrews and Hoskins* [1978], the semigeostrophic frontal equation leads to flattening of the slope of the KE spectrum to −8/3 at the time of frontal collapse. *Boyd* [1992] shows that this slope is ephemeral, and as the front evolves into a discontinuity, the slope of the energy spectra would flatten further to −2, which is characteristic of jump discontinuities. Numerical experiments that are actively being forced, either by buoyancy loss [*Legg and McWilliams*, 2001] or by wind stress [*Capet et al.*, 2008a], also show a −2 slope for EKE, raising questions about the relative contributions of atmospheric forcing, frontogenesis, and APE release by instability in the energetics of submesoscale flows.

Many questions remain open with regard to submesoscales. The numerical experiments described in section 4.2 emphasized the role of wind forcing in driving submesoscale eddy PV fluxes. It remains to be seen if a similar result holds true for submesoscale motions driven by atmospheric buoyancy fluxes. Another question is whether there is a general relation between nonadvective PV fluxes (frictional or diabatic) induced by atmospheric forcing and eddy PV fluxes or whether the relation is specific to the case of down-front winds.

The cumulative effects of submesoscale processes on larger scales are beginning to be assessed in terms of mixed layer restratification (B. Fox-Kemper and R. Ferrari, Parameterization of mixed layer restratification. II: Prognosis and impact, submitted to *Journal of Physical Oceanography*, 2007), although it is not well understood how submesoscale dynamics affect the PV budget at the large scale. A review of the general ocean circulation energetics by *Wunsch and Ferrari* [2004] notes that the energy pathways from mesoscale eddies to internal waves are not well understood and have not been quantified. As it turns out that the limits of balance coincide with the growth of unbalanced instabilities at the submesoscale, it is possible that the energy pathway from mesoscale to internal waves is via unbalanced submesoscale instabilities [*Molemaker et al.*, 2005]. However, it remains to be shown whether this breakdown in balance leads to triggering energy into internal gravity waves or whether direct interactions between the balanced submesoscale flows and a preexisting internal wave field affects energy transfer from large to small scales [*Bühler and McIntyre*, 2005; K. L. Polzin, How Rossby waves break. Results from POLYMODE and the end of the enstrophy cascade, submitted to *Journal of Physical Oceanography*, 2006]. Whatever the mechanism, the downscale cascade of energy from submesoscales could have far-reaching implications on mixing and dissipation. This, plus the implications for vertical transport of buoyancy, momentum, and biogeochemical properties, makes it important to improve our understanding of submesoscale phenomena.

Acknowledgments. We would like to acknowledge support from the National Science Foundation contracts OCE-0549699

(L.T.), OCE-0612058 (L.T.), OCE-0623264 (A.T. & A.M.), OCE-0612154 (A.T.), and the National Oceanic and Atmospheric Administration contract NA05NOS4731206 (A.M.).

REFERENCES

Abraham, E. R. (1998), The generation of plankton patchiness by turbulent stirring, *Nature*, *391*, 577–580.

Allen, J. T., and D. Smeed (1996), Potential vorticity and vertical velocity at the Iceland-Faroes front, *J. Phys. Oceanogr.*, *26*, 2611–2634.

Andrews, D. G., and B. J. Hoskins (1978), Energy spectra predicted by semi-geostrophic theories of frontogenesis, *J. Atmos. Sci.*, *35*, 509–512.

Boccaletti, G., R. Ferrari, and B. Fox-Kemper (2007), Mixed layer instabilities and restratification, *J. Phys. Oceanogr.*, *37*, 2228–2250.

Boyd, J. P. (1992), The energy spectrum of fronts: Time evolution of shocks in Burgers' equation, *J. Atmos. Sci.*, *49*, 128–139.

Bühler, O., and M. E. McIntyre (2005), Wave capture and wave-vortex duality, *J. Fluid Mech.*, *534*, 67–95.

Capet, X., J. C. McWilliams, M. J. Molemaker, and A. F. Shchepetkin (in press), Mesoscale to submesoscale transition in the California Current System: Flow structure, eddy flux, and observational tests, *J. Phys. Oceanogr.*

Capet, X., J. C. McWilliams, M. J. Molemaker, and A. F. Shchepetkin (in press), Mesoscale to submesoscale transition in the California Current System: Frontal processes, *J. Phys. Oceanogr.*

Charney, J. G. (1948), On the scale of atmospheric motions, *Geophys. Publ. Oslo*, *17*(2), 1–17.

Charney, J. G. (1971), Geostrophic turbulence, *J. Atmos. Sci.*, *28*, 1087–1095.

Eliassen, A. (1948), The quasi-static equations of motion, *Geofys. Publikasjoner*, *17*(3).

Eliassen, A. (1951), Slow thermally or frictionally controlled meridional circulation in a circular vortex, *Astrophys. Norv.*, *5*, 19–60.

Emerson, S., P. Quay, D. Karl, C. Winn, L. Tupas, and M. Landry (1997), Experimental determination of the organic carbon flux from open-ocean surface waters, *Nature*, *389*, 951–954.

Flament, P., L. Armi, and L. Washburn (1985), The evolving structure of an upwelling filament, *J. Geophys. Res.*, *90*, 11,765–11,778.

Fox-Kemper, B., R. Ferrari, and R. Hallberg (in press), Parameterization of mixed layer restratification. I: Theory and validation, *J. Phys. Oceanogr.*

Gent, P. R., and J. C. McWilliams (1983), Consistent balanced models in bounded and periodic domains, *Dyn. Atmos. Oceans*, *7*, 67–93.

Gent, P. R., J. C. McWilliams, and C. Snyder (1994), Scaling analysis of curved fronts: validity of the balance equations and semigeostrophy, *J. Atmos. Sci.*, *51*, 160–163.

Greatbatch, R. J. (1998), Exploring the relationship between eddy-induced transport velocity, vertical momentum transfer, and the isopycnal flux of potential vorticity, *J. Phys. Oceanogr.*, *28*, 422–432.

Haine, T. W. N., and J. C. Marshall (1998), Gravitational, symmetric, and baroclinic instability of the ocean mixed layer, *J. Phys. Oceanogr.*, *28*, 634–658.

Hakim, G. J., and D. Keyser (2001), Canonical frontal circulation patterns in terms of Green's functions for the Sawyer-Eliassen equation, *Q. J. R. Meteorol. Soc.*, *127*, 1795–1814.

Hakim, G. J., C. Snyder, and D. J. Muraki (2002), A new surface model for cyclone–anticyclone asymmetry, *J. Atmos. Sci.*, *59*, 2405–2420.

Hallberg, R. (2003), The suitability of large-scale ocean models for adapting parameterizations of boundary mixing and a description of a refined bulk mixed layer model, in *Proceedings of the 2003 'Aha Huliko'a Hawaiian Winter Workshop*, University of Hawaii, pp. 187–203.

Hoskins, B. J. (1975), The geostrophic momentum approximation and the semi-geostrophic equations, *J. Atmos. Sci.*, *32*, 233–242.

Hoskins, B. J., and F. P. Bretherton (1972), Atmospheric frontogenesis models: Mathematical formulation and solution, *J. Atmos. Sci.*, *29*, 11–37.

Hoskins, B. J., and I. Draghici (1977), The forcing of ageostrophic motion according to the semi-geostrophic equations and in an isentropic coordinate model, *J. Atmos. Sci.*, *34*, 1859–1867.

Hoskins, B. J., I. Draghici, and H. C. Davies (1978), A new look at the ω-equation., *Q. J. R. Meteorol. Soc.*, *104*, 31–38.

Jenkins, W. J. (1988), Nitrate flux into the euphotic zone near Bermuda, *Nature*, *331*, 521–523.

Jenkins, W. J., and J. C. Goldman (1985), Seasonal oxygen cycling and primary production in the Sargasso Sea, *J. Mar. Res.*, *43*, 465–491.

Klein P., B. L. Hua, G. Lapeyre, X. Capet, S. Le Gentil, and H. Sasaki (in press), Upper ocean turbulence from high 3-D resolution simulations, *J. Phys. Oceanogr.*

Kunze, E. (2001), Vortical modes, in *Encyclopedia of Ocean Sciences*, edited by S. T. J. Steele and K. Turekian, pp. 3174–3178, Academic, London.

Kunze, E., and T. B. Stanford (1993), Submesoscale dynamics near a seamount, *J. Phys. Oceanogr.*, *23*, 2567–2601.

Lapeyre, G., P. Klein, and B. L. Hua (2006), Oceanic restratification forced by surface frontogenesis, *J. Phys. Oceanogr.*, *36*, 1577–1590.

Lee, C. M., E. A. D'Asaro, and R. Harcourt (2006a), Mixed layer restratification: Early results from the AESOP program, *Eos Trans. AGU*, *87*(52), *Fall Meet. Suppl., Abstract OS51E-04.*

Lee, C. M., L. N. Thomas, and Y. Yoshikawa (2006b), Intermediate water formation at the Japan/East Sea subpolar front, *Oceanography*, *19*, 110–121.

Legg, S., and J. C. McWilliams (2001), Convective modifications of a geostrophic eddy field, *J. Phys. Oceanogr.*, *31*, 874–891.

Legg, S., J. C. McWilliams, and J. Gao (1998), Localization of deep ocean convection by a mesoscale eddy, *J. Phys. Oceanogr.*, *28*, 944–970.

Lévy, M., P. Klein, and A. M. Treguier (2001), Impacts of submesoscale physics on production and subduction of phytoplankton in an oligotrophic regime, *J. Mar. Res.*, *59*, 535–565.

Mahadevan, A. (2006), Modeling vertical motion at ocean fronts:

Are nonhydrostatic effects relevant at submesoscales?, *Ocean Model.*, *14*, 222–240.

Mahadevan, A., and D. Archer (2000), Modeling the impact of fronts and mesoscale circulation on the nutrient supply and biogeochemistry of the upper ocean, *J. Geophys. Res.*, *105* (C1), 1209–1225.

Mahadevan, A., and J. W. Campbell (2002), Biogeochemical patchiness at the sea surface, *Geophys. Res. Lett.*, *29* (19), 1926, doi:10.10292001GL014116.

Mahadevan, A., and A. Tandon (2006), An analysis of mechanisms for submesoscale vertical motion at ocean fronts, *Ocean Model.*, *14 (3-4)*, 241–256.

Maier-Reimer, E. (1993), Geochemical cycles in an ocean general circulation model, *Global Biogeochem. Cycles*, *7*, 645–678.

Marshall, J. C., and A. J. G. Nurser (1992), Fluid dynamics of oceanic thermocline ventilation., *J. Phys. Oceanogr.*, *22*, 583–595.

Martin, A. P., and P. Pondaven (2003), On estimates for the vertical nitrate flux due to eddy-pumping, *J. Geophys. Res.*, *108* (C11), 3359, doi: 10.1029/2003JC001841.

McGillicuddy, D. J., Jr., and A. R. Robinson (1997), Eddy induced nutrient supply and new production in the Sargasso Sea, *Deep-Sea Res.*, *44* (8), 1427–1450.

McGillicuddy, D. J., Jr., A. R. Robinson, D. A. Siegel, H. W. Jannasch, R. Johnson, T. D. Dickey, J. McNeil, A. F. Michaels, and A. H. Knap (1998), Influence of mesoscale eddies on new production in the Sargasso Sea, *Nature*, *394*, 263–266.

McKiver, W. J., and D. G. Dritschel (2008), Balance in non-hydrostatic rotating stratified turbulence, *J. Fluid Mech.*, *596*, 201–219.

McWilliams, J. C. (1985), Submesoscale, coherent vortices in the ocean, *Rev. Geophys.*, *23*, 165–182.

McWilliams, J. C. (2003), Diagnostic force balance and its limits, in *Nonlinear Processes in Geophysical Fluid Dynamics*, edited by O. V. Fuentes, J. Sheinbaum, and J. Ochoa, pp. 287–304, Kluwer, Dordrecht.

McWilliams, J. C., and P. R. Gent (1980), Intermediate models of planetary circulations in the atmosphere and ocean, *J. Atmos. Sci.*, *37*, 1657–1678.

McWilliams, J. C., M. J. Molemaker, and I. Yavneh (2001), From stirring to mixing of momentum: Cascades from balanced flows to dissipation, in *Proceedings of the 12th 'Aha Huliko'a Hawaiian Winter Workshop*, University of Hawaii, pp. 59–66.

Molemaker, M. J., J. C. McWilliams, and I. Yavneh (2005), Baroclinic instability and loss of balance, *J. Phys. Oceanogr.*, *35*, 1505–1517.

Munk, W., L. Armi, K. Fischer, and F. Zachariasen (2000), Spirals on the sea, *Proc. R. Soc. Lond. A*, *456*, 1217–1280.

Nagai, T., A. Tandon, and D. L. Rudnick (2006), Two-dimensional ageostrophic secondary circulation at ocean fronts due to vertical mixing and large-scale deformation, *J. Geophys. Res.*, *111*, C09038, doi:10.1029/2005JC002964.

Najjar, R. G., J. L. Sarmiento, and J. R. Toggweiler (1992), Downward transport and fate of organic matter in the ocean: Simulations with a general circulation model, *Global Biogeochem. Cycles*, *6*, 45–76.

Naveira Garabato, A. C., J. T. Allen, H. Leach, V. H. Strass, and R.

T. Pollard (2001), Mesoscale subduction at the Antarctic Polar Front driven by baroclinic instability, *J. Phys. Oceanogr.*, *31*, 2087–2107.

Niiler, P. P. (1969), On the Ekman divergence in an oceanic jet, *J. Geophys. Res.*, *74*, 7048–7052.

Oschlies, A. (2002a), Can eddies make ocean deserts bloom?, *Glob. Biogeochem. Cycles*, *16*, 1106, doi:10.1029/2001GB001830.

Oschlies, A. (2002b), Improved representation of upper-ocean dynamics and mixed layer depths in a model of the North Atlantic on switching from eddy-permitting to eddy-resolving grid resolution, *J. Phys. Oceanogr.*, *32*, 2277–2298.

Oschlies, A. (2007), Eddies and upper ocean nutrient supply, this volume.

Pallàs Sanz, E., and A. Viúdez (2005), Diagnosing mesoscale vertical motion from horizontal velocity and density data, *J. Phys. Oceanogr.*, *35*, 1744–1762.

Pinot, J. M., J. Tintoré, and D. P. Wang (1996), A study of the omega equation for diagnosing vertical motions at ocean fronts, *J. Mar. Res.*, *54*, 239–259.

Platt, T., and W. G. Harrison (1985), Biogenic fluxes of carbon and oxygen in the ocean, *Nature*, *318*, 55–58.

Plumb, R. A., and R. Ferrari (2005), Transformed Eulerian-mean theory. Part I: Nonquasigeostrophic theory for eddies on a zonal-mean flow, *J. Phys. Oceanogr.*, *35*, 165–174.

Pollard, R. T., and L. Regier (1992), Vorticity and vertical circulation at an ocean front, *J. Phys. Oceanogr.*, *22*, 609–625.

Polzin, K. L., and R. Ferrari (2003), Isopycnal dispersion in NATRE, *J. Phys. Oceanogr.*, *34*, 247–257.

Rudnick, D. L. (1996), Intensive surveys of the Azores front. Inferring the geostrophic and vertical velocity fields, *J. Geophys. Res.*, *101* (C7), 16,291–16,303.

Rudnick, D. L. (2001), On the skewness of vorticity in the upper ocean, *Geophys. Res. Lett.*, *28*, 2045–2048.

Rudnick, D. L., and J. R. Luyten (1996), Intensive surveys of the Azores front 2. Tracers and dynamics, *J. Geophys. Res.*, *101*, 923–939.

Rudnick, D., L. and R. Ferrari (1999), Compensation of horizontal temperature and salinity gradients in the ocean mixed layer, *Science*, *283*, 526–529.

Sawyer, J. S. (1956), The vertical circulation at meteorological fronts and its relation to frontogenesis, *Proc. R. Soc. Lond. A*, *234*, 346–362.

Shay, L. K., T. M. Cook, and P. E. An (2003), Submesoscale coastal ocean flows detected by very high frequency radar and autonomous underwater vehicles, *J. Atmos. Oceanic Technol.*, *20*, 1583–1599.

Shearman, R. K., J. M. Barth, and P. M. Kosro (1999), Diagnosis of three-dimensional circulation associated with mesoscale motion in the California current, *J. Phys. Oceanogr.*, *29*, 651–670.

Stern, M. E. (1965), Interaction of a uniform wind stress with a geostrophic vortex., *Deep-Sea Res.*, *12*, 355–367.

Stone, P. H. (1966), On non-geostrophic baroclinic stability, *J. Atmos. Sci.*, *23*, 390–400.

Stone, P. H. (1970), On non-geostrophic baroclinic stability: Part II, *J. Atmos. Sci.*, *27*, 721–727.

Stone, P. H. (1971), Baroclinic instability under non-hydrostatic conditions, *J. Fluid Mech.*, *45*, 659–671.

Sundermeyer, M. A., and M. P. Lelong (2005), Numerical simulations of lateral dispersion by the relaxation of diapycnal mixing events, *J. Phys. Oceanogr.*, *35*, 2368–2386.

Tandon, A., and C. Garrett (1994), Mixed layer restratification due to a horizontal density gradient, *J. Phys. Oceanogr.*, *24*, 1419–1424.

Thomas, L. N. (2005), Destruction of potential vorticity by winds, *J. Phys. Oceanogr.*, *35*, 2457–2466.

Thomas, L. N. (2007), Formation of intrathermocline eddies at ocean fronts by wind-driven destruction of potential vorticity, *Dyn. Atmos. Oceans*, accepted.

Thomas, L. N., and C. M. Lee (2005), Intensification of ocean fronts by down-front winds, *J. Phys. Oceanogr.*, *35*, 1086–1102.

Thomas, L. N., and P. B. Rhines (2002), Nonlinear stratified spin-up, *J. Fluid Mech.*, *473*, 211–244.

Tulloch, R., and K. S. Smith (2006), A theory for the atmospheric energy spectrum: Depth limited temperature anomalies at the tropopause, *Proc. Natl. Acad. Sci. U.S.A.*, *103* (40), 14,690–14,694.

Vallis, G. K. (2006), *Atmospheric and Oceanic Fluid Dynamics: Fundamentals and Large-scale Circulation*, 745 pp., Cambridge University Press, Cambridge.

Vélez-Belchí, P., M. Vargas-Yáñez, and J. Tintoré (2005), Observation of a western Alborán gyre migration event, *Prog. Oceanogr.*, *66*, 190–210.

Viúdez, A. (2005), The vorticity-velocity gradient cofactor tensor and the material invariant of the semigeostrophic theory, *J. Phys. Oceanogr.*, *62*, 2294–2301.

Viúdez, A., and D. G. Dritschel (2004), Optimal potential vorticity balance of geophysical flows, *J. Fluid Mech.*, *521*, 343–352.

Viúdez, A., J. Tintoré, and R. Haney (1996), Circulation in the Alboran Sea as determined by quasi-synoptic hydrographic observations. Part I: Three-dimensional structures of the two anticyclonic gyres, *J. Phys. Oceanogr.*, *26*, 684–705.

Waite, M. L., and P. Bartello (2006), The transition from geostrophic to stratified turbulence, *J. Fluid Mech.*, *568*, 89–108.

Wunsch, C., and R. Ferrari (2004), Vertical mixing, energy and the general circulation of the oceans, *Annu. Rev. Fluid Mech.*, *36*, 281–314.

Yoshikawa, Y., K. Akitomo, and T. Awaji (2001), Formation process of intermediate water in baroclinic current under cooling, *J. Geophys. Res.*, *106*, 1033–1051.

Amala Madevan, Department of Earth Sciences, Boston University, 685 Commonwealth Avenue, Boston, MA 02215, USA. (amala@bu.edu)

Amit Tandon, Physics Department and Department of Estuarine and Ocean Sciences, University of Massachusetts, Dartmouth, 285 Old Westport Road, North Dartmouth, MA 02747, USA. (atandon@umassd.edu)

Leif N. Thomas, Department of Physical Oceanography, Woods Hole Oceanographic Institution, MS 21, Woods Hole, MA 02543, USA. (lthomas@whoi.edu)

Gulf Stream Separation in Numerical Ocean Models

Center for Ocean-Atmospheric Prediction Studies, Florida State University, Tallahassee, Florida, USA

Atmospheric, Oceanic and Planetary Physics, University of Oxford, Oxford, UK

This chapter summarizes our present knowledge of Gulf Stream separation in numerical ocean models. High horizontal resolution ocean numerical models are now capable of simulating quite realistically the separation and path of the Gulf Stream, and significant advances have been made in the last decade in our understanding of western boundary current separation. However, the Gulf Stream separation in numerical models continues to be a challenge because it remains very sensitive to the choices made for subgrid scale parameterizations.

1. INTRODUCTION

An accurate depiction of currents and associated features, such as mesoscale eddies, fronts, and jets, is essential to any eddy-resolving ocean model when modeling the global ocean circulation. Western boundary currents are an integral part of the wind-driven circulation, and the most conspicuous ones, such as the Gulf Stream and Kuroshio, carry large amounts of warm water from the tropics toward the northern latitudes. After separation, the western boundary currents are subject to intense air–sea interactions that translate into large heat loss (Plate 1).

One of the largest biases in current global climate models is the misrepresentation of the fluxes at the air–sea interface. This is caused by the large sea surface temperature errors (and therefore heat fluxes) that result from western boundary currents that usually overshoot the observed separation latitude [*Kiehl and Gent*, 2004]. The horizontal grid spacing used in the ocean component of these global climate models is quite coarse (on the order of 1°) when compared to state-of-the-art ocean-only global models (on the order of 1/10°).

There is a considerable improvement in the numerical representation of western boundary current separation when the horizontal grid spacing is approximately 10 km or less. As stated by *Bryan et al.* [2007], it has often been argued that this regime shift may result from the fact that the first baroclinic Rossby radius of deformation is actually resolved throughout most of the domain, thereby providing a good representation of baroclinic instability processes, but it may also be at this resolution that a critical Reynolds number is exceeded for separation to occur correctly [*Dengg*, 1993].

Identifying the dynamics responsible for western boundary current separation has been a long-standing challenge. It is fair to say that a proper western boundary current separation in a numerical model is the result of many contributing factors and that the separation mechanism remains very sensitive to choices made in the numerical model for subgrid scale parameterizations. There is yet no single recipe that would guarantee a correct separation of all western boundary currents in a global model. Much of the focus in the literature has been on achieving a proper Gulf Stream separation [e.g., *Dengg et al.*, 1996], but each western boundary current (Gulf Stream, Kuroshio, Brazil/Malvinas, Agulhas, East Australian, etc.) presents its own challenge and will respond differently to the chosen numerical algorithms and forcing functions.

In this chapter, we review our current knowledge of Gulf Stream separation in numerical ocean models. Since the

Ocean Modeling in an Eddying Regime
Geophysical Monograph Series 177
Copyright 2008 by the American Geophysical Union.
10.1029/177GM05

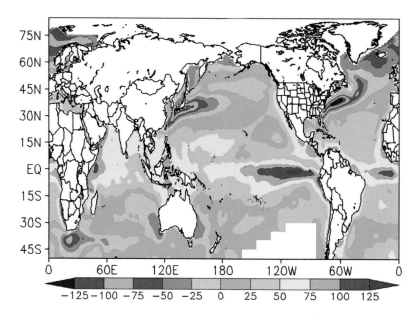

Plate 1. Annual mean ocean to atmosphere surface heat flux from the National Oceanography Centre, Southampton, UK. The units are in W/m². Note the ocean heat losses in the western boundary currents (in blue).

last major review of Gulf Stream separation by *Dengg et al.* [1996], there have been significant advances in our understanding of the influences of coastline geometry, bottom topography, inertia, and the deep western boundary current (DWBC) on the Gulf Stream separation, and also on the influence of boundary conditions as implemented in numerical ocean models. However, the most significant development has been the advent of the higher-resolution numerical ocean models capable of more realistically simulating the separation and path of the Gulf Stream.

The layout of this chapter is as follows. First, in Section 2, a brief overview of the literature on Gulf Stream separation mechanisms is presented with an extensive list of references. In Section 3, the relevance of several of these mechanisms to eddy-resolving models is discussed. Section 4 focuses on the representation of western boundary currents in state-of-the-art eddy-resolving ocean general circulation models (OGCMs) and the difficulties still faced by numerical modelers despite the improvements that result from the use of eddy-resolving grid spacing. Section 5 provides a brief discussion on the validation of the modeled boundary currents against observations.

2. PRIMARY PHYSICAL MECHANISMS

The persistence of unrealistic Gulf Stream separation in numerical models has prompted many theories of possible mechanisms influencing the separation of a western boundary current from the coast. The early linear frictional models [e.g., *Stommel*, 1948; *Munk*, 1950] suggest that separation takes place as a result of the change in sign of the wind stress curl. This theory is further supported by the fact that the observed mean path of the Gulf Stream roughly overlies the zero wind stress curl line (ZWCL). The ZWCL, however, shows considerable seasonal variation [e.g., *Isemer and Hasse*, 1987], while the point of separation shows remarkable consistency. Inclusion of the nonlinear terms and associated boundary conditions (no-slip or free-slip) induces considerable variations in the separation latitude [*Blandford*, 1971; *Moro*, 1988; *Verron and Le Provost*, 1991; *Cessi*, 1991; *Chassignet and Gent*, 1991; *Haidvogel et al.*, 1992; *Verron and Blayo*, 1996; *Adcroft and Marshall*, 1998]. Other mechanisms have been put forth as being important in the separation process, such as potential vorticity crisis [*Kamenkovich*, 1966; *Ierley and Ruehr*, 1986; *Ierley*, 1987, *Cessi et al.*, 1987; *Ierley and Young*, 1988; *Cessi*, 1990; *Kiss*, 2002], a region of adverse pressure gradient [*Haidvogel et al.*, 1992; *Baines and Hugues*, 1996; *Marshall and Tansley*, 2001; *Kiss*, 2002], collision with another western boundary current [*Cessi*, 1991; *Agra and Nof*, 1993], outcropping of isopycnals [*Parsons*, 1969; *Kamenkovich and*

Reznik, 1972; *Veronis*, 1973; *Moore and Niiler*, 1974; *Anderson and Moore*, 1979; *Ou and de Ruijter*, 1986; *Huang*, 1987; *Huang and Flierl*, 1987; *Gangopadhyay et al.*, 1992; *Chassignet and Bleck*, 1993; *Chassignet*, 1995], interaction with the DWBC [*Thompson and Schmitz*, 1989; *Spall*, 1996a, 1996b; *Tansley and Marshall*, 2000], surface cooling [*Veronis*, 1976, 1978; *Pedlosky*, 1987; *Nurser and Williams*, 1990; *Ezer and Mellor*, 1992; *Chassignet et al.*, 1995], and multiple equilibria [*Jiang et al.*, 1995; *Nauw et al.*, 2004]. Most of the cited studies, however, do not include any coastline geometry or bottom topography. Separation can be influenced by a change in coastline orientation or by a change in bottom topography [*Warren*, 1963, *Greenspan*, 1963; *Pedlosky*, 1965; *Kamenkovich and Reznik*, 1972; *Smith and Fandry*, 1976; *Stern and Whitehead*, 1990; *Spitz and Nof*, 1991; *Salmon*, 1992; *Dengg*, 1993; *Salmon*, 1994; *Thompson*, 1995; *Myers et al.*, 1996; *Özgökmen et al.*, 1997; *Stern*, 1998; *Tansley and Marshall*, 2000, 2001; *Munday and Marshall*, 2005]. Eddy–topography interactions have also been surmised to play a role in the separation process [*Holloway*, 1992; *Cherniawsky and Holloway*, 1993; *Eby and Holloway*, 1994; *Hurlburt et al.*, 2008; *Hulburt and Hogan*, 2008].

3. PROCESS STUDY EXPERIMENTS

In this section, we review several process study experiments that are particularly relevant to the discussion of Gulf Stream separation in realistic eddy-resolving ocean models.

3.1. Wind Forcing Boundary Condition

In the ocean, the wind stress acts over the surface Ekman layer, and the wind momentum is distributed within the surface mixed layer. Thus, the depth over which the wind forcing is felt by the ocean is space- and time-dependent. Figure 1 illustrates the sensitivity of the separation latitude to the depth over which the momentum of the wind is distributed by displaying the time average upper nondivergent streamfunction of four eddy-resolving *adiabatic* numerical models configured in a rectangular ocean basin and driven by a meridionally symmetric zonal wind stress.

The approximations made to obtain the quasi-geostrophic equations imply a symmetric response (Figure 1a) when a symmetric forcing is specified. As soon as the latter is relaxed, the midlatitude jet exhibits asymmetries [*Verron and Le Provost*, 1991]. The main difference between the isopycnic- and depth-coordinate models (separation south and north of the ZWCL, respectively; Figures 1c and d) results from the way the wind forcing is applied as a body force over the upper layer or at the first level. In the isopycnic coordinate model, the upper layer interface is shallower in the subpolar

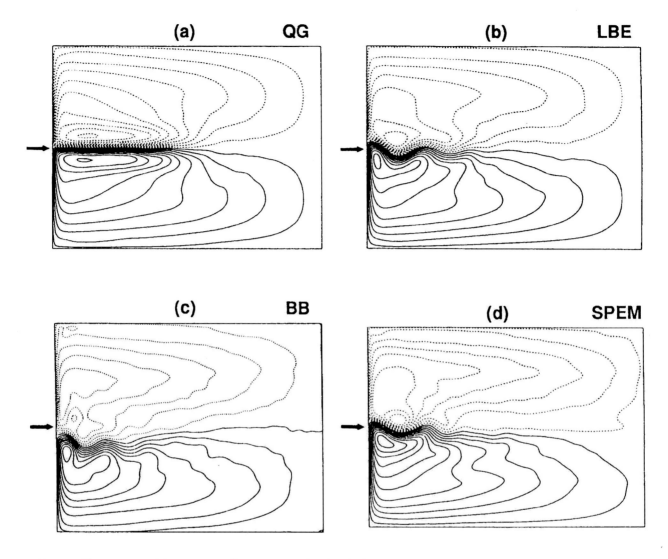

Figure 1. Time average of the upper non-divergent streamfunction for four numerical models: (a) quasi-geostrophic equations, (b) linear balance equations, (c) primitive equations with isopycnic coordinates, and (d) primitive equations with depth coordinates. The *arrow* points to the zero wind stress curl line (ZWCL). For more details, the reader is referred to *Chassignet and Gent* [1991].

gyre and deeper in the subtropical gyre [*Huang*, 1987]. This leads to stronger flows in the subpolar gyre, weaker flows in the subtropical gyre, and therefore the tendency for northward separation due to the Coriolis force (clearly visible in Figures 1b and d for the linear balance equations and depth-coordinate simulations) is then overcome by the tendency for a southward separation (Figure 1c) caused by a stronger subpolar gyre circulation.

Depth variations in the upper layer interface of the isopycnic coordinate model can, in some cases, be large enough for outcropping to occur and move the separation latitude further south [*Parsons*, 1969; *Veronis*, 1973; *Huang*, 1987;

Chassignet and Bleck, 1993]. In the absence of thermodynamics, the amount of fluid in the upper layer is fixed, and this amount, together with the strength of the wind stress, dictates the outcrop location and size. In reality, the mixed layer depth over which the wind stress transfers momentum is deeper in the subpolar gyre than in the subtropical gyre, and a purely adiabatic model cannot capture this distribution. Furthermore, as discussed by *Pedlosky* [1987], *Nurser and Williams* [1990], and *Chassignet et al.* [1995], the presence of diabatic cross-isopycnal flows at the outcrop affects the nature of the solution, and the separation is not required anymore to be south of the ZWCL. In summary, as stated by

Veronis [1978], the intensity of the thermal driving, as well as the wind stress, has to be taken into account in determining the separation latitude of western boundary currents.

3.2. Lateral Boundary Condition

The choice of lateral boundary condition in a numerical model has a profound influence on the boundary current separation in a rectangular ocean basin [*Haidvogel et al.*, 1992; *Verron and Blayo*, 1996]. As stated by *Adcroft and Marshall* [1998], while the appropriate boundary condition for a continuum fluid is no-slip (zero velocity on the boundary), it is less clear that no-slip is appropriate for a finite-resolution ocean model in which the boundary current is barely resolved by the numerical grid. Alternative slippery boundary conditions have been proposed: the free-slip condition is the most widespread, in which the tangential shear at the boundary vanishes; hyper-slip and super-slip conditions have also been proposed to enable advection of vorticity along coastlines (see *Pedlosky* [1996] for a review).

The sensitivity of the separation latitude to lateral boundary conditions where the tangential stress is proportional to the tangential velocity is illustrated in Figure 2 for a quasi-geostrophic numerical model configured in a rectangular ocean basin and driven by a zonally symmetric wind stress. The constant of proportionality (α) has limiting values of zero

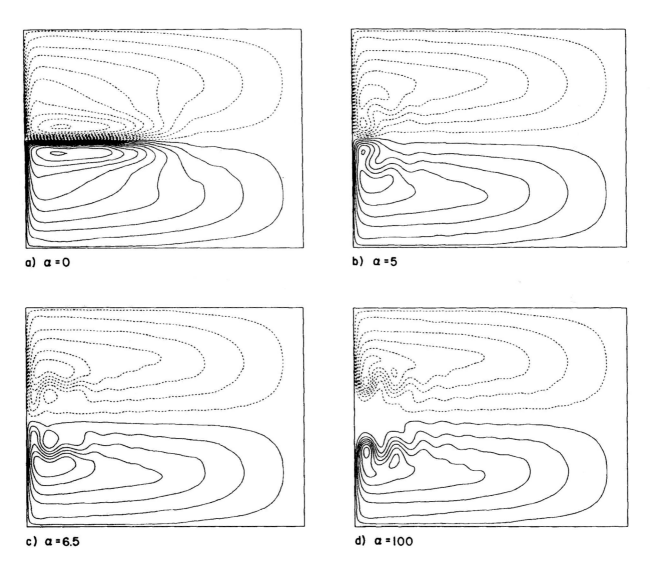

a) $\alpha = 0$

b) $\alpha = 5$

c) $\alpha = 6.5$

d) $\alpha = 100$

Figure 2. Time average of the upper streamfunction with (a) free-slip, (b, c) partial-slip, and (d) no-slip boundary conditions with symmetric wind stress. For more details, the reader is referred to *Haidvogel et al.* [1992].

and infinity, corresponding to free-slip and no-slip conditions, respectively. Significant changes in the time-mean behavior of the solution are observed to occur with increasing α. These changes include a gradual retreat of the separation points of the western boundary currents in the subtropical and subpolar gyres. The difference between free-slip and no-slip boundary conditions is further illustrated in Figure 3 for two experiments identical to Figures 2a and d, except for uniform westerly winds in the northern half of the basin, i.e., zero wind stress curl. The time-mean field for the no-slip experiment (Figure 3b) is virtually identical in the subtropical gyre when compared to the symmetric wind forcing case (Figure 2d). The separation point in the free-slip case (Figure 3a) does move

considerably northward in the absence of a colliding northern gyre boundary current. *Haidvogel et al.* [1992] found the separation to be associated with the occurrence of an adverse value of the higher-order pressure gradient term in the time-mean momentum budget just upstream of the separation point.

The distinction between free-slip and no-slip boundary conditions can become blurred in models when the coastline is irregular and is not oriented north–south as in the above configurations. Coastlines in numerical ocean models are oriented at various finite angles to the model grid and are usually replaced in finite difference models by a piecewise-constant approximation in which the model coastline is everywhere aligned with the model grid. *Adcroft and Marshall* [1998] studied the consequences of this piecewise-constant approximation by rotating the numerical grid at various finite angles to the physical coastline. They demonstrated that piecewise-constant coastlines exert a spurious form stress on model boundary currents, which depends on both the implementation of the lateral boundary condition and the form of the viscous stress tensor. This drag can, in some cases, introduce premature boundary current separation.

The no-slip or free-slip boundary conditions are often implemented in ocean numerical models using "ghost points" located half a grid point outside the model domain (Figure 4a). Along a no-slip boundary, the velocity at the ghost point is set equal and opposite to that of the interior value, whereas along a free-slip boundary, the ghost velocity is set equal to the interior value. There is, however, no need for ghost points when the model variables are staggered in the form of an Arakawa C-grid (used by most eddy-resolving models) and the latter is rotated 45° (Figure 4b). In the latter, the boundary conditions are identical for no-slip and free-slip, which contradicts the use of different boundary conditions. Furthermore, *Adcroft and Marshall* [1998] showed that the net viscous stress is then underestimated by a factor $\sqrt{2}$ for no-slip and finite (instead of zero) for free-slip.

These spurious form stresses are present even if the boundary layer is resolved by several grid points, but the stresses can be reduced (and, in some cases, eliminated) when the boundary condition is directly applied to the viscous stress tensor in a vorticity–divergence form. Similar results are obtained for alternative finite-difference methods (energy conservation, variables staggered on an Arakawa B-grid, etc.). The impacts of different combinations of advection schemes and stress tensors have been studied further by *Dupont et al.* [2003].

Adcroft and Marshall [1998] conclude by stating that ocean modelers need to reconsider the traditional treatment of coastlines and that the ultimate goal should be to develop an ocean model in which the circulation is insensitive to the orientation of the numerical grid. Ultimately, unstructured meshes hold significant promise in this respect [*Piggott et al.*, 2008].

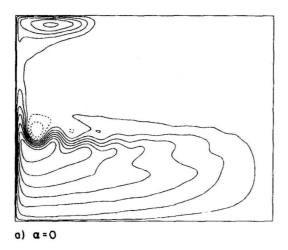

a) $\alpha = 0$

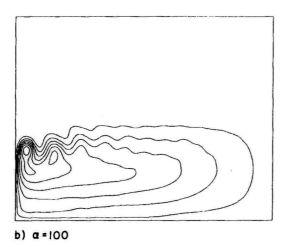

b) $\alpha = 100$

Figure 3. Time average of the upper streamfunction with (a) free-slip and (b) no-slip boundary conditions with asymmetric wind stress. For more details, the reader is referred to *Haidvogel et al.* [1992].

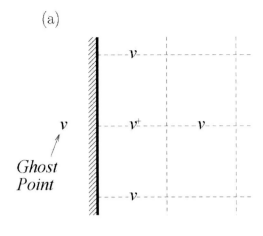

(a)

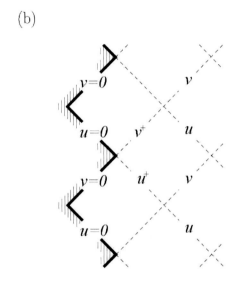

(b)

Figure 4. Schematic showing the distribution of velocity points adjacent to a north–south boundary on (a) an unrotated grid and (b) an Arakawa C-grid rotated at 45°. For more details, the reader is referred to *Adcroft and Marshall* [1998].

3.3. Coastlines, Bottom Topography, and Inertia

The importance of coastline orientation and bottom topography on western boundary current separation has been investigated over the years in some detail, especially for the Gulf Stream. In the North Atlantic, the western boundary current, after flowing over the shallow continental shelf (water depths of less than 1,000 m) between the Florida Straits and Cape Hatteras, separates by crossing the continental rise where the bottom topography drops sharply to 5,000 m within a few hundred kilometers. From a potential vorticity analysis combined with observational considerations, *Warren* [1963]

inferred that the topography of the continental rise was of significant importance in controlling the Gulf Stream's path. In a similar study, *Greenspan* [1963] also argued that bottom topography exerts a considerable influence on the separation of the Gulf Stream from the coastline and on its subsequent meander pattern. In a barotropic numerical study, *Holland* [1967] concluded that the separation of the Gulf Stream, meandering, and transport are strongly related to topographic effects. In a later study, *Holland* [1973] demonstrated that it was essential to include the effect of stratification when investigating the influence of bottom topography, as the bottom torque can be highly modified by the baroclinicity of the flow [see also *Myers et al.*, 1996]. The impact of a continental slope on the dynamics of a western boundary current was also addressed by *Salmon* [1992, 1994] by integrating analytically and numerically the planetary geostrophic equations for a two-layer, double-gyre system. *Salmon* [1994] demonstrated that the low transport values produced by the wind in the subpolar gyre were carried southwestward along the f/h lines to produce a region of southward flow between the coast and the western boundary current of the subtropical gyre. However, the planetary geostrophic equations used by *Salmon* [1992, 1994] do not incorporate the inertial terms, which are of importance in western boundary currents [e.g., *Harrison and Holland*, 1981]. *Thompson* [1995] studied the effect of continental rises in the context of a three-layer quasi-geostrophic model. *Thompson* [1995] considered a symmetrically forced double-gyre circulation with free-slip boundary conditions along the western wall and found that the presence of topography broke the symmetry between the subpolar and subtropical gyres. The mean path of the mid-latitude jet was deflected to the north of the ZWCL, while the separation point remained unaffected.

Dengg [1993] demonstrated in a barotropic flat bottom model that the turn of the coastline away from the western boundary current can induce separation, provided that the current is highly inertial and that no-slip boundary conditions are prescribed. The key process was the production of positive relative vorticity along the coast and its subsequent advection to the separation point via nonlinear terms in the vorticity equation. *Marshall and Tansley* [2001] and *Munday and Marshall* [2005] have related these results to the generation of flow deceleration and an adverse pressure gradient. They show that the β effect acts to accelerate a western boundary current and can suppress its separation. Thus, there is a critical coastline curvature needed to overcome the accelerating influence of the β effect in order for separation to occur. *Özgökmen et al.* [1997] extended the works of *Dengg* [1993] and *Thompson* [1995] to baroclinic flows with bottom topography and coastline orientation by exploring the joint effects of (1) wind forcing, (2) bottom topography, and (3) inertia on the mid-latitude jet separation in a two-layer quasi-geostrophic model. The sensitivity of the separation latitude to the meridional structure of the wind stress distribution was first investigated in a series of no-slip experiments with a simplified wedge-shaped coastline and with a flat bottom. In all cases, the separation latitude was found to be strongly dependent on the position of the maximum wind stress curl (Figure 5).

Inclusion of an idealized bottom topography consisting of a smooth and gradual rise to represent the continental shelf significantly modified the upper-layer flow pattern by forcing the western boundary current to follow the f/h contours and to overshoot the flat-bottom separation latitude (Figure 6a). The influence of bottom topography can be expressed as the difference between the topographic stretching associated with the barotropic flow and a correction term that represents the joint effect of baroclinicity and bottom relief, the so-called JEBAR term [*Sarkisyan and Ivanov*, 1971; *Mertz and Wright*, 1992]. In the absence of direct forcing (i.e., a DWBC or buoyancy forcing), the lower layer can be set in motion only through eddy momentum fluxes generated by instabilities. Topographic stretching is then generated, provided that a mean flow is established in the lower layer by the fluctuations. Feedback to the upper layer will then be es-

tablished via vortex stretching. In other words, topographic effects will be felt by the upper layer in high eddy activity regions. *Özgökmen et al.* [1997] showed that if eddy fluctuations are high at the separation point, then the jet feels the topography strongly and, consequently, is unable to cross the f/h contours. The main factor that was found to minimize the impact of the topographic stress [*Holloway*, 1992] near the separation point is high inertia. Higher inertia facilitates the separation by decoupling the upper layer from the lower layer when the current crosses the f/h contours. The eddy–topography interactions are then minimal over the continental rise and become important only in the meandering jet region (Figure 6b). Contrary to the results of the flat-bottom experiments, once inertial effects dominate, the separation latitude is no longer found to be sensitive to the wind stress distribution. The path of the jet, however, once separated, does depend on the wind stress distribution.

Stern [1998] has also proposed a role for the downstream convergence of the isobaths on boundary current separation; this mechanism may be particularly pertinent for the Gulf Stream, as the continental shelf narrows abruptly at Cape Hatteras. Upstream of this convergence, the topography exerts a stabilizing influence on the boundary current. However, as the isobaths converge, the boundary current is

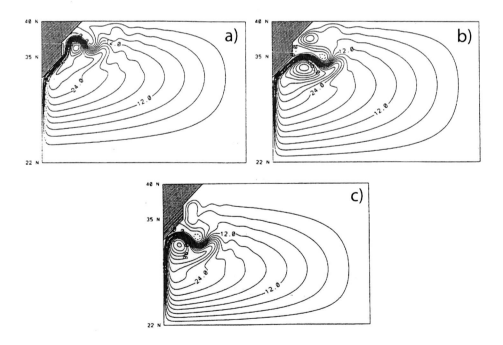

Figure 5. Upper-layer transport streamfunctions (CI = 3 Sv) for the flat bottom experiments with an anticyclonic zonal wind stress of varying meridional structure: (a) intensified to the north, (b) quasi-symmetric, and (c) intensified to the south. For more details, the reader is referred to *Özgökmen et al.* [1997].

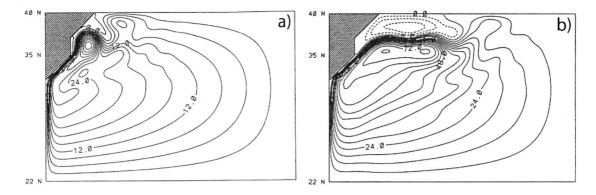

Figure 6. Upper-layer transport streamfunctions for (a) the bottom topography experiment with the same quasi-symmetric anticyclonic zonal wind stress used in the top right panel of Figure 5 (CI = 3 Sv) and (b) the same experiment where the amplitude of the Eckman pumping is doubled, thereby increasing the western boundary current inertia (CI = 6 Sv). For more details, the reader is referred to *Özgökmen et al.* [1997].

forced into deeper water and therefore able to continue as a free jet, shielded from the topographic influence by the stratification beneath. An important aspect of both the *Özgökmen et al.* [1997] and *Stern* [1998] models is that they remind us that in reality, boundary currents such as the Gulf Stream separate from sloping bottom boundaries rather than lateral sidewalls.

Finally, when compared with other western boundary currents, the Gulf Stream separation point is very stable and, in the absence of other credible mechanisms, it seems likely that the geometry of the coastline and/or bottom topography is the major factor responsible for this stability.

3.4. Deep Western Boundary Current

A further factor that may influence separation in the case of the Gulf Stream is its interaction with the DWBC. The path of the DWBC is strongly constrained by bottom topography; at Cape Hatteras, where the continental shelf widens abruptly, the DWBC is forced to move offshore and pass beneath the Gulf Stream. The DWBC may influence Gulf Stream separation through two distinct mechanisms: southward "advection" of the separation point [*Thompson and Schmitz*, 1989] and generation of an adverse pressure gradient within the Gulf Stream through the vortex stretching induced by the descent of the DWBC [*Tansley and Marshall*, 2000].

The first mechanism can be understood by considering a two-layer model of the Gulf Stream and the DWBC. Assuming the flow is in geostrophic balance to leading order, one can show that the depth of the upper layer, and hence the path of the Gulf Stream, is "advected" by the flow in the lower DWBC layer [*Thompson and Schmitz*, 1989]. While this mechanism is inherently transient in nature, *Tansley and*

Marshall [2000] note that the widening of the continental shelf at Cape Hatteras sets a natural limit to the southward advection of the Gulf Stream separation point.

The second mechanism can be understood by considering the vorticity balance of the Gulf Stream. As the DWBC descends beneath the Gulf Stream [*Pickart and Smethie*, 1993], the resultant vortex stretching generates a cyclonic vorticity anomaly within the Gulf Stream, which can be shown to decelerate the flow and induce an adverse pressure gradient [*Tansley and Marshall*, 2000].

Idealized numerical experiments (Figure 7) confirm that the DWBC can shift the Gulf Stream separation point southward [e.g., *Thompson and Schmitz*, 1989]. The extent of this southward shift is increased when both the upper and lower cores of the DWBC are individually resolved due to the increased resolution of the DWBC and the resultant increase in flow speeds [*Spall*, 1996a]. The upper core of the DWBC becomes entrained into the eddy-driven recirculation gyre to the north of the separated Gulf Stream and mixes with water from the gyre interior before returning to the western boundary via the southern recirculation gyre; in contrast, the lower core of the DWBC flows southward along the western boundary and is relatively unaffected by the crossover.

Low-frequency oscillations are also obtained in many parameter regimes, which *Spall* [1996b] attributes to the entrainment of the upper core of the DWBC into the separating Gulf Stream in different phases of the oscillation. In one phase, entrainment of the upper core of the DWBC into the eddy-driven recirculation gyres modifies the meridional potential vorticity gradient and stabilizes the separated Gulf Stream, in turn reducing the eddy potential vorticity fluxes and weakening the eddy-driven recirculation gyres. In the other phase with weakened recirculation gyres, the absence of entrainment

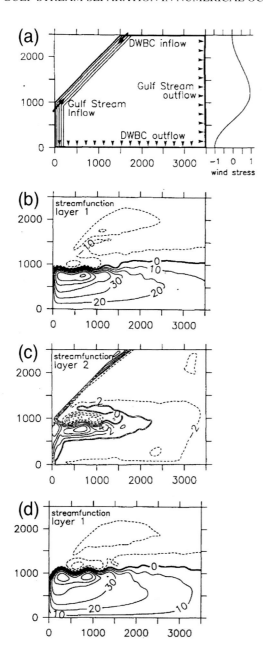

Figure 7. Illustration of the role of the DWBC on Gulf Stream separation in the idealized three-layer model of *Spall* [1996a]: (a) model configuration; streamfunction for the (b) Gulf Stream layer and (c) upper DWBC layer in the control case; (d) streamfunction for the Gulf Stream layer in the case with no DWBC. For more details, the reader is referred to *Spall* [1996a].

of the upper core of the DWBC reverses the meridional potential vorticity gradient, destabilizing the separated Gulf Stream and strengthening the recirculation gyres. *Katsman et al.* [2001] further discuss the stability properties, bifurcation structure, and timescales of these regimes.

4. REPRESENTATION OF GULF STREAM SEPARATION IN EDDY-RESOLVING OGCMS

Recent advances in computer architecture now allow for the numerical integration of state-of-the-art basin models with a grid resolution of 1/10° or higher (routinely on basin scale and less so globally). There are only a handful of results that have been published in the literature on simulations at that resolution, and most of them have focused on the North Atlantic and the Gulf Stream. At that resolution, there is a significant improvement in the separation of the Gulf Stream at Cape Hatteras when compared to coarser resolution experiments [*Bryan et al.*, 2007]. However, there are still substantial differences in the Gulf Stream's path, strength, and variability among the various high-resolution model simulations. In particular, a realistic separation in a 1/10° basin scale simulation of the North Atlantic [*Smith et al.*, 2000] does not guarantee that the separation be as realistic when the identical model is configured globally at the same resolution [*Maltrud and McClean*, 2005].

As already mentioned in the "Introduction", improvement in Gulf Stream separation in the fine mesh (1/10° or higher) simulations when compared to coarser resolution experiments may be due to the fact that the first baroclinic Rossby radius of deformation is resolved throughout most of the domain, thereby providing a good representation of baroclinic instability processes [*Paiva et al.*, 1999; *Smith et al.*, 2000; *Bryan et al.*, 2007], or to the exceeding of some critical Reynolds number [*Dengg*, 1993; *Tansley and Marshall*, 2001], or a combination of both. This is consistent with the results of *Hurlburt and Hogan* [2000] who showed that the Gulf Stream separation and pathways are drastically improved when the resolution is increased from 1/8° to 1/64° in a series of numerical experiments with a hydrodynamic (i.e., no active thermodynamics) primitive equation model.

High resolution appears to be necessary, but it is not necessarily sufficient for a proper Gulf Stream separation. As stated by *Bryan et al.* [2007], substantial uncertainties remain about the robustness of the results obtained at a resolution of 1/10° or higher. The Gulf Stream separation, indeed, turns out to be quite sensitive to a variety of other factors such as subgrid scale parameterization, subpolar gyre strength and water mass properties, DWBC strength, representation of topography, and the choice of model grid (C. Böning, R. Bourdallé-Badie, F. Bryan, M. Hecht, J. McClean, and T. Penduff, personal communication). It is not always clear why certain model configurations lead to a correct western boundary current separation while others do not. For example, Plate 2 shows the mean sea surface height field for a 1/12° North Atlantic basin simulation using the depth-coordinate ocean model NEMO [*Madec*, 2006]. In this simulation, the

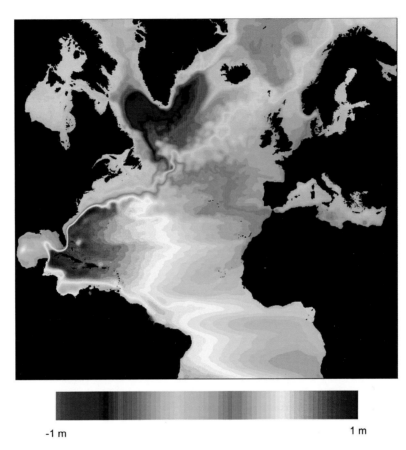

-1 m   1 m

Plate 2. Three-year mean sea surface height from the 1/12° North Atlantic depth coordinate [Nucleus for European Modelling of the Ocean (NEMO); *Madec*, 2006] configured over the North Atlantic between 20°S and 80°N including the Mediterranean Sea (R. Bourdallé-Badie, Y. Drillet, and O. LeGalloudec, personal communication). There are 50 levels in the vertical with an increased vertical resolution near the surface. The bathymetry is a combination of ETOPO2 bathymetry for the deep ocean and GEBCO bathymetry for the continental shelf. The north and south boundaries are buffer zones where the temperature and salinity fields are damped toward a climatological monthly mean [*Levitus et al.*, 1998]. A free surface that filters high-frequency features is used for the surface boundary condition [*Roullet and Madec*, 2000]. An isopycnal Laplacian operator (125 m²/s) is used for the lateral diffusion on the tracers, and a horizontal biharmonic operator (1.25×10^{-10} m²/s²) is used for the lateral viscosity on momentum. A partial slip lateral boundary condition is prescribed along the coast.

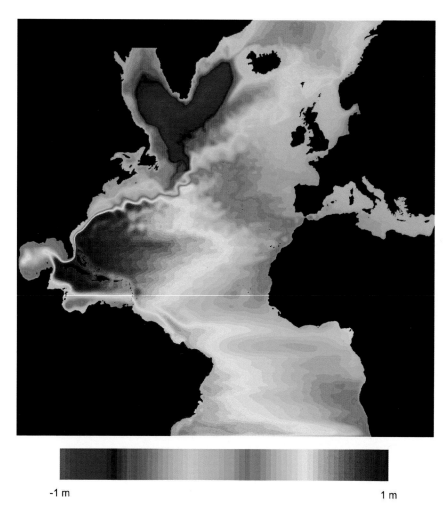

-1 m   **1 m**

Plate 3. Three-year mean sea surface height from the $1/12°$ North Atlantic isopycnic coordinate model [MICOM, *Bleck et al.*, 1992; *Bleck and Chassignet*, 1994] configured over the North Atlantic between 28°S and 70°N including the Mediterranean Sea. There are 26 layers in the vertical with an increased vertical resolution near the surface. The bathymetry is a modified ETOPO5. The north and south boundaries are buffer zones where the temperature and salinity fields are damped toward a climatological monthly mean [*Levitus et al.*, 1998]. The viscosity operator is a combination of biharmonic ($A_4 = V_D \Delta x^3$, with $V_D = 10^{-2}$ m/s) and Laplacian ($A_2 = $ max $[.1\Delta x^2 \times$ deformation tensor, $V_D \Delta x]$, with $V_D = 5 \times 10^{-3}$ m/s) operators. For more details, the reader is referred to *Chassignet and Garraffo* [2001].

Gulf Stream separation is well represented, and the improvements over an earlier simulation [*Drillet et al.*, 2005] are believed to be due to (a) a change in the grid orientation, (b) the use of partial cells to represent topography, (c) an energy- and enstrophy-conserving scheme, and (d) the coupling with the ice model. Indeed, *Barnier et al.* [2006] showed in a series of 1/4° global experiments that the use of partial steps and of the energy- and enstrophy-conserving scheme clearly impact on the Gulf Stream separation. The impact of the energy- and enstrophy-conserving scheme was shown to be greatest at grid cells nearest to a side wall, suggesting a great sensitivity of the momentum advection to the lateral and bottom boundary conditions.

The sensitivity to parameterization choices is clearly illustrated by *Bryan et al.* [2007] in a series of North Atlantic basin simulations with the depth-coordinate POP ocean numerical model [*Dukowicz and Smith*, 1994] with a free surface and a full-cell representation of the topography. *Bryan et al.* [2007] investigated the sensitivity of the modeled circulation to changes in resolution from 0.4° to 0.1° with and without adjustment in horizontal viscosity and diffusivity. When the viscosity is adjusted along with resolution as a function of the grid spacing, the Gulf Stream separation and path improves as the resolution is increased (Figure 8). Both the 0.4° and 0.2° simulations exhibit the traditional stationary eddy north of Cape Hatteras, and the path of the Gulf Stream is to the north of the observed mean path. At 0.1°, the stationary eddy disappears, and the Gulf Stream now separates from the coast. One can also notice the development of strong northern and southern recirculation gyres.

When moving to lower viscosities in the 0.1° experiments, *Bryan et al.* [2007] found that with decreasing viscosity, the Gulf Stream separates at a latitude south of the observed separation point at Cape Hatteras (not illustrated), the Gulf Stream path is more zonal, and the recirculation gyres strengthen. In these experiments, *Bryan et al.* [2007] show that the southern displacement of the separation point is linked to a strengthening of the deep outflow from the Labrador Sea and a substantial increase in the normal component of the velocity of the DWBC at the base of the Gulf Stream near 1,000 m. This is consistent with the mechanism put forward by *Thompson and Schmitz* [1989], i.e., that the Gulf Stream, especially its upper core, can be displaced to the south when the DWBC strength is increased [*Spall*, 1996a].

The Gulf Stream separation is also sensitive to the form of the viscosity operator as illustrated by a series of 1/12° North Atlantic simulations performed with the free surface isopycnic coordinate model MICOM [*Bleck et al.*, 1992; *Bleck and Chassignet*, 1994]. Plate 3 shows the mean sea surface height field for a 1/12° North Atlantic basin simula-

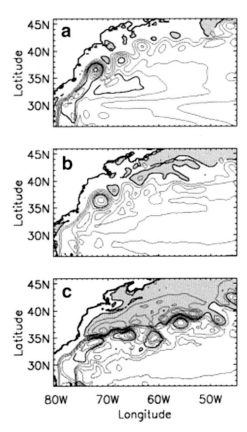

Figure 8. Three-year mean vertically integrated mass transport streamfunction over the Gulf Stream region for a series of North Atlantic basin simulations performed with the depth-coordinate POP ocean numerical model [*Dukowicz and Smith*, 1994] configured over the North Atlantic between 20°S and 73°N including the western Mediterranean Sea: (a) 0.4°, (b) 0.2°, and (c) 0.1°. The contour interval is 5 Sv in (a) and 10 Sv in (b) and (c). For more details, the reader is referred to *Bryan et al.* [2007].

tion using a viscosity operator that is a combination of biharmonic and Laplacian operators.

The mean sea surface height (SSH) of two biharmonic-only simulations are displayed in Figures 9 and 10 for two different magnitudes of the biharmonic viscosity coefficient. When a relatively small value of the biharmonic viscosity coefficient is used (see caption of Figure 9 for details), the western boundary current is seen to separate from the coast early at the Charleston bump before Cape Hatteras (Figure 9). A similar result was observed with the 1/10° POP North Atlantic simulation during the spin-up phase in which both the viscosity and diffusion had to be increased by a factor of 3 to eliminate this feature [*Smith et al.*, 2000]. An increase in the magnitude of the biharmonic viscosity operator did

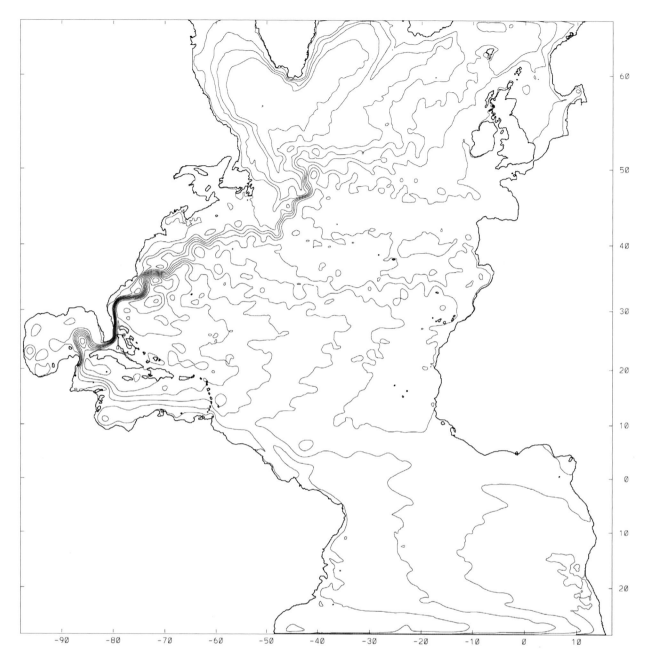

Figure 9. One-year mean sea surface height field from the 1/12° North Atlantic isopycnic coordinate model with a biharmonic viscosity operator (A_4 = max [.1Δ x^4 × deformation tensor, $V_D \Delta x^3$], with $V_D = 10^{-2}$ m/s). For more details, the reader is referred to *Chassignet and Garraffo* [2001].

eliminate the early detachment seen in Figure 9, but it led to the establishment of a permanent eddy north of Cape Hatteras (Figure 10). This eddy is maintained by a series of warm core (anticyclonic) rings that propagate westward, collide with the western boundary, and is only weakly dissipated by the biharmonic viscosity operator. This behavior is reminiscent of other simulations performed with biharmonic dissipation [*Smith et al.*, 2000]. The fact that this permanent eddy only appears in simulations that use biharmonic operators seems to indicate an incorrect representation of the eddy/mean flow and/or of the eddy/topography interactions possibly because of the scale selectiveness of the higher or-

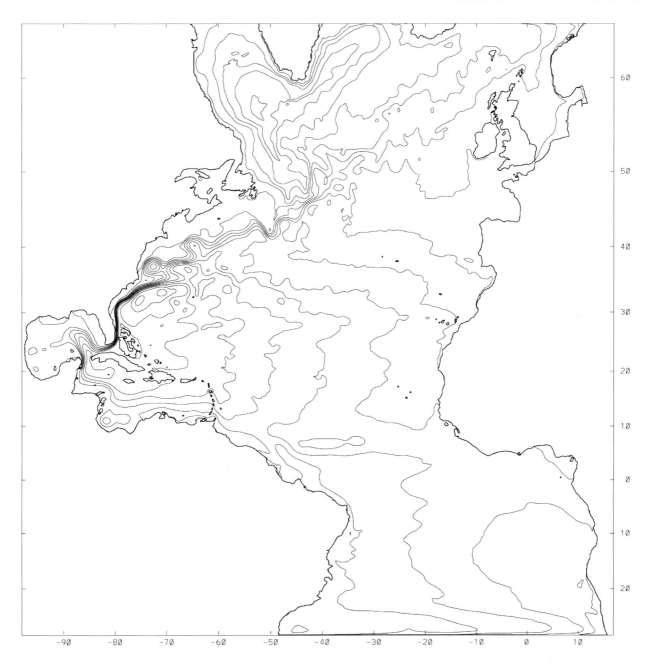

Figure 10. One-year mean sea surface height field from the 1/12° North Atlantic isopycnic coordinate model with a biharmonic viscosity operator ($A_4 = \max [.1 \, \Delta x^4 \times$ deformation tensor, $V_D \, \Delta x^3]$, with $V_D = 2 \times 10^{-2}$ m/s). For more details, the reader is referred to *Chassignet and Garraffo* [2001].

der operator that allows features that are marginally resolved by the grid spacing. In all simulations, the grid spacing is such that both the inertial and the viscous boundary layers are resolved (very well for the inertial and minimally for the viscous).

The mean SSH of the simulation performed with a Laplacian-only viscosity operator is displayed in Figure 11. In that simulation, the Gulf Stream separates well from the coast, but does not penetrate further than the New England Seamounts. The magnitude of the Laplacian viscosity coefficient was the

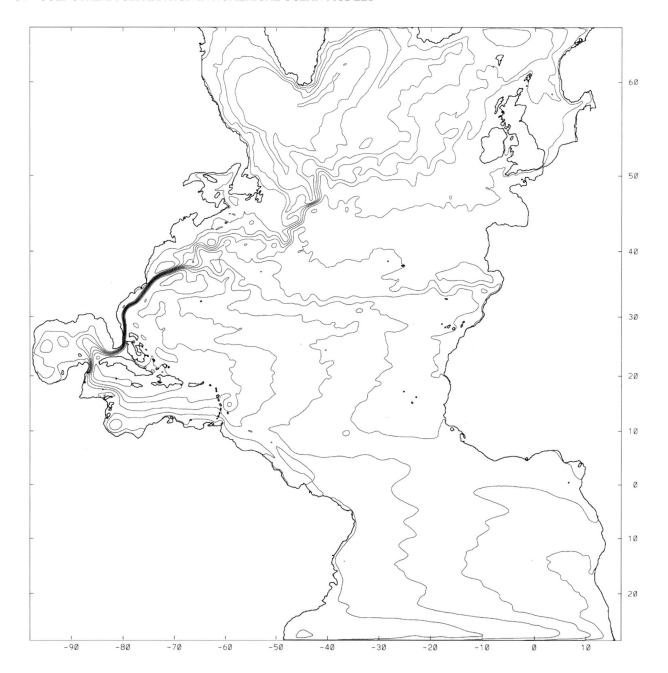

Figure 11. One-year mean sea surface height field from the $1/12°$ North Atlantic isopycnic coordinate model with a Laplacian viscosity operator (A_2 = max [.1 Δx^2 × deformation tensor, $V_D \Delta x$], with $V_D = 10^{-2}$ m/s). For more details, the reader is referred to *Chassignet and Garraffo* [2001].

minimum needed for numerical stability. One can summarize the above results by stating that neither operator (Laplacian or biharmonic) is able to provide satisfactory Gulf Stream separation and penetration. With the biharmonic operator, eddies are found to retain their structure for longer periods than with a Laplacian operator, but with undesirable effects on several features of the large-scale circulation. With the Laplacian operator, the western boundary current and its separation are well represented, but with a weak penetration of the Gulf Stream.

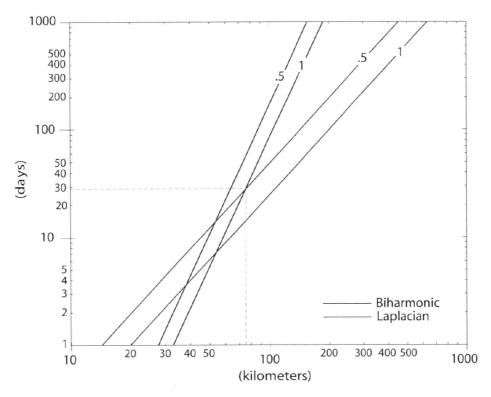

Plate 4. Laplacian and biharmonic decay timescale (in days) as a function of the wavelength k (in km) for values of the diffusive velocity V_D = .5 and 1 cm/s, respectively.

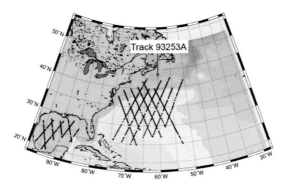

Plate 5. Location of the bathythermographic (BT) data taken during flights under the TOPEX altimeter ground tracks in September 1993. The red arrow points to track 93253A, which is used in Plate 6 to compare various mean dynamic topographies.

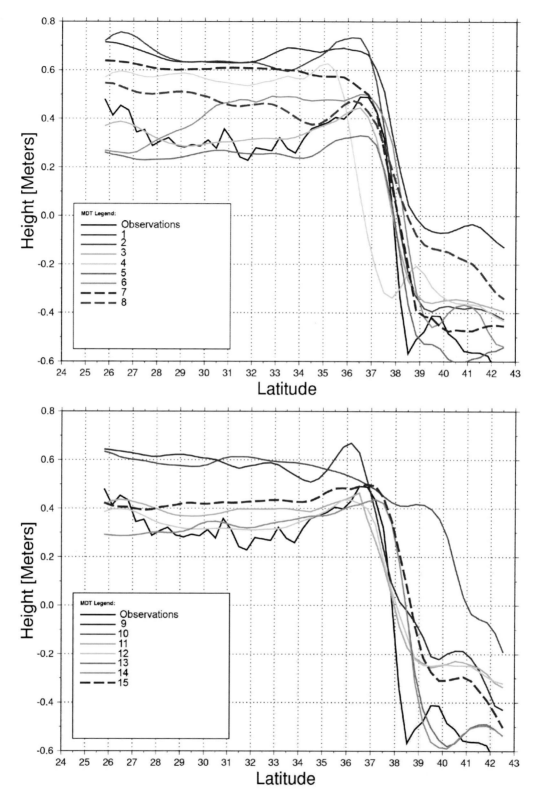

Plate 6. Mean dynamic topography (or sea surface height) in meters from four climatologies (1, 2, 11, and 12) and 11 numerical simulations (3, 4, 5, 6, 7, 8, 9, 10, 13, 14, and 15). The solid line represents the mean dynamic topography derived from the bathythermographic data and altimetry for the track identified in Plate 5.

With a Laplacian (harmonic) dissipation operator, the evolution of a wave $c(t)e^{ikx}$ is damped exponentially with a spin-down time $\tau_2 = A_2^{-1}(2/\Delta x\ \sin(k\Delta x/2))^{-2}$ In the case of a biharmonic operator, the spin-down time is $\tau_4 = A_4^{-1}(2/\Delta x\ \sin(k\Delta x/2))^{-4}$. For comparison purposes, constant harmonic and biharmonic viscosity coefficients can be expressed as a function of a diffusive velocity V_D and the grid spacing Δx as $A_2 = V_D \Delta x$ and $A_4 = V_D \Delta x^3$, respectively. Examples of spin-down times for both operators are given in Plate 4 for the average grid spacing of the MICOM simulations (6 km). For the same diffusive velocity, the biharmonic operator more strongly selects the small scales to dissipate and leaves the large scales relatively untouched.

The Laplacian experiment of Figure 11, when contrasted to the biharmonic experiments of Figures 9 and 10, suggests that some damping of the larger scales is necessary for a reasonable western boundary current behavior. The best separation/penetration results are obtained when the viscosity operator is prescribed as a combination of the biharmonic and Laplacian operators (Plate 3). The main motivation for combining the two operators was to be able to retain the scale selectiveness of the biharmonic operator and to provide some damping at the larger scales [performed in this case by the Laplacian operator for k greater than 80 km (Plate 4)]. This permits the reduction of the magnitude of the Laplacian coefficient A_2 by 50% and, at the same time, ensures numerical stability with an effective damping of the smaller scales by the biharmonic operator (Plate 4). When combined, the individual diffusive velocity V_D specified for each operator is smaller than the minimum value that is needed for numerical stability when only one of the operators is specified.

These results appear to suggest that in a realistic setting, even with such fine grid spacing, the modeled large-scale ocean circulation is strongly dependent on the choices made for the viscosity operators. Furthermore, it appears that the inverse cascade of energy from the small scales to the larger scales may not take place as anticipated and that some large-scale information is needed for a proper representation of the western boundary current. In the experiments described in this paper, the latter is taking place by means of the Laplacian viscosity operator. Hyperviscosity (∇^{2n} operator with $n \geq 2$) is often used in numerical simulations of turbulent flows to extend the range of the inviscid inertial cascade. It has also been argued, however, that hyperviscosity contributes nontrivial spurious dynamics [*Jiménez*, 1994; *Delhez and Deleersnijder*, 2007].

5. VALIDATION

Once a numerical simulation is able to model western boundary currents that separate from the coastline at the correct latitude and with adequate interior penetration, the question remains as to whether the simulation is in agreement with observations. At the scales of interest (tens of kilometers), it is necessary to have the observed means of western boundary ocean currents and associated fronts sharply defined. Global climatologies such as the one derived from surface drifters by *Niiler et al.* [2003] or the one derived from altimetry and direct measurements by *Rio and Hernandez* [2004] can be used to validate the position of the current main axis, but they cannot be used to quantify the width and intensity of the currents because of their coarseness (~0.5–1° horizontal resolution). More specifically, the lack of an accurate geoid on the mesoscale prevents a precise computation of the ocean absolute mean dynamic topography from satellite altimetry. Several satellite missions (GRACE, GOCE) are either underway or planned to try to determine a more accurate geoid, but one will need to have measurements accurate to within a few centimeters on scales down to approximately 30 km so as to be useful for the above validation exercise.

An alternate approach to the validation is to determine an accurate mean dynamic topography using direct in situ measurements. An example of such an approach is to construct the mean dynamic topography along altimetric satellite ground tracks where there exist concurrent and collinear measurements of sea surface height anomalies and dynamic heights (for examples of collinear temperature and dynamic height analyses, see *Carnes et al.* [1990] and *Blaha and Lunde* [1992]). Plate 5 shows the location of the bathythermographic data taken during flights under the TOPEX altimeter ground tracks in September 1993 (courtesy of the Altimetry Data Fusion Center, Naval Oceanographic Office). To compute the mean dynamic topography, one needs to make the assumption that the dynamic topography relative to the geoid can be approximated by the dynamic height relative to a deep pressure surface that is parallel to the geoid (i.e., a level of no motion). Any error in this assumption will lead to different values of mean dynamic topography, but for the spatial scales of interest, as long as the level of no motion is deep enough, choosing a different level just adds or removes a bias. This does not affect the analysis, as we are only interested in the relative difference in mean dynamic topography across the currents (B. Lunde, personal communication).

The mean dynamic topographies displayed in Plate 6 for 4 climatologies and 11 numerical simulations show a wide range in relative difference across the Gulf Stream as well as in sharpness. For the climatologies, the *Niiler et al.* [2003] mean (number 2) clearly outperforms on that particular track the *Rio and Hernandez* [2004] mean (number 11). For the numerical simulations, the 1/32° hydrodynamic (i.e., no thermal forcing) Navy Layered Ocean Model (NLOM) simulation (number 14) of *Hurlburt and Hogan* [2000] performs, best emphasizing again the importance of high horizontal grid spacing. Note that some of the differences between the dynamic topographies in Plate 6 may be attributed to differences in the temporal periods over which the data are compiled in the observations, climatologies, and models.

6. CONCLUDING REMARKS

In summary, identifying the dynamics responsible for Gulf Stream separation continues to be a challenge, and the separation mechanism in numerical models remains very sensitive to choices made in the numerical model for subgrid scale parameterizations. There is yet no single recipe that would guarantee a correct separation of all western boundary currents in a global model. While it can be firmly stated that a resolution on the order of at least 1/10° is a necessary condition for a western boundary current to realistically separate from the coast, it is not yet clear which horizontal resolution is best for a realistic zonal penetration and variability of the separated boundary current. On one hand, a fourfold increase in resolution from 1/16° to 1/64° with the Laplacian operator in the hydrodynamic (i.e., no thermal forcing) NLOM [*Hurlburt and Hogan*, 2000] brought the sea surface height variability to observed levels without altering the pattern of the large-scale circulation. On the other hand, the experiments performed by *Bryan et al.* [2007] with a biharmonic operator show sea surface height variability that exceeds observations (Figure 12 of their paper). While numerical simulations at the above-noted resolutions are becoming more common, they still demand the latest in computing facilities. A fourfold increase in resolution for the thermodynamically forced models cannot be realistically implemented with the present computer resources.

Further evaluation of the impact of dissipation operators on the large-scale circulation also needs to be pursued. For example, the horizontal eddy viscosity implemented in numerical models is usually assumed to be isotropic. There is currently little theoretical or observational evidence to support that assumption, and substantial anisotropy can be generated by the direction of the mean flow in a strong jet [*Smith and McWilliams*, 2003], although more recent theoretical [*Theiss*, 2004] and observational [*Eden*, 2007] analysis suggests that anisotropy can be generated by the β effect. An anisotropic formulation of the viscosity that incorporates the feature of larger along-stream than cross-stream diffusion for zonal jets was first introduced into a coarse-resolution OGCM by *Large et al.* [2001] to improve the simulation of the equatorial currents in the Pacific Ocean. When applied globally, *Doney et al.* [2003] found some improvements in the separation points of western boundary currents, although their results were sensitive to the specific choice of viscous

parameters. *Smith and Gent* [2004], using both anisotropic viscosity and an anisotropic *Gent and McWilliams* [1990] scheme, also found improvements in the separation and path of the Gulf Stream at 0.1° and 0.2° resolutions. The anisotropic *Gent and McWilliams* [1990] scheme led to the greatest overall model improvements especially in the meridional overturning circulation, supporting the argument of *Roberts and Marshall* [1998] for the use of adiabatic tracer diffusion even at eddy-resolving scales.

Finally, we would like to emphasize that a judicious use of data assimilation should increase our physical understanding of western boundary current separation in numerical models. Data assimilation can be used to estimate model parameters by taking into account both the information provided by the model dynamics and by the observations [see *Bennett*, [1992]; *Wunsch* [1996]; *Evensen et al.* [1998]; and *Bennett* [2002] for a review], but there are still many challenges remaining before parameter estimation can be directly applied to eddy-resolving OGCMs. Because of its complexity and nonlinearity, parameter estimation is mostly used in the context of simple dynamical models or idealized configurations. A nice example is the work in progress of A. Wirth and J. Verron in which the outputs of a numerical model that explicitly resolves the entrainment process in gravity currents are used to estimate which frictional parameterization is most adequate in representing the turbulent fluxes that are not resolved in coarser horizontal resolution numerical models.

Acknowledgments. We express our appreciation to Bruce Lunde of the Naval Oceanographic Office and Ole Martin Smedstad of Planning Systems, Inc. for performing the mean dynamic topography analysis presented in Section 5 and for providing Plates 5 and 6. We also would like to thank an anonymous reviewer, L. Bunge, M. Hecht, H. Hurlburt, J. O'Brien, and J. Verron for insightful comments.

REFERENCES

Adcroft, A., and D. Marshall (1998), How slippery are piecewise-constant coastlines in numerical ocean models?, *Tellus, 50A,* 95–108.

Agra, C., and D. Nof (1993), Collision and separation of boundary currents. *Deep Sea Res., 40,* 2259–2282.

Anderson, D. L. T., and D. W. Moore (1979), Cross-equatorial inertial jets with special relevance to very remote forcing of the Somali Current. *Deep Sea Res., 26,* 1–22.

Baines, P. G., and R. L. Hughes (1996), Western boundary current separation: Inferences from a laboratory model. *J. Phys. Oceanogr., 26,* 2576–2588.

Barnier, B., G. Madec, T. Penduff, J.-M. Molines, A.-M. Treguier, J. Le Sommer, A. Beckmann, A. Biastoch, C. Böning, J. Dengg, C. Derval, E. Durand, S. Gulev, E. Remy, C. Talandier, S. Theetten, M. Maltrud, J. McClean, and B. De Cuevas (2006), Impact of partial steps and momentum advection schemes in a global ocean circulation model at eddy-permitting resolution. *Ocean Dyn.*, *56*, 543–567.

Bennett, A. F. (1992), *Inverse Methods in Physical Oceanography*, 346 pp., Cambridge University Press, Cambridge.

Bennett, A. F. (2002), *Inverse Modeling of the Ocean and Atmosphere*, 234 pp., Cambridge University Press, Cambridge.

Blaha, J., and B. Lunde (1992), Calibrating altimetry to geopotential anomaly and isotherm depths in the Western North Atlantic, *J. Geophys. Res.*, *97*, 7465–7477.

Blandford, R. R. (1971), Boundary conditions in homogeneous ocean models. *Deep Sea Res.*, *18*, 739–751.

Bleck, R. (2002), An oceanic general circulation model framed in hybrid isopycnic–Cartesian coordinates, *Ocean Model*, *4*, 55–88.

Bleck, R., and E. P. Chassignet (1994), Simulating the oceanic circulation with isopycnic coordinate models. *The Oceans: Physiochemical Dynamics and Resources*. The Pennsylvania Academy of Science, pp. 17–39.

Bleck, R., C. Rooth, D. Hu, and L. T. Smith (1992), Salinity-driven transients in a wind- and thermohaline-forced isopycnic coordinate model of the North Atlantic. *J. Phys. Oceanogr.*, *22*, 1486–1505.

Bryan, F. O., M. W. Hecht, and R. D. Smith (2007), Resolution convergence and sensitivity studies with North Atlantic circulation models. Part I: The western boundary current system. *Ocean Modelling*, *16*, 141–159.

Carnes, M. R., J. L. Mitchell, and P. W. de Witt (1990), Synthetic temperature profiles derived from GEOSAT altimetry: comparison with air-dropped expendable bathy thermograph profiles. *J. Geophys. Res.*, *95*, 17,979–17,992.

Cessi, P. (1990), Geometrical control of the inertial recirculation. *J. Phys. Oceanogr*, *20*, 1867–1875.

Cessi, P. (1991) Laminar separation of colliding western boundary currents. *J. Mar. Res.*, *49*, 697–717.

Cessi P., G. Ierley, and W. Young (1987), A model of intertial recirculation driven by potential vorticity anomalies. *J. Phys. Oceanogr.*, *17*, 1640–1652.

Cessi, P., R. V. Condie, and W. R. Young (1990), Dissipative dynamics of western boundary currents. *J. Mar. Res.*, *48*, 677–700.

Chassignet, E. P. (1995), Vorticity dissipation by western boundary currents in the presence of outcropping layers. *J. Phys. Oceanogr.*, *25*, 242–255.

Chassignet, E. P., and R. Bleck (1993), The influence of layer outcropping on the separation of boundary currents. Part I: The wind-driven experiments, *J. Phys. Oceanogr.*, *23*, 1485–1507.

Chassignet, E. P., and Z. D. Garraffo (2001), Viscosity parameterization and the Gulf Stream separation, in *From Stirring to Mixing in a Stratified Ocean*, Proceedings 'Aha Huliko'a Hawaiian Winter Workshop January 15–19, 2001, edited by P. Muller and D. Henderson, pp. 37–41, University of Hawaii.

Chassignet, E. P., and P. R. Gent (1991), The influence of boundary conditions on mid-latitude jet separation in ocean numerical models. *J. Phys. Oceanogr.*, *21*, 1290–1299.

Chassignet, E. P., and C. G. H. Rooth (1995), The influence of layer outcropping on the separation of boundary currents. Part II: The wind- and buoyancy-driven experiments. *J. Phys. Oceanogr.*, *25*, 2404–2422.

Cherniawsky, J., and G. Holloway (1991), An upper ocean general circulation model for the North Pacific: Preliminary experiments. *Atmos. Ocean*, *29*, 737–784.

Delhez, E. J. M., and E. Deleersnijder (2007), Overshootings and spurious oscillations caused by biharmonic mixing. *Ocean Modelling*, *17*, 183–198.

Dengg, J. (1993), The problem of Gulf Stream separation: A barotropic approach. *J. Phys. Oceanogr.*, *23*, 2182–2200.

Dengg, J., A. Beckmann, and R. Gerdes (1996), The Gulf Stream separation problem, in *The Warmwatersphere of the North Atlantic*, pp. 253–289, edited by W. Kraus, Gebruder Borntrager, Berlin.

Doney, S. C., S. Yeager, G. Danabasoglu, W. Large, and J. C. McWilliams (2003), Modeling global oceanic interannual variability (1958–1997): Simulation design and model-data evaluation. *NCAR Technical Note*, NCAR/TN-452+STR, 48 pp.

Drillet, Y., R. Bourdallé-Badie, L. Siefridt, and C. Le Provost (2005), Meddies in the Mercator North Atlantic and Mediterranean Sea eddy-resolving model. *J. Geophys. Res.*, *110*, C03016, doi:10.1029/2003JC002170.

Dukowicz, J. K., and R. D. Smith (1994), Implicit free-surface method for the Bryan–Cox–Semtner ocean model. *J. Geophys. Res.*, *99*, 7991–8014.

Dupont, F., D. N. Straub, and C. A. Lin (2003), Influence of a step-like coastline on the basin scale vorticity budget of mid-latitude gyre models. *Tellus*, *55A*, 255–272.

Eby, M., and G. Holloway (1994), Sensitivity of a large scale ocean model to a parameterization of topographic stress. *J. Phys. Oceanogr.* *24*, 2577–2588.

Eden, C. (2007), Eddy length scales in the North Atlantic Ocean. *J. Geophys. Res.*, *112*, C06004, doi:10.1029/2006JC003901.

Evensen G., D. Dee, and J. Schröter (1998), Parameter estimation in dynamical models, in *Ocean Modeling and Parameterization*, NATO Science Series, pp. 373–398, edited by E. P. Chassignet, and J. Verron, Kluwer, New York.

Ezer, T., and G. L. Mellor (1992), A numerical study of the variability and the separation of the Gulf Stream induced by surface atmospheric forcing and lateral boundary flows. *J. Phys. Oceanogr.*, *22*, 660–682.

Gangopadhyay, A., P. Cornillon, and D. R. Watts (1992), A test of the Parsons–Veronis hypothesis related to the separation of the Gulf Stream from the coast. *J. Phys. Oceanogr.*, *22*, 1286–1301.

Gent, P. R., and J. C. McWilliams (1990), Isopycnal mixing in ocean circulation models. *J. Phys. Oceanogr.*, *20*, 150–155.

Greenspan, H. P. (1963), A note concerning topography and inertial currents. *J. Mar. Res.*, *21*, 147–154.

Haidvogel, D., J. McWilliams, and P. Gent (1992), Boundary current separation in a quasigeostrophic, eddy-resolving ocean circulation model. *J. Phys. Oceanogr.*, *22*, 882–902.

Harrison, D. E., and W. R. Holland (1981), Regional eddy vorticity transport and the equilibrium vorticity budgets of a numerical model ocean circulation. *J. Phys. Oceanogr.*, *11*, 190–208.

Holland, W. R. (1967), On the wind-driven circulation in an ocean with bottom topography. *Tellus*, *19*, 582–600.

Holland, W. R. (1973), Baroclinic and topographic influences on the transport in western boundary currents. *Geophys. Fluid Dyn.*, *4*, 187–210.

Holloway, G. (1992), Representing topographic stress for large-scale ocean models. *J. Phys. Oceanogr.*, *22*, 1033–1046.

Huang, R. X. (1987), A three-layer model for wind-driven circulation in a subtropical-subpolar basin. Part I. Model formulation and the subcritical state. *J. Phys. Oceanogr.*, *17*, 664–678.

Huang, R. X., and G. R. Flierl (1987), Two-layer models for the thermocline and current structure in the subtropical/subpolar gyres. *J. Phys. Oceanogr.*, *17*, 872–884.

Hurlburt, H. E., and P. J. Hogan (2000), Impact of 1/8° to 1/64° resolution on Gulf Stream model-data comparisons in basin-scale subtropical Atlantic ocean models. *Dyn. Atmos. Oceans*, *32*, 283–329.

Hulburt, H. E., and P. J. Hogan (2008), The Gulf Stream pathway and the impacts of the eddy-driven abyssal circulation and the deep western boundary current, *Dyn. Atmos. Oceans*, (submitted).

Hurlburt, H. E., E. J. Metzger, P. J. Hogan, C. E. Tillburg, J. F. Shriver, and O. M. Smedstad (2008), Pathways of upper ocean currents and fronts: Steering by topographically-constrained abyssal circulation and the role of flow instabilities, applicability of a two-layer theory to low vs. high vertical resolution models. *Dyn. Atmos. Oceans*, (submitted).

Ierley, G. R. (1987), On the onset of recirculation in barotropic general circulation models. *J. Phys. Oceanogr.*, *17*, 2366–2374.

Ierley, G. R., and O. G. Ruehr (1986), Analytic and numerical solutions of a nonlinear boundary layer problem. *Stud. Appl. Math.*, *75*, 1–36.

Ierley, G., and W. Young (1988), Inertial recirculation in a β-plane corner. *J. Phys. Oceanogr.*, *18*, 683–689.

Isemer, H. J., and L. Hasse (1987), *The Bunker Climate Atlas of the North Atlantic Ocean. Vol. 2, Air–Sea Interactions*, 252 pp., Springer, Berlin.

Jiang, S., Jin, F.-F., and Ghil, M. (1995), Multiple equilibria and aperiodic solutions in a wind-driven double-gyre, shallow-water model, *J. Phys. Oceanogr.*, *25*, 764–786.

Jiménez, J. (1994), Hyperviscous vortices. *J. Fluid Mech.*, *279*, 169–176.

Kamenkovich, V. M. (1966), A contribution to the theory of the inertial-viscous boundary layer in a two-dimensional model of ocean currents. *Isvestiya*, *97*, 781–792.

Kamenkovich, V. M., and G. M. Reznik (1972), A contribution to the theory of stationary wind-driven currents in a two-layer liquid. (English transl.), *Izv. Acad. Sci. USSR Atmos. Oceanic Phys.*, *8*, 238–245.

Katsman, C. A., S. S. Drijfhout, and H. A. Dijkstra (2001), The interaction of a deep western boundary current and the wind-driven gyres as a cause for low-frequency variability. *J. Phys. Oceanogr.*, *31*, 2321–2339.

Kiehl, J. T., and P. R. Gent (2004), The community climate system model, version two. *J. Clim.*, *17*, 3666–3682.

Kiss, A., E. (2002), Potential vorticity "crises", adverse pressure gradients and western boundary current separation. *J. Mar. Res.*, *60*, 779–803.

Large, W.G., G. Danabasoglu, J. C. McWilliams, P. R. Gent, and F. O. Bryan (2001), Equatorial circulation of a global ocean climate model with anisotropic horizontal viscosity. *J. Phys. Oceanogr.*, *31*, 518–536.

Madec G. (2006), *NEMO Reference Manual, Ocean Dynamic Component: NEMO-OPA*, Institut Pierre Simon Laplace (IPSL), 27, ISSN 1288-1619.

Maltrud, M. E., and J. L. McClean (2005), An eddy resolving global 1/10° ocean simulation. *Ocean Modelling*, *8*, 31–54.

Marshall, D. P., and C. E. Tansley (2001), An implicit formula for boundary current separation. *J. Phys. Oceanogr.*, *31*, 1633–1638.

Mertz, G., and D. G. Wright (1992), Interpretations of the JEBAR term. *J. Phys. Oceanogr.*, *22*, 301–305.

Moore, D.W., and P. Niiler (1974), A two-layer model for the separation of inertial boundary currents. *J. Mar. Res.*, *32*, 457–484.

Moro, B., 1988: On the nonlinear Munk model. Part I: Steady flows. *Dyn. Atmos. Oceans*, *12*, 259–287.

Munday, D. R., and D. P. Marshall (2005), On the separation of a barotropic western boundary current from a cape. *J. Phys. Oceanogr.*, *35*, 1726–1743.

Munk, W. H. (1950), On the wind-driven circulation. *J. Meteor.*, *7*, 79–93.

Myers, P. G., A. F Fanning, and A. J. Weaver (1996), JEBAR, bottom pressure torque and Gulf Stream separation. *J. Phys. Oceanogr.*, *26*, 671–683.

Nauw, J. J., H. A. Dijkstra, and E. P. Chassignet (2004), Frictionally-induced asymmetries in wind-driven flows. *J. Phys. Oceanogr.*, *34*, 2057–2072.

Niiler, P. P., N. A. Maximenko, and J. C. McWilliams (2003), Dynamically balanced absolute sea level of the global ocean derived from near-surface velocity observations, *Geophys. Res. Lett.*, *30*(22), 2164, doi:10.1029/2003GL018628.

Nurser, A.J.G. and R.G. Williams (1990), Cooling the Parsons' model of the separated Gulf Stream. *J. Phys. Oceanogr.*, *20*, 1974–1979.

Ou, H. W., and W. P. M. de Ruijter (1986), Separation of an inertial boundary current from a curved coastline. *J. Phys. Oceanogr.*, *16*, 280–289.

Özgökmen, T. M., E. P. Chassignet, and A. M. Paiva (1997), Impact of wind forcing, bottom topography, and inertia on midlatitude jet separation in a quasigeostrophic model. *J. Phys. Oceanogr.*, *27*, 2460–2476.

Paiva, A. M., J. T. Hargrove, E. P. Chassignet, and R. Bleck (1999), Turbulent behavior of a fine-mesh (1/12 degree) numerical simulation of the North Atlantic. *J. Mar. Syst.*, *21*, 307–320.

Parsons, A. T. (1969), A two-layer model of Gulf Stream separation. *J. Fluid Mech.*, *39*, 511–528.

Pedlosky, J. (1965), A necessary condition for the existence of an inertial boundary layer in a baroclinic ocean. *J. Mar. Res.*, *23*, 69–72.

Pedlosky, J. (1987), On Parsons' model of the ocean circulation. *J. Phys. Oceanogr.*, *17*, 1571–1582.

Pedlosky, J. (1996), *Ocean Circulation Theory*, 453 pp., Springer, Berlin.

Pickart R. S., and W. M. Smethie (1993), How does the deep western boundary current cross the Gulf Stream? *J. Phys. Oceanogr.*, 23, 2602–2616.

Piggott, M. D., C. C. Pain, G. J. Gorman, D. P. Marshall, and P. D. Killworth (2008), Unstructured, adaptive meshes for ocean modelling, this volume.

Rio, M.-H., and F. Hernandez (2004), A mean dynamic topography computed over the world ocean from altimetry, in situ measurements, and a geoid model. *J. Geophys. Res., 109*, C12032 doi:10.1029/2003JC002226.

Roberts, M., and D. Marshall (1998), Do we require adiabatic dissipation schemes in eddy-resolving ocean models? *J. Phys. Oceanogr., 28*, 2050–2063.

Salmon, R. (1992), A two-layer Gulf Stream over a continental slope. *J. Mar. Res., 50*, 341–365.

Salmon, R. (1994), Generalized two-layer models of ocean circulation. *J. Mar. Res., 52*, 865–908.

Sarkisyan, A. S., and V. F. Ivanov (1971), Joint effect of baroclinicity and bottom relief as an important factor in the dynamics of sea currents. *Izv. Akad. Nauk* SSSR, *Fiz. Atmos. Okeana, 7*, 173 (Engl. transl., p. 116).

Smith, N. R., and C. B. Fandry (1976), A simple model of Gulf Stream separation. *J. Phys. Oceanogr., 6*, 22–28.

Smith, R. D., and P. R. Gent (2004), Anisotropic Gent–McWilliams parameterization for ocean models. *J. Phys. Oceanogr., 34*, 2541–2564.

Smith, R. D., and J. C. McWilliams (2003) Anisotropic horizontal viscosity for ocean models. *Ocean Modelling, 5*, 129–156.

Smith, R. D., M. E. Maltrud, F. O. Bryan, and M. W. Hecht (2000), Numerical simulation of the North Atlantic Ocean at 1/10°. *J. Phys. Oceanogr., 30*, 1532–1561.

Spall, M. A. (1996a), Dynamics of the Gulf Stream/deep western boundary current crossover. Part I: Entrainment and recirculation. *J. Phys. Oceanogr., 26*, 2152–2168.

Spall, M. A. (1996b), Dynamics of the Gulf Stream/deep western boundary current crossover. Part II: Low frequency internal oscillations. *J. Phys. Oceanogr., 26*, 2169–2182.

Spitz, Y. H., and D. Nof (1991), Separation of boundary currents due to bottom topography. *Deep Sea Res., 38*, 1–20.

Stern, M. E. (1998), Separation of a density current from the bottom of a continental shelf. *J. Phys. Oceanogr., 28*, 2040–2049.

Stern, M. E., and J. A. Whitehead (1990), Separation of a boundary jet in a rotating fluid. *J. Fluid Mech., 217*, 41–69.

Stommel, H. (1948), The westward intensification of wind-driven ocean currents. *Trans. Am. Geophys. Union, 29*, 202–206.

Tansley, C. E., and D. P. Marshall (2000), On the influence of bottom topography and the deep western boundary current on Gulf Stream separation. *J. Mar. Res., 58*, 297–325.

Tansley, C. E., and D. P. Marshall (2001), Flow past a cylinder on a β-plane, with application to Gulf Stream separation and the Antarctic cirumpolar current. *J. Phys. Oceanogr., 31*, 3274–3288.

Theiss, J. (2004), Equatorward energy cascade, critical latitude, and the predominance of cyclonic vortices in geostrophic turbulence. *J. Phys. Oceanogr., 34*, 1663–1678.

Thompson, L. (1995), The effect of continental rises on the wind-driven ocean circulation. *J. Phys. Oceanogr., 15*, 1296–1316.

Thompson, J. D., and W. J. Schmitz, Jr. (1989), A limited area model of the Gulf Stream: Design, initial experiments, and model-data intercomparison. *J. Phys. Oceanogr., 19*, 791–814.

Veronis, G. (1973), Model of world ocean circulation: I. Wind-driven, two-layer. *J. Mar. Res., 31*, 228–288.

Veronis, G. (1976), Model of world ocean circulation: II. Thermally driven, two-layer. *J. Mar. Res., 34*, 199–216.

Veronis, G. (1978), Model of world ocean circulation: III. Thermally and wind driven. *J. Mar. Res., 36*, 1–44.

Verron, J., and E. Blayo (1996), The no-slip condition and separation of western boundary currents. *J. Phys. Oceanogr., 26*, 1938–1951.

Verron, J., and C. Le Provost (1991), Response of eddy-resolved general circulation numerical models to asymmetrical wind forcing. *Dyn. Atmos. Oceans, 15*, 505–533.

Warren, B. A. (1963), Topographic influences on the path of the Gulf Stream. *Tellus, 15*, 167–183.

Wunsch, C. (1996), *The Ocean Circulation Inverse Problem*, 442 pp., Cambridge University Press, Cambridge.

Eddy-Resolving Modeling of Overflows

S. Legg, L. Jackson, and R. W. Hallberg

National Oceanic and Atmospheric Administration, Geophysical Fluid Dynamics Laboratory, Princeton, New Jersey, USA

Dense waters flowing through narrow topographic constrictions or down sloping topography as dense overflows are responsible for generating most of the deep water masses of the ocean following mixing with overlying waters. Overflows involve a variety of different physical processes which together determine the volume, transport, and tracer properties of the dense water mass when it reaches the open ocean. In this paper, we review the current state of eddy-resolving modeling of overflows and understanding of mesoscale eddy processes active in overflows, focusing on models where mesoscale eddies are resolved but small-scale mixing is not. At these resolutions, the significant remaining difficulty is the treatment of diapycnal mixing, and we examine the dependence of this mixing on model parameterizations and numerics in overflow simulations.

1. INTRODUCTION

Dense water, formed in marginal seas or on continental shelves, enters the large-scale ocean circulation in the form of overflows or gravity currents, often flowing through narrow straits and down topographic slopes. Climatologically important overflows include the dense flows through the Denmark Strait, the Faroe Bank Channel, the Gibraltar Strait, and from Red Sea and the Antarctic shelves. The water masses originating in these overflows (e.g., North Atlantic Deep Water, Mediterranean Sea Water, and Antarctic Bottom Water) fill much of the deep ocean. Many different physical processes may be important in controlling the transport of dense water through overflows and its mixing with the ambient water. These include processes such as hydraulic control at sills and straits [*Girton et al.*, 2006; *Kase and Oschlies*, 2000], hydraulic jumps downstream of topographic constrictions [*Holland et al.*, 2002], interactions with narrow canyons, mesoscale eddies generated through instability of the overflow plume [*Bruce*, 1995; *Geyer et al.*, 2006; *Serra and Ambar*, 2002], interactions with overlying water masses, and interactions with tidally driven currents [*Gordon et al.*, 2004]. Other processes are associated with small-scale diapycnal mixing, e.g., shear-driven mixing in the rapid downslope flows [*Mauritzen et al.*, 2005; *Price et al.*, 1993; *Peters et al.*, 2005], bottom friction [*Peters and Johns*, 2006], and diapycnal mixing associated with internal waves and internal tides. While most of these processes are not captured by large-scale climate models, which may not even resolve the scale of the channel from which the overflow originates, mesoscale eddy-resolving models are able to explicitly simulate many of these processes. The exceptions are those associated with small-scale diapycnal mixing. The major challenge therefore for simulations of overflows at eddy-resolving resolution (i.e., high enough to resolve mesoscale eddies) is in the correct representation of the small-scale diapycnal mixing.

In this chapter we will outline the current status of mesoscale eddy-resolving modeling of overflows, review the understanding of some of the above physical processes obtained by previous eddy-resolving simulations of overflows, and discuss the difficulties that remain, particularly in correctly representing the diapycnal mixing processes in eddy-resolving simulations where diapycnal mixing may be both explicitly parameterized and result from numerical advection schemes.

Ocean Modeling in an Eddying Regime
Geophysical Monograph Series 177
10.1029/177GM06

2. PREVIOUS EDDY-RESOLVING STUDIES OF OVERFLOWS

Eddy-resolving simulations of overflows are relatively recent and few in number. As overflows by definition occur in regions of steep topographic slope, may have small widths or thicknesses, and are often associated with small deformation radii as well as strong localized mixing, it is only recently that models have been able to simultaneously cope with all these constraints within realistic regional ocean configurations.

Much of our understanding of the processes acting in overflows therefore comes from idealized simulations and laboratory experiments that allow examination of the sensitivity of mixing to various parameters under controlled conditions. Many of these idealized simulations have represented the overflow as a dense inflow forced through a narrow channel onto a uniform slope, beginning with *Jiang and Garwood* [1996] and continuing with the Dynamics of Overflow Mixing and Entrainment (DOME) collaboration [*Ezer and Mellor*, 2004; *Ezer*, 2005; *Legg et al.*, 2006]. In these simulations, provided resolution is sufficiently high, the dense plume proceeds diagonally down the slope with the path determined by a balance between the Coriolis forces and frictional forces. A large number of eddies often develop on the downslope side and entrain ambient fluid into the plume, altering its characteristics. Both frictional processes and the eddies appear to be important in controlling the downslope flow of the dense fluid. Similar eddy formation has been seen in laboratory experiments [*Lane-Serff and Baines*, 1998; *Cenedese et al.*, 2004] and in observations downstream of the Denmark Strait [*Bruce*, 1995] and Faroe Bank Channel [*Geyer et al.*, 2006]. From these simulations and laboratory experiments, as well as theoretical work, an understanding of the scales and regimes present in overflows has been obtained and is summarized here. Obviously, these idealized simulations and laboratory experiments examine only a few of the overflow processes in isolation to reduce the complexity of the problem. In particular, the DOME scenario focuses on the mesoscale eddies generated through instabilities of the overflow plume and includes both bottom friction and shear-driven mixing but omits most other important physical processes. Nonetheless, it is a useful starting point for understanding overflows where eddying behavior is observed (e.g., the Denmark Strait, Faroe Bank Channel and Gibraltar Strait overflows).

2.1. Processes, Scales, and Regimes in Rotating Overflows

In the absence of friction, a geostrophically balanced dense current would flow along isobaths with a velocity

$$U_N = \frac{\alpha g'}{f}, \tag{1}$$

a velocity scale known as the Nof speed [*Nof*, 1983], where g' is the buoyancy anomaly of the overflow plume, α is the slope of the topography, and f is the Coriolis parameter.

Frictional processes allow flow to cross isobaths so that the angle of the frictionally driven downslope flow is predicted by:

$$tan(\theta) \sim \frac{C_d U_N}{f h_0} = \text{ drag Ekman number} \tag{2}$$

when quadratic bottom drag is the dominant source of friction. Here, C_d is the bottom drag and h_0 is the depth of the overflow layer [*Price and Baringer*, 1994]. *Killworth* [2001] proposes a model for the dense plume in which h_0 is prescribed in terms of g' and U from a Froude number criterion (assuming turbulent bottom boundary layer processes are responsible for determining the depth h_0), which leads to a prediction that the rate of vertical descent ($\alpha tan(\theta) = dD/dl$, where D is the vertical location of the plume and l is the along-plume distance) should be a constant for all overflows (presuming the same turbulent processes apply in all cases).

For linear bottom drag, the angle of descent is:

$$tan(\theta) \sim R_u/f \tag{3}$$

[*Kida*, 2006], where R_u is the linear drag coefficient (with units of s^{-1}). These estimates for the downslope flow angle ignore mixing (which can modify g') and eddies (which provide a further mechanism for downslope transport) and assume the overflow layer is homogeneous so that the bottom drag is applied over the whole dense layer.

If, instead of bottom drag we have Laplacian friction, vertical gradients in velocity in the dense layer will of course be important. However, from simple dimensional arguments, we might predict the angle of descent to again follow an Ekman number-type scaling:

$$tan(\theta) \sim \frac{v_z}{f h_0^2} \sim \text{ frictional Ekman number}, \tag{4}$$

where v_z is the vertical viscosity.

There has been considerable discussion as to the physical mechanism responsible for the eddies seen in laboratory experiments, numerical simulations, and the ocean. *Swaters* [1991] has shown analytically that a lens of dense water on a slope may be unstable to a form of baroclinic instability, which leads to eddy formation on the downslope side of the dense current. The instability theory predicts that these eddies will be anticyclonic in the dense layer of the fluid, with cyclonic vortices in the upper layer. In this theory, the instability growth rate is predicted to scale with the parameter:

$$\mu = \frac{h_0}{\alpha L_{\rho D}} = \frac{h_0 f}{\alpha \sqrt{g'D}}, \tag{5}$$

where $L_{\rho D} = \sqrt{g'D}/f$ is a deformation radius based on the depth of the fluid layer (assumed to be unstratified) above the plume, D.

In contrast, *Lane-Serff and Baines* [1998] (hereafter referred to as LSB98) focus on the vorticity changes induced in the fluid above the overflow. They suggest that the eddies are produced as a result of increased cyclonic vorticity in this layer through potential vorticity (PV) conservation when fluid that is initially high on the slope (i.e., with high absolute PV) is moved to a position lower on the slope (where ambient PV is lower). This instability is therefore a barotropic instability resulting from the generation of a local extremum in potential vorticity by the advection of fluid down the slope. The mechanisms proposed for moving the upper-layer fluid downslope include a coupling with the dense underlying fluid, geostrophic adjustment of the lower layer, and draining of the lower layer by viscous processes. They argue that the eddies are barotropic and cyclonic. LSB98 propose that the relevant parameter to describe this stretching of the upper water column is the parameter

$$\Gamma = \frac{L_{\rho h_0}\alpha}{D} = \frac{\sqrt{g'h_0}\alpha}{fD}, \tag{6}$$

where $L_{\rho h_0} = \sqrt{g'h_0}/f$ is the deformation radius based on the thickness of the overflow layer. LSB98 find that both the strength of the eddies and their frequency increased as Γ increased. They only observed eddies when $\Gamma > 0.05$, perhaps because otherwise, the timescale required for eddies to evolve is too long to be observed. However, it should be noted that they only examined $\Gamma < 0.5$, and one might expect a very different regime, with no eddies, for small f, i.e., large Γ. We can estimate the boundary for the eddy-dominated regime by considering that the overflow becomes geostrophically adjusted after a distance of about $L_{\rho h_0}$. Then, the vertical distance moved down the slope is $\alpha L_{\rho h_0}$. Now, if this vertical distance is greater than the total depth of the fluid, geostrophic adjustment would not occur while the dense overflow is on the slope, and the plume would therefore not be rotationally constrained. Therefore, a plume is not rotationally constrained if

$$\alpha L_{\rho h_0} > D \rightarrow \frac{\alpha\sqrt{g'h_0}}{Df} = \Gamma > 1 \tag{7}$$

Cenedese et al. [2004] indeed find that eddies are not generated for low f.

To compare the baroclinic instability and vortex stretching arguments for eddy formation, one should note that the two key parameters are in fact closely related:

$$\mu = \frac{1}{\Gamma}\left(\frac{h_0}{D}\right)^{3/2} \tag{8}$$

and they are both related to a further parameter,

$$M = \alpha/S = \frac{\alpha}{f}\sqrt{g'/h_0} = \Gamma\frac{D}{h_0} = \frac{1}{\mu}\left(\frac{h_0}{D}\right)^{1/2}, \tag{9}$$

where S is the "geostrophic slope," i.e., the ratio $h_0/L_{\rho h_0}$. However, LSB98 find that their data collapse against Γ, but not against μ or M. Nonetheless, recent experiments by (S. Decamp, and J. Sommeria, Scaling properties for turbulent gravity currents deviated by Coriolis effect on a uniform slope, submitted to *Journal of Fluid Mechanics*, 2007; hereinafter referred to as Decamp and Sommeria, submitted manuscript, 2007) indicate that when D is constant, results do scale with M. The sign of vorticity expected in the eddies in the dense layer is opposite in the two scenarios. *Etling et al.* [2000] suggest from their laboratory experiments that the upper-layer cyclone generation dominates if μ is small, while baroclinic instability dominates if μ is large.

Careful studies with adiabatic (i.e., isopycnal models) have helped to show that both mechanisms may be possible in overflow scenarios. *Kida* [2006] demonstrates instability in a dense current that is already geostrophically balanced as it exits the channel flowing almost along topographic contours. The tendency for upper-layer fluid to be drawn down the slope by coupling with the lower-layer fluid is therefore limited, and generation of cyclonic vorticity through stretching is small. Because a layered model is used, the dense layer does not lose fluid downslope through viscous draining; instead, the whole layer moves downslope. The dense current becomes unstable and breaks up into a series of anticyclonic eddies. The motion in the upper layer consists of alternating cyclones and anticyclones. The instability growth rate and length scale appear to be well predicted by Swaters' instability theory. In particular, as predicted in Swaters' theory, the growth rate is greatly diminished if D, the depth of the upper-layer fluid, is increased. No tendency for cyclone generation is found even at small μ, unlike the results seen in the laboratory experiments of *Etling et al.* [2000] and LSB98.

This suggests that a key difference could be the mechanism for generating the dense current (initializing it moving along slope in geostrophic adjustment) in this study.

Alternatively, *Spall and Price* [1998] (hereafter SP98) show that when the flow exiting the strait has a strong barotropic component, including a component at intermediate densities, this intermediate fluid layer is indeed stretched as the dense current descends the slope. The resulting barotropic instability generates cyclonic eddies with maximum amplitude above the dense layer. This scenario seems to fit the LSB98 model, although the vortex stretching in the intermediate layer is enhanced because of the net flow in that layer. SP98 show that without the flow in the intermediate layer, a weaker instability results—they examine how the coupling between the layers generates a flow from shallow to deep regions in this layer and hence generates cyclonic vorticity, although more weakly than the LSB98 mechanism. The scenario with barotropic flow is especially pertinent to the Denmark Straits overflow where the flow through the strait includes an Artic Intermediate Water layer in addition to the densest deep water. SP98 suggest that this is a reason for more active upper-ocean eddies in the Denmark Strait overflow compared to other overflows.

All the experiments and simulations cited above have been conducted for an ambient fluid that is either unstratified or consists of a few layers. *Lane-Serff and Baines* [2000] (hereafter LSB2000) examine the scenario of a dense overflow entering a region of stable stratification. Then, the vertical coupling between overflow and overlying fluid is limited by stratification to a height

$$d_N = L_{\rho h_0} f / N, \qquad (10)$$

so that the stretching parameter for a stratified fluid becomes

$$\Gamma_N = \frac{L_{\rho h_0} \alpha}{d_N} = \frac{N\alpha}{f}, \qquad (11)$$

which, interestingly, is independent of the buoyancy anomaly and depth of the dense current itself. According to LSB2000, stratification is only expected to significantly affect the evolution of eddies if $d_N < D$, or equivalently $\Gamma_N > \Gamma$. Their laboratory experiments for the stratified case again only find a regime of eddies for $\Gamma_N > 0.05$, but once again only $\Gamma_N > 0.5$ was considered. According to LSB2000, if stratification is important, its effect is always to enhance the barotropic instability because the relative PV gradients in a layer of thickness d_N are greater than those in a layer of thickness D when $d_N < D$. However, it can also be argued that stratification

may suppress the downslope displacement of the overlying fluid (energetically, downslope displacement cannot exceed U/N where U is the velocity scale of the motion). Then, in the presence of strong stratification, the dense overflow layer could become uncoupled from the fluid above it if the dense water descends much further down the slope than a distance d_N. LSB2000 do not consider what is perhaps the most important effect of stratification on the overflow: that the overflow water will not necessarily penetrate all the way down the slope but will rather detrain when the plume reaches its neutral buoyancy level (as shown by *Jiang and Garwood* [1998] and the DOME simulations [*Ezer*, 2005; *Legg et al.*, 2006]). If the neutral buoyancy level were reached before eddy formation occurred, a very different regime might be observed. This regime transition would occur when the depth of the neutral buoyancy level is less than the vertical distance over which geostrophic adjustment occurs, i.e., when

$$\frac{g'}{N^2} < \alpha L_{\rho h_0} \rightarrow \frac{1}{r} < \Gamma_N, \qquad (12)$$

where r is the "relative stratification," $r = N/\sqrt{g'/h_0}$, the ratio of plume stratification to ambient stratification, described by LSB2000. When $r\Gamma_N > 1$, the overflow plume will reach its neutral buoyancy level and detrain into the fluid interior before eddy processes have a chance to develop (note that LSB2000 only considered $r\Gamma_N < 1$).

When eddy processes do develop before the neutral buoyancy level is reached, the eddies appear to play a significant role in this detrainment process, enabling interleaving of plume fluid with ambient fluid at the same density in the ocean interior [*Jiang and Garwood*, 1998; *Legg et al.*, 2006].

Cenedese et al. [2004] and *Smith* [1997] have found that the eddy regime only exists for small Ekman numbers, i.e., $Ek = 2\nu/(fh_0^2) < 0.1$ (and Froude numbers less than unity). For larger Ek, the flow is found to be laminar. SP98 similarly find that large bottom drag prevents the formation of eddies, and they suggest that larger bottom drag or friction causes the overflow layer to descend so rapidly that it decouples from the overlying fluid, and hence prevents the formation of eddies by the vortex stretching mechanism.

The length scale L of the eddies appears to depend on the mechanism responsible for creating them. *Kida* [2006] finds that when baroclinic instability is responsible for creating the eddies, the length scales are well predicted by the wavelength of the fastest growing mode, which in turn is of the order of $L_{\rho D}$, in unstratified ambient fluid. LSB98 find that the length scale of eddies in the vortex-stretching regime is close to $L_{\rho h_0}$. In contrast, SP98 find an eddy length scale that appears to be independent of f. They find that this length scale is well predicted by assuming that the volume of fluid contained within an eddy divided by the eddy generation timescale has to equal the imposed outflow transport so that

$$\frac{\pi L^2 h_0}{T_e} = Q \rightarrow L = \left(\frac{2Q}{\alpha N h_0}\right)^{1/2}, \qquad (13)$$

where Q is the outflow transport. $T_e = \frac{2\pi}{\alpha N}$ is the eddy generation time where the eddy frequency is $\omega = -\alpha N$, as for a short topographic Rossby wave. For SP98, N is determined from the layer model by the density difference between the layers.

LSB98 have a similar estimate for an eddy generation timescale that appears to scale like $T_e \sim 1/(f\Gamma)$. However, their explanation for this timescale is different: if the eddies move at the Nof speed $u_N \sim g'\alpha/f$ (as is seen to some extent in the experiments), and have a radius $L \sim \sqrt{g'h_0}/f$, then the minimum time between eddies must be

$$T_e \sim \frac{2L}{u_N} \sim \frac{1}{\alpha}\sqrt{\frac{h_0}{g'}}. \qquad (14)$$

However, these length, time, and velocity scales do not lead to a total transport that matches the imposed outflow transport, implying that there is some missing physics not accounted for in these scaling arguments.

For a stratified flow, LSB2000 suggest that eddy radii should again scale like $L_{\rho h_0}$. There is a suggestion from comparison of DOME calculations with and without stratification that the eddy radius is reduced by the introduction of stratification [*Legg et al.*, 2006], but whether this can be explained by the change in g' caused by the stratification has not been verified quantitatively (changes in g' leading to changes in eddy radius can also be introduced by changes in mixing due to sill topography and slope [*Ezer*, 2006].)

None of these arguments for eddy scales and regimes takes into account diapycnal mixing. Mixing appears to be most sensitive to the Froude number

$$Fr = U/\sqrt{g'h_0} \qquad (15)$$

or, equivalently, the Richardson number $Ri = 1/Fr^2$, where U is the velocity of the gravity current relative to the environment. This is seen in numerous studies of non-rotating gravity currents [*Ellison and Turner*, 1959; *Özgökmen and Chassignet*, 2002] where mixing is found only when $Fr > 1$. *Cenedese et al.* [2004] find that the Froude number also separates eddying regimes from non-eddying regimes: specifi-

cally, they only found eddies for $Fr < 1$, and when $Fr > 1$, they found a different, wave-dominated regime even with strong rotation. However, these laboratory experiments were at relatively low Reynolds numbers, and it is not certain whether the wave-dominated regime would persist at high Re or transition to a more turbulent flow. *Ezer* [2006] finds a similar supercritical, wave-like regime when $Fr > 1$ in numerical simulations. It should be noted that a single overflow may transition from a $Fr > 1$ regime to a $Fr < 1$ regime by means of a hydraulic jump, and one would expect most of the mixing to occur in the supercritical regime and most of the eddy behavior to occur in the subcritical regime.

In summary, several parameters determine the behavior of eddies and mixing in overflows: the Ekman number (high values leading to laminar overflows), the Froude number (high values leading to mixing), and some form of a stretching parameter Γ, with values of $\Gamma < 1$ leading to eddies. Eddies may be generated by barotropic or baroclinic instability, with the nature of the overflow (the extent to which it is initially ageostrophic and/or barotropic) determining the relative importance of each. Some questions remain concerning the transitions between barotropic and baroclinic instability and the dominant length scales and timescales, particularly for overflows in stratified ambient flow.

2.2. Realistic Eddy-Resolving Simulations of Overflows

While these idealized simulations and laboratory experiments have done much to identify the processes active in dense overflows (namely, baroclinic instability, vortex stretching, and frictional drainage), real overflows are obviously much more complicated than represented in these idealized geometries and include many more active physical processes. Realistic eddy-resolving simulations of overflows are relatively recent due to the high resolution required. Estimates of the length scales of eddies can be made from the scaling arguments given above: For the Denmark Strait, LSB98 predict an eddy length scale of about 18 km, and SP98 find eddies of about 30 km in diameter for Denmark Strait parameters. Other overflows with smaller g' and or h_0 will have smaller horizontal length scales. Hence, to resolve eddies and narrow topography, horizontal resolutions of less than 5 km are necessary. The specification of boundary conditions for the limited area simulations, involving inflows and evolving outflows, can be difficult. Many simulations have used a "dam break" initialization or forcing, whereby the dense water is initially contained entirely within an upstream basin; a dam at the strait or sill is removed instantaneously, and the overflow is allowed to evolve until such time as the downstream boundary conditions begin to influence the solutions. Despite the fact that these simulations are far from the continuous overflows found in nature, the quasi-steady state usually achieved downstream of the sill often has many similarities with the real overflows.

An example of such a dam-break simulation is the Denmark Strait overflow simulation performed by *Käse et al.* [2003] using a 4-km resolution terrain-following model. This simulation shows many of the features noted by SP98 in their more idealized model, namely, generation of barotropic eddies due to vortex stretching at the location where the dense flow begins its descent down the slope. They note, however, that further downstream, the flow is more baroclinic, and baroclinic instability may be more important there. In this downstream regime, the flow has become geostrophically balanced and is moving along the slope as in the idealized study of *Kida* [2006]. Perhaps as suggested earlier, the ambient stratification limits the extent over which the vortex stretching mechanism can operate, and a geostrophic, purely density-driven current appears to be more likely to lead to baroclinic instability. While the simulated eddy behavior appears to agree well with observations, the model entrainment (resulting from diffusion, convective adjustment, and the advection scheme) occurs earlier than seen in observations. More recently, (Tom Haine, private communication) 2-km resolution simulations of the Denmark Strait overflow region using the Massachusetts Institute of Technology general circulation model (MITgcm) with more complex inflow and outflow boundary forcing demonstrate a further likely role for the eddies generated in the surface layer as a result of the overflow: the surface circulation of these eddies seems to entrain cold water from the Greenland Shelf, intermittently flushing it into the open ocean.

The Mediterranean overflow has been simulated at high resolution by several authors [*Jungclaus and Mellor*, 2000; *Papadakis et al.*, 2003; *Xu et al.*, 2007]. *Jungclaus and Mellor* [2000] carry out simulations of the Mediterranean outflow at 5-km resolution using the terrain-following coordinate Princeton Ocean Model (POM). They find instability in the Gulf of Cadiz, which generates eddies on a scale of about 100 km. These eddies are injected into the interior at their neutral buoyancy level and appear similar to the observed "Meddies." The simulated outflow is found to have a double core structure, similar to the observations, which results from the vertical structure in the initial outflow shielding the lower portion of the overflow from mixing and from channeling of the deepest part of the overflow down the slope by a steep canyon.

Papadakis et al. [2003] also find Meddy-like eddy generation in their simulations with the Miami isopycnic model at 1/12° resolution. Their focus is on the dependence of the overflow to the sub-gridscale mixing parameterization. Without explicitly parameterized mixing, the dense overflow in this isopycnal model was able to descend to the bottom of the Atlantic, while with the mixing parameterization, more realistic entrainment was found, with the plume reaching a neutral buoyancy level. More recently, the Mediterranean outflow has been used as a test bed for new entrainment parameterizations [*Xu et al.*, 2007]. *Kida et al.*, 2008, have demonstrated, using highly idealized two-layer simulations, but with the Mediterranean topography, that the overflow and its eddies, through interaction with the topographic slope, can lead to the generation of a topographic beta-plume, which may extend across the ocean basin and may be a model for the Azores Current. The magnitude of this current is predicted to be highly sensitive to the amount of entrainment and to the interaction of eddies with the topography.

The Faroe Bank Channel forms the second major source of deep water in the North Atlantic (after the Denmark Strait). While the channel itself is very narrow, about 100 km downstream of the channel, the dense flow spreads out onto a broad slope. Eddy-like oscillations in transport have been observed [*Geyer et al.*, 2006], as in the Denmark Strait, but unlike the Denmark Straits, there is no barotropic flow through the channel, as the channel only constrains the deepest and densest waters. Recent simulations of this overflow using the MITgcm at 2-km resolution [*Riemenschneider and Legg*, 2007] find that most of the entrainment occurs about 50–100 km downstream of the source at around the location where the dense plume suddenly widens. Ongoing research suggests that this may be a location of a lateral hydraulic jump [*Pratt et al.*, 2007]. The magnitude and location of entrainment are similar to that seen in observations [*Mauritzen et al.*, 2005], but the simulated overflow plume is slightly more diffuse. Downstream of the widening, eddy activity is found. Similar transitions between a supercritical regime downstream of the sill and a eddy-dominated subcritical regime further downstream are found by *Ezer* [2006] using a much more idealized topography in the terrain-following coordinate Princeton Ocean Model at 2.5-km resolution. As in the laboratory experiments of *Cenedese et al.* [2004], the transition between the supercritical regime and the subcritical eddying regime was found to depend on the Froude number of the flow.

Simulations of the Red Sea overflow have been carried out by *Chang et al.* [2007] using the Hybrid Coordinate Ocean Model including parameterization of entrainment. The Red Sea overflow proceeds through two very narrow channels that play an important role in shielding the densest water from mixing. Due to the combination of the low latitude and narrow channel width, the channels are much smaller than the deformation radius so that rotation plays a minimal role. Simulations showed that whereas at a resolution of 0.5 km the overflow structure was well represented compared to observations, at 5-km resolution, the topography could not be resolved, and hence, the overflow evolution was poorly represented.

2.3. The Role of Eddies and Diapycnal Mixing in Overflows

Unlike mesoscale-eddy-resolving simulations of, for example, the Gulf Stream, high-resolution simulations of overflows, while they capture eddy processes, still do not explicitly simulate what is perhaps the dominant process, the overturning and mixing. The influence of diapycnal mixing on the overflow is obvious and direct: more mixing leads to a more diluted overflow product water and a shallower neutral buoyancy depth (in the Mediterranean, for example). The key remaining challenge for mesoscale-eddy-resolving simulations of overflows is therefore the correct representation of sub-gridscale diapycnal mixing.

By comparison, the role of eddies is more subtle and secondary to the mixing. For example, eddies enhance the rate of downslope flow. *Kida* [2006], in particular, shows that the rate of descent is increased (by about a factor of 2) in the region where the instability is developing compared to a case where there is no instability (due to an increase in the total depth of the fluid) and friction alone drives the descent. Even in the eddying simulation, further downstream when the eddy activity has become saturated, the rate of descent reverts to the frictionally driven value given by equation 3. It is not clear how well these results translate to other scenarios (e.g., with ambient stratification) because few simulations have compared eddying and non-eddying rates of descent for the same frictional parameters.

In a stratified ambient fluid, a significant role of eddies is in detraining the overflow fluid away from the topography once the neutral buoyancy level is reached (for example in the Mediterranean overflow). Another role of eddies is in influencing the upper ocean circulation. The upper ocean velocities due to the cyclonic eddies generated by the overflow, for example in the Denmark Strait region, may significantly influence transports and mixing between shelf water and deeper ocean (e.g., the Greenland Shelf "spill jets" seen in simulations by Tom Haine, private communication; see also *Pickart et al.* [2005]). Kida et al. (submitted manuscript, 2007) show that mean flows in the upper ocean are also sensitive to the overflow eddies: in particular, the eddies play an important role in connecting a topographic beta plume to the open ocean in a region of steep slope, allowing a zonal jet (similar to the Azores current) to be established.

Diapycnal mixing and eddies may also influence one another in overflows: eddies certainly lead to stirring and generation of narrow filaments, which, in numerical models, often lead to diapycnal mixing. It is uncertain at present the degree to which such mixing is representative of true physical processes. The extent to which diapycnal mixing influences the evolution of the eddies is also uncertain, as most parameter studies have been done in isopycnal models without mixing or laboratory experiments where the mixing cannot be easily measured.

Given the dominant role that unresolved diapycnal mixing plays in influencing the overflows in mesoscale-eddy-resolving simulations, the focus of the rest of this chapter is therefore on the uncertainties in representation of sub-gridscale mixing processes in these simulations.

3. SENSITIVITY OF OVERFLOWS TO MODEL CLOSURES

Although eddy-resolving calculations are able to explore the rich eddy behavior associated with overflows and have little difficulty capturing the effects of the topography on the flow, the representation of diapycnal mixing in such simulations remains a challenge. Numerical simulations generally include two different types of diapycnal mixing. First, explicit diffusivities may be used (often of the Laplacian or biharmonic form), which may have constant values or be deduced as part of a boundary-layer parameterization scheme. For example, *Ezer and Mellow* [2004] describe simulations in the DOME configuration with constant Laplacian diffusivities in the horizontal and vertical diffusivities deduced from the Mellor-Yamada scheme. Second, numerical diffusivity may result from the smoothing processes inherent in many shape-preserving advection schemes. Models with isopycnal coordinates have no numerical diffusivity, and all diapycnal mixing results from the explicit parameterization only, but both terrain-following and z-coordinate models are subject to numerical diffusivity. In the study of *Legg et al.* [2006] (hereafter LHG06), all simulations with the z-coordinate MITgcm included no explicit diffusivity, so all the (considerable) diapycnal mixing resulted from the advection scheme. The ideal model would have negligible numerical diffusion so that all diapycnal mixing is effected by a well-controlled physical parameterization; however, this is not possible for height and sigma coordinate models. We might therefore expect net diapycnal mixing to depend on the combination of vertical coordinate, advection scheme, and explicit parameterization of diffusivity. The other important consideration for sub-gridscale processes is the model viscosity. Because non-isopycnal coordinate models may include both explicit and implicit diapycnal mixing, it is very important to diagnose and evaluate the total mixing and ascertain its dependence on model numerics.

Studies to date examining the dependence of overflow representation on these factors at eddy-resolving resolution include *Ezer and Mellor* [2004], *Ezer* [2005], LHG06, and

Tseng and Dietrich [2006]. All employ the idealized DOME configuration. Here, we will briefly review the results while focusing only on those simulations at eddy-resolving resolution (i.e., $\Delta x \leq 10$ km). Unfortunately, there is no "truth" against which to compare the mixing in these simulations. Although several laboratory experiments have been carried out in a setup similar to the DOME scenario [*Cenedese et al.*, 2004; *Lane-Serff and Baines*, 1998, 2000; *Etling et al.*, 2000; Decamp and Sommeria, submitted manuscript, 2007], only *Cenedese et al.* [2004] directly measured the entrainment, and then only in a low Reynolds number regime. Furthermore, whereas most of the numerical simulations have been carried out for a regime with ambient stratification, most of the laboratory experiments are for an unstratified ambient fluid. Similarly, many large-eddy simulations and direct numerical simulations of non-rotating gravity currents exist, but none are in this rotating regime. One reason is the enormous resolution necessary to resolve the range of scales which encompasses the mesoscale eddies and the overturning eddies responsible for small-scale mixing. A recommendation for future work therefore would be to carry out detailed measurements of mixing in high Reynolds number, strongly rotating gravity currents in a stratified ambient fluid. However, while we do not have a "truth" against which to compare, it is nonetheless useful to carry out an examination of the sensitivity of model diapycnal mixing to model parameters, as follows.

Ezer and Mellor [2004] compared sigma-grid and z-level grid simulations using a generalized coordinate model [*Mellor et al.*, 2002] that used the same numerics for both grids, the same vertical viscosity and diffusivity (Mellor–Yamada scheme) [*Mellor and Yamada*, 1982], but various values of constant Laplacian diffusion and viscosity. The horizontal diffusivity and viscosity maintain a constant ratio of 1:5, and the diffusivity values vary from 10 to 10^3 m^2 s^{-1}. As would be expected, increasing the diffusion and viscosity in the along-sigma direction leads to a suppression of the eddies in the sigma-coordinate simulations. The z-coordinate simulations perform poorly in terms of moving fluid down the slope, possibly attributable to the excessive mixing generated by the Mellor–Yamada scheme when combined with the stair-step topography or because the lateral diffusion which is applied in the horizontal direction in the z-coordinate model includes a strong diapycnal component. Both vertical coordinate models used a multidimensional positive definite advection transport algorithm, which, according to the authors, necessitates non-negligible explicit diffusivity in the z-coordinate implementation. Despite the relatively poor performance of the stepped topography z-level grid with respect to down-slope plume penetration, increasing the horizontal and vertical resolution in the

z-level model converged the results toward the results of the lower resolution sigma grid.

LHG06 show that z-coordinate simulations can indeed lead to realistic simulations at eddy-resolving resolutions. Their calculations included Laplacian viscosities in both vertical and horizontal, but no explicit diffusivity. All diapycnal mixing was therefore numerical and caused by the Superbee advection scheme used. Nonetheless, at 10-km resolution, simulations with the Hallberg Isopycnal Model (HIM) compared well to the z-coordinate simulations, provided the same value of vertical viscosity was used (so that downslope descent proceeded at the same angle). The HIM simulations used a Smagorinsky horizontal viscosity parameterization and an entrainment parameterization for diapycnal mixing [*Hallberg*, 2000]. These simulations also compare well to the POM sigma coordinate simulations of *Ezer and Mellor* [2004], *Ezer* [2005]. One key result of the LHG06 z-coordinate simulations was that diapycnal mixing was reduced at the intermediate resolutions of $\Delta x = 2.5$ km, $\Delta z = 60$ m and $\Delta x = 10$ km, $\Delta z = 144$ m as compared to the highest resolution simulations at $\Delta x = 500$ m, $\Delta z = 30$ m. However, in addition to changing resolution, the values of both horizontal and vertical viscosities were changed simultaneously (to avoid noise on the grid scale), so the impact of changing resolution on diapycnal mixing cannot be separated from the impact of changing viscosities.

Ezer [2005] examines the impact of the Mellor–Yamada (M–Y) mixing scheme on the overflow simulations, using the sigma-coordinate POM at 10-km horizontal resolution, by comparing the simulation with the full M–Y scheme in the vertical to two special cases in which the vertical diffusivity or vertical viscosity respectively is set to zero. In all three cases, the lateral along-sigma diffusivity has a value of 100 m^2 s^{-1} with a viscosity to diffusivity ratio of 5:1. When vertical diffusivity is set to zero, diapycnal mixing is greatly reduced compared to the full M–Y scheme, the plume descends deeper, and the plume is thinner. Setting the vertical viscosity to zero also leads to a reduction in diapycnal mixing because the diffusivity calculated through the M–Y parameterization changes as a result of alterations in the flow caused by the viscosity. The most undiluted part of the plume again descends deeper than for the full M–Y case, although much of the plume fluid remains near the top of the slope because of the reduction in Ekman drainage.

Tseng and Dietrich [2006] present a series of simulations using the z-coordinate Dietrich Center for Air Sea Technology model. In particular, at a horizontal resolution of 5 km and vertical resolution of 60 m, they vary in horizontal viscosity and diffusivity and background vertical viscosity. A fourth-order centered-difference conservative advection scheme is used for most simulations, with much less numerical diffusion associated with it than the Superbee scheme used in LHG06, and with a much greater level of gridscale noise. Vertical diffusivity and viscosity is parameterized using a *Pacanowski and Philander* [1981] Richardson-number-dependent scheme with a specified background vertical viscosity (such a scheme, which is dependent on dimensional constants, cannot be universally suitable for different regimes with very different velocity and length scales). The control choice of parameters has horizontal viscosity and diffusivity set to 4 m^2 s^{-1} and background vertical viscosity set to 10^{-5} m^2 s^{-1}. Horizontal diffusivity and viscosity are changed simultaneously, and the results show little dependence on these parameters until they become large, i.e., 200 m^2 s^{-1}, when the eddies are suppressed and the plume becomes more sluggish and more diluted. Changing the background vertical viscosity does not have much effect on the plume, presumably because the Richardson-number-dependent part of the vertical mixing dominates and is unchanged. Eliminating vertical diffusivity and viscosity altogether does have an effect; it makes the plume much thicker in the vertical. The authors propose that this increased thickness is caused by greater vertical advection. A simulation with a second-order centered difference scheme is more noisy, with less defined eddies, but largely produces similar results. When these simulations are compared with those in LHG06, the structure of the plumes more closely resembles the highest resolution ($\Delta x = 500m$) simulations of LHG06 in that the plume does not separate from the northern boundary. This may be ascribed to the low values of vertical viscosity in TD06 (the authors state that the maximum background value of 8×10^{-3} m^2 s^{-1} is about the same as the Richardson-number-dependent part of the viscosity, whereas LHG used 5×10^{-2} m^2 s^{-1} at 2.5-km resolution and 2×10^{-2} m^2 s^{-1} at 500 m resolution). The tracer dilution is, however, more like the lower resolution LHG06 (2.5 km) runs, suggesting that the use of the higher-order advection scheme does not help in reducing total diffusion much compared to the Superbee scheme (presumably the explicit diffusion, both horizontal and vertical, which is necessary for stability with the less diffusive scheme is of the same order as the numerical diffusion in the Superbee scheme).

Finally, simulations of the Faroe Bank Channel overflow using the *z*-coordinate MITgcm at 2.5-km resolution [*Riemenschneider and Legg*, 2007] showed a surprising dependence of mixing on the vertical viscosity. As in LHG06, no explicit diffusivity was used so that all diffusion is attributable to the advection scheme. Large values of vertical viscosity were found to suppress vertical mixing so that the plume was less diluted.

In summary, simulated overflow plumes show a complicated dependence on sub-gridscale parameterization, which seems to vary from one model formulation to another. The roles of horizontal viscosity and diffusivity are difficult to distinguish, as most of the studies varied the two together; but provided they are small enough not to suppress the formation of eddies, they have little effect on the model solutions. As they are connected through complex sub-gridscale parameterizations in many models, vertical viscosities and diffusivities are even harder to separate. And finally, while the higher-order scheme of Tseng and Dietrich does not seem to lead to much less mixing than the Superbee scheme used in LHG06, the grid-scale noise it produces is highly undesirable not just in climate simulations but also in regional simulations where spurious tracer extrema may influence surface fluxes and biochemical models. There is therefore a need to examine advection schemes that are less diffusive than Superbee but which still prevent spurious oscillations. When diffusivity can be explicitly prescribed by a parameterization scheme (as in an isopycnal model), there is a need to examine how changing diffusivity parameterizations, independent of the model viscosity parameterization, affects the plume evolution. To this end, and to demonstrate the current state of the art in simulating overflows, here we present several new calculations carried out with the MITgcm and HIM focusing on the role of the sub-gridscale parameterizations in determining the simulated plume behavior.

3.1. Sensitivity of z-Coordinate Model Simulations to Viscosity and Advection Scheme

Here, we will focus on examining the role of viscosity (both horizontal and vertical) and the advection scheme on eddy-resolving simulations of overflows. Unlike the viscosity sensitivity studies discussed above, explicit diffusivity is not simultaneously modified. We will therefore be able to ascertain the influence of viscosity on numerical mixing hinted at by *Riemenschneider and Legg* [2007]. We are not advocating the use of numerical diffusion to replace explicit diffusion parameterizations, but rather, we emphasize that it is necessary to evaluate the level of numerical diffusion so that it can be minimized if an explicit parameterization scheme is implemented. We will examine the overall downslope transport and tracer properties of the simulated overflows at a resolution of 2.5 km in the horizontal and 60 m in the vertical using the MITgcm. All have identical initial conditions and forcing: a dense inflow is forced through the specification of boundary conditions at the entrance to a narrow channel, which then opens onto a uniform slope. The initial conditions consist of uniform stable stratification and no flow. Boundary conditions are no-slip combined with a quadratic bottom drag, as described in LHG06. These specifications correspond to "case 1" of the DOME scenario,

Table 1. Values of Model Parameters in the MITgcm Simulations.

Simulation	$A_z(m^2 s^{-1})$	$A_h(m^2 s^{-1})$	Advection Scheme
Control	5×10^{-2}	5.0	Superbee
Low A_z	2×10^{-4}	5.0	Superbee
High A_z	5×10^{-1}	5.0	Superbee
Low A_h	5×10^{-2}	0.5	Superbee
High A_h	5×10^{-2}	50	Superbee
OS7MP Med A_z	5×10^{-2}	5.0	OS7MP
OS7MP Low A_z	2×10^{-4}	5.0	OS7MP
Prather Med A_z	5×10^{-2}	5.0	Prather
Prather Low A_z	2×10^{-4}	5.0	Prather

described in LHG06. The only differences between the calculations are in the sub-gridscale closures, specifically the choice of vertical and horizontal viscosities and the tracer advection schemes. All the MITgcm calculations use a Laplacian viscosity, with constant coefficients A_z and A_h in the vertical and horizontal, respectively. Three different tracer advection schemes are used and are applied to both buoyancy and a passive tracer. Because our intent is to examine the numerical mixing associated with the advection scheme, no explicit diffusivity is used.

Table 1 gives the values of model parameters for the different runs. We examine results for three different vertical viscosities and two different horizontal viscosities. Three different advection schemes are examined: the Superbee flux-limited advection scheme (used in LHG06); the new OS7MP advection scheme, a seventh-order scheme (Adcroft, personal communication); and the Prather second moment scheme [*Prather*, 1986]. The Prather advection scheme has recently been shown to greatly reduce numerical diffusion in global simulations, while both the OS7MP and Prather schemes have been shown to preserve fronts while limiting diffusion in two-dimensional tests. The calculation labeled "control" is identical to the case 1, 2.5-km resolution case described in LHG06.

In the simulations, a passive tracer is injected into the inflow with a value of unity, compared to zero background value. The inflow has an initial density which matches that at the bottom of the slope. By following the evolution of the tracer field as the dense current descends the slope, we can examine the numerical mixing associated with the overflow through both the dilution of the tracer and the migration of the tracer to lighter density classes.

Plate 1 shows the tracer just above the slope (in color) with buoyancy contours overlain, for runs with different vertical viscosities and advection schemes. The most striking result is the way in which vertical viscosity influences the simulations. Higher vertical viscosities give rise to much less dilution in both tracer and buoyancy; the almost unmixed plume descends the slope at a steep angle, with no evidence of eddies. At intermediate values of vertical viscosity eddies are visible, there is considerable mixing, and the angle of descent is less steep. For the smallest vertical viscosity, there is very rapid mixing, and the dilute plume moves almost along isobaths soon after entering the slope region. Vertical viscosity therefore plays a dominant role in determining the numerical mixing and the angle of descent. This result is independent of the advection scheme used: all three advection schemes show more mixing at lower vertical viscosities. The influence of advection scheme on the results is much more subtle than that of vertical viscosity. Visually, there is little difference between the three runs with different advection schemes at the lowest value of vertical viscosity. For the intermediate vertical viscosity, the Prather advection scheme does appear to produce slightly less dilution and retains somewhat denser fluid in the downslope side of the plume.

The predicted length scale for eddies at the top of the slope $L_\rho = \sqrt{g'h_0}/f$ is about 21 km, while further down the slope, g' is reduced by both mixing with ambient fluid and by approach to neutral stability level. The length scales of the eddies seen in the simulations are considerably larger: about 130 km in diameter at the top of the slope, while in the medium vertical viscosity simulations, smaller diameter eddies are also seen lower down the slope, with diameters of about 80 km. The size of the eddies at the top of the slope where they are generated does not appear to be modified by varying vertical viscosity despite the fact that the level of mixing (and therefore g') is changed.

We can quantify the mixing by calculating the tracer-weighted buoyancy of the plume

$$\overline{b}^\tau(x) = \frac{\int b(x,y,z)\tau(x,y,z)dydz}{\int \tau(x,y,z)dydz} \tag{16}$$

where b is the buoyancy (relative to the surface) and τ is the tracer concentration. Figure 1a shows the weighted plume buoyancy scaled with the initial buoyancy as a function

---

Plate 1. (opposite) Snapshots of the passive tracer field just above the topography (in color) with buoyancy contours overlain. Passive tracer has an initial value of zero in the ambient fluid and is set to 1 in the inflow. Mixing between inflow water and ambient fluid causes dilution of the tracer. The snapshots are for a time 28 days after the beginning of the simulation. All calculations are carried out with the MITgcm with different vertical viscosity and advection schemes.

(a) $A_z = 5 \times 10^{-1}$ m^2 s^{-1}, Superbee

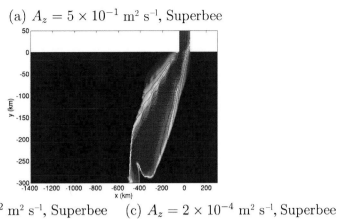

(b) $A_z = 5 \times 10^{-2}$ m^2 s^{-1}, Superbee

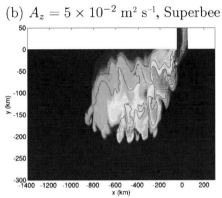

(c) $A_z = 2 \times 10^{-4}$ m^2 s^{-1}, Superbee

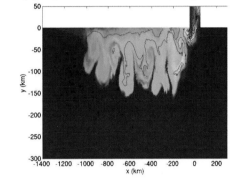

(d) $A_z = 5 \times 10^{-2}$ m^2 s^{-1}, OS7MP

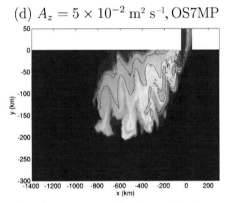

(e) $A_z = 2 \times 10^{-4}$ m^2 s^{-1}, OS7MP

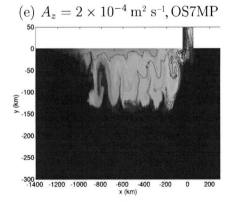

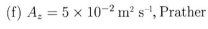

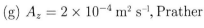

(f) $A_z = 5 \times 10^{-2}$ m^2 s^{-1}, Prather

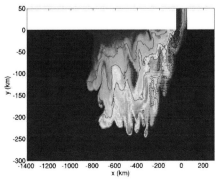

(g) $A_z = 2 \times 10^{-4}$ m^2 s^{-1}, Prather

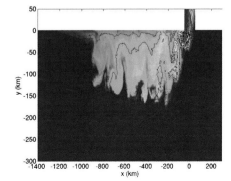

of distance from the channel entrance for all runs with the Superbee advection scheme. For the run with the highest vertical viscosity, the plume buoyancy is about 90% of its initial value, while for the lowest vertical viscosity, the buoyancy is less than 50% of its initial value. Also shown are results for two runs with different horizontal viscosities; the horizontal viscosity appears to play a minor role compared to the vertical viscosity in influencing numerical diffusion.

Figure 1b shows the weighted plume buoyancy for the intermediate vertical viscosity for the three different advection schemes. The OS7MP scheme generates slightly more mixing

(a less dense plume), while the Prather scheme leads to slightly less mixing (a denser plume) just to the left of the inflow (at $x = 0$). However, these differences are small when compared to the changes induced by changing vertical viscosity.

Figure 1c shows results for the lowest vertical viscosity for the three different advection schemes: again, OS7MP leads to slightly more mixing. Now, the Superbee and Prather schemes generate indistinguishable results.

To examine the role of sub-gridscale parameters in influencing the descent of dense fluid down the slope, we calculate the mean position of the plume (determined by the tracer distribution) as a function of along slope distance:

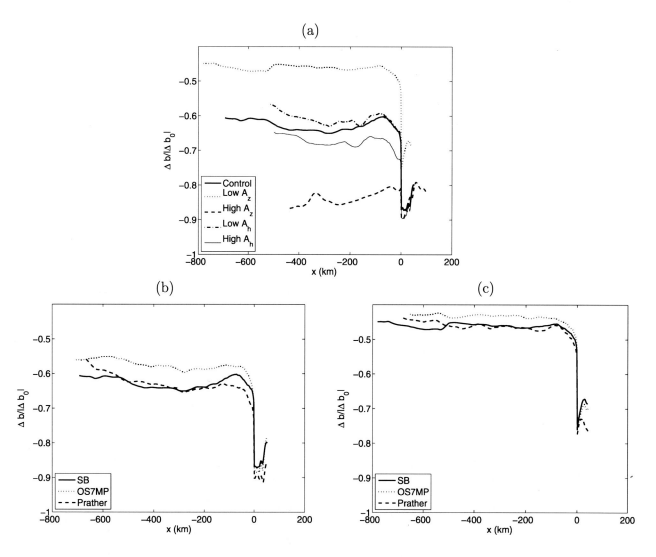

Figure 1. The tracer-weighted plume buoyancy, as a function of along-slope distance from the channel entrance, for the MITgcm calculations. (a) All simulations with the Superbee advection scheme with different horizontal and vertical viscosities. (b) Simulations with $A_z = 5 \times 10^{-2}$ m^2 s^{-1} with different advection schemes. (c) Simulations with $A_z = 2 \times 10^{-4}$ m^2 s^{-1} with different advection schemes.

$$\overline{Y}^\tau(x) = \frac{\int y\tau(x,y,z)dydz}{\int \tau(x,y,z)dydz}. \qquad (17)$$

Shown in Figure 2a is the plume path for all the Super-bee calculations. Higher vertical viscosity leads to a steeper descent, and lower vertical viscosity leads to plumes which travel along isobaths from about 200 km downstream of the inflow entrance. The Ekman numbers corresponding to the three different vertical viscosities are $Ek = 4 \times 10^{-5}$, 10^{-5}, and 0.1 for $A_z = 2 \times 10^{-4}$, 5×10^{-2}, and 5×10^{-1} m^2 s^{-1}, respectively. Note that the Ekman number for the highest vertical viscos-

ity is close to the value at which Cenedese found a transition from eddy-dominated to laminar flow. The initial angles of descent (immediately to the left of the entrance) are $\tan(\theta) =$ cross-slope distance divided by along-slope distance = 0.2, 0.3, and 0.6, respectively. These all greatly exceed the Ekman number, but more closely approach the Ekman number as viscosity increases and the Ekman drainage becomes the dominant process. At lower viscosities, the Ekman drainage is only acting in a fraction of the total plume, and so ageostrophic and eddy processes dominate the initial descent. Again, horizontal viscosity appears to make little difference to the results. Figures 2b and 2c shows the sensitivity of plume path to

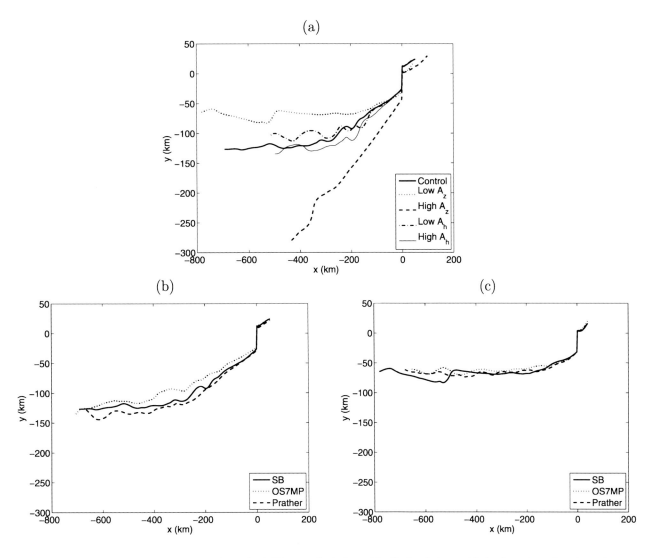

Figure 2. The tracer-weighted plume path for the MITgcm calculations. (a) All simulations with the Superbee advection scheme with different horizontal and vertical viscosities. (b) Simulations with $A_z = 5 \times 10^{-2}$ m^2 s^{-1} with different advection schemes. (c) Simulations with $A_z = 2 \times 10^{-4}$ m^2 s^{-1} with different advection schemes.

advection scheme for two different vertical viscosities: again, the differences are slight. At the intermediate viscosity, the OS7MP scheme leads to a slightly less steep descent, while the Prather scheme leads to a slightly steeper descent.

One way to consider the influence of vertical viscosity is to calculate the Ekman layer thickness $\delta = \sqrt{(2A_z/f)}$ and consider how it varies relative to the total plume thickness and grid spacing. For the three values of $A_z = 2 \times 10^{-4}$, 5×10^{-2}, and 5×10^{-1} m^2 s^{-1}, we have $\delta = 2$, 32, and 100 m, respectively. The actual plume thickness (as defined by the fluid with tracer concentrations of at least 10% of the inflow value) is about 450–1080, 100–500, and 210 m, respectively, with the lower viscosity cases having a large range of thicknesses due to the presence of eddies. It is interesting to note that as A_z is increased, the thickness of the plume is reduced, perhaps because viscous drainage is more effective at carrying fluid down the slope or perhaps because enhanced numerical mixing is responsible for the large thicknesses in the low viscosity cases. In addition, only the highest viscosity case satisfies the *Winton et al.* [1998] criteria $\delta > \alpha\Delta x$, and $\delta < \Delta z$. Hence, if the Ekman layer is responsible for much of the downslope flow, this numerical configuration can only resolve that downslope flow at high values of vertical viscosity. Numerical mixing may therefore be a result of a poorly resolved Ekman layer. This would suggest that sigma-coordinate models would also be subject to the same transition in numerical mixing as a function of vertical viscosity, but would only have to satisfy the vertical constraint $\delta > D\Delta\sigma$, where D is the total depth, and $\Delta\sigma$ is the sigma-coordinate resolution.

To summarize, vertical viscosity plays a very important role in determining the behavior of a simulated dense overflow at high resolution not just by setting the angle of descent (as would be expected) but also by modifying the numerical diffusion. This effect of vertical viscosity on numerical diffusion is independent of the advection scheme used. All three advection schemes employ limiting schemes designed to prevent grid-scale oscillations in the tracer field, and it is through these limiting schemes that numerical diffusion is introduced. The vertical viscosity influences the numerical diffusion by first modifying the small-scale structure of the velocity field: smaller viscosities lead to more small-scale structure in the velocity field and require greater numerical diffusion to eliminate grid-scale noise. Hence, the smaller the vertical viscosity, the higher the numerical diffusivity and the greater the diapycnal mixing. This dependence of mixing on the vertical viscosity agrees with that observed in *Riemenschneider and Legg* [2007]. Note that *Tseng and Dietrich* [2006] see a qualitatively similar effect when their vertical viscosity parameterization is turned off: the plume appears to thicken and become more dilute.

An interesting, although discouraging, result of these simulations is that more sophisticated advection schemes do not appear to dramatically improve results at these resolutions. The Prather scheme does marginally better than the Superbee scheme, but the OS7MP scheme does worse (given that one would like to minimize numerical diffusion so that physically based parameterizations of mixing may be used instead). One cause of the poor performance of the OS7MP scheme may be that the advantage of a seventh-order scheme is lost when simulating a thin layer (less than seven grid points) close to a boundary where the scheme behaves as a lower order scheme. A more sophisticated near-boundary formulation of the scheme could perhaps improve this result. The current implementation of the Prather scheme uses a positive-definite limiter, and a monotonicity-preserving limiter may change the Prather result (Jean-Michel Campin, private communication). For coarse-resolution climate models where fronts are not resolved, the sensitivity of the behavior to advection scheme may be different than seen in these mesoscale-eddy-resolving simulations.

3.2. Sensitivity of Isopycnal Model Simulations to Mixing Parameterization

Having discussed the dependence of numerical diapycnal mixing on advection scheme and viscosity for a *z*-coordinate model, we now examine the influence of explicitly parameterized mixing on the overflow simulations in HIM, a model with no numerical mixing. Three different simulations with HIM configured for the DOME scenario case 1 are shown in Plate 2. Initial conditions and forcing are as described above, one difference being that a sponge condition is used at the western and southern boundaries instead of the radiation condition used in the MITgcm calculations. Also, the horizontal domain is slightly smaller (1,300 km rather than 1,700 km). All three simulations have a resolution of 2.5 km in the horizontal and 25 vertical layers with a total density difference of 2 kg/m^3. A biharmonic Smagorinsky viscosity is used in the horizontal and a constant viscosity of 1×10^{-4} m^2 s^{-1} in the vertical. The three simulations are distinguished by different schemes for parameterization of the mixing between isopycnal layers. In the first simulation, there is no prescribed diapycnal mixing apart from a small background value of 1×10^{-4} m^2 s^{-1}. In the second, diapycnal mixing is prescribed by the Turner scheme (TP) outlined in the work of *Hallberg* [2000] along with the frictional boundary layer mixing described in LHG06. In the third simulation, diapycnal mixing is prescribed by the new Jackson and Hallberg scheme (JH) for shear-driven mixing described in detail by Jackson et al. [2007] along with the frictional boundary layer mixing scheme. The tracer-weighted buoyancy and path of the plume are shown in Figure 3, as defined in equations (16) and (17).

With only a small background diapycnal mixing, the plume stays close to its original density and takes a steep path down the slope. While the steepness of the path is similar to that seen in the z-coordinate case with high vertical viscosity, there is considerable eddy behavior and plume widening in the isopycnal simulation so that the dynamical regime is quite different and not a laminar Ekman flow.

The two simulations with prescribed mixing look very similar so that for these physical parameters, the TP and JH schemes generate a similar amount of mixing, although the former results in slightly more mixing and a lighter plume. This similarity is despite the large difference in the critical Richardson number (Ri_c) at which the turbulent mixing stops: $Ri_c = 0.8$ in TP and $Ri_c = 0.25$ in JH (it should be noted, however, that when the same parameterization schemes are used in the equatorial regions, with no change in parameters, the JH scheme produces much less mixing than the TP scheme and therefore performs much better.) Like the z-coordinate simulations, most of the mixing occurs just after the plume enters the slope region, and the plume forms a series of large eddies through instability; however, with the mixing parameterization schemes, there is also some mixing and a decrease in buoyancy along the slope as seen in Figure 3a. The solutions in Plate 2 for both the JH and TP mixing schemes

are qualitatively most similar to the z-coordinate cases with intermediate vertical viscosity in Plates 1b, 1d, and 1f: the plume is not as dilute as in the low viscosity cases, and although it is mainly flowing along isobaths, it shows some signs of separating from the wall and proceeding downslope (Figure 3b) as in the intermediate viscosity case. Both the final buoyancy of the plume and the angle of descent are in between those of the z-coordinate cases with intermediate and low vertical viscosity. The vertical viscosity in the isopycnal model is close to that in the low vertical viscosity case z-coordinate simulation so that at this value of vertical viscosity, the mixing induced by the numerical scheme in the z-coordinate model exceeds the mixing due to the parameterization scheme in the isopycnal model.

Given that the parameters in the JH diffusivity parameterization have been chosen carefully with reference to direct numerical simulation, this simulation might be regarded as our best estimate of the actual amount of mixing one would expect in this physical scenario in the absence of definitive laboratory measurements of mixing. A z-coordinate scheme without an explicit mixing parameterization gives more mixing for a similar vertical viscosity; hence, until the numerical mixing can be reduced in the z-coordinate simulation, there is little point in focusing attention on implementing sophisticated schemes to parameterize the mixing in such a model. One way to reduce the mixing below the level of parameterized mixing in the isopycnal model is to increase the vertical

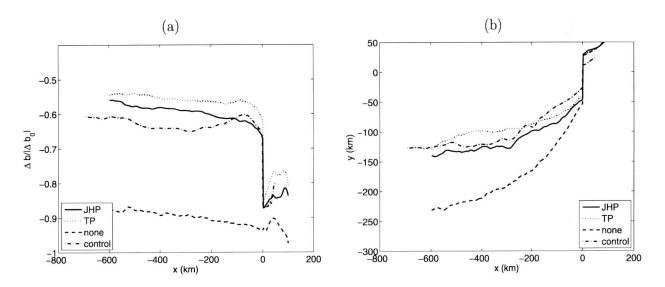

Figure 3. For the Hallberg Isopycnal Model simulations: (a) the tracer-weighted plume buoyancy, as a function of along-slope distance from the channel entrance; (b) the tracer-weighted plume path. Results are shown for three different diapycnal mixing parameterizations: "no mixing" (i.e., only weak background mixing), Jackson–Hallberg parameterization (JHP) and Turner parameterization (TP). For comparison, the MITgcm control simulation is also shown.

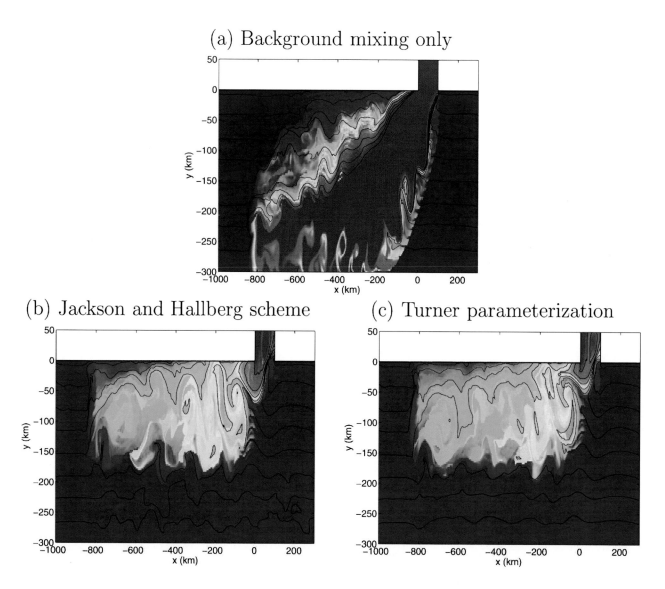

(a) Background mixing only

(b) Jackson and Hallberg scheme

(c) Turner parameterization

Plate 2. Snapshots of the passive tracer field just above topography (in color) with buoyancy contours overlain. The snapshots are for a time 40 days after the beginning of the simulation. All calculations are carried out with the Hallberg Isopycnal Model with different parameterizations of diapycnal mixing.

viscosity; however, this also suppresses eddy behavior and the simulated plume is then governed entirely by Ekman dynamics.

4. DISCUSSION AND CONCLUSIONS

Mesoscale eddies are a feature of many overflows. These eddies may be generated by either barotropic instability when overlying water is transported to deeper depth during the geostrophic adjustment process or by baroclinic instability of the geostrophically adjusted dense fluid on the slope. In fact, examination of the vorticity field in the simulations described above suggests that both processes may operate within the same overflow. Largely barotropic cyclonic vortices are found in the upper layers, close to the coast, generated through barotropic instability, while the dense water below is carried further downslope by more baroclinic eddies. We have reviewed the literature regarding the formation of eddies in overflows and shown that eddies exist when rotation plays an important role in the overflow dynamics so that the overflow comes under the influence of rotation before it reaches its neutral buoyancy level or the bottom of the slope. If frictional effects are large so that the thickness of the overflow plume is close to the Ekman layer thickness, then the plume forms a laminar Ekman layer and eddies are suppressed. Similarly, if Froude numbers are large, the supercritical flow does not form eddies until a transition is made to a subcritical regime.

While mesoscale eddies are easily captured by high-resolution ocean models, the much smaller scale eddies responsible for overturning and mixing cannot be resolved. The diapycnal mixing they are responsible for therefore remains the most significant challenge for overflow simulations and plays a dominant role in determining the dilution of overflow water, setting the tracer properties and volume of the final product water mass. Models are forced to represent this diapycnal mixing through diffusivity parameterizations. Unfortunately, diapycnal mixing also results from numerical diffusion in level coordinate and sigma-coordinate models, complicating matters. We have therefore critically examined the diapycnal mixing in mesoscale-eddy-resolving overflow simulations and examined its dependence on both explicit diffusivity parameterizations and on model numerics. Many past studies are complicated because viscosity and diffusivity are varied simultaneously, and little attempt is made to quantify numerical diffusion. Ideally, one would like a model with negligible numerical diffusion in which all diapycnal mixing is prescribed through an explicit parameterization. To that end, it is necessary to identify model configurations with the minimum numerical diffusion. To complement earlier studies, we have therefore de-

scribed several new simulations focusing on the numerical diffusion in a z-coordinate model (the MITgcm) and its sensitivity to model viscosity and advection scheme and on the influence of diapycnal mixing parameterization in an isopycnal-coordinate model. Numerical diffusion in a z-coordinate model was found to depend strongly, in an inverse sense, on vertical viscosity, as the advection scheme introduces sufficient numerical diffusion to remove any grid-scale noise caused by low values of vertical viscosity. We strongly suspect that a similar dependence would be found in a sigma-coordinate model. As all the numerical diffusion is a result of the advection scheme, we examined the sensitivity to advection scheme and found that while the Prather advection scheme performed a little better than the others, in terms of reducing numerical diffusion, the difference was not large. Finally, while the presence of a diapycnal mixing scheme was found to strongly influence the isopycnal model simulations, the exact nature of that scheme was not found to have a strong influence for this physical scenario (given that parameters have been chosen appropriately in both schemes).

Mesoscale-eddy-resolving models of overflows therefore have to make many choices to correctly capture both the resolved physics and the unresolved diapycnal mixing. These include the vertical coordinate, the advection scheme, and the viscosity and diffusivity parameterization. If sigma-coordinate or z-coordinate models are used, they will inherently have numerical diffusion, and the least diffusive advection scheme would be preferable. However, there is little point choosing a non-diffusive scheme if that scheme generates so much noise as to require large values of explicit diffusivity. The horizontal viscosity should obviously be small enough to permit eddies, but otherwise, this choice is not too important. However, the vertical viscosity does play an important role. For a constant vertical viscosity, it seems appropriate, to prevent too much mixing, to choose a value as large as possible without forcing the plume into the large Ekman number laminar regime. For a variable eddy viscosity parameterization (e.g., Mellor–Yamada), further studies are needed to understand the overflow mixing resulting from the M–Y scheme and how it compares with constant vertical viscosity simulations. Given the complex spatial distribution of vertical viscosity in the gravity current when using the M–Y scheme (e.g., Figure 3 of *Ezer* [2005]), it is unclear if the large sensitivity to viscosity found here will apply to variable eddy viscosity cases.

Isopycnal simulations have the advantage that the advection scheme does not cause diapycnal mixing and that the amount of mixing can be directly controlled by the mixing parameterization completely independently of the vertical viscosity (which also influences the plume path and the eddy dynamics).

Acknowledgments. This paper is a contribution of the Gravity Current Entrainment Climate Process Team, funded by the National Science Foundation and the National Oceanic and Atmospheric Administration as part of the US component of the Climate Variability and Prediction (CLIVAR) program. We would like to thank other team members for sharing their results. We would also like to thank the ocean model development teams at the Massachusetts Institute of Technology (especially Jean-Michel Campin) and at the Geophysical Fluid Dynamics Laboratory (especially Alistair Adcroft) for allowing us to test their recently implemented advection schemes in the overflow configuration. Comments from Anand Gnanadesikan, Matthew Harrison, and two anonymous reviews helped to improve the manuscript. We also thank Matthew Hecht and Hiro Hasumi for inviting us to contribute this chapter.

REFERENCES

Bruce, J. (1995), Eddies southwest of the Denmark Strait, *Deep Sea Res.*, *42*, 13–29.

Cenedese, C., J. A. Whitehead, T. Ascarelli, and M. Ohiwa (2004), A dense current flowing down a sloping botton in a rotating fluid, *J. Phys. Oceanogr.*, *34*, 188–203.

Chang, Y. S., T. Özgökmen, H. Peters, and X. Xu (2008), Numerical simulation of the Red Sea outflow using HYCOM and comparison with REDSOX observations, *J. Phys. Oceanogr.*, *38*, 337–358.

Ellison, T., and J. Turner (1959), Turbulent entrainment in stratified flows, *J. Fluid Mech.*, *6*, 423–448.

Etling, D., F. Gelhardt, U. Schrader, F. Brennecke, G. Kuhn, G. C. d'Hieres, and H. Didelle (2000), Experiments with density currents on a sloping bottom in a rotating fluid, *Dyn. Atmos. Oceans*, *31*, 139–164.

Ezer, T. (2005), Entrainment and diapycnal mixing in three-dimensional bottom gravity simulations using the Mellor-Yamada turbulence scheme, *Ocean Model.*, *9*, 151–168.

Ezer, T. (2006), Topographic influence on overflow dynamics: Idealized numerical simulations and the Faroe Bank Channel overflow, *J. Geophys. Res.*, *111*, C02002 doi:1029/2005JC003195.

Ezer, T., and G. Mellor (2004), A generalized coordinate ocean model and a comparison of the bottom boundary layer dynamics in terrain-following and z-level grids, *Ocean Model.*, *6*, 379–403.

Geyer, F., S. Østerhus, B. Hansen, and D. Quadfasel (2006), Observations of highly regular oscillations in the overflow plume downstream of the Faroe Bank Channel, *J. Geophys. Res.*, *111*, C12020, doi:10.1029/2006JC003693.

Girton, J., L. J. Pratt, D. Sutherland, and J. Price (2006), Is the Faroe Bank Channel overflow hydraulically controlled?, *J. Phys. Oceanogr.*, *36*, 2340–2349.

Gordon, A., E. Zambianchi, A. Orsi, M. Visbeck, C. F. Giulivi, T. Whitworth III, and G. Spezie (2004), Energetic plumes over the western Ross Sea continental slope, *Geophys. Res. Lett.*, *31*, L21302, doi:10.1029/2004GL020785.

Hallberg, R. (2000), Time integration of diapycnal diffusion and Richardson number—dependent mixing in isopycnal coordinate ocean models, *Mon. Wea. Rev.*, *128*, 1402–1419.

Holland, D., R. Rosales, D. Stephanica, and E. Tabak (2002), Internal hydraulic jumps and mixing in two-layer flows, *J. Fluid Mech.*, *470*, 63–83.

Jackson, L., R. Hallberg, and S. Legg (2008), A parametrization of shear-driven turbulence for ocean climate models, *J. Phys. Oceanogr.*, in press.

Jiang, L., and R. Garwood (1996), Three-dimensional simulations of overflows on continental slopes, *J. Phys. Oceanogr.*, *26*, 1214–1233.

Jiang, L., and R. Garwood (1998), Effects of topographic steering and ambient stratification on overflows on continental slopes: A model study, *J. Geophys. Res.*, *103*, 5459–5476.

Jungclaus, J., and G. Mellor (2000), A three-dimensional model study of the Mediterranean outflow, *J. Mar. Sci.*, *24*, 41–66.

Käse, R. H., and A. Oschlies (2000), Flow through Denmark Strait, *J. Geophys. Res.*, *105*, 28,527–28,546.

Käse, R. H., J. B. Girton, and T. B. Sanford (2003), Structure and variability of the Denmark Strait Overflow: Model and observations, *J. Geophys. Res.*, *108*, doi:10.1029/2002JC001548.

Kida, S. (2006), Overflows and upper ocean interaction: a mechanism for the Azores current, Ph.D. thesis, MIT–WHOI Joint Program in Oceanography.

Kida, S., J. Price, J. Yang (2008), The upper oceanic response to overflows: a mechanism for the Azores current, *J. Phys. Oceanogr.* in press.

Killworth, P. (2001), On the rate of descent of overflows, *J. Geophys. Res.*, *106*, 22,267–22,275.

Lane-Serff, G., and P. Baines (1998), Eddy formation by dense flows on slopes in a rotating fluid, *J. Fluid Mech.*, *363*, 229–252.

Lane-Serff, G., and P. Baines (2000), Eddy formation by overflows in stratified water, *J. Phys. Oceanogr.*, *30*, 327–337.

Legg, S., R. Hallberg, and J. Girton (2006), Comparison of entrainment in overflows simulated by z-coordinate, isopycnal and non-hydrostatic models, *Ocean Model.*, *11*, 69–97.

Mauritzen, C., J. Price, T. Sanford, and D. Torres (2005), Circulation and mixing in the Faroese Channels, *Deep Sea Res.*, *52*, 883–913.

Mellor, G. L., and T. Yamada (1982), Development of a turbulence closure model for geophysical fluid problems, *Rev. Geophys.*, *20*, 851–875.

Mellor, G., S. Hakkinen, T. Ezer, and R. Patchen (2002), A generalization of a sigma coordinate ocean model and an intercomparison of model vertical grids, in *Ocean Forecasting: Conceptual Basis and Applications*, edited by N. Pinardi and J. D. Woods, pp. 55–72, Springer, NY.

Nof, D. (1983), The translation of isolated cold eddies on a sloping bottom, *Deep Sea Res.*, *30*, 171–182.

Özgökmen, T., and E. Chassignet (2002), Dynamics of two-dimensional turbulent bottom gravity currents, *J. Phys. Oceanogr.*, *32*, 1460–1478.

Pacanowski, R., and S. Philander (1981), Parameterization of vertical mixing in numerical models of tropical ocean, *J. Phys. Oceanogr.*, *11*, 1443–1451.

Papadakis, M., E. Chassignet, and R. Hallberg (2003), Numerical simulations of the Mediterranean sea outflow: impact of the entrainment parameterization in an isopycnic coordinate ocean model, *Ocean Model.*, *5*, 325–356.

Peters, H., and W. Johns (2006), Bottom layer turbulence in the Red Sea outflow plume, *J. Phys. Oceanogr.*, *36*, 1763–1785.

Peters, H., W. Johns, A. Bower, and D. Fratantoni (2005), Mixing and entrainment in the Red Sea Outflow Plume. Part I. Plume Structure, *J. Phys. Oceanogr.*, *35*, 569–583.

Pickart, R. S., D. J. Torres, and P. S. Fratantoni (2005), The East Greenland Spill Jet, *J. Phys. Oceanogr.*, *35*, 10372013–1053.

Prather, M. J. (1986), Numerical advection by conservation of second-order moments, *J. Geophys. Res.*, *91*, 6671–6681.

Pratt, L., U. Riemenschneider, and K. Helfrich (2007), A transverse hydraulic jump in a model of the Faroe Bank Channel outflow, *Ocean Model.*, *19*, 1–9.

Price, J., and M. Baringer (1994), Overflows and deep water production by marginal seas, *Progr. Oceanogr.*, *33*, 161–200.

Price, J., M. Baringer, R. Lueck, G. Johnson, L. Ambar, G. Parrilla, A. Cantos, M. Kennelly, and T. Sanford (1993), Mediterranean outflow mixing and dynamics, *Science*, *259*, 1277–1282.

Riemenschneider, U., and S. Legg (2007), Regional simulations of the Faroe Bank Channel overflow in a level model, *Ocean Model.*, *17*, 93–122.

Serra, N., and I. Ambar (2002), Eddy generation in the Mediterranean Undercurrent, *Deep Sea Res. II*, *49*, 4225–4243.

Smith, P. (1977), Experiments with viscous source flows in rotating systems, *Dyn. Atmos. Oceans*, *1*, 241–272.

Spall, M., and J. Price (1998), Mesoscale variability in Denmark Strait: The PV outflow hypothesis, *J. Phys. Oceanogr.*, *28*, 1598–1623.

Swaters, G. (1991), On the baroclinic instability of cold-core coupled density fronts on a sloping continental shelf, *J. Fluid. Mech.*, *224*, 361–382.

Tseng, Y.-H., and D. Dietrich (2006), Entrainment and transport in idealized three-dimensional gravity current simulation, *J. Ocean. Atmos. Technol.*, *23*, 1249–1269.

Winton, M., R. Hallberg, and A. Gnanadesikan (1998), Simulation of density-driven frictional downslope flow in *z*-coordinate ocean models, *J. Phys. Oceanogr.*, *28*, 2163–2174.

Xu, X., E. Chassignet, J. Price, T. Özgökmen, and H. Peters (2007), A regional modeling study of the entraining Mediterranean out flow, *J. Geophys. Res.*, *112*, doi:10.1029/2007JC004145.

S. Legg, L. Jackson, and R. W. Hallberg, National Oceanic and Atmospheric Administration, Geophysical Fluid Dynamics Laboratory, Princeton University Forrestal Campus, 201 Forrestal Rd, Princeton, NJ 08540, USA. (Sonya.Legg@noaa.gov)

High-Frequency Winds and Eddy-Resolving Models

Patrice Klein

Laboratoire de Physique des Océans, Ifremer, Plouzané, France

Wind energy input to oceanic near-inertial motions [estimated between 0.5 terawatt (TW) and 0.9 TW] is comparable to the work done by the steady large-scale winds on the general circulation of the ocean (estimated to be 1 TW). This energy input is the largest principally in the regions of atmospheric storm tracks. It is one of the main kinetic energy sources that may sustain the small-scale mixing at depth that is necessary to maintain the deep ocean stratification and therefore to close the total kinetic energy budget. But how much of this wind-driven near-inertial energy penetrates into the deep ocean interior, and where, is still a puzzle. In other words, what is the route to mixing from the surface to the deep interior followed by these motions? In this review, we present different pathways by which near-inertial energy ultimately can reach the deep interior and be available for mixing. The dominant pathway appears to be through the oceanic turbulent eddy fields at mid-latitudes. It makes a nonnegligible part of the near-inertial energy to penetrate into the deep interior with conspicuous maxima at depths as large as 2,500–3,000 m. However, we still lack a precise quantification of the contribution of the different pathways and, in particular, of the part of near-inertial energy used for mixing the upper layers relatively to that used for mixing the deeper layers. Eddy-resolving models at a basin scale (with adequate spatial resolution) forced by high-frequency winds should be able in the close future to explicitly represent the 3-D propagation of the near-inertial waves down to 5,000 m. A few process studies are still needed to parameterize mixing generated by these waves and, in particular, of those in the deep interior.

1. WIND RINGING OF THE OCEAN AT THE NEAR-INERTIAL FREQUENCY

The ocean is a stratified rotating fluid and, as such, supports several classes of high-frequency waves from the Coriolis frequency (f) to the Brunt-Väisälä frequency. At mid-latitude, these frequencies range from 10^{-4} s^{-1} for the Coriolis frequency to 3.10^{-3} s^{-1} for the Brunt-Väisälä frequency. A classical dispersion relation for these waves (in the shallow water framework) is

$$\omega^2 = f^2(1 + k^2 r_d^2) \tag{1}$$

with ω the wave frequency, k their horizontal wave number, and r_d a Rossby radius of deformation[1] that characterizes the oceanic stratification. Atmospheric wind stress can force these different classes of waves if wind energy is present in

Ocean Modeling in an Eddying Regime
Geophysical Monograph Series 177
Copyright 2008 by the American Geophysical Union.
10.1029/177GM07

this large frequency band. However, high-frequency winds are characterized by wavenumber spectra that exhibit energetic horizontal scales of the order of $\mathcal{O}(500–1,000 \text{ km})$, i.e., much larger than the oceanic internal Rossby radii of deformation (<50 km). This means that for oceanic motions forced by such a wind stress, kr_d is much smaller than 1, which explains that the frequency of the resulting wind-driven waves is close to the Coriolis frequency. These near-inertial waves are trapped within the surface mixed layer (50–100 m deep), and therefore, the mixed layer can be considered as an oscillator with the frequency f.

Winds are strongly intermittent in time with periods ranging from 3 to 6 h to much larger timescales. If the wind stress possesses some energy in the inertial frequency band and rotates in the same direction as the near-inertial motions, a systematic increase of the near-inertial kinetic energy and mixing will occur within the mixed layer. Such a resonance mechanism has been well explained by several studies [e.g., *Klein and Coantic*, 1981; *Large and Crawford*, 1995; *Skyllinstad et al.*, 2000].

Figure 1 [from *Klein et al.*, 2004a] can help to explain the impact of the high-frequency winds and the associated resonance mechanism. It shows the time evolution of the wind intensity (in terms of u^{*3}; upper panel), the wind energy flux (middle panel; i.e., $\tau \cdot \mathbf{u}$, where τ is the wind stress vector and $\mathbf{u}$ the surface velocity vector) and the integrated near-inertial energy (lower panel), obtained by using three

Table 1. Statistics over 540 days of the near-inertial energy integrated over the water column (in $m^3\ s^{-2}$) obtained With the 1D model [from *Klein et al.*, 2004a].

Simulations	Near-inertial energy	
	Mean value	RMS value
3-hourly winds	2.03	1.70
6-hourly winds	1.65	1.38
12-hourly winds	0.62	0.56
24-hourly winds	0.28	0.28

hourly meteorological data sampled on the weathership "KILO" located in the North Atlantic. Comparison of the upper and middle panels indicates that not all strong wind events produce a positive wind energy flux and therefore increase the kinetic energy. Actually, a nonnegligible number of strong wind events decreases the kinetic energy, which illustrates the importance of the phase relationship between the wind stress direction and the surface velocity vector. A particular and conspicuous signature of this resonance mechanism is highlighted in tropical hurricane studies [see *Price*, 1981; *Price et al.*, 1994]. They reveal that in the North Hemisphere, this mechanism induces greater cooling on the right side of the hurricane trajectory (where, at a given point, winds rotate clockwise as the hurricane moves across) compared to the left side (where winds rotate anticlockwise).

One interesting feature is that the evolution of the total kinetic energy (shown on the lower panel of Figure 1) is characterized by periods of persisting large amplitude (that last between 10 and 20 days) followed by periods of small amplitude. As found by previous studies [e.g., *Large and Crawford*, 1995; *Skyllinstad et al.*, 2000; *Klein et al.*, 2004a], not only the frequency (with respect to the Coriolis frequency) but also the rotation of the wind fluctuations is important for the resonance mechanism. Thus, another experiment identical to the first one but with the Coriolis parameter having the opposite sign has revealed that although the magnitude of the kinetic energy averaged over 540 days is about the same, its time evolution (not shown) displays strong differences.

To better quantify the impact of the high-frequency wind fluctuations on the ocean ringing, the experiment reported in Figure 1 has been compared with others identical to this one but that use the wind stress averaged over 6, 12, and 24 h. This impact is clearly revealed by Table 1: When the wind stress is averaged over 6, 12, and 24 h, the total kinetic energy is decreased, respectively, by 1.5, 3, and 7! The resonance mechanism mentioned before also explains the near-inertial oscillations with large amplitude that are present in the in situ observations [e.g., *Stockwell et al.*, 2004]. All these results illustrate the necessity to use at least a 3-hourly wind time series to force the ringing of the ocean, i.e., to

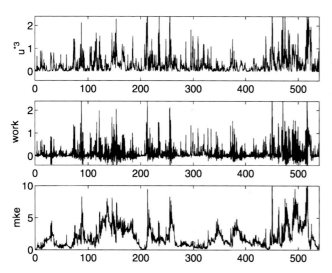

Figure 1. Results [from *Klein et al.*, 2004a] of a 3-hourly simulation using a Mellor and Yamada model and a real wind time series (upper panel) $u^{*3} = |\tau|^{\frac{3}{2}}$ (in $m^3\ s^{-3} \times 1.5 \times 10^5$; middle panel) wind energy flux ($[\tau \cdot \mathbf{u}]$; in $m^3\ s^{-3} \times 10^4$; lower panel) inertial energy integrated over the water column (in $m^3\ s^{-2}$). Abscissa units are days.

correctly estimate the wind energy flux to the ocean in the inertial frequency band.

2. HIGH-FREQUENCY WIND ENERGY FLUX AND THE GLOBAL OCEANIC CIRCULATION

Alford [2001], *Watanabe and Hibiya* [2002], and *Alford* [2003] indicate that the wind kinetic energy input to oceanic near-inertial motions is the largest principally in the regions of atmospheric storm tracks located at mid-latitudes, i.e., where the high-frequency winds are the strongest. These high-frequency winds are essentially linked to the extra-tropical cyclone frequency and intensity and their associated fast traveling atmospheric fronts. Plate 1a [from *Alford*, 2003] illustrates these results during the Northern Hemisphere winter: The excited near-inertial energy is the largest north of 30° latitude. During the Southern Hemisphere winter, the largest energy input regions are south of 30° latitude. The seasonal variation of this wind energy input is much stronger in the Northern Hemisphere than in the Southern Hemisphere. These studies used data reanalysis from Meteorological National Centers (with usually a 2° spatial resolution and a time interval of 6 h). However, *Watanabe and Hibiya* [2002] and *Alford* [2003] have applied correction factors to get a better estimation of the wind-driven near-inertial motions when using such 6-hourly wind data. The ocean part is represented only by a slab mixed-layer model, and a Rayleigh damping is used to mimic the radiation of the near-inertial energy from the mixed layer to the deeper ocean. There is no large or mesoscale oceanic circulation.

Contribution of this wind kinetic energy input to the global oceanic energy budget is discussed by *Wunsch and Ferrari* [2004]. Using *Alford*'s [2003] results, they found that the global energy input due to high-frequency winds (estimated between 0.5 and 0.9 TW) is comparable to the work done by the steady large-scale winds on the general circulation (estimated to be 1 TW). These numbers are probably underestimated because of the low spatial resolution of the meteorological data used (see section 6). However, the importance, for the dynamics of the general oceanic circulation, of this large amount of energy input to the near-inertial waves is clearly invoked by *Munk and Wuncsh* [1998] and *Wunsch and Ferrari* [2004]. They suggest that the resulting wind-driven near-inertial energy could sustain the small-scale mixing in the deep interior that is necessary to resupply the available potential energy removed by mesoscale eddy generation and the Meridional Overtuning Circulation (MOC). Thus, high-frequency winds may provide enough energy to maintain the deep ocean stratification, in particular at mid-latitudes.

Alford [2003] notes that the total high-frequency wind energy input has increased by 28% since 1948. Therefore, assuming that these wind-forced near-inertial motions are a potential source of mechanical energy available for mixing in the deep ocean, then using the arguments raised by *Munk and Wunsch* [1998] and *Wunsch and Ferrari* [2004] a secular increase in extra-tropical atmospheric cyclone frequency and intensity would be accompanied by a strengthened MOC and/or an intensified eddy kinetic energy at mid-latitudes. The consequence of this comment is that the impact of the resulting mixing in the oceanic circulation cannot be parameterized as just a constant global diffusion coefficient. This impact may be considered as a component of the coupled atmosphere-ocean system.

However, how much of this wind-driven near-inertial energy penetrates into the deep ocean interior, and where, is still a real puzzle. We need to better understand the mechanisms through which the near-inertial energy produced at the surface propagates downward. In other words, the question is, what is the detailed route to mixing from the surface to the deep interior followed by these motions? Some clues and suggestions for answering this question are offered in the next sections.

3. DOWNWARD AND EQUATORWARD PROPAGATION OF NEAR-INERTIAL ENERGY DUE TO THE β EFFECT

Near-inertial motions may spread both horizontally and vertically into the interior, but only when their horizontal scale is small enough [e.g., *Gill*, 1984], in fact much smaller than the wind scales. This requires mechanisms to significantly reduce the initially large scales of these wind-driven motions. This is the first condition for the existence of a route to mixing from the surface to the deep interior.

Within this context, one of the promising directions to consider is the β effect [e.g., *D'Asaro*, 1989]. It makes the near-inertial motions generated at mid-latitudes have much smaller horizontal scales than the wind scales, which can be illustrated from the following example. Using $f = f_0 + \beta y$, the near-inertial motions are written as

$$u(y,t) = u_o \sin(ft) = u_o \sin(f_0 t - ly); \qquad (2)$$

with u_o a constant and $l = -\beta t$ [e.g., *Kunze*, 1985]. t is the time. Thus, near-inertial waves are characterized by a meridional wavenumber l that grows with time, which makes them reach a length scale of 100 km within at least a month. Such length scale reduction subsequently allows them to propagate vertically in the interior [e.g., *Gill*, 1984]. Furthermore, because this meridional wavenumber is negative, these waves also freely propagate equatorward [e.g., *D'Asaro*, 1989; *Garrett*, 2001].

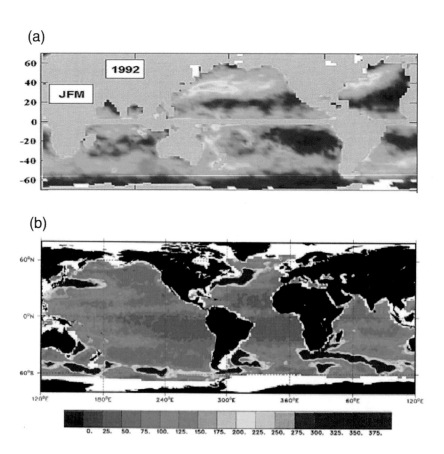

Plate 1. (a) Near-inertial energy input at the ocean surface in the winter season [from *Alford*, 2003] and (b) mesoscale eddy kinetic energy in the upper oceanic layers [from *Stammer and Wunsch*, 1999]. Reproduced from *Zhai et al.* [2005].

When these near-inertial waves propagating equatorward attain a critical latitude (defined as the latitude where their frequencies are twice the local Coriolis frequency), then a large part of their energy may be transferred down to small dissipation scales. Such processes occur through the well-known parametric subharmonic instability (PSI) [e.g., *Staquet and Sommeria*, 2002] that transfers energy across the local internal wave spectrum by nonlinear interactions amongst internal waves. A large amount of the near-inertial energy generated at mid-latitudes by high-frequency winds is then expected to be available for diapycnal mixing at these low latitudes (lower than 30°N). Numerical experiments indicate that PSI is a much more effective mechanism than previously thought (involving a timescale of days instead of weeks or months) for transferring high-frequency energy (inertial and also tidal) to small dissipation scales [e.g., *Hibiya et al.*, 1998, 1999; *Nagasawa et al.*, 2000; *MacKinnon and Winters*, 2005].

Such a direction of investigation that involves the planetary vorticity gradient is appropriate for the global circulation models that do not resolve the mesoscales. However, it should be modified and eventually extended when higher numerical resolutions are considered, such as in eddy-resolving models.

4. DOWNWARD PROPAGATION OF NEAR-INERTIAL ENERGY IN MESOSCALE EDDIES

Since the work of *Weller* [1982], it has been recognized that mesoscale eddies (with a diameter between 50 and 200 km) can distort the near-inertial motions and quickly decrease their length scales down to values smaller than 50 km within a few days [e.g., *Kunze*, 1985; *Van Meurs*, 1998]. *Kunze* [1985] argued that near-inertial waves propagating in geostrophic shear are subject not only to the planetary vorticity gradient but principally to the absolute vorticity gradient, that is, the gradient including the eddy vorticity effects. The eddy vorticity amplitude at mid-latitude can reach root-mean-square (rms) values as large as $0.3 f$ and may involve scales as small as 5–20 km [e.g., *Capet et al.*, 2008a; *Klein et al.*, 2008]. This means that relative vorticity gradients associated with mesoscale eddies can be larger than β and therefore be a much more efficient mechanism to generate small scales. Another major difference with the β effect is that near-inertial motions distorted by mesoscale eddies do not propagate equatorward. Instead they are "polarized" by eddies and more specifically are expelled from cyclonic eddies and trapped within anticyclonic ones [e.g., *Kunze*, 1985; *D'Asaro*, 1995; *Young and Ben Jelloul*, 1997; *Klein et al.*, 2004b].

All these characteristics can be explained by considering the simple situation of Figure 2 that involves a meridional

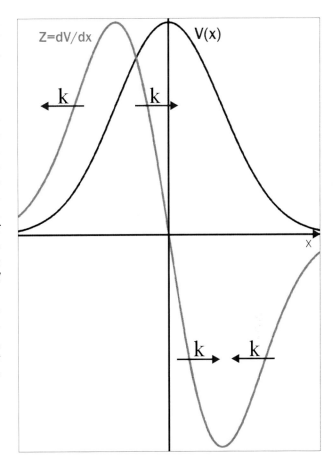

Figure 2. Schematic representation of the propagation of near inertial waves induced by a one dimensional jet $V(x)$. The velocity $V(x)$ and the vorticity $Z = \dfrac{\partial V}{\partial x}$ are shown. The wavenumbers $k = -\dfrac{\partial Z}{\partial x}\dfrac{t}{2}$ are represented by black arrows.

geostrophic jet. Inertial motions feel the rotation of the Earth but also feel the rotation of the mesoscale structures even when the latter is one order of magnitude smaller than the former. Hence, using a first-order Taylor series expansion for the mesoscale vorticity Z (i.e., $Z = Z_o + \nabla Z \cdot x + \ldots$), near-inertial velocity can be written as

$$u(x,t) \approx u_o \sin\left(ft + \frac{Z}{2}t\right) \approx u_o \sin\left(ft + \frac{Z_o}{2}t - k \cdot x\right);$$

(3)

with $k = -\dfrac{t}{2}\nabla Z$, a wavenumber increasing with time [e.g., *Kunze*, 1985]. ∇ is the horizontal gradient operator and $x = [x,y]$ with x and y, respectively, the zonal and meridional coordinates. Thus, as β, mesoscale vorticity gradients reduce the horizontal scales of the near-inertial motions. When

$|\nabla Z| \gg \beta$, this reduction may occur within a scale of a few days instead of months. Furthermore, these motions subsequently propagate but, as illustrated in Figure 2, the sign of k is such that they are expelled from cyclonic structures and are trapped within anticyclonic ones. Consequence of the near-inertial motion distortion by the eddy vorticity and their subsequent horizontal propagation (proportional to the group velocity $C_g = fr_d^2 k/\sqrt{1 + r_d^2 k^2}$ obtained using equation (1)) is that these motions are first trapped within structures of the eddy Laplacian vorticity, i.e., $\nabla^2 Z$, (this is what Figure 2 shows), and then they overshoot this region to spread out within the larger eddy stream function structures. These two limits have been analyzed by *Klein and Tréguier* [1995], *Young and Ben Jelloul* [1997], and *Balmforth et al.* [1998], and are identified as the weak dispersion limit and the strong dispersion limit.

The preceding discussion applies to the horizontal heterogeneity of near-inertial motions associated with one vertical mode. The vertical propagation of these motions within an eddy field can be understood when they involve several vertical normal modes [e.g., *Gill*, 1984]. Indeed, if lower and higher baroclinic modes have the same spatial variability and are initially in phase such that they add up within the mixed layer and cancel out below (with the result of no motion there), then the vertical propagation of the near-inertial motions simply results from the phase decoupling of the different baroclinic modes. Let us introduce the timescale t_{dn} [following *Gill*, 1984] for a 180° phase difference to develop for mode n (using equation (1)),

$$t_{dn} = \frac{\pi}{\omega_n - f} \approx \frac{2\pi}{fk^2 r_{dn}^2}, \qquad (4)$$

with ω_n and r_{dn}, respectively, the frequency of the near-inertial motions and the Rossby radius of deformation, both associated with the mode n. Then, from equation (4), the lower modes (with large r_{dn}) first become out-of-phase with the higher ones (with small r_{dn}), and then the intermediate modes in turn become out-of-phase. The consequence of this phase decoupling is the emergence of nonzero motions at depth. This dynamic of the vertical modes strongly emphasizes the importance of the mechanisms described before (that reduce the length scale of the near-inertial motions) for the propagation of the near-inertial energy into the ocean interior [e.g., *Gill*, 1984].

Most of the related studies have considered isolated geostrophic eddies or jets characterized by only one specific length scale. In this case, a Wentzel-Kramer-Brillouin (WKB) analysis can help to understand the 3-D propagation of the near-inertial motions in the ocean. However, disper-sion properties are quite different for a fully turbulent eddy field (involving strongly interacting eddies characterized by a continuous velocity spectrum). Such properties can be retrieved by using the methodology proposed by *Young and Ben Jelloul* [1997] that discards the WKB approximation and therefore allows any eddy structure, either large or small, to affect the distortion of the near-inertial motions. Extending the results of *Young and Ben Jelloul* [1997], *Klein et al.* [2004b] show that considering a fully turbulent eddy field leads to very different consequences for the behavior of the vertical modes: Lower modes are trapped within structures with relatively large scales, whereas higher modes are trapped within anticyclonic structures with much smaller scales. As a result of these differences in the spatial heterogeneity, the lower modes decouple more quickly from the higher ones and the resulting vertical propagation is therefore much faster and deeper.

5. WHICH PATHWAY PREVAILS?

We may wonder which dynamics principally drive the 3-D propagation, in the global ocean, of the large amount of near-inertial energy produced by the high-frequency winds.

We have mentioned before that atmospheric storm tracks at mid-latitudes are the regions where the wind energy input to the ocean at near-inertial frequencies is the strongest [e.g., *Alford*, 2001; *Watanabe and Hibiya*, 2002; *Alford*, 2003]. In global oceanic models, when mesoscale eddies are not fully resolved, near-inertial waves are usually produced near the surface at mid-latitudes and subsequently propagate equatorward and downward (N. Komori, personal communication, 2007; H. Sasaki, personal communication, 2007). In these simulations, the β effect is the main mechanism that governs the 3-D propagation of these waves toward the Equator [e.g., *Nagasawa et al.*, 2000; *Garrett*, 2001].

However, as noted by *Zhai et al.* [2005], the regions of high-frequency wind energy input are also those where the mesoscale eddy activity is the most energetic. This feature is illustrated by Plate 1, which shows that regions in the North Hemisphere of strong high-frequency wind energy fluxes (Plate 1a) coincide well with the locations of the Gulf Stream and the Kuroshio (Plate 1b). In the Southern Hemisphere, the high-frequency wind energy fluxes during the winter are principally at the same latitude as the Antarctic Circumpolar Current. Such remarkable coincidence indicates that much of the near-inertial energy produced by the high-frequency winds would be trapped within the oceanic mesoscale eddies. The polarization of near-inertial energy by mesoscale eddies should strongly compete with that induced by the β effect.

One question is: Do these energetic eddy regions mostly concern isolated eddies or a fully turbulent eddy field. Considering eddies as just an ensemble of isolated structures was pertinent when observational data indicated a wavenumber velocity spectrum with a slope equal or even steeper than k^{-4} (although they may weakly interact; e.g., *Klein*, 1990). However studies of the last 7 years significantly strengthen the vision of an upper ocean crowded with a large number of strongly interacting eddies. Numerical experiments of *Hurlburt and Hogan* [2000] and *Siegel et al.* [2001] in the North Atlantic basin show that using a horizontal resolution of $1/64°$ leads to an explosion of eddies and to an eddy kinetic energy increase by a factor of 10 compared with classical "eddy-resolving models" (with a $1/6°$ resolution). Furthermore, they reveal that such fully turbulent eddy fields are not only present in well-known "eddy regions" such as the Gulf Stream or the Antarctic Circumpolar Current, but also in a large number of other regions. This corroborates the visual picture provided by the high-resolution satellite images as well as some in situ data [e.g., *Stammer*, 1997; *Wunsch*, 1998; *Rudnick*, 2001; *Assenbaum and Reverdin*, 2005; *Le-Cann et al.*, 2005].

These comments strongly suggest that turbulent eddy fields at mid-latitudes are the prevailing pathways for the 3-D propagation of near-inertial waves. But we still lack a precise quantification of the contribution of these different pathways at a basin scale.

6. THE ROUTE TO MIXING THROUGH EDDY FIELDS: SOME RESULTS

Several numerical studies have described the horizontal and vertical propagation of the near-inertial motions through interacting mesoscale eddies. In a low-resolution ($1/3°$) Southern ocean model, *Zhai et al.* [2005] show that the leakage of the near-inertial energy out of the 50-m-deep mixed layer is strongly enhanced by the presence of eddies. Their results point out a maximum of the horizontal near-inertial energy at a depth of 500 m. They also support the findings

of *Lee and Niiler* [1998] indicating that anticyclonic eddies act as a conduit draining near-inertial energy to the deeper ocean (the "chimney effect"). Using a fully turbulent eddy field and a resolution of $1/15°$, *Danioux et al.* [2008] obtained a shallower penetration of the horizontal near-inertial energy, probably because of the different N^2 profile used. They did not find a clear selective trapping of near-inertial energy within anticyclonic eddies, which is explained by the competition between lower and higher modes when the latter are resolved (see section 4). However, anticyclonic structures still act as a conduit for draining near-inertial energy. As illustrated by Figure 3, the horizontal near-inertial energy penetrates deeper in anticyclonic structures than in cyclonic ones. The horizontal scales involved can be as small as 20 km. More interestingly, their results reveal a conspicuous deeper penetration of the vertical velocity variance associated with the near-inertial motions: The vertical profile of this variance displays a maximum at a depth of 1,700 m!

The competition between the high-frequency wind energy flux through the ocean surface and the near-inertial energy flux leaving the mixed layer to the deeper ocean has strong consequences on the mixing in the surface layers. Indeed, if all the near-inertial energy is trapped within the mixed layer, the excitation of the mixed-layer oscillator by winds with frequencies close to f would force resonance and lead to a significant mixed-layer deepening (as illustrated by the academic results in Figure 4). However, the eddy effects on the near-inertial motion propagation govern the corresponding energy flux at the mixed-layer base, which much reduces the mixed-layer deepening. The decaying timescale of the mixed-layer near-inertial energy, due to its propagation into the deeper layers, is estimated to be on the order of 3–8 days [*D'Asaro*, 1995; *Van Meurs*, 1998]. However, we still lack a precise quantification of this flux at the mixed-layer base, as argued by *Zhai et al.* [2005], and therefore of how much near-inertial energy is used for mixing the upper layers and how much for the deeper layers.

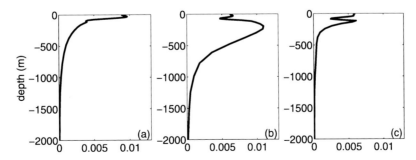

Figure 3. Vertical profiles of the horizontal near-inertial energy (a) averaged over the whole domain, (b) averaged over negative vorticity areas (corresponding to $Z < -2.10^{-5} \ s^{-1}$), and (c) averaged over positive vorticity areas (corresponding to $Z > 2.10^{-5} \ s^{-1}$). From *Danioux et al.* [2008].

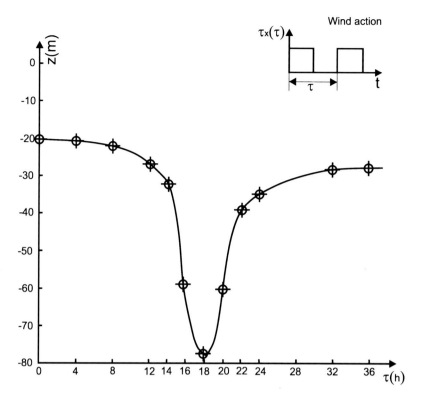

Figure 4. Mixed-layer depth after a 4-day academic simulation in terms of the time period *T* (in hours) of the wind forcing [from *Klein and Coantic*, 1981]. Coriolis frequency is $f = 10^{-4}$ s^{-1} ([18 h]$^{-1}$).

As found by *Danioux et al.* [2008], one important imprint of the high-frequency wind signal in presence of mesoscale eddies is on the vertical velocity field. To our knowledge, very few studies have so far examined this characteristic. Klein et al. (Propagation of high-frequency wind energy in the deep oceanic interior through mesoscale eddies, submitted to *Geophysical Research Letters*, 2008, hereinafter referred to as Klein et al., submitted manuscript, 2008) also analyzed the vertical velocity field in realistic numerical simulations of the North Atlantic Ocean with the Parallel Ocean Program (POP) model at 1/10° and 40 vertical levels [e.g., *Smith et al.*, 2000; *Malone et al.*, 2003]. The oceanic field is forced by, either 6-hourly or daily averaged, winds from Qscat scatterometer (QSCAT) blended with winds from the National Center for Environmental Prediction (NCEP). The vorticity field associated with mesoscale eddies is shown in Plate 2. The signature of the Gulf Stream is clearly visible, but this plate mostly shows that the eddy field is also fully turbulent outside the Gulf Stream. Plate 3 shows the wind energy flux at the ocean surface with the 6-hourly QSCAT/NCEP blended winds (which was the best frequency available). At this time, two significant large-scale atmospheric storms are present in the North Atlantic. But this plate also reveals, superimposed on the signature of the atmospheric storms, much smaller scale features principally associated with the small-scale structures of the eddy field. The vertical velocity fields in two simulations, one with the 6-hourly QSCAT/NCEP blended winds and the other one with the same winds but daily averaged, have been examined. With daily winds, the vertical velocity field (blue curve on Plate 4) is relatively weak. There is an increase near the bottom principally due to the topographic effects. With 6-hourly winds, the vertical velocity field (red curve on Plate 4) is totally different. It displays much larger amplitudes, a very large depth extension, as well as a conspicuous maximum at 2,700 m, that is, well away from the topographic effects. At that depth, the vertical velocity field reaches rms amplitudes on the order of 25 m/d at 2,700 m and involves small-scale patterns (on the order of 40–50 km).

A similar experiment has been repeated using a simulation (performed on the Earth Simulator with the Regional Ocean Modeling System (ROMS) model) of a turbulent eddy field (described by *Klein et al.* [2008]) that results from the baroclinic instability of a large-scale jet in a β-plane channel with no topography. The resolution is much higher (2 km in the horizontal and 100 vertical levels). The ocean is forced by the same wind stress time series as that displayed in Figure 1. The two curves in Plate 5 show the profile of the vertical velocity variance obtained either with the instantaneous

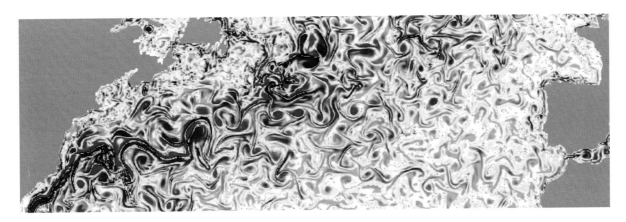

Plate 2. Surface vorticity field in the North Atlantic in January 2002. This field has been obtained from a simulation with the POP model at 1/10° [e.g., *Smith et al.*, 2000].

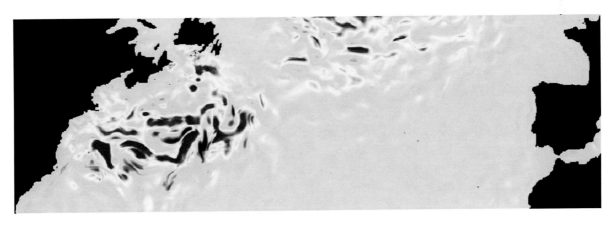

Plate 3. Wind energy input ([τ · **u**]) to the North Atlantic Ocean in January 2002 (simulated with the POP model; see Plate 2) using 6-hourly QSCAT/NCEP blended wind data. From Klein et al. (submitted manuscript, 2008).

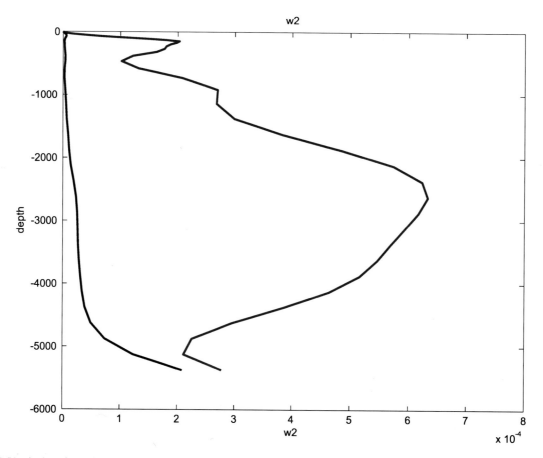

Plate 4. Vertical profiles of the vertical velocity variance in the POP simulation, using a wind forcing with 6-hourly QSCAT/NCEP blended wind data (red curve) and when the wind forcing is averaged over 24 h (blue curve). From Klein et al. (submitted manuscript, 2008).

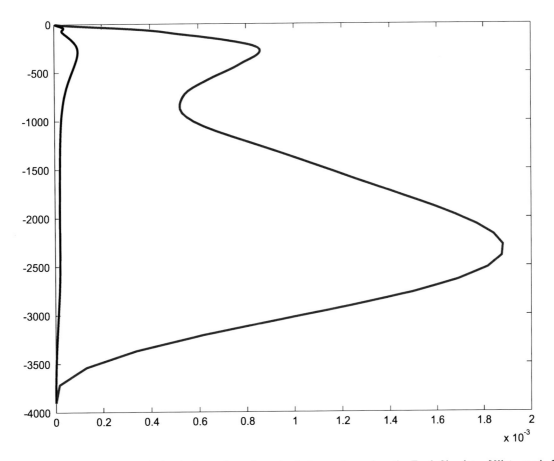

Plate 5. Vertical profiles of the vertical velocity variance in a simulation performed on the Earth Simulator [*Klein et al.*, 2008], using the real 3-hourly wind time series of *Klein et al.* [2004a] (red curve) and when this wind time series is averaged over 24 h (blue curve).

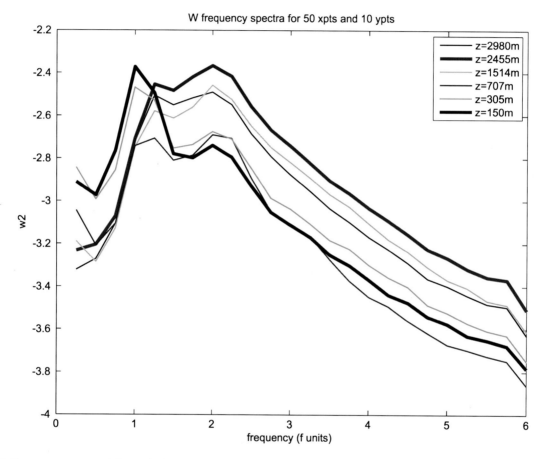

Plate 6. Frequency spectra of the vertical velocity variance at different depths from the Earth Simulator simulation [*Klein et al.*, 2008], using a real 3-hourly wind time series. Frequencies on the horizontal axis are nondimensionalized by f. The dominant frequency at 150 m is f, but it is $2f$ at 2,455 and 2,980 m.

wind forcing (red curve) or with the same wind forcing but daily average (blue curve). Again, the presence of the high-frequency winds leads to a strong increase of the vertical velocity and the emergence of a significant maximum in the deep interior (at a depth of 2,500 m with an rms value of 45 m/d!). On the other hand, when the wind is daily averaged, the vertical velocity is very much weaker with no clear maximum in the deep interior.

The appearance of the deep maximum of the vertical velocity is fully explained by *Danioux et al.* [2008]. It is principally due to the interacting mesoscale eddies that trigger a quick phase decoupling between the lower baroclinic modes of the near-inertial energy and the higher baroclinic modes. As a result, the upper maximum on the red curve of Plate 5 is captured by the higher baroclinic modes, whereas the deeper maximum is captured by the lower modes. The similitude of the deep maximum locations in Plates 4 and 5 is a coincidence, as characteristics of the corresponding experiments are somewhat different. The depth of the maximum in the interior is actually sensitive to the representation of the higher modes and to the existence of vorticity structures at small scale. Indeed, the same experiment as that leading to the results shown on Plate 5, but with a lower resolution (6 km in the horizontal and 50 vertical levels), has produced a maximum at a depth of only 1,700 m. Existence of this deep maximum is strong evidence that the presence of mesoscale eddies and submesoscale structures opens an efficient pathway for the wind energy to go deep into the ocean interior where it could be available for mixing.

Another comment on these results is that in such turbulent eddy fields, the vorticity gradients have a rms value as large as 10^{-10} to 10^{-9} m^{-1} s^{-1}, i.e., 1 to 2 orders of magnitude larger than the β value ($\approx 1.6 \times 10^{-11}$ m^{-1} s^{-1} at mid-latitudes). This means that the dynamic related to the turbulent eddy field strongly dominates that due to the β effect. Therefore, near-inertial motions appear to be mostly trapped in the turbulent eddy field and can weakly propagate equatorward.

One question still to be addressed is: can these near-inertial motions efficiently cascade through the local wave number spectrum down to small dissipation scales? Usually, PSI is the mechanism invoked for such a cascade, but it can only transfer energy from low-vertical mode double-inertial motions to small-scale inertial waves where dissipation occurs [e.g., *Hibiya et al.*, 1998; *Staquet and Sommeria*, 2002]. This means that only near-inertial waves (with frequency very close to f) generated at mid-latitudes that propagate equatorward can cascade down through PSI at latitudes not larger than $\approx 30°$, where the local Coriolis frequency is half that at mid-latitudes. However, it was found recently and confirmed in high-resolution simulations that the dominant frequency of the wind-driven vertical motions present in the deep ocean interior at mid-latitudes (Plates 4 and 5) is not close to f but is close to (and even over) $2f$, that is, twice the Coriolis frequency (see Plate 6). The emergence of such supra-inertial motions (also reported in some observations such as those of *Mori et al.* [2005]) is due to a resonance mechanism described by *Danioux and Klein* [2008]. It therefore makes PSI a likely mechanism to convert vertical kinetic energy into mixing in the deep interior at these mid-latitudes. It is premature to produce any estimation of the corresponding mixing, as PSI cannot be efficiently activated in a primitive-equation model. But this result indicates the existence of a significant source of low-vertical-mode internal waves with frequency equal or over $2f$ at mid-latitudes. Such a result emphasizes the strong potentiality of wind-driven near-inertial motions to significantly contribute to the mixing in the deep ocean interior.

7. EDDY-RESOLVING MODELS AND HIGH-FREQUENCY WINDS: WHAT RESOLUTION IS NEEDED?

Our hypothesis is that eddy-resolving models at a basin scale can be very helpful to address the questions still unanswered about the precise contribution of the different pathways by which near-inertial motions propagate. The eddy-resolving models should allow a particular understanding of how much of the large amount of energy provided by high-frequency winds is available for mixing in both the upper layers and the deep interior and at which latitudes. The prerequisite is to have access to realistic high-frequency winds and to use numerical models at a basin scale with adequate spatial resolution.

High-frequency intermittent winds are usually linked with fast traveling atmospheric fronts associated with storms. Atmospheric fronts are narrow [with a width of $\mathcal{O}(100$ km)] and elongated on a long distance (more than 1,000 km). The atmospheric fronts have a strong signature on the relative vorticity, which explains the strong and rapid variation of both the wind amplitude and direction when they pass over the ocean surface at a given location. NCEP or European Center for Medium Range Weather Forecasting (ECMWF) data cannot represent such features because of their spatial resolution (involving spatial grid equal or larger than 100 km).

However, when these data are blended with data from satellite scatterometers, such as QSCAT and ADEOS-II scatterometers, time variation of the amplitude and rotation of the high-frequency winds are much better represented (e.g., http://www.cora.nwra.com/morzel/). This is illustrated in Plate 7, which shows the atmospheric vorticity field near the surface estimated from only the NCEP data (upper panel) and from blended winds generated with only QSCAT (middle panel) and with both QSCAT and ADEOS-II data (lower panel). Results show that fronts and storms are very much better resolved when NCEP data are blended with scatterometer data. The propagation of fronts and storms, and therefore the intermittency of the winds, are better represented when using two scatterometers instead of one (compare the three panels in Plate 7). With two scatterometers, the blending window can be reduced from 12 to 7.5 h, i.e., every 6-hourly NCEP global field is blended with 7.5 h each of QSCAT and ADEOS-II data. Such data, when they are available, strongly improve the spatial and time intermittency of the high-frequency winds [e.g., *Milliff et al.*, 2004; *Chelton et al.*, 2004], which highlights the strong value and potentiality of high-resolution scatterometers.

Eddy-resolving models at a basin scale should have adequate horizontal and vertical to resolution be able to represent the 3-D propagation of the wind-driven near-inertial energy. This means a horizontal resolution involving a horizontal space grid equal to or less than 3 km and a vertical resolution with at least 100 levels. Two reasons advocate for the use of such spatial resolution.

First, as noted before, the vertical propagation of near-inertial waves down to 4,000-m depth results from the phase decoupling of the lower and higher baroclinic modes. Higher baroclinic modes have a Rossby radius of deformation on the order of a few kilometers. This requires a space grid not larger than 3 km and a vertical resolution involving at least 100 levels on the vertical to resolve these Rossby radii (multiplied by 2π) and to correctly represent the rapid variation with depth of these vertical modes.

The second reason is that the surface dynamics driven by the mesoscale eddy turbulence involves quite small spatial scales or submesoscale structures, with scales as small as 5 km, that need to be explicitly resolved. The energetic character of these submesoscale structures is emphasized by high-resolution simulations of such eddy turbulence [e.g., *McWilliams et al.*, 2004; *Capet et al.*, 2008a; *Klein et al.*, 2008] that exhibit a k^{-2} velocity spectrum in the surface layers. Frontogenesis underlies the physics involved and explains the small-scale structures in Plate 8 with large relative vorticity values. Dynamics of these submesoscale structures are documented by *Mahadevan* [2006], *Mahadevan and Tandon* [2006], and *Thomas et al.* [2008].

As noted before, these submesoscale structures have a significant impact not only on the 3-D propagation of the near-inertial energy (and therefore on the depth of the maximum in the interior), but also on the resulting mixing in both the upper layers and the deep interior. Indeed, they drive a strong restratification in the upper layers, compensated for by a destratification in the ocean interior [e.g., *Lapeyre et al.*, 2006; *Klein et al.*, 2008]. Such restratification competes with the mixed-layer dynamics [e.g., *McWilliams*, 2008].

On the other hand, the wind forcing itself can affect these submesoscale structures as explicitly shown by *Thomas* [2005], *Thomas and Lee* [2005], *Giordani and Caniaux* [2006], *Boccaletti et al.* [2008], and *Thomas et al.* [2008]. All these mechanisms should have a nonnegligeable impact on the resulting mixing by both the near-inertial waves and the submesoscale structures in the upper oceanic layers. At last they have a significant effect on the phase decoupling between lower and higher modes of the near-inertial motions, and therefore on the vertical propagation of these motions, as the higher modes dispersion is strongly sensitive to the presence and amplitude of these submesoscale structures [e.g., *Klein et al.*, 2004b].

Characteristics of the deep maximum of vertical velocity variance observed at 2,500–3,000 m, as found by *Danioux et al.* [2008] and other studies, make the PSI a likely mechanism to convert the vertical kinetic energy into vertical turbulent diffusion [e.g., *Staquet and Sommeria*, 2002; *Koudella and Staquet*, 2006]. As such, the deep maximum of high-frequency motions can potentially participate to the vertical mixing in this region. A very rough scaling analysis (using $K_z \approx <w^2>/f$) indicates that even if only 10% of the near-inertial vertical kinetic energy is converted into energy available for mixing through PSI, the corresponding value (10^{-4} m^2 s^{-1}) would be much larger than the 10^{-6} m^2 s^{-1} molecular value and should be close to that necessary for mixing in the deep interior [e.g., *Wunsch and Ferrari*, 2004]. This comment calls for further studies on these mechanisms to fully evaluate the efficiency of PSI for the characteristics of the supra-inertial energy present in the deep interior. These studies could lead to development of an appropriate parameterization to convert this near-inertial energy at depth into mixing.

Within a climate change scenario, these mechanisms that open a route to mixing from the surface to the deep ocean interior suggest that the atmosphere, through the high-frequency winds, may affect the stratification and the MOC, i.e., the large-scale oceanic circulation. Recent reviews [*Schiermeier*, 2007] and studies [*Sriver and Huber*, 2007; *Emmanuel*, 2001] have already pointed out that tropical cyclones (or hurricanes) mix the upper ocean so well that they are responsible for a staggering amount of heat transfer on a global scale. This means that within an extreme climate scenario, an increase of tropical cyclone power and frequency (due to future changes in tropical sea surface temperature) may accelerate the MOC. This might feed back on climate by redistributing heat poleward and therefore raising global mean temperature [*Sriver and Huber*, 2007]. On the other hand, the mechanisms reviewed in this paper suggest in the same way that an increase of the strength and intensity of the extra-tropical atmospheric storms may be accompanied by a strengthened MOC and/ or an intensified eddy kinetic energy at mid-latitudes with

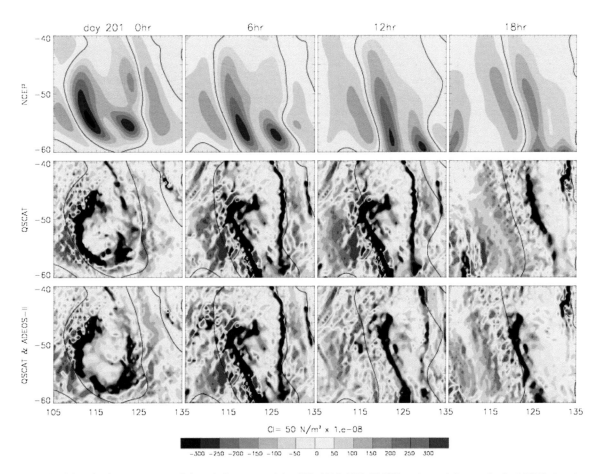

Plate 7. The 6-hourly time sequence of the wind stress curl (at 40°–60°S, 75°–135°E) estimated from only the NCEP data (upper panel), from blended winds generated with QSCAT only (middle panel) and with both QSCAT and ADEOS-II data (lower panel). See http://www.cora.nwra.com/morzel/blendedwinds.qscat.swsa2.html. The dual-blended wind data agree very well with the location in NCEP, whereas the single-blended winds are already 6 h ahead, placing the curl front too far to the East [e.g., *Milliff et al.*, 2004; *Chelton et al.*, 2004].

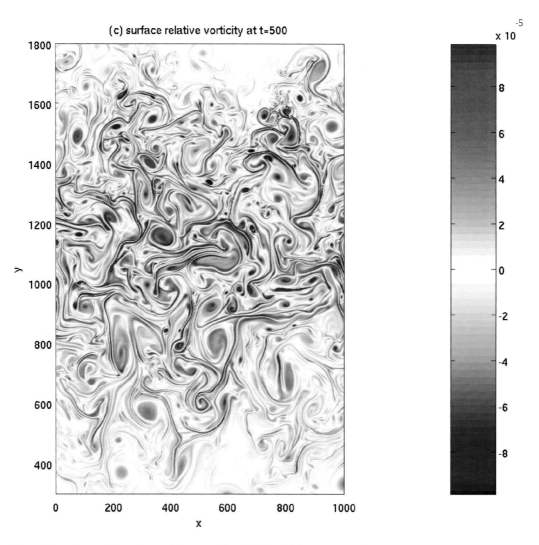

Plate 8. Snapshot of the surface relative vorticity field in a high-resolution simulation [from *Klein et al.*, 2008].

the same climate consequences. However, these effects at mid-latitudes would be efficient on a larger depth than those due to tropical cyclones. All these features, which still have to be quantified, emphasize the need to explicitly take into account the small scales of the atmospheric and oceanic dynamics in order to understand the coupled atmosphere-ocean system at larger scales.

Acknowledgment. Contributions from and discussions with Eric Danioux, Matthew Hecht, Guillaume Lapeyre, Bill Large, and Nobu Komori during the preparation of this short review are greatly acknowledged. This work is supported by the Institut Francais de Recherche pour l'Exploitation de la Mer (IFREMER) and the Centre National de la Recherche Scientifique (CNRS), FRANCE. Some numerical simulations reported here were done on the Earth Simulator (Yokohama, Japan), the access to which has been made possible through a memorandum of understanding between the IFREMER, the CNRS, and the Earth Simulator Center/Japan Agency for Marine-Earth Science and Technology (ESC/JAMSTEC). The support from the French ANR (Agence Nationale pour la Recherche, Contract No. ANR-05-CIGC-010) is also acknowledged. At last, the referees are thanked for their constructive comments.

Note

1. Any motion can be decomposed into vertical normal modes that are solutions of a Sturm-Liouville problem [e.g., *Gill*, 1984], which only requires the knowledge of the vertical stratification. Each mode is associated with a Rossby radius of deformation whose value at mid-latitudes ranges from ≈50 km for the lower modes to ≈2 to 3 km for the higher modes. Therefore, equation (1) can be seen as the dispersion relation associated with one vertical mode.

REFERENCES

Alford, M. H. (2001), Internal swell generation: The spatial distribution of energy flux from the wind to mixed-layer near-inertial motions, *J. Phys. Oceanogr.*, *31*, 2359–2368.

Alford, M. H. (2003), Improved global maps and 54-year history of wind-work on ocean inertial motions, *Geophys. Res. Lett.*, *30*(8), 1424, doi:10.1029/2002GL016614.

Assenbaum, M., and G. Reverdin (2005), Near-real time analyses of the mesoscale circulation during the POMME experiment, *Deep-Sea Res.*, *30*, 2343–2353.

Balmforth, N., S. L. Smith, and W. Young (1998), Enhanced dispersion of near-inertial waves in an idealized geostrophic flow, *J. Mar. Res.*, *56*, 1–40.

Boccaletti, G., R. Ferrari, and B. Fox-Kemper (2008), Mixed layer instabilities, *J. Fluid Mech.*, in press.

Capet, X., J. C. McWilliams, M. Molemaker, and A. Shchepetkin (2008a), Mesoscale to sub-mesoscale transition in the california current system. Part 1: Flow structure and eddy flux, *J. Phys. Oceanogr.*, in press.

Capet, X., J. C. McWilliams, M. Molemaker, and A. Shchepetkin (2008b), Mesoscale to sub-mesoscale transition in the california current system, Part 2: Dynamical processes and observational tests, *J. Phys. Oceanogr.*, in press.

Chelton, D., M. Schlax, M. Freilich, and R. Milliff (2004), Satellite measurements reveal persistent small-scale features in ocean winds, *Science*, *303*, 978–983.

Danioux, E., and P. Klein (2008), A resonance mechanism leading to wind-forced supra-inertial motions with a $2f$ frequency, *J. Phys. Oceanogr.*, in press.

Danioux, E., P. Klein, and P. Rivière (2008), Propagation of wind energy into the deep ocean through mesoscale eddies, *J. Phys. Oceanogr.*, in press.

D'Asaro, E. A. (1989), The decay of wind-forced mixed layer inertial oscillations due to the β effect, *J. Geophys. Res.*, *94*, 2045–2056.

D'Asaro, E. A. (1995), Upper-ocean inertial currents forced by a strong storm. Part iii: Interaction of inertial currents and mesoscale eddies, *J. Phys. Oceanogr.*, *25*, 2953–2958.

Emanuel, K. (2001), Contribution of tropical cyclones to meridional heat transport by the oceans, *J. Geophys. Res.*, *106*(D14), 14,771–14,781.

Garrett, C. (2001), What is the "near inertial" band and why is it different from the rest of the internal wave spectrum?, *J. Phys. Oceanogr.*, *31*, 962–971.

Gill, A. (1984), On the behavior of internal waves in the wakes of storms, *J. Phys. Oceanogr.*, *14*, 1129–1151.

Giordani, L. P., and G. Caniaux (2006), Advanced insights into sources of vertical velocity in the ocean, *Ocean Dyn.*, *56*, 513–524.

Hibiya, T., Y. Niwa, and K. Fujiwara (1998), Numerical experiments of nonlinear energy transfer within the oceanic internal wave spectrum, *J. Geophys. Res.*, *103*, 18,715–18,722.

Hibiya, T., M. Nagasawa, and Y. Niwa (1999), Model predicted distribution of internal wave energy for diapycnal mixing processes in the deep waters of the North Pacific, in *Dynamics of Oceanic Internal Gravity Waves: Proc. AhaHulikoa Hawaiian Winter Workshop*, edited by P. Müller and D. Henderson, pp. 205–215. Hawaii Institute of Geophysics.

Hurlburt, H. E., and P. J. Hogan (2000), Impact of 1/8° to 1/64° resolution on gulf stream model-data comparisons in basin-scale subtropical Atlantic Ocean models, *Dyn. Atmos. Ocean.*, *32*, 283–329.

Klein, P. (1990), Transition to chaos in unstable baroclinic systems: a review, *Fluid Dyn. Res.*, *5*, 235–254.

Klein, P., and M. Coantic (1981), A numerical study of turbulent processes in the marine upper layers, *J. Phys. Oceanogr.*, *11*, 849–863.

Klein, P., and A. Tréguier (1995), Dispersion of wind-induced inertial waves by a barotropic jet, *J. Mar. Res.*, *53*, 1–22.

Klein, P., G. Lapeyre, and W. G. Large (2004a), Wind ringing of the ocean in presence of mesoscale eddies, *Geophys. Res. Lett.*, *31*, L15306, doi:10.1029/2004GL020274.

Klein, P., S. Llewellyn-Smith, and G. Lapeyre (2004b), Spatial organisation of inertial energy by an eddy field, *Q. J. R. Meteorol. Soc.*, *130*, 1153–1166.

Klein, P., B. Hua, G. Lapeyre, X. Capet, S. L. Gentil, and H. Sasaki (2008), Upper ocean turbulence from high 3-D resolution simulations, *J. Phys. Oceanogr.*, in press.

Koudella, C., and C. Staquet (2006), Instability mechanisms, *J. Fluid Mech.*, *548*, 165–196.

Kunze, E. (1985), Near-inertial wave propagation in geostrophic shear, *J. Phys. Oceanogr.*, *15*, 544–565.

Lapeyre, G., P. Klein, and B. L. Hua (2006), Oceanic restratification by surface frontogenesis, *J. Phys. Oceanogr.*, *36*, 1577–1590.

Large, W. G., and G. B. Crawford (1995), Observations and simulations of upper-ocean response to wind events during the ocean storms experiment, *J. Phys. Oceanogr.*, *25*, 2831–2852.

Le Cann, B., M. Assenbaum, J.-C. Gascard, and G. Reverdin (2005), Observed mean and mesoscale upper ocean circulation in the midlatitude northeast Atlantic, *J. Geophys. Res.*, *110*, C07S05, 10.1029/2004JC002768.

Lee, D.-K., and P. P. Niiler (1998), The inertial chimney: The near-inertial energy drainage from the ocean surface to the deep layer, *J. Geophys. Res.*, *103*(C4), 7579–7591.

MacKinnon, J. A., and K. B. Winters (2005), Subtropical catastrophe: Significant loss of low-mode tidal energy at 28.9°, *Geophys. Res. Lett.*, *32*, L15605, doi:10.1029/2005GL023376.

Mahadevan, A. (2006), Modeling vertical motion at ocean fronts: Are non-hydrostatic effects relevant at sub-mesoscales?, *Ocean Modell.*, *14*, 222–240.

Mahadevan, A., and A. Tandon (2006), An analysis of mechanisms for sub-mesoscale vertical motions at ocean fronts, *Ocean Modell.*, *14*, 241–256.

Malone, R., R. Smith, R. Maltrud, and M. Hecht (2003), Eddy-resolving ocean modeling, *Los Alamos Science*, *28*, 223–231.

McWilliams, J. (2008), On the role of eddies, this volume.

McWilliams, J., M. Molemaker, and I. Yavneh (2004), Ageostrophic, anticyclonic instability of a barotropic boundary current, *Phys. Fluids*, *16*, 3720–3725.

Milliff, R., R.F., J. Morzel, D. Chelton, and M. Freilich (2004), Wind stress curl and wind stress divergence biases from rain effects on QSCAT surface wind retrievals, *J. Atmos. Ocean. Technol.*, *21*, 1216–1231.

Mori, K., T. Matsuno, and T. Senjyu (2005), Seasonal/spatial variations of the near-inertial oscillations in the deep water of the Japan Sea, *J. Oceanogr.*, *61*, 761–773.

Munk, W., and C. Wunsch (1998), Abyssal recipes ii: energetics of tidal and wind mixing, *Deep-Sea Res.*, *45*, 1976–2009.

Nagasawa, M., Y. Niwa, and T. Hibiya (2000), Spatial and temporal distribution of wind-induced internal wave energy available for deep water mixing in the North Pacific, *J. Geophys. Res.*, *105*, 13,933–13,943.

Price, J. (1981), Upper ocean response to a hurricane, *J. Phys. Oceanogr.*, *11*, 153–175.

Price, J., T. Sanford, and G. Forristall (1994), Forced stage response to a moving hurricane, *J. Phys. Oceanogr.*, *24*, 233–260.

Rudnick, D. L. (2001), On the skewness of vorticity in the upper ocean, *Geophys. Res. Lett.*, *28*, 2045–2048.

Schiermeier, Q. (2007), Churn, churn, churn, *Nature*, *477*, 522–524.

Siegel, A., J. B. Weiss, J. Toomre, J. C. McWilliams, P. S. Berloff, and I. Yavneh (2001), Eddies and vortices in ocean basin dynamics, *Geophys. Res. Lett.*, *28*, 3183–3186.

Skyllinstad, E., W. Smyth, and G. Crawford (2000), Resonant wind-driven mixing in the ocean boundary layer, *J. Phys. Oceanogr.*, *30*, 1866–1890.

Smith, R., M. Maltrud, F. Bryan, and M. Hecht (2000), Numerical simulation of the North Atlantic Ocean at 1/10°, *J. Phys. Oceanogr.*, *30*, 1532–1561.

Sriver, R., and M. Huber (2007), Observational evidence for an ocean heat pump induced by tropical cyclones, *Nature*, *447*, 577–580.

Stammer, D. (1997), Global characteristics of ocean variability estimated from regional TOPEX/POSEIDON altimeter measurements, *J. Phys. Oceanogr.*, *27*, 1743–1769.

Stammer, D., and C. Wunsch (1999), Temporal changes in eddy energy of the oceans, *Deep-Sea Res.*, *46*, 77–108.

Staquet, C., and J. Sommeria (2002), Internal gravity waves: from instabilities to turbulence, *Annu. Rev. Fluid Mech.*, *34*, 559–593.

Stockwell, R., W. Large, and R. Milliff (2004), Resonant inertial oscillations in moored buoy ocean surface winds, *Tellus*, *56*, 536–547.

Thomas, L. N. (2005), Destruction of potential vorticity by winds, *J. Phys. Oceanogr.*, *35*, 2457–2466.

Thomas, L. N., and C. M. Lee (2005), Intensification of ocean fronts by down-front winds, *J. Phys. Oceanogr.*, *35*, 1086–1102.

Thomas, L. N., A. Tandon, and A. Mahadevan (2008), Submesoscale processes and dynamics, this volume.

Van Meurs, P. (1998), Interactions between near-inertial mixed layer currents and the mesoscale: the importance of spatial variabilities in the vorticity field, *J. Phys. Oceanogr.*, *28*, 1363–1388.

Watanabe, M., and T. Hibiya (2002), Global estimates of the wind-induced energy flux to inertial motions in the surface mixed layer, *Geophys. Res. Lett.*, *29*(8), 1239, doi:10.1029/2001GL014422.

Weller, R. (1982), The relation of near-inertial motions observed in the mixed layer during the JASIN (1978) experiment to the local wind stress and the quasigeostrophic flow field, *J. Phys. Oceanogr.*, *12*, 1122–1136.

Wunsch, C. (1998), The work done by the wind on the oceanic general circulation, *J. Phys. Oceanogr.*, *28*, 2332–2340.

Wunsch, C., and R. Ferrari (2004), Vertical mixing, energy and the general circulation of the ocean, *Annu. Rev. Fluid Mech.*, *36*, 281–314.

Young, W., and M. Ben Jelloul (1997), Propagation of near-inertial oscillations through a geostrophic flow, *J. Mar. Res.*, *55*, 735–766.

Zhai, X., R. J. Greatbach, and J. Zhao (2005), Enhanced vertical propagation of storm-induced near-inertial energy in an eddying ocean channel model, *Geophys. Res. Lett.*, *32*, L18602, doi:1029/2005GL023643.

Patrice Klein, Département d'Océanographie Physique et Spatiale, Ifremer Centre de Brest, BP 70, F29280 Plouzané, France. (pklein@ifremer.fr)

Resolution Dependence of Eddy Fluxes

Yukio Tanaka

Frontier Research Center for Global Change, Japan Agency for Marine-Earth Science and Technology, Yokohama, Japan

Hiroyasu Hasumi

Center for Climate System Research, University of Tokyo, Kashiwa, Japan

The sensitivity of simulation results to horizontal resolution is studied in the Southern Ocean using an ocean model, which is run for three horizontal resolutions of $1/2° \times 1/3°$, $1/4° \times 1/6°$, and $1/8° \times 1/12°$ (longitude $\times$ latitude). Focuses are on eddy fluxes and effects of the eddy fluxes. The temporal residual-mean (TRM) velocity, the location of the Brazil-Malvinas (B/M) Current confluence, and the sea surface height (SSH) variability are investigated as effects of the eddy fluxes. For the low-resolution case, the intensity of the eddy fluxes in the Antarctic Circumpolar Current (ACC), the downward TRM velocity at the B/M confluence, and the SSH variability are relatively small. The B/M confluence is located in the far south compared with the observations. For the medium-resolution case, the intensity of the eddy fluxes in the ACC and the downward TRM velocity at the confluence are about twice as strong as the low-resolution case. The location of the confluence shifts to north, but still is in the south of the observational position. For the high-resolution case, the downward TRM velocity at the confluence becomes about three times as strong as the low-resolution case. The location of the confluence is close to the observations. It is concluded that at least the horizontal resolution of $1/8° \times 1/12°$ is required to simulate the downward TRM velocity at the B/M confluence and the location of the confluence properly.

1. INTRODUCTION

The extent to which we should increase model resolution to properly simulate eddies and their effects on ocean mean states is a large unresolved problem. This problem is investigated in the Southern Ocean, as it is known to be one of the most eddy-rich regions on the globe and its mean state is significantly affected by eddies. Because the Southern Ocean occupies a vast area and eddy activity is high over the whole region, it is the most challenging part in the world ocean in which to study the problem.

The stirring of eddies in the ocean is primarily advective rather than diffusive in character, and this advective transport is expressed as an eddy-induced velocity. Therefore, it is important to know the model resolution at which the eddy-induced velocity can be properly represented. The sum of the eddy-induced and the mean velocity is called a residual-mean velocity. It is this velocity that describes the transport of heat, salinity, and other tracers in the mean state. The residual-mean velocity was first introduced by *Andrews and McIntyre* [1976] to examine atmospheric eddy effects on a zonal mean field. Recently, this residual-mean velocity framework has been applied to oceanic eddies in the Southern Ocean.

Ocean Modeling in an Eddying Regime
Geophysical Monograph Series 177
Copyright 2008 by the American Geophysical Union.
10.1029/177GM08

Marshall [1997] employed the residual-mean velocity to formulate the subduction of a water mass in an eddying ocean. The author formulated the time-mean subduction rate of a water mass in an eddying ocean by the residual-mean velocity. Using an idealized two-dimensional Southern Ocean model, it was shown that the mean and eddy-induced velocity cancel each other at leading order, and the surface buoyancy forcing determines the residual of these two velocities. By assuming plausible patterns of the surface buoyancy forcing, the author obtained subduction of Antarctic Intermediate Water (AAIW) and Antarctic Bottom Water with entrainment of North Atlantic Deep Water in between.

Karsten et al. [2002] diagnosed output data of an eddy-resolving zonal symmetric ocean model and obtained a circulation caused by the residual-mean velocity. The authors showed that the circulation driven by eddies balances the mean wind-driven transport at zeroth order. At the next order, the residual circulation transports buoyancy across the front to balance sea surface buoyancy fluxes.

Radko and Marshall [2006] developed a simple theory for the three-dimensional structure of the Antarctic Circumpolar Current (ACC) and the upper cell of its meridional overturning circulation based on a perturbation expansion about the zonal-average residual-mean model developed by *Marshall and Radko* [2003]. The authors used an observational data to reproduce the residual circulation and found an intensification of the overturning circulation in the Atlantic-Indian sector and reduction in strength in the Pacific Ocean region.

The ocean flow in the ACC has large variability in longitudinal direction so that the residual-mean velocity must be examined in a full three-dimensional framework. *McDougall and McIntosh* [1996, 2001] extended the residual-mean velocity theory to the three-dimensional temporal residual-mean (TRM) velocity. The fundamental difference between the residual mean velocity of *McDougall and McIntosh* [1996, 2001] and that of *Andrews and McIntyre* [1976] is that it takes into account the contribution from the advection of eddy buoyancy variances.

Tanaka and Hasumi (Injection of Antarctic Intermediate Water into the Atlantic subtropical gyre in an eddy resolving ocean model, to appear in *Geophysical Research Letters*, 2008) evaluate the TRM velocity using output data of an eddy-resolving ocean model. The authors find that there is a region with large downward TRM velocity at the Brazil-Malvinas (B/M) confluence. The density and salinity distributions are examined, and it is concluded that this downward TRM velocity causes the injection of AAIW into the Atlantic subtropical gyre. Although a mechanism to cause this TRM velocity is investigated in this study, resolution dependence of the downward TRM velocity has not been studied.

The aim of this work is to study sensitivity of simulation results to the horizontal resolution in the Southern Ocean.

The eddy fluxes, the TRM velocity, the location of the B/M confluence, and the sea surface height (SSH) variability are investigated.

The structure of this paper is as follows. The TRM framework is briefly reviewed in section 2. The ocean general circulation model used in this study is described in section 3. In section 4, the sensitivity of the simulation results to the horizontal resolution is described. The summary and conclusion are in section 5.

2. THE TRM FRAMEWORK

The stirring of eddies in the ocean is primarily advective in character, and this advective transport is expressed by the advection of an eddy-induced velocity. The sum of the usual Eulerian time-averaged velocity and the eddy-induced velocity is called the residual-mean velocity. It is this velocity that *Marshall* [1997] employed to evaluate subduction of water mass in eddy-rich oceans. In this section, we review the residual-mean velocity from the perspective of evaluation using output of the eddy-resolving ocean model.

The steady-state mean buoyancy budget is expressed as

$$\overline{\mathbf{u}} \cdot \nabla \overline{b} + \nabla \cdot \overline{(b'\mathbf{u}')} = \overline{Q}, \tag{1}$$

where b is buoyancy, $\mathbf{u} \equiv (\mathbf{v}, w)$ is velocity, and Q is the source of buoyancy. The overbar represents the temporal mean at fixed height, and the prime represents deviations from the mean. The buoyancy is defined as $b = -g(\rho - \rho_0)/\rho_0$, where g is gravity acceleration, ρ is potential density, and ρ_0 is reference density. The term $\nabla \cdot \overline{(b'\mathbf{u}')}$ represents the effect of eddies on the ocean mean state. Here, we define the eddy component as the deviation from the mean state.

The eddy buoyancy flux in equation (1) is decomposed into an isopycnal component that is along the isopycnals and a vertical (diapycnal) component that is across the isopycnals:

$$\overline{b'\mathbf{u}'} = \overline{(b'\mathbf{u}')}_{\parallel} + \frac{\overline{b'\mathbf{u}'} \cdot \nabla \overline{b}}{\overline{b}_z}\mathbf{k}, \tag{2}$$

where $\mathbf{k}$ is a unit vector of the vertical direction and $\overline{b}_z \equiv \partial \overline{b}/\partial z$ (Figure 1). Note that this decomposition is nonorthogonal one. The advantage of this nonorthogonal decomposition is that mathematical manipulations are easier than with the orthogonal one. The divergence of the isopycnal component can be written as an advection form of the mean buoyancy:

$$\nabla \cdot \overline{(b'\mathbf{u}')}_{\parallel} = \mathbf{u}_1^* \cdot \nabla \overline{b}, \tag{3}$$

$$\mathbf{u}_1^* = \left(\frac{\partial \Psi_1^*}{\partial z}, -\nabla_H \cdot \Psi_1^* \right), \tag{4}$$

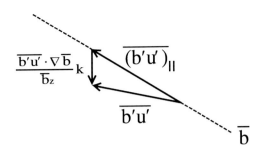

Figure 1. The decomposition method of the eddy buoyancy flux.

$$\Psi_1^* = -\frac{\overline{b'v'}}{\overline{b}_z}, \tag{5}$$

where ∇_H is the horizontal Laplacian (see Appendix A for the derivation). The eddy-induced velocity $\mathbf{u}_1^*$ expresses the advective effect of eddies on the mean state. Then, the eddy term can be written as:

$$\nabla \cdot \overline{(b'\mathbf{u}')} = \mathbf{u}_1^* \cdot \nabla \overline{b} + G_1, \tag{6}$$

$$G_1 = \left[\frac{\overline{b'\mathbf{u}'} \cdot \nabla \overline{b}}{\overline{b}_z}\right]_z, \tag{7}$$

where $[]_z \equiv \frac{\partial}{\partial z} []$. A zonal mean case of equation (5) is first introduced by *Andrews and McIntyre* [1976] to evaluate effect of atmospheric waves on a zonal mean flow field.

Equation (5) has been employed by several authors to evaluate the eddy-induced velocity from the output of eddy-resolving models [*Gille and Davis*, 1999; *Karsten et al.*, 2002; *Henning and Vallis*, 2004, 2005]. However, *Gille and Davis* [1999] demonstrated that the diapycnal term G_1 is not small, and *McDougall and McIntyre* [1996] showed that the term contains an additional advective component (see Appendix B). Therefore, it does not seem appropriate to use equation (5) for evaluating the eddy-induced velocity from the output of an eddy-resolving ocean model.

Eden et al. [2007a] obtained an expansion form of the eddy-induced velocity. For the case of the eddy flux decomposition of equation (2), the expansion form, in which the eddy-induced velocity of *Andrews and McIntyre* [1976] appears as the first term, can be written as:

$$\nabla \cdot \overline{(b'\mathbf{u}')} = \mathbf{u}^* \cdot \nabla \overline{b} - \overline{Q}^*, \tag{8}$$

$$\mathbf{u}^* = \left(\frac{\partial \Psi^*}{\partial z}, -\nabla_H \cdot \Psi^*\right), \tag{9}$$

$$\Psi^* = -\left[\frac{\overline{b'\mathbf{v}'}}{\overline{b}_z}\right] + \frac{1}{\overline{b}_z}\left[\frac{\overline{\phi_2 \mathbf{v}}}{\overline{b}_z}\right]_z - \frac{1}{\overline{b}_z}\left[\frac{1}{\overline{b}_z}\left[\frac{\overline{\phi_3 \mathbf{v}}}{\overline{b}_z}\right]_z\right]_z + \cdots, \tag{10}$$

$$\overline{Q}^* = -\left[\frac{\overline{b'Q'}}{\overline{b}_z}\right]_z + \left[\frac{1}{\overline{b}_z}\left[\frac{\overline{\phi_2 Q}}{\overline{b}_z}\right]_z\right]_z + \cdots, \tag{11}$$

where $\phi_3 = (b'^3/6)$. In this expansion form, the diapycnal term G_1 in equation (6) is separated into higher order eddy-induced velocities and the adiabatic component $\overline{Q}^*$.

The higher order terms in equation (10) contain several differential operators, and it is inevitable to cause numerical errors when evaluating them numerically. Therefore, equation (10) is not used for evaluating the residual-mean velocity.

McDougall and McIntosh [2001] and *Jacobson and Aiki* [2006] show that the horizontal residual-mean velocity can be calculated by temporal averaging of the thickness weighted horizontal velocity. They obtained the following TRM velocity equations:

$$\widetilde{\mathbf{u}}^\sharp \cdot \nabla \widetilde{b} = \widetilde{Q}^\sharp, \tag{12}$$

$$\widetilde{\mathbf{u}}^\sharp = \left(\frac{\partial \widetilde{\Psi}^\sharp}{\partial z}, -\nabla_H \cdot \widetilde{\Psi}^\sharp\right), \tag{13}$$

$$\widetilde{\Psi}^\sharp(z) = \overline{\int_{-H}^{z+z'} \mathbf{v}(z'')dz''}, \tag{14}$$

$$\widetilde{Q}^\sharp(z) = \frac{\partial}{\partial z}\overline{\int_{-H}^{z+z'} Q(z'')dz''}, \tag{15}$$

where $z + z'$ is the instantaneous depth of buoyancy layer and $z = -H$ is the bottom of ocean. The $\widetilde{b}(z)$ represents a temporal averaged buoyancy whose mean depth is z. By expanding equations (14) and (15) and $\widetilde{b}$ with respect to z', *McDougall and McIntosh* [2001] show that equations (12)–(15) are identical to equations (8)–(11) up to the second-order perturbation.

As equation (14) takes into account the higher order terms and seems to be calculated with relatively small numerical errors, we employ equation (14) to evaluate the residual-mean velocity from the output of the eddy-resolving ocean model.

3. MODEL DESCRIPTION

The Center for Climate System Research Ocean Component model [*Hasumi*, 2006] is used. This model solves the primitive equations with the explicit free surface method. The same model was previously applied to eddy-permitting ocean modeling [*Nakano and Hasumi*, 2005; *Sakamoto et al.*, 2005].

The simulation area is the Southern Ocean from 20°S to 75°S. A realistic bathymetry based on *ETOPO2* [2001] is used. Three different horizontal resolutions of $1/2° \times 1/3°$, $1/4° \times 1/6°$, and $1/8° \times 1/12°$ (longitude × latitude) are used and called SO3, SO6, and SO12, respectively. In the vertical, 85 levels are used. The constant grid spacing is 50 m from the sea surface to 2000 m depth, and the grid spacing is linearly increased from 200 to 4000 m depth. From 4000 to 5500 m, a constant 100-m grid spacing is used.

In the tracer equations, a high-accuracy tracer advection scheme [*Hasumi and Suginohara*, 1999] and biharmonic diffusion scheme are employed. The biharmonic diffusivities are 1.6×10^{10}, 10^9, and 5×10^7 m^4 s^{-1} for SO3, SO6, and SO12, respectively. For the vertical diffusivity, a constant of 5×10^5 m^2 s^{-1} is used. In the momentum equations, an enstrophy-conserving scheme [*Ishizaki and Motoi*, 1999] and biharmonic friction with a Smagorinsky-like viscosity [*Griffes and Hallberg*, 2000] are employed. The coefficient of the biharmonic friction depends on the horizontal grid spacing of the models and is controlled by a single parameter. We use the same parameter value of 2.5 for all cases. For the vertical viscosity, a constant value of 10^{-4} m^2 s^{-1} is used.

The model is first spun up for 1 year from a state of rest with constant temperature and salinity. The temperature and salinity fields are relaxed toward annual-mean climatology of the World Ocean Atlas 98 (WOA98) [*Antonov et al.*, 1998; *Boyer et al.*, 1998] with a relaxation time of 30 days throughout the water column during this spin-up period.

From the state created by the spin-up procedure, the model is integrated for 53 years with forcing of the monthly mean climatology of the sea surface wind stress compiled by *Röske* [2001] based on the European Centre for Medium-Range Weather forecasts reanalysis data [*Gibson et al.*, 1997]. The sea surface temperature and salinity are restored to the monthly climatology of WOA98 with a relaxation time of 30 days. The temperature and salinity from 20°S to 23°S and those of the southernmost grid points are restored to the annual mean climatology with a relaxation time of 180 days.

4. RESULTS

The time evolution is monitored using the volume transport through Drake Passage (Figure 2). When reaching a statistically steady state, the time evolution curve of the transport becomes flat on a timescale larger than that of the mesoscale eddies. The data are filtered with a one-year timescale to eliminate mesoscale variability. The curves of the SO03 and SO06

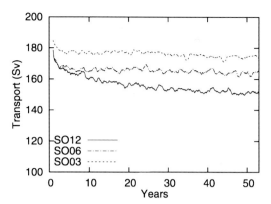

Figure 2. The time evolution of the volume transport through Drake Passage. The transport data are filtered with a 1-year timescale to remove mesoscale eddy variability.

cases are almost flat after a 10-year integration and indicate reaching a statistically steady state. For the SO12 case, the transport decreases continuously up to 45 years and becomes flat after that. Therefore, it is assumed that for these three cases, 50-year integration is enough for reaching the statistically steady state. Consequently, the model is integrated for 53 years, and output data of the last 3 years are collected once every 10 days and used for the following analyses. All variables shown in the following are averaged for the last 3 years.

4.1. SSH Variability

Plate 1 shows the SSH variability evaluated as the root-mean-square of SSH. Plate 2 shows SSH variability calculated by using the combined TOPEX/POSEIDON and European Remote Sensing Satellite altimetry data of year 1993, which is available from the Validation and Interpretation of Satellite Oceanographic data center. For the SO03 case, the SSH variability is weak compared with the observational data. The horizontal mesh size of the SO03 case is about 37 km at a latitude of 54°S, while the deformation radius is about 20 km at this latitude [*Chelton et al.*, 1998]. Therefore, it seems that mesoscale eddies are not resolved well in this resolution. For the SO06 case, the variability increases significantly from the SO03 case. It is high in the equatorward flank of the ACC, especially on the northern boundary of the ACC. The variability is large in the Argentine Basin and the Agulhas Retroflection region. The pattern of the variability agrees well with the observational data. For the SO12 case, the intensity of the variability reaches about 50 cm in the B/M confluence and the Agulhas Retroflection region.

The variability of the SO12 case is systematically larger than that of the observational data. *Pascual et al.* [2006] ar-

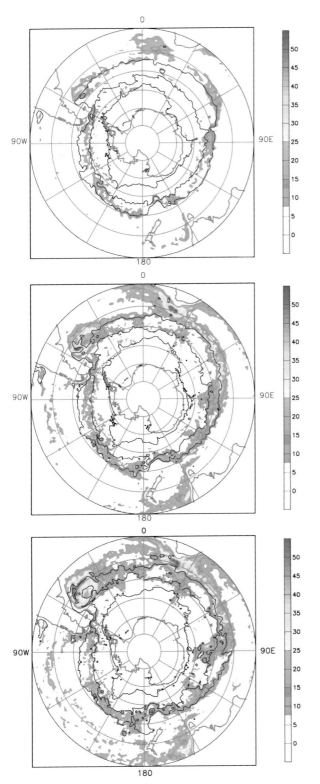

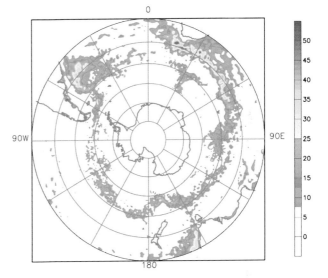

Plate 2. The SSH variability in centimeters obtained by using the combined TOPEX/POSEIDON and ERS altimetry data for the year 1993.

Plate 1. The SSH variability in centimeters. The solid lines show streamlines through Drake Passage. The top, middle, and bottom are the SO03, SO06, and SO12 cases, respectively.

gued that the SSH variability obtained by satellite altimeters is increased by combing with other satellite altimeters. It is anticipated, therefore, that the intensity of the variability of the observational data would be increased and become close to the SO12 case by adding other satellite data.

4.2. Streamlines

The solid lines in Plate 1 show streamlines through Drake Passage obtained by calculating volume transport from the Antarctic Continent. The most southern and northern streamlines mark the zero and maximum eastward transport, respectively. The middle line between the two represents half of the maximum transport. Note that the maximum volume transports are different for the three cases. These are 168, 158, and 144 $\times$ 10^6 m^3 s^{-1} for the SO03, SO06, and SO12 cases, respectively.

When increasing the resolution, the most distinct difference occurs in the location of the B/M confluence. For the SO03 case, the B/M confluence is located at about 45°S and is too far south of observations [e.g., *Legeckis and Gordon*, 1982; *Barré et al.*, 2006]. For the SO06 case, the streamlines shows large meandering in the B/M confluence region, but the location of the confluence is still too far south as compared with observations. For the SO12 case, the location of the B/M confluence is offshore from the Rio de la Plata estuary (38°S and 50°W), and it agrees well with observations.

There is less resolution dependency in the southern boundary of the ACC where the SSH variability is weak. The southern boundary shifts poleward in the Indian Ocean and widens the ACC band. It also shifts poleward sharply in the Pacific-Antarctic Ridge between 145°W and 120°W. This southern boundary stream line broadly agrees with observations [e.g., *Orsi et al.*, 1995].

4.3. Eddy Buoyancy Fluxes

Plate 3 shows northward eddy buoyancy fluxes at 600 m depth. As the resolution increases, the intensity of the fluxes is increased. The regions where the intensity is increased coincide with the ones where the SSH variability is increased. The eddy fluxes in the high eddy active regions appear to change their direction alternately. *Marshall and Shutts* [1981] argued that in the high eddy active regions, a large part of the eddy fluxes, which is represented by a rotational flux, circulates around eddy potential energy contours and has a component up/down the mean buoyancy gradient. *Eden et al.* [2007b] recently confirmed, using an eddy-resolving ocean model of the North Atlantic Ocean, that a large part of the eddy fluxes circulates around the eddy potential energy contours. The eddy fluxes that change direction alternately in Plate 3 seem to contain a large amount of this rotational flux.

Table 1. The Averaged Northward Eddy Buoyancy Flux at 600 m Depth in the ACC (unit, 10^{-6} m^2 s^{-3}).[a]

	SO03	SO06	SO12
Eddy flux	−0.53	−1.12	−1.02

[a]The area over which the fluxes are averaged includes the Argentine Basin (45°S latitude < 30°S; 60°W < longitude < 30°W) as well as the region south of 45°S.

In the ACC, the eddy fluxes transfer buoyancy from the north to south and release potential energy. To represent this transport properly, the eddy fluxes should be well resolved. As the eddy fluxes contain a large rotational component, the fluxes are averaged over the whole ACC area, and the averaged flux is used to evaluate the buoyancy transport across the ACC. Table 1 shows the values of averaged fluxes. As expected, the averaged fluxes are directed southward for all three cases. For the SO03 case, the intensity of the fluxes is relatively weak. For the SO06 case, the intensity is about twice larger than for the SO03 case. For the SO12 case, the value of the intensity is similar to that of the SO06 case.

4.4. The TRM Velocity

Plate 4 shows the vertical component of the TRM velocity calculated by using equations (13) and (14) at 600 m depth. As the calculated velocity contains high-frequency spatial noise, it is smoothed with 1.0° length scale using a method employed by *Bryan et al.* [1999].

To investigate the meridional circulation in the ACC, the vertical TRM velocity is spatially integrated, and the values obtained are shown in Table 2. The areas of integration are defined by dividing the ACC flow regime into four bands of equal eastward volume transport, with the most poleward band identified as B1, the most equatorward band as B4. In the ACC, upwelling flow is expected due to Ekman transport. However, in the regions where eddy activity is strong, eddy-induced transport may cause downwelling flow, as argued by *Marshall* [1997].

For the SO03 case, upwelling transport in the poleward flank of the ACC is relatively weak compared with the higher resolution cases. In the equatorward flank, there is somewhat larger upwelling in B3 band, especially downstream of Drake Passage. In the outermost band of B4, the downward velocity dominates, and the downwelling is large in the B/M confluence. In this resolution case, a main meridional circulation is in the B3 and B4 bands, which are upwelling and downwelling, respectively.

For the SO06 case, the vertical TRM velocity in the poleward flank is somewhat larger than the SO03 case. The downwelling flow dominates in the most northern band (B4).

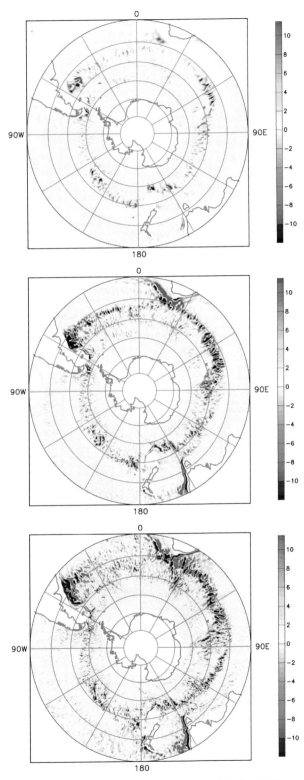

Plate 3. The northward eddy buoyancy fluxes at 600 m depth (unit, 4×10^{-6} m^2 s^{-3}). The top, middle, and bottom are the SO03, SO06, and SO12 cases, respectively.

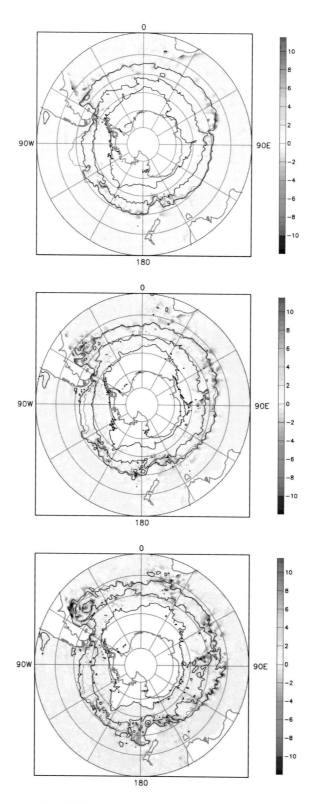

Plate 4. The vertical component of the TRM velocity (unit, 2×10^{-6} m s^{-1}). The top, middle, and bottom are the SO03, SO06, and SO12 cases, respectively.

For the SO12 case, there is a relatively large upwelling flow in the poleward flank of the ACC. The most interesting point of the SO12 case is that there is a very strong downward TRM velocity at the B/M confluence. The strength of the TRM velocity is more than 3×10^{-5} m s^{-1}. The isopycnal line of $\sigma_0 = 27.1$, which is a core density of AAIW [e.g., *Talley*, 1996] in the Atlantic and Indian Oceans, goes through this strong downwelling region.

4.5. Salinity Distribution

The left column of Plate 5 shows the horizontal salinity distribution and the vertical TRM velocity at 600 m depth in the B/M confluence zone. The right column of Plate 5 shows the vertical section of salinity distribution at 30°W. Plate 6 shows that of the WOA98 annual mean salinity at the same longitude.

The Malvinas Current, which contains relatively low salinity water at 600 m depth, flows northward along the Argentine continent and meets the Brazil Current at the confluence point where the TRM velocity is downwelling for all three resolution cases.

For the SO03 case, the strength of the TRM velocity at the confluence point is about 10^{-5} m s^{-1} and does not subduct enough low-salinity water along the $\sigma_0 = 27.1$ surface downstream of the confluence point. The 34.5 salinity water reaches only about 400-m depth at 30°W.

For the SO06 case, the strength of the vertical TRM velocity at the confluence point is increased to 2×10^{-5} m s^{-1}. It subducts the low-salinity water deeper than the SO03 case, and 34.5 salinity water spreads along the $\sigma_0 = 27.1$ surface up to about 500 m depth.

For the SO12 case, the strength of the downward TRM velocity reaches 3×10^{-5} m s^{-1}. The 34.5 salinity water, which is subducted along the $\sigma_0 = 27.1$ surface, reaches more than 1,400 m depth. Although the spreading of the salinity is still somewhat weak even for the SO12 case compared with the climatology salinity distribution, it is much better than the low-resolution cases.

Table 2. The Vertical Transport (unit, 10^6 m^3 s^{-1}) at 600 m depth in the four ACC bands.[a]

Band	SO03	SO06	SO12
B4	−2.7	−1.2	−4.3
B3	6.3	3.3	3.0
B2	2.9	7.7	8.9
B1	1.2	3.4	7.6

[a]The definition of the bands is described in section 4.4. Negative values indicate downwelling.

5. SUMMARY AND CONCLUSION

The resolution dependence of the eddy fluxes and the effects of the eddy fluxes is studied in the Southern Ocean using three different horizontal resolutions of an ocean model.

For the SO03 case, the intensity of the eddy buoyancy fluxes, the SSH variability, and the downward TRM velocity at the B/M confluence are relatively small compared with the higher resolution cases. The B/M confluence is located in far to the south, compared with the observations. The spreading of the low-salinity tongue into the Atlantic gyre is much less than that seen in the observational data.

For the SO06 case, the intensity of the spatially averaged eddy buoyancy flux and the downward TRM velocity at the B/M confluence are about twice as strong as in the SO03 case. The intensity of the SSH variability is increased significantly from the SO03 case. The location of the confluence shifts to north, but remains south of the observed position. The spreading of the low-salinity tongue is increased somewhat.

For the SO12 case, the intensity of the spatially averaged eddy buoyancy flux is similar to that of the SO06 case. The downward TRM velocity at the B/M confluence reaches about three times the strength of the SO03 case. The location of the B/M confluence is close to the observed location. The intensity of SSH variability is increased moderately. The spreading of the low-salinity tongue is increased significantly, becoming close to the observational data.

Therefore, it is concluded that each of the variables has their own resolution dependence. The horizontal resolution of $1/2° \times 1/3°$ is not adequate to simulate all examined variables: the eddy buoyancy flux, the SSH variability, the location of the B/M confluence, and the injection of AAIW. The horizontal resolution of $1/4° \times 1/6°$ may be enough to simulate the eddy buoyancy flux and the SSH variability. However, at least $1/8° \times 1/12°$ seems to be required to properly simulate the location of the B/M confluence and the injection of AAIW.

Acknowledgments. We thank all members of the Global Environment Modeling Research Program, Frontier Research Center for Global Change, Japan Agency for Marine-Earth Science and Technology. We also thank the two anonymous reviewers for providing helpful suggestions to improve the manuscript. The efforts of the editor, Matthew Hecht, are appreciated. We also thank Morrison Bennett, Los Alamos National Laboratory, for editing assistance. This research was supported by CREST, JST. The numerical experiments were performed on the Earth Simulator. The altimeter products were produced by the CLS Space Oceanography Division as part of the Environment and Climate EU ENACT project (EVK2-CT2001-00117) and with support from CNES.

APPENDIX A

The derivation of equation (3) is shown below:

$$\nabla \cdot (\overline{b'\mathbf{u}'})_{\parallel} = \nabla \cdot (\overline{b'\mathbf{u}'}) - \left[\frac{\overline{b'\mathbf{u}'} \cdot \nabla \overline{b}}{\overline{b}_z}\right]_z$$

$$= \nabla_H \cdot (\overline{b'\mathbf{v}'}) - \left[\frac{\overline{b'\mathbf{v}'} \cdot \nabla_H \overline{b}}{\overline{b}_z}\right]_z$$

$$= \nabla_H \cdot \left(\frac{\overline{b'\mathbf{v}'}}{\overline{b}_z}\overline{b}_z\right)$$

$$- \left[\frac{\overline{b'\mathbf{v}'}}{\overline{b}_z}\right]_z \cdot \nabla_H \overline{b} - \left(\frac{\overline{b'\mathbf{v}'}}{\overline{b}_z}\right) \cdot \nabla_H \overline{b}_z$$

$$= \left(\nabla_H \cdot \left(\frac{\overline{b'\mathbf{v}'}}{\overline{b}_z}\right)\right)\overline{b}_z - \left[\frac{\overline{b'\mathbf{v}'}}{\overline{b}_z}\right]_z \cdot \nabla_H \overline{b}.$$

APPENDIX B

The diapycnal term G_1 can be examined using an eddy buoyancy variance budget of the steady state:

$$\nabla \cdot (\overline{\phi_2 \mathbf{u}}) = -\overline{b'\mathbf{u}'} \cdot \nabla \overline{b} + \overline{b'Q'},$$

where $\phi_2 = (b'^2/2)$. The left-hand side term is the advection of the variance by the mean and eddy velocity $\mathbf{u} = \overline{\mathbf{u}} + \mathbf{u}'$. The first term of the right-hand side is a source term of the variance which generates (absorbs) the variance when the eddy fluxes direct downgradient (upgradient). The second term represents diabatic effects. For a weak mean flow, the advection term should be small and the eddy fluxes must be along isopycnals when the diabatic effect is small. In this case, the diapycnal term G_1 must be relatively small, and the effect of eddies is solely represented by equation (3). For a strong mean flow, the advection term is not small and is mainly balanced with the source term. This balance states that a region where the flow goes into the high (low) eddy potential energy region should be identical to a region where the eddy buoyancy flux is directed downgradient (upgradient) [Marshall and Shutts, 1981; McDougall and McIntosh, 1996]. In this case, the diapycnal term G_1 is not negligible.

The variance flux $\overline{\phi_2 \mathbf{u}}$ can be decomposed into an isopycnal component and a vertical component as:

$$\overline{\phi_2 \mathbf{u}} = (\overline{\phi_2 \mathbf{u}})_{\parallel} + \frac{\overline{\phi_2 \mathbf{u}} \cdot \nabla \overline{b}}{\overline{b}_z}\mathbf{k}.$$

The direction of the variance flux $\overline{\phi_2 \mathbf{u}}$ is almost along the mean buoyancy $\overline{b}$, as $\overline{\phi_2 \mathbf{u}} \sim \overline{\phi}_2 \overline{\mathbf{u}}$ and $\overline{\mathbf{u}}$ is almost along the isopycnals $\overline{b}$. Therefore, it is anticipated that the vertical component is relatively small compared with the isopycnal component and the decomposition may be useful to capture a dominant part of the eddy variance flux. The divergence of the isopycnal component can be written as:

$$\nabla \cdot (\overline{\phi_2 \mathbf{u}})_{\parallel} = -\mathbf{F}_2^{\mathrm{rot}} \cdot \nabla \overline{b},$$

$$\mathbf{F}_2^{\mathrm{rot}} = \left(\frac{\partial}{\partial z}\left(\frac{\overline{\phi_2 \mathbf{v}}}{\overline{b}_z}\right), \; -\nabla_H \cdot \left(\frac{\overline{\phi_2 \mathbf{v}}}{\overline{b}_z}\right)\right).$$

Then, the advection of the variance is written as

$$\nabla \cdot (\overline{\phi_2 \mathbf{u}}) = -\mathbf{F}_2^{\mathrm{rot}} \cdot \nabla \overline{b} + \left[\frac{\overline{\phi_2 \mathbf{u}} \cdot \nabla \overline{b}}{\overline{b}_z}\right]_z.$$

The above equation states that the local generation (absorption) of the variance by the advection of the variance is associated with the rotational flux $\mathbf{F}_2^{\mathrm{rot}}$, which is directed downgradient (upgradient) [Marshall and Shutts, 1981; McDougall and McIntosh, 1996]. Bryan et al. [1999] acknowledged that eddy fluxes of a high-resolution ocean model have a large rotational component, and it was an obstacle to compare eddy fluxes of model output with that of parameterizations. This large rotational component seems to correspond to the rotational flux of $\mathbf{F}_2^{\mathrm{rot}}$.

The term associated with the isopycnal component of eddy variance flux can be further modified into an advective form:

$$\left[\frac{\mathbf{F}_2^{\mathrm{rot}} \cdot \nabla \overline{b}}{\overline{b}_z}\right]_z = \left[\frac{1}{\overline{b}_z}\left[\frac{\overline{\phi_2 \mathbf{v}}}{\overline{b}_z}\right]_z \cdot \nabla_H \overline{b} - \nabla_H \cdot \left(\frac{\overline{\phi_2 \mathbf{v}}}{\overline{b}_z}\right)\right]_z$$

$$= \left[\frac{1}{\overline{b}_z}\left[\frac{\overline{\phi_2 \mathbf{v}}}{\overline{b}_z}\right]_z\right]_z \cdot \nabla_H \overline{b} + \frac{1}{\overline{b}_z}\left[\frac{\overline{\phi_2 \mathbf{v}}}{\overline{b}_z}\right]_z \cdot \nabla_H \overline{b}_z$$

$$- \nabla_H \cdot \left[\frac{\overline{\phi_2 \mathbf{v}}}{\overline{b}_z}\right]_z$$

$$= \left[\frac{1}{\overline{b}_z}\left[\frac{\overline{\phi_2 \mathbf{v}}}{\overline{b}_z}\right]_z\right]_z \cdot \nabla_H \overline{b} + \frac{1}{\overline{b}_z}\left[\frac{\overline{\phi_2 \mathbf{v}}}{\overline{b}_z}\right]_z \cdot \nabla_H \overline{b}_z$$

$$- \nabla_H \cdot \left(\left[\frac{1}{\overline{b}_z}\left[\frac{\overline{\phi_2 \mathbf{v}}}{\overline{b}_z}\right]_z\right]\overline{b}_z\right)$$

$$= \left[\frac{1}{\overline{b}_z}\left[\frac{\overline{\phi_2 \mathbf{v}}}{\overline{b}_z}\right]_z\right]_z \cdot \nabla_H \overline{b} - \overline{b}_z \nabla_H \cdot \left(\frac{1}{\overline{b}_z}\left[\frac{\overline{\phi_2 \mathbf{v}}}{\overline{b}_z}\right]_z\right).$$

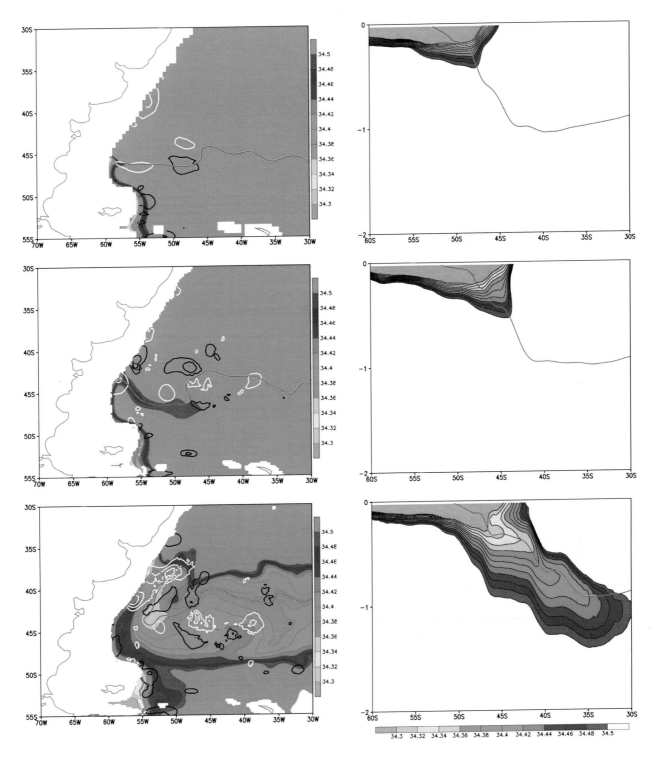

Plate 5. Left: The horizontal salinity distribution at 600-m depth in the B/M confluence zone. Superposed black and white lines represent strength of the vertical TRM velocity. Contour interval is 10^{-5} m s^{-1}, and black and white represent positive and negative values, respectively. The red line is $\sigma_0 = 27.1$ isopycnal. Right: Vertical section of salinity at 30°W . The vertical axis is depth in kilometers. The red line is $\sigma_0 = 27.1$ isopycnal. The top, middle, and bottom rows are the SO03, SO06, and SO12 cases, respectively.

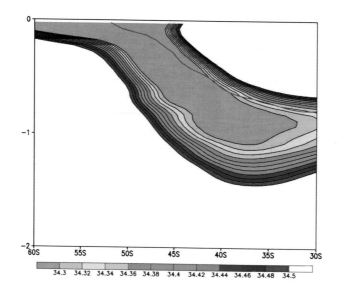

Plate 6. The annual mean climatology salinity of the World Ocean Atlas 98 at 30°W. The vertical axis is depth in kilometers. The contour interval is 0.02. The red line is $\sigma_0 = 27.1$ isopycnal.

Then, the term G_1 is written as:

$$G_1 = \mathbf{u}_2^* \cdot \nabla \bar{b} + \left[\frac{\overline{b'Q'}}{\bar{b}_z} \right]_z - \left[\frac{1}{\bar{b}_z} \left[\frac{\overline{\phi_2 \mathbf{u} \cdot \nabla b}}{\bar{b}_z} \right]_z \right]_z,$$

$$\mathbf{u}_2^* = \left(\frac{\partial \Psi_2^*}{\partial z}, -\nabla_H \cdot \Psi_2^* \right),$$

$$\Psi_2^* = \frac{1}{\bar{b}_z} \left[\frac{\overline{\phi_2 \mathbf{v}}}{\bar{b}_z} \right]_z.$$

Finally, the eddy term can be written as:

$$\nabla \cdot \overline{(b'\mathbf{u}')} = (\mathbf{u}_1^* + \mathbf{u}_2^*) \cdot \nabla \bar{b} - Q_1^* + G_2,$$

$$Q_1^* = - \left[\frac{\overline{b'Q'}}{\bar{b}_z} \right]_z,$$

$$G_2 = \left[\frac{1}{\bar{b}_z} \left[\frac{\overline{\phi_2 \mathbf{u} \cdot \nabla b}}{\bar{b}_z} \right]_z \right]_z.$$

The term $\mathbf{u}_2^* \cdot \nabla \bar{b}$ represents the additional advective component which is first derived by *McDougall and McIntosh* [1996].

REFERENCES

Andrews, D., and M. McIntyre (1976), Planetary waves in horizontal and vertical shear: The generalized Eliassen-Palm relation and the zonal mean acceleration, *J. Atmos. Sci.*, *33*, 2031–2048.

Antonov, J. I., S. Levitus, T. P. Boyer, M. E. Conkright, T. D. O'Brien, and C. Stephens (1998), *World Ocean Atlas 1998, Vol. 1, Temperature of the Atlantic Ocean*, NOAA Atlas NESDIS 27, US Government Printing Office, Washington, DC.

Barré, N., C. Provost, and M. Saraceno (2006), Spatial and temporal scales of the Brazil-Malvinas Current confluence documented by simultaneous MODIS Aqua 1.1-km resolution SST and color images, *Adv. Space Res.*, *37*, 770–786.

Boyer, T. P., S. Levitus, J. I. Antonov, M. E. Conkright, T. D. O'Brien, and C. Stephens (1998), *World Ocean Atlas 1998*, Vol. 4, *Salinity of the Atlantic Ocean*, NOAA Atlas NESDIS 30, US Government Printing Office, Washington, DC.

Bryan, K., J. Dukowicz, and R. Smith (1999), On the mixing coefficient in the parameterization of bolus velocity, *J. Phys. Oceanogr.*, *29*, 2442–2456.

Chelton, D., R. DeSzoeke, and M. Schlax (1998), Geographical variability of the first baroclinic Rossby radius of deformation, *J. Phys. Oceanogr.*, *28*, 433–460.

Eden, C., R. Greatbatch, and D. Olbers (2007a), Interpreting eddy fluxes, *J. Phys. Oceanogr.*, *37*, 1282–1296.

Eden, C., R. Greatbatch, and J. Willebrand (2007b), A diagnosis of thickness fluxes in an eddy-resolving model, *J. Phys. Oceanogr.*, *37*, 727–742.

ETOPO2 (2001), US Department of Commerce, National Oceanic and Atmospheric Administration, National Geophysical Data Center, *2-minute Gridded Global Relief Data*.

Gibson, J., P. Kallberg, S. Uppala, A. Hernandez, A. Nomura, and E. Serrano (1997), ECMWF Re-Analysis, *Project Report Series, 1.ERA Description.* European Centre for Medium-Range Weather Forecasts, Reading, England.

Gille, S., and R. Davis (1999), The influence of mesoscale eddies on coarsely resolved density: An examination of subgrid-scale parameterization, *J. Phys. Oceanogr.*, *29*, 1109–1123.

Griffies, S., and W. Hallberg (2000), Biharmonic friction with a Smagorinsky-like viscosity for use in large-scale eddy-permitting ocean model, *Mon. Weather Rev.*, *128*, 2935–2946.

Hasumi, H. (2006), *CCSR Ocean Compoment Model (COCO) version 4.0*, 103 pp., CCSR Rep. 25, University of Tokyo, Kashiwa, Japan.

Hasumi, H., and N. Suginohara (1999), Sensitivity of a global ocean general circulation model to tracer advection schemes, *J. Phys. Oceanogr.*, *29*, 2730–2740.

Henning, C., and G. Vallis (2004), The effect of mesoscale eddies on the main subtropical thermocline, *J. Phys. Oceanogr.*, *34*, 2428–2443.

Henning, C., and G. Vallis (2005), The effect of mesoscale eddies on the stratification and transport on an ocean with a circumpolar channel, *J. Phys. Oceanogr.*, *35*, 880–896.

Ishizaki, H., and T. Motoi (1999), Reevaluation of the Takano-Onishi scheme for momentum advection on bottom relief in ocean models, *J. Atmos. Oceanic Technol.*, *16*, 1994–2010.

Jacobson, T., and H. Aiki (2006), An exact energy for TRM theory, *J. Phys. Oceanogr.*, *26*, 558–564.

Karsten, R., H. Jones, and J. Marshall (2002), The role of eddy transfer in setting the stratification and transport of a circumpolar current, *J. Phys. Oceanogr.*, *32*, 39–54.

Legeckis, R., and A. Gordon (1982), Satellite observations of the Brazil and Falkland currents—1975 to 1976 and 1978, *Deep-Sea Res.*, *29*, 375–401.

Marshall, D. (1997), Subduction of water masses in an eddying ocean, *J. Mar. Res.*, *55*, 201–222.

Marshall, J., and T. Radko (2003), Residual-mean solutions for the Antarctic Circumpolar Current and its associated overturning circulation, *J. Phys. Oceanogr.*, *33*, 2341–2354.

Marshall, J., and G. Shutts (1981), A note on rotational and divergent eddy fluxes, *J. Phys. Oceanogr.*, *11*, 1677–1680.

McDougall, T., and P. McIntosh (1996), The temporal-residual-mean velocity. Part I: Derivation and the scalar conservation equation, *J. Phys. Oceanogr.*, *26*, 2653–2665.

McDougall, T., and P. McIntosh (2001), The temporal-residual-mean velocity. Part II: Isopycnal interpretation and the tracer and momentum equations, *J. Phys. Oceanogr.*, *31*, 1222–1246.

Nakano, H., and H. Hasumi (2005), A series of zonal jets embedded in the broad zonal flows in the Pacific obtained in eddy-permitting ocean general circulation models, *J. Phys. Oceanogr.*, *35*, 474–488.

Orsi, A., T. Whitworth III, and W. Nowlin Jr (1995), On the meridional extent and fronts of the Antarctic Circumpolar Current, *Deep-Sea Res.*, *42*, 641–673.

Pascual, A., Y. Faugère, G. Larnicol, and P.-Y. L., Traon (2006), Improved description of the ocean mesoscale variability by combining four satellite altimeters, *Geophys. Res. Lett.*, *33*, L02611, doi:10.1029/2005GL024633.

Radko, T., and J. Marshall (2006), The Antarctic Circumpolar Current in three dimensions, *J. Phys. Oceanogr.*, *36*, 651–669.

Röske, F. (2001), An atlas of surface fluxes based on the ECMWF re-analysis—A climatological dataset to force global ocean general circulation models, Max-Planck-Institut für Meteorologie, Hamburg, report No. 323.

Sakamoto, T. T., H. Hasumi, M. Ishii, S. Emori, T. Suzuki, T. Nishimura, and A. Sumi (2005), Responses of the Kuroshio and the Kuroshio Extension to global warming in a high-resolution climate model, *Geophys. Res. Lett.*, *32*, L14617, doi:10.1029/2005GL023384.

Talley, L. (1996), *Antarctic Intermediate Water in the Southern Atlantic.* In The South Atlantic: Present and Past Circulation, edited by G. Wefer, W.H. Berger, G. Siedler and D.J. Webb, pp. 219–238, Springer, Berlin.

Yukio Tanaka, Frontier Research Center for Global Change, Japan Agency for Marine-Earth Science and Technology, 3173-25 Showa-machi, Kanazawa-ku, Yokohama, Japan 236-0001. (ytanaka@jamstec.go.jp)

Eddies and Upper-Ocean Nutrient Supply

A. Oschlies

IFM-GEOMAR, Leibniz-Institut für Meereswissenschaften an der Universität Kiel, Kiel, Germany

The role of mesoscale eddies in biogeochemical cycles is reviewed with a focus on the nutrient supply to the oligotrophic gyres. Previous estimates of the eddy-induced nutrient supply differ by an order of magnitude, which can be related to different assumptions made about timescales for nutrient consumption and subsequent recharging of nutrient-depleted waters. Results from a fully prognostic eddy-resolving ecosystem-circulation model reveal a magnitude of eddy-induced nitrate supply to the North Atlantic subtropical gyre of typically less than 0.05 mol m^{-2} year^{-1}. This number includes the contribution by lateral eddy stirring which, on average, is larger than that by vertical eddy pumping. The results are in line with earlier conservative estimates and suggest that previous high estimates of the nutrient supply by eddy pumping have been overestimates.

1. INTRODUCTION

Observations of mesoscale variability in the color of surface sea water were reported and linked to marine biology already by the great explorers of the eighteenth century, among them Captain James Cook [Bainbridge, 1957]. Nowadays, ubiquitous meso- and sub-mesoscale variations in upper-ocean optical properties are perhaps most spectacularly seen from space, examples including the "line in the sea" recorded on camera by the space shuttle astronauts [Yoder et al., 1994], features associated with equatorial [Strutton et al., 2001] and off-equatorial Rossby waves [Uz et al., 2001; Kawamiya and Oschlies, 2001; Dandonneau et al., 2003] or mid-latitude mesoscale patchiness [Martin, 2003]. In situ studies using towed instruments have also provided in-depth views of mesoscale and sub-mesoscale features [Fasham et al., 1985; Strass, 1992; Washburn et al., 1998].

A global analysis of mesoscale ocean color variability as seen by the sea-viewing wide field-of-view sensor (Sea WiFS) satellite confirmed the global applicability of earlier local and regional findings that small-scale variability

in ocean biology is closely linked to ocean physics [Doney et al., 2003]. Length scales of mesoscale ocean color variability are similar to those estimated from satellite altimeter data and are often related to the first baroclinic Rossby deformation radius. However, the strong west–east asymmetry observed for physical eddy kinetic energy is much less visible in the ocean color data. Significant ocean color variability can also be seen at the sub-mesoscale (<10 km) where scales emerge as a complex interplay of different timescales associated with growth, respiration, and trophic interactions with physical turbulent advection and stirring [Abraham, 1998; Garçon et al., 2001; Mahadevan and Campbell, 2002].

Mesoscale features impact upper-ocean biology through a number of processes, including lateral advection and stirring [Lee and Williams, 2000], frontal instabilities and the associated tilting of isopycnal surfaces [Woods, 1988; Nurser and Zhang, 2000], frontal upwelling [Onken, 1992], and the eddy pumping mechanism according to which the formation and intensification of cyclonic eddies can lift nutrient-replete isopycnals into the light [Jenkins, 1988; Falkowski et al., 1991; McGillicuddy and Robinson, 1997], possibly altered by the interaction of the eddy's surface currents with the wind field [Martin and Richards, 2001; McGillicuddy et al., 2007]. Recent reviews of these processes and the way they influence marine ecosystems include those by Garçon

Ocean Modeling in an Eddying Regime
Geophysical Monograph Series 177
Copyright 2008 by the American Geophysical Union.
10.1029/177GM09

et al. [2001] and *Williams and Follows* [2003]. The present study focuses on the role eddies play in supplying nutrients to the light-lit upper ocean and thereby fueling biological production and driving biogeochemical cycles.

2. WHY BOTHER?

2.1. The Biological Carbon Pump

The biotically mediated export of organic carbon from the surface of the ocean toward its interior is a crucial component of the global carbon cycle. This so-called biological pump [*Volk and Hoffert*, 1985] is responsible for the observed vertical nutrient gradient between the surface ocean, which is generally depleted, and deeper nutrient-rich water. Because of the relatively constant carbon-to-nutrient stoichiometry of marine organic matter, the biological pump is also responsible for a large portion (about three quarters) of the observed vertical gradient of dissolved inorganic carbon. Without the biology acting, the surface concentrations of nutrients and dissolved inorganic carbon would be much larger, resulting in approximately doubled concentrations of atmospheric carbon dioxide (CO_2) [*Maier-Reimer et al.*, 1996].

When linking biological production and the biological pump, a few concepts are relevant. First, only that portion of the total primary production that is not immediately recycled within the upper ocean can contribute to the observed vertical gradients in biogeochemical tracers. This portion is called "new production", which is defined by *Dugdale and Goering* [1967] as the primary production in the euphotic zone that results from nitrogen input from outside the euphotic zone, i.e., inputs from the water below, from land, or from the atmosphere. In a steady state, new production must equal the export of photosynthetically fixed organic matter to the deep ocean [*Eppley and Peterson*, 1979]. This balance does not necessarily hold locally, as lateral advection of nutrients and/or biomass may regionally decouple areas of nutrient supply from areas of biomass export [*Oschlies*, 2002c; *Plattner et al.*, 2005].

Dugdale and Goering [1967] used the euphotic zone (the zone with sufficient light levels for photosynthesis to occur) as the depth range to define new production. Air–sea exchange of CO_2, however, does not depend on light but only on material contact with the air–sea interface. In this respect, the surface mixed layer is the appropriate depth range [*Oschlies and Kähler*, 2004]. As a matter of fact, there is often little difference between the depth ranges of the euphotic zone and the surface mixed layer in the subtropics and summertime mid-latitudes where many of the observational studies have been undertaken and where much of this paper will focus. Nevertheless, in this paper, we will focus mainly on eddy impacts on new production and export production, which both use the base of the euphotic zone as the reference depth. We thus do not explicitly examine the impacts of eddies on mixed-layer depths, although a net eddy-induced shallowing of high-latitude winter mixed-layer depths has been suggested [*Oschlies*, 2002b]. Impacts on new production can also be traced to impacts on primary production and, to some extent, to changes in surface partial pressure of CO_2 (pCO_2) [*Mahadevan et al.*, 2004]. Linking new production to air–sea carbon fluxes is not always straightforward and requires careful consideration of the differences between mixed-layer depth and euphotic-zone depth as well as the different relevant timescales which are typically days to weeks for biological production and months for air–sea gas exchange [*Oschlies and Kähler*, 2004].

2.2. A Case for Eddying Nutrient Supply

Concentrating on the role played by eddies in global biogeochemical cycles, we begin by looking at the annual mean "climatological" nitrate field compiled from a large number of observations by *Conkright et al.* [2002]. A very similar picture exists for phosphate, the other major nutrient essential for biological production. A dominant feature is the depletion of these essential nutrients in the surface waters of the subtropical gyres (Figure 1). A vertical section of the zonally averaged nitrate field shows bowl-shaped nutrient-poor regions in and above the subtropical thermocline.

The presence of closed temporal-mean surfaces of constant nitrate concentration, $NO_3 = const$ (as indicated by closed contours in Figure 1 in the zonal average and in the surface layer), already points at the potential role of eddies in fueling nutrient supply to the nutrient-poor subtropical gyres. Consider some $NO_3 = const$ surface (for example $NO_3 = 5$ mmol m^{-3}) that, in the annual mean, encloses some low-latitude water volume V. Assuming mass (and, for an incompressible fluid, volume) conservation by neglecting any net precipitation-minus-evaporation across the sea surface bounded by the $NO_3 = const$ contours, Gauß's theorem then tells us that the mean advection of the mean nitrate field cannot yield any net transport of nitrate across any closed $NO_3 = const = NO_3^*$ surface:

$$\int_{V(NO_3 \leq NO_3^*)} \nabla \cdot (\mathbf{u}NO_3)d^3x = \int_{NO_3 = NO_3^*} (\mathbf{u}NO_3) \cdot \mathbf{n}dA \quad (1)$$

$$= NO_3^* \int_{NO_3 = NO_3^*} (\mathbf{u} \cdot \mathbf{n})dA \quad (2)$$

$$= NO_3^* \int_V \nabla \cdot \mathbf{u}d^3x = 0 \quad (3)$$

where $\mathbf{u}$ and NO_3 denote annual mean velocity vector and nitrate concentration, respectively. This argument, put forward

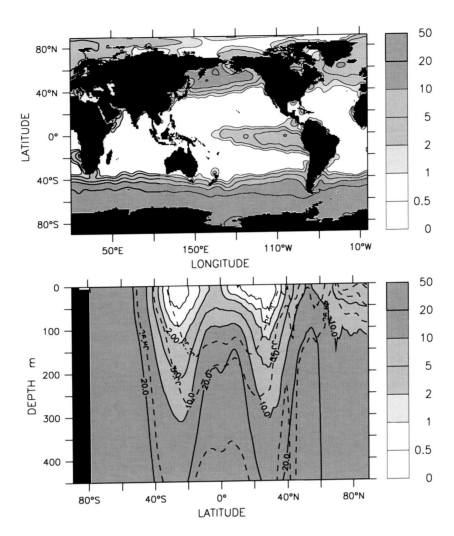

Figure 1. *Top*: Annual-mean surface nitrate concentrations. *Bottom*: Zonally averaged nitrate field. The zonal mean positions of the $\sigma = 25.2$, $\sigma = 26.0$, and $\sigma = 26.8$ isopycnals are indicated by the *dashed lines*. Units are mmol m^{-3}. Data are taken from the World Ocean Atlas 2001 [*Conkright et al.*, 2002].

for ocean heat transport by *Greatbatch et al.* [2007], means that irrespective of biological nitrate sinks and sources, which affect the nitrate field and thus the location and shape of the nitrate iso-surfaces, a steady circulation cannot be a net supplier of nutrients to the tropical and subtropical areas enclosed by a mean NO$_3$=const surface. This does not rule out advective transport across individual areas of this surface, but according to Gauß's theorem, this transport happens in such a way that the total advective transport of nitrate integrated over the closed surface vanishes. Therefore, any export of organic matter out of a volume enclosed by a mean NO$_3$=const surface cannot be fueled by a steady advective nutrient supply.

The following model diagnostics will not be performed on nitrate surfaces but rather on grid levels of the model that, in turn, can be interpreted as the nominal base of the euphotic zone. In this respect, Gauß's theorem has limited practical value, other than demonstrating in a mathematically clean way the general importance of temporal variability and/or mixing for nutrient supply over large parts of the global ocean. Were it not for diffusive transport and temporal variability in nitrate concentrations, in velocities or in both, export production in the subtropical gyres would depend solely on nutrient supply by means of nitrogen fixation, rivers, sediments, or the atmosphere.

2.3. Subtropical Desert Conundrums

The subtropical gyres have been in the center of a number of controversies over the past decades. Over these regions, the surface depletion of the essential nutrients nitrate and phosphorus is particularly pronounced (Figure 1). Ekman downwelling and a stably stratified pycnocline limit re-entrainment of nutrients into the light-lit surface layer and, even in winter, the surface mixed layer does not significantly exceed the depth of the euphotic zone. Accordingly, the subtropical gyres are often referred to as the ocean deserts [*Williams and Follows*, 1998b], and they display the lowest chlorophyll concentrations and clearest surface waters of the world ocean [*Morel et al.*, 2007]. Still, biological production is not zero, and there is ample evidence for substantial new production and export production in these areas. Indirect geochemical estimates of both nutrient supply to the light-lit surface layer, either from oxygen production in the euphotic zone [*Jenkins and Goldman*, 1985; *Spitzer and Jenkins*, 1989], from oxygen consumption in the aphotic zone underneath [*Jenkins*, 1982; *Sarmiento et al.*, 1990], or from a correlation between helium and nitrate (the so-called He flux gauge [*Jenkins*, 1988]), all suggest levels of export production (about 0.4–0.6 mol N m^{-2} year^{-1} if converted to nitrogen using Redfield ratios). A first conundrum is that such values, which are not very different from estimates in mesotrophic and eutrophic areas in mid and high latitudes, are substantially higher than direct biological and physical measurements of nitrate supply and uptake [*Lewis et al.*, 1986] and of particle export [*Lohrenz et al.*, 1992] could account for. A second conundrum addresses the magnitude and even the direction of the biotic contribution to air–sea CO_2 fluxes over the subtropical oceans. A number of observations suggest that respiration exceeds primary production already within the euphotic zone [e.g., *del Giorgio et al.*, 1997; *Williams et al.*, 2004]. Without lateral or atmospheric input of organic carbon, the net respiration would leave no organic carbon that could be exported out of the euphotic zone into the ocean interior.

The mismatch of geochemical tracer-based and direct estimates of nutrient supply and export production in the subtropical gyres has often been attributed to the different time and space scales the different measurements correspond to: the geochemical estimates integrate over hundreds to thousands of kilometers and years to decades, whereas the direct measurements reflect local and instantaneous conditions. Undersampling of episodic events like nutrient injections by mesoscale eddies (see subsequent section) might have led to their under-representation in the direct measurements. Similar arguments may be put forward with respect to the trophic state of the ocean: primary production can be expected to respond faster to changes in environmental conditions and therefore be more sensitive to episodic events than is respiration [*Karl et al.*, 2003]. However, undersampling alone does not systematically bias the results and therefore does not explain the direction of the apparent imbalances between production and export or between production and respira-tion (other than giving each a 50% chance). Statistically, the chances of over-representation of episodic events in any data set are equal to the chances of under-representation. The likelihood of systematic sampling biases should be particularly small for eddy-type features at the long-term time-series sites in the subtropical oceans, the Bermuda Atlantic Time-Series Study (BATS) and the Hawaii ocean time-series with typical sampling intervals (monthly to fortnightly) similar to the lifetime of eddy signals [e.g., *McNeil et al.*, 1999].

It is noteworthy that issues other than episodic events may be relevant for explaining the apparent observational discrepancies. For example, additional nitrogen sources such as nitrogen fixation or atmospheric deposition may significantly contribute to total nutrient input and affect the trophic state of the ecosystem. Erroneous assumptions about stoichiometric relations may be relevant when combining estimates of nutrient supply with concepts of primary production and export production, which are usually formulated in carbon units. Here, we will not further explore these alternative aspects and instead concentrate on the role of episodic events associated with eddies.

3. EDDY PUMPING

A possible scenario whereby eddies may lead to a net increase in nutrient supply to the euphotic zone, and hence biological production, is based on the vertical asymmetry of the light field. Although upward and downward excursions of water parcels due to eddies will average out, nutrients can be more readily taken up when water parcels are moved upward into the light. This concept rests on the fact that within cyclonic eddies and, in parts of the subtropics also within anticyclonic mode water eddies [*McGillicuddy et al.*, 1999], isopycnals are displaced upwards and thereby may move deep and nutrient-rich waters into the euphotic zone. Depending on how long the upwelled water stays in the light, some or all of the upwelled nutrients can be consumed by new production. Within anticyclonic eddies, on the other hand, isopycnal surfaces are displaced downwards. This moves nutrients further away from the euphotic zone, but does not give rise to any major biological response.

A few observations of the process, often termed "eddy pumping," exist but vary with respect to the interpretation of its long-term and basin-scale significance [*Falkowski et al.*, 1991; *McGillicuddy et al.*, 1998; *McNeil et al.*, 1999]. Based on detailed observations of a cyclonic eddy in the subtropical North Pacific, *Falkowski et al.* [1991] inferred some average 20% increase in primary production due to eddies. From observations near Bermuda, *McGillicuddy et al.* [1998] estimated that eddy pumping could increase the nutrient supply by more than 100%. Obviously, extrapolation of the observed individual events to larger space and longer timescales is difficult. Moreover, interpretations and possible extrapolations of local observations will depend on the extent to which the observed features behave more like waves or like eddies. Wave-like features will move water up and down locally. Closed eddies, on the other hand, can displace water vertically during the eddy formation and then move the enclosed water laterally with only a small vertical component due to eddy decay. Observed property differences with respect to the local background may therefore not reflect local processes, but will be affected by the remote history of the observed eddy. Extrapolations from local observations are difficult to achieve, and synoptic satellite observations and numerical models may both help to accomplish this in a consistent way.

3.1. Satellite-Based Estimates

Satellite-based estimates of the contribution of eddy pumping to new production in the Sargasso Sea were presented by *McGillicuddy et al.* [1998] and *Siegel et al.* [1999]. Using satellite altimeter data and statistical regressions derived from in situ observations at the BATS site, they assumed that all negative sea surface height (SSH) anomalies were associated with upward displacements of climatological nitrate profiles and that all nitrate entering the euphotic zone was used up completely during the lifetime of the negative SSH anomaly. The resulting estimates of nutrient supply associated with eddy pumping were 0.19 ± 0.1 and 0.24 ± 0.1 mol N m^{-2} year^{-1} for the two studies, respectively, and thereby account for about half of the new production estimated from geochemical tracers.

As discussed by *Martin and Pondaven* [2003], the *Siegel et al.* [1999] method relies on two assumptions that are mutually exclusive, namely that 100% of the upwelled nitrate is taken up and exported by the biology and that there is sufficient time between successive eddy events to recharge background nitrate to climatological levels. For observed eddy statistics at the BATS site, Martin and Pondaven estimated that not more than about 50% of the nitrate could be taken up locally and that, in addition, recharging of nitrate inventories between successive eddy events cannot be complete. Applying the *Siegel et al.* [1999] method to an eddy-resolving coupled ecosystem-circulation model for estimating the eddy-associated flux in the model, *Oschlies* [2002a] found this estimate to be sixfold larger than the "true" eddy flux calculated from the three-dimensional circulation and biological dynamics.

3.2. Recharging Issues

The issue of recharging becomes more evident when one looks at eddy pumping in an isopycnal framework. An initial upward displacement of an isopycnal into the light may lead to new production and eventually to sinking of particulate organic matter. As sinking is always relative to the ambient water, in a stably stratified pycnocline, sinking is a diapycnal transport. Remineralization will then release the nutrients on denser (and locally deeper) density surfaces than the original one on which the eddy-driven new production had taken place. Recharging the original isopycnal with nutrients will therefore require some diapycnal transport counteracting the diapycnal sinking flux. Such a diapycnal transport could be caused by local diapycnal mixing or by remote diapycnal mixing or water mass transformation in combination with isopycnal transport to the location of interest. The bottom line is that without diapycnal processes, eddies would simply lead to a net deepening of the nutricline by allowing depletion of all isopycnals that can occasionally be moved into the light by eddies. A similar effect could be obtained in a thought experiment by increasing the penetration of light in a non-eddying ocean by a few tens of meters according to the eddy distribution of an eddying ocean. After an initial increase in export production during the first very few deeper light pulses, total new production would again be close to the original levels and be controlled by diapycnal transport of nutrients across the stably stratified nutricline.

The above argument suggests that the one-dimensional concept of eddy pumping cannot work unless there is some extra diapycnal nutrient transport associated with the eddies. There are at least three reasons why eddies may indeed increase diapycnal mixing. First, the additional curvature of the isopycnal surfaces increases the surface area across which turbulent mixing can flux tracers. Second, the kinetic energy of the eddy generates areas of enhanced velocity shear, increasing the likelihood of shear instabilities. Third, by means of interactions with the surface mixed layer, eddies can contribute to water mass modification and thereby change the density class of water parcels [*Radko and Marshall*, 2004]. As these mechanisms include lateral processes in addition to vertical ones, we will now look at fully three-dimensional studies of the impact of eddies on the upper-ocean nutrient supply.

4. EDDY-RESOLVING MODELS

4.1. Regional Models

A number of regional high-resolution models have shown that substantial vertical velocities can be associated with mesoscale variability. For an unstable meandering front. Onken [1992] found alternating vertical velocities of typically 10 m day^{-1}. The additional input of nutrients to the euphotic zone caused by the eddy-induced vertical motions was modelled by Flierl and Davis [1993] for Gulf Stream meanders, suggesting a mean enhancement of phytoplankton biomass by 10–20%. A similar, quasi-geostrophic model setup for the North Equatorial Current by Dadou et al. [1996] showed that biological production within the frontal zone was about twice that of the background values.

An early modeling study of an individual eddy by *Franks et al.* [1986] revealed some additional nutrient supply during the spin-down of an anticyclonic warm-core ring. Eddy–eddy interactions were identified in a model of two warm core rings by *Yoshimori and Kishi* [1994] as the main mechanism responsible for the enhanced biological production simulated at the margins of the eddies. *McGillicuddy et al.* [1995] used a similar approach to study three eddies observed during the Joint Global Ocean Flux Study North Atlantic Bloom Experiment. Employing a quasi-geostrophic model, their findings suggested that the lifting of density surfaces during eddy propagation and interaction with the ambient water and with other eddies was dominating eddy-induced nutrient fluxes compared to minor contributions caused by vortex stretching and interactions with wind-driven motions. A comprehensive study by *Spall and Richards* [2000] elegantly combined Eulerian and Lagrangian modeling to show that eddies shed along a front can locally and temporarily enhance biological production by an order of 100%, but then the net increase is "only" of the order of 10% when averaged over the entire frontal area at the end of their 45-day integration. With a horizontal resolution of 2 km, their isopycnic model was even resolving a substantial portion of the sub-mesoscale spectrum. A similarly high spatial resolution and a similar model setup of an idealized unstable front was used by *Lévy et al.* [2001] for a *z*-level primitive-equation model. Their somewhat shorter model runs (25 days) revealed an increase in biological production by about 100% when mesoscale and sub-mesoscale motions were included.

To obtain a more realistic physical environment, *Mahadevan and Archer* [2000] set up a regional primitive-equation model with sigma-type vertical coordinates. The model was forced at its open boundaries with output taken from a high-resolution global circulation model. By running their nested model for three late summer months over regions in the subtropical North Pacific (near Hawaii) and North Atlantic (near Bermuda), they found that simulated upper-ocean nutrient supply increased by an order of magnitude, for example in the Bermuda region from 0.03–0.05 mol N m^{-2} $year^{-1}$ at 0.4° resolution to 0.30–0.54 mol N m^{-2} $year^{-1}$ at 0.1° resolution. It is interesting to note that their results also showed that modelled surface temperatures were, on average, 0.36°C colder (0.47°C at Hawaii) in the 3-month, 0.1° resolution run compared to the 0.4° resolution run. Assuming this eddy-induced cooling to affect the top 100 m of the ocean, this corresponds to a surface heat–flux difference of 28 W m^{-2} (36 W m^{-2} at Hawaii) between the two runs. Obviously, the enhanced mixing that goes along with higher eddy activity does not only mix nutrients up to the surface but also mixes heat down the thermocline. To the extent that we can limit uncertainties in surface heat fluxes to about 20–30 W m^{-2}, we may be able to use this information to constrain the magnitude of eddy-induced nutrient supply. For a basin-scale eddy-permitting model, *Oschlies* [2002c] revealed a systematic overestimation of subtropical heat uptake by some 25 W m^{-2} in response to simulated surface temperatures being too low by about 0.5°C. This may suggest that there is already too much diapycnal mixing in the model leading to excessive downward transport of heat and likely excessive upward transport of nutrients.

A notorious difficulty of all of the aforementioned regional models is that these cannot usually be run stably for longer than a few months. This makes it difficult to model entire "life cycles" of individual eddies, and the results will always be affected by the initial conditions and boundary conditions set to initiate the eddy field. Even if one gets the eddy-induced mixing intensity right, the mean tracer gradients, which also will determine the eddy-induced tracer transport [*Lévy*, 2003], may not have enough time to adjust to the simulated eddy field in short spin-up or spin-down experiments. Substantially longer simulations require larger domains that reduce the impact of imposed lateral boundary conditions. The quasi-geostrophic model used by *McGillicuddy and Robinson* [1997] covered a 1,000 × 1,000-km area in the Sargasso Sea. Their estimate of a regional eddy-induced nitrate supply of 0.5 mol m^{-2} $year^{-1}$ suggests that eddy pumping is sufficient to explain geochemical estimates of new production. However, although moving the lateral boundaries far away and thereby allowing for longer integration with a fully developed eddy field, this study used a simple nitrate consumption and restoring model in the vertical dimension. In particular, nitrate was restored right below the euphotic zone to climatological values at a timescale of 3 months to parameterize recharging of deep isopycnals by remineralization. This relaxation of vertical nutrient gradients to climatology implies that the mean gradients cannot freely adjust to the eddy field of the model.

4.2. Basin-Scale Models

The first basin-scale model used to investigate the impact of eddies was presented by *Oschlies and Garçon* [1998] who used a 1/3° resolution *z*-level primitive-equation model into which satellite altimeter data were assimilated to ensure more realistic levels of eddy activity. With modeled eddy kinetic energy levels generally being lower than satellite-derived observational estimates by some 30%, and with a fully prognostic biogeochemistry everywhere within the model domain, this study suggested that eddies account for a nitrate supply of about 0.03 mol m^{-2} year^{-1} over the subtropical gyre, which is about the same size as the contribution by small-scale turbulent mixing across the pycnocline. Accordingly, and in contrast to most of the earlier regional models, the conclusion of that study was that the contribution by eddies is not sufficient to explain the apparent discrepancy between low direct measurements and high geochemical estimates of new production.

While the above study was only marginally eddy-resolving and, despite the assimilation of altimeter data did not fully reach observed levels of eddy activity, similar conclusions were reached by a subsequent study using a 1/9° resolution eddy-resolving model [*Oschlies*, 2002a]. In fact, when comparing results of the eddy-permitting, eddy-resolving, and a highly viscous eddy-permitting (to mimic a coarse-resolution model without changing the mean flow too much) model, the large-scale patterns look very similar. This is shown in Plate 1 for the simulated surface chlorophyll. All three model configurations use the same ecosystem model [*Oschlies and Garçon*, 1999], the same initial values, same boundary conditions, and same monthly mean atmospheric forcing fields. Apart from the closed boundaries, there are no restoring terms within the model domain, and for the upper ocean, the solution approaches a self-consistent statistically steady state within a few years. Naturally, the eddy-related conclusions derived from such a fully prognostic model are only as good as the model is able to reproduce at least the statistical moments of the observations. Close agreement with time-series data available at the BATS site was found for the simulated mean nitrate profiles as well as for the variance of vertical displacements of nitrate isosurfaces and their correlations with sea surface height. An investigation of the physical properties showed agreement also for spatial patterns of eddy kinetic energy, eddy length scales, spectral densities of sea-surface height variations, and for winter mixed-layer depths that became generally shallower when eddies were resolved [*Oschlies*, 2002b]. The latter feature was linked to the improved representation of instabilities within the mixed layer [*Nurser and Zhang*, 2000]. This process can actually lead to an eddy-induced reduction in annual nutrient supply in high latitudes where entrainment into the deepening winter mixed layer often dominates the annual budget.

Overall, all three model configurations showed relatively similar patterns of total nitrate input into the upper 126 m of the model, here taken as a proxy for the euphotic zone (Plate 2, Figure 2). Results do not change significantly when the 1% light level is taken as base of the euphotic zone as

done by *Oschlies* [2002a]. Averaged over the entire model domain, nitrate supply varied by less than 2% among the three model configurations (0.204 mol m^{-2} year^{-1} in the viscous experiment, 0.202 mol m^{-2} year^{-1} in the eddy-permitting experiment, and 0.205 mol m^{-2} year^{-1} in the eddy-resolving experiment). This is comparable to the 2–3% increase obtained by switching from monthly mean forcing to daily atmospheric forcing.

One may look at the individual contributions to the total advective nitrate supply from the mean flow and the eddying flow in the following way:

$$\int_{126zm}^{0} \nabla \cdot \overline{(\mathbf{u}NO_3)}\, dz = \overline{wNO_3}\big|_{126m} + \int_{126m}^{0} \nabla_h \cdot \overline{\mathbf{u}_h NO_3}\, dz \tag{4}$$

with

$$\overline{wNO_3}\big|_{126m} = \overline{w}\,\overline{NO_3}\big|_{126m} + \overline{w'NO_3'}\big|_{126m} \tag{5}$$

and

$$\nabla_h \cdot \overline{\mathbf{u}_h NO_3} = \nabla_h \cdot \overline{\mathbf{u}}_h \overline{NO_3} + \nabla_h \cdot \overline{\mathbf{u}_h' NO_3'} \tag{6}$$

where the overbar denotes the 3-year mean, and the prime denotes all deviations from the mean. Possible eddy effects on the mixing terms are not considered in this analysis. The first term on the right-hand side of (5) and (6), respectively, is referred to as the contribution of the mean flow and the second term as contribution of the fluctuating "eddy" flow. Note that the latter term does not only contain fluctuations associated with mesoscale eddies but also includes sub-mesoscale and seasonal variations.

The simulated nitrate supply is broken down into the individual contributions in Plate 3 that also includes the contribution by vertical mixing. The latter is dominant over large parts of the subpolar gyre where winter mixed-layer depths are deeper than the chosen 126-m reference depth for the euphotic zone. Nitrate supply by the mean advective transport (which, according to equations (1) to (3), has to vanish when integrated over volumes enclosed by nitrate isosurfaces) is positive only within relatively narrow bands near the equatorial upwelling and along the margins of the subtropical gyre where Ekman transport fluxes in significant amounts of nutrients across mean streamlines of the horizontal gyre circulation from nutrient-rich areas into the oligotrophic subtropical gyre [*Williams and Follows*, 1998a]. The eddy-induced nitrate supply is positive almost everywhere within the subtropical gyre, reaching highest values of 0.1 mol m^{-2}

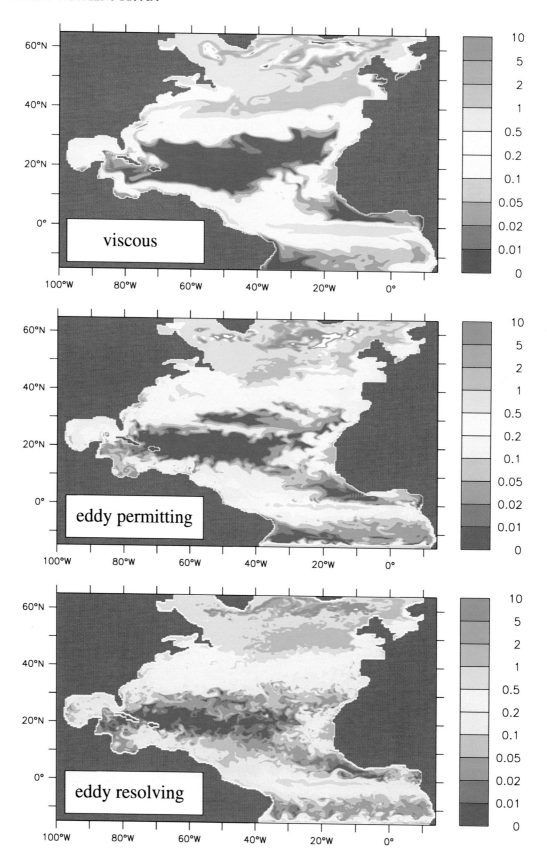

year[-1] right at the margins of the gyre, whereas typical values are smaller than 0.05 mol m[-2] year[-1]. It is interesting to note that the eddy-associated flux is negative over much of the subpolar North Atlantic, along the equator, and along a band extending from the upwelling region off West Africa. Decomposing the eddy contribution into vertical and horizontal components shows that the vertical eddy transport is largest in the Gulf Stream area where eddy formation and associated eddy pumping by lifting of isopycnals during eddy formation is most intense. Contributions of this mechanism reach values of 0.1 mol m[-2] year[-1] only within the Gulf Stream, and in the subtropical gyre, values are usually smaller than 0.05 mol m[-2] year[-1].

Somewhat larger nutrient fluxes are generated by the horizontal component of the eddying nitrate transport, with largest values of 0.2 mol m[-2] year[-1] along the southern margin of the gyre in a narrow band along the North Equatorial Current. Here, the eddy-associated nitrate supply is just north of the input related to the Ekman-related mean flow, suggesting that eddies laterally mix the input by the mean flow further into the oligotrophic gyre. The zonal average of the eddy-induced nutrient fluxes into the upper 126 m of the model is displayed in Figure 3. The zonally averaged nitrate input by vertical eddy motions is positive only in the northern half of the subtropical gyre (north of 25° N), while the zonally averaged input by lateral eddy motions has significant positive contributions also in the southern half. When averaged from 12° N to 40° N, the total eddy-induced nitrate supply amounts to 0.01 mol m[-2] year[-1] in this model. This figure results from an average negative contribution from vertical eddy motion (-0.02 mol m[-2] year[-1]) and a positive contribution from lateral eddy motions (0.03 mol m[-2] year[-1]).

A second basin-scale eddy-resolving modeling study was presented by *McGillicuddy et al.* [2003]. While they used a slightly higher spatial resolution of 0.1° than the 1/9° model of *Oschlies* [2002a], the general architecture of the z-level primitive-equation physical model was the same. The main difference between these two studies lies in the use of artificial restoring: *Oschlies* [2002a] used a nutrient/phytoplankton/zooplankton/detritus-type ecosystem model coupled to the circulation field in fully prognostic mode at all depths without any restoring to climatological nutrient fields, *McGillicuddy et al.* [2003] used a nutrient transport model with nitrate being consumed at rates depending on temperature and light at depths shallower than $z = 104$ m and being restored at depths below $z = 104$ m to density-dependent nitrate fields derived from projecting World Ocean Atlas [*Conkright et al.*, 1998] nitrate data onto the mean density

field of the model. The crucial parameter of this model is the restoring timescale of the "deep" nitrate field. *McGillicuddy et al.* [2003] performed sensitivity experiments using restoring times of 10, 30, and 60 days. However, results were shown only for the 10-day restoring timescale. As pointed out by *Martin and Pondaven* [2003], 10 days is unrealistically short compared to observations at Bermuda [*McNeil et al.*, 1999] and may substantially overestimate eddy-induced nutrient supply by as much as an order of magnitude.

While many features in the two modeling studies turned out to be very similar in both qualitative and quantitative ways, substantial differences were found in the pattern and size of the nutrient input by the eddy-associated vertical advection. The model of *McGillicuddy et al.* [2003] predicted positive values essentially everywhere over the subtropical and tropical Atlantic, in some regions exceeding 0.3 mol N m[-2] year[-1]. The model of *Oschlies* [2002a], on the other hand, restricts positive values of magnitudes typically smaller than 0.05 mol m[-2] year[-1] mostly to the northwestern part of the gyre where eddy activity is largest. It is interesting to note that both models show roughly similar results for the effect of eddies on the lateral nutrient supply while they differ by a factor of 5–10 in the eddy-associated vertical nutrient supply. It is most likely that this difference is caused by the different treatment of nutrient recharging below the euphotic zone.

5. CONCLUDING DISCUSSION

The ocean is full of eddies, and individual eddy events have naturally been observed in biogeochemical observational campaigns. It is difficult to argue that eddies should be under-represented rather than over-represented in the growing observational data base, particularly as this under-representation would seem to be similar for all ocean basins. Given the growing interest in eddies and several dedicated studies with cruise tracks carefully arranged to track eddies detected by remote sensing, one might even ask whether our data bases may over-represent the more "interesting" eddy features. In any case, with an ever-increasing number of observations, the average of the available observations converges to the true average, and there remains less and less chance for a systematic statistical bias needed to explain the apparent observational discrepancies by means of unobserved episodic events. However, given the available observations of eddy-induced nutrient supply [*Falkowski et al.*, 1991; *McNeil et al.*, 1999] and a large number of modeling studies, there is unequivocal agreement that eddies

Plate 1. (Opposite) Simulated surface chlorophyll for 1 May using a chlorophyll-to-nitrogen ratio of 1.59 mg Chl (mmol N)[-1]. *Top*: 1/3 degree model with large lateral viscosity (10[3] m[2] s[-1]). *Middle*: 1/3 degree model. *Bottom*: 1/9 degree model.

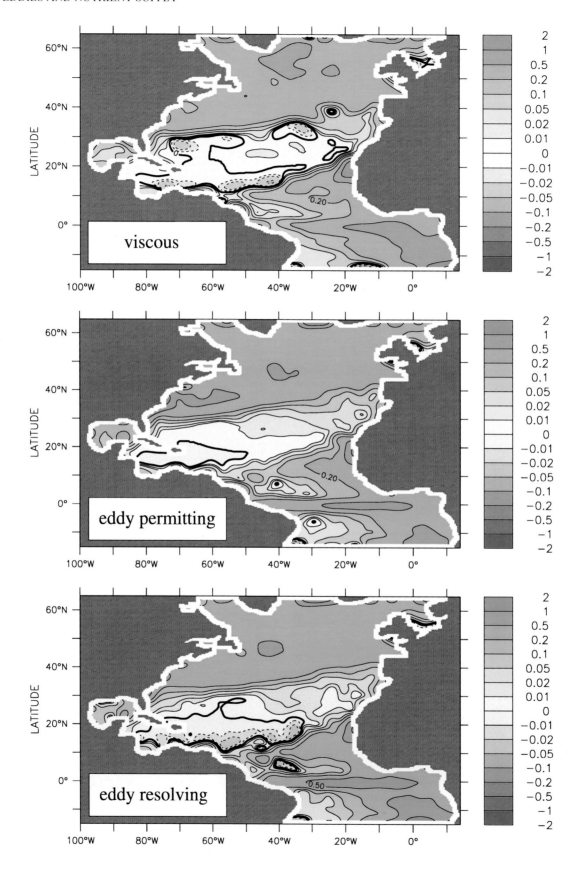

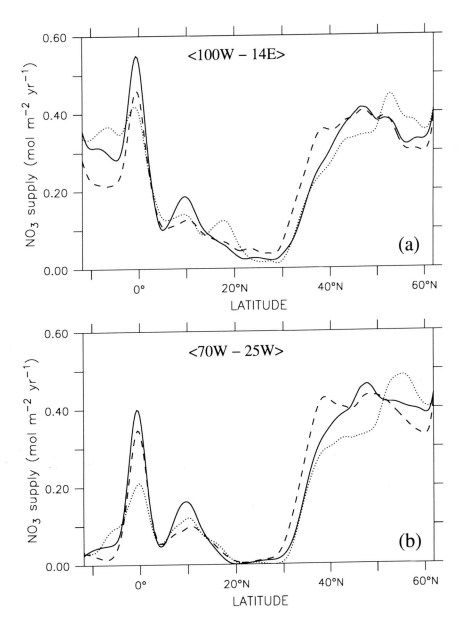

Figure 2. Zonally averaged simulated annual-mean nitrate supply to the upper 126 m of the three models. (**a**) Basin-wide zonal averages. (**b**) Zonal averages between 70°W and 25°W to exclude subtropical coastal upwelling regions. Results are shown for the eddy-resolving 1/9° model (*solid lines*), the eddy-permitting 1/3° model (*dashed lines*), and the viscous 1/3° model (*dotted lines*). Units are mol N m^{-2} year^{-1}.

do play some role in the nutrient supply to the oligotrophic gyres. Extrapolations of individual events to larger areas and longer time spans remain difficult and vary from a mere 20% enhancement of biological production by mesoscale eddies [*Falkowski et al.*, 1991] to a 100–200% enhancement [*McGillicuddy et al.*, 1998].

Modeling studies that can be expected to be particularly useful for inter- and extrapolation of individual observations have also proposed a large variety of values for the average extra nutrient supply generated by eddies, and results seem to fall into two categories: high estimates obtained by models that recharge sub-euphotic zone nutrients by restoring to climatology

Plate 2. (Opposite) Simulated annual-mean nitrate supply to the upper 126 m of the model. *Top*: 1/3 degree model with large lateral viscosity (10^3 m^2 s^{-1}). *Middle*: 1/3 degree model. *Bottom*: 1/9 degree model. Units are mol N m^{-2} year^{-1}.

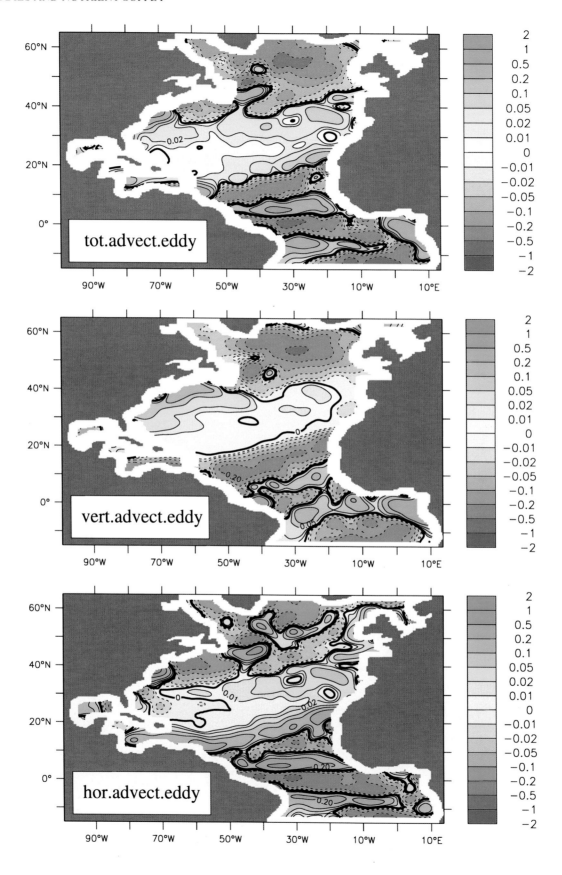

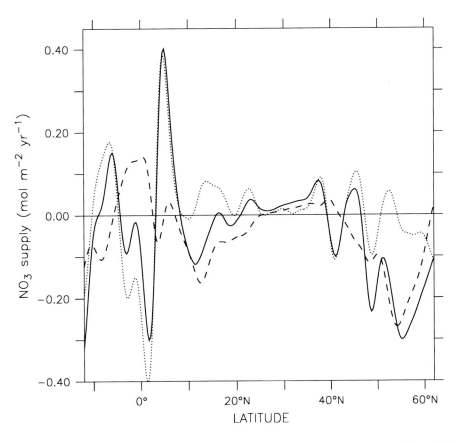

Figure 3. Zonally averaged simulated eddy-induced nitrate supply to the upper 126 m of the eddy-resolving 1/9° model. The *solid line* refers to the total advective supply by eddies, the *dashed line* to the vertical eddy-induced supply, and the *dotted line* to the lateral eddy-induced supply. Units are mol N m⁻² yr⁻¹.

[*McGillicuddy and Robinson*, 1997; *McGillicuddy et al.*, 1998, 2003; *Siegel et al.*, 1999] and models that explicitly compute remineralization and mixing of nutrients [*Oschlies and Garçon*, 1998; *Spall and Richards*, 2000; *Oschlies*, 2002a; *Martin and Pondaven*, 2003]. The later two studies have shown that the "recharging by restoring" method leads, for the restoring timescales employed, to a significant overestimate of the estimated nutrient supply by almost an order of magnitude. Remarkably, the different modeling studies seem to agree upon the role of lateral stirring of nutrients by eddies across the margins of the oligotrophic subtropical gyres [*Lee and Williams*, 2000; *Oschlies*, 2002a; *McGillicuddy et al.*, 2003].

While it remains to be seen whether longer restoring times of the restoring-type models bring results of the two model categories into closer agreement, we do not yet know whether any of the fully prognostic models predict the eddy-induced nutrient flux correctly. Realising that sinking is a diapycnal transport, recharging will have to involve diapycnal transports as well. Such fluxes will not only affect nutrients but also any other tracer such as temperature. To the extent that models tend to underestimate subtropical surface temperatures, diapycnal fluxes are likely too large rather than too small in these models, suggesting that the model-derived eddy-induced nutrient fluxes may be overestimates rather than underestimates. Clearly, more dedicated observational programs will help to put better constraints on our model results. This will have to cover both individual eddies including sub-mesoscale diapycnal transports at the eddies' margins as well as time-series and synoptic remote sensing studies to improve our knowledge about the statistics of episodic nutrient supplies. It may well be that this will eventually confirm the early conservative estimate of *Falkowski et*

Plate 3. (Opposite) Simulated annual-mean nitrate supply to the upper 126 m of the 1/9 degree eddy-resolving model. *Top*: Nitrate supply induced by the eddy transport, $\int_{126m}^{0} \nabla \cdot \overline{(u'NO'_3)}dz$. *Middle*: Nitrate supply associated with the vertical eddy component, $\overline{w'NO'_3}\big|_{126m}$. *Bottom*: Nitrate supply associated with the lateral eddy component, $\int_{126m}^{0} \nabla_h \cdot \overline{(u_h'NO'_3)}dz$. Units are mol N m⁻² year⁻¹.

al. [1991] that puts the eddy-induced average enhancement of biological production into the range of 20%.

A possible impact of eddies on biogeochemical cycles not discussed in this study is their role in structuring pelagic ecosystems. By enclosing water masses for weeks to months, eddies can isolate communities and move them around. The disturbances associated with the eddies may select for certain size classes [*Rodriguez et al.*, 2001] and/or phytoplankton functional types [*Vaillancourt et al.*, 2003; *Benitez-Nelson, et al.*, 2007]. This may well affect fluxes of biogeochemical tracers, but the magnitude and even the direction of any net impact is not yet known.

As it stands, the subtropical desert conundrum, which originally pointed to the possible role of eddies in biogeochemical cycles [*Jenkins*, 1982], has not yet been resolved. Eddies can explain some portion of the apparent observational discrepancy between low nitrate supply and large oxygen consumption at depth (perhaps some 10–20%, about 0.05 mol N m^{-2} year^1 with lateral eddy stirring being at least as important as vertical eddy pumping). There is mounting evidence that in addition to the contribution by eddies, other previously unaccounted transport mechanisms like mixing by double diffusion [*Oschlies et al.*, 2003] and sub-mesoscale upwelling events [*Lévy et al.*, 2001] may play a significant role. However, given that diapycnal mixing will also mix temperature, the constraints imposed by observed surface heat fluxes make it difficult to imagine that we have missed order-of-magnitude increases in near-surface diapycnal mixing that would be required to resolve the conundrum by mixing arguments. Possible alternative explanations include incorrect assumptions about elemental stoichiometry when comparing nitrogen fluxes and oxygen consumption. For example, subduction of dissolved organic carbon can lead to oxygen consumption without requiring any nitrogen supply into the overlying surface waters. Additional nitrogen sources like nitrogen fixation [*Hansell et al.*, 2004] or atmospheric dust supply [*Baker et al.*, 2003] will also help to close the apparent observational discrepancy between low supply of dissolved inorganic nitrogen to the subtropical surface layer and high rates of oxygen consumption at depth.

Acknowledgments. I thank the reviewers for their helpful and constructive comments that helped to improve the text and Matthew Hecht and Hiroyasu Hasumi for giving me the opportunity to participate in this book and for their help and effort in the production process.

REFERENCES

Abraham, E. R. (1998), The generation of plankton patchiness by turbulent stirring, *Nature, 391*, 577–580.

Bainbridge, R. (1957), The size, shape and density of marine phytoplankton concentrations, *Biol. Rev., 32*, 91–115.

Baker, A. R., S. D. Kelly, K. F. Biswas, M. Witt, and T. D. Jickells (2003), Atmospheric deposition of nutrients to the Atlantic Ocean, *Geophys. Res. Lett., 30*(24), 2296, doi:10.1029/GL018518.

Benitez-Nelson, C. R., et al. (2007), Mesoscale eddies drive increased silica export in the subtropical Pacific Ocean, *Science, 316*, 1017–1021.

Conkright, M. E., et al. (1998), World Ocean database 1998, CD-ROM Data Set Documentation. O.C.L., National Oceanographic Data Center Internal Report 14, 111 pp.

Conkright, M. E., H. E. Garcia, T. D. O'Brien, R. A. Locarnini, T. P. Boyer, C. Stephens, and J. I. Antonov (2002), World Ocean Atlas 2001, Volume 4: Nutrients, in *NOAA Atlas NESDIS 54*, edited by S. Levitus, 392 pp , US Government Printing Office, Washington, DC.

Dadou, I., V. Garçon, V. Andersen, G. R. Flierl, and C. S. Davis (1996), Impact of the North Equatorial Current meandering on a pelagic ecosystem: A modeling approach, *J. Mar. Res., 54*, 311–342.

Dandonneau, Y., A. Vega, H. Loisel, Y. du Penhoat, C. Menkes (2003), Oceanic Rossby waves acting as a "Hay Rake" for ecosystem floating by-products, *Science, 302*, 1548–1551.

del Giorgio, P. A., J. J. Cole, and A. Cimbleris (1997), Respiration rates in bacteria production exceed phytoplankton production in unproductive aquatic systems, *Nature, 385*, 148–151.

Doney, S. C., D. M. Glover, S. J. McCue, and M. Fuentes (2003), Mesoscale variability of Sea-viewing Wide Field-of-view Sensor (SeaWiFS) satellite ocean color: Global patterns and spatial scales, *J. Geophys. Res., 108*(C2), 3024, doi:10.1029/2001JC000843.

Dugdale, R. C., and J. J. Goering (1967), Uptake of new and regenerated forms of nitrogen in marine production, *Limnol. Oceanogr., 12*, 196–206.

Eppley, R. W., and B. J. Peterson (1979), Particulate organic matter flux and planktonic new production in the deep ocean, *Nature, 282*, 677–680.

Falkowski, P. G., D. Ziemann, Z. Kolber, and P. K. Bienfang (1991), Role of eddy pumping in enhancing primary production in the ocean, *Nature, 352*, 55–58.

Fasham, M. J. R., T. Platt, B. Irwin, and K. Jones (1985), Factors affecting the spatial pattern of the deep chlorophyll maximum in the region of the Azores front, *Prog. Oceanogr., 14*, 129–165.

Flierl, G. R., and C. S. Davis (1993), Biological effects of Gulf Stream meandering, *J. Mar. Res., 51*, 529–560.

Franks, P. J. S., J. S. Wroblewski, and G. R. Flierl (1986), Prediction of phytoplankton growth in response to the frictional decay of a warm core ring, *J. Geophys. Res., 91*, 7603–7610.

Garçon, V., A. Oschlies, S. Doney, D. McGillicuddy, and J. Waniek (2001), The role of mesoscale variability on plankton dynamics in the North Atlantic, *Deep-Sea Res. II, 48*, 2199–2226.

Greatbatch, R. J., X. Zhai, C. Eden, and D. Olbers (2007), The possible role in the ocean heat budget of eddy-induced mixing due to air-sea interaction, *Geophys. Res. Lett., 34*, L07604, doi:10.1029/2007GL029533.

Hansell, D. A., N. R. Bates, and D. B. Olson (2004), Excess nitrate and nitrogen fixation in the North Atlantic Ocean, *Mar. Chem., 84*, 243–265.

Jenkins, W. J. (1982), Oxygen utilization rates in North Atlantic subtropical gyre and primary production in oligotrophic systems, *Nature, 300*, 246–248.

Jenkins, W. J. (1988), Nitrate flux into the euphotic zone near Bermuda, *Nature, 331*, 521–523.

Jenkins, W. J., and J. Goldman (1985), Seasonal oxygen cycling and primary production in the Sargasso Sea, *J. Mar. Res., 43*, 465–491.

Karl, D. M., E. A. Laws, P. Morris, P. J. leB. Williams, and S. Emerson (2003), Metabolic balance of the open sea, *Nature, 426*, 32.

Kawamiya, M., and A. Oschlies (2001), Formation of a basin-scale surface chlorophyll pattern by Rossby waves, *Geophys. Res. Lett., 28*, 4139–4142.

Lee, M.-M., and R. G. Williams (2000), Examining how eddies affect biological production through the lateral transport and diffusion of nutrients, *J. Mar. Res., 58*, 895–917.

Lévy, M. (2003), Mesoscale variability of phytoplankton and of new production: Impact of the large-scale nutrient distribution, *J. Geophys. Res., 108*(C11), 3358, doi:10.1029/2002JC001577.

Lévy, M., P. Klein, and A.-M. Treguier (2001), Impacts of submesoscale physics on phytoplankton production and subduction, *J. Mar. Res., 59*, 535–565.

Lewis, M. R., W. G. Harrison, N. S. Oakey, D. Herbert, and T. Platt (1986), Vertical nitrate fluxes in the oligotrophic ocea, *Science, 234*, 870–873.

Lohrenz, S. E., G. A. Knauer, V. L. Asper, M. Tuel, A. F. Michaels, and A. H. Knap (1992), Seasonal variability in primary production and particle flux in the northwestern Sargasso Sea: US JGOFS Bermuda Atlantic Time-Series Study, *Deep-Sea Res., 39*, 1373–1391.

Mahadevan, A., and D. Archer (2000), Modeling the impact of fronts and mesoscale circulation on the nutrient supply and biogeochemistry of the upper ocean, *J. Geophys. Res., 105*, 1209–1225.

Mahadevan, A., and J. W. Campbell (2002), Biogeochemical patchiness at the sea surface, *Geophys. Res. Lett., 29*(19), 1926, doi:10.1029/2001GL014116.

Mahadevan, A., M. Lévy, and L. Mémery (2004), Mesoscale variability of sea surface pCO_2: What does it respond to?, *Global Biogeochem. Cycles, 18*, GB1017, doi:10.1029/2003GB002102.

Maier-Reimer, E., U. Mikolajewicz, and A. Winguth (1996), Future ocean uptake of CO_2: Interaction between ocean circulation and biology, *Clim. Dyn., 12*, 711–721.

Martin, A. P. (2003), Phytoplankton patchiness: the role of lateral stirring and mixing, *Prog. Oceanogr., 57*, 125–174.

Martin, A. P., and P. Pondaven (2003), On estimates for the vertical nitrate flux due to eddy pumping, *J. Geophys. Res., 108*(9), 3359, doi:10.1029/2003JC001841.

Martin, A. P., and K. J. Richards (2001), Mechanisms for vertical nutrient transport within a North Atlantic mesoscale eddy, *Deep-Sea Res. II, 48*, 757–773.

McGillicuddy, D. J., Jr., and A. R. Robinson (1997), Eddy-induced nutrient supply and new production in the Sargasso Sea, *Deep-Sea Res. I, 44*, 1427–1450.

McGillicuddy, D. J., Jr., A. R. Robinson, and J. J. McCarthy (1995), Coupled physical and biological modeling of the spring bloom in the North Atlantic (II): Three dimensional bloom and post-bloom processes, *Deep-Sea Res. I, 42*, 1359–1398.

McGillicuddy, D. J., Jr., A. R. Robinson, D. A. Siegel, H. W. Jannasch, R. Johnson, T. D. Dickey, J. McNeil, A. F. Michaels, and A. H. Knap (1998), Influence of mesoscale eddies on new production in the Sargasso Sea, *Nature, 394*, 263–266.

McGillicuddy, D. J., Jr., R. Johnson, D. A. Siegel, A. F. Michaels, N. R. Bates, and A. H. Knap (1999), Mesoscale variations of biogeochemical properties in the Sargasso Sea, *J. Geophys. Res., 104*, 13,381–13,394.

McGillicuddy, D. J., Jr., L. A. Anderson, S. C. Doney, and M. E. Maltrud (2003), Eddy-driven sources and sinks of nutrients in the upper ocean: Results from a 0.1° resolution model of the North Atlantic, *Global Biogeochem. Cycles, 17*(2), 1035, doi:10.1029/2002GB001987.

McGillicuddy, D. J., L. A. Anderson, N. R. Bates, T. Bibby, K. O. Buesseler, C. A. Carlson, C. S. Davis, C. Ewart, P. G. Folkowski, S. A. Goldthwait, D. A. Hansell, W. J. Jenkins, R. Johnson, V. K. Kosnyrev, J. R. Ledwell, Q. P. Li, D. A Siegel, and D. K. Steinberg. (2007), Eddy/wind interactions stimulate extraordinary mid ocean plankton blooms, *Science, 316,* 1021–1026.

McNeil, J. D., H. W. Jannasch, T. Dickey, D. McGillicuddy, M. Brzezinski, and C. M. Sakamoto (1999), New chemical, biooptical and physical observations of upper ocean response to the passage of a mesoscale eddy off Bermuda, *J. Geophys. Res., 104*, 15,537–15,548.

Morel, A., B. Gentili, H. Claustre, M. Babin, A. Bricaud, J. Ras, and F. Tieche (2007), Optical properties of the "clearest" natural waters, *Limnol. Oceanogr., 52*, 217–229.

Nurser, A. J. G., and J. W. Zhang (2000), Eddy-induced mixed layer shallowing and mixed layer/thermocline exchange, *J. Geophys. Res., 105*, 21,851–21,868.

Onken, R. (1992), Mesoscale upwelling and density finestructure in the seasonal thermocline—Adynamical model, *J. Phys. Oceanogr., 22*, 1257–1273.

Oschlies, A. (2002a), Can eddies make ocean deserts bloom? *Global Biogeochem. Cycles, 16*(4), 1106, doi:10.1029/2001GB001830.

Oschlies, A. (2002b), Improved representation of upper ocean dynamics and mixed layer depths in a model of the North Atlantic on switching from eddy-permitting to eddy-resolving grid resolution, *J. Phys. Oceanogr., 32*, 2277–2298.

Oschlies, A. (2002c), Nutrient supply to the surface waters of the North Atlantic: A model study, *J. Geophys. Res., 107*(C5), 3046 doi:10.1029/2000JC000275.

Oschlies, A., and V. Garçon (1998), Eddy-induced enhancement of primary production in a model of the North Atlantic Ocean, *Nature, 394*, 266–269.

Oschlies, A., and V. Garçon (1999), An eddy-permitting coupled physical-biological model of the North Atlantic 1. Sensitivity to advection numerics and mixed layer physics, *Global Biogeochem. Cycles, 13*, 135–160.

Oschlies, A., and P. Kähler (2004), Biotic contribution to air-sea fluxes of CO_2 and O_2 and its relation to new production, export production, and net community production, *Global Biogeochem. Cycles, 18*, GB1015, doi:10.1029/2003GB002094.

Oschlies, A., H. Dietze, and P. Kähler (2003), Salt-finger driven enhancement of upper-ocean nutrient supply, *Geophys. Res. Lett., 30*(23), 2204, doi:10.1029/2003GL018552.

Plattner, G.-K., N. Gruber, H. Frenzel, and J. C. McWilliams (2005), Decoupling marine export production from new production, *Geophys. Res. Lett.*, *32*, L11612, doi:10.1029/2005GL022660.

Radko, T., and J. Marshall (2004), Eddy-induced diapycnal fluxes and their role in the maintenance of the thermocline, *J. Phys. Oceanogr.*, *34*, 372–383,

Rodriguez, J., J. Tintore, J. T. Allen, J. M. Blanco, D. Gomis, A. Reul, J. Ruiz, V. Rodriguez, F. Echevarria, and F. Jimenez-Gomez (2001), Mesoscale vertical motion and the size structure of phytoplankton in the ocean, *Nature*, *410*, 360–363.

Sarmiento, J. L., G. Thiele, R. M. Key, and W. S. Moore (1990), Oxygen and nitrate new production and remineralization in the North Atlantic subtropical gyre, *J. Geophys. Res.*, *95*, 18,303–18,315.

Siegel, D. A., D. J. McGillicuddy, Jr., and E. A. Fields (1999), Mesoscale eddies, satellite altimetry, and new production in the Sargasso Sea, *J. Geophys. Res.*, *104*, 13,359–13,379.

Spall, S. A., and K. J. Richards (2000), A numerical model of mesoscale frontal instabilities and plankton dynamics. 1. Model formulation and initial experiments, *Deep-Sea Res. I*, *47*, 1261–1301.

Spitzer, W. S., and W. J. Jenkins (1989), Rates of vertical mixing, gas exchange and mew production: estimates from seasonal gas cycles in the upper ocean near Bermuda, *J. Mar. Res.*, *47*, 169–196.

Strass, V. H. (1992), Chlorophyll patchiness caused by mesoscale upwelling at fronts, *Deep-Sea Res.*, *39*, 75–96.

Strutton, P. G., J. P. Ryan, and F. P. Chavez (2001), Enhanced chlorophyll associated with tropical instability waves in the equatorial Pacific, *Geophys. Res. Lett.*, *28*, 2005–2008.

Uz, B. M., J. A. Yoder, and V. Osychny (2001), Pumping of nutrients to ocean surface waters by the advection of propagating planetary waves, *Nature*, *409*, 597–600.

Vaillancourt, R. D., J. Marra, M. P. Seki, M. L. Parsons, and R. R. Bidigare (2003), Impact of a cyclonic eddy on phytoplankton community structure and photosynthetic competency in the subtropical North Pacific Ocean, *Deep-Sea Res. I*, *50*, 829–847.

Volk, T., and M. I. Hoffert (1985), Ocean carbon pumps: Analysis of relative strengths and efficiencies in ocean-driven atmospheric CO_2 changes, in *The Carbon Cycle and Atmospheric CO_2: Natural Variations Archean to Present*, Geophys. Monogr. Ser., vol. 32, edited by E.T. Sundquist and W.S. Broecker, pp. 99–110, AGU, Washington, D.C.

Washburn, L., B. M. Emery, B. H. Jones, and D. G. Ondercin (1998), Eddy stirring and phytoplankton patchiness in the subarctic North Atlantic in late summer, *Deep-Sea Res. I*, *45*, 1411–1439.

Williams, P. L. le B., P. J. Morris, and D. M. Karl (2004), Net community production and metabolic balance at the oligotrophic ocean site, station ALOHA, *Deep-Sea Res. I*, *51*, 1563–1578.

Williams, R. G., and M. J. Follows (1998a), The Ekman transfer of nutrients and maintenance of new production over the North Atlantic, *Deep-Sea Res. I*, *45*, 461–489.

Williams, R. G., and M. J. Follows (1998b), Eddies make ocean deserts bloom., *Nature*, *394*, 228–229.

Williams, R. G., and M. J. Follows (2003), Physical transport of nutrients and the maintenance of biological production, in *Ocean Biogeochemistry*, pp. 19–51, edited by M. J. R. Fasham, Global Change IGBP Series, Springer,.

Woods, J. D. (1988), Mesoscale upwelling and primary production, in *Toward a Theory on Biological–Physical Interactions in The World Ocean*, edited by B. J. Rothschild, D. Reidel, 650 pp.

Yoder, J. A., S. G. Ackleson, R. T. Barber, P. Flament, and W. M. Balch (1994), A line in the sea, *Nature*, *371*, 689–692.

Yoshimori, A., and M. J. Kishi (1994), Effects of interaction between two warm-core rings on phytoplankton distribution, *Deep-Sea Res. I*, *41*, 1039–1052.

A. Oschlies, IFM-GEOMAR, Leibniz-Institut für Meereswissenschaften an der Universität Kiel, Düsternbrooker Weg 20, D-24105 Kiel, Germany. (aoschlies@ifmgeomar.de)

Eddies in Eastern Boundary Subtropical Upwelling Systems

X. Capet[1], F. Colas, and J.C. McWilliams

Institute of Geophysics and Planetary Physics, University of California at Los Angeles, Los Angeles, California, USA

P. Penven

Institut de Recherche pour le Developpement, UR097 ECO-UP, Centre IRD de Bretagne, Plouzané, France

P. Marchesiello

Institut de Recherche pour le Developpement, Noumea, New Caledonia

Over the last decade, mesoscale-resolving ocean models of eastern boundary upwelling systems (EBS) have helped improve our understanding of the functioning of EBS and, in particular, assess the role of eddy activity in these systems. We review the main achievements in this regard and highlight remaining issues and challenges. In EBS, eddy activity arises from baroclinic/barotropic instability of the inshore and also offshore currents. Mesoscale eddies play a significant (although not leading) role in shaping the EBS dynamical structure, both directly and through associated submesoscale activity (i.e., primarily frontal). They do so by modifying both momentum and tracer balances in ways that cannot simply be understood in terms of diffusion. The relative degree to which these assertions about eddy activity and eddy role apply to each of the four major EBS (Canary, Benguela, Peru–Chile, and California Current Systems) remains to be established. Besides resolving the eddies, benefits from EBS high-resolution modeling include the possibility of accounting for the fine-scale structures of the nearshore wind, a better representation of the Ekman-driven coastal divergence, and (at resolution $\mathcal{O}$ (1 km) or lower) inclusion of submesoscale (i.e., mainly frontal) processes. Recent numerical experiments suggest that accounting for these various processes in climate models, through resolution increase (possibly locally) or parameterization, would lead to significant basin-scale bias reduction. The mechanisms involved in upscaling from EBS toward the larger scale remain to be fully elucidated.

1. INTRODUCTION

Subtropical oceanic regions off the western coasts of America and Africa are generally referred to as eastern boundary systems (EBS) and share some important characteristics. The subtropical basin-scale circulation and thermohaline structure are characterized by a broad, slow gyre flow directed equatorward and a shallow thermocline. The atmos-

[1]Now at Instituto Oceanografico, University of São Paulo, Brazil.

Ocean Modeling in an Eddying Regime
Geophysical Monograph Series 177
10.1029/177GM10

pheric pressure patterns (highs offshore and lows inland) are conducive to alongshore and equatorward wind regimes and hence to coastal upwelling. Although these two dynamical components (the large-scale circulation and the coastal upwelling) can be considered in isolation, it really is their combination and interplay that lead to the observed oceanographic conditions in the Canary, Benguela, Peru-Chile, and California current systems (respectively denoted by CS, BS, PCS, and CCS). For example, thermocline shallowness is not a mere consequence of the coastal upwelling; on scales of 1,000 km or more, it is also understood in terms of Sverdrup theory, i.e., thermocline deepening toward the west (the direction of Rossby wave propagation). Exemplary features that further illustrate the interplay between the coastal and large-scale dynamics are the upwelling filaments that carry newly upwelled water from the coast to hundreds of kilometers offshore.

Although to different degrees and with significant time variability, EBS have high primary productivity that indirectly sustains intensive fishing activity (more than 20% of the global fish catch takes place within these narrow bands). This is perhaps the main reason why EBS received considerable attention early in the development of modern oceanography and are still a favored place for oceanographic research (e.g., AOSN, *Deep Sea Res.*, special issue; AESOP, http://www.mbari.org/MB2006/AESOP/mb2006-aesop-links.htm; and VOCALS, http://www.eol.ucar.edu/projects/vocals/). In particular, numerical models have been widely applied to these regions, initially for the purpose of understanding processes and increasingly for forecasting on timescales ranging from days to decades. Modeling progress has generally been slow, and we can suggest two explanations. First, eddies and other mesoscale (i.e., with horizontal scales larger than, but comparable to, the first baroclinic deformation radius R_d), or even submesoscale features (e.g., jets, filaments, or fronts that are characterized by at least one horizontal length scale less than R_d), are essential flow components for the dynamical and biogeochemical functioning of the system. In this regard, EBS are similar to many other oceanic regions where mesoscale turbulence is intense. Note, though, that some characteristics of EBS favor energetic submesoscale activity, espe-

cially close to shore [*Durski and Allen*, 2005; *Pedlosky*, 1978; *Barth*, 1994, section 5.1]. Therefore, even eddy-permitting primitive-equation models (i.e., with a couple of grid points to resolve R_d) have difficulty capturing EBS dynamics adequately within a few hundred kilometers of the shore.

Second, the validity of quasi-geostrophic theory breaks down almost by definition in EBS regions because the driving process, i.e., the coastal upwelling, leads to isopycnal outcropping and rather large Froude numbers. Therefore, quasi-geostrophic models have had limited success in reproducing EBS dynamics [*Pares-Sierra et al.*, 1993; *Ikeda and Emery*, 1984].

The aim of this chapter is threefold. In sections 2 and 3, we present some of the main results accumulated over the last decade by mesoscale-resolving EBS numerical studies. These studies have been made almost exclusively using regional models[1] because eddy-resolving global (or even basin-scale) numerical solutions have only become recently affordable. Global models will remain expensive for the years to come, and extensive sensitivity studies, such as those performed using regional models, are still beyond reach. Furthermore, the large-scale modeling community has generally neglected EBS regions because they are considered not of primary importance for the global circulation, as opposed to passage flows and western boundary currents, for example. As modeling errors in the latter regions are being reduced, those related to EBS are beginning to emerge. In section 4, we report on evidence that EBS regions, indeed, exert upscaling effects on the dynamics of their respective oceanic basins. Some important difficulties that hinder further progress in EBS mesoscale modeling are outlined in section 5, and concluding remarks are offered in section 6.

2. GENERATION OF MESOSCALE ACTIVITY IN EBS

Taken as a rough measure of mesoscale activity, the surface eddy kinetic energy (EKE)[2] in EBS is relatively low (lower than 250 cm² s⁻², Plate 1) compared to the most active regions of the world ocean [*Ducet et al.*, 2000]. However, the EBS EKE stands out in the otherwise quiescent subtropical eastern Pacific on global maps of EKE [Plate 8 in *Ducet*

Plate 1. (Opposite) Measured EKE (cm² s⁻²) and sea-surface height (SSH; with 6-cm contour interval; open lines) in the four EBS. SSH contours are also the streamlines for the surface geostrophic current whose direction is indicated by small arrows. EKE is computed from the improved DUACS SSH product for the period 2001–2006 [*Pascual et al.*, 2006]. SSH is computed from Rio05 [*Rio and Hernandez*, 2004] that combines altimetry and gravimetry satellite measurements together with temperature and salinity in situ data (World Ocean Atlas, *Boyer and Levitus* [1998]). Note the good correspondence between high EKE and tight SSH contours at the Azores front, off Central Chile, and off North America.

Plate 2. (Opposite) Top: Same as Plate 1 for a reduced domain. Bottom: EKE and SSH contours from ROMS solutions at 5 km. The EKE is computed using low-pass filtered (6-day averaging and Gaussian spatial filter with 30-km half-width) geostrophic velocities. *Color bars* and SSH intervals are the same for models and observations.

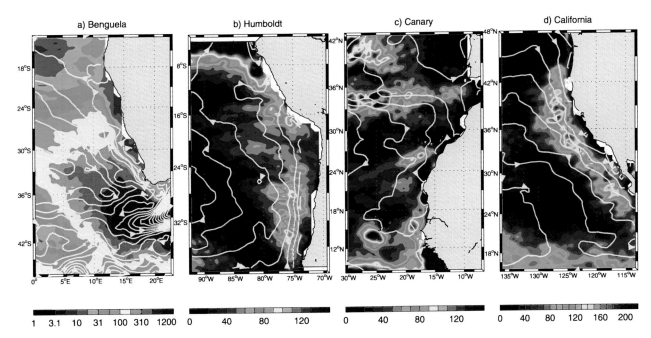

a) Benguela   b) Humboldt   c) Canary   d) California

Plate 1.

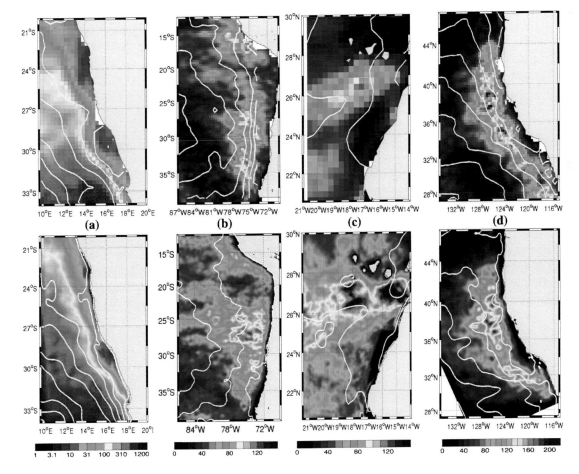

(a)   (b)   (c)   (d)

Plate 2.

et al., 2000]. This is less true in the subtropical eastern Atlantic because mesoscale activity in the CS is weaker than in the other systems for reasons discussed by *Marchesiello and Estrade* [2008] and because the Agulhas Rings overshadow all other sources of mesoscale activity in the southeastern Atlantic.

2.1. Coastal Versus Offshore Generation

We start by reviewing the surface EKE sources in EBS. Early on, idealized studies [*Ikeda et al.*, 1984; *Ikeda and Emery*, 1984] suggested coastal current instabilities for EKE generation, as opposed to wind variability. One of the major achievements of eddy-resolving models in the 1990s is the widely accepted demonstration that mesoscale variability in EBS arises in large part from the instability of coastal currents [*Batteen*, 1997; *Leth and Middleton*, 2004]. This is seemingly at odds with EKE maps derived from altimetric sea-level anomalies that tend to show EKE local minima in a band 100 to 300 km nearshore (Plate 1). EKE coastal minima are found off central Chile, Morocco and Mauritania, and central California. Along the coast of South Africa, EKE levels are also weak, but the situation is peculiar because the offshore region is in the path of the Agulhas Rings. Note that relatively high coastal values indicative of nearshore EKE sources can also be found, e.g., around 36–38°S and 30°S along the Chilean coast; 12–15°N along the coast of Senegal (the high values south of Cape Blanc at 19°N are for the Arguin Bank, a large expanse of water less than 20 m deep where wind-driven variability is probably essential); 38–42°N along the North American West Coast; and around 30°S off South Africa. These regions correspond to well-known upwelling centers.

There are several explanations for the EKE relative minima nearshore. The obvious one is that the sea-surface height (SSH) product, which EKE is derived from, is oblivious to non-geostrophic currents and, more generally, to eddy activity associated with wavelengths smaller than 0.5° [*Ducet et al.*, 2000]. As eddy activity tends to be at small scales nearshore and undergoes an inverse cascade while moving westward, it is progressively more detectable by the altimeter, hence the offshore maximum. A more subtle reason is that mean offshore currents can significantly contribute to eddy variability, however weak and elusive. This might be anticipated on the theoretical grounds for destabilization of meridional currents [*Pedlosky*, 1987; *Spall*, 2000].[3] In the framework of an eastern boundary system, this is confirmed by *Marchesiello et al.* [2003] who simulate a realistic CCS (i.e., the coastal and large-scale circulation, the termination of the West-Wind Drift, and the offshore California Current) and show that eddy activity is due to the intrinsic variability of the system as a whole, with significant EKE sources as far as 500 km off the California-Oregon coast. More specifically, in the offshore region, the main source of energy for eddies is baroclinic conversion (from mean potential to eddy kinetic energy), whereas nearshore, it is more evenly split between baroclinic and barotropic (from mean to eddy kinetic energy) conversion. Thus, EKE patterns may not reflect only advection and propagation of mesoscale structures generated nearshore [*Chaigneau and Pizarro*, 2005; *Hormazabal et al.*, 2004]. The good correspondence between large EKE and intense mean flow in several places in Plate 1 further supports this view.[4] In practice, trying to distinguish eddy activity generated by offshore versus coastal currents may not be very meaningful in this turbulent equilibrium regime: coastal and offshore currents and eddies strongly interact with each other.

Numerical models can now be run at horizontal and vertical resolutions that permit a good representation of eddies in the ocean. However, this does not guarantee a correct EKE field. In fact, given the complexity of EKE generation, surface EKE comparison between a model and data is a stringent test of the upper-oceanic circulation. Using adequate forcing (section 5.2), regional models, such as regional ocean modeling system (ROMS), are able to reproduce general surface EKE patterns with some localized discrepancies and possibly large-scale magnitude differences [e.g., in the southern PCS and CS in particular, EKE is 20 to 30% too high, and in the northern PCS model, EKE is a bit low for reasons discussed by *Penven et al.* 2005]. More rigorous comparisons will be required where numerical solutions are processed in the same way as altimetric observations to be entirely conclusive (in Plate 2, numerical solution EKEs were computed from low-pass filtered 2D sea surface elevations to limit inconsistencies in the estimates).

Although we focus on surface eddy activity, some eddies are subsurface-intensified with little or no hydrographic signature at the surface [*Cornuelle et al.*, 2000]. The general generation mechanism is instability of the nearshore undercurrents [with subsequent insemination of the offshore domain, *Brink et al.* 2000], although offshore generation cannot be ruled out [for the US West Coast (USWC), see *Collins et al.*, 2004]. Numerical models produce subsurface-intensified eddies [*Marchesiello et al.*, 2003], and insofar as their nearshore currents are realistic, they are useful in complementing the generally sparse observational information on the generation, structure, and effects of such eddies.

2.2. Seasonality and Low-Frequency Variability

The average picture described in the previous section is incomplete because mesoscale activity in EBS exhibits considerable variability. A seasonal cycle arises from the wind and heat flux cycles. In the CCS, northern CS, and southern

PCS, coastal surface EKE generation is most pronounced in spring when the upwelling favorable winds pick up, and the current system is concentrated nearshore. The location of the maximum surface EKE then migrates offshore with a significant attenuation over 300 to 500 km. In winter, the surface EKE is relatively weak without any clear maximum. Closer to the equator (northern PCS and BS, southern CS, and CCS), the upwelling is more persistent with a reversed (maximum in fall and winter off Peru) or no EKE seasonal cycle and no clear offshore migration of the EKE signal. In the CCS, the migration and attenuation are both observed in the data [*Kelly et al.*, 1998] and captured in numerical models [*Haney et al.*, 2001; *Marchesiello et al.*, 2003]. These models have played a key role in explaining the underlying dynamics, i.e., barotropization and subduction [*Haney et al.*, 2001] as opposed to an eddy-damping mechanism.

Mesoscale activity can also be modulated on interannual timescales. In the Pacific, the modulation arises primarily from the nearshore circulation changes caused by El Niño Southern Oscillation (ENSO). During El Niño events, currents are intensified poleward. Their destabilization leads to anticyclonic eddy formation especially in the lee of the major capes and headlands, as observed along Peru-Chile and the Mexican, USA, and Canadian West Coasts [*Strub and James*, 2002]. Using an eddy-resolving layered model, *Murray et al.* [2001] demonstrate that anticyclonic eddies shed by the baroclinically unstable coastal currents in the Gulf of Alaska largely control the vorticity budget over the entire region. Similar anticyclones are formed during the 1997–1998 El Niño in a PCS ROMS solution run with realistic forcing (ERS winds and oceanic boundary conditions are derived from a global POP solution; see Colas et al., 1997–1998 El Niño off Peru: A numerical study, submitted to *Progress in Oceanography*, 2007, hereinafter referred to as Colas et al., submitted manuscript, 2007). Although quantitative analyses are needed, these eddies evidently play a role in advecting anomalous water masses and tracer properties (vorticity, heat, and salt) brought from the equatorial region. In the CS, the inter-annual variability is essentially locally forced by the wind variability, as shown by *Roy and Reason* [2001], with the sea-surface temperature (SST) correlated to the ENSO signal with a 4-month phase lag. In the BS, major intrusions of warm saline water of equatorial origin can reach as far as 25°S and are associated with positive coastal sea level anomalies and an increased poleward current. This phenomenon is related to wind anomalies in the western and central Atlantic and is called the Benguela Niño by analogy with its Pacific counterpart [*Shannon et al.*, 1986; *Gammelsrod et al.*, 1998; *Florenchie et al.*, 2004]. The impact of basin-scale interannual variability in the CS and BS has not received significant numerical (or observational) attention.

3. EFFECTS OF THE MESOSCALE IN EBS

3.1. Tracer and Momentum Eddy Fluxes

EBS regions do not have especially large eddy-rectification effects, ignoring the southern part of the Benguela where most of the eddy activity is in fact that of the Agulhas Current that is diverted western boundary current. Bolus velocities[5] are rather small in EBS [*Bryan et al.*, 1999] (but note that they are only computed for density classes that never outcrop at a given location). This should be no surprise considering the fact that isopycnal thickness is quite uniform and the isopycnals rather flat except at the very nearshore (Plate 3); again, this does not apply to the southern Benguela. However, the mean circulation is also weak so that eddy fluxes can potentially exert some influence on EBS dynamical balances [*Marchesiello et al.*, 2003; *Leth and Middleton*, 2004]. Although observations lead to reliable estimates of eddy activity in the ocean (e.g., altimetry outside the nearshore region), they arguably lack spatial and temporal coverage over long enough timescales to allow for comprehensive estimates of eddy effects (fluxes of tracer and momentum and, most importantly, their divergence; *Colbo and Weller*, 2008). Numerical models are a well-suited framework to address these issues provided that they possess a reasonable degree of realism. Depth-integrated vorticity and heat budgets are calculated in *Marchesiello et al.* [2003] [see also *Leth and Middleton*, 2004]. In the CCS, eddy fluxes were found to modify the Sverdrup relation by inducing poleward flow within 150 km from shore and equatorward flow between 150 and 300 km from shore. Comparable budget analyses are lacking for the other systems, but the mechanism is presumably generic. A possible exception would be the northern part of the PCS where the mean offshore circulation is directed poleward rather than equatorward as in all other systems.

Using mesoscale-resolving regional configurations, Capet et al. (Upwelling systems heat balance: Role of the eddies, submitted to *Journal of Marine Research*; hereinafter referred to as Capet et al., submitted manuscript, 2007) examine the upper-oceanic heat budget (1) because of its importance in the coupled atmosphere–ocean system and in particular in the atmospheric boundary layer (section 4).

$$\int_{z_0}^{\zeta} \partial_t \overline{T} dz = -\int_{z_0}^{\zeta} \nabla \cdot \overline{\mathbf{u}T} dz + Q_{as} - \overline{\kappa \partial_z T}|_{z_0}$$

$$T_c = A + Q_{as} + D \tag{1}$$

In (1), **u** is the 3D velocity, T the potential temperature, ζ the sea surface elevation, κ the vertical subgrid-scale

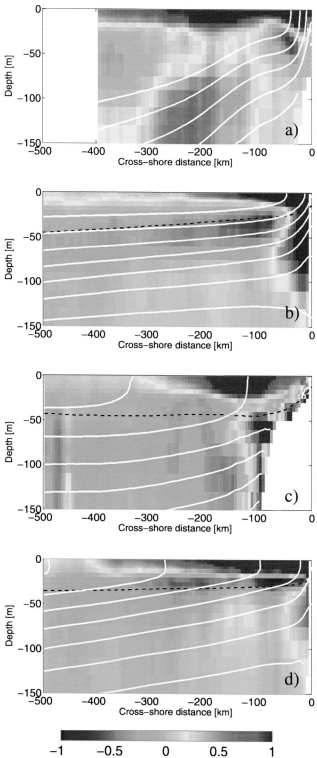

Plate 3. Vertical sections of annual-mean eddy heat flux divergence (W m⁻³) for BS, PCS, CS, and CCS (top to bottom) ROMS solutions. Alongshore averaging is done between, respectively, 25–33°S, 15–25°S, 22–26°N, and 35–43°N. The solid dashed line indicates the mean depth of the surface boundary layer. Open lines represent isotherms (at 1° intervals) that is approximately isopycnals because salinity effects are minor.

diffusivity, and Q_{as} the ocean–atmosphere heat flux including both interface fluxes and penetrating shortwave radiation. $\bar{}$ represents a time-averaging operator. By integrating deep enough (typically $z_0 = 100$ m) and long enough (5 years or more), T_c and D are negligible in EBS. The advection term is further decomposed into mean and eddy contributions presented along with their sum in Plate 4 for the PCS. The mean and eddy terms exhibit significant spatial variability that is absent from their sum. This might be attributed in part to insufficient averaging, especially offshore where convergence as a function of the averaging period seems to take place, albeit slowly, a point that further underlines the difficulty in analyzing heat balances from observations with limited temporal extent.

Nearshore, where spatial variability is most conspicuous, standing eddies play a prominent role in shaping robust structures that have also been observed in nature by their imprint on velocity and temperature (Centurioni, et al., Permanent meanders in the California Current System and comparison of near-surface observations with OGCM solutions, submitted to *Journal of Physical Oceanography*, 2007). This is especially the case in the CCS where standing eddies produce alternating strips of warming and cooling with a magnitude more than 50 W m^{-2} (not shown).

In all systems, the primary balance is between mean advection and atmospheric fluxes, except within roughly 50 to 100 km from shore where eddy fluxes provide a warming contribution that significantly counteracts cooling by the mean advection (in the CS and parts of the BS, the band of large positive eddy flux divergence is displaced farther offshore, but remains centered over the upper-slope and shelfbreak where the upwelling takes place).

Farther offshore, the eddy flux divergence modulates the heat budget with clear signs of cooling over extended regions: off southern Peru (Plate 4), northern California-Oregon, Mauritania-Senegal, and in the Southern California Bight. These regions share the particularity of having a temperature maximum a few hundred kilometers from the coast. As a consequence, mean cross-shore advection in the boundary layer (mainly Ekman) turns into a warming term at some distance away from the coast. This leaves just the eddies to provide the cooling necessary to balance (1). In our solutions, we find an eddy cooling equivalent to −30 W m^{-2}. We emphasize that all these regions are important stratus cloud formation sites for which accurate heat budgets are greatly needed.

3.2. (Sub-)Mesoscale Vertical Fluxes and Subduction

Plate 3 shows vertical sections of eddy heat flux divergence averaged alongshore in the different systems. In accordance with the description in section 3.1, eddy activity in the CCS and PCS tends to cool the offshore domain when averaged over the upper hundred meters, but it also strongly warms the upper half of the boundary layer. The latter effect (and part of the cooling underneath) arises from the vertical component of the eddy heat flux. Ageostrophic secondary circulations that accompany frontogenesis [*Hoskins*, 1982], even at moderately high resolution, act to restratify the near-surface ocean [*Lapeyre et al.*, 2007; *Capet et al.*, 2008b]. The strain induced by the mesoscale eddy field is the key to this process, but the strength of the secondary circulations, and hence the restratification tendency, strongly increases with horizontal resolution, even below 1 km. Around such resolution, the emergence of a baroclinic submesoscale instability can be observed [*Capet et al.*, 2008b] and may contribute to the increased vertical fluxes. Other types of instability may be present as well [*Haine and Marshall*, 1998; *Boccaletti et al.*, 2007].

Traditionally, nearshore upwelled waters are viewed as remaining in the boundary layer where they are carried away by the Ekman flow. During this journey, they are progressively warmed by mixing with surface waters of offshore origin and by atmospheric heat flux. This description, sketched in Plate 5, is consistent with the circulation in oceanic general circulation models (OGCMs) and even mesoscale-resolving EBS models in which restratification is primarily confined to the boundary layer where it is counteracted by mixing arising from wind-induced turbulence. However, at resolutions of around 1 km or below, there are occurrences of subduction events that reach well below the boundary layer (Plate 6) at the edges of mesoscale eddies in the vicinity of surface density fronts. These aspects have been investigated in a submesoscale-resolving (750-m grid size) configuration for an idealized CCS and confirm observational evidence of widespread subduction at depths greater than 100 m [*Kosro et al.*, 1991; *Bograd and Mantyla*, 2005]. Thus, a new element is currently being added to our conceptual view of offshore Ekman transport in EBS. A substantial fraction of newly upwelled water is subducted on its way seaward, while an equivalent amount of upper thermocline water is drawn into the boundary layer (the colored arrows in Plate 5). The importance of subduction within EBS is still unclear, but this process has serious implications for the upper-oceanic heat balance (Capet et al., submitted manuscript, 2007), the transfer of organic material between the euphotic zone and the oceanic interior[6] and possibly EBS thermocline ventilation in general [*Bograd and Mantyla*, 2005]. We expect multi-year, submesoscale-resolving numerical solutions to play a major role in elucidating these questions and providing new insights to be incorporated into OGCMs by means of parameterizations (Fox-Kemper and Ferrari, Parameterization of mixed layer eddies. II: Prognosis and impact, submitted to *Journal of Physical Oceanography*, 2007).

3.3. Mesoscale Activity and Meridional Contrasts in EBS

In EBS, the main frontal feature arises from thermohaline contrasts between nearshore freshly upwelled waters and offshore waters. On average, a front thus runs roughly parallel to the coastline. Other fronts are also present in EBS that stem from meridional gradients in water mass properties; such fronts have a roughly zonal orientation. This is the case for the Benguela-Angola front (at ≈15°S) that separates the Benguela cold waters from the much warmer and more saline northern waters; the Cape Verde front off Mauritania-Senegal and the Ensenada front off Baja California (see Plate 7). Another example is the front between equatorial (25°C and warmer) and subtropical waters (24°C and colder) that develops during El Niño off Peru (Plate 7). Note that this latter front has not been documented by in situ measurements. Model solutions suggest that these fronts are unstable (notice the meanders, filaments, and streamers in the vicinity of the fronts in Plate 7). The instabilities might be key in controlling the cross-front exchange of tracers on a regional scale and, consequently, the water mass that is being upwelled either along the Peruvian coast during El Niño (Colas et al., submitted manuscript, 2007) or along Mauritania-Senegal with some seasonal changes [*Perez-Rodriguez et al.*, 2001]. Further insight into the interplay between the coastal upwelling and such regional fronts would be useful.

4. UPSCALING EFFECTS ON EQUATORIAL CLIMATE

Recent studies indicate that EBS dynamics have nontrivial remote effects and, in particular, exert some influence over the equatorial regions. Using a non-eddy-resolving climate model, *Large and Danabasoglu* [2006] perform sensitivity experiments to evaluate the impact of the temperature warm bias that develops in such coarse coupled models. They compare a baseline solution with a solution that only differs by the fact that temperature within the EBS regions (one or several at a time) is heavily restored toward climatological values. The solutions with restoring exhibit significant improvements of their SST and, subsequently, precipitation patterns in the Pacific and Atlantic equatorial regions. The largest bias reduction is related to the PCS and BS. The impact of the CCS restoring is weak on near-surface properties, but it leads to a 1°–3°C subsurface temperature bias reduction across the entire Pacific between roughly 5°N and 15°N. Atmospheric connections from the EBS toward the equator—thought to occur by locally influencing the extensive stratus cloud deck—might at least partially explain these results. However, subsurface sensitivities suggest the existence of oceanic connections from the EBS toward the oceanic interior (*Large and Danabasoglu* [2006]; see also the analysis of an Atlantic reduced-gravity model by *Lazar et al.* [2002]). The sensitivity experiments presented in Section 5 suggest that eddy-resolving oceanic solutions forced by fine-scale wind fields (or coupled to a high-resolution atmospheric model) will be required to correct equatorial biases in OGCMs without resorting to numerical tricks such as the restoring by *Large and Danabasoglu* [2006].

EBS are also the place where long baroclinic Rossby waves are generated through coastally trapped wave leakage [*Philander and Yoon*, 1982]. Thus, coastal disturbances are transferred thousands of kilometers into the oceanic interior [*White et al.*, 1990; *Vega et al.*, 2003] with consequences that are as yet unclear. Again, the length scale of the processes in play (the baroclinic deformation radii for the cross-shore extension of the coastal trapped waves) implies that mesoscale-resolving oceanic models are needed to assess this issue. Note that another important generation mechanism for long baroclinic Rossby waves in EBS that is wind stress curl [*White and Saur*, 1981], is probably adequately represented in non-eddy-resolving models.

5. MODELING SUBTLETIES

5.1. Resolution Sensitivity

A benefit from high-resolution EBS simulations is better representation of the coastal divergence implied by the Ekman transport. It has been known for some time that mixing processes control the upwelling width, which becomes much less than the offshore deformation radius [*Pedlosky*, 1978], except when the upwelling occurs in regions with wide shelves (because bottom friction is significant over a wide nearshore strip; *Mitchum and Clarke*, [1986]; Estrade et al., Cross-shelf structure of coastal upwelling: A two dimensional expansion of Ekman's theory and a mechanism for innershelf upwelling shut down, submitted to *Journal of Physical Oceanography*, 2007, hereinafter referred to as Estrade at al., submitted manuscript. 2007). This makes upwelling in regions with narrow shelves (e.g., off central California and northern central Chile) very sensitive to horizontal resolution, down to scales hardly attainable with regional models, let alone global ones.[7] As an illustration, we show the summer-mean vertical velocity and temperature cross-sections at different resolutions (Plate 8 for an idealized CCS; *Capet et al.* [2008a]). Heat fluxes are adjusted with a restoring term that is only partly justified on physical grounds [*Barnier et al.*, 1995] so that the long-term impact

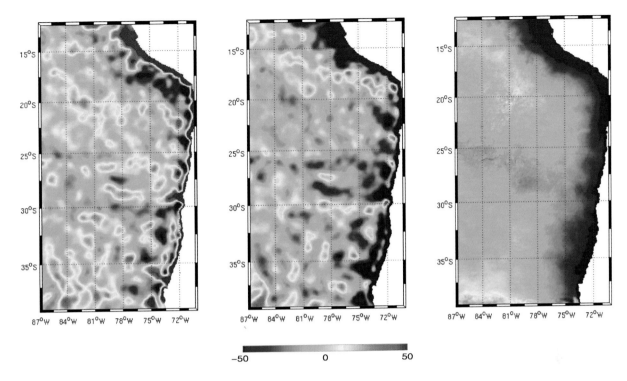

Plate 4. Annual average of vertically integrated (0- to 100-m depth) eddy (left), mean (center), and total (right) heat divergence (W m^{-2}) for a 9-year PCS ROMS solution. Maps of heat flux from the ocean to the atmosphere (i.e., positive upward) are indistinguishable from the panel for total heat flux.

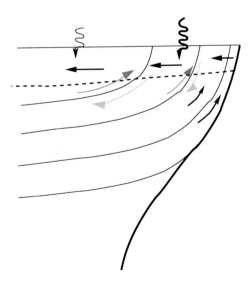

Plate 5. 2D schematic of coastal upwelling processes. Zigzag arrows represent atmospheric heat fluxes. The mixed layer base is delineated by a solid dashed line. Solid arrows indicate mean upwelling and Ekman transport, and the blue and red curved arrows indicate eddy fluxes, mostly along isopycnal surfaces. In the real 3D ocean, the favored locations for subduction or uplifting are related to the instantaneous positions of isopycnal surfaces as in Plate 6 (instead of the mean positions as suggested in this sketch).

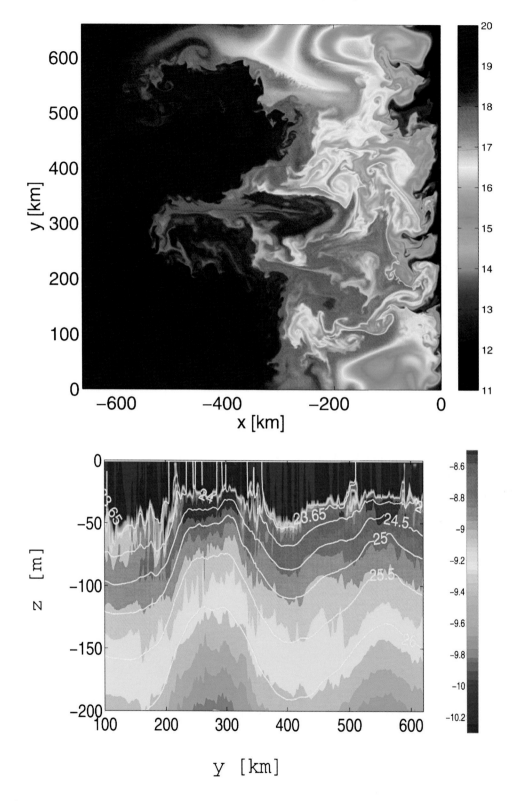

Plate 6. Top: instantaneous SST in an idealized CCS configuration (flat bottom; straight coastline on the right-hand side of the domain; constant upwelling favorable winds) [*Capet et al.*, 2007a]. Bottom: Ertel potential vorticity (PV, in color) and isopycnals (white contours) along the vertical section at $x = -400$ km. The region of low PV near the surface roughly delineates the boundary layer, with an abrupt transition to highly stratified waters (in red, the upper thermocline). Numerous intrusions between low and high PV waters can be seen. They generally coincide with the strongest density (or temperature) fronts at the edges of the main mesoscale structures.

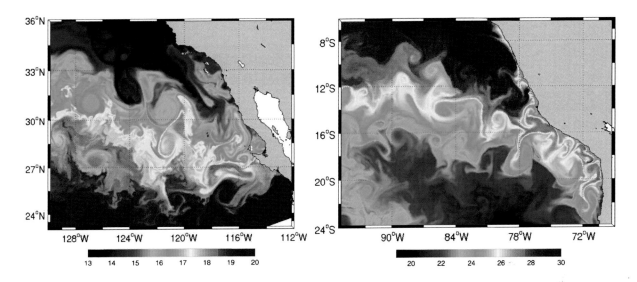

Plate 7. ROMS instantaneous potential temperature at 5 m. Left: during a climatological January off Southern California–Mexico; the unstable behavior of the Ensenada front is revealed by eddies and filaments. Right: in January 1998 shortly after the second major 1997–1998 El Niño pulse; a front separating equatorial from subtropical waters is noticeable as are its accompanying eddies, filaments, and streamers.

of resolution on the upper-oceanic temperature is uncertain. Nonetheless, the nearshore temperatures are almost two degrees colder at 750-m resolution than at 12 km. This difference is related to a tenfold increase in nearshore maximum vertical velocities that reach 50 m/day at the higher resolution, while the vertical flux of water remains roughly constant (because the same Ekman coastal divergence needs to be balanced). This is an important reason for the EBS warm bias present in non-eddy-resolving OGCMs.

5.2. Small Scales in Coastal Winds

It is common that some spatial and temporal uncertainties about atmospheric forcings exist, as analyzed in atmospheric data assimilation models and as simulated with coupled atmosphere-ocean models. The uncertainty is greatest near the shoreline because this is where the wind varies most rapidly and also because measuring or simulating nearshore winds is generally more challenging. QuikSCAT observations are sensitive to land contamination; spectral models such as NCEP suffer Gibbs phenomena nearshore when mountain ranges are present close to the coast that is virtually everywhere along eastern boundaries; nonspectral models still have issues handling the land-ocean transition. The degree to which the uncertainty in the forcing fields affects the oceanic solutions is an important issue. As already noted in section 2.1, upwelling system dynamics can be statistically well simulated without having to include synoptic and intra-seasonal wind fluctuations. In other words, rectification effects due to responses to the high frequencies in the wind are modest. On the other hand, sensitivity studies have demonstrated that mesoscale spatial variability of the nearshore wind has a large impact on upwelling systems. First, expansion fans [*Winant et al.*, 1988; *Enriquez and Friehe*, 1995; *Garreaud and Muñoz*, 2005], in conjunction with topographic irregularities, help the upwelling jet to separate away from the coast [*Castelao and Barth*, 2007] and subsequently generate standing meanders with scales of hundreds of kilometers. The mean nearshore wind profile is also important, and it can be understood in terms of the competition between coastal divergence and Ekman pumping. The occurrence of nearshore wind drop-off (i.e., a tendency for weaker winds toward the shore) favors upward Ekman pumping in a band somewhat distant from the coastline, while stronger nearshore winds (i.e., smaller or no drop-off) favor intense coastal upwelling. In the linear model of *Fennel and Laas* [2006], the outcome of this competition is sensitive to bottom friction. In realistic simulations off Central California and in the Southern California Bight, *Capet et al.* [2004] find that winds with limited drop-off lead to more upwelling, although the uplifting of particles due to Ekman pumping

reaches deeper. The cross-shore wind profile close to shore also plays an important role in determining the alongshore current structure [*Marchesiello et al.*, 2003]. Equatorward winds at the coast force a surface equatorward jet and poleward undercurrent. The depth-averaged poleward currents are related to positive wind curl through Sverdrup balance with some modulation by the Reynolds stress (section 3.1). Large positive wind curl and weak coastal wind favor poleward currents and vice versa. The combination of these effects has important consequences on the system instability and on its resulting mesoscale activity.

To illustrate this, we compare the CCS solution presented in Plates 2d and 3d with that forced by NCEP winds (Plates 9 and 10). The winds for these two solutions differ mainly by the length scale they decrease by toward the coast, around 300 km for NCEP and 50 km for QuikSCAT (see Plate 11). The solution forced by NCEP differs from the one forced by QuikSCAT in a way that is consistent with the discussion above: nearshore poleward currents even near the surface; an offshore doming of the isotherms associated with Ekman pumping; and overall warmer temperatures close to shore. Standing eddies, EKE patterns, and eddy fluxes also differ significantly as evident when comparing Plates 2d and 9 or Plates 3d and 10. With NCEP, the EKE is much reduced nearshore and has its maximum displaced farther offshore (as the California Current itself). The greater stability of prograde currents (i.e., currents that have the coast on the right in the northern hemisphere when facing downstream) is a possible explanation for these differences in EKE. Eddy heat fluxes differ mainly 150 to 300 km offshore where the solution forced with NCEP develops subsurface fluxes that limit the doming of the isotherms.

It seems desirable to reduce the uncertainty in our knowledge of wind profiles; see Plate 11 for a sense of the spread in the case of the CCS. Given that winds in climate models tend to resemble NCEP, this would help determine the role of atmospheric forcing in the EBS warm bias.

6. CONCLUSIONS

Eastern boundary upwelling systems have been largely neglected by the large-scale modeling community because their role in the climate system or the global ocean is less prominent than that of other regions. There are, however, numerous indications that EBS exert some influence on the dynamics of their respective basins, either because they actively contribute to stratus formation over wide areas or because there are oceanic pathways (by advection or through waves) between the EBS and the oceanic basin interior. The role of mesoscale eddies in these climate system connections is far from well established. Upper-oceanic heat

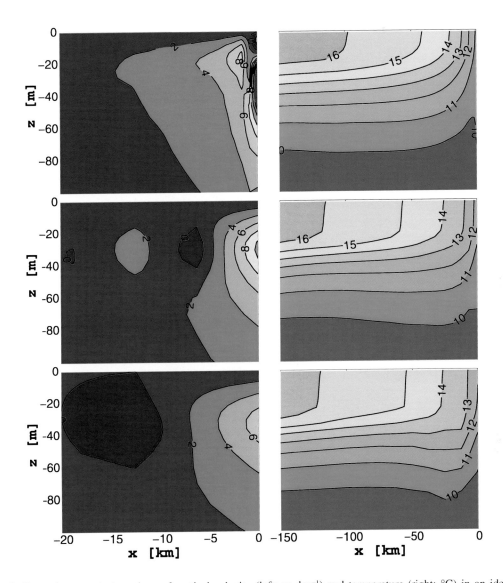

Plate 8. Cross-shore vertical sections of vertical velocity (left; m day^{-1}) and temperature (right; °C) in an idealized California Current System [*Capet et al.*, 2007a] at three different horizontal resolutions (dx = 0.75, 3, and 12 km, top to bottom). Both quantities are averaged alongshore over 600 km. Note the different scales in the cross-shore direction. At 0.75-km resolution, the vertical velocity at the shore reaches 50 m/day^{-1}.

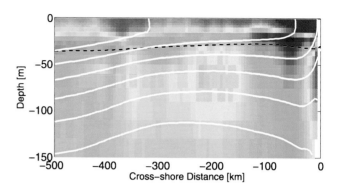

Plate 10. Same as Plate 3d but for a solution forced with NCEP climatological winds.

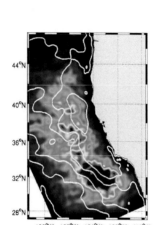

Plate 9. Same as Plate 2d but for a solution forced with NCEP climatological winds.

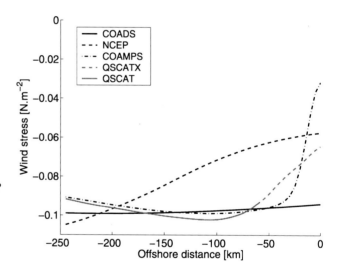

Plate 11. Spring-Summer climatological alongshore wind stress averaged between 34°N and 42°N in the CCS as a function of distance from the coast. COADS [*Da Silva et al.*, 1994], NCEP, COAMPS [*Kindle et al.*, 2002; *Pickett and Paduan*, 2003], and QuikSCAT are represented. Negative wind stress is equatorward. The climatologies are computed over the period 1999–2004, except for COADS (1945–1989). For QuikSCAT, both the actual data (solid gray line) and their near-shore linear extrapolation (dashed gray line) are shown.

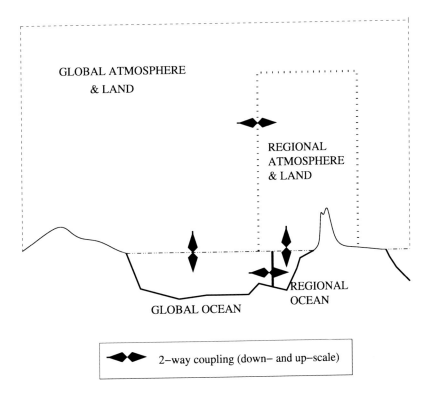

GLOBAL ATMOSPHERE
& LAND

REGIONAL
ATMOSPHERE
& LAND

REGIONAL
OCEAN

GLOBAL OCEAN

◆◆ 2–way coupling (down– and up–scale)

Figure 1. Schematic of an embedded (regional within global) coupled atmosphere-ocean model.

budgets carried out for the four major EBS indicate that eddy heat flux divergences are locally significant and may affect the atmospheric boundary layer functioning, hence also the stratus cloud deck formation [*Colbo and Weller*, 2008]. Explicitly resolving these eddy fluxes, and also better representing coastal processes with scales on the order of the first baroclinic deformation radius or less (coastal upwelling, coastal trapped waves), are the direct main benefits to expect from EBS eddy-resolving models. Indirectly, this may also improve the realism of OGCMs whose EBS biases are quite large at present. Nesting strategies between OGCMs and regional models are currently under development and should limit the computational cost of increasing the resolution for the EBS as well as in other oceanic regions of particular interest (Figure 1). The technical aspects of coupling between such a nested oceanic modeling system and its atmospheric counterpart make this a challenging path, but it would deliver an adequate framework to represent and understand EBS and other regional dynamics in their full complexity, including climate downscaling and upscaling effects as well as ocean-atmosphere feedbacks.

Acknowledgments. The altimeter products were produced by SSALTO/DUACS and distributed by Aviso, with support from CNES. Rio05 was produced by CLS Space Oceanography Division. Alexander Shchepetkin is acknowledged for his unshakable dedication to improving the physical and numerical aspects of ROMS. We acknowledge the Office of Naval Research for support under grant ONR N00014-04-1-0401 and the InterUp (LEFE-IDAO program) for travel support.

Notes

1. Among the various regional models, ROMS [*Shchepetkin and McWilliams*, 2008] is today often preferred when simulating EBS because of its weakly diffusive numerical schemes, a desirable feature given the importance of small-scale processes in upwelling systems.

2. As customary, EKE is defined as $1/2(u'^2 + v'^2)$ where u' and v' are velocity perturbations relative to a long-term time-mean. The bulk of EKE in the ocean is associated with scales around the first-baroclinic deformation radius [*Stammer*, 1997] that is with mesoscale variability, but also see Section 2.1 for a discussion on nearshore EBS regions. Note that EBS EKE estimates differ significantly depending on the type of observations they are based on [*Chereskin et al.*, 2000; *Kelly et al.*, 1998; *Ducet et al.*, 2000; *Pascual et al.*, 2006].

3. These results are expected to apply with the caveat that the intensity of the perturbations should be limited near a continental boundary.

4. Some caution is needed because circulation patterns with scales on the order of 100 km may not be adequately resolved in the Rio05 product

[*Rio and Hernandez*, 2004]. In particular, the standing eddies that are a conspicuous feature of the CCS and, to a lesser extent, the other EBS are nearly absent in Plate 1.

5. Bolus velocities arise from correlations between density (or isopycnal thickness) and velocity [*Gent et al.*, 1995]. In effect, they should be thought of as a mass transport due to eddies.

6. Using mesoscale-resolving solutions for the CCS, Nagai, et al. (Dominant role of eddies in offshore transports in the California Current System, submitted to *Journal of Geophysical Research*, 2007) show that eddy subduction at upper-oceanic fronts takes some of the upwelled nutrients away from the surface, and therefore, primary production tends to decrease when mesoscale activity is resolved.

7. Off Oregon, the scaling of Estrade et al. (submitted manuscript, 2007) for the cross-shore width of an upwelling cell D/s (where D is the Ekman depth and s the slope of the continental shelf) is on the order of 1 km.

REFERENCES

Barnier, B., L. Siefried, and P. Marchesiello (1995), Thermal forcing for a global ocean circulation model using a three-year climatology of ECMWF analyses., *J. Mar. Syst.*, 6, 363–380.

Barth, J. (1994), Short wavelength instabilities on coastal jets and fronts, *J. Geophys. Res.*, 99(C8), 16095–16115.

Batteen, M. (1997), Wind-forced modeling studies of currents, meanders and eddies in the California Current System, *J. Geophys. Res.*, 102(C1), 985–1010.

Boccaletti, G., R. Ferrari, and B. Fox-Kemper (2007), Mixed layer instabilities and restratification, *J. Phys. Oceanogr.*, 37, 2228–2250.

Bograd, S., and A. Mantyla (2005), On the subduction of upwelled waters in the California current, *J. Mar. Res.*, 63, 863–885.

Boyer, T., and S. Levitus (1998), Objective analyses of temperature and salinity for the world ocean on a 1/4° grid. NOAA Atlas NESDIS, Washington, DC.

Brink, K., R. Beardsley, J. Paduan, R. Limeburner, M. Caruso, and J. Sires (2000), A view of the 1993–1994 California current based on surface drifters, floats, and remotely sensed data, *J. Geophys. Res.*, 105(C4), 8575–8604.

Bryan, K., J. Dukowicz, and R. Smith (1999), On the mixing coefficient in the parameterization of bolus velocity, *J. Phys. Oceanogr.*, 29, 2442–2456.

Capet, X. J., P. Marchesiello, and J. C. McWilliams (2004), Upwelling response to coastal wind profiles, *Geophys. Res. Lett.*, 31, L13311.

Capet, X., J. McWilliams, M. Molemaker, and A. Shchepetkin (2008a), Mesoscale to submesoscale transition in the California Current System: Flow structure and eddy flux, *J. Phys. Oceanogr.*, in press.

Capet, X., J. McWilliams, M. Molemaker, and A. Shchepetkin (2008b), Mesoscale to submesoscale transition in the California Current System: Frontal processes, *J. Phys. Oceanogr.*, in press.

Castelao, R., and J. Barth (2007), The role of wind-stress curl in jet separation at a cape, *J. Phys. Oceanogr.*, 37, 2652–2671.

Chaigneau, A., and O. Pizarro (2005), Mean surface circulation and mesoscale turbulent flow characteristics in the eastern South Pacific from satellite tracked drifters, *J. Geophys. Res.*, 110, C05014, doi:10.1029/2004JC002628.

Chereskin, T., M. Morris, P. Niiler, P. Kosro, R. Smith, S. Ramp, C. Collins, and D. Musgrave (2000), Spatial and temporal characteristics of the mesoscale circulation of the California Current from eddy-resolving moored and shipboard measurements, *J. Geophys. Res.*, 105(C1), 1245–1269.

Colbo, K., and R. Weller (2008), The variability and heat budget of the upper ocean under the Chile–Peru stratus, *J. Mar. Res.*, in press.

Collins, C. A., L. M. Ivanov, O. V. Melnichenko, and N. Garfield (2004), California Undercurrent variability and eddy transport estimated from RAFOS float observations, *J. Geophys. Res.*, 109, C05028, doi:10.1029/2003JC002191.

Cornuelle, B., T. Chereskin, P. Niiler, M. Morris, and D. Musgrave (2000), Observations and modeling of a California undercurrent eddy, *J. Geophys. Res.*, 105(C1), 1227–1243.

Da Silva, A., C. Young, and S. Levitus (1994), Atlas of surface marine data 1994, Vol. 1, Algorithms and procedures, *NOAA Atlas NESDIS*, 6, 74 pp.

Ducet, N., P. Y. Le Traon, and G. Reverdin (2000), Global high-resolution mapping of ocean circulation from TOPEX/Poseidon and ERS-1 and -2, *J. Geophys. Res.*, 105(C8), 19477–19498.

Durski, S., and J. Allen (2005), Finite-amplitude evolution of instabilities associated with the coastal upwelling front, *J. Phys. Oceanogr.*, 35, 1606–1628.

Enriquez, A., and C. Friehe (1995), Effect of wind stress and wind stress curl variability on coastal upwelling, *J. Phys. Oceanogr.*, 25, 1651–1671.

Fennel, W., and H. Laas (2006), On the impact of wind curls on coastal currents, *J. Mar. Sys.*, doi:10.1016/j.jmarsys.2006.11.004.

Florenchie, P., C. Reason, J. Lutjeharms, M. Rouault, R. C., and S. Masson (2004), Evolution of interannual warm and cold events in the Southeast Atlantic ocean, *J. Clim.*, 17, 2318–2334.

Gammelsrod, T., C. Bartholomae, D. Boyer, V. Filipe, and M. O'Toole (1998), Intrusion of warm surface water along the Angolan-Namibian coast in February–March 1995: The 1995 Benguela Nino, *S. Afr. J. Mar. Sci.*, 19, 41–56.

Garreaud, R., and R. Muñoz (2005), The low-level jet off the west coast of subtropical South America: structure and variability, *Mon. Weather Rev.*, 133, 2203.

Gent, P., J. Willebrand, T. McDougall, and J. McWilliams (1995), Parameterizing eddy-induced tracer transports in ocean circulation models, *J. Phys. Oceanogr.*, 25, 463–474.

Haine, T., and J. Marshall (1998), Gravitational, symmetric, and baroclinic instability of the ocean mixed layer, *J. Phys. Oceanogr.*, 28, 634–658.

Haney, R., R. Hale, and D. Dietrich (2001), Offshore propagation of eddy kinetic energy in the California Current, *J. Geophys. Res.*, 106(C6), 11709–11717.

Hormazabal, S., G. Shaffer, and O. Leth (2004), Coastal transition zone off Chile, *J. Geophys. Res.*, 109, C01021, doi:10.1029/2003JC001956.

Hoskins, B. (1982), The mathematical theory of frontogenesis, *Annu. Rev. Fluid Mech.*, 14, 131–151.

Ikeda, M., and W. Emery (1984), Satellite observations and modeling of meanders in the California Current System off Oregon and Northern California, *J. Phys. Oceanogr.*, *14*, 1434–1449.

Ikeda, M., L. Mysak, and W. Emery (1984), Observations and modeling of satellite-sensed meanders and eddies off Vancouver Island, *J. Phys. Oceanogr.*, *14*, 3–21.

Kelly, K., R. Beardsley, R. Limeburner, K. Brink, J. Paduan, and T. Chereskin (1998), Variability of the near-surface eddy kinetic energy in the California Current based on altimetric, drifter, and moored current data, *J. Geophys. Res.*, *103*(C6), 13067–13083.

Kindle, J. C., R. M. Hodur, S. deRada, J. D. Paduan, L. K. Rosenfeld, and F. Q. Chavez (2002), A COAMPS™ reanalysis for the Eastern Pacific: Properties of the diurnal sea breeze along the central California coast, *Geophys. Res. Lett.*, *29*(24), 2203, doi:10.1029/2002GL015556.

Kosro, P., et al. (1991), The structure of the transition zone between coastal waters and the open ocean off northern California, Winter and Spring 1987, *J. Geophys. Res.*, *96*(C8), 14707–14730.

Lapeyre, G., P. Klein, and B. Hua (2007), Oceanic restratification forced by surface frontogenesis, *J. Phys. Oceanogr.*, *36*, 1577–1590.

Large, W., and G. Danabasoglu (2006), Attributions and impacts of upper-ocean biases in CCSM3, *J. Clim.*, *19*, 2325–2346.

Lazar, A., T. Inui, P. Malanotte-Rizzoli, A. J. Busalacchi, L. Wang, and R. Murtugudde (2002), Seasonality of the ventilation of the tropical Atlantic thermocline in an ocean general circulation model, *J. Geophys. Res.*, *107*(C8), 3104, doi:10.1029/2000JC000667.

Leth, O., and J. F. Middleton (2004), A mechanism for enhanced upwelling off central Chile: Eddy advection, *J. Geophys. Res.*, *109*, C12020, doi:10.1029/2003JC002129.

Marchesiello, P., and P. Estrade (2008), Eddy activity and mixing in upwelling systems: A comparative study of northwest Africa and California regions, *Int. J. Earth Sci.*, in press.

Marchesiello, P., J. McWilliams, and A. Shchepetkin (2003), Equilibrium structure and dynamics of the California Current System, *J. Phys. Oceanogr.*, pp. 753–783.

Mitchum, G., and A. Clarke (1986), The frictional nearshore response to forcing by synoptic scale winds, *J. Phys. Oceanogr.*, *16*, 934–946.

Murray, C., S. Morey, and J. O'Brien (2001), Interannual variability of upper ocean vorticity balances in the Gulf of Alaska, *J. Geophys. Res.*, *106*(B6), 4479–4491.

Pares-Sierra, A., W. White, and C.-K. Tai (1993), Wind-driven coastal generation of annual mesoscale eddy activity in the California Current, *J. Phys. Oceanogr.*, *23*, 1110–1121.

Pascual, A., Y. Faugère, G. Larnicol, and P.-Y. Le Traon (2006), Improved description of the ocean mesoscale variability by combining four satellite altimeters, *Geophys. Res. Lett.*, *33*, L02611, doi:10.1029/2005GL024633.

Pedlosky, J. (1978), A nonlinear model of the onset of upwelling, *J. Phys. Oceanogr.*, *8*, 178–187.

Pedlosky, J. (1987), *Geophysical Fluid Dynamics*, 710 pp., Springer, Berlin.

Penven, P., V. Echevin, J. Pasapera, F. Colas, and J. Tam (2005), Average circulation, seasonal cycle, and mesoscale dynamics

of the Peru Current System: A modeling approach, *J. Geophys. Res.*, *110*, C10021 doi:10.1029/2005JC002945.

Perez-Rodriguez, P., J. Pelegri, and A. Marrero-Diaz (2001), Dynamical characteristics of the Cape Verde frontal zone, *Sci. Mar.*, *65*(suppl. 1), 241–250.

Philander, S., and J. Yoon (1982), Eastern boundary currents and coastal upwelling, *J. Phys. Oceanogr.*, *12*, 862–879.

Pickett, M. H., and J. D. Paduan (2003), Ekman transport and pumping in the California Current based on the U.S. Navy's high-resolution atmospheric model (COAMPS), *J. Geophys. Res.*, *108*(C10), 3327, doi:10.1029/2003JC001902.

Rio, M.-H., and F. Hernandez (2004), A mean dynamic topography computed over the world ocean from altimetry, in situ measurements, and a geoid model, *J. Geophys. Res.*, *109*, C12032, doi:10.1029/2003JC002226.

Roy, C., and C. Reason (2001), ENSO related modulation of coastal upwelling in the eastern Atlantic, *Prog. Oceanogr.*, *49*, 245–255.

Shannon, L. V., A. Boyd, G. Bundrit, and J. Taunton-Clark (1986), On the existence of an El Nino-type phenomenon in the Benguela system, *J. Mar. Syst.*, *44*, 495–520.

Shchepetkin, A., and J. McWilliams (2007), Computational kernel algorithms for fine-scale, multi-process, long-time oceanic simulations, in *Handbook of Numerical Analysis: Special Volume: Computational Methods for the Atmosphere and the Oceans*, edited by R. Temam and J. Tribbia, Elsevier, in press.

Spall, M. (2000), Generation of strong mesoscale eddies by weak ocean gyres, *J. Mar. Res.*, *58*, 97–116.

Stammer, D. (1997), Global characteristics of ocean variability estimated from regional TOPEX/POSEIDON altimeter measurements, *J. Phys. Oceanogr.*, *27*, 1743–1769.

Strub, P., and C. James (2002), The 1997-1998 oceanic El Nino signal along the southeast and northeast Pacific boundaries—An altimetric view, *Prog. Oceanogr.*, *54*, 439–458.

Vega, A., Y. du-Penhoat, B. Dewitte, and O. Pizarro (2003), Equatorial forcing of interannual Rossby waves in the eastern South Pacific, *Geophys. Res. Lett.*, *30*(5), 1197, doi:10.1029/2002GL015886.

White, W., and J. Saur (1981), A source of annual baroclinic waves in the eastern subtropical North Pacific, *J. Phys. Oceanogr.*, *11*, 1452–1462.

White, W., T. C.K., and J. DiMento (1990), Annual Rossby wave characteristics in the California current region from the GEO-SAT exact repeat mission, *J. Phys. Oceanogr.*, *20*, 1297–1311.

Winant, C., D. Dorman, C. Friehe, and R. Beardsley (1988), The marine layer off Northern California: an example of supercritical channel flow, *J. Atmos. Sci.*, *45*, 3588–3605.

―――――――――

Capet, X. Instituto Oceanográfico, Universidade de São Paulo, Pca. do Oceanografico 191 - Cidade Universitaria, 05508900, Sao Paulo, SP, Brazil. (xcapet@io.usp.br)

The Fidelity of Ocean Models With Explicit Eddies

Julie McClean[1,4], Steven Jayne[2], Mathew Maltrud[3], and Detelina Ivanova[4]

Current practices within the oceanographic community have been reviewed with regard to the use of metrics to assess the realism of the upper-ocean circulation, ventilation processes diagnosed by time-evolving mixed-layer depth and mode water formation, and eddy heat fluxes in large-scale fine resolution ocean model simulations. We have striven to understand the fidelity of these simulations in the context of their potential use in future fine-resolution coupled climate system studies. A variety of methodologies are used to assess the veracity of the numerical simulations. Sea surface height variability and the location of western boundary current paths from altimetry have been used routinely as basic indicators of fine-resolution model performance. Drifters and floats have also been used to provide pseudo-Eulerian measures of the mean and variability of surface and subsurface flows, while statistical comparisons of observed and simulated means have been carried out using James tests. Probability density functions have been used to assess the Gaussian nature of the observed and simulated flows. Length and timescales have been calculated in both Eulerian and Lagrangian frameworks from altimetry and drifters, respectively. Concise measures of multiple model performance have been obtained from Taylor diagrams. The time evolution of the mixed-layer depth at monitoring stations has been compared with simulated time series. Finally, eddy heat fluxes are compared to climatological inferences.

1. INTRODUCTION

The assessment of the fidelity of ocean simulations has become more meaningful and rigorous as ocean models have become more realistic in their representation of ocean processes. Furthermore, data have become available that permit

[1]Scripps Institution of Oceanography, La Jolla, California, USA.
[2]Woods Hole Oceanographic Institution, Woods Hole, Massachusetts, USA.
[3]Los Alamos National Laboratory, Los Alamos, New Mexico, USA.
[4]Lawrence Livermore National Laboratory, Livermore, California, USA.

Ocean Modeling in an Eddying Regime
Geophysical Monograph Series 177

comparisons over a wide range of spatial and temporal scales. The use of increased horizontal and vertical resolution, improved atmospheric surface forcing and bottom bathymetry products, as well as the use of increasingly sophisticated parameterizations of physical processes on temporal and spatial scales not resolved by the model have all contributed to the improved realism of model simulations. The advent of ocean observing satellites and the World Ocean Circulation Experiment in the 1990s have provided unprecedented numbers of observations for diagnostics. These diagnostics can be based on model/data comparisons, statistical measures, budgets, and dynamical balances, as appropriate. While metrics can be used to assess the realism of a model simulation, it should be noted that an ocean model can never be verified or validated, but can only be confirmed to be consistent with the available observations [*Oreskes et al.*, 1994].

The choice of metrics selected to evaluate the performance of an ocean simulation is based on the application and the questions that the model is being used to address. Metrics need to account for the spatial and temporal scales resolved in the simulation along with the representation of processes relevant to the problem being addressed. In coupled climate simulations, these are the processes governing air–sea exchanges and the uptake, transportation, and storage of heat, freshwater, and energy. For synoptic forecasting, high-frequency variability (up to 5 days) of the upper-ocean thermohaline structure and flows is important.

The majority of the kinetic energy in the ocean circulation is generally accounted for by flows with spatial scales of 50 to 500 km, which is referred to as the oceanic mesoscale [*Stammer*, 1997]. Advances in high-performance computing architectures and the increased availability of such platforms for "grand challenge" calculations over the past decade have lead to ocean simulations in which much of this variability is resolved. *Hurlburt and Hogan* [2000] conducted basin-scale

Table 1. Listing of Selected Eddying Models Discussed in this Review

Model	Horizontal Resolution	Vertical Resolution	References
NLOM	1/8°–1/64°	6 layers	*Hurlburt and Hogan* [2000]
POP	1/10°	40 levels	*Smith et al.* [2000] *Le Traon et al.* [2001] *McClean et al.* [2002] *Tokmakian and McClean* [2003] *Brachet et al.* [2004] *Maltrud and McClean* [2005] *Treguier et al.* [2005] *Tokmakian* [2005] *McClean et al.* [2006] *Rainville et al.* [2007]
OFES	1/10°	54 levels	*Masumoto et al.* [2004] *Du et al.* [2005] *Aoki et al.* [2007]
MICOM	1/12°	16 layers	*Paiva et al.* [1999] *Chassignet and Garraffo* [2001] *Garraffo et al.* [2001] *Bracco et al.* [2003] *Treguier et al.* [2005]
HYCOM	1/12°	32 hybrid layers	*Chassignet et al.* [2006] *Kelly et al.* [2007]

NLOM, Navy Layered Ocean Model; POP, Parallel Ocean Program; OFES, Ocean General Circulation Model based on the Modular Ocean Model 3 for the Earth Simulator; MICOM, Miami Isopycnic Coordinate Ocean Model; HYCOM, Hybrid Coordinate Ocean Model.

simulations at 1/8° to 1/64° with five to six layers. Horizontal resolution of 5 to 10 km and higher vertical resolution: 40 to 60 levels [*Smith et al.*, 2000; *Masumoto et al.*, 2004; *Maltrud and McClean*, 2005, hereafter referred to as MM05], 16 layers [*Chassignet and Garraffo*, 2001], and 32 hybrid layers [*Chassignet et al.*, 2006] were used in subsequent basin and global integrations. These simulations are of sufficiently high resolution to allow for the analysis of eddy processes; however, at what horizontal and vertical resolution these models can be considered truly "eddy-resolving" is still to be understood. *Siegel et al.* [2001] suggest that resolutions as high as 1 km are needed for numerical convergence. *Smith et al.* [2000] argued that as the first baroclinic Rossby radius was resolved up to about 50°N in a 0.1° North Atlantic (NA) ocean simulation using the Parallel Ocean Program (POP) and that typical length scales for mesoscale eddies were somewhat larger than the first baroclinic Rossby radius, then eddies would be reasonably well resolved in most of the model domain.

Fine-resolution ocean grids such as those referred to above, in addition to largely resolving eddying processes, produce more realistic depictions of narrow western boundary currents whose dynamics influence the gyre-scale circulation and jets like those in the Antarctic Circumpolar Current (ACC). Their proper representation is essential to the realistic simulation of key climatic quantities such as mass, heat, and salt transports. They also allow for a more realistic representation of ocean bottom bathymetry and coastal geometry, which affects the ocean dynamics and communication between ocean basins and/or marginal seas. In coupled systems, more accurately positioned fronts are important in air–sea interaction processes, as large sea surface temperature errors can result from biases in flow paths.

Centennial global coupled climate simulations are routinely conducted using ocean models whose horizontal resolution is nominally 1°, for example, the Community Climate System Model3 (CCSM3) [*Collins et al.*, 2006]. In these models, ocean eddies are not resolved and are represented using subgrid-scale parameterizations such as *Gent and McWilliams* [1990] and *Ferrari et al.* [2008]. These coupled models have provided depictions of past and present climatic states and are the main tools for projecting future planetary changes. To realistically represent eddies in climate models, it is critical to understand their contribution to the total variability, their transport properties and dynamics, and their impact on the representation of the model mixing on short spatial scales. Measurement of the eddy contribution at basin and global scales, however, is an extremely challenging observational problem. Fine-resolution ocean models, as a type of large eddy simulation that resolve most of the eddy scales, can be used as "truth" to verify the accuracy of eddy parameterizations.

The main objective of this paper was to review fidelity studies of fine-resolution basin and global eddying ocean general circulation models over the past decade. We will concentrate on ocean models whose horizontal resolutions are 1/10th degree and higher. We review a broad base of ocean metric studies; however, the experience of these authors is with POP. Other papers in this monograph review layered models, and some further relevant material may be found within these chapters. The focus here will be on the representation of ocean processes important to climate change and variability as well as the space–timescales resolved by the models in the upper ocean. The duration of the latter, limited to several decades by the capability of present generation computational platforms, is sufficient to allow the flow to mostly adjust dynamically to its initial state ("spin-up" the circulation) but not to bring the model into thermodynamic equilibrium. Below the thermocline in regions away from strong currents or deepwater formation sites, water mass properties do not evolve far from this initial state. Nevertheless, such models can be used for the study of the wind- and buoyancy-driven circulation on timescales up to a decade [*Böning and Semtner*, 2001]. As well, the data upon which to base metrics were primarily sampled in the upper ocean.

We will reflect on the differences between the representation of parameterized eddy heat fluxes in the ocean component of the current generation coupled climate models and the resolved fluxes in the fine-resolution models. It is not possible to do side-by-side comparisons for several reasons. The fine-resolution stand-alone ocean models are typically forced with reanalysis atmospheric surface fluxes from prediction systems. The coupled system is subject to present day greenhouse gas forcing and does not represent a hindcast. It is therefore only possible to discuss statistical representations of the fluxes in such models.

Finally, we will discuss the need to advance the state of metrics to quantitatively assess the fidelity of the relevant ocean processes at both eddy-resolving and non-eddy-resolving scales.

2. OCEAN DATA

The assessment of the fidelity of ocean models has become more quantitative over the past several decades. This, in part, is due to the increasing availability of in situ and satellite data on unprecedented scales that can be used for consistent statistical comparisons with ocean model output. The World Ocean Circulation Experiment (WOCE) provided a collaborative framework for countries all over the world to participate in ocean sampling. Measurement programs already underway were incorporated into WOCE, new measurement techniques

were developed, and complementary satellite missions were launched. Both remotely sensed and in situ measurements have increased since that period using improved technologies. The data sources discussed here are by no means exhaustive; rather, they represent the means by which model performance has been judged to date.

Three data sets, in particular, have spatial coverage on a near-global basis with high temporal resolution: sea surface height anomalies (SSHA) from altimetry [*Wunsch and Stammer*, 1995], blended sea surface temperature (SST) from satellites and in situ sources, and velocities from surface drifting buoys at 15 m [*Pazan and Niiler*, 2001]. The TOPEX/POSEIDON/JASON-1/Ocean Surface Topography Mission is a 15-year time series of SSHA whose spatial coverage is global equatorward of 66°N and 66°S. Along-track samples are spaced approximately 1 s and 6 km apart; the spacing between tracks is about 315 km at the equator, and the global sampling pattern is repeated every 9.92 days. Lagrangian data from the surface drifting buoys are interpolated to create uniform 6-h time series of positions and velocities. The data spans the period 1979–2007; its spatial coverage has progressively approached global extent (apart from the polar oceans) over this period. For more information about these data, see http://www.aoml.noaa.gov/phod/dac/dacdata .html. Very recently, two new high-resolution SST analysis products were developed using optimal interpolation that have a spatial grid resolution of 0.25° and a temporal resolution of 1 day. One product uses advanced very high-resolution radiometer (AVHRR) infrared satellite SST data. The other uses AVHRR and advanced microwave scanning radiometer on the NASA Earth Observing System satellite SST data. Both products also use in situ data from ships and buoys and include a large-scale adjustment of satellite biases with respect to the in situ data [*Reynolds et al.*, 2007]. Other global data sets include the unprecedented numbers of profiles of temperature and salinity over the top 2,000 m of the ocean being collected as part of the Argo program [*Gould et al.*, 2004] and the GRACE satellite mission which is supplying the Earth's time-varying gravity field on a monthly basis from which ocean bottom pressure at a spatial resolution of approximately 500 km can be obtained [*Wahr et al.*, 1998, 2002].

Other data sets are spatially limited, but have provided regional or point-wise time series over long periods. These include the monthly repeat IX1-expendable-bathythermograph (XBT) transect line between Java and Australia [*Meyers*, 1996], which has provided long-term (1987 to the present) monitoring of the Indonesian Throughflow. Profile spacing is about 100 km, and their depth extent is 750 m. In addition, 30 repeat high-resolution (1/2° in the basin interior) XBT transects were collected during 1991–1999 in the North

Pacific at about 22°N [*Roemmich and Gilson*, 2001]. Tide gauge data collected at coastal and island stations across the globe [see http://uhslc.soest.hawaii.edu/] have provided measures of sea level rise and variability over a range of temporal scales. Time series of vertical profiles of upper-ocean temperature data collected at particular geographical locations, the so-called Ocean Time Series are available on a long-term regular basis. The Hawaii Ocean Time-Series (HOT) and the Bermuda Ocean Time-Series Study (BATS), nominally located at 22°N, 158°W and 32°N, 64°W, respectively, have been collected since 1988 to the present on a roughly monthly basis.

Data from several sources have been combined to produce mean volume transports through key inter-basin passages and straits. Using a combination of data from current meters, pressure gauges, and hydrography, *Whitworth and Peterson* [1985] obtained an annual net flow through the Drake Passage of 134 ± 11.2 Sv. *Larsen* [1992], using cable data from 1969 to 1990, obtained a value of 32 ± 3.2 Sv at 27°N for the Florida Current volume transport. From current meter data collected in the southern Indonesian passages, *Gordon* [2005] reports a value of about 10 Sv entering the southern Indian Ocean from current meter data; a better estimate of this value is the object of the International Nusantara Stratification and Transport program [*Sprintall et al.*, 2004]. Meridional sections of upper ocean zonal currents, potential temperature, and salinity are estimated at ten longitudes from 143°E to 95°W using conductivity-temperature-depth and Acoustic Doppler Current Profiler (ADCP) data from 172 synoptic sections taken in the tropical Pacific between 138°E and 86°W, mostly in the 1990s [*Johnson et al.*, 2002].

3. OCEAN METRICS

In this section, we review metrics that have typically been used by the oceanographic community to assess the fidelity of basin- and global-scale eddying ocean models, defined here as having horizontal resolutions of nominally 1/10° and higher. These metrics provide assessments of simulated upper-ocean sea surface height and velocity fields where satellite and surface drifter observations are plentiful. Typically, the paths of simulated western boundary currents and SSH variability from altimetry have provided zero-order indications of model bias. Turbulent statistics of the upper-ocean circulation in these fine-resolution ocean models has been compared with that depicted by near-global-scale observations with temporal resolution of days. Both Eulerian (repeated sampling in time at a particular location such as along a satellite track or by a moored instrument) and Lagrangian (measurements collected in time while following a water particle) statistics have been used as a means of model verification. Lagrangian statis-

tics have long been recognized as a most sensitive statistical means of testing the validity of eddy resolving models [*Krauss and Böning*, 1987]. We also examine the time-varying vertical structure of mixed-layer depth at monitoring stations and surface ventilation processes where temperature and salinity data from hydrography and profiling floats furnish information on water masses. Finally, eddy heat fluxes are compared to climatological inferences.

4. IDENTIFICATION OF PRIMARY OCEAN MODEL BIASES

SSH variability from altimetry has routinely been used as a basic indicator of fine-resolution model performance on both global and basin scales [*Paiva et al.*, 1999; *Hurlburt and Hogan*, 2000; *Smith et al.*, 2000; *Masumoto et al.*, 2004; MM05; *McClean et al.*, 2006; *Chassignet et al.*, 2006]. At these horizontal resolutions, the magnitude of the SSHA variability agrees more closely with that from altimetry than in lower resolution cases (e.g., see MM05; Figure 12); however, significant biases are also seen. In the fully global 0.1°, 40-level POP (MM05; Figure 12a), the 1/12° Hybrid Coordinate Ocean Model (HYCOM) [*Chassignet et al.*, 2006; Figure 3] and near-global 1/10°, 54-level OFES (Ocean General Circulation Model based on the Modular Ocean Model3 for the Earth Simulator) [*Masumoto et al.*, 2004; Figure 3b], the North Atlantic Current is too zonal and it fails to turn to the northwest to form the Northwest Corner off the Grand Banks. On the other hand, these models can be used to understand observational biases such as aliasing arising from data sampling strategies. *Le Traon et al.* [2001] used the 1/10° NA POP to quantify the high-frequency, short wavenumber SSH variability not resolved in the altimeter-derived variability, highlighting the issue of observational sampling biases in model/data comparisons.

Paiva et al. [1999], *Hurlburt and Hogan* [2000], and *Smith et al.* [2000] examined the separation location and accuracy of the path of the Gulf Stream (GS) between Cape Hatteras and the Grand Banks in emerging fine resolution ocean models of the NA. The GS separated at Cape Hatteras in the 1/12°, 16-layer Miami Isopycnic Coordinate Ocean Model (MICOM) and followed an offshore path that lay within the observed northern and southern limits of the GS path as depicted by infrared satellite-derived SST [*Paiva et al.*, 1999; Figure 1]. The 0.1°, 40-level NA POP also showed a realistic separation; however, the GS offshore pathway was displaced to the south by 1° to 1.5° compared with results from moored observations during the Gulf Stream Dynamics and Synoptic Ocean prediction Experiments [*Smith et al.*, 2000; Figure 6]. *Hurlburt and Hogan* [2000] used the mean GS path from satellite-derived SST and SSH altimetry to assess the im-

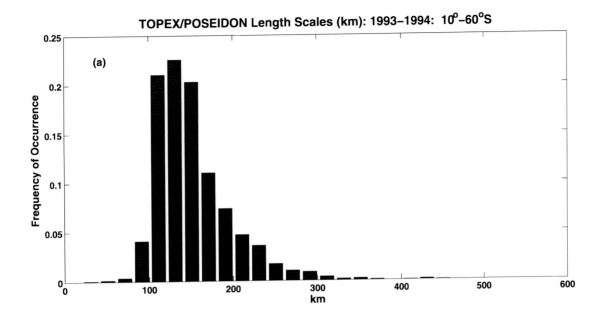

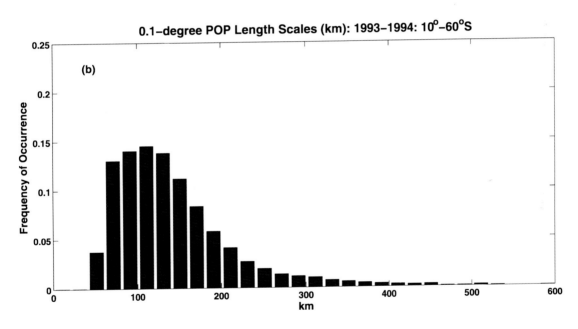

Figure 1. Histograms of length scales (km) from (a) TOPEX/POSEIDON and (b) global 0.1° POP for 1993–1994 in the region 10°–60°S.

provement in the realism of their 1/8°–1/64°, five- to six-layer NLOM simulations with increasing horizontal resolution; they found the best depicted mean GS path in their 1/32° simulation (see their Figure 7). *Tokmakian* [2005] also used observed SST to assess the change in the GS path when forcing the NA 0.1°, 40-level POP with daily fine-resolution (0.25°)

scatterometer winds rather than coarser daily 1.25° European Center for Medium Range Forecasts (ECMWF) winds. All these findings were significant in that the simulated GS path in these fine-resolution simulations was much more realistic than in their lower resolution counterparts where the pathway was typically too northward and/or displayed an anticyclonic

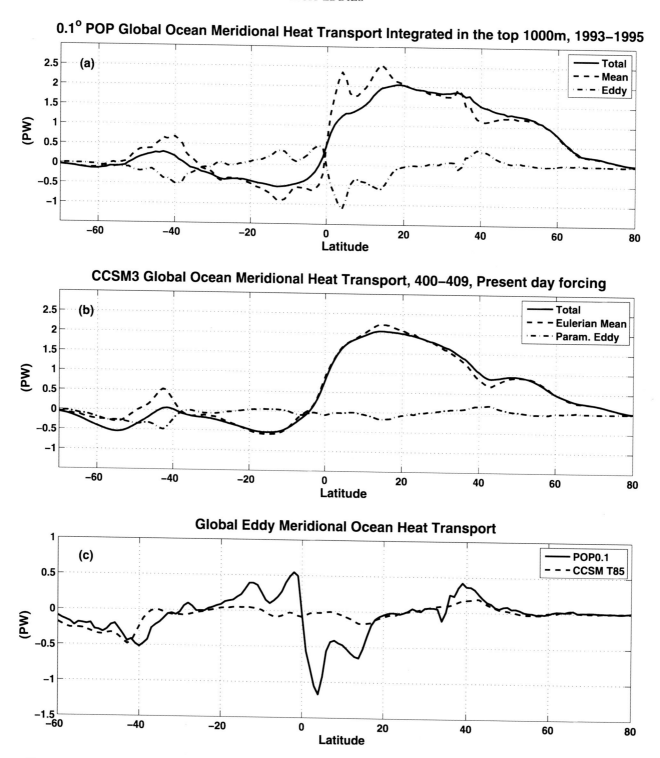

Figure 2. (a) Zonally integrated global eddy heat transport (PW) over the top 1,000 m of 0.1° POP for 1993–1995. May be compared with *Stammer* [1998, Figure 8a]. (b) Zonally integrated global eddy heat transport (PW) from CCSM3 (total depth). The red line is the parameterized eddy heat flux. (c) Zonally averaged global 0.1° POP eddy heat flux and the parameterized eddy heat flux from CCSM3.

circulation on separating from the coast. *Kelly et al.* [2007] advanced beyond path evaluations by developing a hierarchy of metrics to evaluate the Kuroshio Extension (KE) region in a 1/12° Pacific basin HYCOM simulation over a variety of timescales. They compared the simulated KE path and strength and the upper-ocean heat budget with those obtained from altimetric SSHA and observed sea surface temperature. They found that away from the separation region, mean quantities and variations were realistic; however, upstream of 150°E, the KE jet was overly energetic.

Observed values of volume transports through key straits and basins provide another straightforward diagnostic of model performance. MM05 compared observed mean volume transports with those from the 1/10° global POP through key Caribbean and Arctic passages, the Mozambique Channel, the Indonesian Throughflow, the Drake Passage, and from the Agulhas Current. Good agreement with transport estimates of the major current systems was obtained in general, such as 140 and 12 Sv through the Drake Passage and Indonesian Throughflow, respectively. *Masumoto et al.* [2004] reported values of 153 Sv for the Drake Passage and 9 Sv for the Indonesian Throughflow from the 1/10° OFES model. MM05 compared volume transports in surface and intermediate (0–1,250 m), deep (1,250–3,500 m), and bottom waters (below 3,500 m) across latitudinal sections such as 24°N with observational estimates obtained by *Schmitz* [1995] and by *Ganachaud and Wunsch* [2000] from an inverse model. *Du et al.* [2005] compared the annual mean Indonesian Throughflow transport from the monthly repeat IX1-XBT transect line with equivalent 1/10° OFES fields. Readers are referred to these papers for further details of these comparisons.

Du et al. [2005] also validated the simulated mean vertical temperature structure along the IX1 transect and found that the model thermocline was quite a bit tighter than observed in the northern part of the IX1 line, suggesting that the model is underestimating the effects of vertical mixing. MM05 compared the mean zonal POP velocities at 140°W with the observational results from *Johnson et al.* [2002; Figure 2]; they found the model currents to be very close to their correct location with appropriate strengths. They also noted that the strength of the model Equatorial Undercurrent weakened considerably during the height of El Niño events, in agreement with observations.

5. SPACE-TIME DEPICTION OF THE UPPER-OCEAN CIRCULATION

A comparative wavelet analysis of sea surface height variability from NA 0.1° POP, tide gauges, and altimetry was used by *Tokmakian and McClean* [2003] to assess the real-

ism of the model's variability at periods of less than a year. Along the coast, they found the simulated SSH fields to be realistic, as the correlations with tide gauges were on the order of 0.8. Eulerian length scales, representative of the simulated eddy fields, from NA 1/12° MICOM [*Paiva et al.*, 1999] and 1/10° POP [*Smith et al.*, 2000], agreed well with those calculated from altimetry. *Paiva et al.* [1999] used autocorrelation function zero-crossings to define length scales, while *Smith et al.* [2000] used the *Tennekes and Lumley* [1972] microscale length. The microscale length is defined as the square root of the ratio of SSHA to SSHA slope variances:

$$L^2 = \frac{<\hat{h}\hat{h}>}{<\hat{h}'\hat{h}'>}$$

where $\hat{h}$ denotes the SSH residuals, and $\hat{h}'$ represents their slopes in a particular direction.

A comparison of these methodologies can be found in the work of *McClean et al.* [1997]. *Brachet et al.* [2004] focused on the representation of the mesoscale in 0.1° NA POP. They calculated Eulerian sea level height space and timescales and propagation velocities. They found a high level of agreement between the model and the altimeter values for the spatial scales and propagation velocities; however, the POP timescales were significantly longer than those from altimetry in the subtropical regions. They also found good agreement of the high-frequency eddy kinetic energy (EKE) seasonal variations in the Caribbean Sea and at high latitudes despite a phase advance.

Length scales, calculated using the microscale length, are seen in Plate 1c from global 0.1° POP in the Southern Ocean for 1993–1994 at the full horizontal model resolution. The model was forced with synoptic National Center for Environmental Prediction/National Center for Atmospheric Research (NCEP/NCAR) atmospheric surface fluxes of momentum, heat, and freshwater for the period 1979–2003 [MM05]. These length scales can be compared with those from TOPEX/POSEIDON (T/P) for the same period by binning both fields onto 2 × 2 grids (Plates 1a and 1b). The magnitude and spatial distribution of the simulated length scales are generally in good agreement with the altimetric length scales; however, the short scales seen south of 50°S in the model fields are absent in the T/P scales. This is due to T/P not sampling sufficiently often at these high latitudes and thus producing overestimated length scales. Histograms of the distributions of these scales in the region 10°–60°S are shown in Figure 1; the T/P scales occur predominantly in the 100- to 160-km range, while those from the model largely lie between 60 and 160 km.

Garraffo et al. [2001] and *McClean et al.* [2002] compared pseudo-Eulerian and Lagrangian statistics from surface drifting buoys and fine-resolution MICOM and POP basin simulations of the NA, respectively. *Garraffo et al.* [2001] compared NA surface drifter Eulerian statistics and Lagrangian timescales from 1989 to 1998 with those from 1/1° MICOM simulation forced with monthly Comprehensive Ocean-Atmosphere Data climatology. They found that the model underestimated the EKE in the Gulf Stream extension and in the ocean interior. A James test [*Seber*, 1984], which can be used to statistically compare two vector fields with different variances, was used to determine if the differences between the drifter and model mean velocity fields were significant over the domain (see their Figure 4b). Their in situ and modeled Lagrangian timescales were longest in the eastern Atlantic; however, in the interior, the model scales were roughly a factor of two larger than the observed scales. They attributed these results to the lack of a high frequency wind component in the model forcing.

McClean et al. [2002], using a NA 0.1°, 40-level POP simulation with a mixed layer and forced with daily Navy Operational Global Atmospheric Prediction System wind stresses from 1993 to 2000 and climatological heat fluxes, focused on single-particle dispersion of the surface flow. Diffusivities, length, and timescales were calculated from the in situ and simulated velocities. Histograms of drifter and simulated timescales cover a range of 1 to 7 days and 1 to 9 days, respectively; however, most of the scales in both cases are in the 1- to 4-day range. *Bracco et al.* [2003] computed probability density functions of average velocity from numerically simulated surface drifters and 1500-m floats in the NA configuration of MICOM. The distributions and their kurtoses (fourth-order moments) at both depths were non-Gaussian, in agreement with analyses of ocean float trajectories [*Bracco et al.*, 2000].

Treguier et al. [2005] report generally good agreement in the Irminger and Labrador Seas between mean velocities from surface drifters and those from three fine-resolution NA simulations (1/10° POP, 1/12° MICOM, and 1/12° Family of Linked Atlantic Ocean Model Experiments) using multiyear averages when the NAO index was positive. They also compared simulated mean velocities at 700 m with those from floats. Near the East Reykjanes Ridge, the model velocities were seen to more closely follow contours of potential vorticity (*f/h*) than those from the floats, indicating limitations in the model's dynamics.

6. MODEL INTERCOMPARISON METRIC

It is sometimes necessary to represent concisely how a model field compares with a given observational data set, especially if several different models or model variables are being intercompared with the data. The suite of coupled climate simulations conducted for the 4th Intergovernmental Panel on Climate Change is such a case. Single numbers such as the correlation coefficient and the root mean square (RMS) difference provide an overall measure of how well two fields agree, with the former giving information about the structure (or phase) of the two fields and the latter about relative amplitude. In fact, these two measures are related to each other through the following relation:

$$E^2 = 1 + \sigma^2 - 2\sigma R$$

where σ is the ratio of the standard deviation of the model field to the standard deviation of the data, E is the RMS difference between the model and data (normalized by the data standard deviation), and R is the correlation coefficient. By noting that this relationship is simply the triangle law of cosines (with one side of length unity and cos(angle) = R), *Taylor* [2001] introduced a concise way of representing these quantities as a single point on a two-dimensional diagram.

McClean et al. [2006] used a Taylor diagram to consider how well the SSHA variability of the 0.1° global POP simulation compared with AVISO altimetry data and how it related to another global POP simulation that differed only in the horizontal resolution (0.4°) and the magnitude of horizontal friction coefficients (Plate 2). The red/blue circles denote the agreement of the 0.4°/0.1° simulation with the data over the entire globe (from 70°S to 70°N) and three specific oceanic regions representing differing levels of eddy activity: high eddy intensity (Southern Ocean), relatively low intensity (open Pacific Ocean), and mixed (North Atlantic Ocean). In this diagram, the distance from the origin to a given point is equal to σ, the angle between the point and the *x*-axis is related to R, and the distance of the point from unity on the *x*-axis is equal to E. The black semicircle represents the location of perfect agreement (based on these measures) between the simulated field and the data, as such a comparison would yield $\sigma = 1$ (both have the same standard deviation), $R = 1$ (perfect correlation), and $E = 0$ (no RMS difference).

In all four geographical regions, a marked improvement in terms of the agreement with data is seen in the higher resolution simulation. In particular, the standard deviation ratio (σ) for the global domain increases from 0.7 to just over 1, but that is only part of the story, as the correlation has also improved from $R = 0.5$ to $R = 0.65$. Only in the North Atlantic is an improvement in σ not necessarily seen, as the 0.1° run overshoots unity by about the same amount as the 0.4° undershoots. All regions show an increase in correlation (R) and reduction in normalized RMS difference (E). It also appears that much of the apparent improvement in σ for the

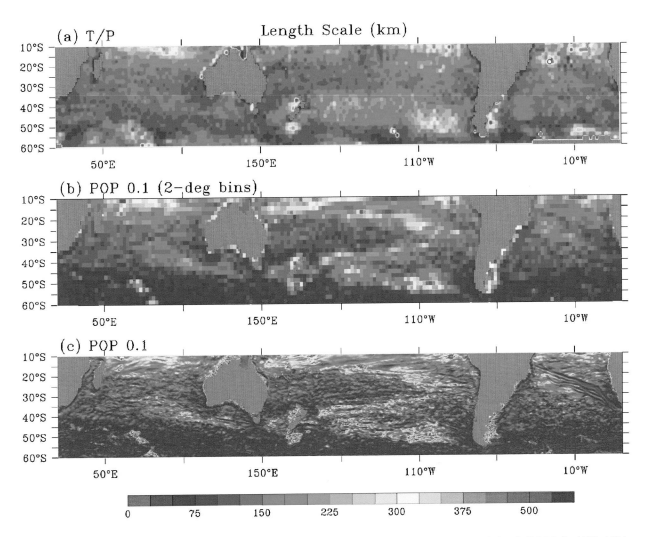

Plate 1. Length scales (km) from (a) TOPEX/POSEIDON, (b) 0.1° POP (2 × 2 bins), and (c) full-resolution 0.1° POP for 1993–1994 in the region 10°–60°S.

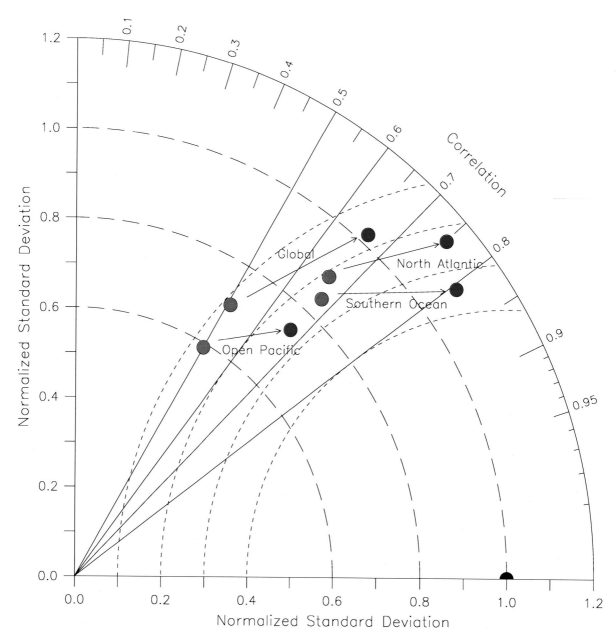

Plate 2. Taylor diagram showing the level of statistical agreement between the 1994–2001 average sea surface height anomaly from the 0.4° (red circles) and 0.1° (blue circles) global POP simulations and AVISO (TOPEX/POSEIDON and ERS 1 and 2) altimetry. The arrows connect the results of both simulations evaluated over the following regions: Global (70°S–70°N), North Atlantic Ocean (20°N–55°N, 100°W–20°W), open Pacific Ocean (30°S–30°N, 150°E–110°W), and Southern Ocean (65°S–40°S). Lines of constant correlation coefficient (*R*) are solid; the long dashed curves denote lines of constant standard deviation ratio (σ); the short dashed curves denote lines of constant RMS difference varying from 0.6 (small radius) to 0.9 by 0.1. The black semicircle represents the location of perfect agreement between the simulation and the comparison data set [after *McClean et al.*, 2006; Figure 5].

global domain is likely due to an overestimate of variability in eddy-active regions, while the more quiescent regions still have too low variability. This diagnostic tool can be used in the future for model inter-comparison studies.

7. VENTILATION PROCESSES

Up to this point, we have discussed the representation of the horizontal circulation of the ocean, but it is also important to investigate the fidelity of the time-varying vertical structure of these fine-resolution simulations. Hydrographic, moored observations, and data from profiling (e.g., PALACE and Argo) floats and ADCPs all can be used for this purpose. As mentioned in the data description section, the Ocean Time Series are monitored long term on a monthly basis, providing a more complete analysis of trends and variability on inter-annual scales at a single point. As the ocean communicates with the atmosphere only at its surface, the fidelity of the model's mixed layer is very important for correctly simulating ventilation processes. *Oschlies* [2002] compared mixed layer depths (MLDs) from three such locations in the subtropical and subpolar NA with spatially co-located time series from two configurations of a NA primitive-equation model: one eddy-permitting (1/3°) and eddy-resolving (1/9°) with both models using identical topographies, monthly climatological forcing fields at the surface, and restoring at the lateral boundaries. He found that the simulated winter MLDs in mid- and high latitudes were systematically shallower by some 50 to 500 m in the higher resolution run and agreed better with observations than those from the lower resolution model. He attributed this improvement to enhanced levels of baroclinic instability leading to a decrease in potential energy and an associated increase in stratification.

McClean et al. [2006] compared simulated co-located time series of MLD from the 0.1° global POP with those from BATS and HOT. The comparison at BATS was extremely good except for winter; during this season, the model was only able to reproduce the maximum depth of the mixed layer. The main reason for the wintertime discrepancy is that the model forcing and output are averaged over a day, while the data are essentially instantaneous samples taken over a day or two for each cluster of observations. This allows measurement of the rapidly evolving mixed layer in different stages of formation, while the model has averaged out such high-frequency motions. At HOT, the density of the model water column was too stably stratified to allow realistically deep penetration of the mixed layer, which was thought to be due to deficiencies in the surface forcing, possibly the use of monthly solar radiation and precipitation climatologies which are part of the NCEP/NCAR reanalysis fluxes used to force this model.

Rainville et al. [2007] analyzed the global 0.1° POP simulation to investigate the formation and destruction of subtropical mode water (STWM) in the Kuroshio Extension region of the North Pacific. The regional mean state as depicted by merged altimetric SSHA measurements and Rio05 absolute dynamical topography [*Rio and Hernandez*, 2004] displayed a smoother jet and a weaker southern recirculation than those simulated by the model, in part a result of the large spatial scale (about 200 km at these latitudes) of the altimetry objective mapping. The observed altimetric variance in the region was in general agreement with that from the model as were the properties of the Kuroshio Current compared to a hydrographic survey of the region. The STMW distribution was found to be highly variable in both space and time, a characteristic often unexplored because of sparse observations or the use of coarse-resolution simulations. Furthermore, its distribution was highly dependent on eddies, as was where it was renewed during the previous winter. Overall, they concluded that eddies were essential to the dynamics of the STWM. *Aoki et al.* [2007] examined the properties of the deep winter mixer layer in the Southern Ocean where Subantarctic Mode Water forms using 1/10° OFES output and Argo float data. General deepening of MLD occurred eastward from 50° to 180°E and from 180° to 80°E, with distinct maxima near bathymetric features. The overall trend and significant peaks of MLD derived from the Argo data were very similar to those from OFES. However, the peaks south of New Zealand, at 130°W–120°W and at 100°W–80°W, were underestimated. This was attributed to sampling irregularity and interannual variability that was not captured in the model due to the use of climatological forcing.

8. EDDY HEAT FLUXES

It is critical to realistically resolve or parameterize eddies in climate models, as they interact with the mean flow and transport heat, fresh water, and momentum. The measurement of the eddy heat contribution at basin and global scales is an extremely challenging observational problem. *Stammer* [1997] used T/P altimeter data and *Levitus et al.* [1994] and *Levitus and Boyer* [1994] climatology together with a mixing length hypothesis to obtain the zonally integrated global eddy heat flux over the top 1,000 m of the water column; these fluxes accounted for up to 20% of the mean meridional fluxes as calculated by *MacDonald and Wunsch* [1996]. Figure 2a shows the same quantity from the 0.1° global POP for the same period (1993–1995). Peak values are found just to the north and south of the equator (−1.1 and 0.5 PW, respectively) and in the ACC (−0.5 PW). The locations of peak values are generally co-located with those of *Stammer* [1998; Figure 8a];

however, the model significantly overestimates the values derived from altimetry and *Levitus et al.* [1994] *Levitus and Boyer* [1994] climatology, which are between ±0.3 PW. Stammer's use of climatology is one likely cause for these different results. *Stammer* [1998] himself commented that he expected his results to represent the lower bound of instantaneous heat flux estimates due to his use of climatology.

Figure 2b shows the total, Eulerian-mean, and eddy heat components from CCSM3. CCSM3 uses the method of *Gent and McWilliams* [1990] to parameterize the eddy heat transport of eddies. It is clear that apart from in the Southern Ocean and north of 45°N, CCSM3 underestimates the eddy heat flux component found in the 0.1° POP (Figure 2c). Close to the equator, the lack of strong signal seen in both the observations and the fine-resolution model is due to the fact that tropical instability waves (TIWs) are not simulated in this nominally one-degree class model. TIWs make a significant contribution to the equatorial mixed-layer budget and hence to the El Niño Southern Oscillation. *Weisberg and Weingartner* [1988] reported that the equatorward heat flux of the TIWs in the upper 50 m of the ocean was comparable to the atmospheric heat flux in the tropics. A dynamical model analysis by *Jochum et al.* [2004] found that their contribution to the mixed-layer heat budget appeared to be smaller than expected; however, it was not unimportant. TIWs also have a dynamical impact on the equatorial circulation as well by removing kinetic energy from the equatorial undercurrent and dissipating it in the equatorial thermocline. Further, TIWs could have an important influence on the phase of the seasonal cycle and the position of the equatorial cold tongue and the intertropical convergence zone [*Jochum et al.*, 2004].

9. DISCUSSION AND CONCLUSIONS

Current practices within the oceanographic community have been reviewed with respect to the use of metrics to assess the realism of the upper-ocean circulation, ventilation processes diagnosed by time-evolving MLD and mode water formation, and eddy heat fluxes in large-scale eddying ocean general circulation model simulations. We have striven to understand the fidelity of these simulations in the context of their potential use in future fine-resolution coupled climate system studies. The availability of data on a near-global basis of sufficiently long duration for statistical analyses posed another constraint leading to the use of both Eulerian and Lagrangian methods for the comparisons of consistent analyses of model output and observations. Overall, these quantitative and semiquantitative metrics provided a gauge of the veracity of the simulated upper-ocean general circulation, indicating that in this regard, these fine resolution models will be useful in future climate simulations.

SSH variability and the location of western boundary current paths from altimetry have been used routinely as zero-order indicators of fine-resolution model performance. *Kelly et al.* [2007] have advanced these metrics by considering the path strength and the upper-ocean heat budget in the KE. Drifters and floats have also been used to provide pseudo-Eulerian measures of the mean and variability of surface and subsurface flows, while statistical comparisons of observed and simulated means have been carried out using James tests. Probability density functions have been used to assess the Gaussian nature of the observed and simulated flows. Length and timescales have been calculated in both Eulerian and Lagrangian frameworks from altimetry and drifters, respectively. Concise measures of multiple model performance have been obtained from Taylor diagrams. The time evolution of the MLD at monitoring stations has been compared with simulated time series.

Impacting these results are the issues of the quality and frequency of the atmospheric surface fluxes used to force ocean models. Daily winds were needed to realistically simulate surface timescales in the North Atlantic. Deficits in the representation of MLDs were attributed to both the frequency and quality of the forcing. Studies reviewed here have shown that simulations with horizontal resolutions of 0.1° and higher more realistically reproduce the mean and variability of the upper-ocean circulation along with the evolution of the MLD. *McClean et al.* [2002] demonstrated that these resolutions more accurately depict Lagrangian intrinsic scales. A comparison of Eulerian length scales from the near-global 0.28° POP simulation [*McClean et al.*, 1997; Plate 6] for the same time period as those in Figure 1 clearly shows that the higher resolution model has reproduced these scales more accurately.

The measurement of the contribution of eddies to global- and basin-scale meridional heat transport is an extremely challenging observational problem. Peak values of the zonally integrated global eddy heat flux from global 0.1° POP were generally co-located but significantly higher than those derived from observations [*Stammer*, 1998]. The discrepancy may be due to the use of climatology in Stammer's calculation or to model biases. Regardless, these fine-resolution eddying ocean models can be used as "truth" when assessing the fidelity of low-resolution ocean components of climate models. CCSM3 underestimates the eddy heat flux component found in the 0.1° global POP except in the Southern Ocean and north of 45°N; particularly, it lacks the strong signal associated with TIWs close to the equator, which are considered important to the equatorial mixed layer heat budget.

The explicit resolution of eddies in ocean models may very well be necessary to improve our projections of climate change. Very recently, *Hallberg and Gnanadesikan* [2006] examined the role played by Southern Ocean winds and eddies in determining the density structure of the global ocean and the magnitude of the Southern Ocean overturning circulation using a suite of Southern hemispheric ocean simulations with realistic geometry at horizontal resolutions ranging from coarse (2°) to eddy-permitting (1/6°). They question the ability of coarse-resolution ocean models to reproduce changes due to eddy activity, although it maybe possible to tune models to make certain aspects of the ocean circulation realistic. They find that the low-resolution models under-represent the eddy response to changes in wind forcing, and they do not think that it is clear that these eddy effects can be effectively parameterized. They recommend that accurate high-resolution ocean models be used to examine the response of the climate system to ensure the validity of climate projections.

We do see the need to advance the state of metrics to quantitatively assess the fidelity of the relevant ocean processes at both eddy-resolving and non-eddy-resolving scales Limited data are always a constraining factor; however, we need to move beyond zero-order comparisons and find novel ways to combine data sets to carry out budget and dynamical balance calculations. The Kuroshio Extension System Study and the CLIvar MOde Water Dynamic Experiment are providing mesoscale measurements in the KE and GS regions, respectively, that can be used as a test bed for the development of more sophisticated metrics.

Acknowledgments. This work was sponsored by the National Science Foundation OCE-0549225 (J.M.) and OCE-0220161 (S.J.), the Office of Naval Research, the Office of Science (BER), US Department of Energy Grant No. DE-FG02-05ER64119, and a Lawrence Livermore National Laboratory/University of California, San Diego intra-university contract. Computer time was provided through the Department of Defense High Performance Computing Modernization Office at the Maui High Performance Computing Center and the Naval Oceanographic Office (NAVO) in Mississippi as part of a Grand Challenge Award. V. Zlotnicki (JPL) calculated the T/P length scales, and the CCSM3 output was made available by F. Bryan (NCAR). The altimeter products used in the Taylor diagram were produced by the CLS Space Oceanography Division as part of the Environment and Climate EU ENACT project (EVK2-CT2001-00117) and with support from CNES. Preparation of this paper was performed using Naval Postgraduate School computers. Discussion of these issues was advanced during an ONR-sponsored workshop at the International Pacific Research Center at the University of Hawaii in February 2006.

REFERENCES

Aoki, S., M. Hariyama, H. Mitsudera, H. Sasaki, and Y. Sasai (2007), Formation regions of Subantarctic Mode Water detected by OFES and Argo profiling floats, *Geophys. Res. Lett.*, *34*, L10606, doi:10.1029/2007GL029828.

Böning, C. W., and A. J. Semtner (2001), High-resolution modeling of the thermohaline and wind-driven circulation, in *Ocean Circulation and Climate, Observing and Modeling the Global Ocean*, edited by G. Siedler, J. Church, and J. Gould, pp. 59–77, Academic, London.

Brachet, S., P. Y. Le Traon, and C. Le Provost (2004), Mesoscale variability from a high-resolution model and from altimeter data in the North Atlantic Ocean, *J. Geophys. Res.*, *109*, C12025, doi:10.1029/2004JC002360.

Bracco, A., J. LaCasce, and A. Provenzale (2000), Velocity probability density functions for oceanic floats, *J. Phys. Oceanogr.*, *30*, 461–474.

Bracco, A., E. P. Chassignet, Z. D. Garraffo, and A. Provenzale (2003), Lagrangian velocity distributions in a high-resolution numerical simulation of the North Atlantic, *J. Atmos. Oceanic Technol.*, *20*, 1212–1220.

Chassignet, E. P., and Z. D. Garraffo (2001), Viscosity parameterization and the Gulf Stream separation, in *From Stirring to Mixing in a Stratified Ocean. Proceedings 'Aha Huliko'a Hawaiian Winter Workshop, January 15–19, 2001*, edited by P. Muller and D. Henderson, pp. 37–41, Univ. of Hawaii at Manoa, Honolulu.

Chassignet, E.P., et al., (2006), Generalized vertical coordinates for eddy-resolving global and coastal ocean forecasts, *Oceanography, 19*, 20–31.

Collins, W. D., et al., (2006), The Community Climate System Model version 3 (CCSM3), *J. Clim.*, *19*, 2122–2143.

Danabasoglu, G., W. G. Large, J. J. Tribbia, P. R. Gent, and B. P. Briegleb (2006), Diurnal coupling in the tropical oceans of CCSM3, *J. Clim.*, *19*, 2347–2365.

Du, Y., T. Qu, G. Meyers, Y. Masumoto, and H. Sasaki (2005), Seasonal heat budget in the mixed layer of the southeastern tropical Indian Ocean in a high-resolution ocean general circulation model, *J. Geophys. Res.*, *110*, C04012, doi: 10.1029/2004JC002845.

Ferrari, R., J. C. McWilliams, V. M. Canuto, and M. Dubovikov (2008), Parameterization of eddy fluxes near oceanic boundaries, *J. Clim.*, in press.

Ganachaud, A., and C. Wunsch (2000), Improved estimates of global ocean circulation, heat transport and mixing from hydrographic data, *Nature, 408*, 453–457.

Garraffo, Z., A. J. Mariano, A. Griffa, C. Veneziani, and E. Chassignet (2001), Lagrangian data in a high resolution numerical simulation of the North Atlantic. I: Comparison with *in-situ* drifter data, *J. Mar. Syst.*, *29*, 157–176.

Gent, P., and J. C. McWilliams (1990), Isopycnal mixing in ocean models, *J. Phys. Oceanogr.*, *20*, 150–155.

Gordon, A. L. (2005) Oceanography of the Indonesian Seas and their throughflow, *Oceanography, 18*(4), 14–27.

Gould, J., D., et al., (2004), Argo profiling floats bring new era of in situ ocean observations, *EoS Trans. AGU*, *85*(19), 185.

Hallberg, R., and A. Gnanadesikan (2006), The role of eddies in determining the structure and response of the wind-driven Southern Hemisphere overturning: Results from the modeling eddies in the Southern Ocean (MESO) Project, *J. Phys. Oceanogr.*, *36*, 2232–2252.

Hurlburt, H. E., and P. J. Hogan (2000), Impact of 1/8° to 1/64° resolution on Gulf Stream model-data comparisons in basin-scale subtropical Atlantic Ocean models, *Dyn. Atmos. Ocean.*, *32*, 283–329.

Jochum, M., P. Malanotte-Rizzoli, and A. Busalacchi (2004), Tropical instability waves in the Atlantic Ocean, *Ocean Modell.*, *7*, 145–163.

Johnson, G. C., B. M. Sloyan, W. S. Kessler, and K. E. McTaggart (2002), Direct measurements of upper ocean currents and water properties across the tropical Pacific during the 1990s, *Prog. Oceanogr.*, *52*, 31–61.

Kelly, K. A., L. Thompson, W. Cheng, and E. J. Metzger (2007), Evaluation of HYCOM in the Kuroshio Extension region using new metrics, *J. Geophys. Res.*, *112*, C01004, doi:10.1029/2006JC003614.

Krauss, W., and C. W. Böning (1987), Lagrangian properties of eddy fields in the northern North Atlantic as deduced from satellite-tracked buoys, *J. Mar. Res.*, *45*, 259–291.

Large, W. G., J. C. McWilliams, and S. C. Doney (1994), Oceanic vertical mixing: A review and a model with a nonlocal boundary layer parameterization, *Rev. Geophys.*, *32*, 363–403.

Larsen, J. C. (1992), Transport and heat flux of the Florida Current at 27°N derived from cross-stream voltages and profiling data: Theory and observations, *Philos. Trans. R. Soc. London, 338 Ser. A.,* 169–236.

Le Traon, P. Y., G. Dibarboure, and N. Ducet (2001), Use of a high-resolution model to analyze the mapping capabilities of multiple-altimeter missions, *J. Atmos. Oceanic Technol.*, *18*, 1277–1287.

Levitus, S., and T. P. Boyer (1994), *World Ocean Atlas 1994*, vol. 4, *Temperature, NOAA Atlas NESDIS 4*, NOAA, Silver Spring, Md.

Levitus, S., R. Burgett, and T. P. Boyer (1994), *World Ocean Atlas 1994*, vol. 3, *Salinity, NOAA Atlas NESDIS 3*, NOAA, Silver Spring, Md.

MacDonald, A. M., and C. Wunsch (1996), An estimate of global ocean circulation and heat fluxes, *Nature*, *382*, 436–439.

Maltrud, M. E., and J. L. McClean (2005), An eddy resolving global 1/10° ocean simulation, *Ocean Modell.*, *8* (1–2), 31–54.

Masumoto, Y., et al., (2004), A fifty-year eddy-resolving simulation of the world ocean—Preliminary outcomes of OFES (OGCM for the Earth Simulator), *J. Earth Simulator*, *1*, 35–56.

McClean, J. L., A. J. Semtner, and V. Zlotnicki (1997), Comparisons of mesoscale variability in the Semtner–Chervin 1/4° model, the Los Alamos 1/6° model, and TOPEX/POSEIDON data, *J. Geophys. Res.*, *102*(C11), 25,203–25,226.

McClean, J. L., P.-M. Poulain, J. W. Pelton, and M. E. Maltrud (2002), Eulerian and Lagrangian statistics from surface drifters and a high-resolution POP simulation in the North Atlantic, *J. Phys. Oceanogr.*, *32*, 2472–2491.

McClean, J. L., M. E. Maltrud, and F.O. Bryan (2006), Measures of the fidelity of eddying ocean models, *Oceanography, 19,* 104–117.

Meyers, G. (1996), Variation of Indonesian Throughflow and the El Niño–Southern Oscillation, *J. Geophys. Res.*, *101*, 12,255–12,263.

Oreskes, N., K. Shrader-Frechette, and K. Belitz (1994), Verification, validation, and confirmation of numerical models in the Earth sciences, *Science*, *263*, 641–646, doi:10.1126/science.263.5147.641.

Oschlies, A. (2002), Improved representation of upper-ocean dynamics and mixed layer depths in a model of the North Atlantic

on switching from eddy-permitting to eddy-resolving grid resolution, *J. Phys. Oceanogr.*, *32*, 2277–2298.

Pacanowski and Gnanadesikan (1998), Transient response in a z-level ocean model that resolves topography with partial-cells, *Mon. Weather Rev.*, *126*, 12.

Paiva, A. M., J. T. Hargrove, E. P. Chassignet, and R. Bleck (1999), Turbulent behavior of a fine mesh (1/12°) numerical simulation of the North Atlantic, *J. Mar. Syst.*, *21*, 307–320.

Pazan, S. E., and P. P. Niiler (2001), Recovery of near-surface velocity from undrogued drifters, *J. Atmos. Oceanic Technol.*, *18*, 476–489.

Rainville, L., S. R. Jayne, J. L. McClean, and M. E. Maltrud (2007), Formation of subtropical mode water in a high-resolution ocean simulation of the Kuroshio Extension region, *Ocean Modell., 17,* 338–356.

Reynolds, R.W., T. M. Smith, C. Liu, D.B. Chelton, K.S. Casey, and M.G. Schlax (2007), Daily high-resolution-blended analyses for sea surface temperature, *J. Clim.*, *20*, 5473–5496.

Rio, M.-H., and F. Hernandez (2004), A mean dynamic topography computed over the world ocean from altimetry, in situ measurements, and a geoid model, *J. Geophys. Res.*, *109*, C12032, doi:10.1029/2003JC002226.

Roemmich, D., and J. Gilson (2001), Eddy transport of heat and thermocline waters in the North Pacific: A key to interannual/decadal climate variability?, *J. Phys. Oceanogr.*, *31*, 675–687.

Schmitz, W. J. (1995), On the interbasin-scale thermohaline circulation, *Rev. Geophys.*, *33*, 151–173.

Seber, G. A. F. (1984), *Multivariate Observations*, 686 pp., John Wiley, New York.

Siegel, A., J. B. Weiss, J. Toomre, J. C. McWilliams, P. S. Berloff, and I. Yavneh (2001), Eddies and vortices in ocean basin dynamics, *Geophys. Res. Lett.*, *28*(16), 3183–3186.

Smith, R. D., M. E. Maltrud, F. O. Bryan, and M. W. Hecht (2000), Numerical simulation of the North Atlantic Ocean at 1/10°, *J. Phys. Oceanogr.*, *30*, 1532–1561.

Smith, W. H. F., and D. T. Sandwell (1997), Global seafloor topography from satellite altimetry and ship depth soundings, *Science*, *277*, 1957–1962.

Sprintall, J., S. Wijffels, A. L. Gordon, A. Ffield, R. Molcard, R. D. Susanto, J. Sopaheluwakan, Y. Surachman, and H. M. van Aken (2004) INSTANTL: A new international array to measure the Indonesian Throughflow, *EoS Trans. AGU*, *85*(39), 369.

Stammer, D. (1997), Global characteristics of ocean variability estimated from regional TOPEX/POSEIDON altimeter measurements, *J. Phys. Oceanogr.*, *27*(8), 1743–1769.

Stammer, D. (1998), On eddy characteristics, eddy transports, and mean flow properties, *J. Phys. Oceanogr.*, *28*, 727–739.

Taylor, K. E. (2001), Summarizing multiple aspects of model performance in a single diagram, *J. Geophys. Res.*, *106*(D7), 7183–7192.

Tennekes, H., and J. L. Lumley (1972), *A First Course in Turbulence*, 300 pp., MIT Press, Cambridge, Mass.

Tokmakian, R. (2005), An ocean model's response to scatterometer winds, *Ocean Modell.*, *9*, 89–103.

Tokmakian, R., and J. L. McClean (2003), How realistic is the high frequency signal of a 0.1° resolution ocean model?, *J. Geophys. Res.*, *108*(C4), 3115, doi:10.1029/2002JC001446.

Treguier, A. M., S. Theetten, E. P. Chassignet, T. Penduff, R. Smith, L. Talley, J. O. Beismann, and C. Böning (2005), The North Atlantic subpolar gyre in four high-resolution models, *J. Phys. Oceanogr.*, *35*, 757–774.

Wahr, J., M. Molenaar, and F. Bryan (1998), Time variability of the Earth's gravity field: Hydrological and oceanic effect and their possible detection using GRACE, *J. Geophys. Res.*, *103*, 30,205–30,229.

Wahr, J. M., S. R. Jayne, and F. O. Bryan (2002), A method of inferring changes in deep ocean currents from satellite measurements of time-variable gravity, *J. Geophys. Res.*, *107*(C12), 3218, doi:10.1029/2001JC001274.

Weisberg, R., and T. Weingartner (1988), Instability waves in the equatorial Atlantic Ocean, *J. Phys. Oceanogr.*, *18*, 1641–1657.

Whitworth, T., and R.G. Peterson (1985), Volume transport of the Antarctic Circumpolar Current from bottom pressure measurements, *J. Phys. Oceanogr.*, *15*, 810–816.

Wunsch, C., and D. Stammer (1995), The global frequency-wavenumber spectrum of oceanic variability estimated from TOPEX/POSEIDON altimetric measurements, *J. Geophys. Res.*, *100*, 24,895–24,910.

Julie McClean, Scripps Institution of Oceanography, La Jolla, California, USA.

Steven Jayne, Woods Hole Oceanographic Institution, Woods Hole, Massachusetts, USA.

Mathew Maltrud, Los Alamos National Laboratory, Los Alamos, New Mexico, USA.

Detelina Ivanova, Lawrence Livermore National Laboratory, Livermore, California, USA.

Common Success and Failure in Simulating the Pacific Surface Currents Shared by Four High-Resolution Ocean Models

Tatsuo Suzuki[1], Hideharu Sasaki[2], Norikazu Nakashiki[3], and Hideyuki Nakano[4]

Using statistical methods, the Pacific Ocean surface current field for the four high-resolution models was described with skill index to estimate the realism against the recent Ocean Surface Currents Analysis-Real time (OSCAR) data. The high-resolution modeling improved the realism of simulated mean surface currents. The improvements are mainly seen in regions with strong mean currents rather than regions with strong eddy activities. The common distinctive aspects of the models are shown. The common features with high skill can be interpreted as successful features in the model, whereas common features with low skill can suggest common problems among the models. Understanding these features as a common modeling problem may help advance high-resolution modeling techniques.

1. INTRODUCTION

In the past decade, the realism of ocean general circulation models (OGCMs) has increased significantly through the various numerical improvements [e.g., *Killworth et al.*, 1991; *Leonard et al.*, 1993], subgrid-scale parameterizations [e.g., *Gent et al.*, 1995; *Mellor and Blumberg*, 2004], and boundary condition quality [e.g., *Röske*, 2001]. High-resolution modeling has especially contributed to the representation of the observed ocean current system. These models show various details of ocean structure that had been difficult to represent within coarser resolution models [e.g., *Masumoto et al.*, 2004; *Nakano and Hasumi*, 2005]. Resolving mesoscale eddies in OGCMs is a significant step

[1]Frontier Research Center for Global Change, Japan Agency for Marine-Earth Science and Technology, Yokohama, Japan

[2]Earth Simulator Center, Japan Agency for Marine-Earth Science and Technology, Yokohama, Japan

[3]Central Research Institute of Electric Power Industry, Abiko, Japan

[4]Meteorological Research Institute, Japan Meteorological Agency, Tsukuba, Japan

Ocean Modeling in an Eddying Regime
Geophysical Monograph Series 177
Copyright 2008 by the American Geophysical Union.
10.1029/177GM12

forward for ocean modeling. The recent advances in computing power have enabled long-term integrations of high-resolution OGCMs with horizontal resolutions of about $1/4°–1/12°$ for global ocean or for a wide basin such as the Pacific or the Atlantic. Some studies refer to $1/4°–1/6°$ models as eddy-permitting and model resolutions higher than about $1/10°$ as eddy-resolving [e.g., *Maltrud and McClean*, 2005; *Bernard et al.*, 2006]. However, even a horizontal resolution of $1/10°$ might not be enough to represent ocean current systems realistically. For example, Kuroshio meandering south of Japan, which is affected by the bottom topography and eddy activity [*Tsujino et al.*, 2006], almost always exists and causes unrealistic separation from the coast south of Kushyu in some model studies [e.g., *Masumoto et al.*, 2005; *Maltrud and McClean*, 2005]. Furthermore, the resolution is not fine enough to resolve the first internal Rossby radius especially at high latitude. Therefore, it is necessary to introduce some sub-grid-scale parameterization, which is different from those of coarser resolution model.

Some studies have started to improve the parameterization for high-resolution model. For example, *Smith and Gent* [2004] introduced an anisotropic Gent–McWilliams (GM) parameterization for eddy-induced tracer transport and diffusion in a high-resolution model, and *Bernard et al.* [2006] showed that a combination of an energy/enstrophy-conserving scheme for momentum advection with a partial step representation of the bottom topography improved the major circulation pattern in an eddy-permitting model. As large computer resources are required to run high-resolution models, efforts toward improvement of parameterizations for high-resolution regimes are not sufficient in comparison with the efforts addressing coarse-resolution models. In addition, 1/4°–1/10° models are expected to be used for the next generation climate modeling. Therefore, further improvement and experiments involving high-resolution models are needed (in this study, 1/4°–1/10° models were referred to as high-resolution model).

In recent years, some experiments using high-resolution global ocean models have been conducted in Japan. As each modeling activity has its own goal and the model is so designed, model configurations such as the boundary conditions, subgrid-scale parameterizations, and model resolutions differ among the models, which have been designed for specific purposes (e.g., global warming projections). These models provide many detailed features. Some of these detailed features are difficult to validate by "observed fact." Furthermore, it is not clear whether the detailed features are consistent among the high-resolution models. Indeed, some models have been produced in different detailed features. It is difficult to explain the cause of these differences because of the many differences in the models' configurations. Additional experiments to investigate the differences may require large computational resources. However, intercomparison of models may help clarify the behavior of high-resolution models and the detailed ocean structure. In this study, we focused on commonalities among models rather than on differences because the common features can suggest common problems and advantages of high-resolution ocean modeling that are not dependent on differences in model configurations.

We specifically examined Pacific surface currents. Numerous observational studies have examined ocean circulation in the Pacific [e.g., *Bograd et al.*, 1999; *Niiler et al.*, 2003; *Isoguchi et al.*, 2004]. However, less is known of the detailed basin-wide features of Pacific surface currents from direct velocity measurements (compared with the Atlantic counterpart [e.g., see Figure 11 of *Rio and Hernandez*, 2004] for example) mainly because Pacific observations have been more confined to specific regions and periods. In particular, the barotropic component is unclear because of limited direct observations of velocity. Although the best description of the real ocean circulation can be provided by the direct measurements, it is worthwhile to examine the common features of the detailed velocity fields resulting from high-resolution models and compare these fields with observations. The recent satellite observation and analysis enabled a multi-year, sustained global monitoring of ocean surface currents processed with eddy-permitting resolution. The recent Ocean Surface Currents Analysis-Real time (OSCAR) project has derived global surface currents using satellite altimeter and scatterometer data in real time [*Bonjean and Lagerloef*,

2002; *Rio and Hernandez*, 2004]. In this study, we assess the realism of the Pacific Ocean surface currents field for four high-resolution global ocean models relative to the OSCAR data and extract the common high-skill or low-skill features among the models. In addition, we also describe the detailed features in specific regions comparing the surface currents features presented by previous studies on the direct measurements such as using current meters and drifters.

2. MODELS AND METHODS

We used four high-resolution global ocean models in this study: the Center for Climate System Research (CCSR) Ocean Component Model (COCO) [*K-1 Model Developers*, 2004], the Ocean General Circulation Model (OGCM) for the Earth Simulator (OFES) [*Masumoto et al.*, 2004], the Parallel Ocean Program (POP) [*Maltrud and McClean*, 2005; *Smith and Gent*, 2004], and the Meteorological Research Institute Community Ocean Model (MRI.COM). These models simulate oceanic mesoscale eddies and solve primitive equations based on the z- (Cartesian) coordinate system (some models also use the sigma-coordinate to calculate sea level in the upper levels). However, some model parameterizations and configurations differ because these models have been developed for specific operations. For high-resolution modeling, an anisotropic Gent–McWilliams (GM) parameterization has been applied in POP, and pseudo-enstrophy preserving scheme [*Ishizaki and Motoi*, 1999] has been applied in COCO and MRI.COM. Table 1 lists model details. The periods for the integration and the output available were different among the models. The models were integrated for more than 19 years under climatologic boundary forcing using an initial dataset based on the World Ocean Atlas [e.g., *Boyar and Levitus*, 1994]. In the analysis, we used the output of the last 5 years for all the models. We also tested the sensitivity to data length by using longer period when available (10 years for COCO and OFES, 6 years for MRI.COM), and the results were almost identical. The outputs were transformed to the same geometrical grids ($0.5° \times 0.5°$) at 10-m depth. In addition, we access a medium-resolution version of COCO (COCO_m) as a typical example for non-eddying ocean models. The horizontal resolution is about 1.4°, except that the meridional resolution is 0.5° near the equator.

We calculated skillA indexes, which were defined by *Holloway and Sou* [1996], to estimate the realism of mean surface current vectors in the high-resolution models against OSCAR data. They introduced the error kinetic energy, $eKE = 0.5(\mathbf{m} - \mathbf{d}) \cdot \mathbf{V}^{-1} \cdot (\mathbf{m} - \mathbf{d})$, where $\mathbf{m}$ and $\mathbf{d}$ are time-averaged velocity of model and observation, respectively, and $\mathbf{V}$ is a diagonal matrix of variance of observed velocity normalized such that its trace

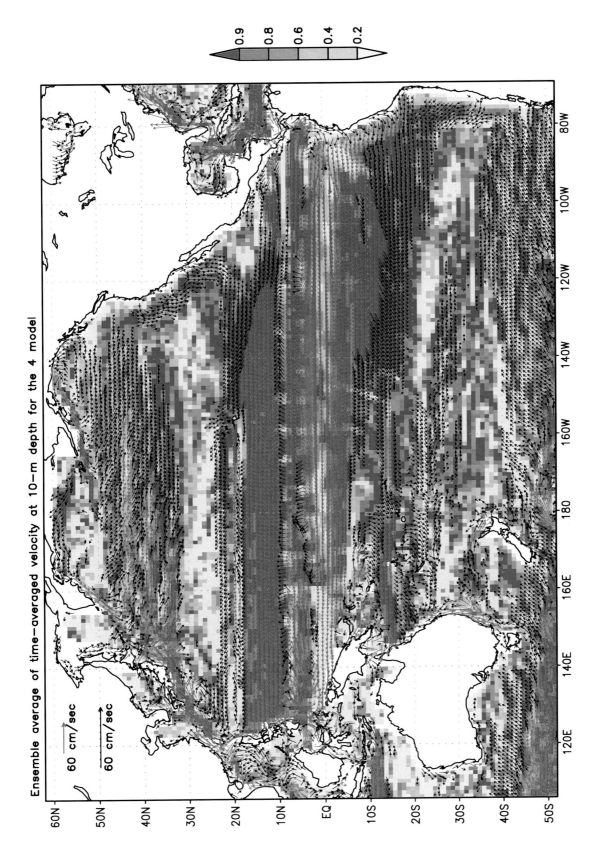

Plate 1. Ensemble-averaged velocity field at 10-m depth of the four models and the skillA index. Velocity vectors above 4 cm/s are shown. Large velocity vectors above 10 cm/s are shown in red and rescaled.

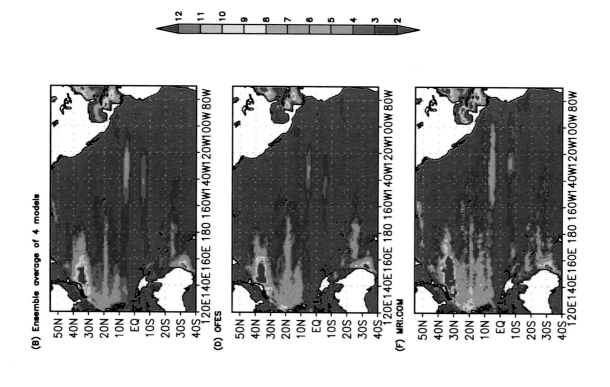

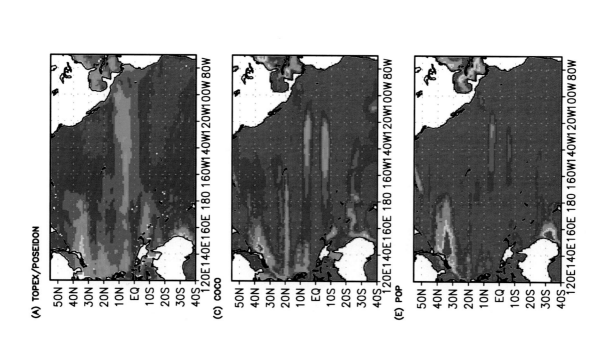

Plate 2. Root-mean-square (rms) of the sea level anomaly from the monthly running mean for each model and the rms of the anomaly from 3-month running mean for TOPEX/POSEIDON (unit: cm).

Table 1. Model Parameterizations and Configurations.

	COCO	OFES	POP	MRI.COM
Resolution	$0.28125° \times 0.1875° \times$ 48 levels	$0.1° \times 0.1° \times$ 54 levels	$0.1° \times 0.1° \times 40$ levels	$0.125° \times 1/12° \times$ 50 levels
Horizontal diffusion	Isopycnal diffusion (1.0E 7 cm²/s) and bi-harmonic diffusion (1.0E 17 cm⁴/s)	Bi-harmonic diffusion (9.0E 17cos³(f) cm⁴/s, f: latitude)	Anisotropic GM (1.5E 6 cm²/s, bolus transport)	Bi-harmonic diffusion (1.0E 16 cm⁴/s)
Horizontal viscosity	Bi-harmonic Smagorinsky	Bi-harmonic (27.0E 17 cos³(f), cm⁴/s, f: latitude)	Anisotropic GM (2.5E 6 cm²/s, parallel to flow direction)	Bi-harmonic Smagorinsky
Mixed layer model	Turbulent closure type [*Noh and Kim*, 1999]	K-Profile parameterization	K-Profile parameterization	Turbulent closure type [*Mellor and Blumberg*, 2004]
Bottom topography	ETOPO (partial cell)	OCCAM 1/30° (partial cell)	ETOPO (full cell)	ETOPO2 + JTOPO30 (18°N–48°N, 120°E–155°E; partial cell)
Boundary condition	*Röske* [2001] (climatology based on ERA15)	NCEP (climatology) and *Rodati and Miyakoda* [1988] bulk formula	NCEP (climatology) and NCEP-bulk formula	NCEP2 (climatology) and *Kara et al.* [2000] bulk formula
Initial condition	WOA94	WOA98	PHC 2.0	WOA98
Integration time	50 model years (last 5 years were used)	50 model years (last 5 years were used)	19 model years (last 5 years were used)	26 model years (last 5 years were used)

becomes unity. They designated skillA = (**d**KE + **m**KE – eKE)/ (**d**KE + **m**KE), where **m**KE = $0.5m \cdot \mathbf{V}^{-1} \cdot m$, **d**KE = $0.5d \cdot \mathbf{V}^{-1} \cdot d$, with properties that a perfect model achieves skillA = 1. We calculated the diagonal matrix **V** and time-averaged velocity **d** from the OSCAR data with 5-day interval for the period 1993–2006 on each horizontal grid point.

3. MEAN FIELD OF PACIFIC OCEAN SURFACE CURRENTS

3.1. General features

Plate 1 shows the ensemble average of the time-averaged surface currents field at 10-m depth for the four high-resolution models with the skillA on the OSCAR data. The surface currents corresponding to major ocean gyres are well represented by the high skillA index. The similar features are also recognized in each model (not shown). The area average of skillA for each high-resolution model is higher than that for COCO_m (Table 2), demonstrating that finer resolution improves the realism of the simulated surface currents. The enhanced meridional

resolution near the equator in COCO_m probably contributes to the high skillA index in the tropical region from 15°S to 15°N. The area-averaged skillA indices of the ensemble for the four high-resolution models are higher than those of each model. The high skill in the ensemble average is the result of some aspects with high skill in each area. If the skill of the aspects in all models is high enough, such distinctive aspects in the ensemble average can be interpreted as a feature of common success among the four models. If the skill in some of the models is significantly low but that in the ensemble average is high enough, on the other hand, it indicates that the models exhibit non-systematic errors affecting the aspect in question. In such a case, we should recognize such low-skill aspects as problems of individual models. On the contrary, if the skill of a distinctive aspect is low in both the ensemble average and in the individual models, then such an aspect can be interpreted as a common failure of all the models.

The root mean square of the surface height variability is shown in Plate 2. To eliminate the seasonal and inter-annual variability, we calculated the anomaly from the monthly mean in each model and from the 3-month running mean in TOPEX/

Table 2. Area Average of SkillA.

	COCO	OFES	POP	MRI	Ave. of Four Models	COCO_m
Area average of SkillA						
100°E–60°W, 60°S–60°N	0.537134	0.556176	0.574815	0.567145	0.617049	0.505486
120°E–100°W, 15°N–55°N	0.483422	0.480064	0.541828	0.499142	0.554385	0.451946
130°E–80°W, 15°S–15°N	0.617010	0.666574	0.661306	0.699103	0.724733	0.610933
140°E–70°W, 50°S–15°S	0.548367	0.548028	0.549480	0.530180	0.603206	0.500479
120°E–180°E, 20°N–50°N	0.350516	0.365729	0.397206	0.400191	0.441565	0.310539
140°E–160°W, 45°S–20°S	0.390987	0.384632	0.396399	0.396769	0.460866	0.312214
Eddy kinetic energy weighted area average of SkillA (EWA-skillA)						
100°E–60°W, 60°S–60°N	0.459135	0.482083	0.502108	0.503830	0.541895	0.424140
120°E–100°W, 15°N–55°N	0.421324	0.495247	0.526157	0.471973	0.551090	0.437455
130°E–80°W, 15°S–15°N	0.510787	0.523573	0.542704	0.563421	0.580047	0.475556
140°E–70°W, 50°S–15°S	0.486659	0.446375	0.447073	0.447281	0.527661	0.359538
120°E–180°E, 20°N–50°N	0.392678	0.464207	0.448716	0.466829	0.545529	0.394283
140°E–160°W, 45°S–20°S	0.383521	0.291901	0.296620	0.329777	0.399355	0.197393
Mean kinetic energy weighted area average of SkillA (MWA-skillA)						
100°E–60°W, 60°S–60°N	0.577989	0.593965	0.599066	0.602378	0.636471	0.483524
120°E–100°W, 15°N–55°N	0.643131	0.630595	0.747360	0.719711	0.786355	0.409157
130°E–80°W, 15°S–15°N	0.682112	0.708537	0.701609	0.711208	0.739615	0.627180
140°E–70°W, 50°S–15°S	0.614760	0.616538	0.614623	0.536100	0.660488	0.537330
120°E–180°E, 20°N–50°N	0.587080	0.555887	0.703622	0.718702	0.778921	0.348000
140°E–160°W, 45°S–20°S	0.480314	0.448460	0.440864	0.414658	0.524047	0.284740

Poseidon data, respectively. Unfortunately, we could not get the surface height variability with uniform quality because the data stored differ. Therefore, we want to examine only the distinctive characteristics. The high-resolution models reproduce the high variability at the Kuroshio Extension, to the east of the Australian coast and to the west and north of Hawaii, as observed. The high-resolution models lead to the significant improvement of the realism of simulated eddy activity and mean surface currents discussed above. Although the better representation of variability is undoubtedly related to the prevalence of mesoscale eddies (i.e., eddy-driven abyssal currents steers upper ocean currents [*Hurlburt et al.*, 1996]), it is not clear if the representation of eddy activity is enough to improve the realism of mean currents in the high-resolution models. We calculated eddy kinetic energy weighted area average and mean kinetic energy weighted area average of skillA index (hereafter EWA-skillA and MWA-skillA, Table 2), respectively. The mean kinetic energy and eddy kinetic energy were estimated from the OSCAR data. The EWA-skillA index is lower than the (no-weighted) area average of skillA (NWA-skillA) index in most regions of the Pacific. On the other hand, the MWA-skillA index was seen to be higher than the NWA-skillA index. These characteristics are shared by the four high-

resolution models despite differences in model configurations. In the case of COCO_m, it is not clear whether these comments apply, as the medium-resolution model is unable to resolve not only the mesoscale eddies but also the narrow strong currents. This consideration of skillA also suggests that the mean surface currents are represented more realistically in regions with strong mean currents, rather than in regions of strong eddy activity, in the high-resolution models.

To assess the robustness of the results, we also tested the area-averaged skillA index of the ensemble average against 100 randomly selected 5-year-long samples of the OSCAR data. Its standard deviation is 0.0135, so we believe that the results are statistically significant.

3.2. The Kuroshio–Oyashio Extension

The mean velocity field of the Kuroshio–Oyashio Extension (KOE) region is very complicated and influenced by narrow western boundary currents forming a boundary between the subtropical and subarctic gyres. Recent studies have argued that air–sea interaction in the KOE affects the decadal variability in the North Pacific [e.g., *Latif and Barnett*, 1994; *Pierce et al.*, 2001; *Schneider et al.*, 2002].

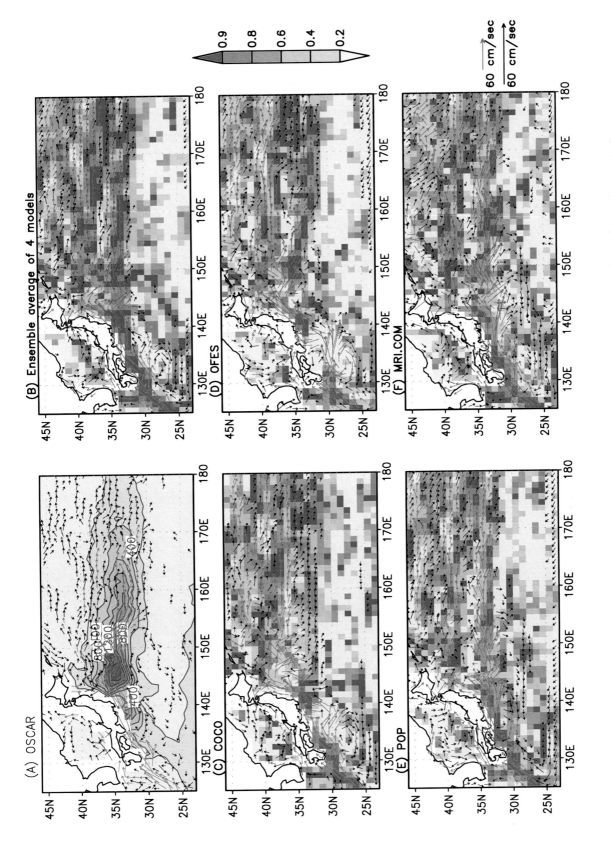

Plate 3. Surface velocity field in the KOE region with skillA index. Velocity vectors above 4 cm/s are shown. Large velocity vectors exceeding 10 cm/s are shown in red and rescaled. Contours in OSCAR indicate eddy kinetic energy (unit: cm²/s²).

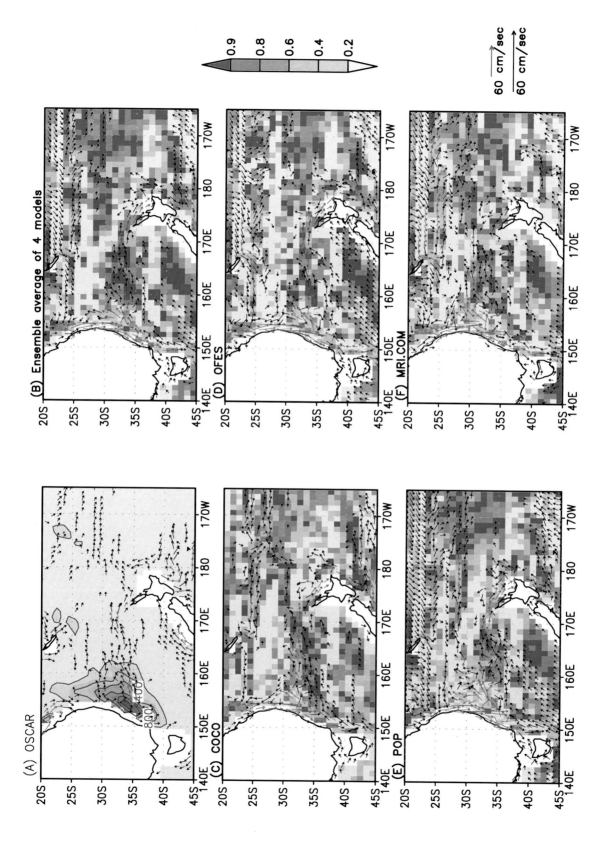

Plate 4. Surface velocity field in the EAC region with skillA index. Velocity vectors above 4 cm/s are shown. Large velocity vectors exceeding 10 cm/s are shown in red and rescaled. Contours in OSCAR indicate eddy kinetic energy (unit: cm²/s²).

Earlier coarse-resolution models, however, cannot resolve the complex features in the KOE region. Thus, understanding the detailed features of high-resolution models is useful for investigating such variability. The Kuroshio and Oyashio appear as narrow currents with detailed features in all the high-resolution models in this study.

Recent observations have indicated that the KOE bifurcates into two eastward jets at the Shatsky Rise around 160°E [e.g., *Niller et al.*, 2003]. Observations by surface drifters place the northern branch at 36°N and the Southern branch at 32°N over the region from 160°E to the International Dateline. The other eastward current has been observed at 38°N east of 155°E. Although some of the features are not clear, these eastward currents may be recognized in the OSCAR data (Plate 3). It is therefore meaningful to estimate the realism of the surface currents against the OSCAR data.

The high-resolution models used in this study present complex features of the surface velocity field in the KOE region. In all the models, three strong eastward jets above 10 cm/s are shown east of 160°E between 30°N and 46°N in the KOE region; the high skillA index suggests that some of these jets correspond with the OSCAR velocity (Plate 3). Although the models have relatively high skill index in the KOE region, three of the four models show the mean latitude of the northern eastward jet around 44°N with low skill index, suggesting that the northern bias of this jet can be interpreted as a common problem among the models. The southern two eastward jets represented in the models appear to be the bifurcation of the Kuroshio Extension (KE). The locations of these branches are consistent, at 34°N and 38°N, for three of the four models with high skill. In a previous numerical study, *Hurlburt et al.* [1996] noted the influence of bottom topography on the surface current path in the KOE region. Although the high-resolution models in this study have enough horizontal resolution to resolve topography such as the Shatsky Rise, some of the paths taken by these two branches of the KE in the high-resolution models are located slightly to the north, as compared with observations including the OSCAR data, but keeping relatively high skill index. Differences in the latitude of the Kuroshio separation among the models may be one reason for this discrepancy.

The pattern of eastward jets is seen to be a common feature of the high-resolution models, for the low skillA northward biases jet of the KOE, as well as for the two more southern jets of the KE, which show much higher skillA. The skillfull common features contribute to the high MWA-skillA indexes in this region, suggesting that fine resolution improves the realism of narrow strong currents (Table 2). Although the realism of the mean currents in the region with strong eddy activity is improved by the enhanced resolution, the increase in skillA is less than that of the regions with higher velocity

currents (Table 2). In the case of COCO, for example, the enhanced resolution did not increase the EWA-skillA index; on the other hand, the MWA-skillA increased from 0.35 to 0.59.

Some anti-cyclonic recirculation regions are evident to the south of the Kuroshio and KE jet in the OSCAR data. Although the location is not consistent between the models, these recirculation regions are seen in each model. However, in the ensemble-averaged velocity field for the four models, the distinctive recirculation was seen from 30°N, 150°E to 25°N, 140°E in agreement with the OSCAR data. Again, the skill of the ensemble average of the models is higher than that of individual high-resolution models.

3.3. The Alaskan Gyre

The Alaskan Gyre (AG) is a partially closed circulation caused by wind and buoyancy forcing [*Van Scoy et al.*, 1991]. The circulation is enclosed by the Subarctic Current (SAC) at the southern flank, the Alaskan Current (AC) at the eastern flank, and the Alaskan Stream (AS) at the northern flank. The AS flows into the Bering Sea through passages or recirculates into the SAC [e.g., *Reed and Stabeno*, 1994; *Bograd et al.*, 1999]. *Bograd et al.* [1999] showed that the AS connects with the SAC around 170°W, completing the circulation of the Alaskan Gyre. The SAC flows eastward along 50°N with a peak in mean speed near 150°W and bifurcates near 48°N, 130–135°W, into northward flow in the AC and southward flow in the California Current (CC) [*Bograd et al.*, 1999].

The AC, the AS, and the bifurcation of the SAC are represented realistically in the ensemble average of the four high-resolution models (Plate 1). These features are also coherent among the models (not shown). However, the northward AC is not seen clearly in the OSCAR data. Furthermore, although the ensemble average shows recirculation of the AS through two or three passages through the Aleutian Chain, around the Dateline and 168°W, to the SAC (Plate 1), as presented by the previous direct observations [e.g., *Reed and Stabeno*, 1994; *Bograd et al.*, 1999], the location of the passages are not consistent among the models and the OSCAR data.

3.4. The East Australian Current

The East Australian Current (EAC) flows southward along the eastern coast of Australia from 18°S to 35°S and accelerates southward following the bifurcation of the South Equatorial Current (SEC). The main portion of the EAC separates from the coast to the south of 32°S and either recirculates northward and connects into the north-eastward Subtropical Counter Current (STCC) or flows eastward across the Tasman Sea. A portion of this eastward flow reattaches at the northern coast of New Zealand and connects to the East Auckland Current (EAUC) [e.g., *Ridgway and Dunn*, 2003]. The rest of the EAC continues southward along the Australian coast as far as Tasmania and then turns westward into the eastern Indian Ocean, having an important impact on global ocean circulation [*Speich et al.*, 2002].

Although the separation of the EAC from the coast of Australia is shown in the OSCAR data (Plate 4), the southward flow along the eastern coast of Australia is not clear probably because the satellite observation is difficult near the coast. Therefore, we could not estimate skill of the models near the coast of Australia. However, the EAC and recirculation are represented as narrow strong currents along the coast by each of the high-resolution models of this study (Plate 4). These features are also coherent among the models. The northward flow connecting to the STCC is not clear in some of the models and is inconsistent among the models in which this flow appears. Each of the models clearly shows the eastward flow across the Tasman Sea. The ensemble average of the four models shows a broad path of eastward flow around 33°N, 162°E with a relatively higher skill index than that for the EAC. The southward portion of the EAC appears in all the models, turning westward south of the Tasman Sea. The ensemble average of the four models shows eastward flow around 45°S, 147°E, above 12 cm/s at 500-m depth (data not shown). The models also resolve some eastward jets which appear further to the east of the Maritime Continent, penetrating into the interior of the Pacific with high skill. One eastward jet appears at 25°S east of 168°E. Another distinctive jet appears at 30°S east of the Dateline. In some models, these jets clearly connect back to the EAC.

The SEC breaks into a series of zonal jets after flowing through island chains, as suggested by recent numerical studies [e.g., *Webb*, 2000; *Kessler and Gourdeau*, 2007]. *Kessler and Gourdeau* [2007] described three permanent jets: the North Vanuatu Jet at 13°S and the North and South Caledonian Jets at 18°S and 24°S, respectively, in the Coral Sea. These jets have almost similar locations in each model. These features are clearer at the subsurface (data not shown).

4. SUMMARY

We have described mean surface current fields from high-resolution global ocean models with skill index to estimate the realism of a model simulation relative to the OSCAR data. The higher resolution models represent the mean surface currents of the major ocean gyres significantly better than a lower resolution model, as shown through a statistical analysis using the skillA index. This improvement with resolution is also seen, to a lesser extent, in regions characterized by both strong eddy activity and strong currents: Although high-resolution models show realistic levels of eddy activity in the KOE, the statistical estimate does not present as much improvement here, relative to that seen in regions with strong currents but weaker eddy activity. It is also likely that some parameterization of unresolved mixing, or use of finer resolution, would be effective in improving reproduction of the surface currents of the KOE region. Investigating the role of mesoscale eddies in the mean currents would be helpful for high-resolution model development, but is beyond the scope of the present study.

Another important aspect of these results relates to the common distinctive aspects among the models. Knowing which features are common among high-resolution models and which common aspects are realistic relative to observations is helpful for advancing ocean modeling. Although our analysis was limited by the lack of OSCAR data in near-coastal regions, this study has, in the context of skill index, revealed some common features. The common aspects with high-skill can be interpreted as common success, and the common aspects with low-skill suggest the common problem in the high-resolution models. In the case of the three eastward jets in the KOE region in this study, we interpret the eastward jet along 44°N as a common problem among the models; on the other hand, we interpret the two more southern jets as relatively successful features in the high-resolution models.

In this study, we compared only the results of *z*-coordinate models, but the comparison with models using different coordinate system, such as sigma-coordinates, might be worthwhile from the point of view that the topography representation influences surface currents.

Acknowledgments. The authors are grateful to Hiroyasu Hasumi, Takashi T. Sakamoto, Takashi Kagimoto, and Masami Nonaka for their valuable discussions and comments. A part of this work has been supported by the Kyousei Project "Project for Sustainable Coexistence of Human, Nature, and Earth," supported by the Ministry of Education, Culture, Sports, Science and Technology of Japan. Some numerical calculations were carried out on the Earth Simulator under support of JAMSTEC. We used datasets from the OSCAR Project. Graphics were produced using Grid Analysis and Display System software.

REFERENCES

Bernard, B., G. Madec, T. Penduff, J.-M. Molines, A.-M. Treguier, J. L. Sommer, A. Beckmann, A. Biastoch, C. Boning, J. Dengg, C. Derval, E. Durand, S. Gulev, E. Remy, C. Talandier, S. Theetten, M. Maltrud, J. McClean, and B. D. Cuevas (2006), Impact of partial steps and momentum advection schemes in a global ocean circulation model at eddy-permitting resolution, *Ocean Dyn.*, 56, 543–567, doi:10.1007/s10236-006-0082-1.

Bograd, S. J., R. E. Thomson, A. B. Rabinovich, and P. H. Leblond (1999), Near-surface circulation of the northeast Pacific Ocean derived from WOCE-SVP satellite-tracked drifters, *Deep-Sea Res., 46*, 2371–2403.

Bonjean F., and G. S. E. Lagerloef (2002), Diagnostic model and analysis of the surface currents in the tropical Pacific Ocean, *J. Phys. Oceanogr., 32*, 2938–2954.

Boyer, T. P., and S. Levitus (1994), Quality Control and Processing of Historical Oceanographic Temperature, Salinity, and Oxygen Data. NOAA Technical Report NESDIS 81, 64 pp.

Gent, P. R., J. Willebrand, T. J. McDougall, and J. C. McWilliams (1995), Parameterizing eddy-induced tracer transports in ocean circulation models, *J. Phys. Oceanogr., 25*, 463–474.

Holloway G., and T. Sou (1996), Measuring skill of a topographic stress parameterization in a large-scale ocean model, *J. Phys. Oceanogr., 26*, 1088–1092.

Hurlburt, H., A. Wallcraft, W. Schmitz Jr., P. Hogan, and E. Metzger (1996), Dynamics of the Kuroshio/Oyashio current system using eddy-resolving models of the North Pacific Ocean, *J. Geophys. Res., 101*(C1), 941–976.

Ishizaki, H., and T. Motoi (1999), Reevaluation of the Takano–Oonishi schemes for momentum advection on bottom relief in ocean models, *J. Atmos. Oceanic Technol., 16*, 1994–2010.

Isoguchi, O., H. Kawamura, and E. Oka (2006), Quasi-stationary jets transporting surface warm waters across the transition zone between the subtropical and the subarctic gyres in the North Pacific, *J. Geophys. Res., 111*, C10003, doi:10.1029/2005JC003402.

K-1 Model Developers (2004), K-1 coupled model (MIROC) description, in *K-1 Tech. Repo. 1.,* edited by H. Hasumi and S. Emori, 34 pp., Center for Climate System Research, University of Tokyo, Tokyo.

Kara, A., P. Rochford, and H. Hurlburt (2000), Efficient and accurate bulk parameterizations of air–sea fluxes for use in general circulation models, *J. Atmos. Oceanic Technol., 17*, 1421–1438.

Kessler W. S., and L. Gourdeau (2007), The annual cycle of circulation of the southwest subtropical Pacific analyzed in an ocean GCM, *J. Phys. Oceanogr., 37*, 1610–1627.

Killworth, P. D., D. Stainforth, D. J. Webb, and S. M. Paterson (1991), The development of a free-surface Bryan–Cox–Semtner ocean model, *J. Phys. Oceanogr., 21*, 1333–1348.

Latif, M., and T. P. Barnett (1994), Causes of decadal variability over the North Pacific and North America, *Science, 266*, 634–637.

Leonard, B. P., M. K. MacVean, and A. P. Lock (1993), Positivity-preserving numerical schemes for multidimensional advection, NASA Tech. Memo. 106055, ICOMP-93-05, 62 pp.

Maltrud, M. E., and J. L. McClean (2005), An eddy resolving model 1/10° ocean simulation, *Ocean Model., 8*(1–2), 31–54.

Masumoto, Y., H. Sasaki, T. Kagimoto, N. Komori, A. Ishida, Y. Sasai, T. Miyama, T. Motoi, H. Mitsudera, K. Takahashi, H. Sakuma, and T. Yamagata (2004), A fifty-year eddy-resolving simulation of the world ocean: Preliminary outcomes of OFES (OGCM for the Earth simulator), *J. Earth Simulator, 1*, 35–56.

Mellor, G., and A. Blumberg (2004), Wave breaking and ocean surface layer thermal response, *J. Phys. Oceanogr., 34*(3), 693–698.

Nakano, H., and H. Hasumi (2005), A series of zonal jets embedded in the broad zonal flows in the Pacific obtained in eddy-permitting ocean general circulation models, *J. Phys. Oceanogr., 35*(4), 474–488.

Niiler, P. P., N. A. Maximenko, G. G. Panteleev, T. Yamagata, and D. B. Olson (2003), Near-surface dynamical structure of the Kuroshio Extension, *J. Geophys. Res., 108*(C6) /3193/ doi:10.1029/2002JC001461.

Noh, Y., and H. Jin Kim (1999), Simulations of temperature and turbulence structure of the oceanic boundary layer with the improved near-surface process, *J. Geophys. Res., 104*(C7), 15621–15634.

Pierce, D. W., T. P. Barnett, N. Schneider, R. Saravanan, D. Dommenget, and M. Latif (2001), The role of ocean dynamics in producing decadal climate variability in the North Pacific, *Clim. Dyn., 18*, 51–70.

Reed, R. K., and P. J. Stabeno (1994), Flow along and across the Aleutian Ridge, *J. Mar. Res., 52*, 639–648.

Ridgway, K., and J. Dunn (2003), Mesoscale structure of the East Australian Current System and its relationship with topography, *Prog. Oceanogr, 56*, 189–222.

Rio, M.-H., and F. Hermandez (2004), A mean dynamic topography computed over the world ocean from altimetry, in situ measurements, and a geoid model, *J. Geophys. Res., 109*, C12032, doi:10.1029/2003JC002226.

Rosati, A., and K. Miyakoda (1998), A general circulation model for upper ocean circulation, *J. Phys. Oceanogr., 18*, 1601–1626.

Roske, F. (2001), An atlas of surface fluxes based on the ECMWF re-analysis—A climatological dataset to force global ocean general circulation models. Max-Planck-Institut für Meteorologie Rep. 323, Max-Planck-Institut für Meteorologie, Hamburg, 31 pp.

Schneider, N., A. J. Miller, and D. W. Pierce (2002), Anatomy of North Pacific decadal variability, *J. Clim., 15*, 586–605.

Smith, R., and P. Gent (2004), Reference manual for the parallel ocean program (POP). Ocean component of the community Climate System Model (CCSM2.0 and 3.0), LAUR-02-2484, Los Alamos National Laboratory, Los Alamos, New Mexico.

Speich S., B. Blanke, P. de Vries, S. Drijfhout, K. Döös, A. Ganachaud, and R. Marsh (2002), Tasman leakage: A new route in the global conveyor belt, *Geophys. Res. Lett., 29*(10), L1416, doi:10.1029/2001GL014586.

Tsujino, H., N. Usui, and H. Nakano (2006), Dynamics of Kuroshio path variations in a high-resolution general circulation model, *J. Geophys. Res., 111*, C11001, doi:10.1029/2005JC003118.

Webb, D. J. (2000), Evidence for shallow zonal jets in the South Equatorial Current region of the Southwest Pacific, *J. Phys. Oceanogr., 30*, 706–720.

T. Suzuki, Frontier Research Center for Global Change, Japan Agency for Marine-Earth Science and Technology, Yokohama, Japan.

H. Sasaki, Earth Simulator Center, Japan Agency for Marine-Earth Science and Technology, Yokohama, Japan.

N. Nakashiki, Central Research Institute of Electric Power Industry, Abiko, Japan.

H. Nakano, Meteorological Research Institute, Japan Meteorological Agency, Tsukuba, Japan.

Eddies in Numerical Models of the Southern Ocean

V.O. Ivchenko[1], S. Danilov, and D. Olbers

Alfred Wegener Institute for Polar and Marine Research Bremerhaven, Germany

Mesoscale eddies play a crucial role in the dynamics of the Antarctic Circumpolar Current (ACC) by facilitating horizontal redistribution and vertical penetration of the momentum, contributing to the meridional heat transport and the budgets of energy and momentum of the ACC. This chapter discusses the dynamics of the ACC based on results of numerical models that permit or resolve the mesoscale eddies. Conclusions are drawn by comparing and contrasting results from different models, including both quasi-geostrophic and primitive equation models.

1. INTRODUCTION

The Southern Ocean is a unique part of the World Ocean. Its essential and most important feature is the Antarctic Circumpolar Current (ACC), the strongest current of the World Ocean transporting about 130 to 150 Sv (1 Sv is 10^6 m³/s) of water. It circumnavigates the Antarctic continent, as there are no meridional barriers blocking it completely. The circumpolar character of the ACC imposes constraints on its dynamics, whereby eddies play an outstanding role in meridional and vertical transfers. Plate 1, adopted from *Hallberg and Gnanadesikan* [2006], illustrates how the complex flow pattern containing numerous small-scale eddies in the ACC emerges as resolution of models increases. We will describe three major ways the eddies influence the ACC dynamics.

The first of them is linked to the fact that the zonally averaged geostrophic meridional velocity is zero in the unconstrained belt of latitudes in the Southern Ocean. Thus, the meridional exchange of water, heat, salt, and other substances between the Southern Ocean, Atlantic, Indian, and Pacific Oceans takes place via the eddy exchanges.

[1]Now at National Oceanography Centre, Southampton, UK.

Ocean Modeling in an Eddying Regime
Geophysical Monograph Series 177
10.1029/177GM13

1. Eddies are a major contributor in meridional exchanges in the Southern Ocean.

As a consequence, they provide an important input into the global thermohaline circulation [*Rintoul et al., 2001; Bryden and Cunningham, 2003*].

The ACC is mainly eastward; however, it deviates from the zonal direction on passing the main topographic features (see Plate 1). A positive zonal component of the mean velocity is observed at all vertical levels from the surface to the bottom. A strong westerly wind above the ACC inputs eastward momentum into the current. The eastward momentum penetrates down to the bottom through the action of interfacial form stress. It occurs in vertically stratified sheared flows with undulating density surfaces in a manner similar to the topographic form stress in flows over topography [*Johnson and Bryden, 1989; Marshall et al., 1993; Ivchenko et al., 1996*]. Eddies displace the isopycnal surfaces from their mean positions and are therefore indispensable in setting the momentum balance. This is the second major way the eddies influence the ACC dynamics:

2. Eddies enable the downward penetration of momentum in the ACC via the interfacial form stress. It is linked to the meridional eddy mass flux.

The ACC is a multijet system with variable positions of the jets in both space and time (see Plate 1). Moreover, the number of jets varies along the path of the ACC. Eddies mix the potential vorticity (PV) in various subdomains along the path of the ACC and create and maintain main fronts and their local branches in the Southern Ocean. The energy budget of the ACC is very different from the energy budget of the main

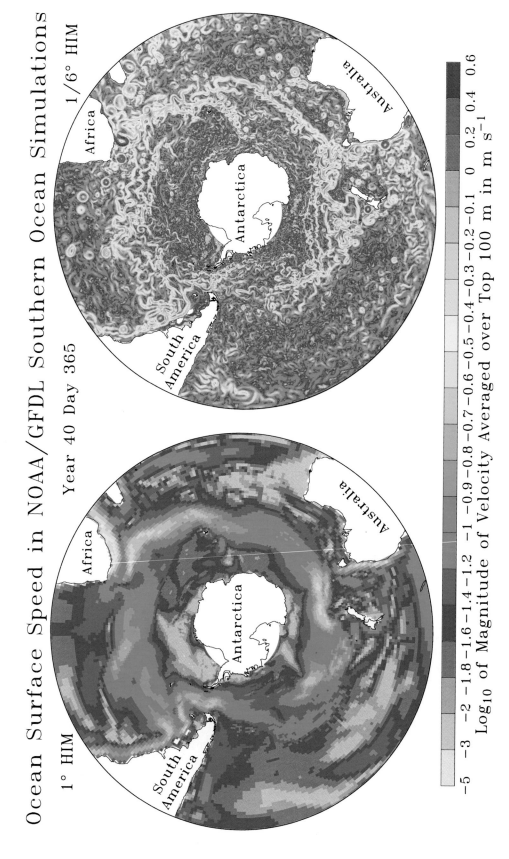

Plate 1. Instantaneous surface speed in the 1° and 1/6° models after 40 years [from *Hallberg and Gnanadesikan, 2006*].

gyres in ocean basins. The eddy contribution is a major constituent in the Southern Ocean. Correspondingly, the third major way the eddies contribute in the ACC dynamics is:

3. *Eddies redistribute momentum, potential vorticity, and energy and also create/maintain the fronts.*

These three issues are described in more detail below. Many results are based on the analysis of the output of eddy-resolving numerical models. The first baroclinic Rossby radius in the Southern Ocean is varying and decreases poleward to values smaller than 10 km because of a weak stratification. This demands high horizontal resolution from a numerical model. For this reason, the first eddy-resolving model of the ACC was a quasi-geostrophic (QG) model set up for an idealized channel with only two layers and reduced zonal length [*McWilliams et al.*, 1978]. This and the other studies based on QG equations illuminated many aspects of the ACC dynamics such as the impact of bottom topography, dependence of the flow on horizontal scales of the channel, and so on [*Treguier and McWilliams*, 1990; *Wolff et al.*, 1991]. In spite of their apparent limitations, they provided a valuable insight into interaction between eddies and mean quasi-zonal flow in "the numerical ACC."

More recent experiments were based on primitive equation models [Semtner and Chervin model, Fine-Resolution Antarctic model (FRAM), Ocean Circulation and Climate Advanced Model (OCCAM), Parallel Ocean Program (POP), Modeling Eddies in the Southern Ocean] working on eddy-permitting or resolving meshes [*Semtner and Chervin*, 1988, 1992; *The FRAM Group*, 1991; *Webb et al.*, 1998; *Maltrud et al.*, 1998; *Maltrud and McClean*, 2005; *Hallberg and Gnanadesikan*, 2006]. All of the models have a good vertical resolution and realistic bottom topography. Forced with realistic wind stresses and surface heat and freshwater fluxes, these models are capable of reproducing the dynamics of the ACC and other processes in the Southern Ocean with a growing degree of realism.

There are many similarities as well as differences in the results of simulations with these models. Comparing and contrasting them allows better understanding of the dynamics of the Southern Ocean and the role played by eddies there.

In this chapter, we are discussing only those aspects of the Southern Ocean dynamics that are directly linked to the eddy activity. Some more indirect effects of eddies hidden in water mass formation and spreading, interaction with the atmosphere, as well as parameterization of the eddy fluxes are not discussed here.

2. QG EDDY-RESOLVING MODELS IN A ZONAL CHANNEL

The progress in early studies of eddy dynamics in zonal flows and eddy interactions with mean flows and topography relies on using QG models, and our current understanding of the ACC dynamics owes much to such studies [*McWilliams et al.*, 1978; *McWilliams and Chow*, 1981; *Wolff and Olbers*, 1989; *Treguier and McWilliams*, 1990; *Wolff et al.*, 1991]. A number of experiments were conducted with flows in rectangular channels of different zonal extent with or without topographic obstacles and driven by different wind stresses. The basic questions answered with QG models concern momentum balance, vertical penetration of momentum, and the convergence of zonal momentum in eastward jets. The total zonal transport as well as the pattern of the mean flow and eddy activity prove to be strongly dependent on the presence and details of the bottom topography.

2.1. Flat Bottom Zonal Channel

In a flat-bottom zonal channel, the momentum imparted by wind to the upper layer is transferred down by interfacial form stress. It can be balanced only by bottom friction if the lateral friction at the side walls is small. This leads to unrealistically high values of the total transport, which is about one order of magnitude higher than the observed ACC transport.

In the absence of forcing and dissipation, the time variation of the depth-integrated zonally averaged zonal momentum is determined by the depth-integrated meridional eddy flux of quasi-geostrophic potential vorticity (QPV) [*Pedlosky*, 1979]. In a two-layer model, this is expressed as

$$\frac{\partial}{\partial t}(H_1\overline{u}_1^x + H_2\overline{u}_2^x) = H_1\overline{v_1'q_1'}^x + H_2\overline{v_2'q_2'}^x, \quad (1)$$

where u_i, v_i are the zonal and meridional components of the horizontal velocity, respectively; subscripts 1 and 2 mark the upper and lower layers whose mean thicknesses H_i are constant. q_i is the QPV,

$$q_i = \nabla^2\psi_i + f + (-1)^i \frac{f_0^2}{g'H_i}(\psi_1 - \psi_2), \quad (2)$$

where ψ_i represents the horizontal geostrophic velocity streamfunctions, $u_i = -\partial\psi_i/\partial y$, $v_i = \partial\psi_i/\partial x$; f is the Coriolis parameter, and f_0 its reference value; $g' = g(\rho_2 - \rho_1)/\rho_0$ is the reduced gravity; ρ_i is the constant density in layer i, and ρ_0 is the reference density; g is the acceleration due to gravity. The overbar with x mark and prime denote the zonal average and eddy component (the deviation from the zonal mean), respectively. It is straightforward to show that the integral over the meridional extent of the channel of the depth-integrated meridional eddy QPV flux is zero for a flat-bottom channel [Bretherton, 1966]:

$$\int_0^L (H_1 \overline{v_1' q_1'}^x + H_2 \overline{v_2' q_2'}^x)\,dy = 0, \qquad (3)$$

where $y = 0$ is the southern boundary, and L is the width of the channel.

Expressions (1) and (3) imply that eddies do not change the total zonal momentum:

$$\frac{d}{dt}\int_0^L (H_1 \overline{u_1}^x + H_2 \overline{u_2}^x)\,dy = 0. \qquad (4)$$

In experiments with a flat bottom, there is no substantial variability in the ACC transport and no standing (stationary in time) eddies because of temporal and zonal invariance. The transient eddies can contain patterns propagating zonally, but are independent (on average) of the zonal coordinate. A strong zonal mean jet forms in each layer with a maximum in the center when the sin-type distribution of zonal wind stress is applied (Figure 1). The upper jet is stronger because the eddy-induced lateral Reynolds stress transfers the eastward momentum to its center, making it narrower and more intense [Held, 1975].

A necessary condition for the baroclinic instability in a two-layer QG model is the difference in signs of the mean meridional QPV gradients [Pedlosky, 1979], given by:

$$\frac{\partial \overline{q_i}^x}{\partial y} = -\frac{\partial^2 \overline{u_i}^x}{\partial y^2} + \beta - (-1)^i \frac{f_0^2}{g' H_i}(\overline{u_1}^x - \overline{u_2}^x), \qquad (5)$$

where $\beta = \partial f / \partial y$. The first term on the right-hand side (rhs) of (5) is the meridional gradient of the relative vorticity that is much smaller than the other two terms [Marshall, 1981]. Thus, the sign of the meridional gradient of the QPV in the upper layer is positive everywhere because $(\overline{u_1}^x - \overline{u_2}^x) > 0$. In the lower layer, $\partial \overline{q_2}^x / \partial y$ can change its sign from negative in the center of the jet to positive at the periphery. The negative value of $\partial \overline{q_2}^x / \partial y$ corresponds to a baroclinic instability and is limited to the central part of the channel where

$$(\overline{u_1}^x - \overline{u_2}^x) > \frac{\beta g' H_2}{f_0^2}. \qquad (6)$$

Note that the criterion of instability for the zonal westward flow is

$$(\overline{u_1}^x - \overline{u_2}^x) < -\frac{\beta g' H_1}{f_0^2}. \qquad (7)$$

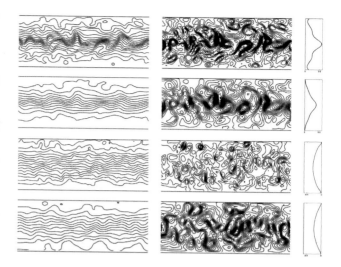

Figure 1. (a) Circulation for eastward wind stress in a flat-bottom QG two-layer model; (b) circulation for westward wind stress [from Olbers, 2005]. In each set the contours of the instantaneous (left upper panel: first layer; left lower panel: second layer) and eddy streamfunction (middle upper panel: first layer; middle lower panel: second layer) are displayed as well as the meridional velocity profiles of the zonal-time mean (right panels). Contour intervals are 2×10^4 m^2 s^{-1} for the instantaneous flow (both layers) and 5×10^3 m^2 s^{-1} for the eddy flow in the first layer and 2.5×10^3 m^2 s^{-1} for the second layer. Velocity is expressed in m s^{-1}.

The upper layer represents the main thermocline with a depth scale of 1,000 m, while the lower one corresponds to the deep ocean with a scale of about 4,000 m, that is, $H_1 < H_2$. The difference in thicknesses implies that the westward flow is more unstable than the eastward flow [Kamenkovich et al., 1986]. Although only the eastward flow simulates the ACC, it is illuminating to compare it with the westward flow, following Olbers [2005], to highlight the role of baroclinic instability in shaping these flows.

The westward flow becomes unstable at smaller amplitudes of the vertical shear, which explains smaller values of available potential energy (APE) of the mean flow and less vigorous eddies. The patterns of the eastward and westward currents differ significantly (Figure 1). The westward current is not a narrow jet, as the eastward flow, but is wide and smooth, with a typical meridional scale close to the meridional scale of the zonal wind stress. In the eastward flow, eddies transport the eastward momentum by Reynolds stress $\overline{u_1' v_1'}^x$ to the jet center both from north and from south (Figure 1a). For the westward flow, they transfer westward momentum to the jet center, but this effect is not large and does not result in a jet concentration [Ivchenko et al., 1997a] (Figure 1b).

The most striking difference is seen in the steady-state APE of the mean flow (671 and 145 m^3 s^{-2} for the eastward and westward flows, respectively). The eddies in the upper layer are more energetic in the eastward flow with the eddy kinetic energy (EKE) 21 m^3 s^{-2} compared to 5 m^3 s^{-2} in the westward flow. The eddy APE in the eastward flow exceeds that of the westward flow almost threefold (28 and 10 m^3 s^{-2}). The quantities reported here are water column or layer-integrated values.

2.2. Zonal Channel With Bottom Topography

A major modification introduced by the bottom topography is that the bottom form stress balances the forcing. This mechanism is inviscid and proves to be very effective in drastically reducing the total zonal transport compared to the flat bottom case. Indeed, the Bretherton theorem (3) in this case is rewritten as [*Ivchenko*, 1987; *Vallis*, 2006]:

$$\int_0^L (H_1 \overline{v_1' q_1'}^x + H_2 \overline{v_2' q_2'}^x)\,dy = f_0 \int_0^L \overline{v_2 b}^x\,dy. \qquad (8)$$

Here, b is the bottom relief measured relative to the unperturbed constant depth of the lower layer H_2. The term under the integral on the rhs of (8) is the topographic form stress, as

$$f_0 \overline{v_2 b}^x = -\overline{p_2 \frac{\partial b}{\partial x}}^x, \qquad (9)$$

where p is the pressure.

Experiments show that even a small zonal variation in b substantially reduces the zonal transport. For example, random depth variability with root mean square (rms) height of about 200 m generates the bottom form stress sufficient to replace the bottom friction in the momentum budget [*Treguier and McWilliams*, 1990]. Isolated features with the same rms height produce even stronger topographic form stress.

In a zonal channel configuration with topography, one separates the eddy field into standing and transient components, with the former representing the time-averaged departure from the zonal mean. Depending on the height and shape of the bottom topography, the standing eddies can play an extraordinary role in the energy budgets and downward penetration of the zonal momentum. The zonal transport in the lower layer can be even negative for some realizations of bottom topography [*Treguier and McWilliams*, 1990; *Wolff et al.*, 1991].

Experiments with the QG models have shed light on the most important processes in the ACC dynamics involving eddies: The vertical penetration and horizontal redistribution of the zonal momentum occur through the interfacial form stress and Reynolds stress created by transient and standing eddies. Topographic form stress mainly balances the input of momentum by wind stress. However, the QG scaling imposes severe limitations that can distort the physics of large-scale dynamics: For example, an absence of outcropping isopycnals does not allow producing correct meridional circulation and meridional tracer and water mass propagation. Such limitations can be lifted using primitive equation models.

3. PRIMITIVE EQUATION MODELS

Numerical models of the Southern Ocean can be either regional, that is, considering only the Southern Ocean, or global. The open boundary of regional models can potentially influence the circulation within the computational domain. An obvious advantage of regional modeling is computational efficiency, as a large part of the World Ocean is excluded.

Among the primitive equation models discussed here, the FRAM [*The FRAM Group*, 1991] is a regional one. The other models are global. They include the OCCAM [*Webb et al.*, 1998] and the POP model [*Maltrud et al.*, 1998].

The horizontal resolution of the FRAM model is 0.5° by 0.25° in zonal and meridional directions, respectively, and it has 32 vertical levels. The vertical grid spacing varies from 20 m at the surface to 230 m close to the bottom. The model covers the region from 78°S to 23°S. Its bottom topography is smoothed at 1°. The model has been run for 16 years and the analysis uses data from the last 6 years.

The OCCAM exists at three different horizontal resolutions of 1°, 1/4°, and 1/12°. The resolution of 1° is too coarse to resolve eddies, 1/4° is "eddy-permitting," and 1/12° is eddy-resolving. The finest version is very demanding with respect to computer resources and was run only over a limited period. The 1/4° model has 36 vertical levels. It was integrated over 12 years, and the analysis was made for the last 4 years.

The results of the POP model presented here were obtained on the Mercator grid with the horizontal size changing between 31.25 km at the equator and 6.8 km at 77°N or S and having 20 vertical layers [*Smith et al.*, 1992; *Dukowicz et al.*, 1993; *Dukowicz and Smith*, 1994; *Maltrud et al.*, 1998]. Such a fine horizontal resolution in high latitudes makes this version a good choice for studies of the Southern Ocean.

Several other numerical studies have appeared recently at horizontal resolution which is superior to that used in the models mentioned above [see *Maltrud and McClean*, 2005; *Hallberg and Gnanadesikan*, 2006]. Yet, our choice here is limited to models and simulations for which thorough analyses of eddy contributions were made.

Discussing the dynamics of the Southern Ocean, we have to specify the domain of analysis. The southern boundary is obvious—it is the Antarctic coast. The northern boundary varies in different studies. Even more care is required in selecting the ACC domain. Both "boundaries" of the ACC are open and vary in time and space. In the Drake Passage, there is a belt of latitudes without continental barriers (we will call it the ACC Belt, or ACCB). The problem is that the ACC is not confined to this belt in many places along its path (see Figure 2). This results in much smaller zonal mean transport of the ACC than the mean transport through the Drake Passage. Also a significant part of kinetic energy of the ACC, both of mean flow and eddies, lies outside the ACCB. In FRAM, 80% of the EKE is generated on the northern flank of the current outside the ACCB [Ivchenko et al., 1996].

Another possibility is to integrate along the time mean path of the ACC transport (we will use the abbreviation ACCP). This approach allows areas such as the northern flank of the ACC to be included in the analysis. The transport in the ACCP coincides with the total transport. A disadvantage of the ACCP approach is that various models produce different positions of streamlines depending on

their topography, surface forcing, and resolution. Hence, the integration along streamlines deals with geographically different locations. Both approaches are valuable, as they give complementary views on the ACC dynamics and should be studied together.

4. MOMENTUM PENETRATION AND THE DEPTH-INTEGRATED BALANCE IN THE ACC

4.1. Zonal Balances

The zonal momentum is imparted to the ACC by strong eastward winds (typical stress values are about 0.1 N/m²). Interfacial form stresses in the ocean are large: Their magnitude is comparable to that of the wind stress [Ferreira et al., 2005]. The sink of the momentum of the depth-averaged flow could be either viscous (by lateral viscosity or bottom friction) or inviscid (by bottom form stress). Estimates show that the balance between the wind stress and any frictional term can be achieved only upon assuming unrealistically high friction. In addition, the eddy meridional flux of the zonal momentum should be too high to provide a balance with the wind stress [Bryden and Heath, 1985]. Munk and Palmen [1951] were the first to formulate the now generally accepted

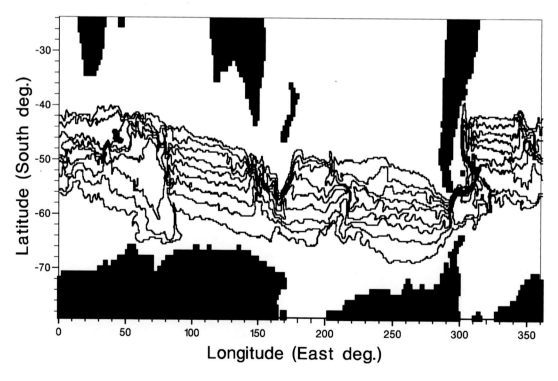

Figure 2. The time-averaged barotropic streamfunction of the ACC from the FRAM. The streamlines range from 10 to 170 Sv with an interval of 10 Sv. The 10-Sv contour is farthest north [from Ivchenko et al., 1996].

view that the topographic form stress $I_{topogr} = \overline{p^b b_x}^{xt}$ balances the wind stress $\overline{\tau_0}^{xt}$:

$$I_{topogr} + \overline{\tau_0}^{xt} = 0, \qquad (10)$$

where the overbar with an xt mark denotes the zonal and time average; p^b and b are the bottom pressure and bottom relief, respectively.

Figure 3 shows the balance taken over the final 2 years of the FRAM run [*Stevens and Ivchenko*, 1997]. Compared to the wind stress or topographic form stress, the poleward momentum-flux divergence and the remaining terms are small. A similar balance is found in the OCCAM and the POP model. However, there are some differences between the balances of these three models [*Grezio et al.*, 2005]. The values of the leading terms in the OCCAM are almost twice as large as in the FRAM and significantly higher then those in the POP model. This is not surprising and is mostly

linked to the wind stress. The eastward component of the wind stress used in the OCCAM is based on the European Center for Medium Range Forcasting climatology, which exceeds that of the Hellerman-Rosenstein climatology used by FRAM by a factor of about 2. The zonal average of the wind stress used in the POP model is lower than that in the OCCAM by about 21% [*Grezio et al.*, 2005]. This difference is caused by the difference in methods of calculating the stress and in data sets.

The vertical penetration of momentum occurs through the action of interfacial form stress. The higher (lower) pressure is found on the upstream (downstream) side of a rise (fall) in the height of a density surface. The expression for the interfacial form stress can be derived by integrating momentum equations along a constant density surface [*Killworth and Nanneh*, 1994]. It is also possible to derive a proxy term for the interfacial form stress for a z-coordinate model [*Johnson and Bryden*, 1989; *Marshall et al.*, 1993;

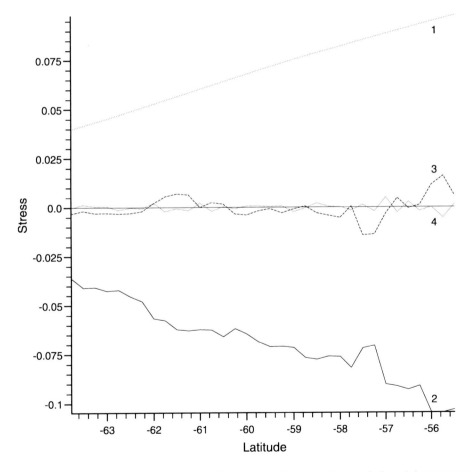

Figure 3. The depth-integrated, time, and zonally averaged momentum balance. Lines 1, 2, 3, and 4 represent the zonal wind stress, bottom form stress, poleward momentum-flux divergence, and remaining (small) terms, respectively [from *Stevens and Ivchenko*, 1997].

Stevens and Ivchenko, 1997]. The interfacial form stress I can be written as

$$I = \overline{p' \frac{\partial \zeta'}{\partial x}}^{xt},$$ (11)

where p and ζ are the pressure and the vertical displacement of a constant density surface, respectively; the prime denotes departure from the zonal and time average. The expression for the interfacial form stress I can be rewritten as

$$I = \overline{p' \frac{\partial \zeta'}{\partial x}}^{xt} = \rho_0 f \frac{\overline{\rho' v'}^{xt}}{\overline{\rho_z}^{xt}},$$ (12)

on assumption of proportionality between the vertical displacement of density surface and the corresponding density variation [*Johnson and Bryden*, 1989]. Note that the contributions from both the transient eddies and "standing eddies"

(deviation from the zonal mean) define the displacements and thus the interfacial form stress.

The Eliassen-Palm theory provides a powerful method for diagnosing the influence of eddies on zonal mean flows [*Eliassen and Palm*, 1961; *Andrews and McIntyre*, 1976; *Edmon et al.*, 1980]. According to this theory, the total eddy influence on the zonal mean flow can be combined in the zonal momentum equation:

$$-f v_r - F = \nabla \cdot \mathbf{E},$$ (13)

where

$$\mathbf{E} = (E^y, E^z) = \left(-\overline{u'v'}^{xt}, f \frac{\overline{v'\rho'}^{xt}}{\overline{\rho_z}^{xt}} \right).$$ (14)

Here, v_r is the meridional component of the "residual velocity" explained below. The term F, including friction and

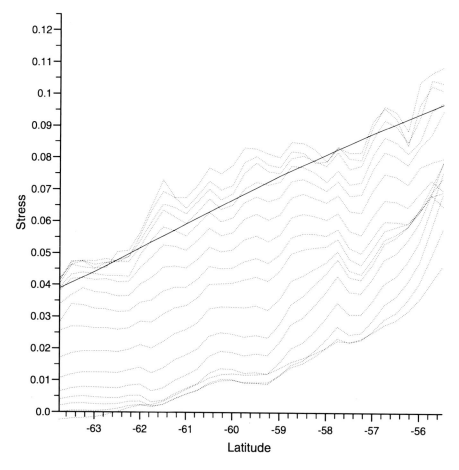

Figure 4. A comparison between the scaled eddy density flux (interfacial form stress E_z) at levels 2–17 of the FRAM (*dotted lines*) and the zonal and time averaged eastward wind stress (*solid line*). The magnitude of the form stress increases with depth. Between levels 13 and 17 (1,000–2,000 m), the form stress approximately equals the wind stress. The units are N/m² [from *Stevens and Ivchenko*, 1997].

advection, is small below the wind-driven surface layer and above the topography. The horizontal component E^y of the Eliassen-Palm vector is the Reynolds stress, while its vertical component E^z is the proxy for the interfacial form stress. A complete set of equations can be written for $\overline{u}^{xt}$ and $\overline{\rho}^{xt}$ and the components of the "residual meridional circulation" v_r, w_r where

$$v_r = \overline{v}^{xt} - \frac{\partial}{\partial z}\left(\frac{\overline{v'\rho'}^{xt}}{\overline{\rho_z}^{xt}}\right), \qquad (15)$$

$$w_r = \overline{w}^{xt} + \frac{\partial}{\partial y}\left(\frac{\overline{v'\rho'}^{xt}}{\overline{\rho_z}^{xt}}\right). \qquad (16)$$

As the Eulerian velocity field ($\overline{v}^{xt}$, $\overline{w}^{xt}$) can be represented in terms of streamfunction ψ_{Eu}, a residual streamfunction can also be introduced, according to (15) and (16). The sec-ond terms in the right-hand side of (15) and (16) represent the eddy-induced components of the residual velocity. The eddies drive the zonal flow by Reynolds stress and interfacial form stress in form of the divergence of the Eliassen-Palm flux. The vertical part dominates in the divergence of the Eliassen-Palm vector [*Killworth and Nanneh*, 1994]. The residual meridional velocity deviates from the Eulerian meridional velocity by this term, which generally is quite substantial, resulting from a strong dependence of the interfacial form stress on depth (Figure 4) [*Ivchenko et al.*, 1996; *Stevens and Ivchenko*, 1997]. The negative divergence of the Eliassen-Palm vector is related to the poleward meridional residual velocity in the Southern Hemisphere and thus supports the ACC. If there are no diabatic sources, the residual circulation is zero [*Ivchenko et al.*, 1996; *Gallego et al.*, 2004; *Olbers*, 2005].

The eddy-induced velocity and components of the Eliassen-Palm vector E^y and E^z can be separated into transient and

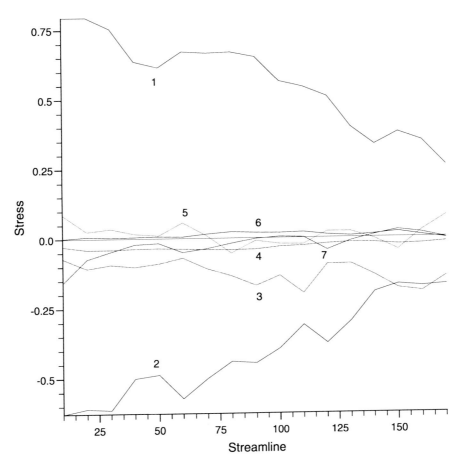

Figure 5. The depth-integrated, time-, and along-streamline-averaged momentum balance in the FRAM. Lines 1, 2, 3, 4, 5, 6, and 7 represent the along-streamline wind stress, bottom form stress, horizontal mixing, bottom friction, horizontal advection of along-stream momentum, vertical advection of along-stream momentum, and remaining (small) terms, respectively. The units are Sv for streamline and dyn/cm² for stress [from *Ivchenko et al.*, 1996].

standing eddy components. They are of comparable magnitude in models of the Southern Ocean. For the ACCB in the FRAM, the magnitude of the standing eddy component exceeds that of the transient eddy component (see Figures 10 and 12 of *Stevens and Ivchenko* [1997]). This can be explained by strong meridional deflection of the ACC path from the zonal direction in many places due to bottom topography. The transient eddy activity in the FRAM is smaller than observed values [*Ivchenko et al.*, 1996, 1997b] because of insufficient horizontal resolution. Refining horizontal resolution leads to increasing the relative importance of the transients in the ACCB.

4.2. Dynamical Balances Along Streamlines

A way of removing the standing component from the analysis is to consider the ACC dynamics along streamlines [*Marshall et al.*, 1993; *Ivchenko et al.*, 1996; *Gille*, 1997]. Some of the results obtained for the ACCB are retained for the ACCP. However, there also are substantial differences between balances in the ACCB and ACCP. The wind stress and topographic form stress are still the largest terms in the quasi-zonal (along streamlines) momentum balance in the ACCP computed from the FRAM output. However, other terms are also significant, in particular the horizontal momentum mixing and bottom friction [*Ivchenko et al.*, 1996]. The lateral friction compared to the topographic form stress is not less than 15%, usually 30–50%, and even larger than the form stress at the southern flank of the ACC (Figure 5).

That this is so is not surprising in hindsight. The main jets of the ACC are located north of the ACCB, so the zonal momentum balance is taken over a large region of sluggish water missing much of the dynamics of the ACC. The along-stream momentum balance follows the main jets, and thus, the horizontal mixing and bottom friction terms are larger, while the magnitude of the wind stress remains the same order of magnitude.

Not surprisingly, the expressions for the vertical penetration of the quasi-zonal momentum are similar to the zonal averaging case. There is very strong eddy-induced quasi-meridional circulation, which results in a strong deviation of an interfacial form stress from the corresponding mean value of the wind stress [*Ivchenko et al.*, 1996]. The EKE substantially varies along streamlines [*Gille*, 1997] with highest values associated with the major topographic obstacles.

5. MERIDIONAL CIRCULATION IN THE SOUTHERN OCEAN

Analysis of meridional circulation addresses the spreading of water masses within the particular ocean domain and the

exchange with the adjacent ocean basins. It is mostly studied in a time-zonal or time-streamline mean and presented by a (meridional) streamfunction describing water exchange in the meridional-vertical plane. Obviously, details of pathways of individual water parcels are lost as well as details of water mass formation. Naturally, the dynamics governing the meridional overturning are identical to the dynamics of the zonal momentum discussed above.

Different types of "zonal" averaging emphasize different properties of the three-dimensional fluid motion. The simplest view on the overturning is obtained from the Eulerian streamfunction $\psi_{Eu}(y,z)$ based on the time and zonal mean on z levels. It is governed by the integrated balance of zonal momentum [*Olbers and Ivchenko*, 2001],

$$- \rho_0 f \psi_{Eu}(y,z) = \overline{\tau_0}^{xt}(y) - \overline{\tau^w}^{xt}(y,z) \\ + I_{\text{topogr}}(y,z) - R(y,z), \quad (17)$$

where $\overline{\tau^w}^{xt}$ is the time-zonal mean of the zonal stress in the water column and R the integrated Reynolds stress divergence (induced by standing and transient eddies). Consequently, with a small $\overline{\tau^w}^{xt}$ below the mixed layer and small Reynolds stresses, ψ_{Eu} generally is dominated by the northward Ekman transport in the surface layers (associated with $\overline{\tau_0}^{xt}$) and a deep geostrophic return flow (associated with I_{topogr}) in the valleys between the highest topography along the particular latitude (see Figure 6a).

This overturning cell is named after Deacon. It does not reflect the more or less adiabatic motion in the ocean interior with transport of active and passive tracers predominantly along isopycnals. This property has more impact on the streamfunction if the zonal average is performed on isopycnals, and indeed the isopycnal streamfunction reveals an essentially different pattern of circulation [see *Döös and Webb*, 1994; *Lee and Coward*, 2003; *Schouten and Matano*, 2006; *Hallberg and Gnanadesikan*, 2006].

The Eulerian view of the overturning can be extended to a closer correspondence with the isopycnal framework using the transformed Eulerian mean (TEM) approach [*Andrews and McIntyre*, 1976]. It acknowledges that the transport (advection) of zonally averaged tracers is performed not only by the time-zonal mean flow with streamfunction ψ_{Eu} but also by eddies. The eddy contribution can easily be inferred by projecting the mean eddy density flux on directions normal and tangent to the mean isopycnals. The total meridional transport of tracer consists then of three terms: Eulerian overturning, standing, and transient eddy-induced components (see section 4.1). The standing eddy-induced term is easily illustrated by an example of gyre circulation carrying warm (cold) water poleward in the western boundary current and returning cold (warm) water on the eastern part of the gyre.

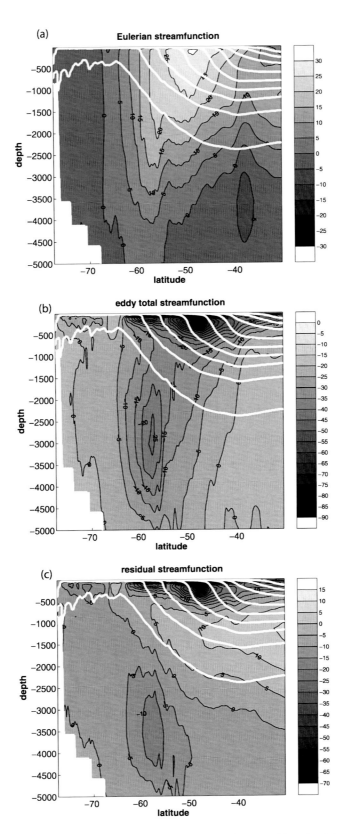

Figure 6. The Eulerian streamfunction ψ_{Eu} (a), eddy-induced streamfunction $\psi'_{ed} + \psi^*_{ed}$ (b), and residual streamfunction ψ_r (c) evaluated from the POP model. The white fat lines represent isopycnals. The units are Sv; CI = 5 Sv [from *Olbers and Ivchenko*, 2001].

For the transient eddies,

$$\overline{\mathbf{u}'\rho'}^{xt} = -K_d' \nabla \overline{\rho}^{xt} - \Psi_{ed}' \nabla_s \overline{\rho}^{xt} , \qquad (18)$$

with $\mathbf{u} = (v, w)$, $\nabla = (\partial/\partial y, \partial/\partial z)$, and $\nabla_s = (-\partial/\partial z, \partial/\partial y)$. Forming the divergence of this expression reveals that K_d' is an eddy-induced diapycnal diffusivity (if positive), and Ψ_{ed}' is an eddy streamfunction advecting the density $\overline{\rho(y,z)}^{xt}$ in addition to Ψ_{Eu}. A corresponding separation can be made for the standing eddies, denoted by the star below, but it is obvious that the construction of a diapycnal diffusivity for this type of eddies is less meaningful. Nevertheless, the advection of $\overline{\rho}^{xt}$ is achieved by the residual streamfunction

$$\psi_r = \psi_{Eu} + \psi_{ed}' + \psi_{ed}^* , \qquad (19)$$

and the diapycnal transport represented by $K_d = K_d' + K_d^*$. All three components of the residual streamfunction ψ_r play an important role in the ACC area. If we discard the eddy components and keep only the Eulerian component, we find a strong Deacon Cell. Taking all three components drastically reduces it.

We should mention that the definition of the TEM streamfunction and diffusivity as given by (18) is not unique. Simpler (but less canonical) forms have been discussed [*Andrews and McIntyre*, 1976; *Held and Schneider*, 1999] and used in the definition (15) and (16) of the residual circulation. These forms partly suffer from incorrect boundary conditions or the a priori assumption of the adiabatic nature of the flow. Indeed, assuming adiabatic conditions, $K_d' = 0$, we find $\psi_{ed}' = \dfrac{\overline{v'\rho'}^{xt}}{\overline{\rho_z}^{xt}}$. More general than (18) is the additional consideration of a rotational flux term which may be used to gauge the eddy streamfunction and diapycnal diffusivity [*Eden et al.*, 2007].

The highest values of transient eddy-induced meridional velocity occur between 65°S and 40°S. This high value is related to the ACC area and strong baroclinic or (and) barotropic instability of the current. The distribution is of a dipole type in the POP model [*Olbers and Ivchenko*, 2001], with the largest values concentrated around two centers. The first cell is located in the upper 200 m, and the second one is centered at about 2,800 m, while its vertical range is between 1,000 and 4,000 m, with maximum transport of 30 Sv. The latter cell mainly appears because of the input from the Southeastern Pacific where the EKE is greatest. These high values are observed in the vicinity of the main core of the ACC because of enhanced instability.

A large component of the standing eddy-induced streamfunction appears in the upper 500 m, which strongly in-creases the total eddy-induced streamfunction (sum of transient and standing; see Figure 6b). This arises from diabatic effects, strong meridional excursions of the main core of the ACC in the area to the north of the ACCB, and also because a rotational part of the eddy fluxes has not been considered. The meridional streamfunction ψ_r in the ACCB and below 500 m is more aligned with the zonal mean isopycnals compared to ψ_{Eu} (Figure 6c). With streamwise averaging, the standing contributions are largely absent.

Hallberg and Gnanadesikan [2006] find two ways the eddies affect the overturning in the Southern Ocean. First, they substantially extend the southward transport of relatively light water across the ACC [see also *Drijfhout*, 2005]. Second, the response to changing wind stresses is systematically smaller than in models with parameterized eddies. The response of the overturning circulation is concentrated on shallower isopycnals (layers).

Recognizing the role played by eddies in setting the meridional circulation, *Marshall and Radko* [2003], *Olbers and Visbeck* [2005], and *Radko and Marshall* [2006] propose simple models to estimate the streamline-averaged residual circulations and the density structure of the ACC. *Gallego et al.* [2004] pursue a similar goal based on a two-layer model. The basic assumption here is the absence of diapycnal fluxes below the mixed layer, in which case, the "zonally" averaged density equation reduces to

$$J(\psi_r, \overline{\rho}^{xt}) = 0. \qquad (20)$$

Here, J denotes the Jacobian operator. This equation implies that the residual streamfunction is a function of only mean density, $\psi_r = F(\overline{\rho}^{xt})$, which is constant on streamlines of the residual velocity [characteristics of the differential equation (20)]. Knowing the density and the functional relation between it and ψ_r just below the mixed layer, the buoyancy structure in the thermocline can be recovered [*Marshall and Radko*, 2003].

Olbers and Visbeck [2005] go a step further. They incorporate a model of mixed layer physics and link the slope of isopycnals on the base of the mixed layer to the surface forcing (wind stress and surface buoyancy flux). They then integrate (20) along characteristics to obtain the mean density structure. *Radko and Marshall* [2006] try to add the gravest mode of zonal variations to this approach.

Common to all these approaches is the parameterization of the eddy-induced velocity in terms of the isopycnal slope (essentially the *Gent-McWilliams* [1990] parameterization), and the assumption that the flow is adiabatic below the mixed layer. *Olbers and Visbeck* [2005] show that given a realistic surface forcing, a realistic density structure can be recovered for the Southern Ocean. These models, although extremely simplified and limited to characteristics exiting from the base of the mixed layer, do emphasize the contribution from eddies to the meridional overturning and propose estimates of eddy-induced and residual circulations.

The total transport of substance in an isopycnal layer consists of the transport by residual velocity (i.e., sum of Eulerian and eddy-induced) and eddy diffusion (see (18), (19), and *Lee et al.* [2007]). The direction of eddy advective transport is governed by the large-scale cross-isopycnal vertical tracer gradient, and the eddy diffusive transport is governed by the large-scale along-isopycnal meridional tracer gradient [see *Lee et al.*, 2007]. They also show that the eddy advective and diffusive heat transports in the Southern Ocean are both poleward in OCCAM (1/12° model). In contrast, eddy advective and diffusive salt transports are equatorward and poleward, respectively. Their results are consistent with those of *Stammer* [1998], where eddy heat and salt transport are inferred from assimilating satellite altimetry and other data. *Henning and Vallis* [2005] show by using a primitive equation high-resolution model with idealized geometry that the residual flow is balanced mainly by the cross-isopycnal eddy flux convergence in the channel region. Eddy buoyancy and heat fluxes can play an important role in the formation of water masses, particularly intermediate waters, by exchanging water between subtropical gyres and the ACC [*Schouten and Matano*, 2006].

6. ENERGY BALANCES AND INSTABILITY

The eddy field simulated in numerical models is a good indicator of model skill, as a growing amount of measurement data is becoming available through the progress in satellite altimetry and new measurements with subsurface floats. The performance of models in this respect is a special issue for the Southern Ocean where eddy contribution is playing a decisive role in meridional transports. The basic questions are:

–What is the mean level of the EKE?
–What is the horizontal and vertical distribution of the EKE?
–What is the main instability mechanism in the Southern Ocean? It is important to have a clear view on the instability processes in models to assess to what extent they are eddy-resolving and whether a further increase in resolution is needed before the Southern Ocean can be modeled realistically. As the Southern Ocean is very inhomogeneous with respect to the EKE, such an assessment can only be made on a regional basis [*Ivchenko et al.*, 1997b].
–What are the main terms in the energy budgets? This question is closely related to the previous one. One can estimate the energy content and energy transfers directly from the model output.

An equation for the EKE can easily be derived from the equation of motion and can be written in the following form:

$$\frac{\partial}{\partial t}\left(\overline{\frac{u_m'^2}{2}}^t\right) = 0 = -\frac{\partial}{\partial x_j}\left(\overline{u_j}^t\,\overline{\frac{u_m'^2}{2}}^t\right)$$
$$-\frac{\partial}{\partial x_j}\overline{u_j'\frac{u_m'^2}{2}}^t - \overline{u_j'u_m'}^t\frac{\partial \overline{u_m}^t}{\partial x_j}$$
$$-\frac{1}{\rho_0}\frac{\partial \overline{u_j'p'}^t}{\partial x_j} - \rho_0^{-1}g\overline{u_3'\rho'}^t + W_H' + W_Z',$$
(21)

where the index $m = 1, 2$ represents the horizontal directions, while the index $j = 1, 2, 3$ represents both horizontal and vertical ($j = 3$) directions; summation over the repeated indices is implied; W_H' and W_Z' represent the viscous terms (horizontal and vertical, respectively) and $\overline{u_m'^2}/2$ is the EKE.

The first two terms on the rhs of (21) are the advection of the EKE by mean and eddy flows, respectively; the third term is the Reynolds stress work. The same term, but with the opposite sign appears in the balance of the kinetic energy of the mean flow, and therefore, it represents the exchange between mean and eddy kinetic energy. The forth term is the work by eddy pressure flux; the fifth term represents an exchange between eddy kinetic and potential energy. Note that the fourth and fifth terms are the parts of the eddy pressure

work $(PW)' = -\overline{\frac{1}{\rho_0}u_m'\frac{\partial p'}{\partial x_m}}^t$:

$$(PW)' = -\frac{1}{\rho_0}\frac{\partial \overline{u_j'p'}^t}{\partial x_j} - \frac{g}{\rho_0}\overline{u_3'\rho'}^t.$$
(22)

The first, second, and fourth terms on the rhs of (21) are written in a divergent form, which means that they redistribute the EKE (first and second) and eddy component of pressure work (fourth) inside the basin. They disappear after integration over a closed basin in a rigid lid approach.

After integration over the volume of a subregion with possible open boundaries, equation (21) becomes:

$$N' + \Pi' + B' + F_H' + F_Z' = 0,$$
(23)

where N' is the nonlinear term, representing the fluxes of the EKE through the open boundaries by the mean flow and eddies, and the Reynolds stress work; $\Pi' = -\rho_0^{-1}\int_{(A)}\overline{u_j'p'}^t dA_j$ is the eddy pressure flux term through boundaries; $B' = -\int_{(V)}\rho_0^{-1}g\overline{u_3'\rho'}^t dV$ is the buoyancy term; $F_H' = \int_{(V)}W_H' dV$ is an integral over volume V of the horizontal friction, and F_z' is an integral over volume V of the vertical friction. The nonlinear term N' can be written as:

$$N' = -\int_{(A)} \left(\overline{u_j}^t \overline{\frac{u_m'^2}{2}}^t + \overline{u_j' \frac{u_m'^2}{2}}^t \right) dA_j$$
$$- \int_{(V)} \overline{u_j' u_m'}^t \frac{\partial \overline{u_m}^t}{\partial x_j} dV, \qquad (24)$$

where A is the boundary of the subregion, and dA_j is the oriented surface element normal to this boundary.

The EKE budget for the ACC area from the FRAM experiment shows a balance between the two biggest terms, the generation of the EKE by buoyancy B', and the horizontal friction F_H' [Ivchenko et al., 1997b]. Observations also give support to the importance of the dissipation of eddies in the circumpolar region [Bryden, 1983]. The other terms, that is, nonlinear transfers, pressure flux through the open boundary, and bottom friction are much smaller. The EKE budget for a wider area of the Southern Ocean from 27.5°S to the Antarctic coast (nearly all FRAM area) is substantially different: The main sources are both the buoyancy term B' and the nonlinear transfer term N'. These two sources are mainly balanced by the lateral friction. The importance of lateral friction arises because several gyres and western boundary currents are included in the domain. High velocities and horizontal velocity shears are observed there. The buoyancy term B' is most important in the ACC. If we consider the FRAM regions where B' is a source of EKE, 92% of generation is confined to the ACC [Ivchenko et al., 1997b]. When the ACC is split into the ACCB and the North ACC domain (the area to the north of the ACCB), the ACCB contributes 24% and the North ACC contributes 76% of the total. This shows that an analysis restricted to the ACCB region misses the largest part of the eddy activity in the ACC.

The ratio $\alpha = N'/B'$ is much higher in the gyre-type domain than in the channel-type domain [Ivchenko et al., 1997b]. For the whole FRAM domain $\alpha = 0.6$, which is to be compared to 0.04, 0.02, and 0.02 for the ACCB, the ACC part outside the Drake Passage latitudes and the whole ACC, respectively. Note that in other models in such regions as the Northern Atlantic, α is often even greater than 1.

The regional analysis of energy budgets and instability could complement the zonal and streamwise analysis. The strongest values of the EKE correspond to the most pronounced topographic features in the Southern Ocean. Measuring the eddy fluxes should be focused on areas downstream of the topography [Best et al., 1999; Hallberg and Gnanadesikan, 2001]. The nonlinear transfer N' is a source of EKE in many

subregions in the FRAM. It may be interpreted as a conversion from the kinetic energy of mean flow to EKE and can be related to the barotropic instability. This conversion is toward the EKE in most regions in the FRAM.

The baroclinic instability can be diagnosed by computing the A' term (introduced by Böning and Budich [1992] and not given here), which is approximately the exchange between potential energy of the mean flow and the eddy potential energy. This term is proportional to the horizontal eddy density flux multiplied by the mean horizontal density gradient and inversely proportional to the vertical gradient of the potential density of the reference state. In the ACC area A' and B' have the same signs and are of the same magnitude. This is consistent with the classical picture of baroclinic instability in which eddy potential and kinetic energies are created out of the potential energy of the mean flow. The ratio of B' of the North ACC to that of the ACCB region is almost equal to the ratio of their EKE densities [Ivchenko et al., 1997b]. This further suggests that the energy levels in these regions are closely linked to the strength of baroclinic instability occurring in each region. For the Antarctic Zone subregion (between the southern boundary of the ACCB and the Antarctic coast), B' is larger than A', which clearly means that the source of EKE is not linked completely to baroclinic instability. Furthermore, the FRAM horizontal grid does not resolve the eddies in this subregion because of weak stratification resulting in a small Rossby radius.

Baroclinic instability can be examined by calculating unstable modes of the zonal flow. The eigenvalue problem is solved in a manner described by Beckmann [1988] for the spatially and time-averaged shear of zonal flow and the corresponding mean density profile. The analysis was applied to a number of dynamically important subregions both inside and outside the ACC based on the FRAM and POP model outputs [Best et al., 1999; Wells et al., 2000]. In all the regions considered, the flow has been found to be baroclinically unstable. In FRAM, the growth rate, that is, the e-folding time of the baroclinic instability, ranges from 3 to 65 days, and in the POP, from 8 to 312 days. In the majority of regions, the most unstable wavelengths are marginally resolved by the zonal grid spacing. Maximum growth rates are found to occur on scales from approximately 1.6 to 3.6 times the first Rossby radius in the FRAM and from 1.3 to 7.1 times the first Rossby radius in the POP analysis.

Instability analysis performed over the ACC jets in the FRAM and POP models showed that baroclinic instability is likely to be the main route for generating EKE. Barotropic instability also contributes across several localized places with especially strong mean velocities and horizontal velocity shears. For example, barotropic instability develops between the Agulhas and Agulhas Return Current [Wells et al.,

2000]. On the northern flank of the Agulhas Current, there is an upgradient momentum flux into the mean flow.

In the FRAM, the upstream flows and flows just to the northeast of the Drake Passage have similar growth rates that are reasonably fast (24 days). A significant difference between the two regions is that the EKE density in the upstream flow is almost six times higher than that of the northeastern flow (58 and 10 cm^2 s^{-2}, respectively). One reason why the flow northeast of Drake Passage remains relatively stable in the FRAM, despite the predicted instability, is that the wavelength at which instability is most likely to occur is only resolved by three zonal grid points. Furthermore, the local first baroclinic Rossby radius is only just over one grid box in length. This strongly suggests that the stability of the flow downstream of the Drake Passage is due to the lack of resolution in FRAM. By way of contrast, the POP gives vigorous eddy distributions downstream of Drake Passage: The EKE density is 39 and 129 cm^2 s^{-2} for the upstream and northeast regions, respectively. Similar vigorous eddy distribution is clearly seen in the TOPEX data in the northeast region. The instability analysis of the POP model in this region, however, gives a very slow growth rate of 312 days. One possible explanation for this is that downstream of Drake Passage barotropic instability is the dominant mechanism. This may be expected since the jet produced by the POP is very tight.

7. ZONAL JETS

The ACC is seen as a broad current in coarse-resolution numerical models. With increasing resolution, the oceanic general circulation models (OGCM) are gaining skill in presenting separate jets and the associated frontal structure as illustrated in Plate 1 [adopted from *Hallberg and Gnanadesikan, 2006*; also see *Maltrud et al., 1998*; *Sinha and Richards, 1999*; *Richards et al., 2006*]. The major fronts in the ACC include the Subantarctic, Polar, and the Southern ACC fronts; yet, a finer frontal structure can be distinguished by closer inspection. Based on the analysis of hydrography at 140°E in the ACC, *Sokolov and Rintoul* [2002] show that the major fronts are split in reality in several branches. There is still some controversy with respect to precise positions of the major fronts [see *Hughes and Ash, 2001*], and even their circumpolar character is not easy to prove in all sectors of the Southern Ocean. The fine frontal structure is variable in time; the jets may appear and disappear by coalescing with each other, and some of them exist only locally. The major fronts are separating water masses with distinct properties. The fine frontal structure is seemingly of dynamic origin.

Despite the progress achieved recently with the fine-resolution OGCM in modeling the Southern Ocean, simulating the observed fine frontal structure still awaits for models with better spatial resolution and perhaps also constrained with observation data. The physical mechanisms of jet formation and the role of eddies in maintaining the jets are not fully understood, and existing evidence is rather controversial [see *Hughes and Ash, 2001*]. The frontal (jet) structure of the ACC is frequently explained by invoking arguments of β-plane turbulence [*Rhines, 1975*]. Indeed, the prediction of jet formation on the Rhines scale agrees generally well with results found in the QG layer models, demonstrating the appearance of multiple jets due to eddies generated by baroclinic instability [see, e.g., *Panetta, 1993*; *Treguier and Panetta, 1994*; *Sinha and Richards, 1999*]. The jet formation or sharpening is linked to the eddy convergence of eastward momentum, and a number of QG studies provide a nice illustration of this fact [*Wolff et al., 1991*; *Panetta, 1993*; *Treguier and Panetta, 1994*; *Olbers, 2005*].

The real situation is much more complicated because bottom topography leads to strong localization of jets downstream of the major topographic features in the Southern Ocean, which influences jet formation and spacing. The simple argument suggested by the theory of two-dimensional turbulence is not necessarily working everywhere in the ACC, yet still remains a plausible departure point.

In this work, we pursue a modest goal of presenting an elementary view on a jet formation mechanism in β-plane turbulence and confronting it with results that follow from existing modeling efforts.

7.1. Phenomenological View on Barotropic β-Plane Turbulence

We begin with barotropic β-plane turbulence driven by small-scale forcing in a flat-bottom box. This is the most elementary system capable of producing multiple jets due to turbulent eddies. It might shed some light on jet formation in the ACC because of its equivalently barotropic character [*Killworth, 1992*; *Killworth and Hughes, 2002*]. According to *Rhines* [1975], barotropic turbulence on a β-plane tends to form a jet-like structure with wavenumber

$$k_{Rh} = (\beta/2U)^{1/2}, \qquad (25)$$

which is simultaneously the scale where the energy cascade toward large scales is arrested. Here, U is the eddy rms velocity. The physical explanation for this tendency is the presence of Rossby wave dispersion that reduces the efficiency of nonlinear transfer involving quasi-zonal wavevectors so that the turbulent energy concentrates at the meridional wavenumber k_{Rh}. Numerous simulations with barotropic models [*Vallis and Maltrud, 1993*; *Danilov and Gurarie, 2002*;

Smith et al., 2002] suggest that the Rhines scale performs well in predicting the observed scale k_{obs} of jets, with a factor k_{obs}/k_{Rh} being between 0.7 and 1.5 in most cases. Similar skill is also seen in layered QG models [*Sinha and Richards*, 1999]. Taking a midlatitude estimate for $\beta = 1.5 \times 10^{-11}$ m^{-1} s^{-1} and an rms velocity of 10 cm/s, one gets a wavelength of 700 km, which is too large compared to the actual distance between the fronts observed in the ACC [see *Sokolov and Rintoul*, 2002]. The influence of topography and baroclinicity can be responsible for this discrepancy [*Sinha and Richards*, 1999].

For a flat-bottom barotropic flow, the Rhines scale can easily be recast in terms of energy production rate ε. If the flow is stabilized at large scales by bottom drag with the inverse timescale λ, the total eddy kinetic energy is then $E = \varepsilon/2\,\lambda$, giving

$$k_{Rh} = \beta^{1/2}(\lambda/4\varepsilon)^{1/4}. \qquad (26)$$

This estimate emphasizes the roles of forcing and dissipation and allows answering the question why zonation is not always observed even if eddies are present.

If the β effect were absent, the inverse energy cascade is stopped by the bottom drag on the friction scale $k_{fr} = (3C_K)^{3/2}$ $(\lambda^3/\varepsilon)^{1/2}$ (with $C_K \approx 6$). Jets are formed only if k_{Rh} exceeds k_{fr} because otherwise, the inverse cascade is arrested before it reaches scales where the Rossby dispersion is important [*Danilov and Gurarie*, 2002; *Smith et al.*, 2002]. The ratio $\gamma = k_{Rh}/k_{fr} = A\beta^{1/2}\varepsilon^{1/4}\lambda^{-5/4} = A\beta^{1/2}U^{1/2}/\lambda$ defines the boundary between the regimes without ($\gamma < 1$) and with ($\gamma > 1$) jets (here, $A = 4^{-1/4}(3C_K)^{-3/2}$). For the ACC taking $\lambda = 0.01$ day^{-1}, one gets $\gamma \approx 0.12$ ($U = 10$ cm s^{-1}), which is too low and no jets are expected. But if λ is reduced to 0.0015 day^{-1}, γ approaches 1, and jets will be produced by small-scale stirring in a barotropic flow.

Clearly, the role of friction is overemphasized in the barotropic approach, and the γ scaling can be applied to the ACC only qualitatively. Yet, it draws attention to the impact of dissipation on the jet formation. Many jets are local features that do not continue over the entire Southern Ocean. The strength of eddies can plausibly depend on the local energy balance between forcing and dissipation. The ability of models to reproduce the jets then depends on their level of dissipation.

Large-scale topography influences barotropic turbulence in an obvious way by redefining the local gradient of unperturbed QPV from β to a combination of $\beta\mathbf{j} + f_0\nabla b/H_0$, where H_0 is the total depth. The second component adds with β on topographic features sloping southwards (in the southern hemisphere) and can easily dominate locally. In such situations, the increase in effective β reduces jet spacing. A slope of 0.001 reduces the wavelength by a factor of around two

when its effect adds with β. *Sinha and Richards* [1999] show that taking the topographic slope into account does indeed make the jet spacing observed in the FRAM and POP consistent with theory if the Rhines scale is computed with the mean effective β.

The phenomenology of β plane turbulence leaves obscure the mechanism of jet formation. It turns out that this mechanism can be understood from a kinematic viewpoint involving the concept of PV mixing by eddies. This concept was used by *Danilov and Gryanik* [2004] and is elaborated from a broader physical perspective by *Dritschel et al.* [2008].

7.2. Multiple Jets From a Kinematic Perspective

Consider first a barotropic flow created by a small-scale stirring within a relatively narrow latitude belt $y_1 \leq y \leq y_2$ on a β plane. The Rossby waves will be generated there and, if friction is moderate, dissipated mostly outside the region of generation. The outgoing Rossby waves are associated with the divergence of the eastward momentum flux and on dissipation leave westward flows outside the stirring belt. If stirring does not impart mean momentum (as is the case with eddies produced by baroclinic instability), a compensating eastward flow occurs in the source region, a situation resembling that at midlatitudes in the atmosphere [see also *Rhines*, 1994; *Vallis*, 2006].

This reasoning is the simplest way of linking the generation of outgoing Rossby waves to the convergence of eastward momentum flux in the stirring region. When a mean background current is present, the Rossby waves are modified, and the waves are existing due to the mean PV gradient. This is assumed everywhere in this section.

Now let us look at this flow pattern from a PV perspective (we assume QG scaling and deal with QPV). Formation of westward flow north of y_2 and eastward flow in some vicinity south of it implies PV mixing around y_2 that tends to locally reduce the QPV gradient there. Mixing is not necessarily perfect and may leave a residual QPV gradient. The same argument applies to the southern source boundary y_1. One gets two zones of partly mixed QPV on flanks of the eastward jet. This mixing is irreversible, as it accompanies dissipation of Rossby waves.

There should be a QPV front joining the zones of mixed QPV within the source region that defines the eastward jet. Thus, the irreversible PV mixing due to Rossby wave dissipation is conductive to formation of PV fronts and eastward jets within the stirring belt. Clearly, this is simply a different language for expressing the well-known phenomena of eastward momentum flux convergence associated with eastward jets (see section 2.1).

If the stirring belt is sufficiently wide, several QPV fronts may form and coexist simultaneously within it, separating zones of partly mixed QPV. The stability of fronts depends on their strength (the QPV jump across the front). Indeed, for a fluid particle to pass across the front, its QPV should be modified by the QPV jump amplitude. Thus, fronts of small amplitude will be destroyed by stirring that leaves wider zones of mixed QPV limited by QPV fronts with higher QPV jumps. Fronts are becoming stronger if zones of mixed QPV are becoming wider (for a fixed degree of mixing). Full mixing is never achieved in practice, as it eliminates the Rossby waves required to redistribute QPV (or momentum) in the meridional direction.

For a given degree of PV mixing, the kinetic energy per unit mass of a zonal flow associated with the mixed zone increases with its width l as l^4. The dissipation of kinetic energy grows accordingly. Yet, it should be balanced with energy generation. This balance sets the size of the zone of mixed QPV, which is clearly the Rhines scale up to a factor of order one.

A limiting case of this scenario is what happens if small-scale stirring is uniform over the entire β plane. The result is a staircase structure of partly mixed QPV zones (or a comb-like structure of the QPV gradient similar to that of *Panetta* [1993]). Writing the gradient of relative vorticity within a zone as $-a\beta$ ($0 \leq a \leq 1$) and requiring the total momentum be zero, one finds the zonal velocity profile

$$u = 0.5a\beta((y - y_c)^2 - l^2/12) \qquad (27)$$

within a zone of width l centered at y_c. Expressing the kinetic energy, one obtains $k_{Rh}l \approx 3.7/a^{1/2}$. This implies that the Rhines scaling is in essence the consequence of QPV mixing in zones between two neighboring fronts, and the success of this scaling in predicting the distance between the jets depends on the degree of mixing inside the zone a.

This view translates to the general case of a baroclinic flow having, like the ACC, a wide zone of high baroclinicity. The interpretation is straightforward for layered flows, and indeed, the PV structure of their upper layers does support the concept of PV mixing (see Figure 4 of *Panetta* [1993] showing a comblike structure for the upper layer QPV gradient and Figure 4a of *MacCready and Rhines* [2001] displaying bands of mixed PV in the upper layer downstream of the topographic ridge). For a continuously stratified fluid, it is the sign of surface or bottom buoyancy gradients that typically determines the baroclinic instability (alternatively, one may introduce surface and bottom QPV sheets). The PV mixing concept then also involves fronts and mixed zones in the surface or bottom buoyancy.

Summing up, the irreversible PV mixing mediated by Rossby waves is a mechanism for multiple jet formation.

7.3. Confronting Theory and Observations

Can the jets in the ACC be explained by the simple theory of QG turbulence? The answer is that it is only partly applicable, and there are two main issues to be mentioned.

First, although *Sinha and Richards* [1999] show that there is a good agreement (on average) between the observed and predicted jet spacing in the FRAM and POP models, closer inspection reveals a difficulty. Figure 7 reproduces their result showing mean zonal velocity and bottom topography at several longitudinal locations in the ACC as derived from the output of the FRAM (left column) and the POP model (right column). Jet spacing is smaller above the southern slope of the ridge centered at 50°S (where effective β is small) than above the northern slope in the upper row, and there is no significant difference in spacing above southern and northern slopes in the lower two panels of Figure 7 (but effective β differs considerably). Therefore, using a local value of effective β would lead to disagreement between observations and theory.

Richards et al. [2006] present a wavelet power spectrum of the zonal component of velocity in the Pacific sector of the ACC simulated by the POP model at 1/10° resolution. They mention discrepancy between the observed scale and the Rhines scale (about a factor of 2).

Clearly, the comparison of observed (simulated) jet scale with the Rhines scale is somewhat ambiguous because no unique sense can be attached to U in a baroclinic flow. Yet, the power spectra for the equatorial belt and the belt of latitudes in the Northern Pacific presented by *Richards et al.* [2006] demonstrate much closer coincidence between the model and theory which supports the view that their estimate for U is consistent.

The observational evidence is also in favor of the possibility of relatively small distances between the jets. *Sokolov and Rintoul* [2002], for example, show the pattern of the sea surface height contours in the 130°–160°E sector that coincide with the fronts identified at section SR3, which are in some places about 1° apart. Even finer frontal structure is indentified by *Sokolov and Rintoul* [2007] based on the analysis of satellite altimetry data. Such a separation between fronts is somewhat too small to be explained by QG turbulence theory (unless the amplitude of U is essentially smaller than the 10 cm/s used here for estimates).

Second, the analyses by *MacCready and Rhines* [2001] based on a two-layer isopycnal model and by *Hughes and Ash* [2001] based on satellite-derived surface geostrophic velocities and eddy fluxes both reveal that eddies do not necessarily accelerate the jets, contrary to the theoretical predictions. Similarly, in the FRAM [*Ivchenko et al.*, 1997b] and the POP model [*Best et al.*, 1999], the integrated effect of eddies is to decelerate the flow.

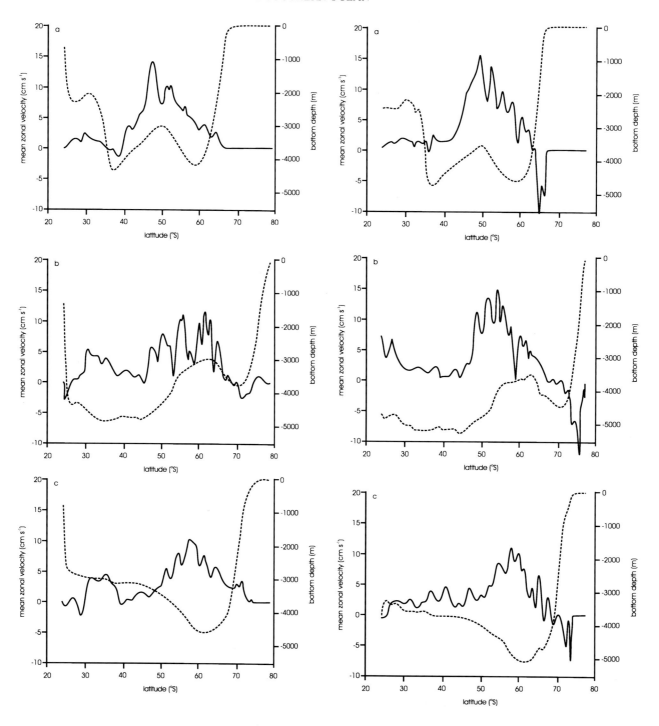

Figure 7. Longitude averaged time mean near-surface (32.5 m) zonal velocity for the FRAM (left column) and POP (right column) models (solid line) for (a) 85°–135°E, (b) 190°–225°E, and (c) 240°–280°E. The dashed lines represent the bottom topography averaged along longitude [from *Sinha and Richards*, 1999].

MacCready and Rhines [2001] consider a flow in a zonally reentrant channel with a topographic ridge and show that the eddy Reynolds stress divergence acts to decelerate jets along lines of constant pressure. Locally, there are regions of acceleration and deceleration, with the latter dominating. However, zonal averaging reveals narrow regions around centers of major jets where the contribution from the Reynolds stress divergence is accelerating. Yet, over the downstream part of the slope where the current is strongest, the contribution from the Reynolds stress divergence is to slow and broaden the jets.

Hughes and Ash [2001] computed the rotational part of the Reynolds stress divergence by processing surface geostrophic velocities derived from TOPEX/Poseidon and ERS altimetry. Their computations show that eddies act to decelerate the jets past main topographic features or act as vorticity sources for other strong jets, consistent with the observation by *MacCready and Rhines* [2001]. However, the sign of eddy force is less clear away from the major topographic features, and one can distinguish contributions of both signs there.

As *Hughes and Ash* [2001] point out, the key factor in explaining this behavior seems to be the bottom topography, which can tend to produce very narrow jets by topographic steering. Broadening of such narrow jets due to eddy stirring is to be expected past topography, and it is equivalent to jet deceleration by eddies.

In flat-bottom experiments, it is only the eddy forcing that maintains narrow jets against dissipation by a mechanism linked to PV mixing. The jets and eddies are in quasi-equilibrium at every zonal location because flow is statistically "invariant" along a zonal coordinate. In the presence of topography, the jet shape, spacing, and baroclinicity of the flow are strongly modified as the jet passes a topographic feature. One can speculate that equilibrium between local generation of turbulence due to baroclinic instability and backward action of turbulent eddies on jets can only be reached further downstream or not reached at all if a new topographic obstacle is encountered. Accordingly, the Rhines scaling can only work qualitatively, while jet intensification by eddies should only be observed locally and presumably away from steep bottom topography. This answer, however, poses new questions as to how the interplay of topography and eddy forcing determines the fine frontal structure downstream of the main topographic features and what maintains this structure in relatively flat regions at scales that are seemingly too small. These questions require additional research and hopefully will be answered at some point in the future.

8. SUMMARY AND DISCUSSION

Substantial progress in understanding the basics of the Southern Ocean dynamics has been achieved during the last three decades. This understanding is linked to the advances in numerical ocean modeling and appearance of models resolving or at least permitting mesoscale eddies. The first step in this direction was made by using quasi-geostrophic models that allowed exploring basic balances, mechanisms of momentum penetration, and horizontal redistribution of momentum and energy [*McWilliams et al.*, 1978; *Treguier and McWilliams*, 1990; *Wolff et al.*, 1991]. However, the intrinsic limitations of these models prevented them from simulating reliable levels of energies, the ACC transport and its variability, as well as many other properties.

The second step involved eddy-resolving experiments carried out with primitive equation models. The output of several such models was employed in a number of studies to investigate the global and local balances mediated by eddies and bringing a quantitative flavor to the mechanisms through which eddies influence ACC dynamics. Substantial progress has been achieved in the studies of generation and propagation of water masses and tracers, in meridional heat transport, and atmospheric-ocean exchange.

There remain important questions concerning the ACC dynamics that are expected to be solved in the near future. They are directly or indirectly linked to eddies that are an indispensable part of the Southern Ocean physics. Theory of the frontal jets in the Southern Ocean, which can describe the fine structures downstream topographic features, is a new challenge for the community. A considerable effort in the modeling of the Southern Ocean is to go into studies of diabatic motion and its role in setting the residual circulation. The eddy dynamics in the Antarctic Zone (to the south of Antarctic Polar Front) is still relatively unexplored because of a weak stratification and a very small Rossby radius. Modeling the ice dynamics and thermodynamics with open boundaries and polynias and convective processes is also a challenge. Convection in the Southern Ocean can be a source of disturbances that quickly propagate from Antarctic to the equatorial ocean [*Ivchenko et al.*, 2004, 2006; *Richardson et al.*, 2005; *Blaker et al.*, 2006]. New mechanisms of strong amplification of the waves coming to the equator were recently found [*Reznik and Zeitlin*, 2006]. The Antarctic sea ice distribution is strongly correlated with the NINO3 index of El Niño Southern Oscillation on timescales of an order of a few months [*Yuan and Martinson*, 2000; *Kwok and Comiso*, 2002]. Correlation occurs with both NINO3 and sea ice leading. Also important is the problem of teleconnections between the Antarctic and equatorial ocean. This problem can be discussed in the context of the role of the Southern Ocean in the global thermohaline circulation and global climate variability.

Acknowledgments. We are grateful to Bob Hallberg for permission to use the figure from Hallberg and Gnanadesikan [2006]. We thank M.-M. Lee for providing her new results, which are included within section 5.

REFERENCES

Andrews, D. G., and M. E. McIntyre (1976), Planetary waves in horizontal and vertical shear: The generalized Eliassen-Palm relation and the mean zonal acceleration, *J. Atmos. Sci.*, *33*, 2031–2048.

Beckmann A. (1988), Vertical structure of mid-latitude mesoscale instabilities, *J. Phys. Oceanogr.*, *18*, 1354–1371.

Best S. E., V. O. Ivchenko, K. J. Richards, R. D. Smith, and R. C. Malone (1999), Eddies in numerical models of the ACC and their influence on the mean flow, *J. Phys. Oceanogr.*, *29*, 328–350.

Blaker, A. T., B. Sinha, V. O. Ivchenko, N. C. Wells, and V. B. Zalesny (2006), Identifying the roles of the ocean and atmosphere in creating a rapid equatorial response to a Southern Ocean anomaly, *Geophys. Res. Lett.*, *33*, L06720, doi:10.1029/2005GL025474.

Böning, C., and R. G. Budich (1992), Eddy dynamics in a primitive equation model: Sensitivity to horizontal resolution and friction, *J. Phys. Oceanogr.*, *22*, 361–381.

Bretherton, F. S. (1966), Critical layer instability in baroclinic flows, *Q. J. R. Meteorol. Soc.*, *92*, 325–334.

Bryden, H. L. (1983), The Southern Ocean, Chapter 14 in *Eddies in Marine Science*, edited by A.R. Robinson, pp. 265–277, Springer, Berlin.

Bryden, H. L., and S. A. Cunningham (2003), How wind-forcing and air-sea heat exchange determine the meridional temperature gradient and stratification for the Antarctic Circumpolar Current, *J. Geophys. Res.*, *108*(C8), 3275, doi:10.1029/2001JC001296.

Bryden, H. L., and R. A. Heath (1985), Energetic eddies at the northern edge of the Antarctic Circumpolar Current in the southwest Pacific, *Prog. Oceanogr.*, *14*, 65–87.

Danilov, S., and V. M. Gryanik (2004), Barotropic beta-plane turbulence in a regime with strong zonal jets revisited, *J. Atmos. Sci.*, *61*, 2283–2295.

Danilov, S., and D. Gurarie (2002), Rhines scale and spectra of the β-plane turbulence with bottom drag, *Phys. Rev. E*, *65*, 067301.

Döös, K., and D. J. Webb (1994), The Deacon Cell and the other meridional cells in the Southern Ocean, *J. Phys. Oceanogr.*, *24*, 429–442.

Drijfhout, S. S. (2005), What sets the surface eddy mass flux in the Southern Ocean? *J. Phys. Oceanogr.*, *35*, 2152–2166.

Dritschel, D. G., and M. E. McIntyre (2008), Multiple jets as PV staircases: The Phillips effect and resilience of eddy-transport barriers, *J. Atmos. Sci.*, *65*, 855–874.

Dukowicz, J. K., and R. D. Smith (1994), Implicit free-surface method for the Bryan-Cox-Semtner ocean model, *J. Geophys. Res.*, *99*(C4), 7991–8014.

Dukowicz, J. K., R. D. Smith, and R. C. Malone (1993), A reformulation and implementation of the Bryan-Cox-Semtner ocean model on the Connection Machine, *J. Atmos. Oceanic Technol.*, *10*, 196–208.

Eden, C., R. J. Greatbatch, and D. Olbers (2007), Interpreting eddy fluxes, *J. Phys. Oceanogr.*, *37*, 1282–1296.

Edmon, H. J., B. J. Hoskins, and M. E. McIntyre (1980), Eliassen-Palm cross sections for the troposphere, *J. Atmos. Sci.*, *37*, 2600–2616.

Eliassen, A., and E. Palm (1961), On the transfer of energy in stationary mountain waves, *Geophys. Publ.*, *22*(3), 1–23.

Ferreira, D., J. Marshall, and P. Heimbach (2005), Estimating eddy stress by fitting dynamics to observations using a residual-mean ocean circulation model and its adjoint, *J. Phys. Oceanogr.*, *35*, 1891–1910.

Gallego, B., P. Cessi, and J. C. McWilliams (2004), The Antarctic Circumpolar Current in equilibrium, *J. Phys. Oceanogr.*, *33*, 1571–1587.

Gent, P, and J. C. McWilliams (1990), Isopycnal mixing in ocean circulation model, *J. Phys. Oceanogr.*, *20*, 150–155.

Gille, S. T. (1997), The Southern Ocean momentum balance: Evidence for topographic effects from numerical model output and altimeter data, *J. Phys. Oceanogr.*, *27*, 2219–2232.

Grezio A., N. C.Wells, V. O. Ivchenko, and B. A. de Cuevas (2005), Dynamical budgets in the Antarctic Circumpolar Current using ocean general circulation models, *Q. J. R. Meteorol. Soc.*, *131*, 833–860.

Hallberg, R., and A. Gnanadesikan (2001), An exploration of the role of transient eddies in determining the transport of a zonally reentrant current, *J. Phys. Oceanogr.*, *31*, 3312–3330.

Hallberg, R., and A. Gnanadesikan (2006), The role of eddies in determining the structure and response of the wind-driven southern hemisphere overturning: Results from the modelling eddies in the Southern Ocean (MESO) project. *J. Phys. Oceanogr.*, *36*, 2232–2252.

Held, I.M. (1975), Momentum transport by quasi-geostrophic eddies, *J. Atmos. Sci.*, *32*, 1494–1497.

Held, I. M., and T. Schneider (1999), The surface branch of the mass transport circulation in the troposphere, *J. Atmos. Sci.*, *56*, 1688–1697.

Henning, C. C., and G. K. Vallis (2005), The effect of mesoscale eddies on the stratification and transport of an ocean with a circumpolar channel, *J. Phys. Oceanogr.*, *35*, 880–896.

Hughes, C. W., and E. R. Ash (2001), Eddy forcing of the mean flow in the Southern Ocean, *J. Geophys. Res.*, *106*, 2713–2722.

Ivchenko, V. O. (1987), The influence of bottom topography on the eddy transfer coefficient, *Izvestiya Acad. Nauk USSR, Atmos. Ocean Phys.*, *21*, 250–260.

Ivchenko, V. O., K. J. Richards, and D. P. Stevens (1996), The dynamics of the Antarctic Circumpolar Current, *J. Phys. Oceanogr.*, *26*, 753–774.

Ivchenko, V. O., K. J. Richards, B. Sinha, and J.-O. Wolff (1997a), Parameterization of mesoscale eddy fluxes in zonal ocean flows, *J. Mar. Res.*, *55*, 1127–1162.

Ivchenko, V. O., A. M. Treguier, and S. E. Best (1997b), A kinetic energy budget and internal instabilities in the Fine Resolution Antarctic Model, *J. Phys. Oceanogr.*, *27*, 5–22.

Ivchenko, V. O., V. B. Zalesny, and M. R. Drinkwater (2004), Can the equatorial ocean quickly respond to Antarctic sea ice/salinity anomalies? *Geophys. Res. Lett.*, *31*, L15310, doi:10.1029/2004GL020472.

Ivchenko V. O., V. B. Zalesny, M. R. Drinkwater, and J. Schröter (2006), A quick response of the equatorial ocean to Antarctic sea ice/salinity anomalies, *J. Geophys. Res.*, *111*, C10018, doi:10.1029/2005JC003061.

Johnson, G. C., and H. L. Bryden (1989), On the size of the Antarctic Circumpolar Current, *Deep Sea Res.*, *36*, 39–53.

Kamenkovich, V. M., M. N. Koshlyakov, and A. S. Monin (1986), *Synoptic Eddies in the Ocean*, D. Reidel Publishing Company, Dordrecht, Holland, 433 pp.

Killworth, P. D. (1992), An equivalent-barotropic mode in the Fine Resolution Antarctic Model, *J. Phys. Oceanogr.*, *22*, 1379–1387.

Killworth, P. D., and C. W. Hughes (2002), The Antarctic Circumpolar Current as a free equivalent-barotropic jet, *J. Mar. Res.*, *60*, 19–45.

Killworth, P. K., and M. M. Nanneh (1994), On the isopycnal momentum budget of the Antarctic Circumpolar current in the Fine Resolution Antarctic Model, *J. Phys. Oceanogr.*, *24*, 1201–1223.

Kwok, R., and J. C. Comiso (2002), Southern Ocean climate and sea ice anomalies associated with the Southern Oscillation, *J. Clim.*, *15*, 487–501.

Lee, M.-M., and A. C. Coward (2003), Eddy mass transport for the Southern Ocean in an eddy-permitting global ocean model, *Ocean Model.*, *5*, 249–266.

Lee, M.-M., A. J. G. Nurser, A. C. Coward, and B. A. de Cuevas (2007), Eddy advective and diffusive transports of heat and salt in the Southern Ocean, *J. Phys. Oceanogr.*, *37*, 1376–1393.

MacCready, P. E., and P. B. Rhines (2001), Meridional transport across a zonal channel: Topographic localization, *J. Phys. Oceanogr.*, *31*, 1427–1439.

Maltrud, M. E., and J. L. McClean (2005), An eddy-resolving global 1/10 degrees ocean simulations, *Ocean Model. 8*, 31–54.

Maltrud, M. E., R. D. Smith, A. J. Semtner, and R. C. Malone (1998), Global eddy-resolving ocean simulations driven by 1985–1994 atmospheric winds, *J. Geophys. Res.*, *103*, 30,825–30,853.

Marshall, J. C. (1981), On the parameterization of geostrophic eddies in the ocean, *J. Phys. Oceanogr.*, *11*, 257–271.

Marshall, J. C., and T. Radko (2003), Residual-mean solutions for the Antarctic Circumpolar Current and its associated overturning circulation, *J. Phys. Oceanogr.*, *33*, 2341–2354.

Marshall, J., D. Olbers, H. Ross, and D. Wolf-Gladrow (1993), Potential vorticity constraints on the dynamics and hydrography of the Southern Ocean, *J. Phys. Oceanogr.*, *23*, 465–487.

McWilliams, J. C., and J. S. Chow (1981), Equilibrium geostrophic turbulence I: A reference solution in a beta plane channel, *J. Phys. Oceanogr.*, *11*, 921–949.

McWilliams, J. C., W. R. Holland, and J. S. Chow (1978), A description of numerical Antarctic Circumpolar Currents, *Dyn. Atmos. Ocean*, *2*, 213–291.

Munk, W. H., and E. Palmén (1951), Note on the dynamics of the Antarctic Circumpolar Current, *Tellus*, *3*, 53–55.

Olbers, D. (2005), On the role of eddy mixing in the transport of zonal ocean currents, in *Marine turbulence. Theories, Observations, and Models*, edited by, H. Baumert, J. Simpson, and J. Sündermann, 630 pp, Cambridge University Press, Cambridge.

Olbers, D., and V. O. Ivchenko (2001), On the meridional circulation and balance of momentum in the Southern Ocean of POP, *Ocean Dyn.*, *52*, 79–93.

Olbers, D., and M. Visbeck (2005), A model of the zonally averaged stratification and overturning in the Southern Ocean, *J. Phys. Oceanogr.*, *35*, 1190–1205.

Panetta, R. L. (1993), Zonal jets in wide baroclinically unstable regions: Persistence and scale selection, *J. Atmos. Sci.*, *50*, 2073–2106.

Pedlosky, J. (1979), *Geophysical Fluid Dynamics*, Springer, New York, Inc., 624 pp.

Radko, T., and J. Marshall (2006), The Antarctic Circumpolar Current in three dimensions, *J. Phys. Oceanogr.*, *36*, 651–669.

Reznik, G. M., and V. Zeitlin (2006), Resonant excitation of Rossby waves in the equatorial waveguide and their nonlinear evolution, *Phys. Rev. Lett.*, 96, 034502, doi:10.1103/PhysRevLett.

Rhines, P. B. (1975), Waves and turbulence on a beta-plane, *J. Fluid Mech.*, *69*, 417–443.

Rhines, P. B. (1994), Jets, *Chaos*, *4*, 313–339.

Richards, K. J., N. A. Maximenko, F. O. Bryan, and H. Sasaki (2006), Zonal jets in the Pacific Ocean, *Geophys. Res. Lett.*, *33*, LO3605, doi:10.1029/2005GL024645.

Richardson, G., M. R. Wadley, K. J. Heywood, D. P. Stevens, and H. T. Banks (2005), Short-term climate response to a freshwater pulse in the Southern Ocean, *Geophys. Res. Lett.*, *32*, L03702.1–L03702, doi:10.1029/2004GL021586.

Rintoul, S. R., C. Hughes, and D. Olbers (2001), The Antarctic Circumpolar Current System, in *Ocean Circulation and Climate*, edited by G. Siedler, J. Church, and J. Gould, pp. 271–302, Academic, New York.

Schouten, M. W., and R. P. Matano (2006), Formation and pathways of intermediate water in the Parallel Ocean Circulation Model's Southern Ocean, *J. Geophys. Res.*, *111*, C06015, doi:10.1029/2004JC002357.

Semtner, A. J., Jr. and R. M. Chervin (1988), A simulation of the Global Ocean circulation with resolved eddies, *J. Geophys. Res.*, *93*, 15,502–15,522.

Semtner, A. J., Jr. and R. M. Chervin (1992), Ocean general circulation from a global eddy-resolving model, *J. Geophys. Res.*, *97*(C4), 5493–5550.

Sinha, B., and K. J. Richards (1999), Jet structure and scaling in Southern Ocean Models, *J. Phys. Oceanogr.*, *29*, 1143–1155.

Smith, K. S., G. Boccaletti, C. C. Henning, I. Marinov, C. Y. Tam, I. M. Held, and G. K. Vallis (2002), Turbulent diffusion in the geostrophic inverse cascade, *J. Fluid Mech.*, *469*, 13–48.

Smith, R. D., J. K. Dukowicz, and R. C. Malone (1992), Parallel ocean general circulation modeling, *Physica D*, *60*, 38–61.

Sokolov, S., and S. R. Rintoul (2002), Structure of Southern Ocean at 140°E., *J. Mar. Syst.*, *37*, 151–184.

Sokolov, S., and S. R. Rintoul (2007), Multiple jets of the Antarctic Circumpolar Current South of Australia, *J. Phys. Oceanogr.*, *37*, 1394–1412.

Stammer, D. (1998), On the eddy mixing and mean flow properties, *J. Phys. Oceanogr.*, *28*, 727–739.

Stevens, D. P., and V. O. Ivchenko (1997), The zonal momentum balance in an eddy-resolving general-circulation model of the Southern Ocean, *Q. J. Royal Meteor. Soc.*, *123*, 929–951.

The FRAM group (1991), Initial results from a fine resolution model of the Southern Ocean, *Eos Trans. AGU*, *72*, 174–175.

Treguier, A. M., and J. C. McWilliams (1990), Topographic influences on wind-driven, stratified flow in a β-plane channel: An idealized model for the Antarctic Circumpolar Current, *J. Phys. Oceanogr.*, *20*, 321–343.

Treguier, A. M., and R. L. Panetta (1994), Multiple zonal jets in a quasigeostrophic model of the Antarctic Circumpolar Current, *J. Phys. Oceanogr.*, *24*, 2263–2277.

Vallis, G. K. (2006), *Atmospheric and Oceanic Fluid Dynamics. Fundamentals and Large-Scale Circulation*, Cambridge University Press, Cambridge, 770 pp.

Vallis, G. K., and M. E. Maltrud (1993), Generation of mean flows and jets on a beta plane and over topography, *J. Phys. Oceanogr.*, *23*, 1346–1362.

Webb, D., B. de Cuevas, and A. Coward (1998), The first main run of the OCCAM global model, *Southampton Oceanography Centre*, Internal Report, No 34. SOC, UK.

Wells, N. C., V. O. Ivchenko, and S. E. Best (2000), Instabilities in the Agulhas Retroflection Current system: A comparative model study, *J. Geophys. Res.*, *105*, 3233–3241.

Wolff, J.-O., and D. J. Olbers (1989), The dynamical balance of the Antarctic Circumpolar Current studied with an eddy-resolving quasi-geostrophic model, in *Mesoscale-Synoptic Coherent Structures in Geophysical Turbulence*, edited by J. C. J. Nihoul and B. M. Jamart, pp. 435–458, Elsevier, Amsterdam.

Wolff, J.-O., E. Maier-Reimer, and D. J. Olbers (1991), Wind-driven flow over topography in a zonal β-plane channel: A quasigeostrophic model of the Antarctic Circumpolar Current, *J. Phys. Oceanogr.*, *21*, 236–264.

Yuan, X., and D. G. Martinson (2000), Antarctic sea ice extent variability and its global connectivity, *J. Clim.*, *13*, 1697–1717.

V.O. Ivchenko, National Oceanography Center, Southamptom, UK. (voi@noc.soton.ac.uk)

High-Resolution Indian Ocean Simulations— Recent Advances and Issues From OFES

Yukio Masumoto,[1,2] Yushi Morioka,[2] and Hideharu Sasaki[3]

Mean circulation and its variability in the Indian Ocean are complex due mainly to unique geographical conditions. Results from an eddy-resolving ocean general circulation model (OGCM) reveal that model performances in simulating Indian Ocean variability are significantly improved. Some examples of the improvements achieved by the model, as well as remaining issues, are described. Magnitude and spatial pattern of root-mean-square variance of the simulated sea surface height variability are nearly identical to the observed ones. However, the skewness of the sea surface height variability in the equatorial region and the regions of large eddy activity show significant differences between the two, suggesting nonlinear aspects of the variability have not yet reproduced appropriately. Simulated meridional heat transport with daily forcing lies within the observed values and is increased by 34% compared to the one forced by climatological winds. Two kinds of origin of westward propagating eddies from the Arabian Sea into the Gulf of Aden are presented, as an example of the mesoscale eddy activities in the Indian Ocean. Most of the eddies are generated in the interior region of the Arabian Sea, while some eddies appearing during the boreal autumn/winter are generated in association with the strong western boundary current through the Socotra Passage. The results also demonstrate that the magnitude of the Indonesian throughflow and its pathways are realistically reproduced due to high resolution representations of the complex topography within the Indonesian Seas, although the simulated water mass characteristics in this region suggest underestimation of the vertical mixing in OGCM for the Earth Simulator (OFES).

1. INTRODUCTION

The Indian Ocean is blocked by the Asian land mass to the north at around 25°N, while its southern part widely opens to the Southern Ocean. This geographical condition makes the Indian Ocean circulations unique compared to those in the other major ocean basins. The equatorial region and the two relatively smaller basins in the northern hemisphere, the Arabian Sea and the Bay of Bengal, are strongly influenced by the Asian monsoon system. The southern hemisphere circulation indicates a structure of a typical subtropical gyre, with additional flow pattern generated by the Indonesian throughflow (ITF), coming into the Indian Ocean through the eastern boundary at around 10°S. Distribution of surface flesh water flux and river runoff determines unique surface

[1]Frontier Research Center for Global Change, Japan Agency for Marine-Earth Science and Technology, Yokohama, Kanagawa, Japan.

[2]Graduate School of Science, University of Tokyo, Tokyo, Japan.

[3]Earth Simulator Center, Japan Agency for Marine-Earth Science and Technology, Yokohama, Kanagawa, Japan.

Ocean Modeling in an Eddying Regime
Geophysical Monograph Series 177
10.1029/177GM14

salinity distribution in the northern Indian Ocean, that is, strong contrast between high salinity water in the Arabian Sea and low salinity water in the Bay of Bengal. Under these external forcings, the variability in the Indian Ocean was thought to be dominated by a seasonal cycle, including a distinct annual and semiannual variations [e.g., *Wyrtki*, 1971, 1973; *McCreary et al.*, 1993]. However, recent observations and modeling results demonstrate that the variability actually spans much wider range of spatial and temporal scales. An extensive review of such circulations in the Indian Ocean can be found in the work of *Schott and McCreary* [2001] and *Tomczak and Godfrey* [1994].

The first attempt to simulate such a circulation in the Indian Ocean goes back to the work by *Lighthill* [1969], in which a generation mechanism of the Somali current is investigated using a rather simple linear wave model. Since this seminal paper of *Lighthill* [1969], there have been many numerical modeling efforts with various kinds of model complexity to clarify dynamics and thermodynamics involved in the variability of the upper ocean circulation until early 1990s, mostly focusing on the variability in the Somali current system [e.g., *Anderson and Moore*, 1979; *Cox*, 1979, 1981; *McCreary and Kundu*, 1988] and equatorial current and wave dynamics [e.g., *McCreary*, 1985; *Kindle and Thompson*, 1989; *Woodberry et al.*, 1989; *Anderson and Carrington*, 1993]. Based on these extensive studies in the past few decades, our understanding on the seasonal variations in the upper ocean has been significantly improved.

In another stream of research on the global circulation, general circulation models (GCMs) are frequently utilized to investigate the three-dimensional thermohaline structure and associated global circulations of the world oceans. The meridional overturning cell in the Indian Ocean is also investigated as an important building block of the meridional heat transport within the oceans [e.g., *Wacongne and Pacanowski*, 1996; *Garternicht and Schott*, 1997; *Lee and Marotzke*, 1998]. Except for a few studies, however, the meridional overturning cell itself and important elements that constitute the cell within the Indian Ocean are usually not the main focus for these studies.

The Indian Ocean exhibits zonally averaged southward heat transport at almost all latitude sections. Southward warm surface water is mainly driven by the Ekman transport associated with the monsoonal winds over the Indian Ocean. Although pathways of subsurface northward return flow of the colder water have not been fully recognized yet, an important role of upwelling regions near the Somali coast and around India on closing the circuit has been suggested [e.g., *Schott and McCreary*, 2001; *Valsala and Ikeda*, 2007].

With recent rapid advances in the computational sciences, high-resolution global GCM simulations can be conducted in several institutions [e.g., *Semtner and Chervin*, 1992; *Coward et al.*, 2002; *Maltrud and McClean*, 2005]. These eddy-permitting or eddy-resolving models provide us with new insights not only into the ocean variability associated with the eddy-scale phenomena themselves, but also into influences of these eddies on the large-scale circulations [*Smith et al.*, 2000; *Hurlburt and Hogan*, 2000; *Chassignet and Garraffo*, 2001].

A couple of eddy-resolving simulation efforts are executing at this moment in several places. Preliminary results from the Ocean Circulation and Climate Advanced Modelling (OCCAM) simulation with horizontal grid spacing of 1/12° are shown by *Coward et al.* [2002], and results from a 15-year integration using the global POP with 0.1° horizontal grid spacing are reported by Maltrud and McClean [2005]. Both results demonstrated quite realistic eddy activities as well as the large-scale circulations, including those in the Indian Ocean sector. In addition to the above efforts, an ocean simulation group at Japan Agency for Marine-Earth Science and Technology (JAMSTEC) has started several years ago a new project, using an ocean general circulation model (OGCM) specifically modified for the use on the Earth Simulator, to conduct a relatively long-term hindcast integration of the world ocean circulations. After the 50-year climatological spin-up integration [*Masumoto et al.*, 2004; *Ohfuchi et al.*, 2005], a hindcast integration from 1950 to 2005 [*Sasaki et al.*, 2008] and a simulation with chlorofluorocarbons (CFCs) tracer have been performed [*Sasai et al.*, 2004]. In this article, as an example of such the eddy-resolving ocean circulation simulations, we highlight some results that can only be shown by the eddy-resolving model, focusing on the variability within the Indian Ocean. At the same time, we will touch on some issues to be solved in future high-resolution models.

The structure of this paper is as follows. The model description is given in section 2. In section 3, statistical representations of the variability in the sea surface height (SSH) anomaly are presented as a general performance of the model. Section 4 introduces the zonally integrated meridional heat transport in the eddy-resolving model. An interesting example of the mesoscale eddy activity in the northern Indian Ocean is described in section 5, and some insights and issues on the Indonesian throughflow obtained in the high-resolution model are shown in section 6.

2. MODEL DESCRIPTION

In the following sections, we show results from particular examples of eddy-resolving simulations to demonstrate achievements of such high-resolution simulations and, at the same time, issues that still remain to be solved. The OGCM

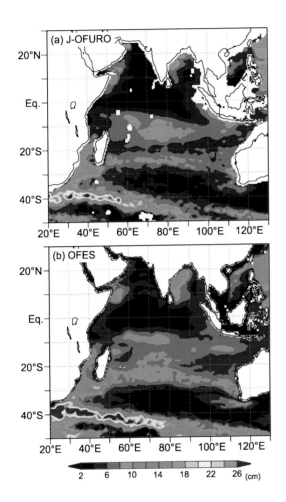

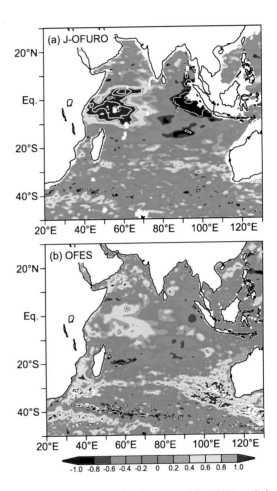

Plate 1. Horizontal distributions of standard deviation of the sea surface height anomaly (SSHA) for (a) satellite derived observation (J-OFURO) and (b) OFES simulation with the QuikScat wind forcing.

Plate 2. Same as in Plate 1, but for skewness of the SSHA variations.

used in this study is based on the Modular Ocean Model version 3 (MOM3) [*Pacanowski and Griffies*, 2000], developed at Geophysical Fluid Dynamics Laboratory/National Oceanic and Atmospheric Administration (GFDL/NOAA). It is highly tuned for the Earth Simulator (ES) and is called as the OGCM for the ES (OFES). OFES is particularly tuned with parallelization and vectorization procedures to attain the best computational performance on the ES. As the result of these optimizations, 1-year model integration using 500 processors (63 nodes) on the ES is completed in 15 h CPU time.

The model solves three-dimensional primitive equations in spherical coordinates under the Boussinesq and hydrostatic approximations, with z-level coordinate in vertical. The model domain covers a quasi-global region from 75°S to 75°N excluding the Arctic Ocean. The horizontal grid spacing and the number of vertical levels are 0.1° and 54, respectively. The vertical grid spacing varies from 5 m at the surface to 330 m at the maximum depth of 6065 m. To represent upper ocean circulation realistically, 20 levels are confined in a layer between the sea surface and the 200-m depth. The model topography is generated using 1/30° bathymetry dataset provided from the OCCAM project at the Southampton Oceanography Centre, and we obtain it through GFDL/NOAA. The partial cell method [*Pacanowski and Gnanadesikan*, 1998] implemented in MOM3, which allows bottom ocean cells to be of different thickness, is employed to represent the topography realistically.

A scale-selective damping of biharmonic operator is utilized for horizontal mixing of momentum and tracers to suppress computational noises with the horizontal scale of the grid spacing. The viscosity and diffusivity coefficients are calculated following *Smith et al.* [2000] with the value of -2.7×10^{10} m^4/s for momentum and -9×10^9 m^4/s for tracers, respectively, at the equator. They vary proportional to the cube of the zonal grid spacing and of *cos(latitude)*. Vertical viscosity and diffusivity are calculated using the K-profile parameterization [*Large et al.*, 1994].

The model ocean is driven by the wind stresses and surface tracer fluxes for the period from 1950 to 2003, from the last condition of the 50-year climatological spin-up integration [*Masumoto et al.*, 2004] as the initial state. The daily mean values obtained from the National Centers for Environmental Prediction/National Center for Atmospheric Research (NCEP/NCAR) reanalysis products [*Kalnay et al.*, 1996] are utilized for the wind stresses and the atmospheric variables that need to be specified for the tracer flux calculation. The bulk formula of *Rosati and Miyakoda* [1988] is adopted for the surface heat flux calculation, while the salinity flux is obtained from precipitation rate of the reanalysis data and evaporation rate derived using the bulk formula. An additional hindcast simulation forced by the QuikScat wind

stress, although other forcings are the same with the NCEP/NCAR wind case, is performed from July 1999 through the end of 2004 [*Sasaki et al.*, 2004].

In addition to this salinity flux, the surface salinity is restored to the climatological monthly mean value of the World Ocean Atlas 1998 (WOA98; *Boyer et al.* [1998a, 1998b, 1998c]), with the restoring timescale of 6 days, so that the effects of the river runoff are included implicitly. In the regions within 3° latitudinal distances from the northern and southern artificial boundaries at 75°N/S, temperature and salinity are restored to the monthly mean climatological values of the WOA98 [*Antonov et al.*, 1998a, 1998b, 1998c; *Boyer et al.*, 1998a, 1998b, 1998c] at all levels, with the restoring timescale that increases linearly from 1 day at the boundaries to 720 days at 3° from the boundaries. This buffer layers suppress the unrealistic wave propagation along the artificial boundaries and can incorporate the effects of sea ice processes. To avoid from being frozen, the surface heat flux is set as zero in OFES when the flux is upward and the sea surface temperature (SST) becomes lower than -1.8°C, a typical freezing temperature of sea water.

General performances of OFES in representing the climatological seasonal variations and the intraseasonal-to-decadal variability are summarized by *Masumoto et al.* [2004] and *Sasaki et al.* [2008], respectively, demonstrating good agreement with the observed variability in the upper oceans. For example, simulated SST anomaly associated with the Indian Ocean Dipole (IOD) events and surface velocity variability at the intraseasonal timescale are well simulated [*Sasaki et al.*, 2008; *Ogata et al.*, 2008]. In addition, the equatorial and coastal waves and currents are reproduced quite realistically, which leads to detailed investigations on a bifurcation mechanism of the coastal currents along the east coast of Sri Lanka [*Vinayachandran et al.*, 2005], on the coastally trapped propagating signals along the eastern boundary of the Indian Ocean [*Iskandar et al.*, 2006] and on heat budget in the eastern tropical Indian Ocean [*Du et al.*, 2005]. All these previous studies demonstrate significant utilities of the OFES outputs and their ability to conduct quantitative analyses on many processes in the oceans.

3. SEA SURFACE HEIGHT VARIABILITY

As one of the major manifestation of the ocean dynamics and because the satellite altimetry observation is available with fine spatial and temporal resolutions with high accuracy, the SSH variability is often utilized to validate the model performance of phenomena with various spatial and temporal scales. It has already been noted in many places that fine horizontal resolution, typically smaller than 0.1° in latitude and longitude, is necessary to reproduce the observed

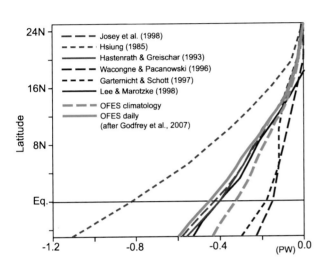

Plate 3. Zonally integrated, annual mean meridional heat transports across each latitude north of 5°S. Both the OFES results forced by the climatological seasonal forcing (green dashed line) and by the daily mean QuikScat wind forcing (green solid line) are shown. Values calculated from the atmospheric flux data [*Hsiung*, 1985; *Hastenrath and Greischar*, 1993; *Josey et al.*, 1999; red lines] and those from the previous numerical model simulations [*Wacongne and Pacanowski*, 1996; *Garternicht and Schott*, 1997; *Lee and Marotzke*, 1998; blue lines] are also shown. [After *Hu and Godfrey*, 2007.]

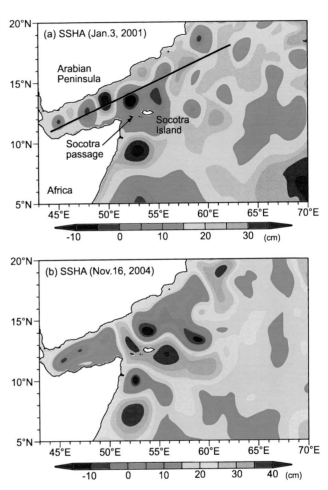

Plate 4. Horizontal SSHA distribution in the northwestern Arabian Sea on (a) January 3, 2001, and (b) November 16, 2004. A black line in (a) denotes a section, for which the distance-time diagram of the SSHA is drawn in Plate 5.

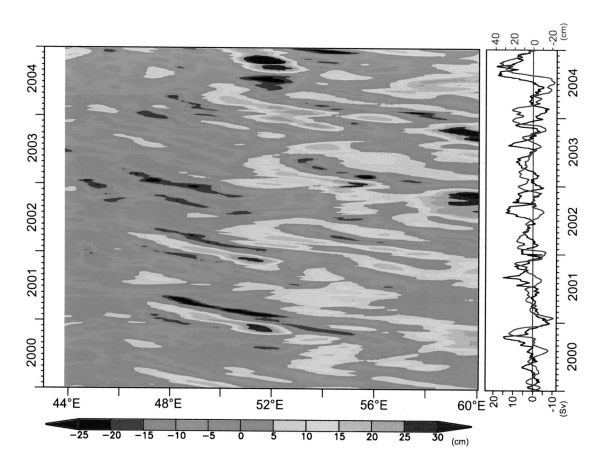

Plate 5. Distance-time diagram of the SSHA along the section in the Gulf of Aden and northeast of the Gulf of Aden, indicated in Plate 4a, from January 2000 to December 2004. Zonally averaged value over the section is subtracted at each time to remove the seasonal signal. A panel in the right-hand side demonstrates time series of the northward volume transport across the Socotra passage (black line) and the SSHA variation at 52°E on the section.

SSH variability appropriately [e.g., *Smith et al.*, 2000]. This is also the case for the Indian Ocean.

The standard deviation (STD) of the SSH anomaly (SSHA) is compared in Plate 1 between the OFES result and the satellite altimeter data, which is obtained from a product of Japanese ocean flux data sets with use of remote sensing observations (J-OFURO) project [*Kubota et al.*, 2002].[1] Here, the SSHA is the deviation from the mean seasonal variations averaged over the period from 1993 to 2004. It is shown that the OFES simulation realistically reproduced the STD of the SSHA field. In some regions with large STD, for example, the Antarctic Circumpolar Current (ACC) region, the Agulhas Current region, and a zonal band along 25°S, the OFES results tend to have larger variability by about a factor of 1.5 to 2.0, while the tropical region and a region along the southern coast off Sumatra and Java islands indicate less variability compared to the observation. The former tendency may be attributed to the biharmonic formulation of the horizontal eddy viscosity, which effectively dumps the grid-scale numerical noises while other signals larger than the scale of a few grids remain unaffected. The latter issue is related to the model performance in reproducing the interannual SSH variability associated with the IOD event, which is a basin-scale air-sea coupled climate mode with typical timescale of several years [*Saji et al.*, 1999]. The simulated amplitude of the IOD events in OFES is relatively small compared to the observation, although the phase of the events is well reproduced [*Sasaki et al.*, 2008].

It is also known that the large SSH variance is associated with the mesoscale eddy activities, and typical examples for such eddy activities can be seen in the Agulhas retroflection and the western half of the ACC in the Indian Ocean section. However, the STD itself cannot indicate any further characteristics of the mesoscale eddies, and the detailed analyses focusing on some specific regions of interests are required to have additional insights on the eddy characteristics. A simple extension of the statistical analysis can provide useful information about the eddy activities that appeared in the SSH variability. Plate 2 shows skewness of the simulated and observed SSHA variability within the Indian Ocean. The skewness is a measure of asymmetry of the probability distribution and defined as the third standardized moment of the SSH anomaly, in the present case. The positive (negative) skewness indicates that the time series demonstrate rather long period of weak negative (positive) SSHA with short period of large positive (negative) SSHA. The model captures the basic horizontal pattern of the skewness. However, its amplitude is significantly smaller in the tropical region and, on the other hand, larger in the area south of 20°S.

Within the tropical region, positive skewness is observed in the western half of the basin and negative values appear off the coast of Sumatra (Plate 2a). This dipole-like structure in skewness is associated with IOD events, indicating that positive IOD events tend to have larger amplitude and occur less frequently compared to negative ones. The small skewness in OFES reflects the fact that the simulated IOD tends to have equal magnitude between positive and negative events. In other words, the nonlinear characteristics of IOD events are not well captured. A possible cause for this discrepancy between the model and observation may be the forcing functions for the model, but further investigation will be needed to verify this.

The skewness in the Agulhas current system, the ACC, and a region west of Australia, on the other hand, indicate significantly large positive values in the OFES results. Again, the STD is of the same order between the simulated and observed SSHA. This indicates that OFES tends to create much stronger anticyclonic eddies in these regions. For example, in the region off southwestern Australia, the anticyclonic mesoscale eddies generated near the southern tip of Tasmania drift into the Indian Ocean along the high skewness band extending from 40°S, 130°E to 30°S, 90°E. The similar isolated anticyclonic eddies can be seen in the satellite altimetry observation, suggesting a possible mechanism for connecting the southern Pacific water into the Indian Ocean, which might be a part of "Super Gyre" reported by *Cai* [2006] and *Speich et al.* [2002, 2007]. Similarly, the anticyclonic eddies generated in the Mozambique Channel and the region south of Madagascar, propagating along the east coast of South Africa, are stronger in OFES compared to the observed values. The negative (positive) skewness to the north (south) of the axis of the ACC is well simulated, but, again, the skewness is larger than the observed values. In the forthcoming eddy-resolving OGCM, not only the STD of the SSHA but also the skewness should be appropriately represented for a better understanding of the role of mesoscale eddies in large-scale circulations and of the mechanisms of a large-scale variability such as IOD events.

4. MERIDIONAL HEAT TRANSPORT IN THE INDIAN OCEAN

Zonally integrated meridional heat transports in the ocean are the key aspects controlling the regional and global climate systems. This issue has been addressed since the advent of large-scale numerical modeling studies. Due to the large uncertainty in the calculation of satellite-based surface fluxes and to the sparseness of in situ observations in the ocean, observed estimates vary widely among studies [e.g., *Hsiung*, 1985; *Hastenrath and Greischar*, 1993; *Josey et al.*, 1998]. The numerical models, naturally, have their own biases and the detailed validation of the simulated meridional heat transport is rather difficult. Several attempts have been conducted to estimate the simulated meridional heat transport in the Indian Ocean [e.g., *Wacongne and Pacanowski*, 1996; *Gartenicht and Schott*, 1997; *Lee and Marotzke*, 1998] and the values are summarized by *Godfrey et al.* [2007].

In general, the simulated meridional heat transports are smaller than the observed estimates (Plate 3). While the smallest observed values by *Hastenrath and Greischar* [1993] is about 0.4 PW of the southward heat transport across the equator, the two examples of the simulated results with relatively coarse resolution indicate about 0.18 PW, which is less than half of the observed estimate. *Lee and Marotzke* [1998] estimated the larger heat transport, which is comparable to the values derived by *Hastenrath and Greischar* [1993]. The OFES results driven by the climatological forcing also show the smaller values compared to the observed estimates at the latitude south of 14°N, and the simulated southward heat transport across the equator is about 0.35 PW. These differences among the model outputs could simply be the model biases. However, another OFES result, forced by the daily mean wind stresses, indicates much larger values with the southward heat transport of 0.47 PW across the equator.

The results of *Lee and Marotzke* [1998] suggest that the meridional heat transport does not change significantly between the seasonal forcing case and the annual mean forcing case. *Godfrey et al.* [2007], however, demonstrate the importance of such a rectification of shorter timescale variability to the mean state. The comparison between the two OFES results demonstrates that inclusion of a much energetic shorter and longer timescale variability can result in the enhancement of the southward heat transport in the Indian Ocean. They attribute this difference to the mixing effect caused by the smaller and shorter scale eddy-like variability simulated in OFES forced by the daily winds. The detailed analyses using results from other eddy-resolving OGCM are required for clear understanding of the role of the eddies.

5. MESOSCALE EDDIES IN THE NORTHWESTERN ARABIAN SEA

One of the interesting examples of the mesoscale eddies in the Indian Ocean and their relation to the western boundary current can be found in a northwestern part of the Arabian Sea, including the Gulf of Aden. Plate 4a shows the horizontal distribution of the simulated SSHA in this region on January 3, 2001, demonstrating a series of positive and negative eddy-like features, with a typical diameter of a few hundred kilometers, aligned along the axis of the Gulf of Aden and the region east of the Gulf. These mesoscale eddies have been observed frequently in the satellite altimetry data and chlorophyll concentration distribution [*Fratantoni et al.*, 2006].

A time-distance section of the SSHA along the axis (indicated by a solid line in Plate 4a) is shown in Plate 5, in which the spatial mean value along the section is subtracted to remove the seasonal signal. The westward propagation of these eddy-like features, generated in the far eastern part of the Arabian Sea and with a phase speed of about 6 cm/s, is clearly

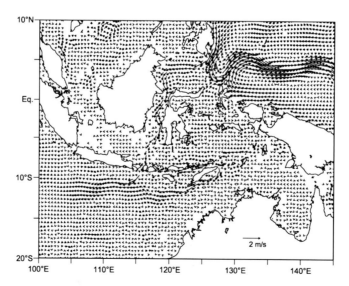

Figure 1. Horizontal distribution of the monthly mean horizontal velocity fields at the depth of 30 m in August, simulated in OFES forced by the climatological seasonal winds.

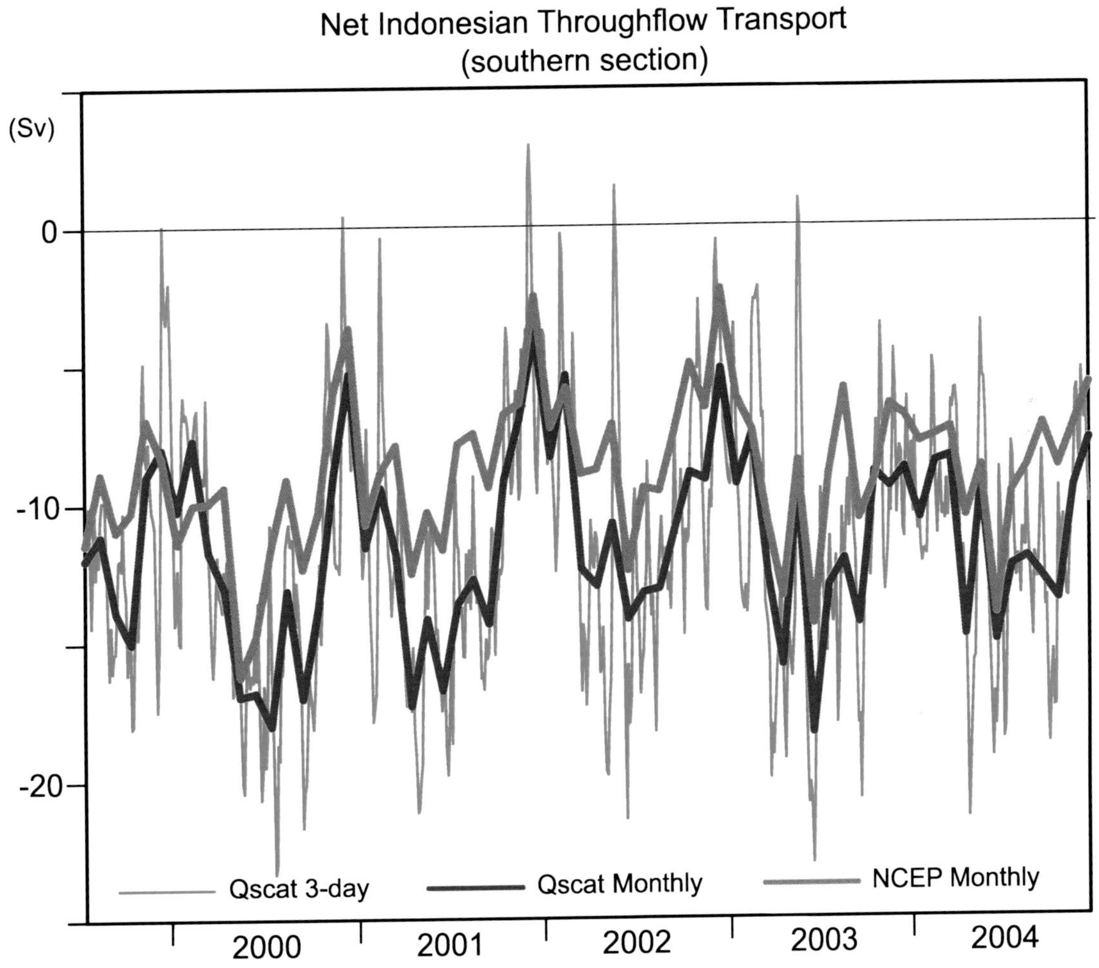

Net Indonesian Throughflow Transport
(southern section)

Plate 6. Time series of the net Indonesian throughflow across the southern boundary of the Indonesian Seas (Malacca, Sunda, Lombok, Sape, Ombai Straits and Timor Sea) from the OFES results forced by the daily mean QuikScat winds (3-day interval snapshot values in the green line, and monthly mean values in the red line) and those forced by the NCEP/NCAR climatological seasonal forcing (monthly mean values in blue line). Negative values indicate southward transport into the Indian Ocean.

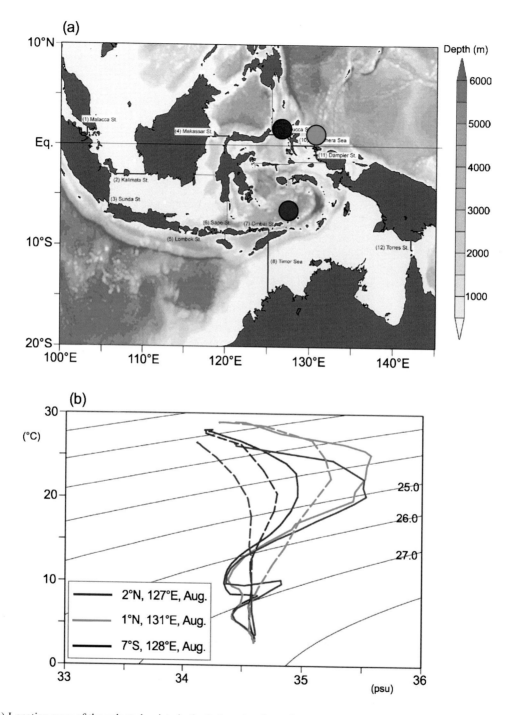

Plate 7. (a) Location map of the selected points in the Indonesian Seas, for which the temperature-salinity distributions are compared between the OFES results and WOA2001 data. (b) Temperature-salinity diagrams for the water column at three locations in August. Solid lines are results from the OFES simulations, and dashed lines are for WOA2001.

observed. The amplitude of the eddies decreases significantly at the entrance to the Gulf of Aden near 51°E. These eddies in the Gulf of Aden seems to affect the circulation pattern within the Gulf and distribution of the Red Sea outflow [*Aiki et al.*, 2006]. Unfortunately, OFES does not capture the realistic outflow from the Red Sea in terms of the transport and water mass properties. Measures to have the realistic Red Sea outflow may be important to simulate the water mass distribution of the northern Indian Ocean appropriately.

In addition to these mesoscale eddies, another type of strong positive disturbances can be seen in the region between 51°E and 52°E during the boreal autumn to winter seasons, especially in 2000/2001 and 2001/2002 winter and in late autumn of 2004. These three events are associated with the migration of the Socotra Gyre to the west of Socotra Island and hence, enhancement of the northward transport across the Socotra passage, between the African continent and Socotra Island. This association can be seen in the time series of the net northward transport above the depth of 1000 m through the Socotra passage and SSHA variation at 52°E (Plate 5). During these events, the large net northward transport through the passage leads the positive SSHA north of the passage by about 1 month.

The horizontal distribution of the SSHA during such a period is shown in Plate 4b for the event in 2004. The positive SSHA, elongated from the Socotra Gyre, intrudes into the region north of the passage, making a large anticyclonic gyre surrounding Socotra Island. In addition, comparison between the time series of the net Socotra passage transport and mean surface velocity within the passage suggests that the variation of the mean surface velocity can be considered as a good measure of the transport variability. This result is consistent with the satellite and in situ hydrographic observations described by *Fratantoni et al.* [2006], and suggests an excellent performance of OFES in reproducing the complex variability in the eddy-rich western boundary region.

6. INDONESIAN THROUGHFLOW

The Indonesian throughflow (ITF), as one of the important boundary conditions for the Indian Ocean, affects the circulations and water mass properties/distributions in the Indian Ocean [e.g., *Hirst and Godfrey*, 1993; *Valsala and Ikeda*, 2007]. There are many attempts to estimate the volume transport of the ITF and its impact on the regional and global ocean circulations and air-sea interactions using global OGCMs and Coupled (ocean-atmosphere) GCMs. Readers can refer to a good review article of *Schiller et al.* [2006] for more details. Although the first-order estimate of the ITF can be obtained with relatively coarse resolution models, a

small grid spacing is essential for the realistic simulation of the ITF, due to many narrow channels and passages within the Indonesian archipelago. OFES can be considered as an ideal tool to evaluate the ITF transport and associated water mass properties in the eddy-resolving OGCM.

The mean pathway of the ITF at the depth of about 30 m in August in the OFES climatological run is shown in Figure 1. OFES reproduces the typical pathway of the upper-layer ITF from the Pacific to Indian Oceans; it mostly originates from the Mindanao Current and passes through Sulawesi Sea, Makassar Strait, and then separates mainly into three branches through Lombok Strait, Ombai Strait, and the Timor Sea to enter into the Indian Ocean.

Plate 6 shows time series of the net ITF transports through the straits across the southern section (Malacca, Sunda, Lombok, Sape, Ombai Straits and Timor Sea) from July 1999 to December 2004, simulated in the OFES hindcast run forced by the QuikScat winds. Both the 3-day interval snapshot values and the monthly mean values are plotted, together with the monthly mean transports obtained in another OFES hindcast run forced by the NCEP/NCAR reanalysis data. Simulated climatological mean net transport forced by the QuikScat winds is 11.5 Sv, with the seasonal amplitude of about 7 Sv taking its maximum (minimum) value in August (February), which is consistent with results from previous observational studies [*Meyers et al.*, 1995; *Wijffels and Meyers*, 2004]. The mean transport, however, decreases by 2 Sv (about 17% of the total values) when forced by NCEP/NCAR reanalysis data. This suggests that uncertainty in the forcing field is one of the important issues for precise determination of the ITF transport.

About 60% of the net transport passes through the Timor Sea, and the Lombok and Ombai Straits share almost the same amount in the remaining transport. This is also the case at the interannual timescale, and therefore the interannual variability in the Timor transport is reflected mostly into the variability in the net ITF transport. Comparison between the 3-day snapshot values and the monthly mean time series indicates existence of large amplitude short-term variability in the ITF. The large intraseasonal variability is observed in the transport through the Lombok and Ombai Straits as well as that through the Timor Sea.

Although the above dynamical aspects of the ITF are simulated rather well in the OFES results, water mass properties in the Indonesian Seas raise an important issue to be addressed in the near future. Plate 7 compares simulated temperature-salinity (T-S) relation of the water column at three locations within the Indonesian archipelago to those from World Ocean Atlas 2001 (WOA2001). Since the WOA2001 data are highly smoothed in time and space, T-S maxima and minima are usually smoothed out significantly. Even so, the comparison demonstrates that the OFES results have unrealistically large subsurface salinity maximum and minimum. For example, at 7°S, 125°E in the Banda Sea, there are salinity maxima at $\sigma = 24.5$ and 26.8, which is around the depth of 120 and 350 m, respectively. At 24.5σ level, the high-salinity water of 35.5 exists and can be traced back to the high-salinity water near the northern coast of Papua New Guinea, suggesting that the water mass originated from the south Pacific Ocean. On the other hand, another high-salinity water of 34.8 at 26.8σ level is originated from the Indian Ocean. It has been known that the vertical mixing process associated with the tidal mixing is strong within the Indonesian archipelago and, therefore, the water mass properties in this area are also significantly modified [*Ffield and Gordon*, 1992; *Schiller et al.*, 1998]. The subsurface high (low)-salinity maxima that appeared in the OFES simulations suggest that the mixing processes within the archipelago in OFES are too weak. It is strongly required to incorporate this additional mixing process indirectly through a parameterization or directly into OFES. This will be an important task to do in the near future.

7. CONCLUDING REMARKS

Results from the eddy-resolving ocean general circulation model, OFES, demonstrate many encouraging outcomes for the mean circulations and their variability in the Indian Ocean. This suggests that OFES can be a promising tool for analyses not only on the basin-scale phenomena but also on the regional variability, including energetic mesoscale eddy activities and narrow, strong boundary currents.

At the same time, however, the results raise several fundamental issues that should be solved to represent more realistic ocean states, and hence, for better understanding of the variability in the Indian Ocean. For example, there is scope to improve the mixed layer processes to represent much appropriate ocean responses as well as the deeper-layer mixing as noted in section 6. To obtain the sea surface temperature properly, surface forcing should also be improved, because there are notable differences in the results forced by NCEP reanalysis data and those forced by Quik-Scat winds. For better representation of the processes associated with fine structures in the coastal areas and narrow frontal regions, much finer grid spacing and/or a nesting-grid application would be inevitable.

Considering the rapid advances in computational sciences, high-resolution ocean circulation modeling is expected to move shortly into the next stage, in which spatial and temporal spectra of the phenomena resolved in the model expand significantly compared to the present models.

Note

1. Data of the J-OFURO product can be accessible at http://dtsv.scc.u-tokai.ac.jp/j-ofuro/.

REFERENCES

Aiki, H., K. Takahashi, and T. Yamagata (2006), The Red Sea outflow regulated by the Indian monsoon, *Cont. Shelf Res.*, 26, 1448–1468.

Anderson, D. L. T., and D. J. Carrington (1993), Modeling interannual variability in the Indian Ocean using momentum fluxes from the operational weather analyses of the United Kingdom Meteorological Office and European Centre for Medium Range Weather Forecasts, *J. Geophys. Res.*, 98, 12, 483–12,499.

Anderson, D. L. T., and D. W. Moore (1979), Cross-equatorial inertial jets with special relevance to very remote forcing of the Somali Current, *Deep-Sea Res.*, 26, 1–22.

Antonov, J. I., S. Levitus, T. P. Boyer, M. E. Conkright, T. O'Brien, and C. Stephens (1998a), World Ocean Atlas 1998, Vol. 1: Temperature of the Atlantic Ocean, NOAA Atlas NESDIS 27. U. S. Government Printing Office, Washington, D. C.

Antonov, J. I., S. Levitus, T. P. Boyer, M. E. Conkright, T. O'Brien, and C. Stephens (1998b), World Ocean Atlas 1998, Vol. 2: Temperature of the Pacific Ocean, NOAA Atlas NESDIS 28. U.S. Government Printing Office, Washington, D. C.

Antonov, J. I., S. Levitus, T. P. Boyer, M. E. Conkright, T. O'Brien, C. Stephens, and T. Trotsenko (1998c), World Ocean Atlas 1998, Vol. 3: Temperature of the Indian Ocean, NOAA Atlas NESDIS 29. U.S. Government Printing Office, Washington, D. C.

Boyer, T. P., S. Levitus, J. I. Antonov, M. E. Conkright, T. O'Brien, and C. Stephens (1998a), World Ocean Atlas 1998, Vol. 4: Salinity of the Atlantic Ocean, NOAA Atlas NESDIS 30, U.S. Government Printing Office, Washington, D. C.

Boyer, T. P., S. Levitus, J. I. Antonov, M. E. Conkright, T. O'Brien, and C. Stephens (1998b), World Ocean Atlas 1998, Vol. 5: Salinity of the Pacific Ocean, NOAA Atlas NESDIS 31. U.S. Government Printing Office, Washington, D. C.

Boyer, T. P., S. Levitus, J. I. Antonov, M. E. Conkright, T. O'Brien, C. Stephens, and B. Trotsenko (1998c), World Ocean Atlas 1998, Vol. 6: Salinity of the Indian Ocean, NOAA Atlas NESDIS 32. U.S. Government Printing Office, Washington, D. C.

Cai, W. (2006), Antarctic ozone depletion causes an intensification of the Southern Ocean super-gyre circulation, *Geophys. Res. Lett.*, 33, L03712, doi:10.1029/2005GL024911.

Chassignet, E. P., and Z. D. Garraffo (2001), Viscosity parameterization and the Gulf Stream separation, in *Aha Huliko'a Hawaiian Winter Workshop*, University of Hawaii, edited by P. Muller and D. Henderson, pp. 37–41.

Coward, A. C., B. A. de Cuevas, and D. J. Webb (2002), Early results from a 1/12° × 1/12° global ocean model. Poster from WOCE Final Conference, http://www.soc.soton.ac.uk/JRD/OCCAM/POSTERS.

Cox, M. D. (1979), A numerical study of Somali Current eddies, *J. Phys. Oceanogr.*, *9*, 311–326.

Cox, M. D. (1981), A numerical study of surface cooling processes during summer in the Arabian Sea, in *Monsoon Dynamics*, edited by M. J. Lighthill and R. P. Pearce, Cambridge University Press, Cambridge.

Du, Y., T. Qu, G. Meyers, Y. Masumoto, and H. Sasaki (2005), Seasonal heat budget in the mixed layer of the southeastern tropical Indian Ocean in a high-resolution ocean general circulation model, *J. Geophys. Res.*, *110*, C04012, doi:10.1029/2004JC002845.

Ffield, A., and A. L. Gordon (1992), Vertical mixing in the Indonesian thermocline, *J. Phys. Oceanogr.*, *22*, 184–195.

Fratantoni, D. M., A. S. Bower, W. E. Johns, and H. Peters (2006), Somali Current rings in the eastern Gulf of Aden, *J. Geophys. Res.*, *111*, C09039, doi:10.1029/2005JC003338.

Gartenicht, U., and F. Schott (1997), Heat fluxes of the Indian Ocean from a global eddy-resolving model, *J. Geophys. Res.*, *102*, 21,147–21,159.

Godfrey, J. S., R.-J. Hu, A. Schiller, and R. Fiedler (2007), Explorations of the annual mean heat budget of the tropical Indian Ocean. Part I: Studies with an idealized model, *J. Clim.*, *20*, 3210–3228.

Hastenrath, S., and L. Greischar (1993), The monsoonal heat budget of the hydrosphere-atmosphere system in the Indian Ocean sector, *J. Geophys. Res.*, *98*, 6869–6881.

Hirst, A. C., and J. S. Godfrey (1993), The role of Indonesian throughflow in a global OGCM, *J. Phys. Oceanogr.*, *23*, 1057–1086.

Hsiung, J. (1985), Estimates of global oceanic meridional heat transport, *J. Phys. Oceanogr.*, *15*, 1405–1413.

Hu, R.-J., and J. S. Godfrey (2007), Explorations of the annual mean heat budget of the tropical Indian Ocean. Part II: Studies with a simplified ocean general circulation model, *J. Clim.*, *20*, 3229–3248.

Hurlburt, H. E., and P. J. Hogan (2000), Impact of 1/8° to 1/64° resolution on Gulf Stream model-data comparisons in basin-scale subtropical Atlantic Ocean models, *Dyn. Atmos. Oceans, 32*, 283–329.

Iskandar, I., T. Tozuka, H. Sasaki, Y. Masumoto, and T. Yamagata (2006), Intraseasonal variations of surface and subsurface currents off Java as simulated in a high-resolution ocean general circulation model, *J. Geophys. Res.*, *111*, C12015, doi:10.1029/2006JC003486.

Josey, S. A., E. C. Kent, and P. K. Taylor (1999), New insights into the ocean heat budget closure problem from analysis of the SOC air-sea flux climatology, *J. Clim.*, *12*, 2856–2880.

Kalnay, E., M. Kanamitsu, R. Kistler, W. Collins, D. Deaven, L. Gandin, M. Iredell, S. Saha, G. White, J. Woollen, Y. Zhu, M. Chelliah, W. Ebisuzaki, W. Higgins, J. Janowiak, K. C. Mo, C. Ropelewski, A. Leetmaa, R. Reynolds, and R. Jenne (1996), The NCEP/NCAR 40-year reanalysis project, *Bull. Am. Meteorol. Soc.*, *77*, 437–471.

Kindle, J. C., and J. D. Thompson (1989), The 26- and 50-day oscillations in the western Indian Ocean: Model results, *J. Geophys. Res.*, *94*, 4721–4736.

Kubota, M., N. Iwasaka, S. Kizu, M. Konda, and K. Kutsuwada (2002), Japanese ocean flux data sets with use of remote sensing observations (J-OFURO), *J. Oceanogr.*, *58*, 213–225.

Large, W. G., J. C. McWilliams, and S. C. Doney (1994), Oceanic vertical mixing: A review and a model with a nonlocal boundary layer parameterization, *Rev. Geophys.*, *32*, 363–403.

Lee, T., and J. Marotzke (1998), Seasonal cycles of meridional overturning and heat transport of the Indian Ocean, *J. Phys. Oceanogr.*, *28*, 923–943.

Lighthill, M. J. (1969), Dynamic response of the Indian Ocean to the onset of the southwest monsoon, *Philos. Trans. R. Meteorol. Soc. A, A265*, 45–92.

Maltrud, M. E., and J. L. McClean (2005), An eddy resolving global 1/10° ocean simulation, *Ocean Modell., 8*, 31–54.

Masumoto, Y., H. Sasaki, T. Kagimoto, N. Komori, A. Ishida, Y. Sasai, T. Miyama, T. Motoi, H. Mitsudera, K. Takahashi, H. Sakuma, and T. Yamagata (2004), A fifty-year eddy-resolving simulation of the world ocean: Preliminary Outcomes of OFES (OGCM for the Earth Simulator), *J. Earth Simul.*, *1*, 35–56.

McCreary, J. P. (1985), Modeling equatorial ocean circulation, *Annu. Rev. Fluid Mech.*, *17*, 359–409.

McCreary, J. P., and P. K. Kundu (1988), A numerical investigation of the Somali Current during the Southwest Monsoon, *J. Mar. Res.*, *46*, 25–58.

McCreary, J. P., P. K. Kundu, and R. L. Molinari (1993), A numerical investigation of dynamics, thermodynamics and mixed-layer processes in the Indian Ocean, *Prog. Oceanogr.*, *31*, 181–244.

Meyers, G., Bailey, R. J., and A. P. Worby (1995), Geostrophic transport of Indonesian throughflow, *Deep-Sea Res.*, *42*, 1163–1174.

Ogata, T., H. Sasaki, V. S. N. Murty, M. S. S. Sarma, and Y. Masumoto (2008), Intraseasonal meridional current variability in the eastern equatorial Indian Ocean, *J. Geophys. Res.*, doi:10.1029/2007JC004331, in press.

Ohfuchi, W., H. Sasaki, Y. Masumoto, and H. Nakamura (2005), Mesoscale resolving simulations of global atmosphere and ocean on the Earth Simulator, *Eos Trans. AGU*, *86*(5), 45–46.

Pacanowski, R. C., and A. Gnanadesikan (1998), Transit response in a z-level ocean model that resolves topography with partial-cells, *Mon. Weather Rev.*, *126*, 3248–3270.

Pacanowski, R. C., and S. M. Griffies (2000), *MOM 3.0 Manual*, Geophysical Fluid Dynamics Laboratory/National Oceanic and Atmospheric Administration, 680 pp.

Rosati, A., and K. Miyakoda (1988), A general circulation model for upper ocean circulation, *J. Phys. Oceanogr.*, *18*, 1601–1626.

Saji, N. H., B. N. Goswami, P. N. Vinayachandran, and T. Yamagata (1999), A dipole mode in the tropical Indian Ocean, *Nature, 401*, 360–363.

Sasai, Y., A. Ishida, Y. Yamanaka, and H. Sasaki (2004), Chlorofluorocarbons in a global ocean eddy-resolving OGCM: Pathway and formation of Antarctic Bottom Water, *Geophys. Res. Lett.*, *31*, L12305, doi:10.1029/2004GL019895.

Sasaki, H., Y. Sasai, M. Nonaka, Y. Masumoto, and S. Kawahara (2006), An eddy-resolving simulation of the quasi-global ocean driven by satellite-observed wind field—Preliminary outcomes from physical and biological fields, *J. Earth Simul.*, *6*, 35–49.

Sasaki, H., M. Nonaka, Y. Masumoto, Y. Sasai, H. Uehara, and H. Sakuma (2008), An eddy-resolving hindcast simulation of the quasiglobal ocean from 1950 to 2003 on the Earth Simulator, in *High Resolution Numerical Modelling of the Atmosphere and Ocean*, edited by K. Hamilton and W. Ohfuchi, Chapter 10, pp. 157–185, Springer, New York.

Schiller, A., J. S. Godfrey, P. C. McIntosh, G. Meyers, and S. Wijffels (1998), Seasonal near-surface dynamics and thermodynamics of the Indian Ocean and Indonesian throughflow in a global ocean general circulation model, *J. Phys. Oceanogr., 28,* 2288–2312.

Schiller, A., S. Wijffels, and J. Sprintall (2006), Variability of the Indonesian Throughflow: A review and model-to-data comparison, in *Global Climate Change and Response of Carbon Cycle in the Equatorial Pacific and Indian Oceans and Adjacent Landmasses,* Elsevier Oceanogr. Ser., vol. 73, edited by H. Kawahata and Y. Awaya, pp. 175–209, Elsevier, Amsterdam.

Schott, F., and J. P. McCreary (2001), The monsoon circulation of the Indian Ocean, *Prog. Oceanogr., 51,* 1–123.

Semtner, A. J., Jr., and R. M. Chervin (1992), Ocean general circulation from a global eddy-resolving model, *J. Geophys. Res., 97,* 5493–5550.

Smith, R. D., M. E. Maltrud, F. O. Bryan, and M. W. Hecht (2000), Numerical simulation of the North Atlantic Ocean at 1/10°, *J. Phys. Oceanogr., 30,* 1532–1561.

Speich, S., B. Blanke, P. de Vries, S. Drijfhout, K. Doos, A. Ganachaud, and R. Marsh (2002), Tasman leakage: A new route in the global ocean conveyor belt, *Geophys. Res. Lett., 29,* doi:10.1029/2001GL014586.

Speich, S., B. Blanke, and W. Cai (2007), Atlantic meridional overturning circulation and the Southern Hemisphere supergyre, *Geophys. Res. Lett., 34,* L23614, doi:10.1029/2007GL031583.

Tomczak, M., and J. S. Godfrey (1994), *Regional Oceanography: An Introduction,* Pergamon Press, Oxford.

Valsala, K. V., and M. Ikeda (2007), Pathways and effects of the Indonesian Throughflow water in the Indian Ocean using Particle trajectory and Tracers in an OGCM, *J. Clim., 20,* 2994–3017.

Vinayachandran, P. N., T. Kagimoto, Y. Masumoto, P. Chauhan, S. R. Nayak, and T. Yamagata (2005), Bifurcation of the East India Coastal Current east of Sri Lanka, *Geophys. Res. Lett., 32*(15), L15606, doi:10.1029/2005GL022864.

Wacongne, S., and R. Pacanowski (1996), Seasonal heat transport in a primitive equations model of the tropical Indian Ocean, *J. Phys. Oceanogr., 26,* 2666–2699.

Wijffels, S., and G. Meyers (2004), An intersection of oceanic waveguides: Variability in the Indonesian throughflow region, *J. Phys. Oceanogr., 34,* 1232–1253.

Woodberry, K. E., M. E. Luther, and J. J. O'Brien (1989), The wind-driven seasonal circulation in the southern tropical Indian Ocean, *J. Geophys. Res., 94,* 17,985–18,002.

Wyrtki, K. (1971), *Oceanographic Atlas of the International Indian Ocean Expedition,* National Science Foundation, Washington, D. C., 531 pp.

Wyrtki, K. (1973), An equatorial jet in the Indian Ocean, *Science, 181,* 262–264.

Y. Masumoto, Frontier Research Center for Global Change, Japan Agency for Marine-Earth Science and Technology, Yokohama, Kanagawa, Japan.

Y. Masumoto and Y. Morioka, Graduate School of Science, University of Tokyo, Tokyo, Japan.

H. Sasaki, Earth Simulator Center, Japan Agency for Marine-Earth Science and Technology, Yokohama, Kanagawa, Japan.

Toward a Physical Understanding of the North Atlantic: A Review of Model Studies in an Eddying Regime

Matthew W. Hecht

Computational Physics Group, CCS Division, Los Alamos National Laboratory, Los Alamos, New Mexico, USA

Richard D. Smith

Los Alamos, National Laboratory, Los Alamos, New Mexico, USA

We survey progress in modeling of the North Atlantic Ocean since a set of reviews last appeared around a dozen years ago. Predictions that a threshold in resolution was yet to be crossed proved to be true: 1/10° or higher-resolution models are now capable of producing Gulf Stream separation at the observed location of Cape Hatteras, and other features of the Gulf Stream/North Atlantic Current system are much more realistically represented. The surface expression of eddy kinetic energy agrees well between models and observations. Mesoscale eddies also invigorate the deep circulation, which in turn improves the mean circulation via enhanced current-topography interactions. The successful generation of these prominent features of the circulation, however, is found to depend sensitively on model configuration issues that are not yet fully understood. More generally, the North Atlantic basin remains a particularly well-observed region in which high-resolution models can be used to explore fundamental issues in physical oceanography, such as the manner in which the turbulent variability feeds back on the large scale mean circulation, and the related issue of the range of spatial and temporal scales required to adequately describe the circulation.

1. INTRODUCTION

Ocean modeling of the North Atlantic has crossed over a threshold into a regime in which the variability of the circulation is comparable to that which is observed. Long-standing biases in the mean circulation have to a great degree been corrected, suggesting a critical role of variability and nonlinearities in determining the mean flow. The last set of review papers addressing modeling of the basin were

published just more than a decade ago [*Stammer and Böning*, 1996; *Böning and Bryan*, 1996; *Dengg et al.*, 1996] as models were approaching that threshold. Our aim here is to survey more recent developments that have contributed to our emerging understanding of the complex circulation of the North Atlantic.

Early theoretical work was naturally focused on the mean circulation, in particular the Gulf Stream. An early map showing the Gulf Stream was drawn by the most important physical scientist of his day, Benjamin Franklin [*Stommel*, 1965]; awareness of this strong current presented an advantage to American captains attempting to slip through the British naval blockade during the American Revolution. The elegant work of *Stommel* [1948] explained the intensification of the Gulf Stream and other western boundary currents as arising

Ocean Modeling in an Eddying Regime
Geophysical Monograph Series 177
10.1029/177GM15

not from just rotation but from rotation of a spherical planet (the β effect), bringing a first and still critical level of understanding to that portion of the gyre not addressed by the theory of *Sverdrup* [1947]. Following on Stommel's insight into the intensification of western boundary currents, *Munk* [1950] presented an understanding of the viscous forces that regulate that intensification. These earlier theoretical works were linear and are therefore limited by the neglect of the advection of relative vorticity in the boundary current.

The modern recognition of the tremendous degree of oceanic variability required more extensive observations. Early maps of eddy energy were derived from ship drift data [*Wyrtki et al.*, 1976], while XBT data sets indicated the generation of eddies in regions of strong shear [*Dantzler*, 1976, 1977].

Much of the initial study of the influence of mesoscale variability was focused on the North Atlantic, as it was the best sampled ocean basin, with a strong source of shear instability in the Gulf Stream. Other characteristics of the basin include its extensive span in latitude, with exchange of waters across the North Atlantic Sill. It is relatively compact in longitude, raising the influence of continental and mid-basin topography. The North Atlantic Current forms the most poleward penetration of any of the subtropical gyres of the World Ocean, at least in its modern configuration of the Holocene; evidence from ocean sediment cores suggests that it followed an eastward drift to the Mid-Atlantic Ridge at the Last Glacial Maximum [*Robinson et al.*, 1995]. A much discussed feature of the northern basin on climate timescales, which may be related at some level to changes in the path of the North Atlantic Current, is the variability in deep water formation and meridional overturning [*Broecker et al.*, 1985; *Rahmstorf*, 2002]. The basin not only presents a context in which to develop our understanding of physical oceanography but also presents a challenge to the understanding of prominent modes of climate variability, a topic that we touch on through discussion of the North Atlantic Current. We expect the relevance of mesoscale eddy feedback on mean North Atlantic circulation, and the oceanic response to changes in forcing, to be the subject of research over the coming years.

Early model-based research into the role of mesoscale eddy variability in Atlantic-like domains made use of idealized geometries. Using a flat-bottomed, rectangular domain with two layers on a β-plane, *Holland and Lin* [1975] found strong feedback of mesoscale eddies on the mean circulation, and in particular on the deep flow. Whereas *Holland and Lin* [1975] primarily studied eddy generation through baroclinic instability, *Robinson et al.* [1977] focused on barotropic instability in a five level primitive equation model at 40-km resolution, again with an idealized flat-bottomed domain.

Semtner and Mintz [1977] added an idealized continental slope to the problem, reporting on sensitivity to dissipation (and establishing an enduring preference for biharmonic closure over Laplacian at high spatial resolution), identifying the process of baroclinic-to-barotropic energy conversion as dominant in the western boundary region, and also producing the diagnosis of an equivalent eddy diffusivity.

The 1/3°, 18 level study of *Cox* [1985] was set in an idealized Atlantic domain. With the inclusion of bottom topography, the first realistic primitive equation model of the North Atlantic in an eddying regime was presented by *Böning* [1989], followed shortly by a factor of two horizontal refinement by *Böning and Budich* [1992]. At this point, in the early 1990s, ocean general circulation modeling had become a useful tool for the investigation of physical oceanography involving mesoscale variability, along with idealized process studies and physical theory.

As ocean general circulation modeling in an eddying regime became established, focus on the North Atlantic was maintained with the Community Modeling Effort (CME) [*Bryan and Holland*, 1989]. This period also saw groundbreaking research on the Southern Ocean, with the Fine Resolution Antarctic Model, *FRAM Group* [1991], and the nearly global study of *Semtner and Chervin* [1988] at 0.5° resolution with 20 levels, prognostic in the thermocline, with climatological restoring below. Careful analysis of the CME by *Treguier* [1992] confirmed and quantified the predominance of baroclinic-to-barotropic energy conversion.

Mean circulation was better simulated in some respects at resolutions as high as 1/6°, but in the Northwest Atlantic, comparison with observations, even those from as early as *Iselin* [1936], still showed first-order biases in the North Atlantic Current and late separation of the Gulf Stream from the continental slope. These biases persisted even as *Böning et al.* [1995] showed that spurious cross-frontal mixing (the "Veronis effect"; *Veronis* [1975]), associated with explicit horizontal diffusion, gave rise to a short circuiting of the meridional heat transport in models. Use of an isopycnal parameterization of eddy-driven tracer mixing [*Gent and McWilliams*, 1990] improved northward heat transport but did not result in correction of the Gulf Stream or North Atlantic Current biases.

In the review of *Böning and Bryan* [1996], which was largely based on the CME simulations, it was demonstrated that despite persistence of low eddy energy levels over broad areas, models were capable of producing useful estimates of northward heat transport and amenable to detailed analysis of the underlying mechanisms. Their analysis of the heat transport showed that the thermohaline-driven meridional overturning dominates at low latitudes, while the wind-driven gyre provides the stronger contribution at high latitudes. The details of this decomposition were found to depend on horizontal resolution, with resolution-dependent differences in the severity of the Veronis effect and path of the Gulf Stream Extension. In the context of idealized coupled experimentation, with ocean resolutions ranging from 4° down to 0.25°, *Fanning and Weaver* [1997] also discussed the dependence of oceanic poleward heat transport on resolution, emphasizing the need to improve representation of the Gulf Stream.

New insight was also being gained from observational data, both on a climate timescale and in a near-instantaneous sense. The insight of *Broecker et al.* [1985] that the rate of deep water formation in the North Atlantic had undergone periods of dramatic and sudden change was based on the Greenland ice core record. On the shorter end of the temporal spectrum, satellite-based altimetry became established in the 1990s as one of the most important sources of data in oceanography with the very accurate measurements of the TOPEX/Poseidon mission (see *Wunsch and Stammer* [1995] for early spectral analysis of this altimetry).

The TOPEX/Poseidon mission produced a slightly revised estimate, relative to the earlier GEOSAT mission, for the apparently linear relationship between first internal Rossby radius R_1 and eddy length scale L_o, with $L_o = 1.7R_1 + 86$km [*Stammer and Böning*, 1996]. It is interesting to note that the 1/3° CME model produced realistic eddy length scales over the low-to-mid latitudes [*Stammer and Böning*, 1992], and the 1/6° version of the model produced fairly realistic eddy length scales even at higher latitudes [*Beckmann et al.*, 1994], and nevertheless further increases in resolution were required to finally produce a realistic Gulf Stream/North Atlantic Current system.

Beyond this slight revision of eddy length scale, the TOPEX/Poseidon mission provided far more accurate estimates of variability over the less active regions of the ocean and allowed for greater discrimination of model fidelity. *Fu and Smith* [1996] made use of the data in the first detailed comparison of a global eddy-admitting ocean model with satellite altimetry, considering mean circulation, mesoscale variability and amplitude and phase of the annual cycle. The model variability in sea surface height was identified as being small by a factor of two (the model resolution was at 0.28°), with the deficiency attributed to insufficient resolution of the first internal Rossby radius.

The remote sensing data also provided further detail on the long-recognized problem of Gulf Stream separation and the path of the North Atlantic Current, allowing for quantitative comparison of model-produced variability with the surface expression of that variability in the sea surface height field over the entire extent of the current system. Another component of the current system included in this new and greatly clarified bird's eye view was the Azores Current, which also

branches from the Gulf Stream around the Southeast Newfoundland Rise [*Klein and Siedler*, 1989]. The more precise the instruments, the better quantified are the levels of eddy kinetic energy in more quiescent regions, providing sufficient data from which to characterize the turbulent cascade, with frequency-dependent spectral decays largely confirming quasigeostrophic theory and significant anisotropy appearing only at the longer timescales associated with Rossby waves [*Stammer and Böning*, 1996]. More recent analysis of *Eden* [2007] indicates the anisotropy to be restricted to latitudes below 30°, while the issue of turbulent cascade has also been revisited. The paper of *Scott and Wang* [2005] explores the question of why an inverse cascade is observed in surface altimetry, in accord with two-dimensional theory, despite the expectation that one should see a strong signature of the first baroclinic mode.

While other model deficiencies mentioned above were recognized, the greatest attention in the mid-1990s was paid to the unsolved problem of Gulf Stream separation. *Dengg et al.* [1996] provided an excellent review of the subject, including a description of the Gulf Stream itself, describing the very stable location of separation at Cape Hatteras, the recirculation gyres and downstream development, and the vertical and horizontal profiles of the Stream (see Figure 1 for the geography of the region). Modeling and theoretical work seemed to indicate that separation was not simply a function of one dominant physical process but a more vexing balance of a number of important processes including the effects of advective nonlinearity, bathymetry (including bottom topography and coastline), and stratification (the possible role of the interaction of the Gulf Stream with the Deep Western Boundary Current was commented on but underappreciated).

Stammer and Böning [1996] and *Böning and Bryan* [1996] together provide a review of what we would now call "eddy-admitting" modeling of the North Atlantic. We proceed to survey progress made since that time.

2. GULF STREAM SEPARATION

As discussed by *Dengg et al.* [1996], various hypotheses based on a number of physical mechanism were presented to explain Gulf Stream separation in idealized circumstances. When idealization and limiting assumptions were relaxed, however, no single hypothesis seemed adequate to explain the ocean modeling failure of the Stream to separate at the observed location.

Even while refinement from 1° to 1/3° to 1/6° in the CME model failed to solve the Gulf Stream separation problem, producing instead a more pronounced anticyclonic stationary meander as the Stream overshot the observed separation point

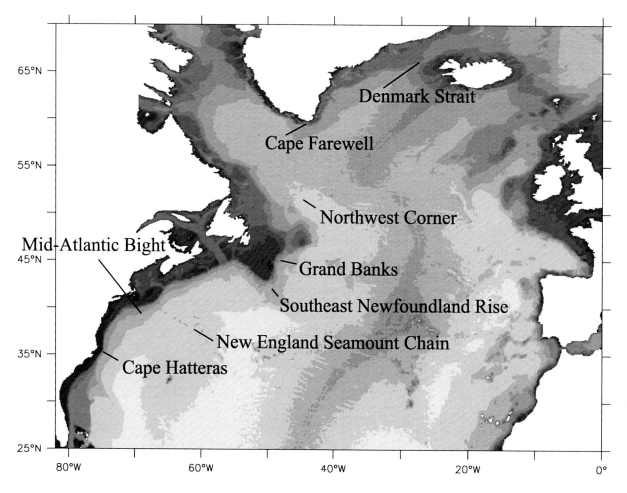

Figure 1. Topography of the North Atlantic with labeling of a number of the features referred to in the text.

of Cape Hatteras in which much of the Stream's kinetic energy was dissipated. *Böning and Bryan* [1996] insightfully raised the likelihood that a modeling threshold was yet to be crossed as the first internal Rossby radius came within resolution.

This insight that the crossing of a threshold might be imminent was strongly suggested in the 1/6° study of *Chao et al.* [1996] and then proven out in the North Atlantic regional study of *Smith et al.* [2000]. The simulation was based on the same Parallel Ocean Program (POP) [*Smith et al.*, 1992; *Dukowicz and Smith* 1994] used in the earlier 0.28° global simulation of *Maltrud et al.* [1998], but with a 0.1° Mercator projection and 40 vertical levels spanning the Atlantic from approximately 20°S to somewhat beyond the North Atlantic Sill. Lateral boundary conditions were provided through restoring of hydrography to climatology within buffer zones; forcing was based on climatological means for heat flux and salinity with daily reanalysis winds. Within five or so years

of spin-up, kinetic energies had largely equilibrated, and the Gulf Stream was seen to separate at the correct location of Cape Hatteras without evidence of an anticyclonic meander at its separation from the coast, as shown in Plate 1 (taken from the newsletter piece of *Bryan and Smith* [1998]); see also Plate 2. The early 1/12° isopycnal model simulation of *Paiva et al.* [1999] also showed Gulf Stream separation at Cape Hatteras, providing a base state for the refinement of *Chassignet and Garraffo* [2001]. The results of *Paiva et al.* [1999] provided support for the importance of achieving a sufficiently inertial regime as a prerequisite for separation to occur at Cape Hatteras.

A zonal-average of the model's first internal Rossby radius was shown in Figure 1 of *Smith et al.* [2000], reproduced here as our Figure 2. The act of "resolving" the Rossby radius is not as simple as maintaining a grid spacing less than or equal to the first internal Rossby radius. A span of

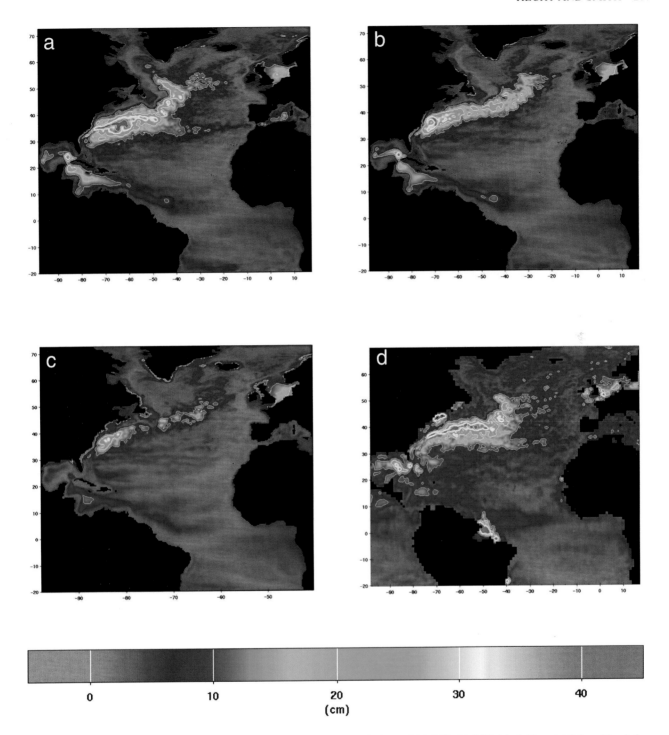

Plate 1. Sea surface height variability of North Atlantic models at resolutions of (a) 0.1°, (b) 0.2°, (c), 0.4°, and (d) for a blended ERS-TOPEX/POSEIDON altimetric product [*Le Traon and Ogor*, 1998]. Figure is taken from *Bryan and Smith* [1998].

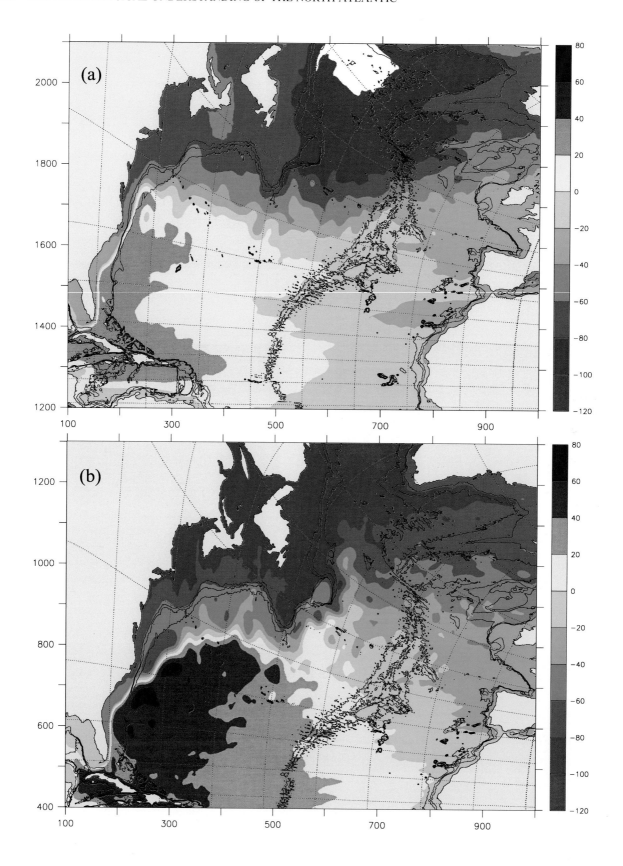

several grid lengths is required if a numerical ocean model is to be capable of representing a feature. The eddy length scale is considerably larger than the Rossby radius, as mentioned above in reference to the paper of *Stammer and Böning* [1996], although it exhibits a linear dependence on the Rossby radius. It was not clear then, a priori, that 0.1° would prove sufficient to cross this threshold. Later investigation showed 0.1° as barely adequate, with strong sensitivities to model configuration [*Chassignet and Garraffo*, 2001; *Eden and Böning*, 2002; *Maltrud and McClean*, 2005; *Bryan et al.*, 2007], as discussed further in the next section. It is important to note that all of these works solve the same basic equations of fluid flow as in the now-classic paper of *Bryan* [1969], even if questions of numerical implementation remain consequential.

3. MODEL SENSITIVITIES

Sensitivity studies in a strongly eddying regime have most often been conducted in a North Atlantic regional context, with sensitivity to lateral dissipation most thoroughly explored.

Most of our discussion in this section addresses modeling studies that include thermohaline as well as wind forcing, with dozens of levels or layers in the vertical. A satisfactory convergence study has yet to be attempted with such a model but has been performed with an isopycnal model with up to six layers without heat and fresh water forcings: The study of *Hurlburt and Hogan* [2000] focused on the Gulf Stream region of the North Atlantic and found great improvement in the pathway of the Gulf Stream and in the strength of abyssal flows when increasing resolution from 1/8° to 1/6°. Further refinement to 1/32° brought additional moderate improvement in these features. The authors see evidence of convergence at 1/64° resolution, in some regions, with more substantial dependence on resolution remaining evident in the region of the Grand Banks.

3.1. Sensitivity to Lateral Dissipation

Despite (or perhaps because of) the relatively thorough consideration that sensitivity to lateral dissipation has received, we address the question only briefly here. The topic is taken up in greater depth by *Hecht et al.* [2008], in this volume, and is also discussed in section 5, as the North Atlantic Current and its penetration into the region of the Northwest Corner show particularly strong dependence on model configuration.

The work of *Chassignet and Garraffo* [2001], however, deserves mention here as well as in the report of *Hecht et al.* [2008]: They demonstrate an effective approach of combining Laplacian and biharmonic forms of dissipation at 1/12° in the layered Miami Isopycnal Coordinate Ocean Model in response to unsatisfactory results in their model with biharmonic dissipation alone. Also deserving mention is the work of *Smith and Gent* [2004], where they demonstrate that adequate levels of energy can be obtained with Laplacian viscosity if the anisotropic form of *Smith and McWilliams* [2003] is implemented with appropriate coefficients; they find improvement in water mass properties and meridional heat transport with the first use of an anisotropic form of the Gent-McWilliams isopycnal tracer mixing scheme [*Gent and McWilliams*, 1990]. *Bryan et al.* [2007] identify a relatively narrow range of acceptable values of biharmonic lateral mixing in the *z*-coordinate POP model and further clarifies the roles of model resolution and lateral mixing with a limited study of grid convergence.

The range of acceptable values of lateral mixing coefficients is determined largely by the circulation the model produces, underlining the point that model configuration retains an aspect of tuning even at the higher resolutions discussed in this paper. The models have not converged at 1/10° or 1/12°. The sensitivity to parameter values, particularly evident in the path of the Gulf Stream, is consistent with the findings of *Hurlburt and Hogan* [2000] in a slightly simplified context where they reported needing resolutions of 1/16° and higher to produce a robust path.

3.2. Sensitivity to Topography

Barnier et al. [2006] attributed a part of their success in achieving some penetration of the North Atlantic Current into the Northwest Corner in their 1/4° global model to the use of a partial-step representation of topography (as explained by *Adcroft et al.* [1997]), along with the use of a more accurate advection scheme. The partial-step representation of topography has also been credited with significant improvement in Labrador Sea circulation, as discussed in section 5.

Plate 2. (on page 7) Time-averaged North Atlantic sea surface height from a global simulation on the 0.1° "displaced pole" grid of *Maltrud and McClean* [2005] (a), and on a specially configured North Atlantic regional version of the same grid designed to explore the sensitivity to lateral boundary conditions (b). Imposition of restoring lateral boundary conditions is sufficient to correct the separation of the Gulf Stream and the path of the North Atlantic Current. Notice that the map projection is in terms of model grid indices (*i, j*) showing the distortion associated with the grid. Lines of longitude and latitude are overlain in black.

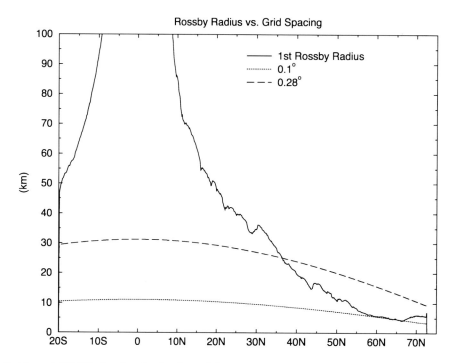

Figure 2. From *Smith et al.* [2000], showing the first baroclinic Rossby radius, temporally and zonally averaged from their 0.1° North Atlantic model, along with grid spacings of the 0.1° model and the 0.28° model of *Maltrud et al.* [1998].

Further analysis of the model presented by *Barnier et al.* [2006] identifies the free-slip boundary condition as being essential to bringing out the improvements attributed to the formulation of advection, as those improvements were lost with reversion to the more commonly used no-slip boundary condition [*Penduff et al.*, 2007].

The use of partial steps allows for a smoother representation of bottom topography in regions of shallow slope. Explicit smoothing of topography was also noted by *Barnier et al.* [2006], motivated by the suggestive work of *Penduff et al.* [2002]. The impact of smoothing of topography probably merits further investigation in strongly eddying ocean modeling where eddy-mean flow excitation of the deep flow would appear to raise the importance of more accurate representation of topographic interaction.

Horizontal grid discretization of course determines where the bottom depth is to be defined, and so is intertwined with topographic sensitivity. This point is touched upon in the following subsection.

3.3. Sensitivity to Domain

Following on the experience of *Smith et al.* [2000], a fully global 0.1° configuration of the POP model was developed by *Maltrud and McClean* [2005] with the expectation that the North Atlantic sector of the global model would look much as it did in the earlier regional simulation. It did not, and subsequent sensitivity studies were performed to understand and correct the somewhat late Gulf Stream separation and lack of penetration of the North Atlantic Current into the Northwest Corner within the global simulation (see figures 7 and 12 of Maltrud and McClean [2005]). These subsequent studies identified sensitivities to dissipation and to the use of partial cells, although neither was sufficient to fully correct the Gulf Stream/North Atlantic Current System (Bryan et al., personal communication).

The lateral boundary conditions differed, of course, between regional and global configurations, as well as the discretization of the grid. The grid used by *Smith et al.* [2000] was a Mercator projection (uniform aspect ratio, $dx/dy = 1$) of a latitude/longitude grid, whereas the global grid of *Maltrud and McClean* [2005] used a displaced pole in the northern hemisphere [see *Smith and Gent*, 2002, and references therein], with the pole located in Hudson's Bay. The gridding of the southern hemisphere remained Mercator, while the gridding over the Northwest Atlantic departed greatly from that simpler projection, with grid cell aspect ratios departing significantly from one.

Lateral boundary conditions in the regional model were set through restoring of potential temperature and salinity within buffer zones, whereas in the global configuration, they were provided by the model itself in a more natural but less constrained fashion. A regional subdomain of the global model was extracted to evaluate the influence of lateral boundary conditions, with the location of lateral boundaries where restoring conditions were applied corresponding approximately to those of *Smith et al.* [2000]. As shown by *Hecht et al.* [2006], the constraint of restoring boundary conditions at the edges of the regional domain was sufficient to correct the Gulf Stream separation and send the North Atlantic Current into the Northwest Corner. This result is presented in our Plate 2, which also indicates the orientation of the underlying model grid lines of the displaced-pole grid.

Ongoing work in the 0.1° POP global model suggests the more satisfying result that the combined use of a more regular tripolar grid [*Smith and Gent*, 2002] and partial bottom cells [*Adcroft et al.*, 1997] is sufficient to correct the Gulf Stream separation and Northwest Corner, even using the full global domain (Maltrud et al., personal communication).

A number of the studies we discuss were performed with nested models [*Zhai et al.*, 2004; *Dietrich et al.*, 2004; *Greatbatch and Zhai*, 2006; *Chanut et al.*, 2008], and the paper of *Capet et al.* [2008] in this same volume also present studies performed with nesting. This approach has emerged from the domain of research and development, serving now to focus resolution on the region of interest in a particular study while allowing the lateral boundary conditions of that highly resolved domain to evolve with time, and, in the case of two-way nesting, to adjust to the state of the more highly resolved subdomain.

3.4. Numerical Formulation

The requirement for resolution may, to some extent, be reduced by the use of improved numerical schemes. In particular, *Dietrich et al.* [2004] make a case for the advantage of higher order numerical treatments in the course of their exploration of the sensitivity of Gulf Stream separation to viscosity. Using the DieCAST model [*Dietrich et al.*, 1997], which has fourth-order accurate numerics (as contrasted with the second-order accurate numerics of most other models), they demonstrate separation at Cape Hatteras with sufficiently low dissipation even with relatively moderate resolution of 1/6° in the separation (with boundary conditions provided by a 1/2° version of the model). They take particular note of the sensitivity of the Deep Western Boundary Current to viscosity, arguing for the importance of the mechanism of *Thompson and Schmitz* [1989]. It should be noted that DieCAST also differs from the other primitive equation models we discuss through its spatial discretization and by the choice to allow static instability to form (the model is run without convective adjustment).

Barnier et al. [2006] provide support for the advantage of numerical schemes with physically based conservation properties using the energy and estrophy-conserving scheme of *Arakawa and Lamb* [1981] in the momentum equations. The improvements they find in the Gulf Stream/North Atlantic Current within a 1/4° global model are jointly attributed to the transport scheme and the use of partial bottom cells (discussed above).

It should also be noted here that whereas conventional wisdom would suggest that advective errors become less of a problem with increasing resolution, this only applies in a straightforward way to laminar flow: *Griffies et al.* [2000] point out that increased diapycnal mixing of a spurious nature will be associated with increased eddy activity. The extent to which these advective errors may compromise eddy-resolving applications has never been directly addressed and remains an important question.

Questions of numerical formulation tend to bring one back to the fundamental choice of model. Most prominent among questions to consider with respect to model is the vertical discretization, a subject that is treated extensively elsewhere, notably in the paper of *Griffies and Adcroft* [2008] within this volume. The choice that many modeling teams consider between level and isopycnal vertical coordinates determines the way in which diapycnal mixing is done. The result of *Chassignet and Garraffo* [2001], in which their isopycnal coordinate ocean model required a different form of lateral viscosity, relative to that of a *z*-coordinate model, to produce satisfactory results, represents one example of the consequences of choice of model.

4. EDDY MIXING AND PARAMETERIZATION

The North Atlantic has been studied to explore fundamental questions of physical oceanography as well as to understand the oceanography of the basin. Studies are relatively well constrained due to good observational data coverage, and the regional models require an order of magnitude fewer grid points than does a global model at comparable resolution. The North Atlantic offers particularly challenging tests of theoretical understanding and of the application of that theoretical understanding to model formulation in determining the balance of forces which result in the observed structure and path of the Gulf Stream and North Atlantic Current.

The rapidly increasing level of eddy activity with resolution would appear to be a factor in obtaining Gulf Stream separation at the observed location of Cape Hatteras, as discussed in the preceding section, and in successful reproduction of many of the features discussed in the sections that follow. The rise of domain-averaged kinetic energy with resolutions ranging from 0.4° to 0.1° is illustrated with simulations of the POP model in Figure 3; comparison of sea surface height variability with observations derived from

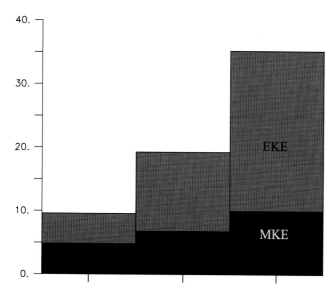

Figure 3. Kinetic energies, in units of cm²/s², from 3-year time averages over the entire domain, from POP North Atlantic model simulations at horizontal grid resolutions of 0.4°, 0.2°, and 0.1° as indicated on the abscissa. The separation into mean kinetic energy (black) and eddy kinetic energy (gray) is done simply in terms of overall time mean and departure from that time mean.

satellite-based altimetry, shown in Plate 1, strongly suggests the higher kinetic energy of the 0.1° case to be more realistic. The time-mean kinetic energy, indicated in black on Figure 3, rises modestly with resolution; the strong increase in total kinetic energy with resolution comes primarily from eddy kinetic energy. Spectral analysis of sea surface height variability, as in Figure 19 of *Smith et al.* [2000], supports the notion that a threshold has been crossed.

Attempts have been made over the past decade to directly diagnose eddy fluxes to better parameterize the role of eddies in non-eddy-resolving models, including those used for climate projection. We focus here on one recent paper, that of *Eden et al.* [2007], as an example of work on this important front in theoretical development. It also serves to illustrate the use of eddy-resolving models of the North Atlantic in the development of a theoretical understanding of the feedback of mesoscale variability on the mean flow and in the parameterization of that interaction.

The so-called Gent-McWilliams parameterization of eddy mixing has proven to be one of the most significant advances not just in ocean modeling but also in climate modeling. The breakthrough of *Gent and McWilliams* [1990] was the recognition that eddies not only mix tracers along isopycnal surfaces, as had already been parameterized by *Redi* [1982], but that they also act on layer thickness, as in the release of potential energy associated with baroclinic instability, where

downgradient diffusion of thickness occurs. The two diffusive coefficients, one for tracers and another for thickness, have most often been taken to be identical. In the adiabatic interior, the thickness diffusion can be recast as a transport term, bringing an eddy-induced component to the transport equations through what is referred to as the bolus velocity.

Diagnosis of a sign-definite isopycnal diffusivity was found to be problematic [*Rix and Willebrand*, 1996; *Treguier*, 1999]. *Bryan et al.* [1999] determined that the bolus velocity contains a significant rotational component, presenting a complication to the diagnosis of a downgradient thickness diffusivity from eddy flux data (either synthetic or observed).

To understand this statement regarding a rotational component of the eddy flux field, at least at a superficial level, some explanation is in order. Mesoscale eddy variability would be defined in terms of deviations from a temporal mean. Accordingly, eddy fluxes would appear in a tendency equation for the time-mean buoyancy, following the notation of *Eden et al.* [2007], as

$$\bar{b}_t = -\bar{\mathbf{u}} \cdot \nabla \bar{b} - \nabla \cdot \mathbf{F} + \bar{Q} \tag{1}$$

where b is the buoyancy, Q is any diabatic forcing (an overbar indicates time-averaging), and $\mathbf{F}$ is the eddy flux in question.

Equation (1) is unchanged by the substitution of $\mathbf{F} + \nabla \times \theta$ in place of $\mathbf{F}$, where θ is an arbitrary vector field, as the divergence of the curl of a vector field is zero. This invariance of the tendency equation to inclusion of a rotational component is referred to as a "gauge" invariance, borrowing from the language of field theory. The consequence of choice of gauge, while irrelevant to the mean buoyancy budget, is, however, consequential to the diagnosis of a thickness diffusivity from eddy fluxes. *Fox-Kemper et al.* [2003] discuss the indeterminacy of a decomposition into rotational and divergent components and also identify diagnostics which remain invariant to the choice of decomposition.

Working in the 1/12° model of *Dengler et al.* [2004], *Eden et al.* [2007] confirm earlier findings of a very ambiguous sign of thickness diffusivity in the North Atlantic before consideration of rotational fluxes (their Figure 3). A positive sign of thickness diffusivity would be expected under conditions of baroclinic instability, with release of available potential energy (APE). This release of APE can also be diagnosed directly from the correlation term $\overline{w'b'}$, which shows a pattern of APE release over most of the domain (their Figure 4), with the Gulf Stream region near and just downstream of Cape Hatteras being the one region in the mid-to-high latitudes in which the eddy buoyancy flux contributes broadly to

(a) 0.1 (3/91−2/94)

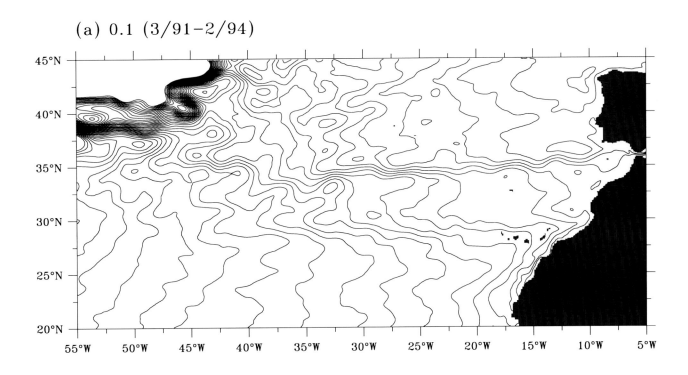

(b) 0.1 (12/92)

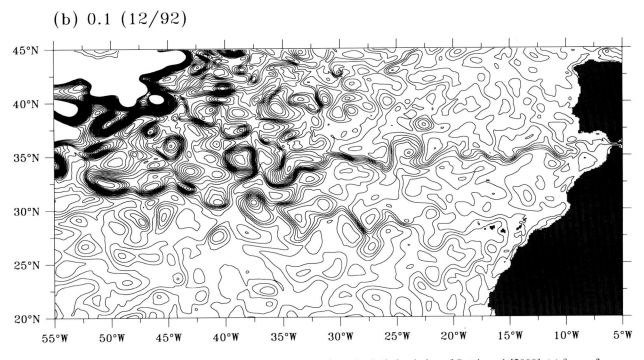

Figure 4. Sea surface height in the region of the Azores Current, from the 0.1° simulation of *Smith et al.* [2000], (a) from a 3-year time average and (b) instantaneous.

an increase, rather than a decrease, of APE. This observation implies that existing isopycnal mixing parameterizations cannot be expected to adequately model the subgrid-scale in this region.

The inconsistency between the patterns of $\overline{w'b'}$ and of thickness diffusivity diagnosed from eddy fluxes is associated with the gauge degree of freedom discussed above, and hence to the parameterization of that eddy flux as a downgradient diffusion. Whereas the usual bolus velocity of *Gent and McWilliams* [1990] is associated with the flux of buoyancy across contours of mean buoyancy, the rotational component of the eddy flux (the choice of which "fixes" the gauge) produces a bolus velocity along those contours of mean buoyancy, breaking the isotropy of the lateral mixing coefficient. The discrepancy between the sign of $\overline{w'b'}$ and that of thickness diffusivity diagnosed from eddy fluxes by *Eden et al.* [2007] is indicative of a need for some degree of spatially dependent anisotropy in the isopycnal mixing formulation. This isotropy is found to be associated with westward advection, confirming zonal propagation of Rossby waves as a physical source of anisotropy in ocean mixing [*Smith and McWilliams*, 2003].

Considerable work has been done on the role of eddies in setting ocean stratification, particularly in the Southern Ocean (see, e.g., *Karsten et al.* [2002]). The very recent work of *Greatbatch et al.* [2007] builds on that of *Eden et al.* [2007], addressing the role of eddies in the determination of surface heat fluxes, with their analysis focused on the Gulf Stream region.

5. THE NORTHWEST ATLANTIC

Mesoscale eddies mediate energy conversion between baroclinic and barotropic modes, and nowhere is this more evident than in the Northwest Atlantic. The degree to which the deeper flows in a model are energized through the inclusion of eddies appears to be critical in determining the character of the circulation in this region, with the biases of non-eddy-resolving models particularly extreme here.

5.1. The North Atlantic Current

The North Atlantic Current (NAC) has been described by *Rossby* [1996] as existing at a crossroads. The northward turn of the NAC around the Grand Banks and into the so-called Northwest Corner (see *Lazier* [1994] for a regional review) has already been mentioned at several points above, as it is a particularly elusive feature in ocean models. This remarkable northward excursion in fact represents the most poleward penetration of the near-surface waters of any of the subtropical gyres, and the absence of this northward penetration in non-eddy-resolving ocean models is responsible for some of the largest biases in sea surface temperature in ocean climate models, of the order of 5° to 10°C; this point is taken up in section 6.3.

Simulation of the North Atlantic Current depends, to a great extent, on the simulation of the Gulf Stream. It was understood as early as *Beckmann et al.* [1994; see the summary of *Dengg et al.*, 1996] that the bias of a model Gulf Stream toward a strong anticyclonic meander north of Cape Hatteras results in excessive dissipation of the kinetic energy of the Stream. Models that exhibit this spurious, dissipative anticyclonic meander at separation also produce a poor description of total transport and structure of the Gulf Stream in the free jet region. *Bryan et al.* [2007] commented that they never saw penetration of the NAC into the Northwest Corner without Gulf Stream separation at Cape Hatteras in their sensitivity study.

Failure to separate at the observed location tends to correlate with a more surface-trapped Stream without the deep penetration of kinetic energy that would appear to be required for reattachment of the Gulf Stream at the Grand Banks and for the subsequent northward turn of the North Atlantic Current. This requirement for deep penetration of kinetic energy at the Grand Banks contrasts with the situation at Cape Hatteras where the Deep Western Boundary Current serves to isolate the Gulf Stream from the influence of the underlying topography. The importance of barotropization of the Stream, of deep penetration of kinetic energy, can be taken as confirmation of the relevance of the early work of *Holland and Lin* [1975] and others on baroclinic instability in eddying ocean models. Eddy fluxes of potential vorticity can be related to a vertical transport of horizontal momentum, as shown by *Rhines and Holland* [1979], and as one might expect of a process that causes the circulation to become less variable with depth.

Penduff et al. [2006] noted insufficient vertical penetration of momentum in the CLIPPER model of the North Atlantic at 1/6° resolution. More recently, it has come to be understood by *Penduff et al.* [2007] that improved treatment of advection and the use of free-slip lateral boundary conditions are sufficient, when taken together, to raise levels of deep kinetic energy.

The importance of eddies in the establishment of the deep flow, and in the generation of pressure gradient torques that support the large-scale mean circulation, will be addressed further in sections 6 and 7.

5.2. Labrador Sea

Käse et al. [2001] show that to obtain a realistic mid-depth circulation, as elucidated by the observational analysis of *Lavender et al.* [2000], it is important to resolve the topography well. They make use of a previously mentioned partial step representation of topography [*Adcroft et al.*, 1997] to improve the resolution of the topographic slope, working at a resolution of 1/6°. *Myers and Deacu* [2004] also explored the impact of partial cells on the circulation of the Labrador Sea, and on the fresh water budget in particular, although at a lower resolution of 1/3°, finding that the Labrador Sea Water became saltier with partial bottom cells due to a strengthened Labrador Sea Counter-Current, with greater import of salty Atlantic waters and enhanced export of Labrador Sea waters to the Irminger Basin.

The large-scale cyclonic circulation of the Labrador Sea causes isopycnal surfaces to dome in the interior of the basin, creating conditions under which deep convection can occur. Eddies generated through a number of mechanisms act to reduce that doming and to restore stratification.

Spall [2004] considered the role of rim current eddies generated through baroclinic instability, commenting on the significant modification that occurs with departure from a flat-bottomed topography. *Eden and Böning* [2002] discussed the generation of Irminger rings from Cape Desolation, off Southwest Greenland. Working in a realistic model at 1/12° horizontal resolution and diagnosing energy transfer at depths between 100 and 400 m, they attributed ring generation to barotropic instability. *Katsman et al.* [2004], in work related to that of *Spall* [2004], studied this problem of restratification through boundary current-generated eddy fluxes in an idealized Labrador Sea at 7.5-km resolution and with 15 vertical levels. They confirm the finding of barotropic instability in the upper 400 m off Cape Desolation but also find baroclinic instability below this depth, raising the likelihood that Irminger rings are generated through a mixed instability. A third type of eddy in the Labrador Sea is generated at the edges of the convective patches themselves. These convective eddies may be thought of as submesoscale, if rather deep-reaching, occurring within regions of very weak stratification and hence being of smaller size than what one would ordinarily refer to as a mesoscale eddy. Convective eddies and restratification following open-ocean convection have been studied by a number of authors, including *Visbeck et al.* [1996] and *Jones and Marshall* [1997]; these studies on convective eddies form the background for the current work on submesoscale eddies discussed in two other papers within this volume [*Thomas et al.*, 2008; *Fox-Kemper and Smyth*, 2008].

The three types of eddies that we consider in the problem of Labrador Sea restratification are:

1. Rim current eddies generated through baroclinic instability,
2. Irminger rings generated most likely through a

mixed barotropic-baroclinic instability at Cape Desolation, and
3. Convective eddies.

The role of all three types of eddies in restratification of the Labrador Sea has been examined by *Chanut et al.* [2008] in a realistic model with 4-km discretization of the Labrador Sea embedded within a 1/3° North Atlantic model. They resolve, at least marginally, even convective eddies, if only at the larger end of the spectrum.

In the region of deep convection, *Chanut et al.* [2008] found that rim current eddies flux enough heat laterally to explain the containment of deep convection in their model, setting up the large-scale conditions under which convective eddies can work to restratify the water column immediately following deep convection, restricting the duration of events and the volume of water produced. They find Irminger rings only in the north of the basin, primarily poleward of about 60°N, although they have been observed to the south in the region of deep convection [*Lilly et al.*, 2003] where they can be expected to further contribute to restratification.

Consideration of the role of eddies in limiting deep convection in the Labrador Sea points to a limitation of Gent-McWilliams style parameterization of isopycnal tracer mixing. The generation of eddies may be described by local processes, but eddies propagate, and so their impacts do not necessarily remain local. A secondary issue is the limitation of the approach to the parameterization of baroclinic processes, with eddy generation through barotropic instability remaining unaddressed. While Gent-McWilliams isopycnal tracer mixing has been very successful on balance, we presume that these deficiencies which have yet to be addressed explain much of the benefit of explicit inclusion of the eddy field.

Waters formed in and exported from the Labrador Sea have observed densities in the range of $27.68 < \sigma_0 < 27.80$ [*Schott et al.*, 2004] and are particularly distinguished as being relatively fresh, with salinities of 34.95 psu and below. Treguier's analysis of model biases in Labrador Sea Water composition focuses in part on the underestimate of the salinity minimum shared by all four high-resolution models [*Treguier et al.*, 2005], arguing it to be the result of biases in horizontal transport, involving either transport from the eastern side of the basin or from much closer range where the fresh water flux into the East Greenland Current is poorly represented. It bears mention that the salinity bias in question is nearly an order of magnitude less severe than that produced in low-resolution, non-eddying models.

An important source of freshwater for the Labrador Sea is the throughflow from the Canadian Archipelago, which may require special attention to reproduce correctly, or even to represent at all, owing to the intricacy of the pathway. *Komuro and Hasumi* [2005] show that allowing for this throughflow makes for stronger Atlantic deep circulation, with the overflow waters from the GIN Sea spared from unrealistic dilution. *Oka and Hasumi* [2006] investigate the dependence of throughflow on resolution, which is understandably high; *Oka and Hasumi* [2004] explore the dependence of the North Atlantic deep circulation on freshwater forcing, finding the specification of river runoff at high northern latitudes to represent another critical factor. Freshwater forcing at high latitudes and the pathways taken by those waters are essential to the determination of meridional overturning, and hence to climate study, and yet they remain poorly known.

The Irminger Basin is the other place in the North Atlantic, south of the sills, where renewal of deep water may occur. In contrast to the situation in the Labrador Sea where deep water formation is well documented, deep water formation in the Irminger Basin has never actually been observed, but is inferred from circumstantial evidence [*Pickart et al.*, 2003]. The properties of this presumed Irminger Sea Deep Water are close to those produced in the Labrador Sea. *Pickart et al.* [2003] argue that what has been considered to be LSW must be derived in part from the Irminger Basin.

Patterns of deep water formation in ocean general circulation models are infamously sensitive to the details of model configuration. This can be understood in part as due to the discontinuous threshold behavior of the phenomenon: Small biases in wind or buoyancy forcing, parameterizations of mixing, or representation of transport may be sufficient to trigger deep convection or alternatively to erroneously suppress it. A certain degree of sensitivity to model configuration persists in the wintertime convective activity of eddy-resolving ocean models, as seen in Plate 3. Partial step topography and anisotropic parameterizations of lateral mixing both tend to preserve the density of waters from the Denmark Strait overflow (the influence of the anisotropic lateral mixing parameterizations was documented by *Smith and Gent* [2004]). Evidently, this tendency to preserve the density of the overflow waters that subsequently encircle the slope of the Northwest Atlantic translates into enhanced convection where that water of greater density becomes entrained in the strongly mixed portion of the water column. The extent of mixing in the Irminger Basin seen in panels (b) and (c) of Plate 3 is almost certainly unrealistic. A very high resolution modeling study, in the spirit of *Chanut et al.* [2008], could yet be applied to the broader region, perhaps bringing the analysis of the freshwater budget of *Treguier et al.* [2005] together with the analysis of overflow and entrainment of *Legg et al.* [2008] so as to better understand deep water formation in the Northwest Atlantic.

5.3. *The Mann Eddy*

If the North Atlantic Current exists at a crossroads, then the Mann Eddy is not far from the center of that crossroads. Located at approximately 42°N and 43°W, *Rossby* [1996] notes that the anticylonic eddy contains the warmest waters at 1,000 m in the entire North Atlantic, a remarkable fact in light of its location at what would otherwise be the poleward limit of subtropical gyre waters.

The structure of the Eddy itself is not entirely clear from instantaneous inspection but emerges from a time-averaging, as commented by *Smith et al.* [2000]. It is located toward the southern end of the set of topographically constrained and more persistent meanders that the NAC forms as it flows northward around the Grand Banks [*Kearns and Paldor*, 2000] where some of the flow of NAC continues its eastward drift, but much of it makes the northward turn, with some of that northward-tending water doubling back to the west as it forms the southern flank of the Eddy. Slightly to the south, closer to the Southeast Newfoundland Rise, the Azores Current, addressed in section 6, also branches off from the Gulf Stream and NAC.

The temporal variability of the Mann Eddy presents a challenge to the observational description of the region; note that the section line analyzed by *Schott et al.* [2004] has its offshore end extending into the region occupied by the western flank of the eddy. More complete description of the Mann Eddy could be accomplished through synthesis of model results and observations where the temporal and spatial coverage of the model can contribute to the understanding of this remarkable feature found at the intersection of several major currents of the North Atlantic.

6. GYRE STRUCTURE AND BASINWIDE TRANSPORT

The James test [*Seber*, 1984] provides a method by which to judge the level of agreement between model velocity fields and observational drifter data. In their study, *McClean et al.* [2002] (where the method is explained in an Appendix) apply the test to observed drifter data and Lagrangian floats in the 0.1° model of *Smith et al.* [2000] (but with the US Navy's daily NOGAPS wind forcing product and the K-profile parameterization of vertical mixing; *Large et al.* [1994]). The test was used to indicate agreement between model and observations at the 95% significance level. Agreement throughout the Labrador Sea was impressive, although the resolution of spatial binning and near-surface depth of inspection would not bring out the more intricate recirculations discussed above in section 5. Discrepancies between the model circulation and observations were identified in

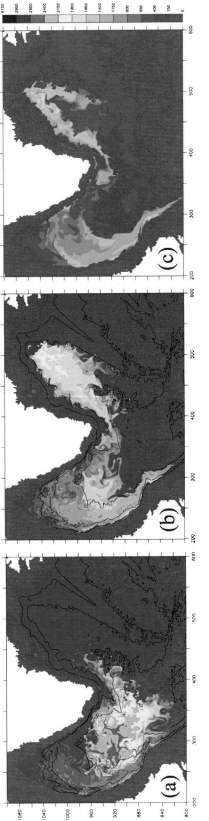

Plate 3. Mixed layer depths in the Labrador Sea and Irminger Basin in late March, 2000 for 0.1° configurations of the POP model, as in the study of *Smith et al.* [2000] and *Bryan et al.* [2007], from cases (a) with biharmonic lateral dissipation and full step representation of topography, (b) with biharmonic lateral dissipation and partial step representation of topography, and (c) with partial steps and anisotropic formulations of isopycnal tracer mixing and viscosity, as in the work of *Smith and Gent* [2004].

the Mid-Atlantic Bight, parts of the Gulf Stream where the model produces an overly zonal flow, the North Atlantic Drift, where the North Atlantic Current flows after leaving the Northwest Corner, and in the Canary Current (off the northwest coast of Africa). In the more recent paper of *Mc-Clean et al.* [2006], the same test is applied to the North Atlantic regional model under daily surface heat fluxes (as opposed to the monthly heat flux climatology used by *Smith et al.* [2000]), with the most notable improvement being that of the Mid-Atlantic Bight.

While much of the more recent comparison of turbulent statistics from model and observations has made use of the surface expression through altimetry, *McClean et al.* [2006] also compared dispersion rates of numerical floats in the model of *Smith et al.* [2000] to those of the EUROFLOAT deployment in which 21 floats followed the circulation at 1,750-m depth in the northeast Atlantic [*Speer et al.*, 1999]. In the region covered by the floats, over the eastern half of the basin and extending into subpolar and subtropical gyres, one sees an initial period of linear dispersion of floats (average distance from center of mass of deployment proportional to elapsed time), as one would expect when the floats are following their initial outward trajectories. At later times, after the float trajectories become decorrelated, the dispersion rate is slower, showing the square root dependence on time characteristic of a random walk (average distance from center of mass of deployment growing with $\sqrt{t}$). For both model and observations, the transition between these dispersion regimes occurred between 10 and 20 days, a timescale characteristic of the mesoscale.

6.1. Azores Current

The Azores Current is the more southerly part of the Gulf Stream Extension. *Klein and Siedler* [1989] place the Current's origin off the Southeast Newfoundland Rise, near the Gulf Stream/North Atlantic Current juncture.

Smith et al. [2000] provided the first satisfactory comparison of the Azores Current from an ocean model with observations. One of their Azores plots is reproduced here as Figure 4. It is evident from this comparison of time-mean and instantaneous sea surface height fields that the variability of the flow is not necessarily small relative to the mean, posing a difficulty for any theoretical derivation relying on the ratio of u' to u as an expansion parameter.

From a study with three different models, *New et al.* [2001] tell us that the Azores Current presents a front that tends to limit the southward extent of ventilated Eastern North Atlantic Water. Realistic strength of the Current may be important to the determination, in turn, of the potential vorticity and density structure of subtropical mode waters.

The intercomparison upon which the work of *New et al.* [2001] was based deserves mention as a careful consideration of the impact of model formulation. Known as the Dynamo experiment, models based on z, isopycnal, and sigma coordinates were configured at 1/3° eddy-permitting resolution. *Willebrand et al.* [2001] reported significant differences in large-scale circulation that were attributed largely to differences in the overflow of the North Atlantic Sill.

An argument was also developed by *Jia* [2000], based on dynamo results, for the importance of Mediterranean overflow and entrainment to development of the Azores Current. This argument has yet to be reexamined in a more energetic eddy-resolving context.

Stammer and Böning [1996], in concluding remarks, commented on weak or nonexistent Azores Front as among the most egregious deficiencies in model circulation. A number of authors, including *Böning and Bryan* [1996] and *Fu and Smith* [1996], had raised the possibility that crossing over into a regime with more adequate resolution of first internal Rossby radius would bring out such features. This speculation was confirmed when *Smith et al.* [2000] demonstrated that a strong Azores Current indeed emerges from such refinement. The establishment of a realistic Azores Current is found to depend sensitively on model configuration as well as on resolution, as with the previously described sensitivity of the Gulf Stream and North Atlantic Current; that the Azores Current shows a similar sensitivity should come as no surprise, knowing that its origin can be traced back to the "crossroads" of *Rossby* [1996] to the point at which the Gulf Stream rides over the Southeast Newfoundland Rise.

6.2. Recirculation Gyres

The barotropic stream function plots of *Bryan et al.* [2007], reproduced here as Figure 5, not only illustrate the sensitivity of Gulf Stream separation to resolution and lateral dissipation but also illustrate the establishment of a strong and extensive Northern Recirculation Gyre with separation. *Zhang and Vallis* [2007] argue that vorticity production associated with downslope flow of the deep western boundary current gives rise to the Northern Recirculation Gyre, then causing Gulf Stream separation at the observed location of Cape Hatteras.

Zhai et al. [2004] present a "semidiagnostic" method by which the influence of eddies may be ascertained while correcting for biases in large-scale mean fields (their method is rooted in the "semi-prognostic" method of *Eden et al.* [2004], but with modification to avoid damping of eddies). They explore this method in the context of the northwest Atlantic, finding that the eddies are directly responsible for more than half of the transport of the Northern Recirculation Gyre even

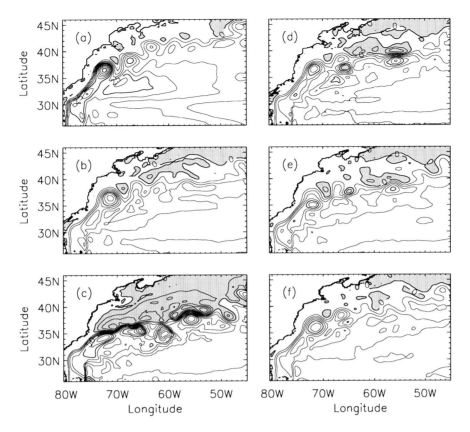

Figure 5. Barotropic stream function averaged over the period 1989–1991. Contour interval is 10 Sv (5 Sv in panel a), negative values are shaded. Panel (c) is from the 0.1° North Atlantic modeling study of *Smith et al.* [2000], whereas panels (b) and (a) are coarsened 0.2° and 0.4° configurations of the model, with lateral dissipation scaled as grid cell area to the 3/2 power (see *Hecht et al.* [2008] in this volume, for a discussion of scaling of dissipation with grid cell area). Panel (d) is the 0.2° but with the lower values of lateral dissipation used in (c); panel (e) is the compliment of this, the 0.1° model but with the higher values of lateral dissipation used in (b). Panel (f) is as per panel (e), but with the coarser 0.2° resolution of bathymetry, although the model is otherwise configured at 0.1°. Notice that the Gulf Stream only separates at Cape Hatteras, as observed in (c), with the highest resolution of 0.1° and with appropriately low value of viscosity. This figure is reprinted from *Bryan et al.* [2007] (their Figure 4).

at a model resolution of only 1/3°, apparently contributing to the establishment of the large-scale bottom pressure torque that supports the gyre. The Northern Recirculation Gyre in turn is critical not only to the establishment of the offshore location of the free jet but strongly reinforces the mass transport of the Stream. In their analysis, they report that the eddies alter the mean flow by fluxing momentum up-gradient. Up-gradient eddy fluxes were also diagnosed by *Eden et al.* [2007] where they find the eddies to contribute to the building, rather than release, of available potential energy over much of the free-jet region of the Gulf Stream. Further research on subgrid-scale parameterizations and numerical schemes is required if such up-gradient turbulent fluxes are to be accounted for at coarser resolution.

This understanding of the dominance of bottom pressure torque in generating the northern recirculation, and of the role of eddies in establishing the bottom pressure field, is reinforced in the closely related paper of *Greatbatch and Zhai* [2006]. In this later work, the authors allow the mean circulation to be unconstrained while assimilating eddies from the study of *Zhai et al.* [2004], finding that the assimilated eddies contribute more weakly toward the establishment of a northern recirculation when the hydrography is allowed to drift. They also find support for the assertion of *Hogg and Stommel* [1985] of constraint within continuous f/H contours, which, in the case of the anticyclonic circulation of the Northwest Corner, are limited by the depth of the Gibbs Fracture Zone.

6.3. Northern Limit of the Subtropical Gyre

It has already been noted that the North Atlantic Current's penetration into the region of the Northwest Corner represents the most poleward penetration of any of the subtropical gyres and that this has been an even more elusive feature in models than the more widely recognized problem of Gulf Stream separation. The penetration of the NAC into the Northwest Corner would also appear to be a historically variable feature: There is evidence that the Current instead followed an eastward drift at the time of the Last Glacial Maximum, as surmised from ocean bottom sediment cores by *Robinson et al.* [1995]. *Rossby* [2003] has presented a possible mechanism for the rapid switching of the Current from LGM conditions to modern involving topographic Rossby waves.

As has also been mentioned above, ocean models used in climate general circulation models fail to produce North Atlantic Current penetration into the Northwest Corner, producing instead a gyre boundary more like that of the Northwest Pacific or of the Northwest Atlantic at the time of the LGM. The question of whether these large biases in sea surface temperature matter to atmospheric model circulation was explored by *Weese and Bryan* [2006], who used the method of *Eden et al.* [2004] (which also provided *Zhai et al.* [2004] the basis for their variant of that method) to adiabatically correct the ocean model circulation in the Northwest Atlantic, finding that correction of the sea surface temperature errors resulted in an improvement in the Icelandic Low of the atmospheric model.

It is quite unknown, however, the degree to which the path of the North Atlantic Current matters to the stability of climate in the twenty-first century, a line of questioning raised by *Hecht et al.* [2006]. Eddy-resolving ocean models will soon provide us with estimates of the stability of Atlantic northward heat transport based on a circulation that more correctly represents the modern penetration of subtropical gyre waters into the Northwest Corner. A component of the water masses that eventually participate in deep water formation are conditioned along this Gulf Stream/North Atlantic Current path. Eddy-resolving ocean models will make a significant contribution to climate science by settling this question of the degree to which our estimates of twenty-first century climate stability are compromised, or not, by this bias of non-eddy-resolving ocean models.

6.4. Resolution Dependence of Northward transports

Tremendous interest has been shown toward the estimation of Atlantic meridional overturning circulation and heat transport in recent years, as seen in the concerted effort of the Rapid Climate Change Programme of the UK's National Environment Research Council, and yet, there is much still to be learned about the resolution dependence of these transports in models, as alluded to just above.

A preliminary analysis of the resolution dependence of northward heat transport in the Atlantic was presented by *Bryan and Smith* [1998]. Extending on the work of *Böning et al.* [1995], *Böning and Bryan* [1996], and *Fanning and Weaver* [1997], *Bryan and Smith* [1998] analyzed transports at resolutions of 4/10°, 2/10°, and 1/10° with consistent model configuration across this range of resolution. They found a maximum in northward heat transport somewhere between 25°N and 30°N in all cases. That maximum value increased by 50% from lowest to highest resolution, as shown in Figure 6, with the heat transport in the highest resolution case falling near the central value of the observational estimate of *Macdonald and Wunsch* [1996]. Nearly all of the increase in the total transport was accounted for by changes in the time mean flow rather than by direct contribution from eddies.

One paper that explains a role for model resolution in the variability of the meridional overturning circulation is that of *Getzlaff et al.* [2005]. Building on earlier observational and more idealized modeling work, they confirm the correlation of meridional overturning circulation (MOC) with the North Atlantic Oscillation (NAO), with the forcing leading by 2 to 3 years, and then go on to identify a rapid speed of dynamical signal propagation in an eddy-admitting (1/3°) version of their model, with the MOC response to changes in Labrador Sea deep water communicated to lower latitudes through this rapid communication.

The fast dynamical signals they see are in the form of Kelvin waves, triggered by the NAO-driven anomalies in the thickness of Labrador Sea Waters. The Kelvin waves propagate along the western boundary to the equator where they cross the basin in the equatorial wave guide. In contrast, the non-eddy-resolving version of their model shows a far more significant lag in response time with latitude, with the signal propagation occurring at the much slower advective rate, illustrating the linkage between processes occurring at small spatial and fast temporal scales.

7. DEEP FLOW

The flow of the upper ocean, and the dependence of that flow on mesoscale variability, has been a focus of much discussion. The tremendous spatial and temporal coverage of remotely sensed altimetry only tends to bring more of our attention toward the surface. The response of the deep ocean to the mesoscale eddy variability, however, is every bit as remarkable as that of the upper ocean. Here, we briefly

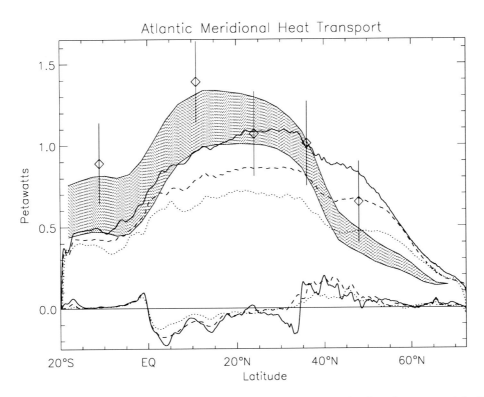

Figure 6. Net meridional heat transport (upper curves) and transport by the time varying flow (lower curves) for North Atlantic models at resolutions of 0.1° (solid), 0.2° (dashed), and 0.4° (dotted). The estimate of *Trenberth* [1998] is indicated by the shaded region, and the inverse model results of *Macdonald and Wunsch* [1996] are indicated by the diamond symbols and vertical bars. Figure is reprinted from *Bryan and Smith* [1998].

describe recent developments in our understanding of the structure of the deep flow of the Northwest Atlantic and then touch on the importance of the interaction between the Gulf Stream and the deep flow.

7.1. Export from the Subpolar Gyre

Tracer distributions imply that a significant fraction of the Deep Western Boundary Current emerges from the Labrador Sea and circulates around the Grand Banks [*Talley and McCartney*, 1982; *Pickart and Smethie*, 1998; *Smethie et al.*, 2000], passing from subpolar to subtropical gyres. In marked contrast, PALACE and SOLO floats, which have been used in an attempt to trace out this circulation, fail to follow such a path around the Grand Banks, detraining, instead, into the counter-flowing deep waters of the North Atlantic Current [*Lavender et al.*, 2000; *Fischer and Schott*, 2002].

This contradiction is addressed by *Getzlaff et al.* [2006]. They work with an offline transport model based on the flow field from a 1/12° Mercator grid z-coordinate model with the bottom boundary layer scheme of *Beckmann and Döscher*

[1997] (as documented by *Dengg et al.* [1999]; this is largely the same model as used in the analysis of *Eden et al.* [2007], discussed above).

The picture that *Getzlaff et al.* [2006] find is consistent with the tracer observations. Of the fraction of numerical floats that make it from a release point of 53°N to the western boundary at 32°N, south of Cape Hatteras (around one in six floats), they find that 60% round the Grand Banks, with the remainder diverted through a more circuitous path, flowing eastward to the mid-Atlantic Ridge before resuming a southward course (this interior path is consistent with the tracer-based observations presented by *Rhein et al.* [2002]). The interior flow follows the Ridge until around 38°N where it takes a largely westward course to the point where it rejoins the western boundary flow. That point where the two principal paths rejoin was found to lie in the vicinity of the crossover point, near Cape Hatteras.

The initial deflection of a sizable fraction of the DWBC is reported by *Getzlaff et al.* [2006] to occur as these waters interact with the quasi-stationary meanders of the NAC. The strength of these meanders, in turn, can be expected to depend sensitively on the ability of a model to reproduce a strong North Atlantic Current with sufficient depth penetration, as touched on by *Bryan et al.* [2007]. The existence of this detour through which much of the flow of the DWBC passes would appear to owe its existence, therefore, to the presence of eddies.

The explanation for the failure of floats embedded in the DWBC to trace out the main branch of the Current emerges when *Getzlaff et al.* [2006] transport numerical floats that are made to surface at 10-day intervals, spending 12 h at the surface, as for SOLO floats. As with the real-world floats, they find that few of their numerical floats follow the main branch of the DWBC around the Grand Banks. The disturbance associated with remaining in surface waters for 12 h is simply too great; an alternative strategy involving surfacing for 2-h periods, which is expected to be technically feasible, was found to compromise the float trajectories to a much more tolerable extent.

In the North Atlantic model of *Smith and Gent* [2004], which uses anisotropic forms of horizontal viscosity and isopycnal mixing, there is some evidence for a splitting of the deep transport into western boundary and interior branches at the Grand Banks, but with a great deal of recirculation seen in between the two branches, as shown in Figure 7 (their case A′). These deep recirculations are associated with the intensification of the Gulf Stream transport downstream of Cape Hatteras; the dominance of the mesoscale is seen even in the time-average view of Figure 7. It is also evident from the figure that waters in the denser and colder 2°–3°C class of potential temperature have their origin in the overflow of the Denmark Strait (upper panel), whereas the warmer 4°–5°C class (generally higher in the column) traces out waters from the Iceland-Faroe-Scotland Ridge.

As explained by *Smith and Gent* [2004], anisotropic forms of dissipation are particularly effective in maintaining the density of overflow waters from the Denmark Strait. The subject of entrainment within the early and intense mixing of the overflow, and the role of model formulation and resolution, is an important one, and is discussed by *Legg et al.* [2008] in this volume.

7.2. Crossover

A significant fraction of the deep flow crosses under the Gulf Stream at Cape Hatteras within the Deep Western Boundary Current. *Tansley and Marshall* [2000] argue that crossover occurs here due to the steepening of topography associated with the Cape. *Bryan et al.* [2007] find evidence that the underlying deep current serves to isolate the Stream from bathymetry such that the potential vorticity change associated with vertical stretching of cross-slope flow might be lessened.

The importance of the Deep Western Boundary Current, and the fact that steepening of the continental shelf brings the Deep Current beneath the Stream at Cape Hatteras, had been explored by *Thompson and Schmitz* [1989]. They never saw separation at Cape Hatteras in their two-layer, 0.2° model, but saw southward migration of the offshore Stream (after late separation). Now, with the benefit of more recent findings from models combining realistic configuration and a more realistic mesoscale eddy spectrum [*Bryan et al.*, 2007], it appears that the argument of *Tansley and Marshall* [2000] that the point of separation is locally fixed through topographic constraint is correct. Furthermore, findings to date, as surveyed above, strongly suggest that a realistically intense interaction between the Stream and Deep Western Boundary Current is required for separation to occur in a model; this requirement for a realistically strong Deep Western Boundary Current in turn requires eddy-mediated processes of energy conversion to invigorate the deep flow.

A point that is not widely appreciated is the effect of the interaction between the two currents on the deep flow at the crossover where much of the deep flow is swept along with the Stream to the east. This is particularly clear in the 3°–4°C potential temperature class, shown in Figure 8, where the eastward excursion of the deep flow is only turned back at the New England Seamount Chain. A strong, well-defined Deep Western Boundary Current is reestablished shortly before reaching The Bahamas (see Figure 14 of *Smith et al.* [2000] for a model section of velocity off Abaco).

This split of the deep western boundary current, with the upper layers being carried to the east along with the Gulf Stream, was identified from hydrographic survey data of *Pickart and Smithie* [1993] and later confirmed with the Lagrangian float study of *Bower and Hunt* [2000]. The split was first studied within a model by *Spall* [1996a] and the implications for eddy-mean flow influence on the southern recirculation gyre explored further by *Spall* [1996b], where the Gulf Stream and upper core of the deep western boundary current were seen to form a low-frequency oscillator, which may present a source of decadal variability.

8. SUMMARY

Since the last set of published reviews a dozen years ago [*Stammer and Böning*, 1996; *Böning and Bryan*, 1996; *Dengg et al.*, 1996], modeling of the North Atlantic Ocean has moved from what would be called the eddy-admitting regime into the eddy-resolving regime: Eddy kinetic energies are now in much better agreement with observations. It has become clear that the variability in the circulation is generally stronger than the mean, apart from the regions occupied by mean jets.

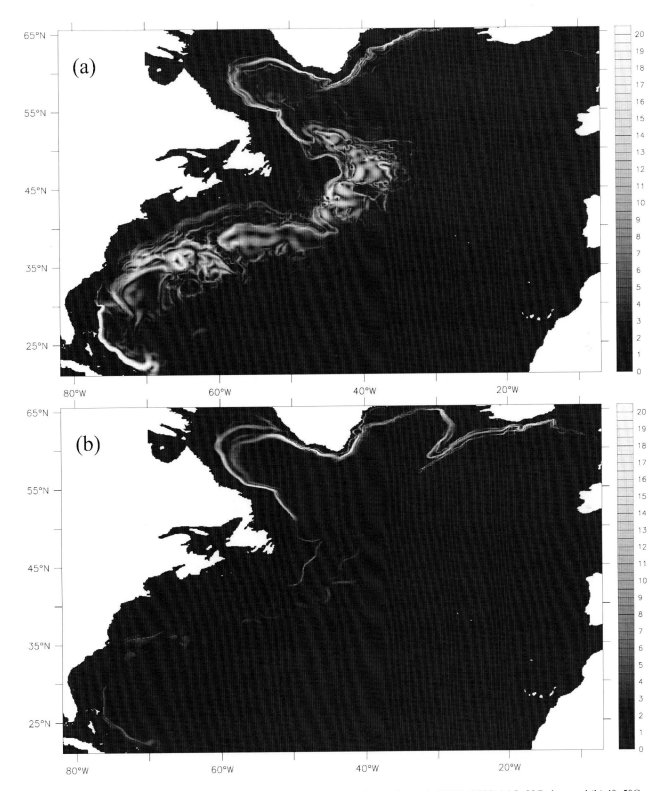

Figure 7. Deep transport within potential temperature classes, averaged over the period 1998–2000. (a) 2–3°C class and (b) 4°–5°C class. Units are in 10^{-4} m²/s, such that integration over the width of any section produces the net volume transport across that section within the potential temperature class. The 2,500-m depth contour is indicated in black.

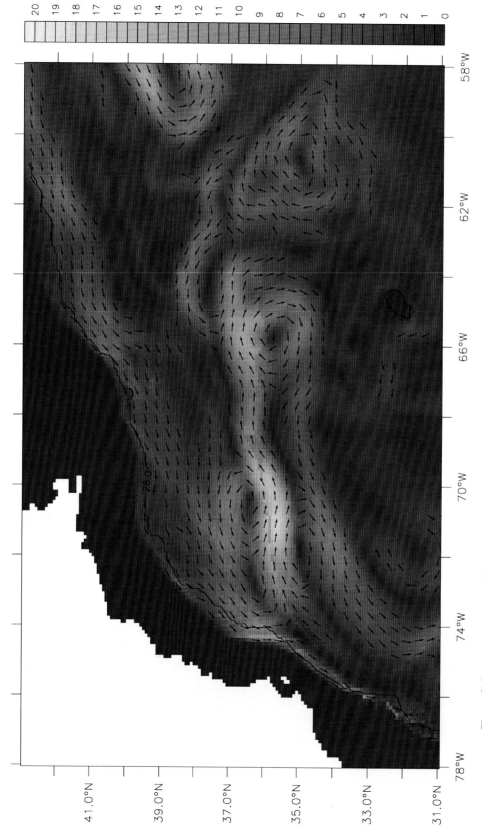

Figure 8. Deep transport, as in Figure 7, but for the 3°–4°C class over the Gulf Stream region from Cape Hatteras to the New England Seamounts. Vectors of uniform length have been added, where the transport exceeds a threshold, to trace out the sense of the flow.

The Gulf Stream separates at the observed location of Cape Hatteras in these eddy-resolving models, although results remain very sensitive to model configuration. Major improvements are seen downstream in the Gulf Stream system as well, with a North Atlantic Current that rounds the Grand Banks correctly forming a remarkable poleward penetration of subtropical gyre waters into the region of the Northwest Corner. The Azores Current also branches off of the Gulf Stream in the vicinity of the Southeast Newfoundland Rise in these models, forming a front that presents an important constraint on the spreading of mode waters.

It is not entirely clear at this point where increased resolution is working to better resolve eddies and where the greater impact of resolution may perhaps lie in finer resolution of the topography, although *Bryan et al.* [2007] addressed this question in a limited sense. There is evidence supporting the use of even higher resolutions to adequately resolve processes within the mixed layer: the 1/20° study of *Paci et al.* [2007] provides evidence for the importance not only of the mesoscale but of the submesoscale. The relevance of the submesoscale is discussed further by *Thomas et al.* [2008] in this volume. Analysis of high-resolution simulations will contribute to further development of parameterizations, particularly in terms restratification of the mixed layer, as in the work of *Fox-Kemper et al.* [2008].

An issue that has only been lightly addressed is that of appropriate forcing for strongly eddying ocean models. The mismatch between high-resolution ocean and low-resolution reanalysis is one that has been discussed informally for years. Performing initial fully coupled simulations with high-resolution ocean and relatively low-resolution atmosphere is attractive as a means to minimize the number of variables changed and perhaps maximize understanding, but the extent to which this may compromise the effect of air-sea fluxes remains to be determined.

Further progress on all of these issues could of course be enhanced by the continued development and adoption of metrics. This is a developing area where strongly eddying simulations figure prominently because of their more satisfactory comparison with observations [see *McClean et al.*, 2002, 2006, 2008].

Even acknowledging gaps in our understanding, it does appear that improvements in the deep circulation are as meaningful as those seen in the upper ocean. Repeatedly through the years, comment has been made on the invigoration of the deep flow with the inclusion of eddies, as eddy mediation of energy conversion between baroclinic and barotropic modes works to couple the circulation of the upper ocean to that of the depths and to bring the influence of topographic torque on the large-scale circulation to a more realistic level. It appears that this eddy-mediated coupling of circulations in the deep and upper ocean is fundamental to the establishment of the Northern Recirculation Gyre and to the North Atlantic Current's northward excursion around the Grand Banks.

Density differences are small enough in the Northwest Atlantic that much of the steering must be produced by topographic torque, as pointed out by *Treguier et al.* [2005]. In turn, then, topography presents a major constraint on layer thickness—on the potential vorticity of the flow—and hence eddies and topography together play a major role in determining the large-scale dynamics of the North Atlantic.

This coupling of upper ocean and depths also appears to be critical to the determination of the degree to which the Deep Western Boundary Current and Gulf Stream interact at Cape Hatteras. At this location, the steep topography of the shelf at Cape Hatteras causes the two currents to cross, with sufficient strength of interaction required for the separation to occur, along with a sufficiently inertial character of the Stream. This is how we best understand the process of Gulf Stream separation in our most realistic, eddy-resolving models today.

The North Atlantic has also provided a context for theoretical development, most notably with respect to the effects of eddies and the parameterization of those effects in lower resolution models. Even while our ability to parameterize the effect of eddies is developing, however, we are presented now with the possibility to reevaluate the stability of Atlantic meridional overturning circulation and the stability of twenty-first century climate, with models that produce a very much more realistic path of the North Atlantic Current, and hence corrected preconditioning of those waters that subsequently participate in the formation of North Atlantic Deep Water.

When the first coupled climate simulations of twenty-first century climate are performed, initialized with the more realistic mean ocean circulation that may be attained with a strongly eddying ocean, the question of whether the stability of the meridional overturning is different than in today's models will be one of the important questions to be posed. If the stability of the overturning is found to be substantially different than with non-eddying models, then we anticipate that further numerical experimentation will be required to understand the robustness of the results with confidence.

Acknowledgments. Mike Spall and an anonymous reviewer provided thoughtful reviews that led to numerous improvements. M. Hecht gratefully acknowledges the support of the Climate Change Prediction Program within the Department of Energy's Office of Science. Support for R. Smith's participation in the preparation of this manuscript was provided by NSF grant 065422 to the National Center for Atmospheric Research. Together, we thank our program managers who have supported our cross-agency research in ocean circulation and climate over the decade covered in this review. LANL is operated by Los Alamos National Security, LLC for the Department of Energy.

REFERENCES

Adcroft, A., C. Hill, and J. Marshall (1997), Representation of topography by shaved cells in a height coordinate ocean model, *Mon. Weather Rev.*, *125*, 2293–2315.

Arakawa, A., and V. R. Lamb (1981), A potential enstrophy and energy conserving scheme for the shallow-water equations, *Mon. Weather Rev.*, *109*(1), 18–36.

Barnier, B., et al. (2006), Impact of partial steps and momentum advection schemes in a global ocean circulation model at eddy-permitting resolution, *Ocean Dyn.*, *56*, 543–567, doi:10.1007/s10236-006-0082-1.

Beckmann, A., and R. Döscher (1997), A method for improved representation of dense water spreading over topography in geo-potential-coordinate models, *J. Phys. Oceanogr.*, *27*, 581–591.

Beckmann, A., C. W. Böning, C. Koberle, and J. Willebrand (1994), Effects of increased horizontal resolution in a simulation of the North Atlantic Ocean, *J. Phys. Oceanogr.*, *24*(2), 326–44.

Böning, C., and F. Bryan (1996), Large-scale transport processes in high-resolution circulation models, in *The Warmwatersphere of the North Atlantic Ocean*, edited by W. Krauss, pp. 91–128, Gebrüder Borntraeger, Berlin.

Böning, C. W. (1989), Influences of a rough bottom topography on flow kinematics in an eddy-resolving circulation model, *J. Phys. Oceanogr.*, *19*(1), 77–97, doi:10.1175/1520-0485(1989)019<0077:IOARBT>2.0.CO;2.

Böning, C. W., and R. G. Budich (1992), Eddy dynamics in a primitive equation model: Sensitivity to horizontal resolution and friction, *J. Phys. Oceanogr.*, *22*(4), 361–381, doi:10.1175/1520-0485(1992)022<0361:EDIAPE>2.0.CO;2.

Böning, C. W., W. R. Holland, F. O. Bryan, G. Danabasoglu, and J. C. McWilliams (1995), An overlooked problem in model simulations of the thermohaline circulation and heat-transport in the Atlantic Ocean, *J. Clim.*, *8*(3), 515–523.

Bower, A. S., and H. D. Hunt (2000), Lagrangian observations of the deep western boundary current in the North Atlantic Ocean. Part II: The Gulf Stream-deep western boundary current crossover, *J. Phys. Oceanogr.*, *30*(5), 784–804, doi:10.1175/1520-0485(2000)030<0784:LOOTDW>2.0.CO;2.

Broecker, W. S., D. M. Peteet, and D. Rind (1985), Does the ocean–atmosphere system have more than one stable mode of operation?, *Nature*, *315*(6014), 21–26.

Bryan, F., and W. Holland (1989), A high-resolution simulation of the wind- and thermohaline-driven circulation in the North Atlantic Ocean, in *Proceedings of the 'Aha Huliko,' a Hawaiian Winter Workshop*, edited by P. Muller and D. Henderson, pp. 99–115, Univ. of Hawaii at Manoa, Honolulu.

Bryan, F. O., and R. D. Smith (1998), Modeling the North Atlantic circulation: From eddy-permitting to eddy-resolving, *Int. WOCE Newsl.*, *33*, 12–14.

Bryan, F. O., M. W. Hecht, and R. D. Smith (2007), Resolution convergence and sensitivity studies with North Atlantic circulation models. Part I: The western boundary current system, *Ocean Modell.*, *16*(3–4), 141–159, doi:10.1016/j.ocemod.2006.08.005.

Bryan, K. (1969), A numerical method for the study of the circulation of the World Ocean, *J. Comput. Phys.*, *4*, 347–376.

Bryan, K., J. K. Dukowicz, and R. D. Smith (1999), On the mixing coefficient in the parameterization of bolus velocity, *J. Phys. Oceanogr.*, *29*(9), 2442–56, doi:10.1175/1520-0485(1999)029<2442:OTMCIT>2.0.CO;2.

Capet, X., F. Colas, P. Penven, and J. C. McWilliams (2008), Eddies in eastern-boundary subtropical upwelling systems, this volume.

Chanut, J., B. Barnier, W. Large, L. Debreu, T. Penduff, and J. Molines (2008), Mesoscale eddies in the Labrador Sea and their contribution to convection and restratification, *J. Phys. Oceanogr.*, in press.

Chao, Y., A. Gangopadhyay, F. O. Bryan, and W. R. Holland (1996), Modeling the Gulf Stream system: How far from reality?, *Geophys. Res. Lett.*, *23*(22), 3155–3158.

Chassignet, E. P., and Z. D. Garraffo (2001), Viscosity parameterization and Gulf Stream separation, in *Proceedings of the 'Aha Huliko'a Hawaiian Winter Workshop*, edited by P. Muller and D. Henderson, pp. 367–374, Univ. of Hawaii at Manoa, Honolulu.

Cox, M. D. (1985), An eddy resolving numerical model of the ventilated thermocline, *J. Phys. Oceanogr.*, *15*(10), 1312–1324.

Dantzler, H. L. (1976), Geographic variations in intensity of the North Atlantic and North Pacific oceanic eddy fields, *Deep Sea Res.*, *23*(9), 783–794.

Dantzler, H. L. (1977), Potential energy maxima in the tropical and subtropical North Atlantic, *J. Phys. Oceanogr.*, *7*(4), 512–519.

Dengg, J., A. Beckmann, and R. Gerdes (1996), The Gulf Stream separation problem, in *The Warmwatersphere of the North Atlantic Ocean*, edited by W. Krauss, chap. 9, pp. 253–290, Gebrüder Borntraeger, Berlin.

Dengg, J., C. Böning, U. Ernst, R. Redler, and A. Beckmann (1999), Effects of an improved model representation of overflow water on the Subpolar North Atlantic, *Int. WOCE Newsl.*, *37*, 10–14.

Dengler, M., F. A. Schott, C. Eden, P. Brandt, J. Fischer, and R. J. Zantopp (2004), Break-up of the Atlantic deep western boundary current into eddies at 8° S, *Nature*, *432*(7020), 1018–1020.

Dietrich, D. E., C. A. Lin, A. Mestas-Nunez, and D. S. Ko (1997), A high resolution numerical study of Gulf of Mexico fronts and eddies, *Meteorol. Atmos. Phys.*, *64*(3–4), 187–201.

Dietrich, D. E., A. Mehra, R. L. Haney, M. J. Bowman, and Y. H. Tseng (2004), Dissipation effects in North Atlantic Ocean modeling, *Geophys. Res. Lett.*, *31*, L05302, doi:10.1029/2003GL019015.

Dukowicz, J. K., and R. D. Smith (1994), Implicit free-surface method for the Bryan–Cox–Semtner ocean model, *J. Geophys. Res.*, *99*, 7991–8014, doi:10.1029/93JC03455.

Eden, C. (2007), Eddy length scales in the North Atlantic Ocean, *J. Geophys. Res.*, *112*, C06004, doi:10.1029/2006JC003901.

Eden, C., and C. Böning (2002), Sources of eddy kinetic energy in the Labrador Sea, *J. Phys. Oceanogr.*, *32*, 3346–3363.

Eden, C., R. J. Greatbatch, and C. W. Böning (2004), Adiabatically correcting an eddy-permitting model using large-scale hydrographic data: Application to the Gulf Stream and the North Atlantic Current, *J. Phys. Oceanogr.*, *34*, 701–719.

Eden, C., R. J. Greatbatch, and J. Willebrand (2007), A diagnosis of thickness fluxes in an eddy-resolving model, *J. Phys. Oceanogr.*, *37*(3), 727–742, doi:10.1175/JPO2987.1.

Fanning, A. F., and A. J. Weaver (1997), A horizontal resolution and parameter sensitivity study of heat transport in an idealized coupled climate model, *J. Clim.*, *10*(10), 2469–2478, doi:10.1175/1520-0442(1997)010<2469:AHRAPS>2.0.CO;2.

Fischer, J., and F. A. Schott (2002), Labrador Sea Water tracked by profiling floats—From the boundary current into the open North Atlantic, *J. Phys. Oceanogr.*, *32*(2), 573–84.

Fox-Kemper, B., and B. Smyth (2008), Can large eddy simulation techniques improve mesoscale rich ocean models?, this volume.

Fox-Kemper, B., R. Ferrari, and J. Pedlosky (2003), On the indeterminancy of rotational and divergent eddy fluxes, *J. Phys. Oceanogr.*, *33*(2), 478–483, doi:10.1175/1520-0485(2003)033<0478:OTIORA>2.0.CO;2.

Fox-Kemper, B., R. Ferrari, and R. Hallberg (2008), Parameterization of mixed layer eddies. I: Theory and diagnosis, *J. Phys. Oceanogr.*, in press.

FRAM Group (1991), An eddy-resolving model of the Southern Ocean, *Eos Trans. AGU*, *72*, 169, 174–175.

Fu, L. L., and R. D. Smith (1996), Global ocean circulation from satellite altimetry and high-resolution computer simulation, *Bull. Am. Meteorol. Soc.*, *11*(77), 2625–2636.

Gent, P. R., and J. C. McWilliams (1990), Isopycnal mixing in ocean circulation models, *J. Phys. Oceanogr.*, *20*, 150–155.

Getzlaff, J., C. W. Böning, C. Eden, and A. Biastoch (2005), Signal propagation related to the North Atlantic overturning, *Geophys. Res. Lett.*, *32*, L09602, doi:10.1029/2004GL021002.

Getzlaff, J., C. W. Böning, and J. Dengg (2006), Lagrangian perspectives of deep water export from the subpolar North Atlantic, *Geophys. Res. Lett.*, *33*, L21S08, doi:10.1029/2006GL26470.

Greatbatch, R. J., and X. Zhai (2006), Influence of assimilated eddies on the large-scale circulation in a model of the northwest Atlantic Ocean, *Geophys. Res. Lett.*, *33*, L02614, doi:10.1029/2005GL025139.

Greatbatch, R. J., X. Zhai, C. Eden, and D. Olbers (2007), The possible role in the ocean heat budget of eddy-induced mixing due to air-sea interaction, *Geophys. Res. Lett.*, *34*, L07604, doi:10.1029/2007GL029533.

Griffies, S., R. Pacanowski, and R. Hallberg (2000), Spurious diapycnal mixing associated with advection in a z-coordinate ocean model, *Mon. Weather Rev.*, *128*, 538–564.

Griffies, S. M., and A. J. Adcroft (2008), Formulating the equations of ocean models, this volume.

Hecht, M. W., F. O. Bryan, M. Maltrud, and R. D. Smith (2006), The northwest corner of the Atlantic and rapid climate change, *Rep. LA-UR-06-7346*, Los Alamos Natl. Lab., Los Alamos, N.M. (Available at http://www.noc.soton.ac.uk/rapid/rapid2006/ic06pres.php.)

Hecht, M. W., E. Hunke, M. E. Maltrud, M. R. Petersen, and B. A. Wingate (2008), Lateral mixing in the eddying regime and a new broad-ranging formulation, this volume.

Hogg, N. G., and H. Stommel (1985), On the relation between the deep circulation and the Gulf Stream, *Deep Sea Res., Part A*, *32*(10), 1181–1193.

Holland, W. R., and L. B. Lin (1975), On the generation of mesoscale eddies and their contribution to the oceanic general circulation. I. A preliminary numerical experiment, *J. Phys. Oceanogr.*, *5*(4), 642–57.

Hurlburt, H. E., and P. J. Hogan (2000), Impact of 1/8° to 1/64° resolution on Gulf Stream model—Data comparisons in basin-scale subtropical Atlantic Ocean models, *Dyn. Atmos. Ocean*, *32*(3–4), 283–329.

Iselin, C. O. (1936), A study of the circulation of the western North Atlantic, *Pap. Phys. Oceanogr. Meteorol.*, *4*, 101.

Jia, Y. (2000), Formation of an Azores Current due to Mediterranean overflow in a modeling study of the North Atlantic, *J. Phys. Oceanogr.*, *30*(9), 2342–2358.

Jones, H., and J. Marshall (1997), Restratification after deep convection, *J. Phys. Oceanogr.*, *27*(10), 2276–2287.

Karsten, R., H. Jones, and J. Marshall (2002), The role of eddy transfer in setting the stratification and transport of a circumpolar current, *J. Phys. Oceanogr.*, *40*(1), 39–54.

Käse, R. H., A. Biastoch, and D. B. Stammer (2001), On the mid-depth circulation in the Labrador and Irminger Seas, *Geophys. Res. Lett.*, *28*, 3433–3436.

Katsman, C. A., M. A. Spall, and R. S. Pickart (2004), Boundary current eddies and their role in the restratification of the Labrador Sea, *J. Phys. Oceanogr.*, *34*(9), 1967–1983, doi:10.1175/1520-0485(2004)034<1967:BCEATR>2.0.CO;2.

Kearns, E. J., and N. Paldor (2000), Why are the meanders of the North Atlantic Current stable and stationary?, *Geophys. Res. Lett.*, *27*, 1029–1032.

Klein, B., and G. Siedler (1989), On the origin of the Azores Current, *J. Geophys. Res.*, *94*, 6159–6168.

Komuro, Y., and H. Hasumi (2005), Intensification of the Atlantic deep circulation by the Canadian Archipelago throughflow, *J. Phys. Oceanogr.*, *35*(5), 775–789, doi:10.1175/JPO2709.1.

Large, W. G., J. C. McWilliams, and S. C. Doney (1994), Oceanic vertical mixing: A review and a model with a nonlocal boundary layer parameterization, *Rev. Geophys.*, *32*, 363–403.

Lavender, K. L., R. E. Davis, and W. B. Owens (2000), Mid-depth recirculation observed in the interior Labrador and Irminger Seas by direct velocity measurements, *Nature*, *407*, 66–69.

Lazier, J. R. (1994), Observations in the northwest corner of the North Atlantic Current, *J. Phys. Oceanogr.*, *24*(7), 1449–1463, doi:10.1175/1520-0485(1994)024<1449:OITNCO>2.0.CO;2.

Legg, S., L. Jackson, and R. W. Hallberg (2008), Eddy-resolving modeling of overflows, this volume.

Le Traon, P. Y., and F. Ogor (1998), ERS-1/2 orbit improvement using TOPEX/POSEIDON: The 2 cm challenge, *J. Geophys. Res.*, *103*(C4), 8045–8057.

Lilly, J. M., P. B. Rhines, F. Schott, K. Lavender, J. Lazier, U. Send, and E. D'Asaro (2003), Observations of the Labrador Sea eddy field, *Prog. Oceanogr.*, *59*(1), 75–176.

Macdonald, A. M., and C. Wunsch (1996), An estimate of global ocean circulation and heat fluxes, *Nature*, *382*(6590), 436–439, doi:10.1038/382436a0.

Maltrud, M. E., and J. L. McClean (2005), An eddy resolving 1/10° ocean simulation, *Ocean Modell.*, 8(1–2), 31–54.

Maltrud, M. E., R. D. Smith, A. J. Semtner, and R. C. Malone (1998), Global eddy-resolving ocean simulations driven by 1985–1994 atmospheric winds, *J. Geophys. Res.*, 103, 30,825–30,853.

McClean, J., S. Jayne, M. Maltrud, and D. Ivanov (2008), The fidelity of ocean models with explicit eddies, this volume.

McClean, J. L., P.-M. Poulain, J. W. Pelton, and M. E. Maltrud (2002), Eulerian and Lagrangian statistics from surface drifters and a high-resolution POP simulation in the North Atlantic, *J. Phys. Oceanogr.*, 32, 2472–2491.

McClean, J. L., M. E. Maltrud, and F. O. Bryan (2006), Measures of the fidelity of eddying ocean models, *Oceanography*, 19, 104–117.

Munk, W. H. (1950), On the wind-driven ocean circulation, *J. Meteorol.*, 7(2), 79–93.

Myers, P., and D. Deacu (2004), Labrador Sea freshwater content in a model with a partial cell topographic representation, *Ocean Modell.*, 6(3–4), 359–377.

New, A. L., Y. Jia, M. Coulibaly, and J. Dengg (2001), On the role of the Azores Current in the ventilation of the North Atlantic Ocean, *Prog. Oceanogr.*, 48(2–3), 163–194.

Oka, A., and H. Hasumi (2004), Effects of freshwater forcing on the Atlantic deep circulation: A study with an OGCM forced by two different surface freshwater flux datasets, *J. Clim.*, 17(11), 2180–2194, doi:10.1175/1520-0442(2004)017<2180:EOFFOT>2.0.CO;2.

Oka, A., and H. Hasumi (2006), Effects of model resolution on salt transport through northern high-latitude passages and Atlantic meridional overturning circulation, *Ocean Modell.*, 13(2), 126–147.

Paci, A., G. Caniaux, H. Giordani, M. Lévy, L. Prieur, and G. Reverdin (2007), A high-resolution simulation of the ocean during the POMME experiment: Mesoscale variability and near surface processes, *J. Geophys. Res.*, 112, C04007, doi:10.1029/2005JC003389.

Paiva, A. M., J. T. Hargrove, E. P. Chassignet, and R. Bleck (1999), Turbulent behavior of a fine mesh (1/12°) numerical simulation of the North Atlantic, *J. Mar. Syst.*, 21, 307–320.

Penduff, T., B. Barnier, M.-A. Kerbiriou, and J. Verron (2002), How topographic smoothing contributes to differences between the eddy flows simulated by sigma- and geopotential-coordinate models, *J. Phys. Oceanogr.*, 32(1), 122–137, doi:10.1175/1520-0485(2002)032<0122:HTSCTD>2.0.CO;2.

Penduff, T., B. Barnier, J. M. Molines, and G. Madec (2006), On the use of current meter data to assess the realism of ocean model simulations, *Ocean Modell.*, 11(3–4), 399–416, doi:10.1016/j.ocemod.2005.02.001.

Penduff, T., J. Le Sommer, B. Barnier, A.-M. Treguier, J.-M. Molines, and G. Madec (2007), Influence of numerical schemes on current-topography interactions in 1/4° global ocean simulations, *Ocean Sci. Discuss.*, 4(3), 491–528.

Pickart, R. S., and W. M. Smethie (1993), How does the deep western boundary current cross the Gulf Stream?, *J. Phys. Oceanogr.*, 23(12), 2602–2616.

Pickart, R. S., and W. M. Smethie (1998), Temporal evolution of the deep Western Boundary Current where it enters the subtropical domain, *Deep Sea Res., Part I*, 45(7), 1053–1083.

Pickart, R. S., F. Straneo, and G. W. K. Moore (2003), Is Labrador Sea water formed in the Irminger basin, *Deep Sea Res.*, 50(1), 23–52.

Rahmstorf, S. (2002), Ocean circulation and climate during the past 120,000 years, *Nature*, 419(6903), 207–214.

Redi, M. H. (1982), Oceanic isopycnal mixing by coordinate rotation, *J. Phys. Oceanogr.*, 12(10), 1154–1158.

Rhein, M., J. Fischer, W. M. Smethie, D. Smythe-Wright, R. F. Weiss, C. Mertens, D.-H. Min, U. Fleischmann, and A. Putzka (2002), Labrador Sea water: Pathways, CFC inventory, and formation rates, *J. Phys. Oceanogr.*, 32(2), 648–665.

Rhines, P. B., and W. R. Holland (1979), Theoretical discussion of eddy-driven mean flows, *Dyn. Atmos. Oceans*, 3(2–4), 289–325.

Rix, N. H., and J. Willebrand (1996), Parameterization of mesoscale eddies as inferred from a high-resolution circulation model, *J. Phys. Oceanogr.*, 26(10), 2281–2285, doi:10.1175/1520-0485(1996)026<2281:POMEAI>2.0.CO;2.

Robinson, A. R., D. E. Harrison, Y. Mintz, and A. J. Semtner (1977), Eddies and the general circulation of an idealized oceanic gyre: A wind and thermally driven primitive equation numerical experiment, *J. Phys. Oceanogr.*, 7(2), 182–207.

Robinson, S. G., M. A. Maslin, and I. N. McCave (1995), Magnetic susceptibility variations in upper Pleistocene deep-sea sediments of the NE Atlantic: Implications for ice rafting and paleocirculation at the Last Glacial Maximum, *Paleoceanography*, 10(2), 221–250.

Rossby, T. (1996), The North Atlantic Current and surrounding waters: At the crossroads, *Rev. Geophys.*, 34(4), 463–481.

Rossby, T., and J. Nillson, (2003), Current switching as the cause of rapid warming at the end of the Last Glacial Maximum and Younger Dryas, *Geophys. Res. Lett.*, 30(2), 1051, doi:10.1029/2002GL015423.

Schott, F. A., R. Zantopp, L. Stramma, M. Dengler, J. Fischer, and M. Wibaux (2004), Circulation and deep-water export at the western exit of the subpolar North Atlantic, *J. Phys. Oceanogr.*, 34(4), 817–843.

Scott, R. B., and F. Wang (2005), Direct evidence of an oceanic inverse kinetic energy cascade from satellite altimetry, *J. Phys. Oceanogr.*, 35(9), 1650–1666, doi:10.1175/JPO2771.1.

Seber, G. A. F. (1984), *Multivariate Observations*, 686 pp., John Wiley, Hoboken, N.J.

Semtner, A. J., Jr., and R. M. Chervin (1988), A simulation of the global ocean circulation with resolved eddies, *J. Geophys. Res.*, 93(C12), 15,502–15,522.

Semtner, A. J., and Y. Mintz (1977), Numerical simulation of the Gulf Stream and mid-ocean eddies, *J. Phys. Oceanogr.*, 7(2), 208–230.

Smethie, W. M., Jr., R. A. Fine, A. Futzka, and E. P. Jones (2000), Tracing the flow of North Atlantic Deep Water using chlorofluorocarbons, *J. Geophys. Res.*, 105, 14,297–14,323.

Smith, R. D., and P. Gent (2002), Reference manual of the Parallel Ocean Program (POP), *Rep. LA-UR-02-2484*, Los Alamos Natl. Lab., Los Alamos, N.M.

Smith, R. D., and P. R. Gent (2004), Anisotropic GM parameterization for ocean models, *J. Phys. Oceanogr.*, 34, 2541–2564, doi:10.1175/JPO2613.1.

Smith, R. D., and J. C. McWilliams (2003), Anisotropic horizontal viscosity for ocean models, *Ocean Modell.*, *5*, 129–156.

Smith, R. D., J. K. Dukowicz, and R. C. Malone (1992), Parallel ocean general circulation modeling, *Physica D*, *60*, 38–61.

Smith, R. D., M. E. Maltrud, F. O. Bryan, and M. W. Hecht (2000), Numerical simulation of the North Atlantic Ocean at 1/10°, *J. Phys. Oceanogr.*, *30*, 1532–1561.

Spall, M. A. (1996a), Dynamics of the Gulf Stream/deep western boundary current crossover. Part I: Entrainment and recirculation, *J. Phys. Oceanogr.*, *26*, 2152–2168.

Spall, M. A. (1996b), Dynamics of the Gulf Stream/deep western boundary current crossover. Part II: Low-frequency internal oscillations, *J. Phys. Oceanogr.*, *26*, 2169–2182.

Spall, M. A. (2004), Boundary currents and watermass transformation in marginal seas, *J. Phys. Oceanogr.*, *34*(5), 1197–1213, doi:10.1175/1520-0485(2004)034<1197:BCAWTI>2.0.CO;2.

Speer, K. G., J. Gould, and J. LaCasce (1999), Year-long float trajectories in the Labrador Sea Water of the eastern North Atlantic Ocean, *Deep Sea Res., Part II*, *46*(1–2), 165–179.

Stammer, D., and C. W. Böning (1992), Mesoscale variability in the Atlantic Ocean from Geosat altimetry and WOCE high-resolution numerical modeling, *J. Phys. Oceanogr.*, *22*(7), 732–752.

Stammer, D., and C. Böning (1996), Generation and distribution of mesoscale eddies in the North Atlantic Ocean, in *The Warmwatersphere of the North Atlantic Ocean*, edited by W. Krauss, pp. 159–193, Gebrüder Borntraeger, Berlin.

Stommel, H. (1948), The westward intensification of wind-driven ocean currents, *Trans., AGU*, *29*, 202–206.

Stommel, H. (1965), *The Gulf Stream*, Univ. of Calif. Press, Los Angeles.

Sverdrup, H. U. (1947), Wind-driven currents in a baroclinic ocean; with application to the equatorial currents of the eastern Pacific, *Proc. Natl. Acad. Sci., U. S. A. 33*, 318–326.

Talley, L. D., and M. S. McCartney (1982), Distribution and circulation of Labrador Sea Water, *J. Phys. Oceanogr.*, *12*(11), 1189–1205.

Tansley, C. E., and D. P. Marshall (2000), On the influence of bottom topography and the deep western boundary current on Gulf Stream separation, *J. Mar. Res.*, *58*, 297–325.

Thomas, L., A. Tandon, and A. Mahadevan (2008), Submesoscale processes and dynamics, this volume.

Thompson, J. D., and W. J. Schmitz (1989), A limited-area model of the Gulf Stream: Design, initial experiments, and model-data intercomparison, *J. Phys. Oceanogr.*, *19*(6), 791–814.

Treguier, A. M. (1992), Kinetic energy analysis of an eddy resolving, primitive equation model of the North Atlantic, *J. Geophys. Res.*, *97*(C1), 687–701.

Treguier, A. M. (1999), Evaluating eddy mixing coefficients from eddy-resolving ocean models: A case study, *J. Mar. Res.*, *57*(1), 89–108, doi:10.1357/002224099765038571.

Treguier, A. M., S. Theetten, E. P. Chassignet, T. Penduff, R. Smith, L. Talley, J. O. Beismann, and C. Boning (2005), The North Atlantic subpolar gyre in four high-resolution models, *J. Phys. Oceanogr.*, *35*(5), 757–774.

Trenberth, K. E. (1998), The heat budget of the atmosphere and ocean, in *Proceedings of the First International Conference on Reanalysis*, *WMO/TD 876*, pp. 17–20, World Meteorol. Organ., Geneva, Switzerland.

Veronis, G. (1975), The role of models in tracer studies, in *Numerical Models of Ocean Circulation*, pp. 133–146, Natl. Acad. of Sci., Washington, D.C.

Visbeck, M., J. Marshall, and H. Jones (1996), Dynamics of isolated convective regions in the ocean, *J. Phys. Oceanogr.*, *26*(9), 1721–1734.

Weese, S. R., and F. O. Bryan (2006), Climate impacts of systematic errors in the simulation of the path of the North Atlantic Current, *Geophys. Res. Lett.*, *33*, L19708, doi:10.1029/2006GL027669.

Willebrand, J., B. Barnier, C. W. Böning, C. Dieterich, P. D. Killworth, C. LeProvost, Y. Jia, J. M. Molines, and A. L. New (2001), Circulation characteristics in three eddy-permitting models of the North Atlantic, *Prog. Oceanogr.*, *48*, 123–161.

Wunsch, C., and D. Stammer (1995), The global frequency-wavenumber spectrum of oceanic variability estimated from TOPEX/POSEIDON altimetric measurements, *J. Geophys. Res.*, *100*(C12), 24,895–24,910.

Wyrtki, K., L. Magaard, and J. Hager (1976), Eddy energy in the oceans, *J. Geophys. Res.*, *81*(15), 2641–2646.

Zhai, X., R. J. Greatbatch, and J. Sheng (2004), Diagnosing the role of eddies in driving the circulation of the northwest Atlantic Ocean, *Geophys. Res. Lett.*, *31*, L23304, doi:10.1029/2004GL021146.

Zhang, R., and G. K. Vallis (2007), The role of bottom vortex stretching on the path of the North Atlantic Western Boundary Current and on the Northern Recirculation Gyre, *J. Phys. Oceanogr.*, *37*(8), 2053, doi:10.1175/JPO3102.1.

M. W. Hecht, Computational Physics Group, CCS Division, Los Alamos National Laboratory, Los Alamos, NM, 87545, USA. (e-mail: mhecht@lanl.gov)

R.D. Smith, P. O. Box 1342, Los Alamos, NM 87544, USA. (e-mail: rds@ncar.ucar.edu)

Towards Eddy-Resolving Models of the Arctic Ocean

Wieslaw Maslowski and Jaclyn Clement Kinney

Department of Oceanography, Naval Postgraduate School, Monterey, California, USA

Douglas C. Marble

Office of Naval Research, Code 322, Arlington, Virginia, USA

Jaromir Jakacki

Institute of Oceanology, Polish Academy of Sciences, Sopot, Poland

High-resolution model results from two regional grid configurations, 1/6° and 1/12°, for the pan-Arctic domain are intercompared and validated against limited observational data to examine the main characteristics and distribution of simulated eddies and to determine limitations of the employed spatial resolution. Several regions within the larger domain are examined in particular: the Gulf of Alaska, Bering Sea, Chukchi Sea, Labrador Sea, Nordic Seas, and Barents Sea. These regions are selected either because of their known high-level eddy activity and/or because some data exist there for model validation. We find that doubling horizontal resolution roughly from 18 km to 9 km increases mean eddy kinetic energy (EKE) by an order of magnitude or more. In some regions, such as the southern Labrador Sea and the Gulf of Alaska, EKE distribution from the 1/12° model compares well, both in magnitude and spatial distribution, with estimates from observations. Model and altimeter estimates of EKE in the sub-Arctic Pacific both show high values associated with the Alaskan Stream and low values in the western Gulf of Alaska and deep Bering Sea. The model EKE values are less than the altimeter EKE values along the shelf break in the central Bering Sea. Spectral analysis of model and altimeter-measured sea surface topography suggest that the 9-km grid is not sufficient to fully resolve eddy features with wavelengths shorter than 100–150 km. Presented regional analyses imply that spatial resolution of order few kilometers is needed to fully represent eddy energetics in the Arctic Ocean.

1. INTRODUCTION

The primary focus of this chapter is to analyze the simulated eddy kinetic energy (EKE) in increasingly high-resolution model configurations of the pan-Arctic region and to illustrate the gains realized from doubling model resolution from 18 km to 9 km. The comparison is between two regional coupled ice ocean models with many similar characteristics (Figure 1, Table 1).

Ocean Modeling in an Eddying Regime
Geophysical Monograph Series 177
10.1029/177GM16

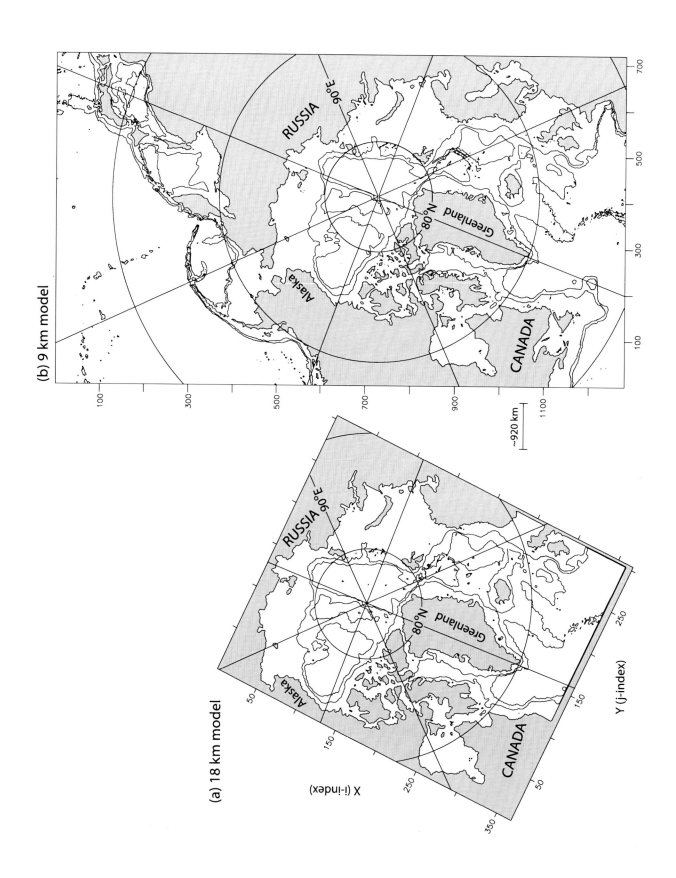

(b) 9 km model

(a) 18 km model

Table 1. The 18-km (PCAP) and 9-km (NAME) Model Configurations and Key Parameters. Bdry-boundary, rpt-repeat, Sal-Salinity, pe-processing elements, PHC-Polarscience Center Hydrographic Climatology, ASRC-Arctic Region Supercomputing Center, ETOPO5-5-minute Gridded Global Relief Data from the National Oceanic and Atmospheric Administration.

Parameter		18-km Model	9-km Model
ocean model		LANL POP, free surface	LANL POP, free surface
ice model		Hibler (1979)	Hibler (1979)
horizontal grid		368×304	1280×720
vertical levels		30	45
bathymetry		IBCAO + ETOPO5	modified ETOPO5/IBCAO
initialization fields		PHC 2.0	PHC 1.0
atmospheric forcing		ECMWF	ECMWF
restoring fields	surface	PHC 2.0 monthly mean	PHC 2.0 monthly mean
	Lat. Bdry	PHC 2.0 annual mean	PHC 1.0 annual mean
restoring timescale	surface	Temp (365 d), Sal (120 d) (0–20m)	Temp/Sal (30 d) (0–5m)
	Lat. Bdry	30 d	10 d
timestep	ocean	20 min	8 min
	ice	120 min	48 min
horizontal diffusion coefficients	tracer	-4.00×10^{18}	-5.00×10^{17}
	momentum	-1.00×10^{19}	-1.25×10^{18}
vertical diffusion coefficients	background diffusion	0.1	0.05
	background viscosity	1	0.2
spinup integration completed		10-yr rpt 1979–1993 mean, 3 × 1979–1981 cycle (9 yr)	27-yr rpt 1979–1993 mn, 6 yr rpt 1979, 3 × 1979–1981 cycle (9 yr)
diagnostic integration completed		1979–1998	1979–2004
approximate integration time		~28 h/yr on 64 pe, ARSC T3E-900	~168 h/yr on 128 pe, ARSC T3E-900

The initial question: "Is the 9-km model better at simulating ocean and sea-ice dynamics and distribution of properties than the 18-km model?" can be answered with a qualified yes. Yes, in that better representation of the bathymetry has an immediate positive impact on the modeled circulation in several regions of the Arctic Ocean and its adjacent seas, where bathymetry is the controlling factor, such as the Barents Sea, the Chukchi Sea, and the Canadian Arctic Archipelago (CAA). Another yes, in that several narrow boundary currents that were absent or poorly represented in the 18-km model appear or become stronger and better defined in the 9-km model. An example of this will be shown in the Labrador Sea. Another yes, in that volume transports through Bering Strait, Fram Strait, and the Barents Sea are approaching observed values [*Clement et al.*, 2005; *Maslowski et al.*, 2004], with the acknowledgment that even at ~9-km resolution, transport through the CAA is still possibly underrepresented. Finally, the ability of the 9-

km model to reasonably represent eddy kinetic energy in the southern Labrador Sea and the Gulf of Alaska and to decrease the gap between modeled and observed values in the northern Labrador Sea from a factor of 10 to a factor of 1.22 indicates greater skill at simulating ocean energy levels, which in many regions determine observed conditions.

The representation of certain aspects of Arctic Ocean and sea ice characteristics are significantly improved in the 9-km model [NPS Arctic Modeling Effort or (NAME)] when compared with output from a previous 18-km model [Parallel Coupled Arctic Program or (PCAP)]. However, there are unknowns and shortcomings in the simulation of certain features that the 9-km model shares with the 18-km version, such as the representation of bottom boundary layer or tidally induced regional mixing and currents, verification of which is made difficult by a lack of observational data. The focus in this chapter will be to illustrate how representation

Figure 1. The 18-km PCAP model domain (a) and 9-km NAME model domain (b). The 18-km model image has been rotated 26°. Approximate distance scale is equivalent to one hundred 9-km model grid points and fifty 18-km model grid points. The 500- and 2500-m contours are shown in black.

of EKE in the pan-Arctic region has been improved because of the resolution increase.

The most significant finding is that inadequately resolving the basic large scale circulation can adversely impact a model's ability to properly represent water mass formation, distribution, and interactions due to advection, which will then impact heat, salt, and possibly mass balances. Underrepresentation of eddies and eddy kinetic energy will also impact the transport and mixing of water mass properties. In essence, improvements gained through perfecting subgrid-scale parameterizations will not help coarser models because improper representation of the circulation will inhibit or prevent water mass interactions. As shown by *Siegel et al.* [2001], eddy kinetic energy and the generation of eddies increase with increases in resolution, and at higher resolutions the rate of increase slows somewhat. Their highest resolution experiment, a 1.56-km resolution wind-driven, closed-basin, quasi-geostrophic ocean model, is approximately six times the resolution of the 9-km model discussed in this report. This indicates further increases in resolution should continue to result in improvements in the representation of mesoscale and smaller scale processes in the Arctic Ocean.

The 18-km model used here is similar to that used by *Maslowski et al.* [2000, 2001] except for the bathymetry dataset, which was derived from the International Bathymetric Chart of the Arctic Ocean (IBCAO) [*Jakobsson et al.*, 2000] and interpolated onto the model grid [*Marble*, 2001].

The remainder of this chapter will first briefly discuss improvements gained through the increase in resolution in the following order: (1) representation of the bathymetry and circulation in regions where bathymetry is the major controlling factor; (2) the representation of shelf break and coastal boundary currents; and (3) the increase in eddies and eddy kinetic energy noted in the 9-km model. Analyses of modeled eddies and EKE are organized in three separate regions: subpolar North Atlantic, subpolar North Pacific, and the Arctic Ocean. The choice of model output used was made based on its availability at the time of particular regional analyses. Every effort has been made to compare equivalent depths and regions. The NAME model domain is shifted ~26° to the west when compared to that of PCAP. Therefore, 18-km model images, when placed next to 9-km model images, have been rotated 26°.

2. SUBPOLAR NORTH ATLANTIC

2.1. Labrador, Nordic, and Barents Seas

2.1.1. Bathymetry Impacts. It is obvious that increased resolution will result in improved representation of the geography and bathymetry within a model domain. Higher sampling along a coastline or a vertical section will capture higher frequency variations in that profile. Some smoothing will occur and many small features will still be missed, but as resolution increases, features attain a more realistic shape. What is less clear is what resolution will ensure enough of the horizontal and vertical variations have been captured to accurately simulate the circulation and mass balances.

As discussed by *Maslowski et al.* [2004], the 9-km model exhibits considerable skill in representing the mean circulation in the Barents Sea, a region where bathymetry is a major controlling factor in the circulation [e.g., *Pfirman et al.*, 1994]. Circulation in the PCAP Barents Sea (Plate 1a) exhibits many similar features to the 9-km model. Yet due to poorer vertical resolution and the resultant representation of the bathymetry in the Barents Sea, specifically Great Bank and Central Bank, PCAP displays pathways that are not in agreement with the observed circulation in the Barents Sea [*Ozhigin et al.*, 2000] nor with the modeled circulation from the NAME model (Plate 1b). There appear to be two paths the warm Atlantic Water follows through the Barents Sea, with larger velocities and more concentrated currents along the northern path (Plate 1a). This dynamical difference has significant physical implications on volume and property fluxes and on the nature of the water mass transformation that takes place in the Barents Sea. The placement of the primary pathway of warm Atlantic Water through the Barents Sea farther to the north also impacts the oceanic frontal structure as well as the ice edge, which in turn force the region's ecosystem.

2.1.2. Boundary Current. Boundary currents are crucial in the transport of volume, heat, salt, and freshwater throughout the Arctic Ocean and the subpolar seas [*Rudels*, 1987; *Aagaard*, 1989, *Rudels et al.*, 1994]. The Labrador Sea plays a key role in the global thermohaline circulation as one of the few deepwater formation regions in the world [*Broecker*, 1991; *Skyllingstad et al.*, 1991]. Furthermore, variations in the transport of cold, fresh Arctic Ocean outflow, as well as sea-ice and icebergs, by boundary currents such as the East and West Greenland Currents, the Baffin Current, and the Labrador Current precondition the Greenland Sea and the Labrador Sea for deep convection and deep water formation. Large freshwater outflows can alter deep convection in either area [*Dickson et al.*, 1988, 1996; *Visbeck et al.*, 1995]; therefore, proper representation of these currents is crucial.

Labrador Sea boundary currents are narrower and stronger in the 9-km model as represented by the mean velocity in the top 180 m when compared to the 18-km model (Plate 2). Improved representation of northward flow through Davis Strait is apparent as well as bathymetric effects on the

Plate 1. Distribution of 1980 annual mean velocity (cm/s) in the Barents Sea. (a) PCAP, 0–225 m (model levels 1–7), every vector is plotted; (b) NAME, 0–223 m (model levels 1–15), every other vector is plotted. Background shading is current speed (cm/s) and is the same for both model images. Note the differing vector scales.

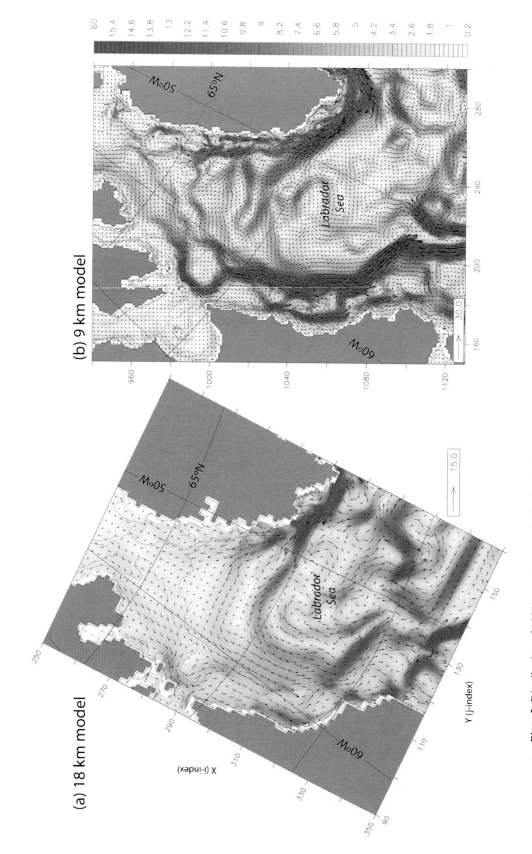

Plate 2. Distribution of 1980 annual mean velocity (cm/s) in the Labrador Sea. (a) PCAP, 0–180 m (model levels 1–6); (b) NAME, 0–183 m (model levels 1–14). Every other vector is plotted. Background shading as in Plate 1. Note the differing vector scales.

southerly flowing Baffin Land/Labrador Current. The width of the western branch of the Baffin Land/Labrador Current traveling around the 500-m contour at the mouth of Hudson Strait is ~50 km in the 9-km model versus ~100 km in the 18-km model. The separation of the West Greenland Current into several branches of westward flow across the Labrador Sea [*Cuny et al.*, 2002] is more distinct, as are the interactions between the current along the Labrador coast and the shelf break current. The proximity of the 18-km model boundary and modifications made to the coastline must be taken into account in the behavior of the coastal current south of 55°N and representation of the northward intrusion of North Atlantic Current meanders into the Labrador Sea (Plate 2a).

2.1.3. Eddies and Eddy Kinetic Energy. Eddies play a significant role in oceanic circulation in that they can result in the propagation of significantly different water masses outside of their place of origination [*Gent et al.*, 1995] and their dissipation transfers momentum between length scales and transfers properties between water masses. Their size is governed by the scale of the Rossby deformation radius defined as:

$$R = (g'D)^{1/2}/f \qquad (1)$$

where $g' = g\Delta\rho/\rho$ is the reduced gravitational acceleration, g is the acceleration due to gravity, $\Delta\rho/\rho$ is a characteristic density difference ratio for the fluid over the vertical scale of motion D, and f is the Coriolis acceleration [*Pedlosky*, 1987]. In the Arctic Ocean, due to high latitude (i.e., large f) and relatively weak vertical stratification, this scale, hence the size of eddies, is on the order of 0 (10 km) or less [*Manley and Hunkins*, 1985; *D'Asaro*, 1988; *Plueddemann et al.*, 1998; *Pickart et al.*, 2005]. In the sub-Arctic regions, the eddy radius increases to 20 km or more in the Labrador Sea [e.g., *Pickart et al.*, 1996] and 25–100 km in the Bering Sea/Gulf of Alaska [*Okkonen*, 1993; *Schumacher and Stabeno*, 1994; *Stabeno and Reid*, 1994; *Crawford et al.*, 2000; *Ladd et al.*, 2007].

Proper eddy parameterization in coarser resolution models is found to be important insofar as eddy-topography interactions contribute to propelling and sustaining narrow boundary currents [*Nazarenko et al.*, 1997, 1998]. *Semtner and Chervin* [1992] presented results from the first eddy-resolving global ocean circulation model and since then, further improvements in the efficiency of model codes and supercomputer capabilities have allowed continued resolution increases in many applications, which have broadened our view of the intensity and distribution of eddies in the oceans. These resolution increases do come at considerable computational cost (cf. Table 1 for this limited regional example). There is a vigorous and continuing discussion about finding the proper balance between resolution increases and parameterizations that will allow long integrations to accurately simulate circulation at multiple length scales and maintain mass and property balances.

Because the high spatial resolution of the 9-km model allows analysis of eddies theoretically down to a scale of ~30 km, we investigate the distribution of eddy kinetic energy (defined as EKE = $(u'^2 + v'^2)/2$, following *Stevens and Kilworth* [1992]). These calculations are done for monthly and daily fluctuations, referred to the annual mean. The monthly average data sets used in EKE calculations represent the average of all timesteps in that month (roughly 5400 timesteps), whereas a daily snapshot represents the last timestep out of the approximately 180 timesteps during one day. In addition, due to storage and integration time considerations, the daily snapshots contain only surface velocity components among other limited variables.

The annual mean velocity components, $\bar{u}$ and $\bar{v}$, were subtracted from the daily velocity values or monthly average velocity values to obtain u' and v'. EKE was calculated for the region including the northern Norwegian, Greenland, and Barents Seas. The annual mean surface EKE for the 9-km model, calculated from monthly average output, was found to represent approximately 15% of the annual mean surface EKE calculated from daily snapshots (8.40 cm²/s² versus 54.53 cm²/s²; with standard deviations about the means of 12.6 cm²/s² versus 70.2 cm²/s², respectively).

A comparison of the EKE calculated in a similar region using output from the 18-km resolution model [*Maslowski et al.*, 2000] shows area-averaged monthly mean EKE in the 9-km model exceeds EKE in the 18-km model by a factor of 5.

A comparison of model 1993–1997 mean surface EKE with similar multiyear mean EKE values deduced from observed surface drifter data [*Cuny et al.*, 2002] indicates that the 9-km model is able to reproduce the geographic distribution of EKE maxima similar to that observed (Plate 3). Model EKE values match those observed in a northern extension of the North Atlantic Current, the lower right corner of Plates 3a and 3b. Farther north, however, the model EKE values are roughly 20–30% lower than the observed values. The maximum daily surface EKE in the region in Plate 3a is ~1542 cm²/s², the annual mean ~113 cm²/s² with a standard deviation about the mean of 168 cm²/s².

The difference in EKE in the northern Labrador Sea may be due to differences in the atmospheric forcing between daily-averaged, relatively low-resolution, and spatially smoothed European Center for Medium-Range Weather Forecasting (ECMWF) data prescribed in the model and the real conditions experienced by surface drifters. Another potential cause of the difference may be due to model resolution. As latitude increases and vertical density stratification decreases, the internal Rossby radius decreases. The North Atlantic Current is well stratified in comparison to the nearly vertically homogeneous waters of the northern Labrador Sea, thus the Rossby radius within the North Atlantic Current is larger and the model is able to reproduce the observed eddy kinetic energy. That the 9-km model is able to reproduce the observed spatial distribution and magnitudes of EKE is quite significant, especially when compared to EKE values calculated in a similar region using output from a coarser resolution ~18-km model. The 9-km model values are 10 times larger than the 18-km model values.

Results presented above indicate that the 9-km model was able to accurately represent the mean levels of EKE observed in the southern Labrador Sea. Whether the difference between modeled and observed EKE in the northern Labrador Sea is a function of latitude and vertical stratification or the significant differences between the model and actual atmospheric forcing adds uncertainty to any conclusions drawn about 9-km model performance in this region. However, comparison of snapshots of EKE in the top ~45 m of the Labrador Sea clearly indicates that EKE is significantly underrepresented in the 18-km model (Plate 4). The southern maximum in EKE in the 18-km model (Plate 4a) corresponds to energy within a branch of the North Atlantic Current that extends to the north, then east. Geographic distribution of EKE maxima in the 9-km model are in reasonable agreement with the observed means (Plate 4b). The underrepresentation of EKE in the 18-km model may cause insufficient mixing of water from the West Greenland Current with surface water in the central Labrador Sea. This would in turn impact the degree of stratification, which would affect overturning and deepwater formation.

Dramatic differences exist in the distribution of 0- to 45-m eddy kinetic energy in the Nordic Seas between the two models (Plate 5). When viewed as an aggregate, the concentration of increased EKE in the 9-km model appears to define the path of the North Atlantic Current passing south of Iceland, and the Norwegian Atlantic Current as it travels north, along the west coast of Norway (Plate 5b). Indication of such a pathway is absent in the distribution of 18-km model EKE (Plate 5a). There is an order-of-magnitude difference in EKE statistics computed for the regional snapshots discussed previously (Table 2) as well as in the statistics computed for similar regional annual mean surface (0–20 m, PCAP level 1; 0–5 m, NAME level 1) EKE values (not shown).

The 9-km EKE in the model is still underrepresented due to inadequate resolution, the use of mean atmospheric forcing to drive the model during spinup, and the monthly mean output used in many of these calculations. The internal Rossby radius of deformation is about 6 km in the St. Anna Trough (U. Schauer, personal communication) and a model would require a grid cell spacing on the order of 1–2 km to be able to reasonably resolve features of that scale.

Considerable seasonal variability is noted in the 9-km model EKE distribution in the Labrador Sea (Plate 6) in qualitative agreement with limited observational estimates [*White and Heywood*, 1995]. The winter EKE maximum in the Labrador Sea, the majority of which remains ice-free year round, is due to an increase in storm activity during the winter months. The daily EKE maximum in the Labrador Sea in wintertime is roughly 25% larger than the summer maximum, 2400.1 versus 1919.8 cm²/s², respectively. The Labrador Sea winter mean EKE, 135.6 cm²/s², is ~1.7 times the summer mean of 80.2 cm²/s⁻² and there is more variability in the values in the wintertime, with winter standard deviation about the mean ~205 cm²/s² and summer standard deviation about the mean ~150 cm²/s².

3. SUBPOLAR NORTH PACIFIC

3.1. Gulf of Alaska

The 18-km (PCAP) model domain did not include the subpolar North Pacific (Figure 1), therefore we analyze and present results from the 9-km (NAME) model in this section on the Gulf of Alaska, and the following section will focus on the Bering Sea. We examine individual eddies and their frequency, along with associated impacts on the surrounding water masses.

Observations and models have identified the region south of the Aleutian Island Arc, in the vicinity of the Alaskan Stream, as an area of high eddy activity [*Maslowski et al.*, 2008; *Crawford et al.*, 2000; *Reed and Stabeno*, 1999; *Okkonen*, 1992]. Large (mean diameter of 160 km) and long-lived (1–3 years) anticyclonic eddies have been observed propagating along the Alaskan Stream at an average speed of 2.5 km d⁻¹ [*Crawford et al.*, 2000]. Sea surface height anomalies (SSHA) approached 72 cm. Similarly, model results show anticyclonic eddies with a mean diameter of 168 km and SSHA of 62 cm [*Maslowski et al.*, 2008]. Using satellite altimetry, *Crawford et al.* [2000] recorded the presence of six anticyclonic eddies during a 6-year period (September 1992–September 1998). Over this same time frame, modeled SSHA also show six eddies propagating westward south of the Aleutian Island Arc (Figure 2).

Model simulations over a 25-year period (1979–2003) showed a total of 20 anticyclonic eddies crossed a north-south line (~173.5°W) in the Alaskan Stream just east and south of Amukta Pass [*Maslowski et al.*, 2008]. These eddies, with SSHA greater than 30 cm, make for an average of 0.8 per year. Twelve of the 20 eddies had SSHA greater than 50 cm for an average of 0.5 per year.

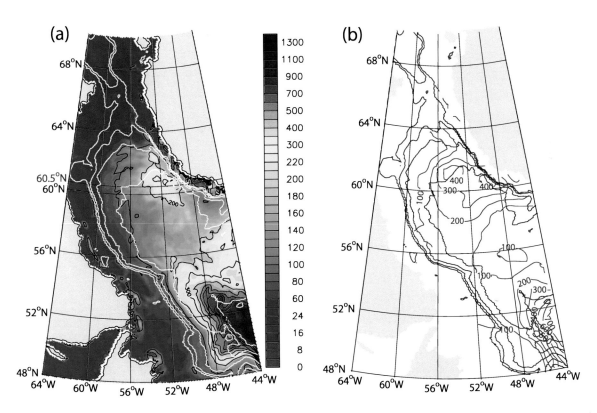

Plate 3. Horizontal distribution of eddy kinetic energy (cm²/s²) in the Labrador Sea. (a) 1993–1997 annual mean, 0–5 m (model level 1) calculated from daily model output. EKE contours 100, 200, 300, 400, 500, 600, 700, and 800 cm²/s² in black. (b) Eddy kinetic energy deduced from surface drifter data released in North Atlantic Ocean and Labrador Sea in during 1993–1997. (After Figure 7 from *Cuny et al.* [2002].)

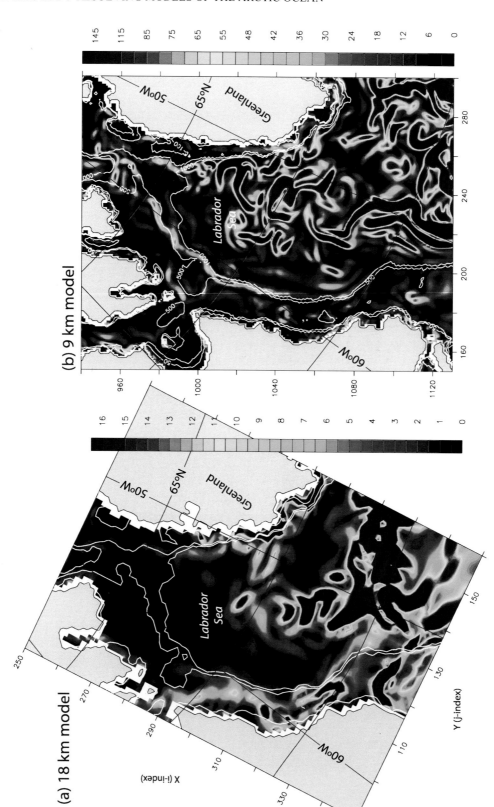

Plate 4. August 1980 snapshot of surface layer eddy kinetic energy (cm²/s²) in the Labrador Sea: (a) PCAP, 0–45 m (model levels 1–2); (b) NAME, 0–43 m (model levels 1–7). Note the different shading scales.

The propagation of one of these large eddies affects the flow of the Alaskan Stream by shifting the westward velocity core offshore by 70–155 km, depending on the size (or diameter) of the eddy and the particular location along the Aleutian Island Arc. As an example, Figure 3 shows the offshore displacement of the Alaskan Stream just east and south of Amukta Pass (~173.5°W) in December 1993 due to an eddy. (This eddy is shown when it was further upstream in April 1993 in Figure 2 as eddy 2.) The 25-year mean velocity structure shows speeds up to 70 cm/s at the surface, with speeds exceeding 5 cm/s as deep as 2000 m. The velocity core is typically located near the slope; however, as Figure 3 shows, an eddy can cause a large horizontal displacement, in this case ~155 km offshore. During the eddy passage, the maximum surface speed slows to 55 cm/s. However, the total westward volume transport through this cross section is actually higher during eddy passage (60 Sv) compared to the long-term mean (53 Sv), due to the widening and deepening of the velocity core. In fact, speeds of more than 5 cm/s reach the bottom at a depth of 5000 m during eddy passage.

This eddy, which we use as an example, was in the vicinity of Amukta Pass from October 1993 to February 1994 and caused significant changes in the salinity structure of the region over this period. During the eddy's approach, low salinity waters that typically lie over the shelf are drawn offshore to the south by more than 200 km. In addition, the eddy is responsible for bringing relatively salty water up the slope and into Amukta Pass. During the passage of the eddy, monthly mean volume transport and salt flux anomalies through Amukta Pass reached 1.25 Sv and 40 million kg/s, respectively. Below the euphotic zone, salinity is positively correlated with nutrient concentrations (P. Stabeno, personal communication). This means that eddy-induced upwelling along the slope has the potential to supply the Aleutian shelves with nutrient-rich water.

3.2. Bering Sea

Eddies in the Bering Sea are primarily found in the basin and along the slope in the Bering Slope Current (BSC) region. Our model results and observational data [Okkonen, 1993] indicate that the lifetimes of eddies in the Bering Sea are typically a few months to a year and tend to be shorter than those in the Gulf of Alaska, which range from 1 to 3 years. Another difference between these two regions, based on model results, is the presence of both cyclonic and anticyclonic eddies in the Bering Sea basin, whereas the Alaskan Stream is more prone to large anticyclonic features.

To better understand the ability of the model to represent realistic energies, we compare TOPEX altimeter observations along orbital ground track D79 (located along the shelf break region of the central Bering Sea) with model results for 1994–2001. Figure 4 shows power spectra of observed SSHA along this TOPEX ground track and modeled SSHA interpolated at points corresponding to the TOPEX ground track locations. At longer wavelengths (greater than ~200 km), the observed and modeled spectra exhibit similar energies. However, the modeled energy drops off more rapidly than the observations at shorter wavelengths, causing the spectra to diverge. Peaks at ~40 and ~20 km can be seen in the model spectrum and an increasing amount of noise shows up at shorter wavelengths in the altimeter spectrum. Interpolating 9-km gridded model output to the 6.2-km TOPEX sample spacing likely caused spatial aliasing, which shows up as peaks at the shortest wavelengths in the model spectrum.

The BSC is a predominant feature of the Bering Sea; however, this current has been described as more of a system of eddies rather than a continuous current [Okkonen, 1993]. Model results show eddies with diameters ranging from 90 to 325 km are frequently found in the Bering Sea basin, with a high concentration along the downstream portion of the Bering Slope Current region to the northwest [Clement Kinney et al., 2008]. For example, in November 1993, a cyclonic eddy with a diameter of 145 km is located just south of Zhemchug Canyon (ZC) (Figure 5). We can also see an anticyclonic feature adjacent to the cyclonic eddy, suggesting dipole formation due to the presence of the canyon.

The eddy's presence in the canyon strongly affects the local distribution of salinity. Vertical sections of salinity anomalies at BSC (not shown) and ZC (Plate 7) show high salinity water being upwelled and pushed onto the shelf when the eddy is present, during November 1993. Monthly mean salinity anomalies approach 0.12 in the surface water and are even higher near bottom (up to 0.18). Velocities are relatively strong (monthly mean up to 12 cm/s) and directed on-shelf.

It has been suggested [Paluszkiewicz and Niebauer, 1984; Okkonen, 1993; Mizobata et al., 2006] that topographic planetary waves generated by baroclinic instabilities are responsible for eddy generation along the Bering slope. The modeled eddy described above had a diameter of 145 km and period of ~90 days, which is similar to observations made by Paluszkiewicz and Niebauer [1984; 140 km/84 days] and Okkonen [1993; 184 km/72 days]. Based on these results and the assumption that deep Bering Sea water is high in nutrient concentrations [Clement et al., 2005], it can be inferred that mesoscale eddies are responsible for providing the Bering Sea shelf with nutrient-rich water from the basin. The location, duration, and frequency of eddy activity would then be highly important to ecosystems on the shelf. As seen in the model results, the Zhemchug Canyon region appears to be a prime location for eddy activity and subsequent upwelling of salty, nutrient-rich water.

4. ARCTIC

4.1. Chukchi Sea, Beaufort Sea, and Canada Basin

The Arctic Ice Dynamics Joint Experiment (AIOJEX) of 1972 was the first study to identify subsurface mesoscale eddies in the Arctic Ocean [*Manley and Hunkins*, 1985]. Over the 14-month time frame, a total of 127 eddies were identified between 50 and 300 m depth. They ranged in diameter from 10 to 20 km and were predominantly anticyclonic with an anticyclonic/cyclonic eddies ratio of 5:2. Later work by *D'Asaro* [1988] revealed the presence of vertical pairs of counterrotating eddies in the Beaufort Sea. Although poorly sampled, it was observed twice that anticyclonic eddies in the upper water column (but below 50 m) were associated with deep cyclonic features. *Plueddemann et al.* [1998] were able to clearly observe 10 eddies in 23 months of data collected by a drifting buoy equipped with an ADCP in the Beaufort Gyre. Again, a predominance of anticyclonic eddies was noted (9 of 10); however, no observations of a deep counterrotating eddy were made, possibly due to a lack of samples in deep enough water. Modeling work by *Chao and Shaw* [1996] showed how a pair of counterrotating eddies might occur in the Beaufort Sea. A pulse of either lower or higher density water could create a shallow eddy with a deep counterrotating eddy; however, the shallow one is more quickly dissipated due to surface friction exerted by sea ice. *Pickart et al.* [2005] suggest two general mechanisms for formation of eddies found in the Canada Basin: current-topographic interactions (primarily in Barrow Canyon) and hydrodynamic instability. In this section, we will first examine the energetics of the Chukchi and Beaufort shelves and then later describe eddies found in deeper waters of the slope and basin. We use results from the 9-km model instead of the 18-km model, due to its ability to better represent smaller features and the small Rossby radius of deformation found in the Arctic Ocean, as noted earlier.

Modeled eddy kinetic energy at 20- to 26-m depth on the Chukchi and Beaufort shelves from the 9-km model is highest during the summer and autumn (Plate 8). Values of more than 500 cm²/s² are found in shallow water (<50 m deep) in Bering Strait and near Point Barrow. However, these high values of EKE are likely due in large part to fluctuations in current speed and also current reversals, rather than actual mesoscale eddies.

More interestingly, EKE values up to 10 cm²/s² are found along the Beaufort shelf break and in the Canada Basin throughout the year. In an attempt to more closely examine these features, we created animations of monthly "snapshots" of EKE for 5 years (1985–1989). "Snapshots" represent the last timestep of each month. Five-year mean (1985–1989) values of the horizontal components of velocity were used as the "long-term" mean in calculating the EKE. We completed the same exercise for several depth levels from 20–26 to 395–478 m. A typical EKE distribution is shown in Plate 9 for July 31, 1986, at 268–326 m. At least five anticyclonic and three cyclonic eddies are visible, with most present in the Western Beaufort Sea, especially near the Barrow Canyon outflow. Eddies that are located along the Beaufort slope appear to be associated with the Barrow Canyon, and are an important means of communication between the shelf and basin [*Pickart et al.*, 2005]. In addition, there is a large anticyclone off the northeast corner of the Chukchi Cap. Eddies were commonly found inside the black box (Plate 9); therefore we examine the vertical structure of one of these eddies in the following paragraph.

After careful examination of the vertical structure of EKE over the 5-year period and over the entire water column, we found that it is sometimes possible to have a vertical pair of counterrotating eddies. Plate 10 shows EKE and velocity on November 30, 1985 (left), at three representative depth levels. The shallower, anticyclonic eddy is at ~15- to 100-m depth, with speeds up to 5 cm/s⁻¹ and a diameter of ~75 km (Plate 10a). The lower, cyclonic eddy (Plate 10c) has a larger diameter (~110 km) and is found at depths from 150 to well below 400 m. Rotational speeds are slightly higher for the cyclonic eddy, reaching 5.5 cm/s⁻¹. Weak and unorganized flow exists at depths ranging from 100 to 150 m between the two eddies. Six months later (on May 31, 1986), the shallow, anticyclonic eddy has dissipated, but the deeper cyclone remains intact (Plate 10). In fact, this eddy has a lifetime of 18 months versus only 6 months for the upper one, likely due to surface wind forcing and friction from the overlying sea ice [*Chao and Shaw*, 1996].

A vertical cross section of salinity and density (σ_θ) through these counterrotating eddies (along the black line shown in Plate 10) is shown in Plate 11. On November 30, 1985, in the upper 50 m, the doming of the isopycnals (and isohalines, as

Table 2. Eddy Kinetic Energy (cm/s) Statistics for the 0- to 45-m Regional Snapshots Presented in Plates 4 and 5

Model	Labrador Sea EKE			Nordic Seas EKE		
	Maximum	Mean	Std Dev	Maximum	Mean	Std Dev
PCAP58	132.50	4.90	9.20	269.40	4.70	13.30
PIPS	3998.00	70.40	203.70	4959.00	43.50	142.70

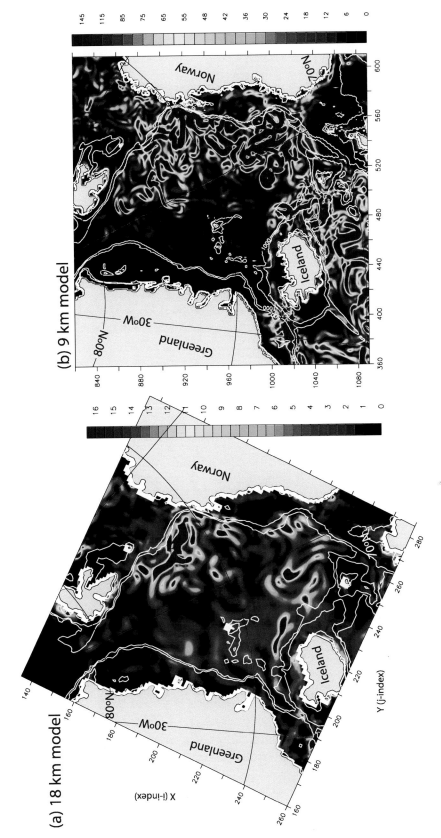

Plate 5. August 1980 snapshot of surface layer eddy kinetic energy (cm²/s²) in the Nordic (Greenland, Iceland, and Norwegian) Seas: (a) PCAP, 0–45 m (model levels 1–2); (b) NAME, 0–43 m (model levels 1–7). Note the different shading scales.

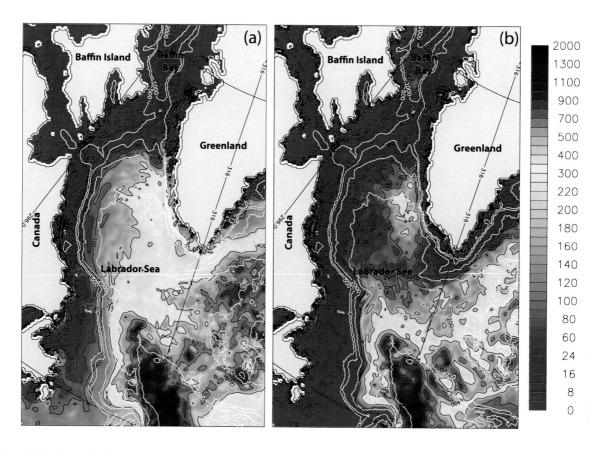

Plate 6. Horizontal distribution of surface (0–5 m, model level 1) 1993–1997 eddy kinetic energy (cm²/s²) in the Labrador Sea for different seasons: (a) winter (J-F-M) average; (b) summer (J-A-S) average. EKE contours: 8, 18, 32, 50, 72, 100, 200, 400, 600, and 800 cm²/s².

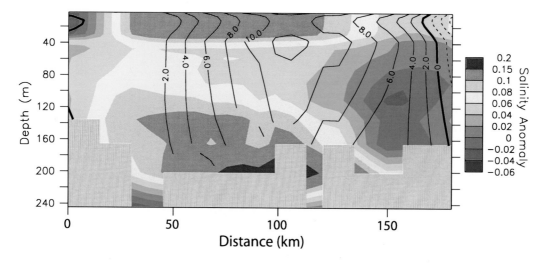

Plate 7. Vertical section of salinity anomaly across Zhemchug Canyon (section location shown in Figure 5) during November 1993. Contour lines indicate the across-slope velocity (positive onto the shelf; cm/s). Modified from *Clement Kinney et al.* [2008].

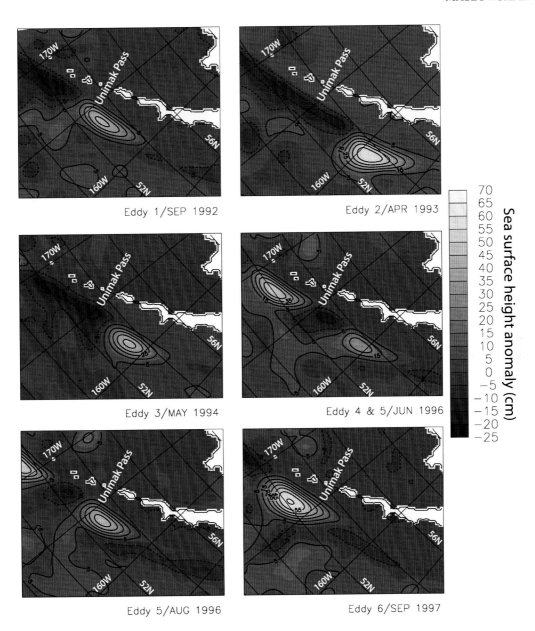

Figure 2. Modeled monthly mean sea surface height anomaly associated with six eddies propagating along the Alaskan Stream during the period September 1992 through September 1998. Shading represents the total SSHA (cm). Modified from *Maslowski, et al.* [2008].

well) is readily apparent, with vertical displacements up to ~20 m. In addition, sea surface height shows a local high centered over the eddy (not shown). A depression of the isopycnals and isohalines occurs below 50 m depth (likely due to the accumulation of water above) and shows even greater vertical displacements (up to 30 m). After the shallow eddy has dissipated on May 31, 1986, the upper water column is characterized by much flatter isopycnals and isohalines (Plate

11, right). However, below 150 m the depression of the isopycnals is still present, indicative of the cyclonic circulation that is seen in Plate 10f. The black dashed boxes in Plate 11 correspond to the three depth levels shown in Plate 10.

Modeled eddies occurred frequently throughout the 5-year period (1985–1989). However the diameters of these eddies were typically 80–150 km, which is significantly larger than observations (10–20 km) [*Manley and Hunkins*, 1985;

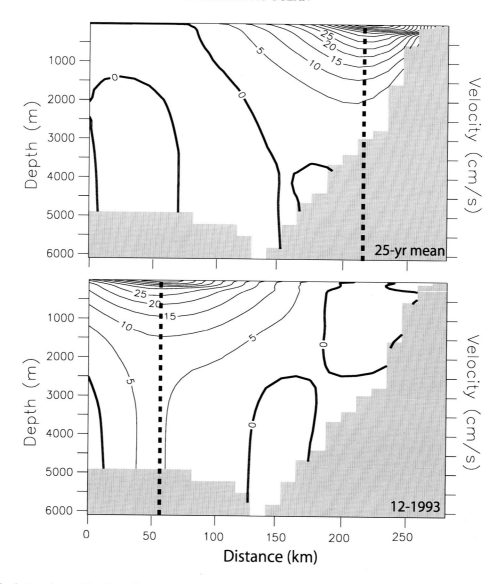

Figure 3. Vertical sections of the along-shore velocity during 1979–2003 mean (upper) and December 1993 (lower) along a line just east and south of Amukta Pass (~173.5°W) in the Alaskan Stream. The dashed lines indicate the position of the velocity cores.

D'Asaro, 1988]. As mentioned earlier, the 9-km horizontal grid cell spacing limits the model's ability to properly represent small eddies in this region that is characterized by such a small Rossby radius of deformation. On the other hand, the model simulated several eddies (both cyclonic and anticyclonic) in the Canada Basin and was even able to represent a pair of counterrotating eddies comparable to observations by D'Asaro [1988] and simulations by Chao and Shaw [1996]. Similar to predictions by Chao and Shaw [1996], the upper anticyclone dissipated first, in May 1986, and the lower cyclone lasted until May 1987.

5. SUMMARY AND CONCLUSIONS

Significant improvements in model skill are realized through a doubling of resolution, from 18 to 9 km. Increasingly realistic bathymetry due to higher resolution both in the horizontal and vertical direction, improves the simulation of topographically steered flows, which in the case of the Barents Sea and other coastal areas can change the representation of ocean circulation as well as regional distribution and transformation of water masses. As a result, the long-term circulation patterns in the Barents Sea agree

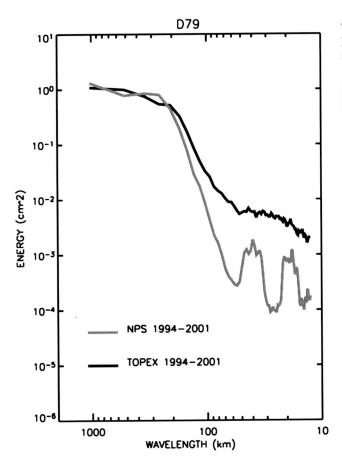

Figure 4. Eight-year mean (1994–2001) power spectra for model (gray line) and TOPEX altimeter (black line) sea surface height anomalies along TOPEX orbital ground track D79 (located along the shelf break region of the central Bering Sea).

well with published maps [*Maslowski et al.*, 2004; *Ozhigin et al.*, 2000]. The Atlantic Water flow through the Barents Sea, from the merging of portions of the North Cape Current and the Norwegian Coastal Current south of the Central Bank to the St. Anna Trough outflow, is more realistically defined. Calculations of EKE indicate a response to seasonal variability as well as areas of increased activity apparently associated with bathymetric features. Reducing the model grid spacing, from 18 to 9 km, has resulted in an increase in EKE by a factor of 5 in this region. It is expected that further increases in resolution will increase model EKE levels in the Barents Sea. Five numerical simulations using a closed basin quasi-geostrophic ocean model, varying only in horizontal resolution (from 25 to 1.56 km) and viscosity coefficients, and therefore Reynolds number (*Re*), indicated a continual increase in EKE with increased resolution [*Seigel et al.*, 2001]. *Seigel et al.* [2001] also found that the rate

of increase slows somewhat at the highest *Re*, indicating the possibility of a regime where eddy variability becomes insensitive to further increases in *Re*.

Boundary currents become narrower and stronger at 9-km resolution and the appearance of opposing boundary currents demonstrates a significant increase in horizontal and vertical shear. The representation of circulation within and around the model Labrador Sea is significantly improved. Better boundary current representation results in improvements in water mass transport around the perimeter of the Arctic Ocean as well as outflow through the Norwegian and Labrador Seas.

Comparison of the distribution of 0- to 45-m eddy kinetic energy in the Nordic/Labrador Seas between the two models shows dramatic differences (Plates 4 and 5). When viewed as an aggregate, the concentration of increased EKE in the 9-km model appears to define the path of the North Atlantic Current passing south of Iceland, and the Norwegian Atlantic Current as it travels north, along the west coast of Norway (Plate 5b). Indication of such a pathway is absent in the distribution of 18-km model EKE (Plate 5a). Based on EKE statistics computed for the regional daily snapshots (Table 2), there is an order-of-magnitude difference in magnitude between the two models. This increase in eddy kinetic energy between the 18-km model and the 9-km model has resulted in simulated values matching observed values in the southern Labrador Sea. However, 8–10° to the north, modeled values are about 20–30% lower than the observed. A definitive judgment on model skill in representing EKE in this region is difficult to make due to the differences in the atmospheric forcing between that used in the model and actual conditions. Another factor may be model resolution being just at the threshold of capturing features of a size on the order of the internal Rossby radius of deformation in the southern Labrador Sea and too coarse farther north.

Similarly in the Gulf of Alaska, EKE distribution from the 9-km model compares well, both in magnitude and spatial distribution, with estimates from observations. Model and altimeter estimates of EKE in the sub-Arctic Pacific both show high values associated with the Alaskan Stream and low values in the western Gulf of Alaska and deep Bering Sea. The model EKE values are less than the altimeter EKE values along the shelf break in the central Bering Sea. Spectral analyses of model and altimeter-measured sea surface topography suggest that the 9-km grid is not sufficiently fine to fully represent eddy energy levels at wavelengths shorter than 100–150 km.

Eddies are commonly simulated in the central Arctic Ocean in the 9-km model in similar regions as reported from observations. However, their typical size (i.e., diameter) ranges from 80 to 150 km compared to the 10- to 20-km range of observed eddies [e.g., *D'Asaro*, 1988]. We

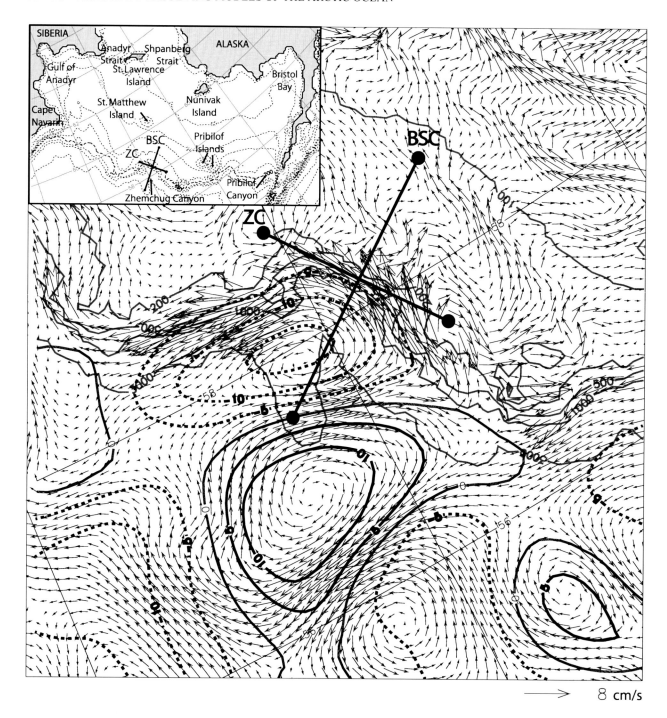

Figure 5. Velocity vectors (cm/s) and sea surface height anomaly contours (cm) during November 1993 in the vicinity of Zhemchug Canyon. Straight black lines indicate the positions of two cross sections (ZC, Zhemchug Canyon; BSC, Bering Slope Current). Solid gray lines represent bathymetry (m). Inset shows the geographic location.

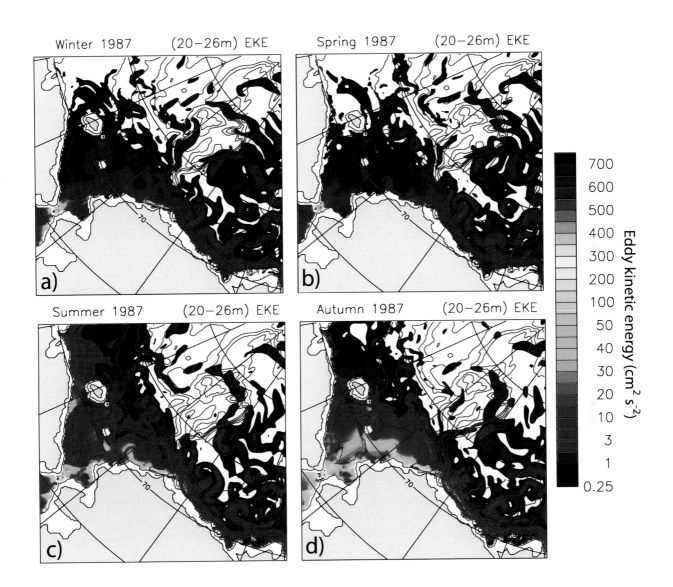

Plate 8. Eddy kinetic energy during 1987 at a depth of 20–26 m from the 9-km model for (a) winter (J-F-M), (b) spring (A-M-J), (c) summer (J-A-S), and autumn (O-N-D). Red lines indicate bathymetry.

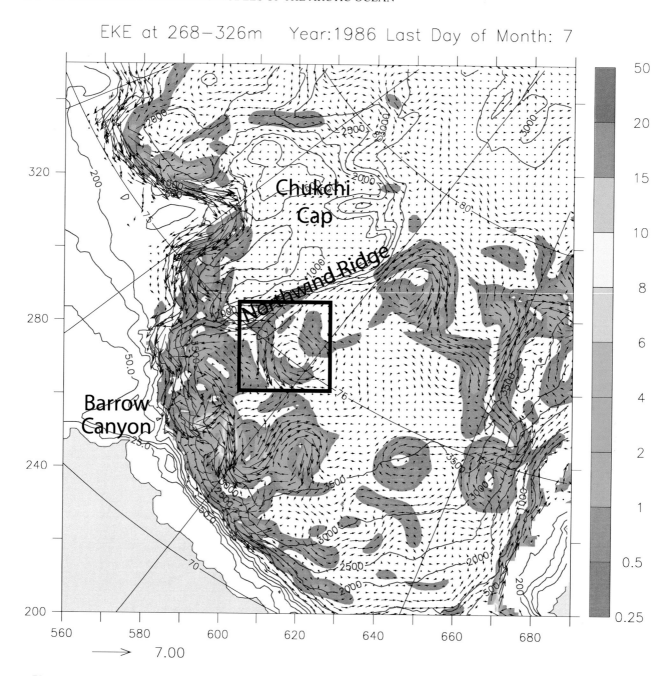

Plate 9. Modeled eddy kinetic energy (cm²/s²; color shading) and velocity (cm/s; vectors) in the Western Arctic from the 9-km model on July 31, 1986, at a depth of 268–326 m. Red lines indicate bathymetry. The black box indicates the region of interest for Plate 10.

Plate 10. Modeled eddy kinetic energy (cm²/s²; color shading) and velocity (cm/s; vectors) at (a, d) 42–53 m, (b, e) 99–122 m, and (c, f) 326–395 m from the 9-km model on November 30, 1985 (left) and May 31, 1986 (right). The black line shows the location of a vertical cross section shown in Plate 11.

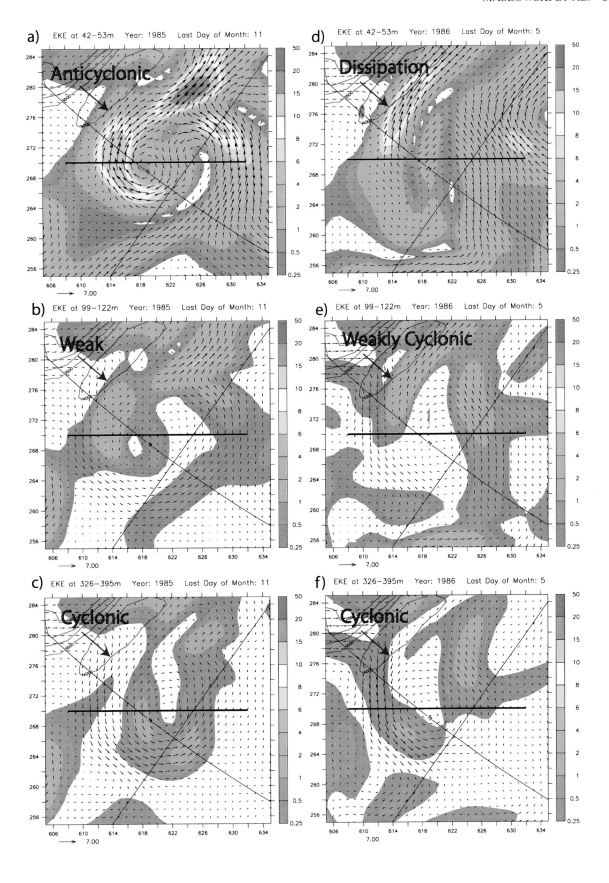

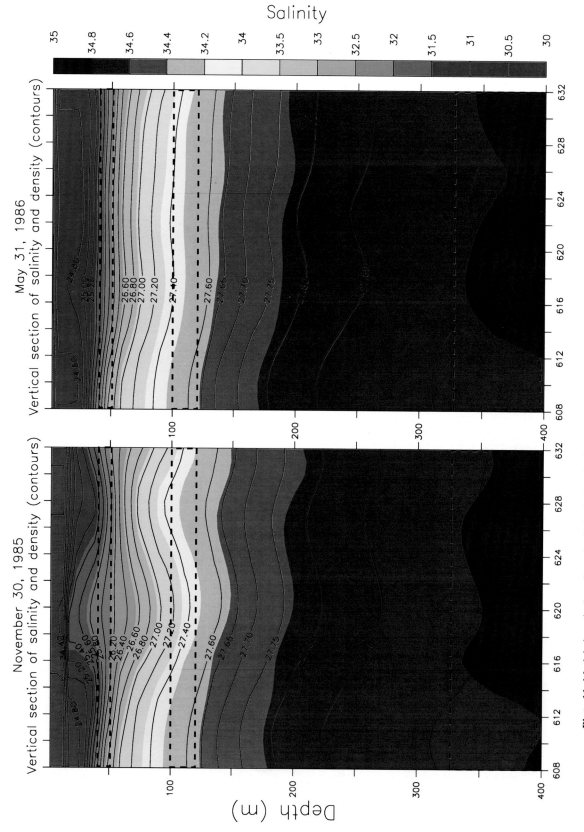

Plate 11. Modeled vertical section of salinity (color shading) and σ_θ (contours) through two counterrotating eddies from the 9-km model on November 30, 1985 (left) and May 31, 1986 (right). The black dashed boxes correspond to the three depth levels shown in Plate 10. The section location is shown in Plate 10.

hypothesize that given the limiting resolution of 9 km the model attempts to represent integrated energy of multiple smaller eddies averaged over a larger area of a modeled eddy. It is clear that significant increases in model resolution beyond 9 km are necessary to fully resolve eddies in the central Arctic. In addition, more observational data are needed to allow verification of modeled EKE distribution over larger areas.

Presented regional analyses imply that spatial resolution on the order of a few kilometers is needed to fully represent eddy energetics, in the Arctic Ocean. In addition, validation of other model results against observations (not shown) indicate that a similar horizontal resolution is required to properly represent buoyancy-driven narrow coastal currents (e.g., Alaska Coastal Current or Norwegian Coastal Current; *Maslowski and Walczowski* [2002]), small-scale bathymetry (e.g., Barrow Canyon), and land futures (e.g., passages through the CAA). Finally, more realistic and higher temporal and spatial resolution atmospheric forcing is needed for eddy-resolving ocean simulation. All these features might play a critical role in the large scale ocean dynamics as well as affect sea ice cover variability and freshwater advection and export into the North Atlantic.

Acknowledgments. Funding for the development and integration of the 18- and 9-km coupled ice-ocean models was provided by the Arctic System Science Program of the National Science Foundation, the Climate Change Prediction Program of the Department of Energy, and the Office of Naval Research. A portion of this work was toward partial completion of the requirements for a Ph.D. in physical oceanography (D.M.). Computer resources were provided by the Arctic Region Supercomputing Center (ARSC), Fairbanks, Alaska, through the Department of Defense High Performance Computer Modernization Program (DOD/HPCMP). Finally, we would like to thank the two anonymous reviewers for their comments, which helped improve the manuscript.

REFERENCES

Aagaard, K. (1989), A synthesis of the Arctic Ocean circulation, *Rapp P.-V. Reun. Cons. Int. Explor. Mer.*, *188*, pp. 11–22.

Broecker, W. S. (1991), The great ocean conveyor, *Oceanography*, *4*(2), 79–89.

Chao, S. Y., and P. T. Shaw (1996), Initialization, asymmetry, and spindown of Arctic eddies, *J. Phys. Oceanogr.*, *26*, 2076–2092.

Clement, J. L., W. Maslowski, L. Cooper, J. Grebmeier, and W. Walczowski (2005). Ocean circulation and exchanges through the northern Bering Sea—1979–2001 model results, *Deep-Sea Res. II*, *52*, 3509–3540, doi: 10.1016/j.dsr2.2005.09.010.

Clement Kinney, J., W. Maslowski, and S. Okkonen (2008), On the processes controlling shelf-basin exchange and outer shelf dynamics in the Bering Sea, *Deep-Sea Res. II*, in press.

Crawford, W. R., J. Y. Cherniawsky, and M. G. G. Foremann (2000), Multi-year meanders and eddies in the Alaskan Stream as observed by TOPEX/Poseidon altimeter, *Geophys. Res. Lett.*, *27*(7), 1025–1028.

Cuny, J., P. B. Rhines, P. P. Niiler, and S. Bacon (2002), Labrador Sea boundary currents and the fate of the Irminger sea water, *J. Phys. Oceanogr.*, *32*, 627–647.

D'Asaro, E. A. (1988), Observations of small eddies in the Beaufort Sea, *J. Geophys. Res.*, *93* (C6), 6669–6684.

Dickson, R. R., J. Meincke, S.-A. Malmberg, and A. J. Lee (1988), The "Great Salinity Anomaly" in the northern North Atlantic 1968–1982, *Progr. Oceanogr.*, *20*, 103–151.

Dickson, R. R., J. Lazier, J. Meincke, P. Rhines, and J. Swift (1996), Long-term coordinated changes in the convective activity of the North Atlantic, *Progr. Oceanogr.*, *38*, 241–295.

Gent, P. R., J. Willebrand, T. McDougall, and J. C. McWilliams (1995), Parameterizing eddy-induced tracer transports in ocean circulation models, *J. Phys. Oceanogr.*, *25*, 463–474.

Jakobsson, M., N. Z. Cherkis, J. Woodward, R. Macnab, and B. Coakley (2000), New grid of Arctic bathymetry aids scientists and mapmakers, *Eos Trans. AGU*, *81*(9), 89.

Ladd, C. A., C. W. Mordy, N. B. Kachel, and P. J. Stabeno (2007), Northern Gulf of Alaska eddies and associated anomalies, *Deep-Sea Res. I*, *54*, 487–509.

Manley, T. O., and K. Hunkins (1985), Mesoscale eddies of the Arctic Ocean, *J. Geophys. Res.*, *90*(C3), 4911–4930.

Marble, D. C. (2001), Simulated annual and seasonal Arctic Ocean and sea-ice variability from a high resolution coupled ice-ocean model," Ph.D. dissertation, Naval Postgraduate School, September 2001.

Maslowski, W., and W. Walczowski (2002), On the circulation of the Baltic Sea and its connection to the Pan-Arctic region—A large scale and high-resolution modeling approach, *Boreal Environ. Res.*, *7*, 319–325.

Maslowski, W., B. Newton, P. Schlosser, A. J. Semtner, and D. Martinson (2000), Modeling recent climate variability in the Arctic Ocean, *Geophys. Res. Lett.*, *27*(22), 3743–3746.

Maslowski, W., D. C. Marble, W. Walczowski, and A. J. Semtner (2001), On large-scale shifts in the Arctic Ocean and Sea Ice Conditions during 1979–98, *Ann. Glaciol.*, *33*, 319–325.

Maslowski, W., D. Marble, W. Walczowski, U. Schauer, J. L. Clement, and A. J. Semtner (2004), On climatological mass, heat, and salt transports through the Barents Sea and Fram Strait from a pan-Arctic coupled ice-ocean model simulation, *J. Geophys. Res.*, *109*, C03032, doi:10.1029/2001JC001039.

Maslowski, W., R. Roman, and J. Clement Kinney (2008), Effects of mesoscale eddies on the flow of the Alaskan Stream, *J. Geophys. Res.*, in press.

Mizobata, K., J. Wang, and S. Saitoh (2006), Eddy-induced cross-slope exchange maintaining summer high productivity of the Bering Sea shelf break, *J. Geophys. Res.*, *111*, C10017, doi:10.1029/2005JC003335.

Nazarenko, L., T. Sou, M. Eby, and G. Holloway (1997), The Arctic ocean-ice system studied by contamination modelling, *Ann. Glaciol.*, *25*, 17–21.

Nazarenko, L., G. Holloway, and N. Tausnev (1998), Dynamics of transport of "Atlantic signature" in the Arctic Ocean, *J. Geophys. Res.*, *103*(C13), 31,003–31,015.

Okkonen, S. R. (1992), The shedding of an anticyclonic eddy from the Alaskan Stream as observed by the Geosat altimeter, *Geophys. Res. Lett., 19*(24), 2397–2400.

Okkonen, S. R. (1993), Observations of topographic planetary waves in the Bering Slope Current using Geosat altimeter, *J. Geophys. Res. 98*(C12) 22,603–22,613.

Ozhigin, V. K., A. G. Trofimov, and V. A. Ivshin (2000), The Eastern Basin Water and currents in the Barents Sea, in *ICES Annual Science Conference 2000*.

Pfirman, S. L., D. Bauch, and T. Gammelsrod (1994), The Northern Barents Sea: Water Mass Distribution and Modification, in *The Polar Oceans and Their Role in Shaping the Global Environment: The Nansen Centennial Volume, Geophys. Monogr. Ser.,* vol. 85, edited by O. M. Johannesen, R. D. Muench, and J. E. Overland, pp. 77–94, AGU, Washington, D. C.

Pickart, R. S., T. J. Weingartner, L. J. Pratt, S. Zimmermann, D. J. Torres (2005), Flow of winter-transformed Pacific water into the Western Arctic, *Deep-Sea Res., 52,* 3175–3198, doi: 10.1016/j.dsr2.2005.10.009.

Plueddemann, A. J., R. Krishfield, T. Takizawa, K. Hatakeyama, and S. Honjo (1998), Upper ocean velocities in the Beaufort Gyre, *Geophys. Res. Lett., 25*(2), 183–186.

Reed, R. K., and P. J. Stabeno (1999), A recent full-depth survey of the Alaskan Stream, *J. Oceanogr., 55,* 79–85.

Rudels, B. (1987), On the mass balance of the Polar Ocean, with special emphasis on the Fram Strait, *Norsk Polarinst. Skr., 188,* 53 pp.

Rudels, B., E. P. Jones, L. G. Anderson, and G. Kattner (1994), On the intermediate depth waters of the Arctic Ocean, in *The Polar Oceans and Their Role in Shaping the Global Environment: The Nansen Centennial Volume, Geophys. Monogr. Ser.,* vol. 85, edited by O. M. Johannesen, R. D. Muench, and J. E. Overland, pp. 33–46, AGU, Washington, D. C.

Schumacher, J. D., and P. J. Stabeno (1994), Ubiquitous eddies of the Eastern Bering Sea and their coincidence with concentrations of larval pollock, *Fish. Oceanogr., 3,* 182–190.

Seigel, A., J. B. Weiss, J. Toomre, J. C. McWilliams, P. S. Berloff, and I. Yavneh (2001), Eddies and vortices in ocean basin dynamics, *Geophys. Res. Lett., 28*(16), 3183–3186.

Semtner, A. J., Jr., and R. M. Chervin (1992), Ocean general circulation from a global eddy-resolving model, *J. Geophys. Res., 97*(C4), 5493–5550.

Skyllingstad, E. D., D. W. Denbo, and J. Downing (1991), Convection in the Labrador Sea: Community Modeling Effort (CME) results, in *Deep Convection and Deep Water Formation in the Oceans*, edited by P. C. Chu and J. C. Gascard, Elsevier, New York, NY.

Stabeno, P., and R. Reed (1994), Circulation in the Bering Sea Basin Observed by Satellite-Tracked Drifters: 1986–1993. *J. Phys. Oceanogr., 24,* 848–854.

Stevens, D. P., and P. D. Kilworth (1992), The distribution of kinetic energy on the Southern Ocean: A comparison between observations and an eddy resolving general circulation model, *Philos. Trans. Biol. Sci., 338*(1285), Antarctica and Environmental Change, pp. 251–257.

Visbeck, M., J. Fischer, and F. Schott (1995), Preconditioning the Greenland Sea for deep convection: Ice formation and ice drift, *J. Geophys. Res., 100*(C9), 18,489–18,502.

White, M. A., and K. J. Heywood (1995), Seasonal and interannual changes in the North Atlantic subpolar gyre from Geosat and TOPEX/POSEIDON altimetry, *J. Geophys. Res., 100*(C12), 24,931–24,941.

W. Maslowski, Department of Oceanography, Naval Postgraduate School, Monterey, California, USA.

J. Clement Kinney, Department of Oceanography, Naval Postgraduate School, Monterey, California, USA.

D. C. Marble, Office of Naval Research, Code 322, Arlington, Virginia, USA.

J. Jakacki, Institute of Oceanography, Polish Academy of Sciences, Sopot, Poland.

Pacific Upper Ocean Response to Global Warming—Climate Modeling in an Eddying Ocean Regime

Takashi T. Sakamoto

Frontier Research Center for Global Change, Japan Agency for Marine-Earth Science and Technology, Yokohama, Kanagawa, Japan

Hiroyasu Hasumi

Center for Climate System Research, the University of Tokyo, Kashiwa, Chiba, Japan

Responses of upper ocean currents to global warming are described using the results of global warming experiments conducted by coupled climate models with eddy-permitting and coarse-resolution ocean models. Difference between the two models is discussed by focusing on the Pacific, especially on the Kuroshio and subtropical cells. Changes of these currents also affect the spatial patterns of sea surface temperature (SST) rise. The Kuroshio is accelerated under global warming in the climate model with an eddy-permitting ocean model due to intensification of the Westerlies, and it leads to large SST rise to the south and east of Japan. On the other hand, global warming weakens the Trades, and it results in slowdown of the subtropical cells and tropical cells. Relatively large SST rise is found in the tropical Pacific, which seems to be associated with slowdown of subtropical and tropical cells. The climate model with an eddy-permitting ocean model exhibits a larger SST rise compared with the climate model with a coarse-resolution ocean model. The difference is attributable to the different response of the tropical cells to change of zonal winds around the equator.

1. INTRODUCTION

Since coupled atmosphere–ocean general circulation models (CGCMs) came into practical use for projecting future climate changes, a lot of efforts have been made to improve models and to reduce uncertainties in projection. Increase of resolution is one of the main foci of such efforts. Most of the CGCMs contributing to the second assessment report of the Intergovernmental Panel on Climate Change (IPCC) [*IPCC*,

1995] adopt horizontal resolution of several hundreds of kilometers both for the atmosphere and the ocean, whereas typical horizontal resolution of 100–200 km has been attained among those contributing to the IPCC fourth assessment report [AR4; *IPCC*, 2007]. Further increase of resolution is still required for both uncertainty reduction and practical application. As for the latter, people's interest in future climate changes has shifted from global views to regional aspects and extreme events such as frequency of hurricanes or torrential rainfall, which necessitate an order-of-magnitude higher atmospheric horizontal resolution.

For ocean models, horizontal resolution of ~100 km is unsatisfactory even when looking at large-scale features. Oceanic large-scale circulation is made up of a number of narrow swift currents. For example, western boundary currents, such

Ocean Modeling in an Eddying Regime
Geophysical Monograph Series 177
Copyright 2008 by the American Geophysical Union.
10.1029/177GM17

as the Kuroshio and the Gulf Stream, occupy only a small portion of the ocean (~100 km width for each), while they account for an essential part of oceanic poleward heat transport. Proper modeling of such western boundary currents requires, at least, the so-called eddy-permitting horizontal resolution, which is typically 20–30 km at mid-latitudes. Their future behavior and/or their influence on future climate changes can never be appropriately discussed without CGCMs comprised of eddy-permitting ocean models. Such attempts have just started [*Roberts et al.*, 2004; *Suzuki et al.*, 2005a]. Upper ocean currents are primarily wind-driven, and recent studies indicate that eddy-permitting ocean resolution, not just an increase of atmospheric resolution, leads to significant improvement of surface winds in CGCMs [*Roberts et al.*, 2004; *Suzuki et al.*, 2005b]. In this respect, too, high-resolution ocean models are required for reliable future projection of the ocean.

Warming climate accompanies changes of surface winds. Most CGCMs project acceleration of the Westerlies under global warming [*Yamaguchi and Noda*, 2006; *Fyfe and Saenko*, 2006]. Such changes could significantly affect wind-driven upper ocean currents. Indeed, *Sakamoto et al.* [2005] show that the Kuroshio and the Kuroshio Extension are significantly accelerated in response to surface wind changes based on their future climate projection using a CGCM with an eddy-permitting ocean model. Changes of such large-heat-carrying currents could influence sea surface temperature (SST), and its effect could then feed back to the atmosphere. It is of interest how ocean model resolution affects climate simulation and projection by CGCMs incorporating such feedbacks.

Using the results of twentieth century simulations conducted by 14 models contributing to IPCC AR4 [*IPCC*, 2007], *Zhang and McPhaden* [2006] examine CGCM-simulated changes of the subtropical cells (STCs) which are shallow meridional overturning circulation extending from the equator to the subtropics [e.g., *McCreary and Lu*, 1994; *Liu et al.*, 1994; *Schott et al.*, 2004] during the last few decades of the twentieth century. Based on observations, STC convergence in the Pacific has a downward trend during the last three decades of the twentieth century, and the variation in tropical SST is negatively correlated with the STC convergence [*McPhaden and Zhang*, 2002]. Only 1 in 14 models examined reasonably reproduces such trends, and it employs an eddy-permitting ocean model. They suggest that it is due to a success in reproducing the changes of winds not only in the tropics but also in the off-equatorial tropics, as the Trades are thought to be the primary factor inducing STC variability [*Nonaka et al.*, 2002; *Klinger et al.*, 2002]. The success in reproducing the change of winds by a climate model with an eddy-permitting ocean model is another indication of the importance of ocean model resolution in properly simulating surface wind changes and oceanic responses under changing climate.

In this paper, we describe how the ocean responds to surface wind changes under global warming and how ocean model resolution matters in such responses based on global warming experiments by CGCMs with eddy-permitting and coarse-resolution ocean models. The focus is on the Pacific, especially on the Kuroshio and STCs. The paper is organized as follows. The model and experiments are documented in section 2. The results are described in section 3 where overviews of the projected climate are first presented, and oceanic response to surface wind changes and influence of ocean model resolution therein are shown. Section 4 gives a summary and discussion.

2. MODEL AND EXPERIMENTAL DESIGN

The CGCM used in the present study is the Model for Interdisciplinary Research on Climate (MIROC) version 3.2, which was developed by the Center for Climate System Research (CCSR), the University of Tokyo; the National Institute for Environmental Studies (NIES); and the Frontier Research Center for Global Change (FRCGC), Japan Agency for Marine-Earth Science and Technology [*K-1 Model Developers*, 2004]. Its atmospheric component is a global spectral general circulation model (GCM), CCSR/NIES/FRCGC Atmospheric General Circulation Model (AGCM) 5.7b [*Numaguti et al.*, 1997]. The land-surface and river-routing components are minimal advanced treatments of surface interaction and runoff, [*Takata et al.*, 2003] and the total runoff integrating pathways [*Oki and Sud*, 1998], respectively. The oceanic component is a *z*-coordinate ocean GCM, CCSR Ocean Component Model [*Hasumi*, 2006], and the sea ice component is based on zero-layer thermodynamics [*Semtner*, 1976], two-category thickness representation [*Hibler*, 1979], and elastic–viscous–plastic rheology [*Hunke and Dukowicz*, 1997].

MIROC3.2 has contributed to IPCC AR4 [*IPCC*, 2007] by using two different resolution setups. In a high-resolution setup (MIROC3.2_hires), the atmospheric resolution is T106 with 56 vertical levels, and the horizontal resolution of the land and river components is 0.56° and 0.5°, respectively. The oceanic and sea ice components are "eddy-permitting," with zonal and meridional resolution of 0.28125° and 0.1875°, respectively, and 48 vertical levels. In a medium-resolution setup (MIROC3.2_medres), the same horizontal resolution, T42, is employed for the atmosphere, land, and river, with 20 vertical levels for the atmosphere. For the ocean and sea ice, the zonal resolution is 1.4°, and the meridional resolution is 1.4° at latitudes higher than 65°, 0.56° at latitudes lower than 8°, with smooth changes in between. The ocean model has 44 vertical levels.

In the present study, two setups of MIROC3.2 are used. One is the same as the high-resolution setup described above and is hereafter denoted by "hAhO" (high-resolution atmospheric component coupled to high-resolution oceanic component). The other employs the high-resolution setup for the atmospheric, land, and river components and the medium-resolution setup for the oceanic and sea ice components. Hereafter, this setup is denoted by "hAmO" (high-resolution atmospheric component coupled to medium-resolution oceanic component).

In the oceanic and sea ice components of these two setups, the same parameterizations are applied except that resolution-dependent parameters are adjusted for each. Note that the equilibrium climate sensitivity of the atmospheric component coupled to a slab ocean model, responding to CO_2 doubling, is 4.3 K.

Two experiments are conducted with each of the above setups. One is a simulation for a control climate state (hereafter control run), which is carried out by fixing the external forcing at the year 1900 (pre-industrial condition) in terms of solar and volcanic forcing, greenhouse gas concentration, various aerosol emissions, and land use. The other is a global warming experiment (hereafter CO_2 run) where the atmospheric CO_2 concentration is increased at the rate of 1% per year from the pre-industrial condition (295.9 ppm at the year 1900). Both the control and CO_2 runs are initiated by a spin-up calculation of more than 100 years under pre-industrial condition for each setup and are integrated without any flux correction for 100 and 90 years, respectively.

3. RESULTS

3.1. Global Features of the Control Experiments

Figure 1 shows the time series of globally averaged 2-m height air temperature (T2) of each experiment. In the control experiments of both hAmO and hAhO, the model climate does not exhibit a long-term drift. The mean temperature of the model climate over the entire period is slightly higher in hAhO (14.0°C) than in hAmO (13.6°C).

Biases in modeled T2, as compared with the average of the ECMWF re-analysis ERA40 from 1971 to 2000 [*Simmons and Gibson*, 2000], are generally improved by the increase of ocean model resolution (Plate 1). The standard deviations of the mean T2 in the models from the re-analysis are about 1.5 K in hAhO and about 2.2 K in hAmO. Areas of improved biases are found over the low-latitude oceans, the North Atlantic mid-latitudes, and the Southern Ocean. On the other hand, little improvement is found over the Arctic Ocean and the North Pacific high latitudes. The large errors in the Barents Sea and the Southern Ocean are associated with the position of the sea ice edge. In hAmO, the sea ice cover around Antarctica is small, one of the reasons for which is that deep convection is taking place in the open ocean off the Weddell Sea. The Barents Sea, a significant part of which is ice-free even in winter because of intrusion of the Norwegian Atlantic Current, is mostly covered by sea ice in hAmO. On the other hand, sea ice cover in the Barents Sea is too small in hAhO. The difference of the sea ice cover induces the large negative T2 bias over the Barents Sea in hAmO and the large positive bias in hAhO.

There is a significant positive T2 bias over the entire Arctic Ocean in hAhO, while this is not the case in hAmO. This difference in T2 bias is due to the fact that the Arctic sea ice is thin overall in hAhO (figure not shown). The central Arctic Ocean is covered by sea ice all the year round, so there is no such bias in SST. Except for the ice-covered areas, the biases in SST are very similar to those in T2.

3.2. Global Features of the Global Warming Experiments

Change in T2 during the 90 years of the CO_2 run is 3.7 K in hAhO and 3.3 K in hAmO (Figure 1). Ocean model resolution

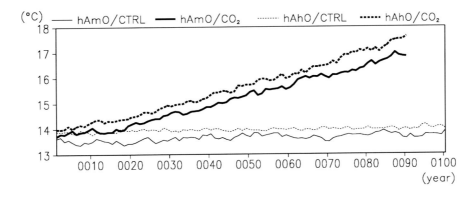

Figure 1. Time series of globally averaged annual-mean T2 in the control runs (thin lines) and CO_2 runs (bold lines) of hAmO (solid lines) and hAhO (dashed lines).

seems to matter in this different response. *Yokohata et al.* [2007] compare the difference of climate sensitivity between MIROC3.2_hires (identical to hAhO) and MIROC3.2_medres (where the ocean resolution is the same as in hAmO and the atmospheric resolution is lower). Sensitivity is higher in the MIROC3.2_hires, and the difference is attributed to ice-albedo feedback and ocean heat uptake. Based on their analysis, the difference in the effect of ice-albedo feedback is caused by the difference of sea ice extent between the control experiments of the two models. The sea ice extent in hAmO resembles that of MIROC3.2_medres; the explanation of *Yokohata et al.* [2007] carries over to the present study.

Both hAmO and hAhO exhibit a large T2 rise in the Northern Hemisphere and a small rise in the Southern Ocean (figures not shown). Such features are also recognized in the SST change (Plate 2). The large SST rise in the Northern Hemisphere high latitudes is associated with decrease of sea ice cover. A large SST rise is also found over the Kuroshio Extension (KE) region and the tropical Pacific, especially in hAhO.

Surface winds also change under global warming (Plate 3). In both hAmO and hAhO, the Westerlies are intensified, which is also recognized as strengthening of the Aleutian Low over the North Pacific, and the Trades are weakened over the western and central tropical Pacific. These changes of winds are likely to modify the North Pacific upper ocean circulation, such as the Kuroshio, KE, and STCs, and seem to be related to the aforementioned regions of large SST rise. Details of these features are described in the following subsections.

3.3. Responses of the Kuroshio to Global Warming

Responses of the Kuroshio to global warming are described in this section mostly along the line of discussion made by *Sakamoto et al.* [2005]. Only the results of hAhO are presented. The Kuroshio in hAmO is represented as an unrealistically wide (~300 km) and slow (~0.4 m s⁻¹) current overshooting to the north compared with the actual point of separation from the western boundary, so discussion of its responses makes little sense.

In the mean state of the control run (Figure 2a), it is well captured that the Kuroshio has two paths to the south of Japan: one is a straight path along the southern coast of Japan and the other is a meandering path flowing off the southern coast between 135° and 140°E. This reflects the well-known bimodality of the path of the Kuroshio [*Shoji*, 1972; *Kawabe*, 1995]. In the warming climate, the upper-ocean velocity of the Kuroshio and KE to the west of 155°E obviously increases in comparison with the control climate, while the

positions of the currents change little (Figure 2). The acceleration of the current speed is large in the KE region, especially in the area of 35°–37°N and 145°–155°E where the increase of the current speed reaches up to 0.3 m s⁻¹. This acceleration is reflected to the rise of dynamic surface height to the south of the Kuroshio and KE, which also appears as a region of enhanced rise in actual sea level [*Suzuki et al.*, 2005a]. Compared with standard deviation of decadal variability in the control run, which is calculated by 9-year running mean of annual averages, the change of dynamic height in the aforementioned region under global warming is three times as large and, thus, is significant (Figure 2c). The change is less significant (against natural decadal variability) in some part of the region of the large KE acceleration. In this region, variance of dynamic height due to decadal variability is large (figure not shown), and it is caused by meridional fluctuation of the Kuroshio separation point. If the velocity change in the KE region were caused by the change of the KE position, it could be seen in Figures 2a and b.

Spinning up of the subtropical gyre due to surface wind changes in the North Pacific is a possible cause of the acceleration. However, the whole of the subtropical gyre is not spun up, only its sub-region to the west of the date line is (Plate 4). Therefore, a local change of wind stress is likely to be the reason why the Kuroshio and KE are accelerated. Change of wind stress curl is negative in the northwestern North Pacific, while it is positive in the southwestern and northeastern regions (Plate 3b). As mentioned before, the former is associated with intensification of the Aleutian Low, and the latter indicates weakening of the northeasterly Trades. Caused by the negative change of wind stress curl in the northwestern region, together with a narrow zonal band of negative change over 30°–38°N extending from 140°E to 150°W, the subtropical gyre is partially spun up, as indicated by the change of Sverdrup transport in Plate 4. The magnitude of this change of Sverdrup transport is three times larger than the standard deviation of decadal variability, and thus is significant relative to decadal variability, along most of the Kuroshio and KE near Japan where we are now focusing. This intensified Sverdrup transport accompanies strengthening of its compensating currents, i.e., the Kuroshio and KE.

Although the linear Sverdrup theory supports the acceleration of the Kuroshio and KE, the change of circulation, as depicted by the dynamic sea surface height response (Plate 4), differs from what is expected from the linear theory. The vertically integrated volume transport of the KE averaged over 145°–155°E, where the acceleration of the current is large, increases by about 29 Sverdrup (Sv; Sv = 10⁶ m³ s⁻¹), while the increase of Sverdrup transport calculated from the wind stress change is 5–6 Sv. This discrepancy is accounted for by spin-up of recirculation, which is induced

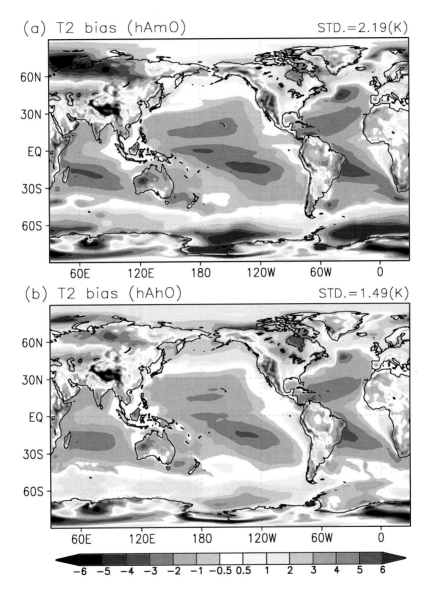

Plate 1. Spatial distribution of time-averaged T2 bias (in the unit of K) from ERA40 in the control runs of (a) hAmO and (b) hAhO. The time average is taken over the entire period of the experiments.

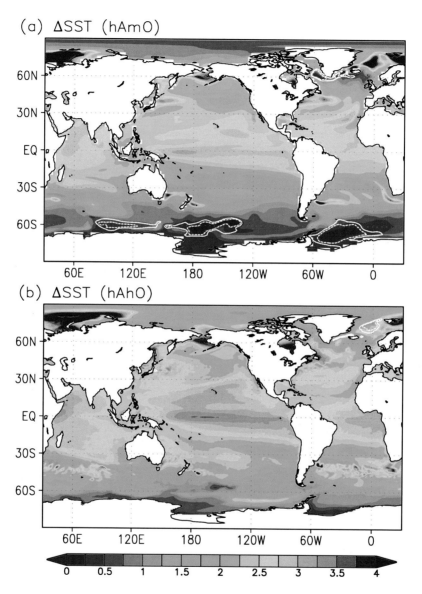

Plate 2. SST difference (in the unit of K) between the CO_2 run and the control run (former minus latter) for (a) hAmO and (b) hAhO. Average is taken over the years 71–90 for the CO_2 runs and over the entire 100-year period for the control runs. White dashed and solid contours indicate places where the ΔSST is two and three times, respectively, larger than standard deviation of SST decadal variability in the 100-year control runs. SST decadal variability is extracted using a 9-year moving average.

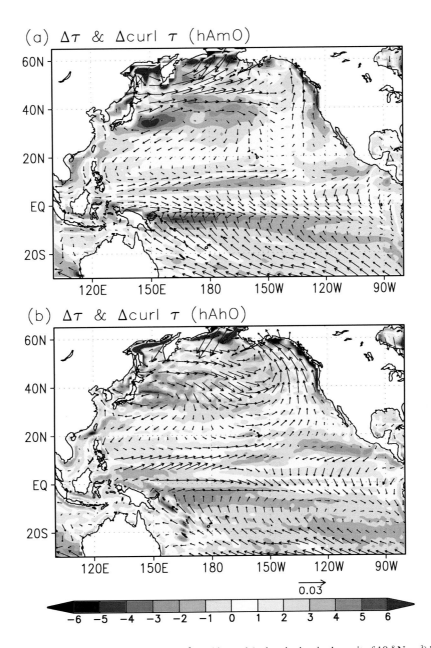

Plate 3. Difference of wind stress (vectors, in the unit of N m^{-2}) and its curl (color shades, in the unit of 10^{-8} N m^{-3}) between the CO$_2$ run and the control run (former minus latter) for (a) hAmO and (b) hAhO. The average period is the same as in Plate 2.

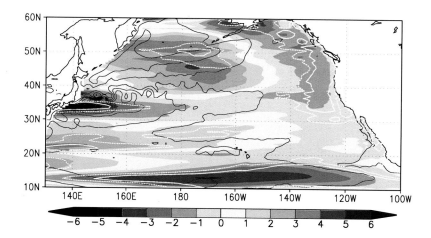

Plate 4. Difference of mean dynamic sea surface height relative to 2,048-m depth (black contours, in the unit of m) and Sverdrup transport stream function (color shades, in the unit of Sv) between the CO_2 run and the control run (former minus latter) in the North Pacific. Contour interval is 0.05 m. White dashed and solid contours mean the same as in Plate 2 but for Sverdrup transport stream function.

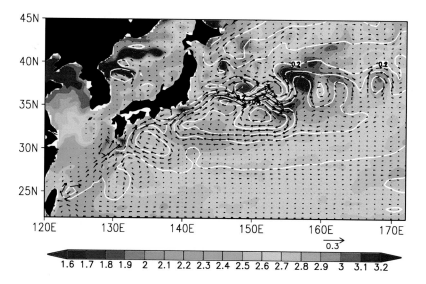

Plate 5. Sea surface temperature change around Japan (shades, enlargement of Plate 2b), superimposed on dynamic sea surface height change (white contours) and 100-m depth velocity change (vectors) shown in Figure 2c. Note that the color scale is different from Plate 2.

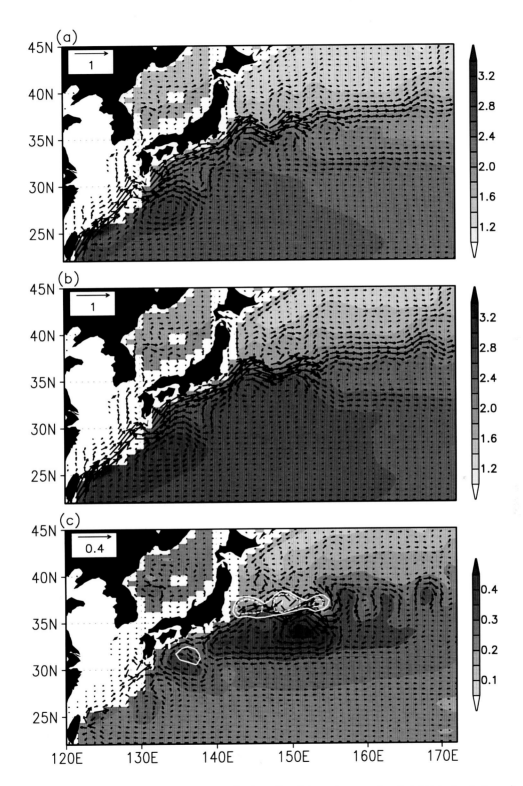

Figure 2. Mean velocity at 100-m depth (vectors, in the unit of m s^{-1}) and dynamic sea surface height (shades, in the unit of m) relative to 2,048-m depth for (a) the control run, (b) the CO_2 run, and (c) their difference (latter minus former). The average period in each run is the same as in Plate 2. White dashed and solid contours mean the same as in Plate 2 but for dynamic sea surface height.

by a nonlinear effect associated with the acceleration of the Kuroshio and KE itself [*Taguchi et al.*, 2005]. Note that intensification of the eastward volume transport excluding the contribution from the recirculation is about 5.8 Sv when averaged over 145°–155°E, which is consistent with the change of Sverdrup transport.

A large SST rise is found along the Kuroshio and KE (Plate 5). Increase of volume transport by the Kuroshio and KE accompanies the increase of heat transport, which could account for this large SST rise. Another region of large SST rise is found to the north of the KE. The region of elevated SST is also likely due to the acceleration downstream within the KE, as it mainly occurs along the northern branch of the KE bifurcation (Figure 2c).

3.4. Response of STC to Global Warming

Some of the previous studies investigating the change of the STC at the end of the twentieth century [*McPhaden and Zhang*, 2002, 2004; *Zhang and McPhaden*, 2006] define STC transport convergence as the equatorward volume transport convergence in the upper pycnocline ($\sigma_\theta = 22$–26 and 22.5–26.2 for the Northern and Southern Hemispheres,

respectively) zonally integrated at 9°S and 9°N. Here, it is rather simply defined by the difference between the maximum and minimum of the stream function of the zonally integrated meridional overturning circulation in the Pacific between 9°S and 9°N and above 300 m depth (see Figure 3). The STC transport convergence thus defined gradually decreases during the course of global warming in both hAhO and hAmO (Figure 4a).

Slowdown of the STCs is linked to a rise in tropical SST, as the STC transports cooler water from the extratropical subsurface to the tropical surface [e.g., *McPhaden and Zhang*, 2002, 2004; *Merryfield and Boer*, 2005; *Zhang and McPhaden*, 2006]. Both hAhO and hAmO exhibit slowdown of the STCs and a region of enhanced SST rise in the tropical Pacific (Plate 2). However, although the STC slowdown is similar between hAhO and hAmO (Figure 4a), the tropical SST rise is larger in hAhO than in hAmO. Its maximum in the equatorial Pacific is about 3.5 K in hAhO, while it is about 3 K in hAmO.

In the STCs, there are recirculating cells approximately between 5°S and 5°N (Figure 3). Here, these recirculating cells are referred to as tropical cells (TCs), as in previous studies [e.g., *Liu et al.*, 1994; *McCreary and Lu*, 1994; *Schott*

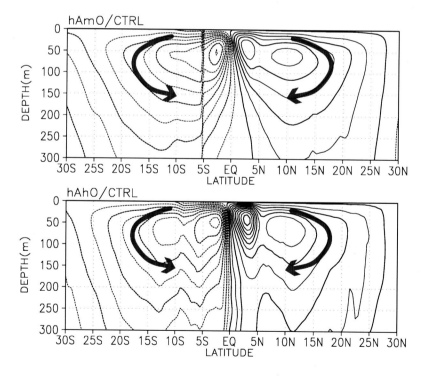

Figure 3. Stream function of the 100-year averaged meridional overturning circulation in the upper 300 m of the Pacific for the control runs of (top) hAmO and (bottom) hAhO. Contour interval is 5 Sv, and dashed contours indicate negative values. Gray arrows show the direction of circulation. Discontinuity at 5°S in (top) is caused by the Indonesian Through Flow. Such discontinuity is not found in (bottom) because representation of straits is significantly different between hAmO and hAhO.

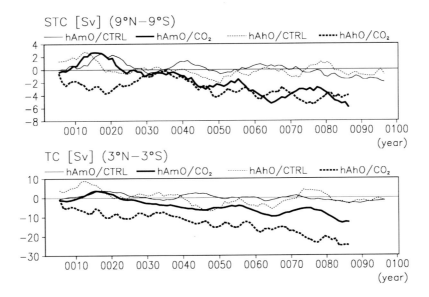

Figure 4. Time series of transport convergence (in the unit of Sv = 10^6 m^3 s^{-1}) of (a) STC and (b) TC for each experiment. Nine-year moving averaged anomalies from the 100-year mean of the corresponding control run are shown.

et al., 2004], and TC transport convergence is defined by the difference between the maximum and minimum of the meridional overturning stream function in the Pacific between 3°S and 3°N and above 300 m depth.

The TCs also weaken during the course of global warming, and its slowdown is larger in hAhO than in hAmO (Figure 4b). The larger slowdown in hAhO is consistent with the larger tropical SST rise. The TCs are closely associated with zonal winds around the equator [*Philander et al.*, 1987], and the TCs and tropical SST are highly anti-correlated to each other [*McPhaden and Zhang*, 2002; *Lohmann and Latif*, 2005; *Merryfield and Boer*, 2005]. The Trades are weakened under global warming, as described before (Plate 3), and the change of zonal winds at the equator is almost the same between hAhO and hAmO (Figure 5). Therefore, the response

of TCs is significantly different between hAhO and hAmO, contrasting with a similar change of equatorial winds in the two cases. This difference in TCs' response may cause the dissimilarity in the tropical SST rise. The different response of the TCs is likely to be caused by the difference in ocean model resolution.

The Trades and TCs are highly related to each other as mentioned above, and they are also closely associated with the North Equatorial Countercurrent (NECC). The TCs consist of Ekman upwelling over the equatorial Pacific, which is induced by the Trades, and Ekman convergence and downwelling over the NECC region [e.g., *Wyrtki and Kilonsky*, 1984; *Johnson et al.*, 2001; *Schott et al.*, 2004]. This downwelling influences the thermocline depth and strength of the NECC [*Philander et al.*, 1987; *Lohmann and Latif*, 2005].

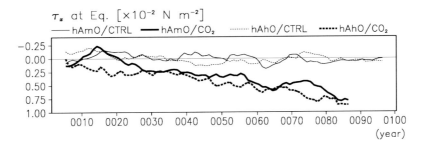

Figure 5. Time series of zonally averaged (140°E–110°W) zonal wind stress at the equator for each experiment. Nine-year moving averaged anomalies from the 100-year mean of the corresponding control run are shown. Note that the y-axis is reversed.

Therefore, the modeled response of the TCs to global warming depends critically on how realistically the NECC is reproduced and responds to the global warming. In comparison with observed features [cf. *Johnson et al.*, 2001], the NECC is far better reproduced in hAhO than in hAmO (Figure 6): the NECC is very weak (maximum velocity is about 0.3 m s⁻¹) and broad in hAmO. During the course of global warming, the NECC slows down, and the deceleration is larger in hAhO than in hAmO (Figure 7), which is consistent with the response of TCs (Figure 4b). The slowdown of the NECC and TCs under global warming is consistent with the results of *Vecchi and Soden* [2007], who conducted a multimodel ensemble analysis using the model projection datasets compiled for the IPCC AR4 [*IPCC*, 2007].

The strength and position of the NECC vary with the seasonality of the Trades, which is associated with the meridional migration of the Intertropical Convergence Zone [*Philander et al.*, 1987]. In the control runs of hAmO and hAhO, climatological seasonal variability of the Trades over the equatorial Pacific is very similar, but the amplitude of the NECC velocity variability is larger in hAhO than in hAmO (figure not shown). The difference in the seasonal response of the NECC to the Trades implies that ocean model resolution affects the sensitivity of the NECC to temporal variations of

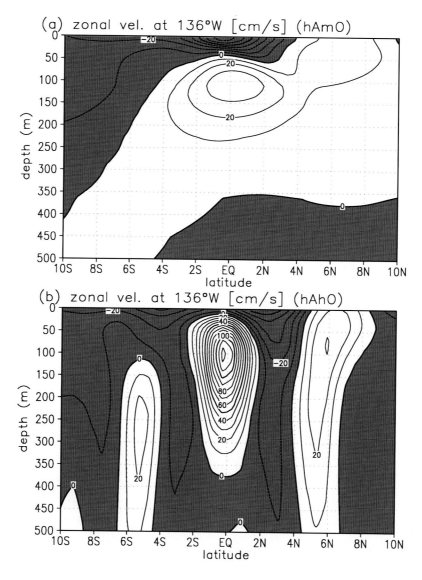

Figure 6. Mean zonal velocity at 136°W (in the unit of cm s⁻¹) in the control runs of (a) hAmO and (b) hAhO. Negative (westward) velocity is shaded.

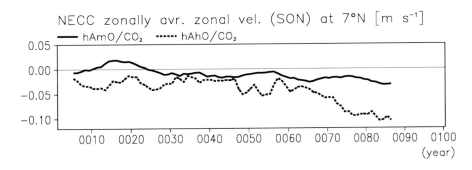

Figure 7. Zonally averaged (140°E–110°W) zonal velocity at 7°N (in the unit of m s^{-1}) in the CO_2 runs of hAmO (54 m depth) and hAhO (50 m depth). Average in each model year is taken from September to November when the NECC is relatively strong, and 9-year moving averaged anomalies from the 100-year mean of the corresponding control run are shown.

the Trades. This could account for the differences in response of the NECC and TCs in hAmO and hAhO to the weakened Trades under global warming.

These differences in the response of TCs in hAhO and hAmO affect meridional heat transport accounted for by overturning circulation. In hAhO, poleward overturning heat transport by meridional overturning circulation between 5°S and 5°N in the CO_2 run is less than that in the control run, although it is only a little less in hAmO (Figure 8b). On the other hand, the change of total meridional heat transport in hAhO is small (Figure 8a). This is because equatorward heat transport induced by eddies related to the tropical instability waves (TIWs) [e.g., *Wang and McPhaden*, 1999; *Kessler*, 2006; *Willett et al.*, 2006], which occur in the eastern Tropical Pacific, is less vigorous in the CO_2 run of hAhO

than in its control run (figure not shown). This reduction of TIW activity in the CO_2 run is related to the "El Nino-like" trends in the CO_2 run [*Contreras*, 2002; *Wu and Bowman*, 2007]. Weakened Trades bring about slowdown of the South Equatorial Current (SEC), the Equatorial Undercurrent (EUC), and NECC. The slowdown causes reduction of shear between the SEC and the NECC and between the SEC and the EUC, which are considered to cause TIWs through barotropic and baroclinic instabilities [*Philander*, 1978; *Contreras*, 2002; *Willet et al.*, 2006]. Needless to say, eddies induced by TIWs are not resolved in hAmO, so eddy heat transport is very small both in the control and CO_2 runs of hAmO (Figure 8c).

Some previous studies point out that TCs do not significantly influence meridional heat transport because they are

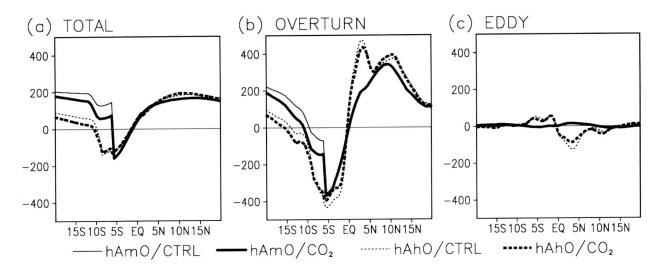

Figure 8. (a) Total meridional heat transport and (b) its overturning and (c) eddy components (expressed in K Sv) for the Pacific basin. Discontinuity at 5°S in (a) and (b) is due to the same reason as in Figure 3.

confined within the thermocline [*Hazeleger et al.*, 2001; *Schott et al.*, 2004]. On the other hand, *Lohmann and Latif* [2005] show that the TCs are spun up in their global warming simulation using a CGCM, although the STCs are slowed down. Further investigation is needed.

4. SUMMARY AND DISCUSSION

Based on global warming experiments conducted with CGCMs with eddy-permitting and coarse-resolution ocean models, responses of upper ocean currents to surface wind changes under global warming are described, and their dependence on ocean model resolution is discussed. The focus is on the Pacific, especially on the Kuroshio and STCs. Changes in these currents seem to be associated with the spatial pattern of SST rise. Global warming intensifies the Westerlies, which accelerates the Kuroshio in the CGCM with an eddy-permitting ocean model. Acceleration of the Kuroshio leads to large SST rises to the south and east of Japan. Global warming also brings about weakening of the Trades, which induces slowdown of the STCs and TCs. Relatively large SST rises in the tropical Pacific seem to be associated with slowdown of the STCs and TCs. The CGCM with an eddy-permitting ocean model exhibits a larger SST rise in the tropical Pacific compared with a CGCM with a coarse-resolution ocean model, and the SST rise is attributable to the different responses of the TCs to changes of zonal winds around the equator.

It is still to be investigated how such changes of upper ocean currents and their influence on SST feed back to the atmosphere. *Wu et al.* [2007] suggest that some part of the decadal variability of the tropical Pacific climate is controlled by the decadal variability in the extra-tropics through a coupled wind-evaporative-SST feedback. According to this mechanism, the Aleutian Low, SST in the tropics, winds over extra-tropics, and STCs are closely related. Both the intensified Aleutian Low and the enhanced SST rise in the tropics are persistent features of the global warming experiments described herein, so such a feedback mechanism is worth examining. In this regard, the change of the Kuroshio might be linked to that of the STCs and TCs.

Schneider and Cornuelle [2005] show that ocean circulation anomalies in the Kuroshio-Oyashio Extension affect the North Pacific climate at decadal timescales. This indicates that the accelerated Kuroshio and KE may affect the atmosphere over the subtropical North Pacific. When the mechanisms suggested by *Schneider and Cornuelle* [2005] and *Wu et al.* [2007] are combined, there is a possibility that responses of the atmosphere and the ocean over the tropical and extratropical North Pacific are very intricately connected.

Oceanic major currents and heat transport as well as their changes under global warming can only be appropriately modeled by adopting eddy-resolving resolution. Progress of computational technology has made possible the modeling of long-term climate changes with such high-resolution ocean models. Such a direction is necessary and promising both from the standpoints of scientific understanding and practical application of climate projection results.

Acknowledgments. The authors gratefully acknowledge valuable discussions with and comments from Dr. Masami Nonaka and Dr. Tatsuo Suzuki. This work is supported by the first subject of the Kyousei Project (Project for Sustainable Coexistence of Human, Nature, and the Earth), which is established by the Ministry of Education, Culture, Sports, Science, and Technology of Japan. All the experiments presented in this study were carried out on the Earth Simulator.

REFERENCES

Contreras, R. F. (2002), Long-term observations of tropical instability waves, *J. Phys. Oceanogr.*, 32, 2715–2722.

Fyfe, J. C., and O. A. Saenko (2006), Simulated changes in the extratropical Southern Hemisphere winds and currents, *Geophys. Res. Lett.*, 33, L06701, doi:10.1029/2005GL025332.

Hasumi, H. (2006), *CCSR Ocean Component Model (COCO) version 4.0*, CCSR Rep. 25, 103 pp., University of Tokyo, Tokyo.

Hazeleger, W., P. de Vries, and G. J. van Oldenborgh (2001), Do tropical cells ventilate the Indo-Pacific equatorial thermocline?, *Geophys. Res. Lett.*, 28, 1763–1766.

Hibler, W. D., III (1979), A dynamic thermodynamic sea ice model, *J. Phys. Oceanogr.*, 9, 815–846.

Hunke, E. C., and J. K. Dukowicz (1997), An elastic–viscous–plastic model for sea ice dynamics, *J. Phys. Oceanogr.*, 27, 1849–1867.

IPCC (1995), *Climate Change 1995: The Science of Climate Change*, edited by J. T. Houghton, L. G. Meira Filho, B. A. Callender, N. Harris, A. Kattenberg, and K. Maskell, 572 pp., Cambridge University Press, Cambridge.

IPCC (2007), *Climate Change 2007—The Physical Science Basis*, contribution of Working Group I to the Fourth Assessment Report of the IPCC, edited by S. Solomon, D. Qin, and M. Manning, 996 pp., Cambridge University Press, Cambridge.

Johnson, G. C., M. J. McPhaden, and E. Firing (2001), Equatorial Pacific Ocean horizontal velocity, divergence and upwelling, *J. Phys. Oceanogr.*, 31, 839–849.

K-1 Model Developers (2004), *K-1 Coupled Model (MIROC) Description*, K-1 Tech. Rep., 1, edited by H. Hasumi and S. Emori, 34 pp., Center for Climate System Research, University of Tokyo, Tokyo.

Kawabe, M. (1995), Variations of current path, velocity, and volume transport of the Kuroshio in relation with the large meander, *J. Phys. Oceanogr.*, 25, 3103–3117.

Kessler, W. S. (2006), The circulation of the eastern tropical Pacific: A review, *Prog. Oceanogr.*, 69, 181–217.

Klinger, B. A., J. P. McCreary Jr., and R. Kleeman (2002), The relationship between oscillating subtropical wind stress and equatorial temperature, *J. Phys. Oceanogr.*, *32*, 1507–1521.

Liu, Z., S. G. H. Philander, and R. C. Pacanowski (1994), A GCM study of tropical–subtropical upper-ocean water exchange, *J. Phys. Oceanogr.*, *24*, 2606–2623.

Lohmann, K., and M. Latif (2005), Tropical Pacific decadal variability and the subtropical–tropical cells, *J. Clim.*, *18*, 5163–5178.

McCreary, J. P., Jr., and P. Lu (1994), Interaction between the subtropical and equatorial ocean circulations: The subtropical cell, *J. Phys. Oceanogr.*, *24*, 466–497.

McPhaden, M. J., and D. Zhang (2002), Slowdown of the meridional overturning circulation in the upper Pacific Ocean, *Nature*, *415*, 603–608.

McPhaden, M. J., and D. Zhang (2004), Pacific Ocean circulation rebounds, *Geophys. Res. Lett.*, *31*, L18301, doi:10.1029/2004GL020727.

Merryfield, W. J., and G. J. Boer (2005), Variability of upper Pacific Ocean overturning in a coupled climate model, *J. Clim.*, *18*, 666–683.

Nonaka, M., S.-P. Xie, and J. P. McCreary (2002), Decadal variations in the subtropical cells and equatorial Pacific SST, *Geophys. Res. Lett.*, *29*(7), doi:10.1029/2001GL013717.

Numaguti, A., M. Takahashi, T. Nakajima, and A. Sumi (1997), Description of CCSR/NIES atmospheric general circulation model, in *CGER's Supercomputer Monograph Report 3*, pp. 1–48, Center for Global Environmental Research, National Institute for Environmental Studies, Ibaraki, Japan.

Oki, T., and Y. C. Sud (1998), Design of total runoff integrating pathways (TRIP)—A global river channel network, *Earth Interact.*, *2*, 1–37.

Philander, S. G. H. (1978), Instabilities of zonal equatorial currents, 2, *J. Geophys. Res.*, *83*, 3679–3682.

Philander, S. G. H., W. J. Hurlin, and A. D. Seigel (1987), Simulation of the seasonal cycle of the tropical Pacific Ocean, *J. Phys. Oceanogr.*, *17*, 1986–2002.

Roberts, M. J., H. Banks, N. Gendney, J. Gregory, R. Hill, S. Mullerworth, A. Pardaens, G. Rickard, R. Thorpe, and R. Wood (2004), Impact of an eddy-permitting ocean resolution on control and climate change simulations with a global coupled GCM, *J. Clim.*, *17*, 3–20.

Sakamoto, T. T., H. Hasumi, M. Ishii, S. Emori, T. Suzuki, T, Nishimura, and A. Sumi (2005), Responses of the Kuroshio and the Kuroshio Extension to global warming in a high-resolution climate model, *Geophys. Res. Lett.*, *32*, L14617, doi:10.1029/2005GL023384.

Schneider, N., and B. D. Cornuelle (2005), The forcing of the Pacific decadal oscillation, *J. Clim.*, *18*, 4355–4373.

Schott, F. A., J. P. McCreary Jr., and G. C. Johnson (2004), Shallow overturning circulations of the tropical-subtropical oceans in *Earth's Climate: The Ocean-Atmosphere Interaction, Geophys. Monogr. Ser*, vol. 147, edited by C. Wang, S.-P. Xie, and J. A. Carton, pp. 261–304, AGU, Washington, D.C.

Semtner, A. J., Jr. (1976), A model for the thermodynamics growth of sea ice in numerical investigations of climate, *J. Phys. Oceanogr.*, *6*, 379–389.

Shoji, D. (1972), *Time Variation of the Kuroshio South of Japan, Kuroshio—Its Physical Aspects*, edited by H. Stommel and K. Yoshida, pp. 217–234, University of Tokyo Press, Tokyo.

Simmons, A. J., and J. K. Gibson, (2000), The ERA-40 Project Plan, *ERA-40 Project Report Series*, *1*, 62 pp. European Centre for Medium-Range Weather Forecasts, Reading, UK.

Suzuki, T., H. Hasumi, T. T. Sakamoto, T. Nishimura, A. Abe-Ouchi, T. Segawa, N. Okada, A. Oka, and S. Emori (2005a), Projection of future sea level and its variability in a high-resolution climate model: Ocean processes and Greenland and Antarctic ice-melt contributions, *Geophys. Res. Lett.*, *32*, L19706, doi:10.1029/2005GL023677.

Suzuki, T., T. T. Sakamoto, T. Nishimura, N. Okada, S. Emori, A. Oka, and H. Hasumi (2005b), Seasonal cycle of the Mindanao Dome in the CCSR/NIES/FRCGC atmosphere–ocean coupled model, *Geophys. Res. Lett.*, *32*, L17604 doi:10.1029/2005GL023666.

Taguchi, B., S.-P. Xie, H. Mitsudera, and A. Kubokawa (2005), Response of the Kuroshio Extension to Rossby waves associated with the 1970s climate regime shift in a high-resolution ocean model, *J. Clim.*, *18*, 2979–2995.

Takata, K., S. Emori, and T. Watanabe (2003), Development of the minimal advanced treatments of surface interaction and runoff, *Glob. Planet. Change*, *38*, 209–222.

Vecchi, G. A., and B. J. Soden (2007), Global warming and the weakening of the tropical circulation, *J. Clim.*, *20*, 4316–4340.

Wang, W., and M. J. McPhaden (1999), The surface-layer heat balance in the equatorial Pacific Ocean. Part I: Mean seasonal cycle, *J. Phys. Oceanogr.*, *29*, 1812–1831.

Willet, C. S., R. R. Leben, and M. F. Lavín (2006), Eddies and tropical instability waves in the eastern tropical Pacific: A review, *Prog. Oceanogr.*, *69*, 218–238.

Wu, L., Z. Liu, C. Li, and Y. Sun (2007), Extratropical control of recent tropical Pacific decadal climate variability: A relay teleconnection, *Clim. Dyn.*, *28*, 99–112, doi:10.1007/s00382-006-0198-5.

Wu, Q., and K. P. Bowman (2007), Interannual variations of tropical instability waves observed by the Tropical Rainfall Measuring Mission, *Geophys. Res. Lett.*, *34*, L09701, doi:10.1029/2007GL029719.

Wyrtki, K., and B. Kilonsky (1984), Mean water and current structure during the Hawaii-to-Tahiti Shuttle Experiment, *J. Phys. Oceanogr.*, *14*, 242–254.

Yamaguchi, K., and A. Noda (2006), Global warming patterns over the North Pacific: ENSO versus AO, *J. Meteorol. Soc. Jpn*, *84*, 221–241.

Yokohata, T., et al. (2007), Different transient climate responses of two versions of an atmosphere-ocean coupled general circulation model, *Geophys. Res. Lett.*, *34*, L02707, doi:10.1029/2006GL027966.

Zhang, D., and M. J. McPhaden (2006), Decadal variability of the shallow Pacific meridional overturning circulation: Relation to tropical sea surface temperature in observations and climate change models, *Ocean Modell.*, *15*, 250–273.

Formulating the Equations of Ocean Models

Stephen M. Griffies

*National Oceanic and Atmospheric Administration (NOAA) Geophysical Fluid Dynamics Laboratory,
Princeton, New Jersey, USA*

Alistair J. Adcroft

*Princeton University Atmospheric and Oceanic Sciences Program,
Princeton, New Jersey, USA*

We formulate mathematical equations describing the thermo-hydrodynamics of
the ocean and introduce certain numerical methods employed by models used for
ocean simulations.

1. INTRODUCTION

The purpose of this chapter is to formulate the equations of ocean models and to outline solution methods. Global ocean climate models, including those representing mesoscale eddies, are traditionally based on the hydrostatic primitive equations. We nonetheless discuss extensions to the more fundamental non-hydrostatic equations, which are used in certain fine resolution process studies, such as for convection and mixing, and increasingly for coastal and regional modeling. The target audience for this chapter includes students and researchers interested in fundamental physical and numerical aspects of ocean models. We thus aim to present a reasonably concise yet thorough accounting of the rationalization required to pose the problem of ocean modeling. We take a first principles perspective to allow readers with little background in ocean fluid mechanics to follow the full development. This goal necessitates starting from the basics as we develop the model equations and methods. For this purpose, much material was culled from various research papers and textbooks, such as *Gill* [1982], *Pedlosky* [1987], *Lion et al.* [1992], *Marshall et al.* [1997], *Haidvogel and Beck-*

mann [1999], *Griffies et al.* [2000a], *Griffies* [2004, 2005], *Vallis* [2006], *Higdon* [2006], and *Müller* [2006].

Our presentation focuses on developing the fluid mechanics of the ocean and weaves into this discussion elements appropriate for the formulation of ocean models. We begin with a discussion of ocean fluid kinematics in Section 2 where we introduce mass conservation as well as the notions of dia-surface transport. Section 3 then focuses on tracer budgets, which follow quite naturally from mass budgets, only with the introduction of possible nontrivial fluxes of tracer which occur in the absence of mass fluxes. Section 4 introduces a dynamical description that arises from the use of Newton's Second Law of Motion applied to continuous fluid parcels. Section 5 presents the equation of state for density and discusses the material evolution of density. Section 6 derives some energetic properties of the equations of motion, with energetics providing a guiding principle for developing certain numerical solution methods. Section 7 introduces notions of non-equilibrium thermodynamics, a subject which forms the basis for establishing budgets of heat within the ocean. Section 8 linearizes the dynamical equations to deduce various linear modes of motion fundamental to ocean dynamics. These motions also have direct relevance to the development of methods used to solve the ocean equations. They furthermore motivate certain approximations or filters used to simplify the supported dynamical modes, with certain approximations described in Section 9. Section 10 presents an overview of vertical coordinates. The

Ocean Modeling in an Eddying Regime
Geophysical Monograph Series 177
Copyright 2008 by the American Geophysical Union.
10.1029/177GM18

choice of vertical coordinate is fundamental to the numerical algorithms of an ocean model. Section 11 presents a general discussion of solution methods used for numerical models of the ocean. Section 12 closes this chapter with a brief summary and discussion of certain features of ocean modeling that present a barrier between what is desired theoretically and what is realizable in practice.

2. KINEMATICS

Kinematics is the study of intrinsic properties of motion without concern for dynamical laws. Fluid kinematics is concerned with establishing constraints on fluid motion due to interactions with geometrical boundaries of the domain, such as the land–sea, ice–sea, and air–sea boundaries of an ocean basin. A fundamental element of kinematics is the set of coordinates used to describe motion. For fluid motion, we are led to notions of generalized vertical coordinates, which are a critical element in theoretical and numerical models of the ocean. Although not strictly a kinematic issue, fluid kinematics also concerns itself with establishing the balances of mass for infinitesimal fluid parcels, as well as for finite regions.

It is convenient and conventional to formulate the mechanics of a continuous fluid by focusing on infinitesimal mass conserving parcels [e.g., *Batchelor*, 1967]. Choosing to do so allows many notions from classical particle mechanics to transfer over to continuum mechanics of fluids, especially when describing fluid motion from a Lagrangian perspective. Mass conservation is also a fundamental property of the ocean, with the mass of the ocean changing only through boundary input.

2.1. Parcel Kinematics

Consider an infinitesimal parcel of seawater contained in a volume $dV = dx\,dy\,dz$ with a mass $dM = \rho\,dV$, where ρ is the in situ mass density of the parcel and $\mathbf{x} = (x, y, z)$ is the Cartesian coordinate of the parcel with respect to an arbitrary origin. Conservation of mass for this parcel implies that dM is materially constant, i.e., $d/dt\,(dM) = 0$. For convenience, we write mass conservation as

$$\frac{d}{dt}\ln(dM) = 0. \tag{1}$$

Mass conservation is realized as the parcel volume and density change in complementary manners where the volume of a fluid parcel changes according to the divergence of the velocity field

$$\frac{d}{dt}\ln(dV) = \nabla \cdot \mathbf{v}, \tag{2}$$

and the density changes according to the convergence

$$\frac{d}{dt}(\ln\rho) = -\nabla \cdot \mathbf{v}. \tag{3}$$

Hence, parcel volume increases when moving through a diverging velocity field, while the density decreases.

The mass budget, given equivalently by equations (1)–(3), describes fluid motion from the perspective of a moving fluid parcel. This perspective provides a Lagrangian description of fluid motion. The complementary Eulerian perspective measures fluid properties from a fixed-space frame. Time tendencies in the two reference frames are related by the coordinate transformation

$$\frac{d}{dt} = \partial_t + \mathbf{v} \cdot \nabla, \tag{4}$$

where ∂_t measures Eulerian time tendencies from a fixed-space point. The advection term $\mathbf{v} \cdot \nabla$ reveals the fundamentally nonlinear character of fluid dynamics, with the parcel's velocity $\mathbf{v} = d\mathbf{x}/dt$ measuring the time changes of its position.[1] Use of relation (4) converts the Lagrangian statement of mass conservation given by equation (1) to the Eulerian form

$$\rho_{,t} + \nabla \cdot (\rho\,\mathbf{v}) = 0. \tag{5}$$

This equation is termed the mass continuity equation. Note that we introduced a comma as shorthand for the partial time derivative taken at a fixed point in space

$$\rho_{,t} = \partial\rho/\partial t. \tag{6}$$

We use an analogous notation for other partial derivatives.[2] A useful relation used throughout this chapter follows by combining the material time derivative in equation (4) with mass continuity in equation (5) to render

$$\rho \frac{d\Psi}{dt} = (\rho\Psi)_{,t} + \nabla \cdot (\rho\,\mathbf{v}\Psi), \tag{7}$$

where Ψ is any scalar field.

It is common in fluid mechanics to move between the Lagrangian and Eulerian descriptions, as they offer useful complementary insights. Certain ocean models likewise exploit the advantages of these two descriptions. For example, the vertical coordinate in isopycnal models moves with the motions of an adiabatic fluid parcel. It is therefore a Lagrangian vertical coordinate. In contrast, geopotential vertical coordinate models retain a fixed vertical position as determined by the static depth of a grid cell, and so this is an Eulerian vertical coordinate. Horizontal coordinates in most ocean models remain fixed in space, and so are Eulerian.

2.2. Dia-Surface Transport

In providing a mechanistic description of ocean budgets, it is often useful to measure the material or momentum transfer through a surface. This transport is termed the dia-surface transport. We are particularly interested in the transport through three surfaces, with the following general discussion relevant for each.

The first surface is the ocean free surface. Here, water and tracer penetrate this surface through precipitation, evaporation, river runoff (when applied as an upper ocean boundary condition), and sea ice melt. Momentum exchange arises from stresses between the ocean and atmosphere or ice. The ocean free surface can be represented mathematically by the identity $z - \eta(x, y, t) = 0$. For mathematical expediency, we assume that the surface height η is smooth and contains no overturns at the scales of interest. That is, we assume that breaking surface waves are filtered or averaged.

Second, we may describe the solid Earth lower boundary mathematically by using the time-independent expression $z + H(x, y,) = 0$. It is typically assumed that there is no fluid mass transport through the solid Earth. However, in the case of geothermal heating, we may consider an exchange of heat between the ocean and the solid Earth. Momentum exchange through the action of stresses occur between the solid Earth and ocean fluid.

Third, within the ocean interior, transport across surfaces of constant generalized vertical coordinate $s = s(x, y, z, t)$ constitutes the dia-surface transport affecting budgets of mass, tracer, and momentum within layers bounded by two generalized vertical coordinate surfaces. A canonical example is provided by isopycnal layers formed by surfaces of constant potential density, as used in isopycnal ocean models as well as theoretical descriptions of adiabatic ocean dynamics. A surface of constant generalized vertical coordinate can be successfully used to partition the vertical so long as the transformation between the generalized vertical coordinate and the geopotential is invertible. The Jacobian of transformation is given by $z_{,s}$, which must then be single signed for useful vertical coordinates. This constraint means that we do not allow the surfaces to overturn, which is the same assumption made about the ocean surface $z = \eta(x, y, t)$. This restriction places a limitation on the ability of isopycnal models to describe non-hydrostatic processes, such as overturning, common in Kelvin–Helmholz billows or vertical convection. We refer to the Jacobian $z_{,s}$ as the specific thickness, with this name motivated by noting that the vertical thickness of a layer of coordinate thickness ds is given by $dz = z_{,s} ds$.

To develop the mathematical description of dia-surface fluid transport, we note that at an arbitrary point on a surface

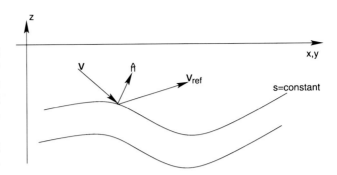

Figure 1. Surfaces of constant generalized vertical coordinate living interior to the ocean. An upward normal direction $\hat{\mathbf{n}}$ is indicated on one of the surfaces. Also shown is the orientation of the velocity of a fluid parcel $\mathbf{v}$ and the velocity $\mathbf{v}^{(\mathrm{ref})}$ of a reference point living on the surface.

of constant generalized vertical coordinate (see Figure 1), the rate at which fluid moves in the direction normal to the surface is given by

$$\text{rate of flow in direction } \hat{\mathbf{n}} = \mathbf{v} \cdot \hat{\mathbf{n}}. \tag{8}$$

In this equation, $\hat{\mathbf{n}} = \nabla s \, |\nabla s|^{-1}$ is the surface unit normal direction. If we are working with the free surface, then the unit normal takes the form $\hat{\mathbf{n}} = \nabla(z - \eta)/|\nabla(z - \eta)|$, whereas at the solid Earth bottom, $\hat{\mathbf{n}} = -\nabla(z + H)/|\nabla(z + H)|$. Introducing the material time derivative $ds/dt = s_{,t} + \mathbf{v} \cdot \nabla s$ to equation (8) leads to the equivalent expression

$$\mathbf{v} \cdot \hat{\mathbf{n}} = |\nabla s|^{-1}(d/dt - \partial_t)s. \tag{9}$$

That is, the normal component to the velocity of a fluid parcel is proportional to the difference between the material time derivative of the surface and its partial time derivative.

As the surface is generally moving (except the solid Earth lower boundary), the net flux of seawater penetrating the surface is obtained by subtracting the velocity of the surface $\mathbf{v}^{(\mathrm{ref})}$ in the $\hat{\mathbf{n}}$ direction from the velocity component $\mathbf{v} \cdot \hat{\mathbf{n}}$ of the fluid parcels

$$\begin{array}{c} \text{rate of relative normal flow} \\ \text{across surface} = \hat{\mathbf{n}} \cdot (\mathbf{v} - \mathbf{v}^{(\mathbf{ref})}). \end{array} \tag{10}$$

The velocity $\mathbf{v}^{(\mathrm{ref})} = \mathbf{u}^{(\mathrm{ref})} + w^{(\mathrm{ref})} \hat{\mathbf{z}}$ is that of a reference point fixed on the surface. Correspondingly, the material time derivative of the surface, taken with respect to the reference velocity, vanishes: $d^{(\mathrm{ref})}s/dt = 0$. This result allows us to write the reference vertical velocity component $w^{(\mathrm{ref})} = dz^{(\mathrm{ref})}/dt$ as $w^{(\mathrm{ref})} = -z_{,s}(\partial_t + \mathbf{u}^{(\mathrm{ref})} \cdot \nabla_z)s$, thus rendering

$$\hat{\mathbf{n}} \cdot \mathbf{v}^{(\mathrm{ref})} = -s_{,t} |\nabla s|^{-1}. \tag{11}$$

Hence, the normal component to the velocity of the surface vanishes when the surface is static, as may be expected. When interpreting the dia-surface velocity component below, we find it useful to note that relation (11) leads to

$$z_{,s} \nabla s \cdot \mathbf{v}^{(\mathrm{ref})} = z_{,t}. \tag{12}$$

To reach this result, we used the identity $s_{,t} z_{,s} = -z_{,t}$, with $z_{,t}$ the time tendency for the depth of a particular constant s surface.

Using expression (11) in equation (10) for the net flux of seawater crossing the surface leads to

$$\hat{\mathbf{n}} \cdot (\mathbf{v} - \mathbf{v}^{(\mathrm{ref})}) = |\nabla s|^{-1} ds/dt. \tag{13}$$

The material time derivative of the generalized surface thus vanishes if and only if no water parcels cross it. This is a very important result that is used throughout ocean theory and modeling. It provides an expression for the volume of seawater crossing a generalized surface, per time, per area. The area normalizing the volume flux is that area $dA_{(\hat{\mathbf{n}})}$ of an infinitesimal patch on the surface of constant generalized vertical coordinate with outward unit normal $\hat{\mathbf{n}}$. This area can generally be written $dA_{(\hat{\mathbf{n}})} = |z_{,s} \nabla s| dA$, where $dA = dxdy$ is the area of the surface projected onto the horizontal plane formed by surfaces of constant depth. Hence, the volume per time of fluid passing through the generalized surface $\hat{\mathbf{n}} \cdot (\mathbf{v} - \mathbf{v}^{(\mathrm{ref})}) dA_{(\hat{\mathbf{n}})}$ is equivalent to $|z_{,s}|(ds/dt)dA$. This result motivates us to introduce the dia-surface velocity component

$$w^{(s)} = z_{,s} \frac{ds}{dt}, \tag{14}$$

which measures the volume of fluid passing through the surface, per unit horizontal area, per unit time. That is,

$$w^{(s)} \equiv \frac{\hat{\mathbf{n}} \cdot (\mathbf{v} - \mathbf{v}^{(\mathrm{ref})}) dA_{(\hat{\mathbf{n}})}}{dA} \tag{15}$$

$$= \frac{(\mathrm{volume}/\mathrm{time})\mathrm{fluid\ through\ surface}}{\mathrm{horizontal\ area\ of\ surface}}. \tag{16}$$

The dia-surface velocity component can be written in the following equivalent forms

$$w^{(s)} = z_{,s} ds/dt$$

$$= z_{,s} \nabla s \cdot (\mathbf{v} - \mathbf{v}^{(\mathrm{ref})})$$

$$= (\hat{\mathbf{z}} - \nabla_s z) \cdot \mathbf{v} - z_{,t}$$

$$= w - (\partial_t + \mathbf{u} \cdot \nabla_s)z \tag{17-20}$$

where $\nabla_s z = -z_{,s} \nabla_z s$ is the slope of the s surface projected onto the horizontal directions, and the penultimate step follows from the identity (12). When the surface is static, then the dia-surface velocity component reduces to $w^{(s)} = w - \mathbf{u} \cdot \nabla_s z$. If the surface is flat, then the dia-surface velocity component measures the flux of fluid moving vertically relative to the motion of the generalized surface. Finally, if the surface is flat and static, the dia-surface velocity component becomes the vertical velocity component $w = dz/dt$ used in geopotential coordinate models.

The expression (14) for $w^{(s)}$ brings the material time derivative (4) into the following equivalent forms

$$\frac{d}{dt} = \left(\frac{\partial}{\partial t}\right)_z + \mathbf{u} \cdot \nabla_z + w \left(\frac{\partial}{\partial t}\right) \tag{21}$$

$$= \left(\frac{\partial}{\partial t}\right)_s + \mathbf{u} \cdot \nabla_s + \frac{ds}{dt} \left(\frac{\partial}{\partial s}\right) \tag{22}$$

$$= \left(\frac{\partial}{\partial t}\right)_s + \mathbf{u} \cdot \nabla_s + w^{(s)} \left(\frac{\partial}{\partial z}\right), \tag{23}$$

where $\partial_s = z_{,s} \partial_z$ provides a relationship between the vertical coordinate partial derivatives. Note that the subscripts in these expressions denote variables held fixed for the partial derivatives. We highlight the special case of no fluid parcels crossing the generalized surface. This occurs in the case of adiabatic flows with $s = \rho$ being an isopycnal coordinate. For adiabatic flow, the material time derivative in equation (23) only has a horizontal two-dimensional advective component $\mathbf{u} \cdot \nabla_\rho$. This result should not be interpreted to mean that the velocity of a fluid parcel in an adiabatic flow is strictly horizontal. Indeed, it generally is not, as the form given by equation (21) makes clear. Rather, it means that the advective transport of fluid properties occurs along surfaces of constant ρ, and such transport is measured by the convergence of horizontal advective fluxes as measured along surfaces of constant ρ.

2.3. Kinematic Boundary Conditions

The discussion so far of dia-surface transport focused on a surface with a constant generalized vertical coordinate within the ocean interior. These results can also be applied to the ocean free surface (Figure 2) and solid Earth lower boundary to derive kinematic boundary conditions. For the lower boundary, again assuming no material transport through the boundary, we have the trivial result

$$w^{(s)} = 0 \quad \text{at} \quad s = s_{\mathrm{bot}}, \tag{24}$$

$$\rho \left(\frac{d(z - \eta)}{dt} \right) = -\rho_w q_w \qquad \text{at } z = \eta. \qquad (28)$$

The identity $dz/dt = z_{,s} \, ds/dt$ leads to the kinematic boundary condition in generalized vertical coordinates

$$\rho \, z_{,s} \left(\frac{d(s - s_{\text{top}})}{dt} \right) = -\rho_w q_w \qquad \text{at } s = s_{\text{top}}, \qquad (29)$$

where $s_{\text{top}} = s(x, y, z = \eta, t)$ is the value of the generalized vertical coordinate at the ocean surface.

These material statements of the kinematic boundary condition can also be derived by considering the mass budget over either an infinitesimal region near the upper ocean surface, or the budget over a full column of water extending from a static ocean bottom at $z = -H(x, y)$ to a dynamic ocean surface at $z = \eta(x, y, t)$. We present the column mass budget approach because it has application for later considerations. The total mass per horizontal area of fluid inside the column is given by the integral $\int_{-H}^{\eta} \rho \, dz$. Conservation of mass for this column implies that mass changes in time through imbalances in fluxes crossing the ocean free surface and convergence of advective mass transport through the vertical sides of the column.[4] These considerations lead to the balance

$$\partial_t \left(\int_{-H}^{\eta} dz \, \rho \right) + \nabla \cdot \mathbf{U}^\rho = q_w \rho_w, \qquad (30)$$

where

$$\mathbf{U}^\rho = \int_{-H}^{\eta} dz \, \rho \, \mathbf{u} \qquad (31)$$

is a shorthand notation for the vertically integrated horizontal momentum per volume. Now, to derive the surface kinematic boundary condition, perform the derivative operations on the integrals in the mass budget expressed in equation (30), use the no-flux lower boundary condition, and use the Eulerian mass conservation relation (5) to render

$$\rho \left(\partial_t + \mathbf{u} \cdot \nabla \right) \eta = \rho_w q_w + \rho \, w \qquad \text{at } z = \eta. \qquad (32)$$

This is an Eulerian version of the material kinematic boundary condition of equation (28).

3. TRACER BUDGET

The tracer concentration C is defined to be the mass of tracer per mass of seawater for material tracers such as salt or biogeochemical tracers. Hence, the total tracer mass within a finite region of seawater is given by the integral $\int C \rho \, dV$. The

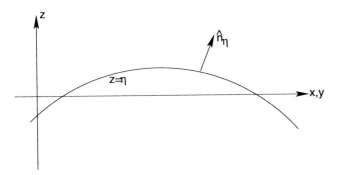

Figure 2. Schematic of the ocean's upper surface with a smoothed undulating surface at $z = \eta(x, y, t)$ and outward normal direction $\hat{n}_\eta$. Undulations of the surface height are on the order of a few meters due to tidal fluctuations in the open ocean and order 10–20m in certain embayments (e.g., Bay of Fundy in Nova Scotia). When imposing the weight of sea ice onto the ocean surface, the surface height can be depressed even further, on the order of 5–10m, with larger values possible in some cases. It is important for simulations to employ numerical schemes facilitating such wide surface height undulations.

which is equivalent to the no-normal-flow boundary condition $\mathbf{v} \cdot \hat{\mathbf{n}} = 0$.

At the ocean surface, mass transport arises from the passage of water across the penetrable free surface. We define this transport as

$$\text{(mass/time) through surface} = Q_w^\eta \, dA_\eta, \qquad (25)$$

with Q_w^η the mass flux through the ocean surface, normalized by the area element dA_η on the surface. We next exploit the assumption that the surface interface of ocean models has no overturns, in which case we can introduce the horizontal area dA to rewrite the mass flux as

$$\text{(mass/time) through surface} \equiv Q_w \, dA. \qquad (26)$$

Q_w is the mass flux used in ocean models, with some models defining

$$Q_w = \rho_w q_w \qquad (27)$$

where ρ_w is the density of the water crossing the ocean surface, and q_w is the fresh water flux (with units of velocity).[3]

To develop the surface kinematic boundary condition (Figure 2), we note that the free surface, defined at $z - \eta(x, y, t) = 0$, materially evolves according to the flux of mass crossing it so that material evolution of tracer mass within a Lagrangian parcel of mass conserving fluid is given by

$$\rho \frac{dC}{dt} = -\nabla \cdot \mathbf{J} + \rho \mathcal{S}, \tag{33}$$

where S is a tracer source that cannot be written as the convergence of a flux. There are many biogeochemical tracers that have a non-trivial S. The tracer flux $\mathbf{J}$ arises from subgrid-scale (SGS) transport of tracer occurring in the absence of mass transport. Such transport consists of SGS diffusion and advection. Use of the identity (7) allows us to bring the Lagrangian parcel tracer budget (33) into the following Eulerian flux form

$$(\rho C)_{,t} + \nabla \cdot (\rho \mathbf{v} C + \rho \mathbf{F}) = \rho \mathcal{S}, \tag{34}$$

where $\mathbf{J} = \rho \mathbf{F}$ introduces the tracer concentration flux $\mathbf{F}$, with dimensions velocity × tracer concentration.

As the tracer flux $\mathbf{J}$ and tracer source S are not associated with mass transport or mass sources, they both vanish when the tracer concentration is uniform, in which case the tracer budget of equation (34) reduces to the mass budget of equation (5). This compatibility relation between mass and tracer budgets follows trivially from the definition of tracer concentration. It forms an important guiding principle that a numerical algorithm must maintain in order for the simulation to conserve tracer. Not all ocean models satisfy this constraint, in which case they suffer from local or global tracer non-conservation [Griffies et al., 2001; Campin et al., 2004; White et al., 2007].

In a manner analogous to the definition of a dia-surface velocity component in Section 2.2, it is useful to identify the amount of tracer transported through a surface from the effects of SGS processes as follows:

$$\frac{(\text{SGS tracer mass through surface})}{\text{time}} = dA_{(\hat{n})} \, \hat{\mathbf{n}} \cdot \mathbf{J}. \tag{35}$$

For this purpose, we are led to introduce the dia-surface SGS tracer flux

$$J^{(s)} \equiv \frac{dA_{(\hat{n})} \, \hat{\mathbf{n}} \cdot \mathbf{J}}{dA} \tag{36}$$

$$= z_{,s} \nabla s \cdot \mathbf{J} \tag{37}$$

$$= (\hat{\mathbf{z}} - \nabla_s z) \cdot \mathbf{J}, \tag{38}$$

where $\nabla_s z$ is the slope vector for the generalized surface introduced following equation (20). In words, $J^{(s)}$ is the tracer mass per time per horizontal area penetrating surfaces of constant generalized vertical coordinate by processes that

are unresolved by the dia-surface velocity component $w^{(s)}$. At the ocean boundaries, $J^{(s)}$ embodies the transport of tracer into the ocean from other components of the climate system.

4. LINEAR MOMENTUM BUDGET

The linear momentum of a fluid parcel is given by $\mathbf{v}\rho dV$. Through Newton's Second Law of Motion, momentum changes in time due to the influence of forces acting on the parcel. There are two external (or body) forces and two internal (or contact) forces acting on a fluid parcel that concern ocean modelers. Body forces act throughout the fluid media, with gravitational and Coriolis forces of concern.[5] Contact forces act on the volume of a continuous media by acting on the boundaries of the media. Pressure and friction are the two contact forces of concern here. Through the Green–Gauss theorem of vector calculus, the contact forces are transformed into body forces, which provides a means to formulate the equations of motion for an infinitesimal fluid parcel.

4.1. Gravitational Force and Spherical Geometry

The effective gravitational force is noncentral due to the Earth's rotation and due to inhomogeneities in the Earth's mass distribution. Hence, if the Earth were an ideal fluid, matter would flow from the poles toward the equator, thus ensuring that the Earth's surface would everywhere be perpendicular to the effective gravitational acceleration, $\mathbf{g}$. Indeed, the Earth does exhibit a slight equatorial bulge. However, inhomogeneities in the Earth's composition and surface loading by continents, glaciers, and seawater make its shape differ from the ideal case. For purposes of global ocean modeling, we ignore the inhomogeneities, but we do not ignore the equatorial bulge.

Veronis [1973], *Phillips* [1973], and *Gill* [1982] discuss how the Earth's geometry can be well approximated by an oblate spheroid, with the equatorial radius larger than the polar due to centrifugal effects. With this geometry, surfaces of constant geopotential are represented by surfaces with a constant oblate spheroid radial coordinate [page 662 of *Morse and Feshbach*, 1953]. However, the oblate spheroidal metric functions, which determine how to measure distances between points on the spheroid, are less convenient to use than the more familiar spherical metric functions. To provide a simpler form of the equations of motion on the Earth, *Veronis* [1973] and *Gill* [1982] (see in particular page 91 of Gill) indicate that it is possible, within a high level of accuracy, to maintain the best of both situations. That is, surfaces of constant r are interpreted as best fit oblate spheroidal geopotentials, yet the metric functions used to measure distance between points in the surface are approximated as spherical. As the metric functions determine the geometry of the surface, and hence the form of the equations of motion, the equations are exactly those which result when using spherical coordinates on a sphere. Hence, throughout this chapter, the geometry of the Earth is spherical, yet the radial position r represents a surface of constant geopotential, which is approximated by an oblate spheroid.

In summary, the gravitational field most convenient for ocean modeling is an effective gravitational field, which incorporates the effects from the centrifugal force. The effective gravitational field is conservative so that the gravitational acceleration of a fluid parcel can be represented as the gradient of a scalar,

$$g = -\nabla \Phi, \tag{39}$$

with Φ the geopotential. As $(\rho dV)\Phi$ is the gravitational potential energy of a parcel, Φ is also the gravitational potential energy per mass. In most ocean modeling applications, the local vertical direction is denoted by z, with $z = 0$ the surface of a resting ocean, in which case

$$\Phi \approx gz, \tag{40}$$

with $g \approx 9.8 \mathrm{m\ s^{-2}}$ the acceleration due to gravity, which is generally assumed constant for ocean climate modeling. This assumption is not fundamental and can be readily jettisoned, as indeed is important for accurate ocean tide calculations.

4.2. Coriolis Force

Ocean models generally are written in the reference frame of an observer at a fixed lateral position on the rotating Earth. This moving reference frame then leads to a Coriolis force per mass, which is written [Marion and Thornton, 1988; Gill, 1982]

$$\mathbf{F}_c = -2\mathbf{\Omega} \wedge \mathbf{v}. \tag{41}$$

The Earth's rotational vector $\mathbf{\Omega}$ points outward through the north pole, with the Earth's rotation counterclockwise if looking down onto the north pole.

The Earth's angular velocity is comprised of two main contributions: the spin of the Earth about its axis and the orbit of the Earth about the Sun. Other astronomical motions can be neglected for ocean modeling. Therefore, in the course of a single period of 24 h, or $24 \times 3600 = 86400$ s, the Earth experiences an angular rotation of $(2\pi + 2\pi/365.24)$ radians, in which case the angular velocity of the Earth is approximated by

$$\begin{aligned}
\Omega &= \left(\frac{2\pi + 2\pi/365.24}{86400\mathrm{s}} \right) \\
&= \left(\frac{\pi}{43082} \right) \mathrm{s}^{-1} \\
&= 7.2921 \times 10^{-5} \mathrm{s}^{-1}.
\end{aligned} \tag{42}$$

For the purposes of ocean modeling, this angular velocity can be assumed constant in time.

4.3. Stresses From Pressure and Friction

When parcels exchange momentum with other parcels and/or boundaries, this exchange can be represented by the components of a symmetric stress tensor whose elements have units of a force per area. There are two types of stress of concern for ocean fluid dynamics: diagonal stresses associated with pressure p and stresses associated with friction organized into the components of a symmetric and trace-free frictional stress tensor τ. The frictional stress tensor is also known as the deviatoric stress tensor [e.g., Aris, 1962; Batchelor, 1967] because it represents deviations from the static case when stress is due solely to pressure.

The contact force from friction and pressure acting on the boundaries of a fluid region can generally be written

$$\mathbf{F}_{\mathrm{stress}} = \int (\tau \cdot \hat{\mathbf{n}} - p\hat{\mathbf{n}}) \, dA_{(\hat{\mathbf{n}})}. \tag{43}$$

The surface integral is taken over the bounding surface of the domain whose outward normal is $\hat{\mathbf{n}}$. Pressure acts on a surface in the direction opposite to the outward normal, and so always acts in a compressive manner (Figure 3). Deviatoric stresses create more general forces on the surface, which can have compressive, expansive, and/or shearing characteristics. It is notable that the mechanical pressure considered here is the same as the pressure used for equilibrium and non-equlibrium thermodynamical considerations (Section 7).

The Green–Gauss theorem of vector calculus can be used to convert the area integral in equation (43) to a volume integral so that

$$\mathbf{F}_{\mathrm{stress}} = \int \nabla \cdot (\tau - \mathbf{I}p) \, dV, \tag{44}$$

where $\mathbf{I}$ is the identity tensor. This is a fundamental result of practical relevance in the formulation of pressure forces in ocean models. That is, one may choose to formulate the

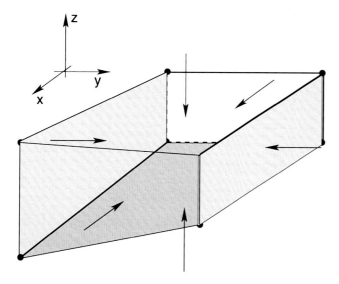

Figure 3. Schematic of a grid cell bounded at its top and bottom in general by sloped surfaces and vertical side walls. The top and bottom surfaces can represent linear piecewise approximations to surfaces of constant generalized vertical coordinates, with $s = s_1$ at the top surface and $s = s_2$ at the bottom surface. They could also represent the ocean surface (for the top face) or the ocean bottom (for the bottom face). The *arrows* represent the pressure contact forces which act in a compressive manner along the boundaries of the grid cell and in a direction normal to the boundaries. These forces arise from contact between the shown fluid volume and adjacent regions. Due to Newton's Third Law, the pressure acting on an arbitrary fluid parcel **A** due to contact with a parcel **B** is equal and opposite to the pressure acting on parcel **B** due to contact with parcel **A**. If coded according to finite volume budgets, as in *Lin* [1997], this law extends to the pressure forces acting between grid cells in an ocean model.

pressure force as the gradient of pressure integrated over the volume of the cell, as in equation (44), or as the accumulation of pressure forces acting on the boundary of the cell, as in equation (43). Both formulations are equivalent in the continuum. However, certain discrete formulations break the symmetry. For example, the finite difference approach of *Bryan* [1969a] uses an energetically consistent formulation of the pressure gradient, yet the energy-consistent method is not equivalent to a contact force formulation. In contrast, *Lin* [1997] proposes to use a finite volume formulation [e.g., chapter 6 of *Hirsch*, 1988] in which the contact force formulation (written as a closed contour integral) is constructed to be equivalent to a finite volume formulation of the pressure gradient.

The frictional stresses in a fluid arise from strains acting in the horizontal and vertical directions which, through the assumptions of a Newtonian fluid, are directly proportional to stress. The proportionality is in the form of a viscous tensor.

The stress tensor is symmetric, reflecting the inability of internal stresses to impart a net angular momentum on a fluid.

Anticipating the kinetic energy discussion in Section 6.1, we note that stresses arising from molecular viscosity dissipate kinetic energy. This result places a constraint on the form of the viscous tensor, and it motivates the name frictional stress tensor, as friction generally dissipates mechanical energy. To illustrate this property mathematically, consider friction arising from molecular viscosity to be represented in a Laplacian form and assume a planar geometry to simplify the tensor analysis. In this case, the inner product of velocity and the friction vector, which appears in the kinetic energy budget [equation (67)], can be written

$$\mathbf{v} \cdot \nabla \cdot \tau = v_m (\gamma v_{m,n})_{,n}$$
$$= \nabla \cdot (\gamma \nabla K) - \gamma \mathbf{v}_{,m} \cdot \mathbf{v}_{,m}. \quad (45)$$

In this equation, a comma represents a partial derivative, and repeated indices are summed over their range from 1,2,3. The strength of the Laplacian friction operator is scaled by the nonnegative number γ, which is the molecular dynamic viscosity for water. Typical values are around $\gamma \approx 10^{-3}$ kg m^{-1} s^{-1} [*Gill*, 1982]. More commonly considered in applications is the kinematic viscosity

$$\nu = \gamma/\rho, \quad (46)$$

whose values for water are around $\nu \approx 10^{-6}$ m^2 s^{-1}. The negative semi-definite term in equation (45) thus represents a kinetic energy sink associated with local viscous dissipation. It is termed Joule heating, as it represents a conversion of mechanical energy to heat (Sections 6.3 and 7.1). It is commonly written as

$$\epsilon = \nu \mathbf{v}_{,m} \cdot \mathbf{v}_{,m} \geq 0. \quad (47)$$

As noted by *McDougall* [2003], frictional dissipation in the ocean interior associated with molecular viscosity is on the order[6]

$$\epsilon \approx 10^{-9} \, \text{W kg}^{-1}. \quad (48)$$

The number 10^{-9} W kg^{-1} sounds small, as indeed it is. To put it into perspective, using the heat capacity of seawater $C_p \approx$ 3989 Joules kg^{-1} K^{-1}, frictional dissipation through molecular viscosity warms seawater at a rate of less than 10^{-3} °K per hundred years. This is a negligible amount of heating from a large-scale ocean circulation perspective, and so it is commonly neglected in large-scale models.

4.4. Momentum Budget for a Parcel

With the above considerations, the equation for linear momentum of a fluid parcel takes the form

$$\rho \frac{d\mathbf{v}}{dt} + 2\mathbf{\Omega} \wedge \rho\,\mathbf{v} = -\rho\,\nabla\Phi + \nabla \cdot (\tau - \mathbf{I}p). \tag{49}$$

The left-hand side of this equation is the time tendency for the linear momentum per volume of a parcel, along with the Coriolis force, and the right-hand side is the sum of the gravitational, pressure, and frictional forces. The momentum equation (49) is a form of Cauchy's equation with the diagonal pressure force split from the stress tensor. Cauchy's equation becomes the Navier–Stokes equation when assuming the frictional stress is linearly proportional to the fluid strain according to a Newtonian fluid [*Aris*, 1962; *Batchelor*, 1967].

There are two general forms that the linear momentum equation (49) appears in ocean models: the advective form and the vector invariant form. The two differ by how the material time derivative is translated into an Eulerian form. The advective form exploits the identity

$$\rho \frac{d\mathbf{v}}{dt} = (\rho\,\mathbf{v})_{,t} + \nabla \cdot (\rho\,\mathbf{v}\mathbf{v}) + M\,(\hat{\mathbf{z}} \wedge \rho\,\mathbf{v}), \tag{50}$$

where $M = v\partial_x (\ln dy) - u\partial_y (\ln dx)$ defines an advective metric frequency. Its form given here assumes that the lateral directions are described by locally orthogonal coordinates, which is the typical case for ocean fluid mechanics. For example, in spherical coordinates[7] (r, λ, ϕ), the grid cell increments are given by $dx = (r\cos\phi)\,d\lambda$, $dy = r\,d\phi$, in which case the advective metric frequency is given by $M = (u/r)\tan\phi$. Use of equation (50) in the linear momentum balance of equation (49) leads to the Eulerian budget

$$(\rho\,\mathbf{v})_{,t} + \nabla \cdot (\rho\,\mathbf{v}\mathbf{v}) + (2\mathbf{\Omega} + \hat{\mathbf{z}}M) \wedge \rho\,\mathbf{v}$$
$$= -\rho\,\nabla\Phi + \nabla \cdot (\tau - \mathbf{I}p). \tag{51}$$

The vector invariant form exploits the identity

$$\rho \frac{d\mathbf{v}}{dt} = \rho\,(\partial_t + \omega\wedge)\mathbf{v} + \rho\,\nabla\mathcal{K}, \tag{52}$$

where

$$\omega = \nabla \wedge \mathbf{v} \tag{53}$$

is the three-dimensional vorticity, and

$$\mathcal{K} = \mathbf{v} \cdot \mathbf{v}/2 \tag{54}$$

is the kinetic energy per mass of a fluid parcel. Use of equation (52) in the linear momentum balance of equation (49) leads to the prognostic equation for the linear momentum per mass; i.e., the velocity $\mathbf{v}$

$$[\partial_t + (2\mathbf{\Omega} + \omega) \wedge]\,\mathbf{v} = -\nabla\mathcal{E} + \rho^{-1}\nabla \cdot (\tau - \mathbf{I}p), \tag{55}$$

where

$$\mathcal{E} = \Phi + \mathcal{K} \tag{56}$$

is the total mechanical energy per mass of a fluid parcel. The vector invariant velocity equation (55) exposes vorticity and mechanical energy per unit mass, whereas the linear momentum equation (51) focuses on nonlinear self-advection along with the coordinate-dependent advection metric frequency.

4.5. Vorticity and Potential Vorticity

Vorticity is one of the most important dynamical variables in fluid mechanics. Furthermore, the associated potential vorticity scalar is key to understanding and predicting aspects of geophysical fluid flows. This section introduces these vorticities, with more complete discussions available in such places as *Gill* [1982], *Pedlosky* [1987], *Müller* [1995], *Salmon* [1998], and *Vallis* [2006].

To derive the vorticity equation, take the curl of the vector-invariant form of the velocity equation (55) to lead to the material evolution of absolute vorticity $\omega_a = \omega + 2\mathbf{\Omega}$

$$\frac{d\omega_a}{dt} = \underbrace{-\omega_a\,(\nabla \cdot \mathbf{v})}_{\text{vortex stretching}} + \underbrace{(\omega_a \cdot \nabla)\mathbf{v}}_{\text{vortex tilting}}$$
$$+ \underbrace{\rho^{-2}\,(\nabla\rho \wedge \nabla p)}_{\text{baroclinicity}} + \underbrace{\nabla \wedge \mathbf{F}^{(v)}}_{\text{friction}}, \tag{57}$$

where we wrote the friction vector in the form $\nabla \cdot \tau = \rho\mathbf{F}^{(v)}$. The four terms on the right-hand side represent various manners whereby the absolute vorticity of a parcel is modified. The names associated with these terms represent the mechanisms under which vorticity is affected. A discussion of the physics of these mechanisms is outside the scope of the present considerations. Instead, Chapter 2 of *Pedlosky* [1987] is highly recommended for garnering a physical understanding.

Ertel [1942] determined that the potential vorticity

$$\Pi = \rho^{-1} \omega_{\mathrm{a}} \cdot \nabla \chi \tag{58}$$

is materially conserved so long as the scalar χ is materially conserved and representable as just a function of density and pressure. *Müller* [1995] and *Vallis* [2006] discuss oceanographically relevant examples of χ. Ertel's potential vorticity theorem generalizes all vorticity theorems of fluid mechanics. Furthermore, the theorem provides a practical means for determining constraints on the fluid motion. In particular, potential vorticity plays a fundamental role in hydrostatic isopycnal models. *Salmon* [1998] discusses the connection of Ertel's potential vorticity conservation to the relabelling symmetry possessed by fluid parcels. In the presence of a two-component equation of state, as in the ocean, there is no materially conserved potential vorticity [*Müller*, 1995]. Nonetheless, oceanographers have made great use of approximate forms of potential vorticity, and it therefore remains of fundamental importance in modeling.

5. DENSITY

The density of seawater is an important variable to measure in the ocean and to accurately compute in an ocean model. In particular, density variations, by means of their effects on the pressure field, provide one of the most important driving forces for large-scale circulation.

Density at a point in the ocean (the in situ density) is generally a function of temperature, salinity, and pressure,

$$\rho = \rho(\theta, S, p), \tag{59}$$

where we choose to use either potential or conservative temperature (Section 7.2) in the functional relation. This choice is more convenient than the alternative in situ temperature, as ocean models generally carry the more conservative θ as a prognostic variable rather than in situ temperature (see Section 7 for a discussion).

Equation (59) is known as the equation of state. Its precise form is determined empirically. The most accurate equation of state appropriate for ocean models has been given by *Jackett et al.* [2006]. This work is based on that of *Feistel* [1993], *Feistel and Hagen* [1995], and *Feistel* [2003]. Most ocean models are now switching to such accurate equations of state, as the earlier approximate forms, such as *Bryan and Cox* [1972], maintain a relatively narrow range of salinity variations over which the equation is valid. With ocean mod-

els of refined grid resolution and realistic fresh water forcing, it is desirable to remove such limitations, as model salinity can vary quite widely, especially near river mouths and sea ice.

The equation of state (59) is often approximated by replacing the pressure dependence with a depth dependence

$$\rho(\theta, S, p) \rightarrow \rho(\theta, S, p_o(z)), \tag{60}$$

where $p_o(z)$ is a predefined reference pressure profile generally set as the hydrostatic pressure arising from the initial density profile (Section 9.2). Converting pressure dependence to depth dependence produces an infinite acoustic speed, which removes acoustic modes from the system (Section 8.1). A time-dependent mass conservation is retained with the approximate density of equation (60), and this density is used to define the seawater parcel mass, the tracer mass, and the linear momentum. We may choose, however, to employ a more accurate expression for density in computing pressure if the hydrostatic approximation is used (Section 9.2). The resulting fluid is termed pseudo-incompressible [*Durran*, 1999] or quasi-non-Boussinesq [*Greatbatch et al.*, 2001].

The functional relation $\rho = \rho(\theta, S, p)$ allows us to develop the material time derivative of in situ density

$$\frac{d \ln \rho}{dt} = \frac{1}{\rho c_s^2} \frac{dp}{dt} + \beta_S \frac{ds}{dt} - \alpha_\theta \frac{d\theta}{dt}. \tag{61}$$

In this equation, we introduced the thermal expansion and saline contraction coefficients

$$\alpha_\theta = -\left(\frac{\partial \ln \rho}{\partial \theta} \right)_{p,S} \tag{62}$$

$$\beta_S = \left(\frac{\partial \ln \rho}{\partial S} \right)_{p,\theta} \tag{63}$$

as well as the squared sound speed

$$c_s^2 = \left(\frac{\partial p}{\partial \rho} \right)_{S,\theta}. \tag{64}$$

It is interesting to note that when parcels mix as they are materially transported, e.g., from molecular diffusion with diffusivities κ_θ and κ_S, the potential temperature and salinity terms in equation (61) become

$$\frac{1}{\rho c_s^2} \frac{dp}{dt} + \beta_S \nabla \cdot (\kappa_S \nabla S) - \alpha_\theta \nabla \cdot (\kappa_\theta \nabla \theta) = -\nabla \cdot \mathbf{v},$$

(65)

where mass conservation in the form of equation (3) was used to replace the density derivative with velocity convergence. This equation indicates that in addition to material changes in pressure, the mixing of salinity and potential temperature act, by means of mass continuity, to balance changes in the volume of a fluid parcel. For example, absent salinity and pressure effects, raising the potential temperature of a mass conserving fluid parcel by molecular diffusion ($\nabla \cdot (\kappa_\theta \nabla \theta) > 0$) causes an increase in the volume of a parcel ($d \ln(dV)/dt = \nabla \cdot \mathbf{v} > 0$) when the thermal expansion coefficient α_θ is positive. We caution that this is a deceptively simple thought experiment, as expansion of a region of fluid by heating is actually mediated by pressure fluctuations occuring as acoustic modes, a subject we consider in Section 8.1.

6. ENERGETIC BUDGETS

There are fundamental symmetries that the momentum equation (49) respects, and these symmetries lead to conservation laws for certain combinations of dynamical variables, such as kinetic energy and total energy. In the construction of numerical models, it is often beneficial to build analogous symmetries and conservation laws into the discrete equations. Such practices have been demonstrated to yield robust algorithms and physically realizable solutions. In general, it is desirable to be able to manipulate the discretized model equations in an analogous fashion to the manipulations used in obtaining the conservation laws in the continuum. Unfortunately, it is not always possible to maintain the exact conservation laws and symmetries in the discrete equations.

6.1. Kinetic Energy Budget

Energy is a useful scalar currency in physics because the total energy of a closed system is conserved. The ocean is not closed, but instead is a forced dissipative system. Nonetheless, the governing equations are energetically self-consistent, and so it is useful to consider the energetic budgets in numerical models. We start by considering the kinetic energy of a fluid parcel, which is given by $(\rho dV)\mathbf{v} \cdot \mathbf{v}/2 = (\rho dV)\mathcal{K}$. Bounds on this quadratic quantity can provide indirect constraints with which to develop numerical algorithms for the momentum equation. These constraints are useful, as linear momentum is not conserved on a sphere [see Section 4.11 of *Griffies*, 2004].

The kinetic energy budget is obtained by taking the inner product of $\mathbf{v}$ with the linear momentum equation (49) to find

$$\rho \frac{d\mathcal{K}}{dt} = -\rho \mathbf{v} \cdot \nabla \Phi - \mathbf{v} \cdot \nabla p + \mathbf{v} \cdot \nabla \cdot \tau.$$

(66)

Use of the identity (7) renders the Eulerian budget

$$(\rho \mathcal{K})_{,t} + \nabla \cdot (\rho \mathbf{v}\mathcal{K} + \mathbf{v}p) = p\nabla \cdot \mathbf{v} - \rho \mathbf{v} \cdot \nabla \Phi + \mathbf{v} \cdot \nabla \cdot \tau.$$

(67)

Note that we could have obtained this budget by working with either of the Eulerian forms: the advective form of the momentum budget (51) or the vector invariant velocity equation (55). On the discrete lattice, it is often more convenient to work with the vector invariant form, such as commonly used with the Arakawa C-grid models [*Mesinger and Arakawa*, 1976].

Terms on the right-hand side of the kinetic energy budget (67) represent energy conversion processes, whereby kinetic energy is exchanged for other forms of energy. Recall from the discussion of mass conservation in Section 2.1, the volume of a fluid parcel expands in a diverging velocity field according to equation (2). We thus identify $p\nabla \cdot \mathbf{v}$ as a pressure work term: as pressure works to compress a fluid parcel ($p\nabla \cdot \mathbf{v} < 0$), the internal energy of the parcel increases at the cost of decreasing its kinetic energy. The term $-\rho \mathbf{v} \cdot \nabla \Phi$ represents an exchange of kinetic energy for gravitational potential energy arising from vertical motions. That is, as parcels move up the gravitational field gradient ($\rho \mathbf{v} \cdot \nabla \Phi > 0$), kinetic energy decreases as potential energy increases (see Section 6.2 for more on gravitational potential energy). Note that in the special case, common in ocean models, where the geopotential is aligned according to the local vertical, then $\nabla \Phi = g\hat{\mathbf{z}}$, and so $\rho \mathbf{v} \cdot \nabla \Phi = \rho wg$. Finally, the frictional stress term $\mathbf{v} \cdot \nabla \cdot \tau$ was discussed in Section 4.3 where we noted that equation (45) provides the form for this term due to molecular viscous effects.

Bringing these results together leads to the Eulerian budget for kinetic energy per volume of a fluid parcel

$$(\rho \mathcal{K})_{,t} + \nabla \cdot (\rho \mathbf{v}\mathcal{K} + \mathbf{v}p - \mathbf{v} \cdot \tau)$$
$$= p\nabla \cdot \mathbf{v} - \rho \mathbf{v} \cdot \nabla \Phi - \rho \epsilon.$$

(68)

In this relation, we reintroduced the more general form of the viscous transport $\mathbf{v} \cdot \tau$, which allows us to identify a kinetic energy flux

$$\mathbf{J}_{\mathcal{K}} = \rho\,\mathbf{v}\mathcal{K} + \mathbf{v}p - \mathbf{v}\cdot\tau. \tag{69}$$

The terms in this flux alter kinetic energy locally but integrate to boundary terms when considering a global budget.

In the derivation of the local kinetic energy budget (68), the Coriolis term has exactly zero contribution to the energetics. This result follows, as the Coriolis force arises from our choice to describe motion in a moving reference frame at a point on the rotating Earth. Such an arbitrary choice of reference frame can have no impact on the energy of a parcel. This result suggests a desirable property for the discretized form of the momentum equation: that there be no local kinetic energy source due to the Coriolis force. Unfortunately, this property is difficult to achieve on the discrete lattice if components of discrete velocity reside on a staggered grid and are not co-located. The Arakawa B-grid has both horizontal velocity components co-located, whereas the Arakawa C-grid places them on adjacent cell faces [*Arakawa*, 1966]. When local conservation is unobtainable, it may still be useful to satisfy the global kinetic energy budget

$$\partial_t \int \mathcal{K}\rho\,\mathrm{d}V = \int \mathrm{d}V\,(p\nabla\cdot\mathbf{v} - \rho\mathbf{v}\cdot\nabla\Phi - \rho\epsilon), \tag{70}$$

in which we dropped boundary terms for brevity. In the case of the Coriolis force on an Arakawa C-grid, there are some discretizations that ensure there is no net global spurious source or sink of kinetic energy associated with these terms [*Sadourny*, 1975; *Arakawa and Lamb*, 1981; *Arakawa and Hsu*, 1990].

6.2. Gravitational Potential Energy Budget

The evolution of gravitational potential energy $(\rho\,\mathrm{d}V)\Phi$ for a parcel follows trivially by use of mass conservation

$$(\rho\,\Phi)_{,t} + \nabla\cdot(\rho\,\mathbf{v}\Phi) = \rho\,(\partial_{,t} + \mathbf{v}\cdot\nabla)\,\Phi. \tag{71}$$

Time dependence can arise for the geopotential through tidal effects. In this case, the energetic balances for total energy of the terrestrial ocean fluid (Section 6.3) includes a source term representing input of potential energy from the astronomical bodies affecting the tidal forcing. In addition to time-dependent effects, the potential energy of a parcel is affected by motions through the gravitational field. Namely, motions up the geopotential gradient $(\rho\mathbf{v}\cdot\nabla\Phi > 0)$ increase gravitational potential energy. This mechanical energy conversion term is equal and opposite to the corresponding conversion term in the kinetic energy budget (68).

Notably, mixing processes, which affect internal energy (Section 6.3), are absent on the right-hand side of the gravitational potential energy budget (71). Therefore, the connection between potential energy and internal energy is indirect. That is, mixing leads to local density modifications, which then lead to divergent flow through mass conservation [equation (3)]. This then leads to work being done on the fluid, which converts internal energy to kinetic energy. Then, through an adiabatic adjustment process, motions through the gravitational field are realized so as to modify potential energy. This adiabatic adjustment process is carried out very rapidly by acoustic modes (Section 8.1).

The total mechanical energy, $\mathcal{E} = \mathcal{K} + \Phi$, is the sum of kinetic and gravitational potential energies, and it evolves according to

$$(\rho\,\mathcal{E})_{,t} + \nabla\cdot(\rho\mathbf{v}\mathcal{E} + \mathbf{v}p - \mathbf{v}\cdot\tau) = \rho\,\Phi_{,t} + p\nabla\cdot\mathbf{v} - \rho\,\epsilon \tag{72}$$

On the left side is the divergence of the mechanical energy flux, with this flux acting to transport mechanical energy throughout the fluid. On the right side are source terms that represent time-dependent gravitational effects $(\rho\Phi_{,t})$ and the conversion between kinetic energy and internal energy.

6.3. Mechanical Plus Internal Energies

In the previous discussion, we have inferred the existence of internal energy based on the conversion of mechanical energy into a non-mechanical form. This inference is founded on an assumption that total energy of a fluid parcel is conserved. Quite generally, internal energy represents the energy of the molecular degrees of freedom that are averaged out when formulating a continuum description of a fluid. That is, the internal energy per mass, I, embodies the energy of molecular thermal agitation and molecular interactions, with details of this energy unavailable with a continuum description. Another source of energy introduced in our discussion of gravitational potential energy (Section 6.2) arises from time-dependent gravitational fields $\rho\Phi_{,t}$, which represent tidal forcing. We consider time-dependent gravitational fields as a source of energy external to the terrestrial ocean fluid.

The total energy per mass, T (specific energy), of a fluid is therefore written as the sum of the mechanical plus internal energies

$$T = \mathcal{K} + \Phi + I, \tag{73}$$

where, again, $\mathcal{K} = \mathbf{v}^2/2$ is the kinetic energy per mass of a fluid parcel, Φ is the gravitational potential energy per mass, and I is the internal energy per mass. Energy conservation for a fluid parcel means that total energy per mass evolves according to the conservation law

$$(\rho\,T)_{,t} + \nabla \cdot \mathbf{J}_T = \rho\,\Phi_{,t} \qquad (74)$$

for some flux of energy $\mathbf{J}_T$. Nonzero normal flux components arise for open fluid systems such as the ocean. Again, the nonzero source term $\rho\Phi_{,t}$ arises from gravitational effects external to the terrestrial ocean system.

Based on considerations of mechanical energy flux for a parcel affected by friction, we define the flux of total energy as

$$\mathbf{J}_T = \rho\,\mathbf{v}\mathcal{E} + \mathbf{v}\,p - \mathbf{v}\cdot\tau + \mathbf{J}_q. \qquad (75)$$

We introduced here the heat flux $\mathbf{J}_q$, which is generally a function of temperature as well as tracer concentration [for discussions, see *Fofonoff*, 1962; *Gregg*, 1984; *Landau and Lifshitz*, 1987; *Davis*, 1994; *McDougall*, 2003]. Subtracting the mechanical energy budget (72) from the total energy budget (74) leads to the internal energy budget for a fluid parcel

$$\rho\,\frac{d\mathcal{I}}{dt} = -\nabla\cdot\mathbf{J}_q - p\,\nabla\cdot\mathbf{v} + \rho\,\epsilon. \qquad (76)$$

Internal energy of a parcel is thus affected by the convergence of heat fluxes and sources due to pressure work and frictional dissipation. Notice how internal energy is increased by pressure work acting to compress the fluid. In the absence of irreversible effects due to heat transport and friction, internal energy is affected only by pressure work. Notably, equation (76) is a reflection of the First Law of Thermodynamics applied to a moving fluid parcel assumed to be locally in thermodynamic equilibrium, but non-locally to be out of equilibrium. We have more to say on such applications of linear irreversible thermodynamics to a moving fluid in Section 7.

7. BASIC NON-EQUILIBRIUM THERMODYNAMICS

The equations of an ocean model embody Newton's Laws of motion applied to a continuum fluid. Additionally, they employ results from linear irreversible, or non-equilibrium, thermodynamics, which is the subject of this section. In particular, it is useful to work with a thermodynamic variable that is readily measured, provides information about the heat of a fluid parcel, and is conservatively transported through the fluid. However, there is no strictly conservative thermodynamic scalar that measures heat, as there are always sources, such as from frictional dissipation or heat of mixing. The purpose of this section is to introduce some

basic notions of non-equilibrium thermodynamics and, in the process, expose a few details about useful temperature variables.

7.1. Budgets for Entropy and In Situ Temperature

We start the discussion with the fundamental thermodynamic relation [see, for example, Section 5.2.4 of *Griffies*, 2004]

$$d\mathcal{I} = T\,d\zeta - p\,d\rho^{-1} + \mu_S\,dS, \qquad (77)$$

where T is the in situ temperature, ζ is the entropy per mass, $1000\mu_S = \mu_{salt} - \mu_{water}$ is the relative chemical potential between salt and fresh water.[8] This relation holds between infinitesimal changes in thermodynamical state functions. Hence, although derived for quasi-static processes from the First Law of Thermodynamics using connections to work and heat, equation (77) holds for arbitrary infinitesimal changes; its connection to the First Law of Thermodynamics holds only for quasi-static processes.

Now assume that each fluid parcel is in local thermodynamic equilibrium yet allow the full ocean system to be out of equilibrium. These assumptions yield the following internal energy time evolution

$$\rho\,\frac{d\mathcal{I}}{dt} = \rho\,T\,\frac{d\zeta}{dt} - p\,\nabla\cdot\mathbf{v} + \rho\,\mu_S\,\frac{dS}{dt}, \qquad (78)$$

where we used the mass balance (3) to relate material changes in density to the velocity convergence. The result (78) allows one to transfer the methods of equilibrium thermodynamics to the non-equilibrium or linear irreversible thermodynamics of moving fluid parcels. The term linear in this name refers to an assumption that the system is close to thermodynamic equilibrium. In this case, the dissipative thermodynamic fluxes are linear functions of the gradients of the thermodynamic state variables. Nonlinear effects are not absent, however, as there are nonlinear effects from advective transport, nonlinear source terms, a nonlinear equation of state, and nonlinear dependence of the transport coefficients. *DeGroot and Mazur* [1984] provide a thorough accounting of this subject, and *Gregg* [1984] and *Davis* [1994] apply these methods to small-scale mixing in the ocean. Slightly different formulations can be found in *Batchelor* [1967] and *Landau and Lifshitz* [1987], and their approaches are preferred in the following.

Using equation (76) for the evolution of internal energy in equation (78) leads to the expression for evolution of entropy in a seawater parcel

$$T \rho \frac{d\zeta}{dt} = -\nabla \cdot \mathbf{J}_q + \rho \varepsilon - \rho \mu_S \frac{dS}{dt}. \tag{79}$$

This equation implies that entropy of a fluid parcel evolves by three irreversible mixing processes: (1) convergence of heat fluxes; (2) frictional dissipation sources that increase a parcel's heat content by Joule heating; and (3) salinity mixing. Correspondingly, a parcel generally maintains constant entropy if processes associated with its evolution are adiabatic, frictionless, and isohaline. As the friction source is very small in the ocean, adiabatic isohaline transport is very nearly isentropic. Indeed, when ocean modelers refer to adiabatic and isohaline processes, they typically assume this to be synonymous with isentropic.[9]

We now expose a few steps along the path toward developing a scalar field whose evolution is approximately conservative and which provides a measure of heat in the ocean. For this purpose, we develop an equation for the evolution of in situ temperature. To start, note that specific entropy ζ can be considered a function of pressure, temperature, and salinity $\zeta(p, T, S)$. Consequently, its incremental change is given by

$$d\zeta = \zeta_{,p}\, dp + \zeta_{,T}\, dT + \zeta_{,S}\, dS, \tag{80}$$

with each of the partial derivatives taken with the other independent variables held fixed. Use of the following Maxwell thermodynamic relations

$$\rho \left(\frac{\partial \zeta}{\partial p} \right)_{T,S} = -\alpha_T \tag{81}$$

$$\left(\frac{\partial \zeta}{\partial S} \right)_{T,p} = -\frac{\partial \mu_S}{\partial T} \tag{82}$$

leads to

$$\rho T\, d\zeta = -T \alpha_T\, dp + \rho C_{pS}\, dT - \rho T \frac{\partial \mu_S}{\partial T}\, dS, \tag{83}$$

where we introduced the following thermodynamic response functions

$$C_{pS} = T \left(\frac{\partial \zeta}{\partial T} \right)_{p,S} \tag{84}$$

$$\alpha_T = - \left(\frac{\partial \ln \rho}{\partial T} \right)_{p,S}, \tag{85}$$

with C_{pS} the specific heat with constant pressure and salinity and α_T the thermal expansion coefficient for in situ temperature [in contrast to that defined for potential temperature or

conservative temperature used in equation (62)]. As for deriving the internal energy equation (78), assume local thermodynamic equilibrium for parcels, thus allowing relation (83) to hold for material parcels moving through the fluid, in which case

$$\rho C_{pS} \frac{dT}{dt} = \rho T \frac{d\zeta}{dt} + T \alpha_T \frac{dp}{dt} - \rho T \frac{\partial \mu_S}{\partial T} \frac{dS}{dt}. \tag{86}$$

Now employ the relation (79) for entropy evolution to render

$$\rho C_{pS} \frac{dT}{dt} = T \alpha_T \frac{dp}{dt} + \rho \left(\mu_S - T \frac{\partial \mu_S}{\partial T} \right) \frac{dS}{dt} + \rho \epsilon - \nabla \cdot \mathbf{J}_q. \tag{87}$$

Temperature of a seawater parcel is thus affected by the following processes: (1) adiabatic pressure effects which alter the temperature by expansion or contraction of the parcel, (2) material changes in salinity, (3) Joule heating from frictional dissipation, and (4) the convergence of heat fluxes. It is possible to remove the adiabatic compression effects by introducing potential temperature (Section 7.2). However, the remaining source terms cannot in general be absorbed into another scalar function.

7.2. Potential and Conservative Temperatures

Vertical motion made without changes to entropy or salinity change the hydrostatic pressure of a fluid parcel, which causes its in situ temperature to change according to [see equation (83)]

$$dT = \Gamma dp, \tag{88}$$

where $\Gamma = (T\alpha_T)/(\rho C_{pS})$ is the adiabatic lapse rate. Consequently, in situ temperature is not a conservative thermodynamic variable to label water parcels of common origin, as it changes even in the absence of mixing or heating. This observation leads one to consider removing adiabatic pressure effects from in situ temperature.

Potential temperature is defined as the in situ temperature that a water parcel of fixed composition would have if isentropically transported from its in situ pressure to a reference pressure p_r, with the reference pressure typically taken at the ocean surface. Mathematically, the potential temperature θ is the reference temperature obtained by integration of $dT = \Gamma dp$ for an isentropic and isohaline in situ temperature change with respect to pressure [e.g., *Apel, 1987*]:

$$\theta(S, T, p; p_r) = T + \int_p^{pr} \Gamma(S, \theta, p') \, dp'. \qquad (89)$$

By definition, the in situ temperature, T, equals the potential temperature, θ, at the reference pressure $p = p_r$. Elsewhere, these two temperature fields differ by an amount determined by the adiabatic lapse rate. The potential temperature of a parcel is constant when the entropy and material composition of the parcel are constant. Mathematically, this result follows by noting that when entropy changes at a fixed pressure and composition, $p = p_r$, then in situ temperature equals potential temperature. Equation (83) then leads to

$$d\zeta = C_{pS} \, d\ln\theta, \qquad (90)$$

implying $d\zeta = 0$ if and only if $d\theta = 0$.

Potential temperature has proven useful for many oceanographic purposes. However, we have yet to ask whether it is a convenient variable to mark the heat content in a parcel of seawater. Traditionally, the potential temperature multiplied by the heat capacity is used for this purpose. *Bacon and Fofonoff* [1996] provide a review with suggestions for this approach. In contrast, *McDougall* [2003] argues that potential temperature multiplied by heat capacity is less precise, by some two orders of magnitude, than an alternative thermodynamic tracer called potential enthalpy. Indeed, enthalpy is what is more commonly associated with heat in thermodynamics [*Fofonoff*, 1962], so it is sensible that ocean models should be carrying an enthalpy variable to represent heat content.

At present, most ocean models consider their heat variable to be potential temperature, and this variable is assumed to be conservative. This assumption has implications for the equation of state (59) and the calculation of heat fluxes at the ocean boundaries. *McDougall* [2003] notes that if we reinterpret the conservative heat variable in a model to be proportional to potential enthalpy, then the conservation equation

$$\rho \, C_p^o \frac{d\Theta}{dt} = -\nabla \cdot \mathbf{J}_q \qquad (91)$$

is an approximate statement of the first law of thermodynamics for the ocean. In this equation, Θ is the conservative temperature variable, and C_p^o is an appropriately chosen constant heat capacity. The alternative to equation (91), whereby Θ is replaced by potential temperature θ, is commonly used by ocean modelers. It is roughly 100 times less accurate and can lead to sea surface temperature differences upwards of 1°C in regions of large salinity deviation, such as river mouths. The National Aeronautics and Space Administration Goddard Institute for Space Studies ocean model [*Russell et al.*,

1995] uses potential enthalpy for its heat variable, and the new version of the Modular Ocean Model [*Griffies*, 2007] provides an option for using conservative temperature. For the remainder of this document, we use potential temperature θ as the heat scalar, noting that the equations remain the same if using the more accurate Θ.

8. LINEAR MODES OF MOTION

Having now developed the fundamental equations of the ocean, we move on to the task of introducing the linear dynamical modes admitted by these equations. This analysis initiates an exploration of the multitude of dynamical processes active in the ocean. More central to our purposes, the analysis provides us with guidance toward which numerical methods are needed to integrate the equations. The results are generally used to motivate certain approximations, so some material here anticipates approximation methods discussed in Section 9.

We are particularly interested here in the speed of various linear dynamical modes. This then allows us to determine a guide for the time step required to explicitly represent a particular mode by making use of the Courant–Friedrichs–Lewy (CFL) constraint [e.g., *Haltiner and Williams*, 1980; *Durran*, 1999]. Depending on details of space and time discretization, this constraint says that when simulating a propagating signal on a discrete lattice, $U\Delta t/\Delta$ must remain less than a number on the order of unity. Here, U is the speed of the mode, Δ is the discrete grid spacing, and Δt is the discrete time step. The CFL constraint says that as motions increase in speed, the numerical model must reduce its time step to represent these motions. Finer grid spacings also require smaller time steps. If the model fails to satisfy the CFL constraint for a particular mode, the model will likely go unstable, and it generally will do so quite rapidly.

8.1. Acoustic Waves

Linear acoustic fluctuations arise from small amplitude adiabatic, frictionless, and isohaline motion [e.g., *Apel*, 1987]. Such motions lead, through the equation of state (59), to the equation for pressure fluctuations in terms of density fluctuations [see equation (61)]

$$\frac{dp}{dt} = \rho \, c_s^2 \frac{d\ln\rho}{dt}. \qquad (92)$$

Noting the approximate form of mass conservation in Section 9, we write mass conservation in the form

$$\epsilon_{nb} \frac{d \ln \rho}{dt} = -\nabla \cdot \mathbf{v}, \qquad (93)$$

where we introduced the non-dimensional parameter ϵ_{nb}, with $\epsilon_{nb} = 1$ with mass conserving non-Boussinesq kinematics and $\epsilon_{nb} = 0$ for incompressible flow. Likewise, we write the linearized velocity equations, in the absence of the Coriolis force, friction, and gravity force, in the form

$$\rho \mathbf{u}_{,t} = -\nabla_z p \qquad (94)$$

$$\epsilon_{nh} \rho w_{,t} = -p_{,z} \qquad (95)$$

where ∇_z is the horizontal gradient operator, and we introduced the non-dimensional parameter ϵ_{nh}, which is unity for non-hydrostatic dynamics and vanishes for hydrostatic dynamics. Use of these relations in a linearized version of the pressure equation (92) leads to the wave equation for linear pressure fluctuations[10]

$$[\epsilon_{nb} \epsilon_{nh} \partial_{tt} - c_s^2 (\epsilon_{nh} \nabla_z^2 + \partial_{zz})] p = 0. \qquad (96)$$

Considering a single Fourier mode with space–time dependence of the form $\exp[i(\omega t - kx - ly - mz)]$, an approximate modal analysis of the above system yields the dispersion relation involving two of the four modes to be of the form

$$\left(\frac{\epsilon_{nh} \epsilon_{nb}}{c_s^2} \right) \omega^2 = \epsilon_{nh} (k^2 + l^2) + m^2. \qquad (97)$$

The unapproximated system (with $\epsilon_{nb} = \epsilon_{nh} = 1$) has non-dispersive modes that travel at speed c_s in three dimensions. The phase speed $c_s \sim 1500 \, ms^{-1}$ in the ocean is roughly two orders of magnitude faster than motions of interest in most climate and regional applications. If these modes were explicitly represented in models, then the time step would be very small, making the model prohibitively expensive.

There are three distinct ways in which the acoustic modes can be "filtered" from the system.

1. Make the equation of state (59) independent of pressure (incompressible), in which case $1/c_s \to 0$. This approach has the advantage that only the equation of state is modified. It has the disadvantage that it is inappropriate to neglect the effect of pressure on density at global scales [e.g., *Dewar et al.*, 1998].

2. Constrain the flow to be incompressible by setting $\epsilon_{nb} = 0$. Here, sound waves are prohibited because the acoustic mode propagation requires divergent flow to drive density and pressure anomalies. This approach, used alone, renders the system elliptic in pressure. It is the approach used in the

Massachusetts Institute of Technology general circulation model (MITgcm) when integrating the Boussinesq non-hydrostatic equations [*Marshall et al.*, 1997].

3. Assume hydrostatic (or quasi-hydrostatic) balance in the vertical momentum equation (set $\epsilon_{nh} = 0$). In this case, only the $m = 0$ mode satisfies the dispersion relation (97). This is the traditional approach in meteorology, which filters vertically propagating sound waves but retains an external acoustic mode known as the Lamb wave.

In oceanography, the traditional filters used are the second and third in conjunction. This approach filters out all acoustic modes and converts the elliptic problem for pressure into the local one-dimensional hydrostatic balance. Recently, however, non-Boussinesq ocean models are becoming the norm. In these models, only hydrostatic balance is used to filter acoustic modes, thus retaining the Lamb wave. The Lamb wave has not yet presented itself as a cause for concern in the stability of non-Boussinesq hydrostatic ocean models probably because the time-implicit or split-explicit treatment of the external mode is sufficient to damp or resolve this mode [see comment at the end of *DeSzoeke and Samelson*, 2002].

A fourth approach to numerically handling acoustic modes has been used in regional models. Here, rather than filtering the modes, the models slow them down so that they can be explicitly resolved [*Browning et al.*, 1990]. As for the first method above, this approach is likely to be inappropriate for global scale modeling.

8.2. Inertia–Gravity Waves

After the acoustic modes, the next fastest linear modes are the inertia–gravity waves. These are rotationally modified gravity waves which exist as external modes as well as an infinite range of internal modes. The external mode can be analyzed in the context of the depth-integrated Boussinesq equations or equivalently by considering a homogenous layer of constant density fluid [e.g., Sections 5.6 and 8.2 of *Gill*, 1982]. These equations are often referred to as the shallow water equations, which we write in their linear form as

$$(\partial_t + f \hat{\mathbf{z}} \wedge)\mathbf{u} = -g \nabla \eta \qquad (98)$$

$$\eta_{,t} = -H \nabla \cdot \mathbf{u}, \qquad (99)$$

where $\mathbf{u}$ is the horizontal velocity field in the homogeneous layer, η is the surface height fluctuation with respect to a resting fluid at $z = 0$, and H is the depth of the resting fluid, which is assumed constant for present purposes. We assume the Coriolis parameter f to be constant, which defines the f-plane approximation.

Introducing a space–time dependence of the form $\exp[i(\omega t - kx - ly)]$ leads to three linear eigenmodes. The first occurs with zero frequency $\omega = 0$, which is the geostrophic mode where the Coriolis force balances pressure $f\hat{\mathbf{z}} \wedge \mathbf{u} = -g\nabla\eta$. The geostrophic mode is a stationary mode of variability. It therefore places no time step constraint on the simulation. However, it is a critical element determining the large-scale structure of the ocean circulation. The nonzero frequency modes satisfy the dispersion relation

$$\omega^2 = f^2 + gH(k^2 + l^2). \tag{100}$$

The waves satisfying this relation are a pair of dispersive inertia–gravity or Poincaré waves. These waves provide the mechanism by which a fluid adjusts to an imbalance which then leads to geostrophic balance.

The inability of a numerical simulation to adequately adjust by inertia–gravity waves is very often the cause of grid-scale noise. For example, models built on the Arakawa B-grid can exhibit a checkerboard mode in the surface height field, and this is a direct consequence of the grid scale gravity waves exhibiting a null mode (spurious zero frequency numerical mode) [Mesinger, 1973, Killworth et al., 1991], that is, certain of the numerical gravity waves are spuriously static, rather than propagating. Similarly, coarse-resolution models built on the Arakawa C-grid exhibit longitudinal or latitudinal coherent noise which is a direct consequence of a null-mode associated with the numerical representation of the Coriolis force [see Adcroft et al., 1999, for a review of this issue].

Regardless of the spatial treatment of the inertia–gravity modes, permitting these modes in a simulation introduces a limitation on the model time step if they are to be treated explicitly. For short waves, the phase and group speed are approximately that of surface gravity waves. In the deep open ocean, this speed is of order $\sqrt{gH} \approx 200$ ms^{-1}. Satisfying the CFL condition for these waves in a model with $\Delta = 100$ km horizontal grid spacing (roughly 1° resolution) means the time step must satisfy $\Delta t = \Delta/U \approx 500$ s. Although longer than the time step required to admit acoustic waves, this time step is far smaller than practical when considering the needs of global ocean modeling, given the present power of computers. Other approaches must be used to avoid this limitation for the full model equations (see Section 11.6).

In contrast to the short waves, long inertia–gravity waves are dominated by rotation ($w^2 \sim f^2$). In this case, we are led to a time step limitation as a function of the Coriolis parameter. The most stringent limitation arising from these inertial waves occurs at the pole, where $1/(2\Omega) \sim 1.9$ h.

We now consider internal modes in which stratification is relevant. For this purpose, consider the following linear Boussinesq non-hydrostatic system

$$\rho_o(\partial_t + f\hat{\mathbf{z}} \wedge)\mathbf{u} = -\nabla_h p \tag{101}$$

$$\epsilon_{nh} w_{,t} + g + p_{,z}/\rho = 0 \tag{102}$$

$$\nabla \cdot \mathbf{u} + w_{,z} = 0 \tag{103}$$

$$\rho_{,t} - (N^2\rho_o/g)w = 0, \tag{104}$$

where ρ_0 is the constant Boussinesq reference density, and we again introduce a non-dimensional parameter ϵ_{nh} to monitor non-hydrostatic effects. We ignore horizontal density variations because our focus is on effects of vertical stratification as represented by the squared buoyancy frequency

$$N^2 = -(g/\rho_o)\rho_{,z}, \tag{105}$$

which is assumed constant for present purposes. A linear modal analysis assuming a space–time dependence of the form $\exp[i(\omega t - kx - ly - mz)]$ leads to both the geostrophic mode ($\omega = 0$) and the internal inertia-gravity wave dispersion relation

$$\left(\epsilon_{nh}(k^2 + l^2) + m^2\right)\omega^2 = m^2f^2 + (k^2 + l^2)N^2. \tag{106}$$

Non-hydrostatic effects are generally relevent only for regimes where the aspect ratio (ratio of vertical to horizontal scales) is order unity, meaning the horizontal wave numbers are on the order of the vertical: $k^2 + l^2 \approx m^2$. These modes are responsible for allowing the fluid to adjust toward geostrophic balance as well as to adjust to hydrostatic balance in the case of non-hydrostatic models. For hydrostatic inertia–gravity waves, the long waves are dominated by rotation, as were the external waves, while short waves have phase speed approaching the internal wave speed, N/m.

8.3. Rossby Waves

Rossby waves represent a slowly evolving, nearly geostrophic fluctuation. They arise from the gradient of the Coriolis parameter [see equation (121) for definition of Coriolis parameter] with respect to latitude

$$\begin{aligned}\beta &= f_{,y} \\ &= (2\Omega/R)\cos\phi. \end{aligned} \tag{107}$$

To develop the dispersion relation for Rossby waves, reconsider the linear shallow water system of equations (98) and (99), only now, let the Coriolis parameter be given by a linear function of latitude $f = f_o + \beta y$, with f_o and β constant. Assuming a space–time dependence of the form

$\exp[i(\omega t - kx - ly)]$ leads to the dispersion relation satisfied for this beta plane shallow water system [e.g., Section 6.4 of *Cushman-Roisin*, 1994]

$$\omega = -\beta L_{\mathrm{d}}^2 \left(\frac{l}{1 + L_{\mathrm{d}}^2 (k^2 + l^2)} \right), \qquad (108)$$

where $L_{\mathrm{d}}^2 f_o^2 = gH$ defines the Rossby radius of deformation.

Spurious behavior of numerical Rossby waves can often be associated with spurious behavior of inertia–gravity waves. This arises because spurious behavior of inertia–gravity waves implies a spurious gravitational adjustment process, which in turn leads to a poor representation of the geostrophic balance for some modes. Short-scale Rossby waves are generally dissipated locally before they can propagate far. Indeed, this, and a preferred westward drift for large-scale waves, is the mechanism of western enhancement of boundary currents [*Pedlosky*, 1987]. A common numerical problem associated with Rossby waves is due to insufficient dissipation necessary to trap eastward propagating, short-scale Rossby waves. Removing this problem requires enhancing horizontal friction sufficiently to resolve the Munk boundary layer [*Munk*, 1950; *Griffies and Hallberg*, 2000; *Large et al.*, 2001]. Once inertial boundary currents are resolved, eddy–mean flow interactions and other nonlinear interactions tend to be sufficient. Further discussion of numerical representation of Rossby waves can be found in *Wajsowicz* [1986] and *Fox-Rabinovitz* [1991].

8.4. Implications for Stability of Numerical Models

The stability of numerical models depends on the choice of numerical time-integration method, the spatial discretization, and the permitted modes in the equations. Ocean models generally do not permit acoustic modes that would otherwise be prohibitive: a grid spacing of 100km would require a time step of order less than 1 min if the model admitted acoustic modes. The next fastest modes are the external gravity waves, with speeds exceeding 200m s^{-1} in the deep ocean. As discussed in Section 11.6, these modes are usually treated separately from the full three-dimensional fluctuations. We thus do not consider external modes in this section. The remaining processes may cause a numerical model to be unstable either through a direct numerical instability or through the generation or admission of excessive grid-scale noise.

A process may be directly numerically unstable in the von Neuman sense [*Durran*, 1999] if it is treated explicitly, and the shortest characteristic timescale of that process is not resolved by the model time step. Time-implicit treatment

of a process often yields unconditional numerical stability, although other considerations such as accuracy may lead to constraints on the model time step.

The simplest example of a process with an identifiable term in the equations is the inertial oscillation, for which the Coriolis term is responsible. The characteristic timescale is f^{-1}, which is shortest at the poles: $(4\pi/1\mathrm{day})^{-1} \sim 1.9$ h. To resolve inertial oscillations, a time-explicit integration scheme requires that

$$f\Delta t < \gamma, \qquad (109)$$

where γ is a number that depends on the details of the numerical integration scheme. We choose $\gamma = 1/2$ as a representative number. Thus, the maximum time step allowed to integrate inertial oscillations stably is $\Delta t_{\mathrm{max}}^f = \gamma/f$. For reference, $1/(2f) \sim 57$ min is plotted in Figure 5.

Advection is characterized by a velocity scale, U. The shortest advective charactistic timescale in a numerical model is Δ/U, where Δ is the spatial grid scale and U is representative of the largest characteristic velocity. The CFL number is the ratio of this characteristic timescale to the model time step,

$$C_u = \frac{U \Delta t}{\Delta}. \qquad (110)$$

This dimensionless ratio is often known as the Courant number. Most time-explicit schemes for advection require that C_u be less than a number on the order of unity, with the constraint more restrictive in higher dimensions due to the possibility of propagation diagonal to the discrete grid lines. That is, the largest time step that can support numerical stability for a given flow and grid spacing scales is

$$\Delta t_{\mathrm{max}}^u \propto \frac{\Delta}{U}. \qquad (111)$$

This result applies in all spatial directions for which advection is explicit. In practice, this constraint can be most restrictive for regions of fine vertical resolution with strong surface wind stress curls. Furthermore, the CFL criteria may be either additive or independent, again depending on the algorithm details.

The flow speed U is a result of the forcing and balances in a model simulation. It is also a function of resolution, particularly for low-resolution models. The transport of ocean boundary currents is determined by the basin-wide forcing, and numerical models respect this transport even at coarse resolution. However, if the boundary current is not resolved, such as when there is only one cell in the current, then U becomes inversely proportional to Δ to maintain the proper transport. For the purposes of this discussion, we have chosen the profile for $U(\Delta)$ depicted in Figure 4. The corresponding limitation on time step (dashed line in Figure 5) has a Δ^1 dependence at fine resolution and a Δ^2 dependence at coarse resolution. These two resolution regimes for U and Δt^U are indicated in Table 1.

The gravest internal gravity wave (lowest vertical eigenmode) propagates with a characteristic speed $c_g \propto NH$. These waves have a grid scale characteristic time of Δ/c_g, which in turn leads to a stability constraint that Δt must be smaller than

$$\Delta t_{\max}^{c_g} \propto \frac{\Delta}{c_g}. \tag{112}$$

Unlike advection, the fastest internal wave speed is independent of resolution.

A friction operator is typically used to control noise in numerical models and to maintain a finite grid Reynolds number to keep the solution stable. Typical causes of noise include a high grid Reynolds number (i.e., velocity advection dominates accelerations) or insufficient dissipation to damp short eastward propagating Rossby waves near boundaries. As shown below, at coarse resolutions, the boundary resolution criteria dominates the need for viscosity, whereas the grid Reynolds number criteria dominates at fine resolution. Additionally, biharmonic friction operators are favored at fine resolution due to their enhanced scale selectivity, thus increasing the energetics of the flow while, ideally, maintaining a sufficient level of dissipation at the grid scale [e.g., *Semtner and Mintz*, 1977; *Griffies and Hallberg*, 2000]. Notably, a lack of sufficient friction may not immediately translate into a catastrophic model instability (i.e., model blow-up). Instead, depending on grid resolution, forcing, and numerical methods, it is possible for models to run stably, albeit with unphysically huge levels of grid noise, using very small, if not zero, interior friction.

We now consider the time step constraints introduced by the Laplacian and biharmonic friction operators. Viscous dissipation terms have a grid-scale characteristic time of Δ^2/A_2 and Δ^4/A_4, for harmonic and biharmonic viscosities, respectively. The explicit stability criteria require that the time step be smaller than

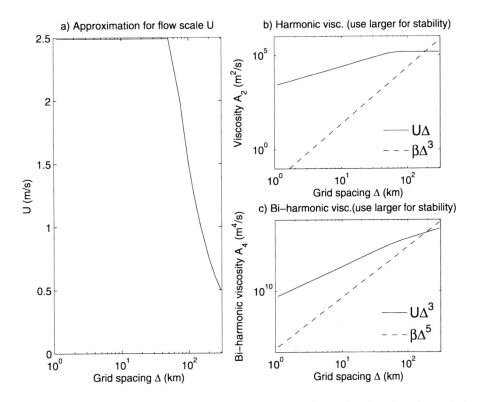

Figure 4. (a) Scaling for the maximum speed U seen in global numerical models as a function of spatial resolution. (b) The harmonic viscosity required to maintain a finite grid Reynolds number or that required to ensure the Munk boundary layer is resolved. (c) As for Figure 4b, but for a biharmonic viscosity. The spatial resolution is assumed to be isotropic and relatively uniform. This assumption is not generally the case for many global model grids.

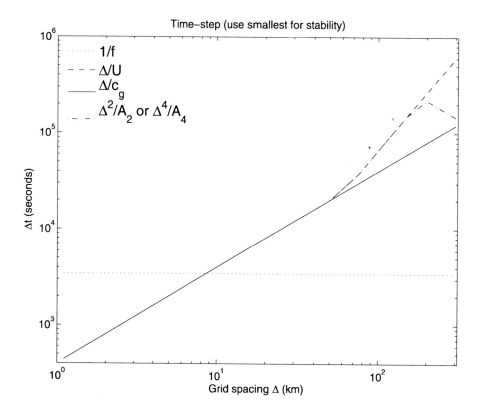

Figure 5. Using the scaling for U and viscosities shown in Figure 4, the maximum Δt allowed by each process: Coriolis term, $1/f$; advection Δ/U; internal gravity waves Δ/c_g; viscosity Δ^2/A_2 or Δ^4/A_4. Note that the timescales of processes have arbitrarily been equalized at the grid-scale, as we have neglected the details of the numerial discretizations. The curves shown can not be compared to each other, as they should each be scaled by appropriate factors to reflect the numerical details.

$$\Delta t_{\max}^{A_2} \propto \frac{\Delta^2}{A_2} \quad \text{or} \quad \Delta t_{\max}^{A_4} \propto \frac{\Delta^4}{A_4}, \tag{113}$$

respectively. These appear to have higher power dependence on Δ than the advective ($\Delta t_{\max}^u$) and gravity wave ($\Delta t_{\max}^{c_g}$) constraints. But this result is only true for given viscosity parameters (A_2 or A_4). In practice, the viscosity applied in ocean models is a strong function of resolution, and there are two distinct criteria for setting the viscosity. At very coarse resolution, a significant source of noise can occur when the viscous Munk boundary layer is not resolved [*Munk*, 1950]. The boundary layer scale, L_b, scales as

$$L_b \propto \begin{cases} \left(\dfrac{A_2}{\beta}\right)^{1/3} & \text{harmonic friction} \\[2ex] \left(\dfrac{A_4}{\beta}\right)^{1/5} & \text{biharmonic friction,} \end{cases} \tag{114}$$

where $\beta \sim 2 \times 10^{-11}$ m^{-1} s^{-1} is the planetary vorticity gradient at the equator. If we chose the viscosities to be sufficient to create a Munk boundary layer wide enough to be resolved by a grid spacing, Δ, then the viscosities will scale as

$$A_2 \propto \beta \Delta^3 \quad \text{or} \quad A_4 \propto \beta \Delta^5, \tag{115}$$

respectively.[11] This scaling determines the slope of the curves for A_2 and A_4 in the coarse grid spacing regime of Figure 4. Thus, the effective characteristic timescales [equation (113)] for viscosity becomes $(\beta \Delta)^{-1}$ for both harmonic and biharmonic forms. This scaling is seen in the far right column of Table 1.

The second criteria for setting the viscosity applies to finer resolution where nonlinearity in the momentum equation is sufficient to form an inertial boundary layer (which will be thinner than the frictional boundary layer of coarse-resolution models). In this regime, the role of the viscous terms is to dissipate grid-scale energy and noise. The required viscosity can be estimated by requiring the grid Reynolds number to be finite so that

$$A_2 \propto U_b \Delta \quad \text{or} \quad A_4 \propto U_b \Delta^3, \tag{116}$$

where U_b is the scale of the boundary speed, which itself is also a function of grid resolution. At very fine resolution, the grid-length characteristic timescale [equation (113)] of either form of viscosity becomes

$$\Delta t_{\max}^{A_{2,4}} \propto \frac{\Delta}{U_b}. \tag{117}$$

There is a marginal resolution where the viscous boundary layer is resolved, but the inertial boundary layer may be marginally resolved so that the maximum realized velocity is still a function of resolution. In this narrow regime, the characteristic timescale for viscosities becomes

$$\Delta t_{\max}^{A_{2,4}} \propto \frac{\Delta^2}{U_b L_b}, \tag{118}$$

where L_b is now the realistic inertial boundary layer scale. The three regimes for scaling behavior of the viscous limitation on maximum time step are listed in Table 1.

Figure 5 schematically shows the scaling of maximum time step discussed above and tabulated in Table 1. The absolute values are not necessarily appropriate to any specific model because discretization modifies the numerical coefficients that have been ignored for these curves. Of note, though, is that the scaling behavior of the largest stable time step allowed in a model is very complicated at coarse and marginal resolutions, more so than is indicated by the curves shown. In contrast, the scaling becomes simple at fine resolutions. In practice, the numerical details of the Laplacian and biharmonic friction operators will make the maximum allowed time step shorter. At coarse resolution, we typically find $\Delta t_{\max}^{A_4} < \Delta t_{\max}^{A_2} < \Delta t_{\max}^u$. This result is not apparent in Figure 5, as we have arbitrarily set all timescales to be equal at the grid scale. That is, at coarse resolution where the boundary currents are not resolved, the time step is usually limited by viscosity.

Besides grid-scale-dependent viscosities, which are now common in ocean models, *Smagorinsky* [1963, 1993] proposed that the Laplacian viscosity due to unresolved scales should be proportional to the resolved horizontal deformation rate times the squared grid spacing. In effect, the Smagorinsky viscosity tailors the local dissipation to both the local flow state and the local grid resolution using only a single non-dimensional adjustable parameter. If this parameter is properly chosen, the resulting viscosity ensures that the flow respects the numerical stability properties previously discussed even when simulating multiple flow and grid regimes such as occur in realistic ocean simulations. It is for these pragmatic reasons that the Smagorinsky viscosity has found notable use in large-scale ocean models [e.g., *Blumberg and Mellor*,

Table 1. Table of scaling relations for maximum time step permitted by each process.

Process	Timescale	$\Delta < L_b$	$\Delta > L_b$	
Advection, U $\Delta t_{\max}^u \propto$	$\dfrac{\Delta}{U}$	$U \sim U_b$ $\dfrac{\Delta}{U_b}$	$U \sim U_b L_b / \Delta$ $\dfrac{\Delta^2}{U_b L_b}$	
Gravity wave, C_g $\Delta t_{\max}^{c_g} \propto$	$\dfrac{\Delta}{c_g}$	$\dfrac{\Delta}{c_g}$		
Harmonic, $A_2 \propto$ $\Delta t_{\max}^{A_2} \propto$	$\dfrac{\Delta^2}{A_2}$	ΔU_b $\dfrac{\Delta}{U_b}$	$L_b U_b$ $\dfrac{\Delta^2}{U_b L_b}$	$\beta \Delta^3$ $\dfrac{1}{\beta \Delta}$
Biharmonic, $A_4 \propto$ $\Delta t_{\max}^{A_4} \propto$	$\dfrac{\Delta^4}{A_4}$	$\Delta^3 U_b$ $\dfrac{\Delta}{U_b}$	$L_b \Delta^2 U_b$ $\dfrac{\Delta^2}{U_b L_b}$	$\beta \Delta^5$ $\dfrac{1}{\beta \Delta}$

U is the flow speed realized in the model, U_b is maximum oceanic flow speed observed in boundary currents, c_g is the speed of the gravest internal gravity waves, Δ is the smallest grid spacing, L_b is a boundary layer scale, β is the planetary vorticity gradient at the Equator and A_2 and A_4 are the Laplacian and biharmonic viscosities, respectively.

1987; *Rosati and Miyakoda*, 1988; *Bleck and Boudra*, 1981; *Bleck et al.*, 1992], with *Griffies and Hallberg* [2000] also arguing for its utility with a biharmonic operator.

In Figure 5, we have also shown the Δt^f timescale arising from inertial oscillations. This timescale is independent of spatial resolution. The Δt^{C_g} timescale arises from internal gravity waves, and it has a scaling of Δ^1, with $c_g \sim NH$ invariant with resolution (Section 8.2). Vertical friction also has stability criteria, but is somewhat easier to treat implicitly in time because of the non-periodic nature of the vertical direction and the small aspect ratio of the computational grid. Notably, there are typically many fewer degrees of freedom in the vertical than the horizontal directions.

9. APPROXIMATIONS

We denote the set of equations developed thus far the unapproximated ocean equations. There have indeed been approximations made in deriving these equations: the fluid is approximated by a continuum, geopotentials are approximated by surfaces of constant oblate spheroid radius (Section 4.1), and the angular rotation rate of the Earth is assumed constant. Nonetheless, the suite of phenomena described by these equations is immense, with space scales and timescales ranging from millimeter and seconds to global and millenia. Various methods have been used to filter the equations to focus on particular subranges of this spectrum. From a modeling perspective, filtering, or approximating, the equations helps to reduce the cost of the resulting simulation. The previous discussion of linear modes anticipated some of the approximations commonly made in physical oceanography. We more formally review these approximations in this section.

9.1. Shallow Ocean Approximation

In the shallow ocean approximation,[12] the metric functions measuring horizontal distances on the Earth are dependent only on the lateral coordinates. Radial dependence of the metric functions is reduced to the constant radial factor $R = 6.367 \times 10^6$ m. This radius corresponds to the ellipsoid of best fit to the sea level geopotential. This is the appropriate value for the "Earth's radius" of use in ocean models. Note that R in ocean models is often taken as the slightly larger value $R = 6.371 \times 10^6$ m. This value corresponds to the radius of a sphere with the same volume as the Earth [*Gill*, 1982, page 597].

The shallow ocean approximation is motivated by noting the relatively small thickness of the ocean relative to the Earth's radius. Within this approximation, distances used to compute partial derivatives, covariant derivatives, areas, and

volumes are determined by a metric tensor whose components are functions only of the lateral position on the sphere. Additionally, assumptions regarding the metric function dependence, as well as assumptions about the smallness of vertical accelerations associated with the hydrostatic approximation (Section 9.2), have implications toward the energy and angular momentum conservation laws. In particular, the angular momentum about the Earth's center is computed with a moment-arm that has a fixed radius $r = R$. Hence, motion in the vertical direction does not alter angular momentum in the shallow ocean approximation.

9.2. Hydrostatic Approximation

The hydrostatic approximation exploits the large disparity between horizontal motions, occurring over scales of many tens to hundreds of kilometers, and vertical motions, occurring over scales of tens to hundreds of meters. In this case, it is quite accurate to assume the moving fluid maintains the hydrostatic balance, whereby the vertical momentum equation takes the form

$$p_{,z} = -\rho\, g. \tag{119}$$

Because the vertical momentum budget has been reduced to the hydrostatic balance, the Coriolis force per mass must be given by

$$\mathbf{F}_c = -f\,\hat{\mathbf{z}} \wedge \mathbf{v}, \tag{120}$$

where

$$f = 2\Omega\,\sin\phi \tag{121}$$

is the Coriolis parameter and ϕ is the latitude. That is, we drop the nonradial component of the Earth's angular rotation vector when computing the Coriolis force in a hydrostatic fluid.[13]

By truncating, or filtering, the vertical momentum budget to the inviscid hydrostatic balance, we are obliged to parameterize strong vertical motions occurring in convective regions, as hydrostatic equations cannot explicitly represent these motions. Such has led to various convective parameterizations in use by ocean models [*Killworth*, 1989; *Marshall and Schott*, 1999]. These parameterizations are essential for the models to accurately simulate various deep water formation processes, especially those occurring in the open ocean due to strong buoyancy fluxes.

The kinetic energy density for a hydrostatic fluid involves only the horizontal motions [e.g., *Bokhove*, 2000] so that

$$\mathcal{K} = \frac{1}{2}\mathbf{u} \cdot \mathbf{u}. \qquad (122)$$

No other change is required for the energetic relations established in Section 6 to follow through for the hydrostatic fluid. This result is self-consistent with the scaling implicit in the hydrostatic balance that $w \ll |\mathbf{u}|$. Correspondingly, the hydrostatic relative vorticity vector is

$$\boldsymbol{\omega} = \nabla \wedge \mathbf{u}, \qquad (123)$$

where $\mathbf{u}$ is the horizontal velocity vector.

Making these three changes in the non-hydrostatic velocity equation (55) leads to the hydrostatic vector invariant velocity equation

$$[\partial_t + (f\hat{\mathbf{z}} + \boldsymbol{\omega}) \wedge]\mathbf{v} = -\nabla \mathcal{E} + \rho^{-1} \nabla \cdot (\tau - \mathbf{I}p), \quad (124)$$

where, again, $\mathcal{E} = \mathbf{u} \cdot \mathbf{u}/2 + \Phi$. Note that the vertical component of equation (124) reduces to the hydrostatic balance upon setting the time derivative of the vertical velocity component to zero and by noting that the hydrostatic form of $\mathcal{K}$ and $\boldsymbol{\omega}$ mean that $\nabla\mathcal{K} + \boldsymbol{\omega} \wedge \mathbf{v}$ has a zero vertical component.

In hydrostatic ocean models, the effects of horizontal stresses are usefully split from vertical stresses when discussing friction. Friction from vertical stresses are generally parameterized by downgradient diffusion of momentum

$$(\nabla \cdot \tau)_{\text{vertical strain}} = \partial_z (\kappa_{\text{model}}\mathbf{u}_{,z}), \qquad (125)$$

where $\kappa_{\text{model}} > 0$ is the vertical viscosity used in the model. The vertical (more generally dianeutral) viscosity is generally assumed to be equal to, or more often roughly ten times greater than, the vertical or dianeutral diffusivity employed for tracer. This vertical Prandtl number (ratio of viscosity to diffusivity) is not well measured in the ocean, leaving modelers to tune this parameter based on simulation integrity.

Vertically integrating the hydrostatic balance (119) over the full depth of the ocean fluid leads to

$$p_b - p_a = g \int_{-H}^{\eta} \rho \, \mathrm{d}z, \qquad (126)$$

with p_b the hydrostatic pressure at the ocean bottom and p_a the pressure at the ocean surface applied from the overlying atmosphere or ice. Use of this result in the mass budget (30) then leads to

$$\partial_t (p_b - p_a) = -\nabla \cdot \mathbf{U}^p + q_w \rho_w. \qquad (127)$$

Assuming knowledge of the tendency for the applied surface pressure p_a, this budget is isomorphic to that for the Boussinesq surface height [equation (143)].

9.3. Oceanic Boussinesq Approximation

The Boussinesq approximation is an attempt to simplify the appearance of density in the ocean equations. In situ density in the large-scale ocean varies by a relatively small amount, with a 5% variation over the full ocean column at the upper end of the range, and most of this variation due to compressibility has no dynamical consequence. Furthermore, the dynamically relevent horizontal density variations, $\Delta\rho$, are on the order of 0.1%. Thus, it is justifiable to make approximations to the density in certain terms within the ocean equations, as discussed in this section.

There are two distinct steps to the Boussinesq approximation. We refer to these two steps in conjunction as the oceanic Boussinesq approximation. The first step of the Boussinesq approximation applies a linearization to the velocity equation (55) by removing the nonlinear product of density times velocity

$$[\partial_t + (f\hat{\mathbf{z}} + \boldsymbol{\omega})\wedge]\rho_o \mathbf{v} + \rho_o \nabla_z \mathcal{K} + \nabla_z p + \rho \nabla_z \Phi = \nabla \cdot \tau. \qquad (128)$$

To obtain this equation, the product $\rho\mathbf{v}$ was replaced by $\rho_o\mathbf{v}$, where ρ_o is a constant Boussinesq reference density.[14] Importantly, one retains the in situ density dependence of the gravitational potential energy, and, correspondingly, it is retained for computing the hydrostatic pressure. It is through pressure that variations in density create critical dynamical effects. It is notable that as shown by equation (174), hydrostatic non-Boussinesq models based on pressure as the vertical coordinate naturally eliminate the nonlinear product $\rho\mathbf{v}$, thus removing the need to make any approximations [Huang et al., 2001; DeSzoeke and Samelson, 2002; Marshall et al., 2004; Losch et al., 2004]. Interest in removing these nonlinear products arises in hydrostatic ocean modeling using depth-based vertical coordinates. The associated Boussinesq kinetic energy budget is given by

$$(\rho_o\mathcal{K})_{,t} + \nabla \cdot (\rho_o\mathcal{K}\mathbf{v} + p\mathbf{v} - \mathbf{v} \cdot \tau) = p\nabla \cdot \mathbf{v} - \rho\mathbf{v} \cdot \nabla\Phi - \rho_o\epsilon. \qquad (129)$$

The second step in the oceanic Boussinesq approximation considers the mass continuity equation (5) where it is noted that to leading order, the three-dimensional flow is incompressible

$$\nabla \cdot \mathbf{v} \approx 0. \tag{130}$$

It is this step in the approximation which filters acoustic modes, if they are not already filtered by the hydrostatic approximation (Section 8.1). More formally, the nearly incompressible observation manifests in the following scaling

$$\frac{d \ln \rho}{dt} \ll \nabla \cdot \mathbf{v}. \tag{131}$$

This scaling follows, as

$$d \ln \rho / dt \sim (U/L) \Delta \rho / \rho_o, \tag{132}$$

whereas each term of $\nabla \cdot \mathbf{v}$ scales as U/L or W/H. In this equation, (U,W) are horizontal and vertical velocity scales, and (L,H) are horizontal and vertical length scales. With $\nabla \cdot \mathbf{v} = 0$, mass conserving kinematics of the non-Boussinesq system are translated into volume conserving kinematics, in which case the mass of a parcel is approximated by $dM = \rho_o dV$, and the tracer mass in a parcel is approximated by $(\rho_o dV)C$.

There are some confusing points that arise when considering the Boussinesq approximation. Namely, volume conservation for a parcel, through the mass budget (2), means that the three-dimensional velocity field $\mathbf{v}$ is non-divergent. A non-divergent velocity field cannot support acoustic modes (Section 8.1), and this is useful for purposes of large-scale modeling. A non-divergent velocity field also cannot support material evolution of in situ density [equation (3)]. Furthermore, through equation (61), a non-divergent velocity only supports, in general, adiabatic and isohaline motions. These motions are of interest for ideal incompressible fluid mechanics. They are, however, insufficient for describing the ocean circulation where mixing and heating are critical.[15]

How does the oceanic Boussinesq approximation work in ocean models? The oceanic Boussinesq approximation assumes the resolved flow to be incompressible, in which case acoustic modes are not supported. This approximation furthermore retains the dependence of density on pressure, heating, and salinity mixing, thus avoiding any assumption regarding the fluid properties. To support a nontrivial material evolution of density, as arises through pressure changes, mixing, and heating, requires a divergent velocity field, which is unresolved in oceanic Boussinesq models: the effect of this divergent velocity field manifests through nontrivial density evolution.

To illustrate how a Boussinesq model can support nontrivial density evolution, write the velocity as the sum of divergent and non-divergent components

$$\mathbf{v} = \mathbf{v}_{nd} + \mathbf{v}_d. \tag{133}$$

The divergent velocity $\mathbf{v}_d$ is associated with the acoustic modes. Although we do not present a formal asymptotic analysis here, the acoustic fluctuations are of small amplitude and high frequency with respect to the oceanic flows of interest, which are embodied in $\mathbf{v}_{nd}$. That is,

$$|\mathbf{v}_d| \ll |\mathbf{v}_{nd}|. \tag{134}$$

By construction, the continuity equation can now be split into the following two parts:

$$\nabla \cdot \mathbf{v}_{nd} = 0 \tag{135}$$
$$\nabla \cdot \mathbf{v}_d = -d \ln \rho / dt. \tag{136}$$

Given the scaling noted above, the non-divergent velocity contributes to leading order in the material time derivative on the right-hand side of equation (137) so that

$$\nabla \cdot \mathbf{v}_d \approx -\frac{d^{nd} \ln \rho}{dt}, \tag{137}$$

where $d^{nd}/dt = \partial_t + \mathbf{v}^{nd} \cdot \nabla$. The divergent velocity is seen, through mass conservation (3), to support a nonzero material evolution of density. This evolution, through equation (61), is affected by pressure fluctuations, salinity mixing, and heating

$$\frac{1}{\rho c_s^2} \frac{dp}{dt} + \beta_S \frac{dS}{dt} - \alpha_\theta \frac{d\theta}{dt} = -\nabla \cdot \mathbf{v}_d. \tag{138}$$

The oceanic Boussinesq approximation considers the resolved prognostic velocity field to be the non-divergent velocity $\mathbf{v}_{nd}$, and this maintains an incompressible prognostic flow field that does not support acoustic modes. It is this velocity which is time stepped by using the Boussinesq momentum equation, and it is this velocity which transports tracer through advection. The divergent velocity $\mathbf{v}_d$ does not vanish, however, as each term on the left-hand side of equation (138) is generally nonzero for the oceanic Boussinesq approximation. Instead, its divergence can, in principle, be diagnosed by evaluating the terms in equation (138).[16] Again, it is the existence of $\mathbf{v}_d$ which allows the oceanic Boussinesq system to self-consistently employ a realistic equation of state in which density is a function of pressure, temperature, and salinity, thus supporting nonzero material time variations of the in situ density. These variations are critical for representing the thermohaline-induced variations in density which are key drivers of the large-scale ocean circulation.

9.3.1 Implications for gravitational potential energy. To obtain the gravitational potential energy equation, multiply the approximate mass budget in equation (137), involving the divergent velocity, by the geopotential Φ, to render the Boussinesq gravitational potential energy equation

$$(\rho\,\Phi)_{,t} + \nabla \cdot (\rho\,\Phi\,\mathbf{v}_{\mathrm{nd}}) = \rho\,(\partial_t + \mathbf{v}_{\mathrm{nd}} \cdot \nabla)\Phi - \Phi\rho\,\nabla \cdot \mathbf{v}_{\mathrm{d}}.$$

(139)

Hence, from the perspective of the Boussinesq ocean model, which time steps the non-divergent velocity $\mathbf{v}_{\mathrm{nd}}$, there is a new term affecting potential energy relative to the unapproximated budget (71). This term is given by

$$-\Phi\rho\,\nabla \cdot \mathbf{v}_{\mathrm{d}} = \Phi\frac{d\rho}{dt}$$

(140)

$$= \Phi\rho\left(\frac{1}{\rho\,c_s^2}\frac{dp}{dt} + \beta_S\frac{dS}{dt} - \alpha_\theta\frac{d\theta}{dt}\right),$$

(141)

where we used equations (137) and (138) to replace the divergence $\nabla \cdot \mathbf{v}_{\mathrm{d}}$ with the material changes in pressure, temperature, and salinity and where the material time derivative is taken with the resolved non-divergent velocity $\mathbf{v}_{\mathrm{nd}}$. This source is affected by fluctuations in the pressure field, heating, and salinity mixing. Importantly, these three processes are coupled, with heating and mixing, for example, affecting pressure and pressure affecting dynamics. In Boussinesq models that replace the pressure dependence of density with depth dependence, as in equation (60), the source takes the form

$$-\Phi\rho\,\nabla \cdot \mathbf{v}_{\mathrm{d}} = \Phi\rho\left(\frac{w}{\rho\,c_s^2}\frac{\partial\rho}{\partial p}\frac{\partial p_o}{\partial z} + \beta_S\frac{dS}{dt} - \alpha_\theta\frac{d\theta}{dt}\right).$$

(142)

Contrary to the more general form of equation (141), the pressure contribution is more readily diagnosed in an ocean model using this approximated equation of state. Further discussion of energetics of Boussinesq equations using the simpler equation of state can be found in *Vallis* [2006].

9.3.2 Implications for sea level height. We now ask how well the Boussinesq ocean model approximates the surface height relative to the non-Boussinesq model. The surface height in a Boussinesq ocean model satisfies the approximate balance of volume conservation for the column

$$\eta_{,t}^{\mathrm{Bouss}} = -\nabla \cdot \mathbf{U} + q_{\mathrm{w}}\rho_{\mathrm{w}}/\rho_o,$$

(143)

where $\mathbf{U} = \int_{-H}^{\eta} dz\,\mathbf{u}$ is the vertically integrated horizontal velocity. This equation approximates the more exact result for a mass conserving fluid which is obtained by vertically integrating the mass conservation equation (3) over a column of seawater, using the bottom kinematic boundary condition (24) and surface kinematic boundary condition (32), to find

$$\eta_{,t} = -\nabla \cdot \mathbf{U} + \frac{q_{\mathrm{w}}\rho_{\mathrm{w}}}{\rho(\eta)} - \int_{-H}^{\eta} dz\frac{d\ln\rho}{dt},$$

(144)

where $\rho(\eta)$ is the density at the ocean surface. The missing term in the Boussinesq surface height equation (143) arises from stretching and compressing a vertical column associated with changes in the ocean hydrography within a fluid column. The absence of this steric effect represents a limitation of Boussinesq ocean models for prognostically simulating, for example, effects of anthropogenic climate changes on sea level [*Greatbatch*, 1994; *Mellor and Ezer*, 1995].[17]

10. ELEMENTS OF VERTICAL COORDINATES

A key characteristic of rotating and stratified fluids, such as the ocean, is the dominance of lateral over vertical transport. Hence, it is traditional in ocean modeling to orient the two horizontal coordinates orthogonal to the local vertical direction as determined by gravity. The more difficult choice is how to specify the vertical coordinate and the associated transport across surfaces of constant vertical coordinate. Indeed, the choice of vertical coordinate is arguably the single most important aspect in the design of an ocean model. The main reason it is crucial is that practical issues of representation and parameterization are often directly linked to the vertical coordinate choice, and these issues enter at a level fundamental to developing the model algorithms.

10.1. Three Flow Regimes

Currently, there are three main vertical coordinates in use by ocean modelers, and they arose from applications focusing on complementary dynamical regimes. The following characterizes these regimes and provides a qualitative assessment of the abilities of the three coordinates. This assessment is subject to modifications due to algorithmic improvements, continually being developed, which push the envelope of applicability for the various vertical coordinates.

- *Upper ocean mixed layer.* This is a generally turbulent region dominated by transfers of momentum, heat, freshwater, and tracers with the overlying atmosphere, sea ice, rivers, etc. The mixed layer is of prime importance for climate system modeling and operational oceanography. It is typically very well mixed in the vertical through three-dimensional turbulent processes. These processes involve non-hydrostatic physics, which requires very fine horizontal and vertical resolution (i.e., a vertical-to-horizontal grid aspect ratio near unity) to explicitly represent. In this region, it is useful to employ a vertical co-ordinate that facilitates the representation and pa-rameterization of these highly turbulent processes. Geopotential and pressure coordinates, or their rela-tives, are the most commonly used coordinates, as they facilitate the use of very refined vertical grid spacing, which can be essential to simulate the strong exchanges between the ocean and atmosphere, rivers, and ice. These coordinates, in particular geopotential coordinates, have been the dominant choice of model-ers focusing on global climate.

- *Ocean interior*. Tracer transport processes in the ocean interior predominantly occur along neutral di-rections [*McDougall*, 1987]. The transport is domi-nated by large-scale currents and mesoscale eddy fluctuations. Water mass properties in the interior thus tend to be preserved over large space scales and timescales (e.g., basin and century scales). It is critical to represent this property of the ocean inte-rior in a numerical simulation of ocean climate. An isopycnal coordinate framework is well suited to this task, whereas geopotential and sigma models have problems associated with numerical truncation errors. As discussed by *Griffies et al.* [2000b], the problem can become more egregious as the model resolution is refined due to the enhanced levels of eddy activity that pumps tracer variance to the grid scale. Quasi-adiabatic dissipation of this variance is difficult to maintain in non-isopycnal models. We have more to say on this spurious mixing problem in Section 12.1.

- *Solid Earth boundary*. The solid Earth topography directly influences the ocean currents. In an un-stratified ocean, the balanced flow generally fol-lows lines of constant f/H, where f is the Coriolis parameter and H is the ocean depth. Additionally, there are several regions where density driven cur-rents (overflows) and turbulent bottom boundary layer processes act as strong determinants of wa-ter mass characteristics. Many such processes are crucial for the formation of deep water properties in the World Ocean and for representing coastal processes in regional models. It is for this reason that terrain following sigma models have been developed over the past few decades, with their

dominant application focused on the coastal and estuarine problem.

As reviewed by *Griffies et al.* [2000a], the geopotential, isopycnal, and sigma models each focus on one of the above regimes. Each do quite well within the confines of the sepa-rate regimes. It is in the overlap where problems arise. Be-cause the ocean involves all of the regimes, there remain problems applying one particular coordinate choice for simulating the global ocean climate system. It is not clear whether these problems are insurmoutable. Indeed, much progress continues to be made at addressing various weak-nesses. Nonetheless, the problems have motivated some ef-fort to develop generalized vertical coordinates,[18] whereby the model algorithms determine the vertical coordinate ac-cording to the physical flow regime, e.g., pressure near the surface, isopycnal in the interior, and terrain following next to the solid Earth [*Bleck*, 2002]. We have more to say on such approaches when discussing solution methods in Sec-tion 11.

10.2. Depth and Pressure Isomorphism

A natural set of vertical coordinates of use for describing Boussinesq ocean models is based on the depth, or geopo-tential, vertical coordinate, as depth measures the volume per area above a point in a fluid column. Depth-based ocean models are the oldest of those models used for studying cli-mate, with classical references for this first generation of ocean climate models being *Bryan and Cox* [1967]; *Bryan* [1969a, b]; *Bryan et al.* [1975]; *Bryan and Lewis* [1979]; *Cox* [1984]. The work of *Huang et al.* [2001], *DeSzoeke and Samelson* [2002], *Marshall et al.* [2004], *Losch et al.* [2004] highlights an isomorphism between depth-based Boussinesq mechanics and pressure-based non-Boussinesq mechanics (see Section 11.4 for details). This isomorphism has allowed for a straightforward evolution of depth-based models to the pressure-based models more commonly considered in recently developed ocean climate models. Pressure-based vertical coordinates are naturally used to describe non-Boussinesq hydrostatic fluids, as pressure in a hydrostatic fluid measures the mass per area above a point in a fluid column (Section 9.2).

10.3. Non-Orthogonality

The generalized vertical coordinates used in ocean mod-eling are not orthogonal, which contrasts with many other applications in mathematical physics.[19] Hence, it is useful to keep in mind the following properties that may seem odd on initial encounter.

The horizontal velocity in ocean models measures motions in the horizontal plane, perpendicular to the local gravitational field. That is, horizontal velocity is mathematically the same regardless the vertical coordinate, be it geopotential, isopycnal, pressure, or terrain following. The key motivation for maintaining the same horizontal velocity component is that the hydrostatic and geostrophic balances are dominant in the large-scale ocean. Use of an alternative quasi-horizontal velocity, for example one oriented parallel to the generalized surface, would lead to unacceptable numerical errors. Correspondingly, the vertical direction is anti-parallel to the gravitational force in all of the coordinate systems. We do not choose the alternative of a quasi-vertical direction oriented normal to the surface of a constant generalized vertical coordinate.

It is the method used to measure transport across the generalized vertical coordinate surfaces which differs between the vertical coordinate choices. That is, computation of the dia-surface velocity component detailed in Section 2.2 represents the fundamental distinction between the various coordinates. In some models, such as geopotential, pressure, and terrain following, this transport is typically diagnosed from volume or mass conservation. In other models, such as isopycnal layered models, this transport is prescribed based on assumptions about the physical processes producing a flux across the layer interfaces. We return to this key point in Section 11 when discussing solution methods.

11. SOLUTION METHODS

The purpose of this section is to introduce some of the steps needed to develop an algorithm for solving the ocean equations.

11.1. Finite Volumes

In formulating the budgets for an ocean model, it is typical to focus on mass, tracer, and momentum budgets for a finite domain or control volume, such as that of an ocean model grid cell. The budget for a grid cell is distinct from the budget for infinitesimal mass conserving Lagrangian fluid parcels moving with the fluid. Mass conserving fluid parcels form the fundamental system for which the budgets of mass, tracer, momentum, and energy are generally formulated from first principles. Grid cell budgets are then derived from the fundamental parcel budgets. Formulating budgets over finite-sized regions is an important first step toward developing a numerical algorithm. In particular, it is an essential step when working with a finite volume formulation [e.g., chapter 6 of *Hirsch*, 1988] such as with the MITgcm [*Marshall et al.*, 1997; *Adcroft et al.*, 1997].

The grid cells of concern for typical ocean models have vertical sides fixed in space–time, but with the top and bottom generally moving (Figure 6). In particular, the top and bottom either represent the ocean top, ocean bottom, or a surface of constant generalized vertical coordinate. As before, we assume that at no place in the fluid do the top or bottom surfaces of the grid cell become vertical. This assumption allows for a one-to-one relation to exist between geopotential depth z and a generalized vertical coordinate s (i.e., the relation is invertible).

To establish the grid cell budget, we integrate the budget for mass conserving fluid parcels over the cell volume. A first step is to take a differential relation of the form (7) and transform it to a finite domain relation by integrating over a region such as that for the grid cell shown in Figure 6. The following finite domain result follows by using standard vector calculus:

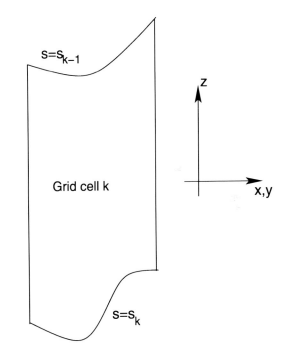

Figure 6. Schematic of an ocean grid cell labeled by the vertical integer k. Its sides are vertical and oriented according to $\hat{\mathbf{x}}$ and $\hat{\mathbf{y}}$, and its horizontal position is fixed in time. The top and bottom surfaces are determined by constant generalized vertical coordinates s_{k-1} and s_k, respectively. Furthermore, the top and bottom are assumed to always have an outward normal with a nonzero component in the vertical direction $\hat{\mathbf{z}}$. That is, the top and bottom are never vertical. We take the convention that the discrete vertical label k increases as moving downward in the column, and grid cell k is bounded at its upper face by $s = s_{k-1}$ and lower face by $s = s_k$.

$$\int \rho \frac{d\Psi}{dt} dV = \partial_t \left(\int \rho \Psi dV \right)$$
$$+ \int dA_{(\hat{n})} \hat{n} \cdot (\mathbf{v} - \mathbf{v}^{(ref)}) (\rho \Psi). \quad (145)$$

Hence, the mass weighted integral of the material time derivative of a field Ψ is given by the time derivative of the mass weighted field integrated over the domain, plus a boundary term that accounts for the transport across the domain boundaries, with allowance made for moving domain boundaries. Applying this result to the parcel tracer budget (33) leads to the finite domain tracer budget

$$\partial_t \left(\int C \rho dV \right) = \int S^{(C)} \rho dV$$
$$- \int dA_{(\hat{n})} \hat{n} \cdot [(\mathbf{v} - \mathbf{v}^{ref}) \rho C + \mathbf{J}]. \quad (146)$$

Again, the left-hand side of this equation is the time tendency for tracer mass within the finite-sized grid cell region. When the tracer concentration is uniform, the SGS flux and source vanish, in which case the tracer budget (146) reduces to the finite domain mass budget

$$\partial_t \left(\int \rho dV \right) = - \int dA_{(\hat{n})} \hat{n} \cdot [(\mathbf{v} - \mathbf{v}^{ref}) \rho]. \quad (147)$$

Further work leads to similar domain statements for the momentum budget.

11.2. Reynolds Averaging

The finite volume budgets provide a first step along a particular avenue toward discretizing the ocean equations. The next step explicitly considers the shape and size of the grid cells, and approximates these geometric details as well as fields within these cells, given information resolved by the discrete model. This step exposes our lack of information about the scales smaller than the grid scale, and in so doing, introduces the SGS parameterization problem.

From the finite volume perspective, the SGS parameterization problem arises when a particular form of the resolved model variables is assumed, e.g., the model variables represent a mass weighted average of the continuous variables such as

$$\Psi_{model} \equiv \frac{\int \Psi \rho dV}{\int \rho dV}. \quad (148)$$

The averaging described here is one form of a more generic Reynolds averaging procedure required to specify the ocean equations appropriate for the chosen discretization. Averaging of the form (148), or any other with more sophisticated weighting, introduces correlation terms between nonlinear products of SGS fields. A prescription for the correlations depends on the model grid and the unresolved physical processes. It also depends on whether the average is performed at a constant point in space (Eulerian average), on a moving surface such as an isopycnal (quasi-Lagrangian average), on a pressure surface, or another surface.

Regardless of the details of the Reynolds averaging, averaging over the subgrid scales appropriate for ocean models (e.g., scales smaller than 10 to 100 km) produces correlation terms that are many orders larger than the effects from molecular processes (e.g., molecular tracer diffusion and molecular friction). Hence, for all purposes of large-scale ocean modeling, the SGS flux $\mathbf{J}_c$ for tracer C is just that from Reynolds averaging, as is the momentum friction tensor τ.

Although we can, in principle, formulate a Reynolds averaging procedure for the ocean equations, there has not been a satisfying first principles closure for these equations relevant at the scale of global ocean models. Hence, the Reynolds averaged ocean equations are closed by introducing ad hoc steps that are unsatisfying both in principle and practice. We have more to say on this point regarding the lateral friction used in ocean models in Section 12.2.

11.3 General Comments on Solution Algorithms

The numerical procedures required to solve the ocean equations are dependent on details of the approximations or filters applied to the equations. For example, the non-Boussinesq and non-hydrostatic ocean equations,

$$[\partial_t + (2\Omega + \omega) \wedge] \mathbf{v} = -\nabla \mathcal{E} + \rho^{-1} \nabla \cdot (\tau - \mathbf{I} p) \quad (149)$$

$$\rho_{,t} + \nabla \cdot (\rho \mathbf{v}) = 0 \quad (150)$$

$$(\rho \theta)_{,t} + \nabla \cdot (\rho \mathbf{v} \theta + \mathbf{J}_\theta) = \rho S_\theta \quad (151)$$

$$(\rho S)_{,t} + \nabla \cdot (\rho \mathbf{v} S + \mathbf{J}_S) = \rho S_S \quad (152)$$

$$\rho = \rho(\theta, S, p) \quad (153)$$

permit acoustics modes (Section 8.1; recall that $\mathcal{E} = \mathcal{K} + \Phi$ is the total mechanical energy per mass of a fluid parcel). An algorithm to solve these equations would significantly differ from an algorithm developed to solve the oceanic Boussinesq equations

$$[\partial_t + (2\Omega + \omega)\wedge]\rho_o\mathbf{v} = -\rho_o\nabla\mathcal{K} - \rho\nabla\Phi + \nabla\cdot(\tau - \mathbf{I}p) \tag{154}$$

$$\nabla\cdot\mathbf{v} = 0 \tag{155}$$

$$\theta_{,t} + \nabla\cdot(\mathbf{v}\theta + \mathbf{F}_\theta) = S_\theta \tag{156}$$

$$S_{,t} + \nabla\cdot(\mathbf{v}S + \mathbf{F}_S) = S_S \tag{157}$$

$$\rho = \rho(\theta, S, p) \tag{158}$$

which do not support acoustic modes. The choice of vertical coordinate also has a critical impact on the solution algorithm. The major distinction here is between Eulerian and Lagrangian viewpoints [*Adcroft and Hallberg*, 2006]. We briefly describe the procedures resulting from these two viewpoints for hydrostatic models, which can be applied to both the Boussinesq and non-Boussinesq momentum equations.

11.4. Sample Eulerian Algorithms

We start our consideration of solution algorithms with the Boussinesq hydrostatic equations written in geopotential (or depth) vertical coordinates. We also consider the shallow ocean approximation (Section 9.1). The resulting Boussinesq hydrostatic primitive equations are given by

$$[\partial_t + (f\hat{\mathbf{z}} + \omega)\wedge]\mathbf{v} + \nabla_z\mathcal{E} + \rho_o^{-1}\nabla_z p = \rho_o^{-1}\nabla\cdot\tau \tag{159}$$

$$\Phi_{,z} + \rho^{-1}p_{,z} = 0 \tag{160}$$

$$\nabla_z\cdot\mathbf{u} + w_{,z} = 0 \tag{161}$$

$$\theta_{,t} + \nabla_z\cdot(\mathbf{u}\theta) + (w\theta)_{,z} = S_\theta \tag{162}$$

$$S_{,t} + \nabla_z\cdot(\mathbf{u}S) + (wS)_{,z} = S_s \tag{163}$$

$$\rho = \rho(\theta, S, p). \tag{164}$$

Notice that we allow for the general pressure dependence of the equation of state according to the discussion in Section 9.3. In these equations, ω, $\mathcal{E}$, and τ each assume their hydrostatic forms given in Section 9.2

$$\omega = \nabla\wedge\mathbf{u} \tag{165}$$

$$\mathcal{E} = \mathbf{u}\cdot\mathbf{u}/2 + \Phi \tag{166}$$

$$\nabla\cdot\tau = \partial_z(\kappa_{model}\mathbf{u}_{,z}) + \nabla\cdot\tau_{horizontal\ strain}. \tag{167}$$

Equations (159)–(164) are seven equations with seven unknowns (u, v, w, p, θ, S, ρ). There are four predictive or prognostic equations: two components of the horizontal velocity equation (159) and the potential temperature and salinity equations (162) and (163). There are three diagnostic equations: the hydrostatic balance (160), the Boussinesq continuity equation (161), and the equation of state (164). Predictive equations are used to update in time (i.e., predict) the corresponding model variable, while the remaining variables are determined by the diagnostic relations.

The system of equations (159)–(164) can be solved using explicit-in-time algorithms. For example, given an initial hydrographic specification of θ and S and p, we can proceed as follows:

1. The in situ density ρ can be diagnosed from the equation of state (164) using pressure at a lagged time step [e.g., *Dewar et al.*, 1998; *Griffies et al.*, 2001];
2. Hydrostatic pressure p can be diagnosed by vertical integration of the hydrostatic equation (160);
3. Horizontal velocity $\mathbf{u}$ can be predicted using the velocity equation (159);
4. The vertical velocity component w can be diagnosed by vertical integration of the continuity equation (161);
5. Tracers θ and S can be predicted using the tracer equations (162) and (163);
6. Returning to step 1 repeats the cycle for a subsequent time step.

This algorithm will work using fully explicit time-integration methods. However, the time step will be limited by the fastest mode, which, in this case, is the external gravity wave which can exceed speeds of 200 ms^{-1} in the deep ocean. As noted in Section 8.2, resolving external gravity waves in a global ocean model is prohibitively expensive. An alternative treatment of the fast external modes is discussed in Section 11.6.

The hydrostatic Boussinesq equations (159)–(164) permit four linear internal modes of variability: two inertia–gravity, one geostrophic, and one thermohaline mode.[20] Furthermore, the equations are written in a transparent form for developing solution algorithms, as there is no ambiguity about the way to use each equation, and there is only one obvious candidate for which predictive equation to use to predict a particular variable. As we now discuss, changing the nature of an equation within the system can potentially lead to inconsistencies. To illustrate this point, consider the non-Boussinesq and hydrostatic primitive equations written with a general vertical coordinate s:

$$[\partial_t + (f\,\hat{\mathbf{z}} + \boldsymbol{\omega})\wedge]\mathbf{v} + \nabla_s \mathcal{E} + \rho^{-1}\nabla_s p = \rho^{-1}\nabla\cdot\boldsymbol{\tau} \quad (168)$$

$$\partial_s \Phi + \rho^{-1}\partial_s p = 0 \quad (169)$$

$$(\rho\, z_{,s})_{,t} + \nabla_s\cdot(\rho\,\mathbf{u}\,z_{,s}) + (\rho\,w^{(s)})_{,s} = 0 \quad (170)$$

$$(\rho\, z_{,s}\,\theta)_{,t} + \nabla_s\cdot(\rho\,\mathbf{u}\,z_{,s}\,\theta) + (\rho\,w^{(s)}\,\theta)_{,s} = \rho\, z_{,s} S_\theta \quad (171)$$

$$(\rho\, z_{,s}\,S)_{,t} + \nabla_s\cdot(\rho\,\mathbf{u}\,z_{,s}\,S) + (\rho\,w^{(s)}\,S)_{,s} = \rho\, z_{,s} S_S \quad (172)$$

$$\rho = \rho(\theta, S, p) \quad (173)$$

where, again, $w^{(s)} = z_{,s}\,\mathrm{d}s/\mathrm{d}t$ is the dia-surface velocity component introduced in Section 2.2. Equations (168)–(173) also permit four linear internal modes of variability, but the equations now contain five time derivatives. Hence, there are now five predictive equations and only two diagnostic relations. Consider a situation where $z_{,s}$ is known or prescribed (for instance, with geopotential vertical coordinates $z_{,s} = 1$); there are then two possible equations that might be used to determine the in situ density ρ: either equation (170) or equation (173). It is not immediately obvious which is the proper equation to use. If we use the equation of state (173), then mass conservation (170) might be violated, but if we use mass conservation, then the equation of state may not be satisfied. A judicious choice of vertical coordinate removes this ambiguity. For example, setting $s = p$ and noting that $g\rho z_{,p} = -1$ brings equations (168)–(173) into the form

$$[\partial_t + (f\,\hat{\mathbf{z}} + \boldsymbol{\omega})\wedge]\mathbf{u} + \nabla_p \mathcal{E} = \rho^{-1}\nabla\cdot\boldsymbol{\tau} \quad (174)$$

$$\partial_p \Phi + \rho^{-1} = 0 \quad (175)$$

$$\nabla_p\cdot\mathbf{u} + \partial_p(\dot{p}) = 0 \quad (176)$$

$$\theta_{,t} + \nabla_p\cdot(\mathbf{u}\,\theta) + (\theta\,\dot{p})_{,p} = S_\theta \quad (177)$$

$$S_{,t} + \nabla_p\cdot(\mathbf{u}\,S) + (S\,\dot{p})_{,p} = S_S \quad (178)$$

$$\rho = \rho(\theta, s, p), \quad (179)$$

where $\dot{p} = \mathrm{d}p/\mathrm{d}t$ is the material time derivative of pressure. These equations are, term for term, isomorphic to the Boussinesq hydrostatic geopotential coordinate equations (159)–(164). This isomorphism, already mentioned in Section 10.2, has been exploited in meteorology for many years [e.g., *Haltiner and Williams*, 1980] and more recently has been brought into use for non-Boussinesq ocean models [*Huang et al.*, 2001; *DeSzoeke and Samelson*, 2002; *Marshall et al.*, 2004; *Losch et al.*, 2004]. A solution procedure for the compressible pressure-coordinate equations (174)–(179) is identical to that for the incompressible equations in height coordinates (159)–(164), except solving for Φ instead of p

and without the need to temporally lag pressure to evaluate the equation of state.

Returning to the general coordinate non-Boussinesq ocean equations (168)–(173), we note that there are two distinct approaches to solving these equations. The approaches are distinguishable by their treatment of the continuity equation (170). The first method adopts an Eulerian perspective, used in the preceding algorithms, where continuity is used to diagnostically determine transport across coordinate surfaces (vertically integrating equation (170) for the dia-surface velocity component $w^{(s)}$). To achieve this, the time derivative $(\rho z_{,s})_{,t}$ must be prescribed, or more specifically, $\rho z_{,s}$ must be functionally related to other dependent or independent model variables. This is the case for the terrain-following coordinates commonly referred to as σ-coordinates of which a simple example is the *Phillips* [1957] sigma-coordinate $s = \sigma = (p - p_a)/(p_b - p_a)$, where p_b is the pressure at the solid-Earth boundary, and p_a is the pressure applied at the top of the water column. For this vertical coordinate, the factor $\rho z_{,s} = -(p_b - p_a)/g$ is minus the mass per area of the water column, and this two dimensional field is predicted by the external mode equations (Section 11.6).

11.5. Sample Lagrangian Algorithms

The second approach to solving the non-Boussinesq ocean equations (168)–(173) adopts the Lagrangian perspective, common to most layer (isopycnal or stacked shallow water) models. For this approach, the cross-coordinate flow, $w^{(s)}$, is prescribed. We focus here on isopyncal models, in which case $w^{(s)}$ is set to zero in the adiabatic limit. The continuity equation (170) is then used prognostically to predict the layer mass per area

$$\rho\,\mathrm{d}z = \rho\, z_{,s}\,\mathrm{d}s. \quad (180)$$

Most isopycnal algorithms [e.g., *Bleck et al.*, 1992] assume the potential density of each layer to be constant in time and space. With zero dia-surface flow, the continuity equation (170) leads to

$$(\rho\, z_{,s})_{,t} + \nabla_s\cdot(\rho\,\mathbf{u}\,z_{,s}) = 0, \quad (181)$$

which predicts layer mass per area. A difficulty with this approach is that θ and S are predicted independently. Hence, there is no guarantee that the diagnosed potential density $\rho_{ref} = \rho(\theta, S, p_{ref})$ will correspond to the assumed potential density of the layer. There are various approaches to correcting for this evolution of layer density, such as wrapping the layer remapping together with dianeutral processes like cabelling, as proposed by *Oberhuber* [1993], *McDougall and Dewar*

[1997], *Hallberg* [2000]. We have more to say on this point in Section 12.1.

A more general approach has been proposed [*Hirt et al.*, 1974; *Bleck*, 2002] in which remapping is used to reallocate mass so as to bring the layer densities back to their targets. This extra step is the foundation of the arbitrary Lagrangian–Eulerian method (ALE), which allows for the remapping step to remap to any coordinate, not necessarily isopycnal. The ALE method has become the preferred approach for general coordinate modeling because it is quite flexible. The ALE-based procedure for solving the general-coordinate hydrostatic non-Boussinesq equations is generally as follows:

1. Given the mass per area $\rho z_{,s} ds$ of a layer, pressure can be found by vertically integrating the hydrostatic equation (169).
2. Given the pressure p, potential temperature θ, salinity S, the in situ density can be calculated from the equation of state (173).
3. The horizontal velocity **u** can be predicted using the velocity equation (168).
4. The mass thickness $\rho z_{,s} ds$, θ, and S can be predicted, assuming the generalized s surfaces are material ($\dot{s} = 0$), by using equations (170)–(172).
5. A remapping step redistributes mass, θ, and S so as to bring the surfaces of s to the desired position. This step introduces irreversible numerical mixing processes (Section 12.1), although it is sometimes hidden within the operators describing physical mixing processes [*Hallberg*, 2000].

11.6. Solving for the External Mode

The ocean external modes are significantly faster than the internal modes. Treating the full three-dimensional system explicitly with a time step short enough to resolve external motions would be prohibitively expensive. There are three methods that have been used for avoiding this limitation: the rigid lid method, the split-explicit method, and the time-implicit method.

Ocean model algorithms approximate the external mode by using solutions to the vertically integrated continuity and momentum equations. These equations are coordinate-independent and take the form

$$\partial_t \mathbf{U}^\rho + g(H + \eta)\nabla p_b + \mathbf{F}(\mathbf{U}^\rho) = \mathbf{S} \tag{182}$$

$$\partial_t p_b + \nabla \cdot \mathbf{U}^\rho = q_w \rho_w + \partial_t p_a \tag{183}$$

where $\mathbf{U}^\rho = \int_{-H}^{\eta} \rho \mathbf{u} dz$ [equation (31)] was introduced when discussing mass conservation for a vertical column. We have grouped various terms into fast (**F**) and slow (**S**) vectors. The choice of which terms are placed in the fast vector (**F**) is a matter of taste, and it varies from model to model. For the present discussion, we ignore these terms, yet note that many models include the Coriolis force in this vector, whereas others also include horizontal friction or bottom drag. The linear forms of equations (182) and (183) describe plane waves with phase and group speed of $c_g \sim \sqrt{gH}$. These are the external gravity waves discussed in Section 8.2. The essential role of these equations is to allow the external mode to adjust the system to an imbalance of divergent mass transport. This adjustment occurs on timescales that are short compared to the baroclinic evolution of the model.

The first generation of ocean climate models, based on the algorithm of *Bryan* [1969a], employed the rigid lid approximation with the oceanic Boussinesq approximation. This approach sets the time tendency of the surface height in equation (143) to zero and drops the surface water fluxes. Consequently, the vertically integrated horizontal velocity is non-divergent, thus allowing for the introduction of a streamfunction for the vertically integrated velocity in the solution algorithm. This approach was revolutionary for its time, as it facilitated efficient time stepping, whereby fast barotropic gravity waves are absent from the algorithm. Nonetheless, it is physically unsatisfying due to the inability to represent tides and through the lack of a direct water forcing [*Huang*, 1993]. Additionally, it is computationally awkward, as the resulting elliptic problem is very difficult to solve accurately in realistic ocean geometries. Hence, practitioners often halted the elliptic solver searches after a maximum number of steps regardless of the remaining distance to convergence.

The split-explicit approach solves the depth averaged equations explicitly with a short external mode time step. In contrast, the baroclinic portion of the model is time stepped with a larger time step determined by the slower baroclinic processes. Investigations have revealed that the split-explicit methods, such as those from *Blumberg and Mellor* [1987], *Bleck and Smith* [1990], *Killworth et al.* [1991], *Dukowicz and Smith* [1994], *Griffies et al.* [2001], can be just as efficient computationally as the rigid lid method, yet without sacrificing tides, direct surface water forcing, or compromising the realism of the ocean geometry. Hence, the rigid lid methods are largely obsolete in the more recent (e.g., post-2000) models used for global climate.

There are many variants on the implementation details of the split-explicit approach, with essentially two broad strategies taken. In Eulerian models, the flow is partitioned as

$$\mathbf{u} = \hat{\mathbf{u}} + \bar{\mathbf{u}}, \qquad (184)$$

with the barotropic component determined by the following depth and mass-weighted average:

$$\bar{\mathbf{u}} = \frac{\mathbf{U}^\rho}{\int_{-H}^{\eta} \rho \, dz}. \qquad (185)$$

The baraclinic component is the remainder: $\hat{\mathbf{u}} = \mathbf{u} - \bar{\mathbf{u}}$. The full momentum equations may be rewritten to predict only the baroclinic component in a manner analogous to the rigid lid approach of *Bryan* [1969a; e.g., *Blumberg and Mellor*, 1987; *Killworth et al.*, 1991; *Griffies et al.*, 2001]. Alternatively, the full flow may be corrected by replacing the barotropic component with the results of the barotropic solver. In Lagrangian vertical coordinate models, the nonlinearity of the continuity equation and, in particular, the nonlinearity of the advection schemes used to guarantee positive definiteness of thickness, make a decomposition of the flow according to equation (184) troublesome. The alternative approach used is to make adjustments to either the layer transports or the barotropic transports to make them consistent [e.g., *Bleck and Smith*, 1990]. Earlier models allowed these two estimates of barotropic transport to remain inconsistent and used weak coupling to drive one toward the other [*Hallberg*, 1997; *Higdon*, 2005].

The time-implicit treatment of the external mode requires that the fast barotropic terms, principally the pressure gradient force due to the bottom pressure and the depth-integrated mass divergence, to be evaluated in the future part of the time step (i.e., solved implicitly). The analogous depth average equations can be rearranged into a two-dimensional elliptic equation which closely resembles the wave equation in structure. This wave equation takes the form

$$\left[\nabla \cdot ((H + \eta)\nabla) - \frac{\Gamma}{\Delta t^2} \right] p_b^{n+1} = \text{r.h.s.} \qquad (186)$$

where the details of the right-hand side terms (r.h.s.) and Γ are dependent on the choice of time discretization. Notably, this elliptic operator is better conditioned than that arising in the rigid lid approach [*Dukowicz and Smith*, 1994], and so convergence is more rapid. Such is fortunate, as this elliptic equation must be solved fairly accurately to ensure that the residual mass divergence does not grow during the integration of the baroclinic model. The elliptic equation should strictly be nonlinear. However, it is typically linearized by lagging the nonlinear terms in time. For deep ocean calculations, this linearization is justifiable, but it becomes less appropriate in shallow regions. The time-implicit approach can

recover the rigid lid approximation by dropping the Helmholtz term ($\Gamma/\Delta t^2 \to 0$). The physical interpretation of this limit is that we are finding p_b so as to adjust the barotropic flow to be exactly non-divergent.

11.7. Non-Hydrostatic Methods

Non-hydrostatic effects are significant only for large aspect ratio flows (e.g., Kelvin–Helmholtz instability) but can lead to systematic differences at small aspect ratios. Non-hydrostatic global models generally are too expensive to be in routine use. However, as spatial resolution is refined and the sub-mesoscale is resolved, non-hydrostatic effects need to be included. Relaxing the hydrostatic approximation in the non-Boussinesq equations permits acoustic modes, and so most non-hydrostatic models of the ocean assume the oceanic Boussinesq approximation. The methods of *McDougall et al.* [2003] might well be able to incorporate quasi non-Boussinesq effects in non-hydrostatic models.

The non-hydrostatic algorithms used in ocean models use derivatives of the projection method [*Chorin*, 1968]. Here, the problem is posed as follows: what (non-hydrostatic) pressure gradients are required to correct the flow to make the flow exactly non-divergent? By summarizing the full three-dimensional momentum equations and continuity as

$$\rho_o \mathbf{v}^{n+1} + \Delta t \nabla p = \rho_o \mathbf{v}^n + \Delta t \, G \qquad (187)$$

$$\nabla \cdot \mathbf{v}^{n+1} = 0 \qquad (188)$$

where $\mathbf{G}$ is the vector of all explicit forces, a rearrangement and substitution of variables yields the three-dimensional elliptic equation

$$\nabla \cdot \nabla p = \nabla \cdot \left(\frac{\rho_o}{\Delta t} \mathbf{v}^n \right) + \nabla \cdot \mathbf{G}. \qquad (189)$$

The first term on the right-hand side is important for implementation; otherwise, a residual divergence in the system may accumulate and lead to numerical instability. The three-dimensional elliptic equation for pressure replaces the vertical integration for hydrostatic pressure in the hydrostatic algorithms discussed in Section 9.2.

In the atmosphere, non-hydrostatic models do not make the Boussinesq approximation, and such models therefore permit acoustic modes. However, atmospheric acoustic waves are slow enough to be resolvable explicitly in the horizontal directions, whereas an implicit treatment in the vertical direction is sufficient to render a stable algorithm. Thus, non-hydrostatic atmospheric models are essentially hyperbolic, while oceanic non-hydrostaic models are elliptic in pressure.

12. SUMMARY AND OUTSTANDING PROBLEMS

We have aimed in this chapter to present a compendium of the scientific rationale for the equations used in physical ocean models. This rationale is independent of model resolution, and so provides a basis for both coarse-resolution global models as well as very fine-resolution regional and coastal models. Differences in applications largely reflect on the approximations made to the equations as well as choices for SGS parameterizations.

There remain many outstanding problems with ocean models. A large number can be alleviated by refining the resolution, thus allowing the simulation to rely less on the often ad hoc SGS parameterizations and depending more on resolved dynamical features. This reliance on enhanced resolution has a cost in computational expense. It also is limited by numerical algorithmic integrity. Namely, for enhanced resolution to accurately capture the dynamics, the numerical methods must respect the dynamics. This statement is relevant at all resolutions. However, large levels of viscous dissipation generally used at coarse resolution can sometimes hide problems revealed only upon refining the grid and reducing viscosity. Our goal in this final section is to discuss two issues which present a limitation on ocean simulations and which do not appear to be remedied upon refining the resolution: (a) the spurious mixing problem and (b) the specification of horizontal friction.

12.1. The Spurious Mixing Problem

Direct measurements of tracer diffusivity in the ocean were pioneered with the purposeful passive tracer release experiments of *Ledwell et al.* [1993]. These measurements indicate that on the large scales (on the order of hundreds of kilometers), the associated neutral to dianeutral anisotropy in mixing can be as high as 10^8 in the ocean interior, with smaller anisotropies in regions of strong dianeutral mixing such as within boundary layers or above rough topography. Another method for determining the dianeutral diffusivity uses the indirect approach suggested by *Osborn* [1980] and reviewed by *Gregg* [1987] and *Davis* [1994]. In this indirect approach, momentum dissipation at small scales is directly measured, and dianeutral diffusivity is inferred based on a theoretical connection between buoyancy mixing and momentum dissipation. These microstructure techniques likewise indicate that the levels of interior dissipation are very small, in general agreement with the direct tracer release measurements. They each indicate that the level of interior dianeutral mixing corresponds to a diffusivity on the order 10^{-5} m^{2}s^{-1}, with 10 to 100 times smaller values suggested at the equator by measurements from *Gregg et al.* [2003]. This is a strong statement regarding the level to which the ocean interior respects the neutral orientation of transport. Ocean models, especially those used for purposes of climate simulations, must respect this anisotropy.

For a numerical simulation to respect these small levels of mixing requires a tremendous level of integrity for the tracer transport algorithms. There are two main approaches that modelers have taken in this regard. First, modelers intent on respecting this level of mixing have tended to work with an isopycnal vertical coordinate. The advantage isopycnal models have over alternative vertical coordinates is that their advective transport operator is the sum of a two-dimensional lateral operator which acts independently of the dia-surface transport operator, with the dia-surface transport set, ideally, according to physical processes leading to mixing between density classes. Although suitable for the ideal adiabatic simulations, this is not sufficient for the real ocean. Here, a nonlinear equation of state introduces new physical sources of mixing, and the independent transport of two active tracers (temperature and salinity) requires remapping algorithms to retain fields within pre-specified density classes. These details introduce levels of dianeutral mixing which have yet to be systematically documented. They have nonetheless generally been considered negligible, with remapping often wrapped together with diapycnal processes [*Oberhuber*, 1993; *McDougall and Dewar*, 1997; *Hallberg*, 2000], thus using physical mixing processes in lieu of numerical mixing.

The second approach focuses on improving numerical transport methods. As discussed in *Griffies et al.* [2000b], the difficulties of maintaining small levels of spurious mixing are enhanced when moving to an eddying regime where the quasi-geostrophic cascade pumps tracer variance toward the grid scale. Dissipating this variance is required to damp unphysical grid scale features. Dissipation methods include the addition of an operator acting on the small scales (e.g., Laplacian or biharmonic), or dissipation inherent in the advection scheme (e.g., odd order schemes and/or flux limiters). However, most dissipation methods remain ignorant of the constraints based on spurious dianeutral transport. Notable exceptions include the adiabatic dissipation arising from the Laplacian operator of *Gent and McWilliams* [1990] and *Gent et al.* [1995], which attempts to parameterize physical processes, or the operators of *Smith* and *Gent* [2004] and *Roberts and Marshall* [1998], which are motivated from numerical considerations. For advection operators and dissipation operators, the question remains whether they can be constructed so that in practice, their numerical truncation errors are comparable with, or ideally less than, the tiny levels of physical mixing seen in the ocean interior. This problem remains at the forefront of ocean modeling practice, especially as eddying simulations for global climate become the norm.

12.2. Frictional Stresses in Ocean Models

As discussed in Section 8.4, modelers generally must set the strength of lateral fricional stresses to satisfy requirements of certain numerical constraints.[21] For example, in the presence of solid-Earth boundaries, friction must be sufficient to maintain a nontrivial side boundary layer. Even in the absence of boundaries, horizontal friction must be sufficient to maintain a finite grid Reynolds number $U\Delta/\nu_{model}$, with ν_{model} the model horizontal viscosity. Otherwise, the simulation may go unstable, or at best, it will produce unphysical noise-like features. This constraint on the numerical simulation is unfortunate, as it generally requires a model viscosity many orders of magnitude larger than the molecular viscosity relevant for the ocean. The reason is that model grid sizes, even in mesoscale eddy-permitting simulations, are much larger than the Kolmogorov scale ($\approx 10^{-3}$m) where molecular friction acts.

Various methods have been engineered to employ the minimal level of horizontal friction required to meet the numerical constraints [e.g., *Griffies and Hallberg*, 2000; *Large et al.*, 2001; *Smith and McWilliams*, 2003]. Notably, modelers generally assume that the frictional stresses are isotropic in the horizontal direction, with anisotropy only between the horizontal and vertical stresses. However, as noted by *Large et al.* [2001] and *Smith and McWilliams* [2003], we may choose to allow one more degree of freedom by breaking horizontal isotropy. Doing so provides a practical avenue toward reducing the overall dissipation, and it can have a nontrivial advantage for simulating certain features such as the equatorial currents.

Quite generally, methods for selecting model horizontal friction are ad hoc. They furthermore lead to some of the most unsatisfying elements in ocean model practice, as details of friction can strongly influence the simulation. Unfortunately, it appears that this sensitivity remains as resolution is refined [*Chassignet and Garraffo*, 2001]. One compelling approach to resolve this problem with ad hoc friction operators is to remove horizontal friction operators from the models altogether. In their place, one allows dissipation to occur within the momentum transport operators. This approach, formally termed implicit large eddy simulation, holds some promise [*Margolin et al.*, 2006]. It is analogous to the trend for handling the tracer equation in eddying simulations, whereby lateral SGS operators are removed or rendered far subdominant to the resolved advection process. The hope is that numerical methods for the resolved transport can be designed that are smarter and more robust than the suite of SGS operators engineered thus far.

Acknowledgments. We thank Matthew Hecht and Hiroyasu Hasumi for inviting us to write this chapter and for editing this volume. We also thank Riccardo Farnetti, Bob Hallberg, Hyun-Chul Lee, and Geoff Vallis for comments on earlier drafts.

Notes

1. The three dimensional velocity vector is written $\mathbf{v} = (\mathbf{u}, w)$, with $\mathbf{u} = (u, v)$ the horizontal components and w the vertical component.
2. This notation is standard in theoretical physics. It serves our purposes by distinguishing between a partial derivative and a tensor label.
3. Water crossing the ocean surface is typically quite fresh, such as for precipitation or evaporation. However, rivers and ice melt can generally contain a nonzero salinity.
4. We assume there to be no mass flux through the solid Earth boundary.
5. Electromagnetic effects are generally ignored in physical ocean models.
6. Note that $Wkg^{-1} = m^2 s^{-3}$.
7. For spherical coordinates, r is the radial position, λ is the longitude, and ϕ is the latitude.
8. We consider seawater to be a binary system of fresh water and salt. The factor of 1000 accounts for the use of salinity in parts per thousand rather than salt concentration.
9. This meaning for isentropic ocean models is consistent with the models not including frictional heating.
10. Note that when $\epsilon_{nh} = 0$, we require the buoyancy force to recover hydrostatic balance. We ignore this force here to focus on the linear dynamical modes arising without buoyancy.
11. As noted by *Griffies et al.* [2000b], it is prudent to admit at least two grid points in a Munk boundary layer to minimize spurious levels of mixing associated with advection truncation errors.
12. This approximation is distinct from the shallow water approximation.
13. The quasi-hydrostatic approximation discussed in *Marshall et al.* [1997] keeps the full Coriolis terms, but still integrates a balance equation for pressure as in Section 11.4.
14. Some ocean models choose the Boussinesq density to be $\rho_o = 1000$kg m^{-3} [e.g., *Cox*, 1984], which is roughly the density of freshwater at standard conditions, whereas others [e.g., *Griffies et al.*, 2004] choose $\rho_0 = 1035$kg m^{-3}, which is roughly the mean density of seawater in the World Ocean [page 47 of *Gill*, 1982].
15. See *Veronis* [1973] for a thorough discussion, with critique, of the Boussinesq approximation in which salinity, temperature, and potential density are materially conserved.
16. The existence of a nonzero $\mathbf{v}_d$ in the oceanic Boussinesq approximation is analogous to the presence of a nonzero ageostrophic flow in quasi-geostrophic models in which the ageostrophic flow is not directly computed but can be diagnosed.
17. For diagnostic purposes in a non-Boussinesq model, it is of interest to determine the local effects from column stretching on sea level. The material time derivative in equation (144) is difficult to diagnose. Hence, we offer the alternative expression, derived from the column integrated mass budget (30): $\eta_{,t} = (1/\overline{\rho})(-\nabla \cdot \mathbf{U}^\rho + q_w\rho_w - D\partial_t \ln\overline{\rho})$, with $\overline{\rho} = (H + \eta)^{-1} \int_{-H}^{\eta} dz\rho$ the vertically averaged density in a column. It is now clear that $-D\partial_t \ln \overline{\rho}$ represents a positive contribution to the surface height when the vertically averaged in situ density within a column decreases.

18. So far as we know, generalized vertical coordinates are synonymous with hybrid vertical coordinates.

19. It is nonetheless notable that a similar set of generalized vertical coordinates has found use in other areas of theoretical physics. In particular, condensed matter physicists and biophysicists studying the dynamics of fluctuating membranes use these coordinates, where the coordinates go by the name Monge gauge. Their mathematical aspects are lucidly described in Section 10.4 of *Chaikin and Lubensky* [1995].

20. See *Müller and J. Willebrand* [1986] for a discussion of a thermohaline mode arising from the presence of two tracers, temperature and salinity, affecting in situ density.

21. There is a notable exception to the discussion here, where *Holloway* [1992] argues that unresolved stresses associated with interactions between the ocean and the solid-Earth boundary impart a net momentum to the fluid.

REFERENCES

Adcroft, A., and R. W. Hallberg (2006), On methods for solving the oceanic equations of motion in generalized vertical coordinates, *Ocean Model.*, *11*, 224–233.

Adcroft, A., C. Hill, and J. Marshall (1997), Representation of topography by shaved cells in a height coordinate ocean model, *Mon. Weather Rev.*, *125*, 2293–2315.

Adcroft, A., C. Hill, and J. Marshall (1999), A new treatment of the Coriolis terms in c-grid models at both high and low resolutions, *Mon. Weather Rev.*, *127*, 1928–1936.

Apel, J. R. (1987), *Principles of Ocean Physics*, International Geophysics Series, vol. 38, Academic Press, London.

Arakawa, A. (1966), Computational design for long-term numerical integration of the equations of fluid motion: Two-dimensional incompressible flow. Part 1, *J. Comput. Phys.*, *1*, 119–143.

Arakawa, A., and Y.-J. Hsu (1990), Energy conserving and potential enstrophy dissipating schemes for the shallow water equations, *Mon. Weather Rev.*, *118*, 1960–1969.

Arakawa, A., and V. Lamb (1981), A potential enstrophy and energy conserving scheme for the shallow water equations, *Mon. Weather Rev.*, *109*, 18–36.

Aris, R. (1962), *Vectors, Tensors and the Basic Equations of Fluid Mechanics*, Dover Publishing, New York.

Bacon, S., and N. P. Fofonoff (1996), Oceanic heat flux calculation, *J. Atmos. Ocean. Technol.*, *13*, 1327–1329.

Batchelor, G. K. (1967), *An Introduction to Fluid Dynamics*, Cambridge University Press, Cambridge, England.

Bleck, R. (2002), An oceanic general circulation model framed in hybrid isopycnic-Cartesian coordinates, *Ocean Model.*, *4*, 55–88.

Bleck, R., and D. B. Boudra (1981), Initial testing of a numerical ocean circulation model using a hybrid- (quasi-isopycnic) vertical coordinate, *J. Phys. Oceanogr.*, *11*, 755–770.

Bleck, R., and L. Smith (1990), A wind-driven isopycnic coordinate model of the north and equatorial Atlantic Ocean. 1. Model development and supporting experiments, *J. Geophys. Res.*, *95*(C3), 3273–3285.

Bleck, R., C. Rooth, D. Hu, and L. T. Smith (1992), Salinity-driven thermocline transients in a wind and thermohaline forced isopycnic coordinate model of the North Atlantic, *J. Phys. Oceanogr.*, *22*, 1486–1505.

Blumberg, A. F., and G. L. Mellor (1987), A description of a three-dimensional coastal ocean circulation model, in *Three-Dimensional Coastal Ocean Models, Coastal and Estuarine Series*, vol. 4, edited by N. Heaps, AGU, Washington, DC.

Bokhove, O. (2000), On hydrostatic flows in isentropic coordinates, *J. Fluid Mech.*, *402*, 291–310.

Browning, G., W. Holland, H. Kreiss, and S. Worley (1990), An accurate hyperbolic system for approximately hydrostatic and incompressible oceanographic flows, *Dyn. Atmos. Oceans*, *14*, 303–332.

Bryan, K. (1969a), A numerical method for the study of the circulation of the world ocean, *J. Comput. Phys.*, *4*, 347–376.

Bryan, K. (1969b), Climate and the ocean circulation III: The ocean model, *Mon. Weather Rev.*, *97*, 806–824.

Bryan, K., and M. D. Cox (1967), A numerical investigation of the oceanic general circulation, *Tellus, XIX*, 54–80.

Bryan, K., and M. D. Cox (1972), An approximate equation of state for numerical models of the ocean circulation, *J. Phys. Oceanogr.*, *4*, 510–514.

Bryan, K., and L. Lewis (1979), A water mass model of the world ocean, *J. Geophys. Res.*, *84*(C5), 2503–2517.

Bryan, K., S. Manabe, and R. C. Pacanowski (1975), A global ocean atmosphere climate model. Part II. The oceanic circulation, *J. Phys. Oceanogr.*, *5*, 30–46.

Campin, J.-M., A. Adcroft, C. Hill, and J. Marshall (2004), Conservation of properties in a free-surface model, *Ocean Model.*, *6*, 221–244.

Chaikin, P. M., and T. C. Lubensky (1995), *Principles of Condensed Matter Physics*, Cambridge University Press, Cambridge, United Kingdom.

Chassignet, E. P., and Z. Garraffo (2001), Viscosity parameterization and the Gulf Stream separation, in *From Stirring to Mixing in a Stratified Ocean*, edited by P. Müller and D. Henderson, Proceedings of the 12th 'Aha Huliko'a Hawaiian Winter Workshop, pp. 37–41, University of Hawaii at Manoa.

Chorin, A. (1968), Numerical solution of the Navier–Stokes equations, *Mathematical Computations*, 22, 745–762.

Cox, M. D. (1984), *A Primitive Equation, 3-Dimensional Model of the Ocean*, NOAA/Geophysical Fluid Dynamics Laboratory, Princeton, USA.

Cushman-Roisin, B. (1994), *Introduction to Geophysical Fluid Dynamics*, Prentice-Hall, Englewood Cliffs, USA, 320 + xv pp.

Davis, R. E. (1994), Diapycnal mixing in the ocean: equations for large-scale budgets, *J. Phys. Oceanogr.*, *24*, 777–800.

DeGroot, S. R., and P. Mazur (1984), *Non-Equilibrium Thermodynamics*, Dover Publications, New York, 510.

DeSzoeke, R. A., and R. M. Samelson (2002), The duality between the Boussinesq and non-Boussinesq hydrostatic equations of motion, *J. Phys. Oceanogr.*, *32*, 2194–2203.

Dewar, W. K., Y. Hsueh, T. J. McDougall, and D. Yuan (1998), Calculation of pressure in ocean simulations, *J. Phys. Oceanogr.*, *28*, 577–588.

Dukowicz, J., and R. Smith (1994), Implicit free-surface method for the Bryan–Cox–Semtner ocean model, *J. Geophys. Res.*, *99*(C4), 7991–8014.

Durran, D. R. (1999), *Numerical Methods for Wave Equations in Geophysical Fluid Dynamics*, Springer, Berlin, 470 pp.

Ertel, H. (1942), Ein neuer hydrodynamicscher wirbelsatz, *Meteorol. Z.*, *59*, 271–281.

Feistel, R. (1993), Equilibrium thermodynamics of seawater revisited, *Progr. Oceanogr.*, *31*, 101–179.

Feistel, R. (2003), A new extended Gibbs thermodynamic potential of seawater, *Progr. Oceanogr.*, *58*, 43–114.

Feistel, R., and E. Hagen (1995), On the Gibbs thermodynamic potential of seawater, *Progr. Oceanogr.*, *36*, 249–327.

Fofonoff, N. P. (1962), Physical properties of seawater, in *The Sea*, vol. 1, edited by M. N. Hill, pp. 3–30, Wiley-Interscience, New York.

Fox-Rabinovitz, M. (1991), Computational dispersion of horizontal staggered grids for atmospheric and ocean models, *Mon. Weather Rev.*, *119*, 1624–1639.

Gent, P. R., and J. C. McWilliams (1990), Isopycnal mixing in ocean circulation models, *J. Phys. Oceanogr.*, *20*, 150–155.

Gent, P. R., J. Willebrand, T. J. McDougall, and J. C. McWilliams (1995), Parameterizing eddy-induced tracer transports in ocean circulation models, *J. Phys. Oceanogr.*, *25*, 463–474.

Gill, A. (1982), *Atmosphere-Ocean Dynamics, International Geophysics Series*, vol. 30, Academic, London, 662 + xv pp.

Greatbatch, R. J. (1994), A note on the representation of steric sea level in models that conserve volume rather than mass, *J. Geophys. Res.*, *99*(C6), 12,767–12,771.

Greatbatch, R. J., Y. Lu, and Y. Cai (2001), Relaxing the Boussinesq approximation in ocean circulation models, *J. Atmos. Ocean. Technol.*, *18*, 1911–1923.

Gregg, M., T. Sanford, and D. Winkel (2003), Reduced mixing from the breaking of internal waves in equatorial waters, *Nature*, *422*, 513–515.

Gregg, M. C. (1984), Entropy generation in the ocean by small-scale mixing, *J. Phys. Oceanogr.*, *14*, 688–711.

Gregg, M. (1987), Diapycnal mixing in the thermocline: A review, *J. Geophys. Res.*, *92*(C5), 5249–5286.

Griffies, S. M. (2004), *Fundamentals of Ocean Climate Models*, Princeton University Press, Princeton, USA, 518 + xxxiv pages.

Griffies, S. M. (2005), Some ocean model fundamentals, in *Ocean Weather Forecasting: an Integrated View of Oceanography*, vol. 577, edited by E. P. Chassignet, and J. Verron, pp. 19–73, Springer, Berlin.

Griffies, S. M. (2007), *Elements of mom4p1*, NOAA/Geophysical Fluid Dynamics Laboratory, Princeton, USA, 346 pp.

Griffies, S. M., and R. W. Hallberg (2000), Biharmonic friction with a Smagorinsky viscosity for use in large-scale eddy-permitting ocean models, *Mon. Weather Rev.*, *128*, 2935–2946.

Griffies, S. M., C. Böning, F. O. Bryan, E. P. Chassignet, R. Gerdes, H. Hasumi, A. Hirst, A.-M. Treguier, and D. Webb (2000a), Developments in ocean climate modelling, *Ocean Model.*, *2*, 123–192.

Griffies, S. M., R. C. Pacanowski, and R. W. Hallberg (2000b), Spurious diapycnal mixing associated with advection in a *z*-coordinate ocean model, *Mon. Weather Rev.*, *128*, 538–564.

Griffies, S. M., R. Pacanowski, M. Schmidt, and V. Balaji (2001), Tracer conservation with an explicit free surface method for *z*-coordinate ocean models, *Mon. Weather Rev.*, *129*, 1081–1098.

Griffies, S. M., M. J. Harrison, R. C. Pacanowski, and A. Rosati (2004), *A Technical Guide to MOM4*, NOAA/Geophysical Fluid Dynamics Laboratory, Princeton, USA, 337 pp.

Haidvogel, D. B., and A. Beckmann (1999), *Numerical Ocean Circulation Modeling*, Imperial College Press, London.

Hallberg, R. W. (1997), Stable split time stepping schemes for large-scale ocean modeling, *J. Comput. Phys.*, *135*, 54–65.

Hallberg, R. W. (2000), Time integration of diapycnal diffusion and Richardson number-dependent mixing in isopycnal coordinate ocean models., *Mon. Weather Rev.*, *128*, 1402–1419.

Haltiner, G. T., and R. T. Williams (1980), *Numerical Prediction and Dynamic Meteorology*, Wiley, New York, USA.

Higdon, R. (2005), A two-level time-stepping method for layered ocean circulation models: further development and testing, *J. Comput. Phys.*, *206*, 463–504.

Higdon, R. (2006), Numerical modelling of ocean circulation, *Acta Numer.*, pp. 385–470.

Hirsch, C. (1988), *Numerical Computation of Internal and External Flows*, Wiley, New York.

Hirt, C., A. Amsden, and J. Cook (1974), An arbitrary Lagrangian–Eulerian computing method for all flow speeds, *J. Comput. Phys.*, *14*, 227–253.

Holloway, G. (1992), Representing topographic stress for large-scale ocean models, *J. Phys. Oceanogr.*, *22*, 1033–1046.

Huang, R. X. (1993), Real freshwater flux as a natural boundary condition for the salinity balance and thermohaline circulation forced by evaporation and precipitation, *J. Phys. Oceanogr.*, *23*, 2428–2446.

Huang, R. X., X. Jin, and X. Zhang (2001), An oceanic general circulation model in pressure coordinates, *Advances in Atmospheric Physics*, *18*, 1–22.

Jackett, D. R., T. J. McDougall, R. Feistel, D. G. Wright, and S. M. Griffies (2006), Algorithms for density, potential temperature, conservative temperature, and freezing temperature of seawater, *J. Atmos. Ocean. Technol.*, *23*, 1709–1728.

Killworth, P. (1989), On the parameterisation of deep convection in ocean models, in *Parameterizing Small Scale Processes in the Ocean*, edited by P. Müller and G. Holloway, Proceeding of the 5th 'Aha Huliko'a Hawaiian Winter Workshop, pp. 59–74, University of Hawaii at Manoa.

Killworth, P. D., D. Stainforth, D. J. Webb, and S. M. Paterson (1991), The development of a free-surface Bryan-Cox-Semtner ocean model, *J. Phys. Oceanogr.*, *21*, 1333–1348.

Landau, L. D., and E. M. Lifshitz (1987), *Fluid Mechanics*, Pergamon Press, Oxford, UK, 539 pp.

Large, W. G., G. Danabasoglu, J. C. McWilliams, P. R. Gent, and F. O. Bryan (2001), Equatorial circulation of a global ocean climate model with anisotropic horizontal viscosity, *J. Phys. Oceanogr.*, *31*, 518–536.

Ledwell, J. R., A. J. Watson, and C. S. Law (1993), Evidence for slow mixing across the pycnocline from an open-ocean tracer-release experiment, *Nature*, *364*, 701–703.

Lin, S. J. (1997), A finite volume integration method for computing pressure gradient force in general vertical coordinates, *Q. J. R. Meteorol. Soc.*, *123*, 1749–1762.

Lion, J., R. Temam, and S. Wang (1992), On the equations of the largescale ocean, *Nonlinearity*, *5*, 1007–1053.

Losch, M., A. Adcroft, and J.-M. Campin (2004), How sensitive are coarse general circulation models to fundamental approximations in the equations of motion?, *J. Phys. Oceanogr.*, *34*, 306–319.

Margolin, L., W. Rider, and F. Grinstein (2006), Modeling turbulent flow with implicit LES, *J. Turbul.*, *7*, 1–27.

Marion, J. B., and S. T. Thornton (1988), *Classical Dynamics of Particles and Systems*, Harcourt Brace Jovanovich, San Diego, USA, 602 pp.

Marshall, J., and F. Schott (1999), Open-ocean convection: observations, theory, and models, *Rev. Geophys.*, *37*, 1–64.

Marshall, J., C. Hill, L. Perelman, and A. Adcroft (1997), Hydrostatic, quasi-hydrostatic, and nonhydrostatic ocean modeling, *J. Geophys. Res.*, *102*(C3), 5733–5752.

Marshall, J., A. Adcroft, J.-M. Campin, and C. Hill (2004), Atmosphere–ocean modeling exploiting fluid isomorphisms, *Mon. Weather Rev.*, *132*, 2882–2894.

McDougall, T., and W. Dewar (1997), Vertical mixing and cabbeling in layered models, *J. Phys. Oceanogr.*, *28*, 1458–1480.

McDougall, T. J. (1987), Neutral surfaces, *J. Phys. Oceanogr.*, *17*, 1950–1967.

McDougall, T. J. (2003), Potential enthalpy: A conservative oceanic variable for evaluating heat content and heat fluxes., *J. Phys. Oceanogr.*, *33*, 945–963.

McDougall, T. J., R. J. Greatbatch, and Y. Lu (2003), On conservation equations in oceanography: How accurate are Boussinesq ocean models?, *J. Phys. Oceanogr.*, *32*, 1574–1584.

Mellor, G. L., and T. Ezer (1995), Sea level variations induced by heating and cooling: an evaluation of the Boussinesq approximation in ocean models, *J. Geophys. Res.*, *100*(C10), 20,565–20,577.

Mesinger, F. (1973), A method for construction of second-order accurate difference schemes permitting no false two-grid-interval waves in the height field., *Tellus*, *25*, 444–457.

Mesinger, F., and A. Arakawa (1976), Numerical methods used in atmospheric models, in *GARP Publication Series*, vol. 1, p. 66 pages.

Morse, P. M., and H. Feshbach (1953), *Methods of Theoretical Physics Part I and II*, McGraw-Hill Book Company, New York.

Müller, P. (1995), Ertel's potential vorticity theorem in physical oceanography, *Rev. Geophys.*, *33*, 67–97.

Müller, P. (2006), *The Equations of Oceanic Motions*, 1st ed., Cambridge University Press, Cambridge, 302 pp.

Müller and J. Willebrand, P. (1986), Compressibility effects in the thermohaline circulation: A manifestation of the temperature salinity mode, *Deep Sea Res.*, *33*, 559–571.

Munk, W. H. (1950), On the wind-driven ocean circulation, *Journal of Meteorology*, *7*, 79–93.

Oberhuber, J. (1993), Simulation of the atlantic circulation with a coupled sea ice-mixed layer-isopycnal general circulation model. Part I: Model description, *J. Phys. Oceanogr.*, *23*, 808–829.

Osborn, T. R. (1980), Estimates of the local rate of vertical diffusion from dissipation measurements, *J. Phys. Oceanogr.*, *10*, 83–89.

Pedlosky, J. (1987), *Geophysical Fluid Dynamics*, 2nd ed., Springer, Berlin, 710 + xv pp.

Phillips, N. (1973), Principles of large-scale numerical weather prediction, in *Dynamic Meteorology*, edited by P. Morel, pp. 1–96, Riedel, Boston.

Phillips, N. A. (1957), A coordinate system having some special advantages for numerical forecasting, *J. Meteorol.*, *14*, 184–185.

Roberts, M. J., and D. Marshall (1998), Do we require adiabatic dissipation schemes in eddy-resolving ocean models?, *J. Phys. Oceanogr.*, *28*, 2050–2063.

Rosati, A., and K. Miyakoda (1988), A general circulation model for upper ocean simulation, *J. Phys. Oceanogr.*, *18*, 1601–1626.

Russell, G. L., J. R. Miller, and D. Rind (1995), A coupled atmosphere ocean model for transient climate change studies, *Atmosphere-Ocean*, *33*, 683–730.

Sadourny, R. (1975), The dynamics of finite-difference models of the shallow-water equations, *J. Atmos. Sci.*, *32*, 680–689.

Salmon, R. (1998), *Lectures on Geophysical Fluid Dynamics*, Oxford University Press, Oxford, England, 378 + xiii pp.

Semtner, A. J., and Y. Mintz (1977), Numerical simulation of the Gulf Stream and mid-ocean eddies, *J. Phys. Oceanogr.*, *7*, 208–230.

Smagorinsky, J. (1963), General circulation experiments with the primitive equations: I. The basic experiment, *Mon. Weather Rev.*, *91*, 99–164.

Smagorinsky, J. (1993), Some historical remarks on the use of nonlinear viscosities, in *Large Eddy Simulation of Complex Engineering and Geophysical Flows*, edited by B. Galperin and S. A. Orszag, pp. 3–36, Cambridge University Press, Cambridge.

Smith, R. D., and P. R. Gent (2004), Anisotropic Gent–McWilliams parameterization for ocean models, *J. Phys. Oceanogr.*, *34*, 2541–2564.

Smith, R. D., and J. C. McWilliams (2003), Anisotropic horizontal viscosity for ocean models, *Ocean Model.*, *5*, 129–156.

Vallis, G. K. (2006), *Atmospheric and Oceanic Fluid Dynamics: Fundamentals and Large-scale Circulation*, 1st ed., Cambridge University Press, Cambridge, 745 + xxv pp.

Veronis, G. (1973), Large scale ocean circulation, *Adv. Appl. Mech.*, *13*, 2–92.

Wajsowicz, R. C. (1986), Adjustment of the ocean under buoyancy forces. Part II: The role of planetary waves, *J. Phys. Oceanogr.*, *32*, 2115–2136.

White, L., V. Legat, and E. Deleersnijder (2008), Tracer conservation for three-dimensional, finite element, free-surface, ocean modeling on moving prismatic meshes, *Mon. Weather Rev.*, (in press), doi: 10.1175/2007MWR2137.1.

Stephen M. Griffies, NOAA Geophysical Fluid Dynamics Laboratory, Princeton Forrestal Campus Rte. 1, 201 Forrestal Road, Princeton, NJ 08542, USA. (Stephen.Griffies@noaa.gov)

Can Large Eddy Simulation Techniques Improve Mesoscale Rich Ocean Models?

B. Fox-Kemper

Cooperative Institute for Research in the Environmental Sciences and Department of Atmospheric and Oceanic Sciences, University of Colorado, Boulder, Colorado, USA

D. Menemenlis

Jet Propulsion Laboratory, California Institute of Technology, Pasadena, California, USA

The necessary adaptations to large eddy simulation (LES) methods so that they can be used in mesoscale ocean large eddy simulations (MOLES)—where the gridscale lies just below the Rossby deformation radius—are presented. For MOLES, the Smagorinsky model is inappropriate and the similar Leith model is appropriate. The latter is preferable, as the gridscale lies nearer to a potential enstrophy cascade than an energy cascade. A trivial modification of the Leith scheme is shown to improve numerical stability in a global eddy-permitting simulation. Dynamical adjustment of subgrid-scale models—where online diagnosis of eddy terms are used to improve the subgrid-scale model—is introduced and recommended for future investigation.

1. INTRODUCTION

Large-eddy simulations (LES) differ from other fluid flow computations in that the largest eddies are explicitly resolved and the smaller eddies are modeled (as engineers call it) or parameterized (as oceanographers call it). Modeling the ocean and atmosphere inspired the first large-eddy simulations, but true large-eddy simulations of the ocean and atmosphere are surprisingly rare.

Ocean models almost always have eddy parameterizations, be they just larger than natural values of constant diffusivities and viscosities, perhaps oriented along isopycnal surfaces [*Redi*, 1982], or more sophisticated models such as the eddy extraction of potential energy by along-isopycnal bolus flux model of *Gent and McWilliams* [1990], perhaps with nonlinear transfer coefficient scaling [*Visbeck et al.*, 1997;

Ocean Modeling in an Eddying Regime
Geophysical Monograph Series 177
Copyright 2008 by the American Geophysical Union.
10.1029/177GM19

Held and Larichev, 1996], the "Neptune" effect parameterizing eddy-topography interaction [*Holloway*, 1993], or parameterizations based on linear instability analysis [*Stone*, 1972; *Branscome*, 1983; *Killworth*, 2005]. All of these parameterizations are designed for situations where eddies are not present or weak, and, as a result, the model resolution does not appear explicitly. The Herculean—perhaps Sisyphean—task of parameterizing the important effects of eddies in ocean models is done by offline theoretical analysis in lieu of resolved eddies. It is no wonder that there is enormous model sensitivity to the parameterization choices made [e.g., *Steiner et al.*, 2004]. Leaving all to theory seems very unlikely to succeed, as eddies are sensitive to difficult to quantify parameters, such as interaction with unresolved topography [*Holloway*, 1993], regional variations of hydrodynamic instability [*Killworth*, 2005], bottom drag [*Arbic and Flierl*, 2004a; *Thompson and Young*, 2006], and subtle sensitivity to the Coriolis parameter variation [*Arbic and Flierl*, 2004b; *Thompson and Young*, 2007].

Subgrid-scale parameterizations for LES are quite different in character because they attempt only to represent the effects of an *unresolved cascade of smaller eddies* to complement the resolved large eddies. Subgrid-scale parameterizations for LES depend explicitly on the resolution of the model at hand, and their goal is to anticipate higher resolution results at any given resolution. Fortunately, as will be argued quantitatively below, the largest eddies are the ones that do most of the work in stirring tracers both in engineering scale flows and in ocean mesoscale flows. Thus, in LES modeling, results are expected to be less sensitive to subgrid-scale parameterizations than in coarse-resolution modeling. However, even high-resolution ocean models are sensitive to subgrid-scale parameterizations (a frequent topic during sessions like the one on which this volume is based), so parameterization improvement is important. Of course, the range of scales in the ocean is vast and diverse, so phenomena that are substantially smaller than the gridscale—such as submesoscale eddies and microstructure turbulence—will continue to require purely theoretical parameterizations in large-scale models for the next few decades at least [e.g., *Fox-Kemper et al.*, 2008; *Fox-Kemper and Ferrari*, 2008].

This chapter will first review briefly the traditional subgrid-scale model of engineering-scale 3D LES: the *Smagorinsky* [1963] nonlinear viscosity. The necessary adaptations appropriate for large-scale ocean models will then be presented. The emphasis will be on ocean models in a larger domain than traditional ocean boundary layer LES [e.g., *McWilliams et al.*, 1997; *Wang et al.*, 1998; *Skyllingstad et al.*, 1999], but high-resolution enough to resolve at least the first baroclinic deformation radius. Simulations at this scale will be called mesoscale ocean large-eddy simulations (MOLES). In MOLES, both the tail of the mesoscale eddy spectrum and smaller scale turbulence closures must be provided. The emphasis here is on the former, and it is argued that the mesoscale spectrum can be sensibly represented with a nonlinear diffusivity and viscosity combination adapted from LES methods. However, the Smagorinsky viscosity is not useful for MOLES or in coarser resolution simulations because it relies on the gridscale being in the inertial range of a *Kolmogorov* [1941] (forward) energy cascade. Smagorinsky's viscosity might be useful in submesoscale or boundary layer ocean models, but the large-scale high-resolution ocean models that are the primary subject of this book are more likely to have their gridscale in an inertial potential enstrophy cascade of effectively quasi-geostrophic (QG) eddies (if there is an inertial range at all). Models coarser than MOLES tend to not have eddies at all or to have only a weak inverse energy cascade; in these cases, no cascade-based parameterization is useful.

The second section reviews a more ambitious pursuit of subgrid-scale modelers using LES at engineering scales, to use the resolved eddies to improve the subgrid-scale model itself, a process known as dynamical adjustment. Dynamical adjustment has made LES simulations more accurate and reliable with fewer ad hoc assumptions and unknown parameters, beginning with *Bardina et al.* [1980] and *Germano et al.* [1991] and reviewed by *Meneveau and Katz* [2000]. An advantage relevant to MOLES is that dynamical adjustment allows the subgrid-scale parameterization to recognize flow heterogeneity—spatial variation in the natural instabilities of the resolved flow leads to spatial variation in the parameterization as well. For example, dynamical adjustment would allow for a low-viscosity Gulf Stream where it is attached to the coast, and enhanced subgrid-scale effects in the separated boundary current and North Atlantic drift where resolved eddy effects are strong. There is a critical resolution threshold needed for dynamical adjustment: more than one scale of eddies needs to be permitted by the model resolution. Mesoscale eddies at the deformation radius are well represented at O(10 km) resolution, so dynamical adjustment will be possible once O(5 km) resolution is possible. It is argued here that using the adaptations for subgrid-scale parameterization relevant to mesoscale eddies, dynamical adjustment may prove to be a powerful approach in improving subgrid-scale parameterizations for the near future of O(5 km) resolution.

The goal here is to introduce simply the concepts of nonlinear viscosity, nonlinear diffusivity, and dynamical adjustment to an oceanographic audience and to show the adaptations necessary for potential use in future MOLES. Of course, there is much more to LES than just nonlinear viscosities and dynamical adjustment [see, e.g., *Sagaut*, 2005; *Lesieur et al.*, 2005], and in some circumstances, alternative subgrid-scale parameterizations may be superior [*Geurts and Holm*, 2003]. Adding a prognostic model for additional unresolved quantities, such as turbulent eddy kinetic energy, has been shown to be useful in LES [*Deardorff*, 1980; *Mellor and Yamada*, 1982; *Moeng*, 1984; *Burchard et al.*, 1998], but implementation of these methods for MOLES is just beginning (A. Adcroft, personal communication). For brevity, the focus here is a presentation we believe will be most palatable to oceanographers and most readily applied to MOLES.

2. NONLINEAR VISCOSITIES AND DIFFUSIVITIES

2.1. The Cascades

The basic principle underlying all of the subgrid-scale parameterizations discussed here is the concept of a cascade through an inertial range. The idea is due to *Kolmogorov* [1941], who proposes such a scenario for isotropic, homogeneous, 3D turbulence. Subsequent work has shown that similar ideas generalize to isotropic, homogeneous 2D, and QG turbulence as well.

Consider the mean kinetic energy per unit mass (angle brackets are volume mean):

$$\langle E \rangle = \frac{1}{V} \iiint \frac{1}{2} (\mathbf{u} \cdot \mathbf{u}) dV = \int_0^\infty E(k) dk. \quad (1)$$

V is the domain volume, $\mathbf{u}$ is the velocity, $E(k)$ is the kinetic energy spectrum (assuming homogeneous, isotropic turbulent flow), and k is the wavenumber. The energy spectrum obeys

$$\frac{\partial E(k)}{\partial t} = -\frac{\partial F_E}{\partial k} - \nu k^2 E(k) + S_E(k, t), \quad (2)$$

where F_E is the energy flux through a given wavenumber due to nonlinear turbulent interactions, ν is the molecular viscosity of the fluid, and $S_E(k, t)$ represents sources of kinetic energy from conversion of potential energy by fluid instabilities or from external forces.

Kolmogorov [1941] supposes that in some flows, a quasi-steady state might arise in between the forcing and dissipation, that is, over a range of wavenumber s higher than where $S_E(k)$ is active and lower than where $\nu k^2 E$ is active. If a wavenumber k_* is assumed to be in this range, then the energy flux through that wavenumber is constant in k, as all of the other terms in (2) vanish except $\partial F_E / \partial k$. If the constant value of the flux is denoted by ε, then the assumptions made amount to

$$\varepsilon = \int_0^{k_*} S_E(k, t) dk = \int_{k_*}^\infty \nu k^2 E(k) dk = F_E(k_*). \quad (3)$$

Dimensional analysis yields:

$$[E] = L^3 T^{-2}, \quad [k] = L^{-1}, \quad [\varepsilon] = L^2 T^{-3},$$
$$E(k) \propto \varepsilon^{2/3} k^{-5/3}, k_d = \varepsilon^{1/4} \nu^{-3/4}. \quad (4)$$

(Brackets are used to denote the dimensions of a quantity.) The wavenumber k_d is the energy dissipation wavenumber (sometimes called the Kolmogorov wavenumber), which is large enough for frictional effects to be important.

In 2D flows, there is a second conserved quantity in addition to energy, the enstrophy:

$$\langle G \rangle = \frac{1}{V} \iiint \frac{1}{2} (\nabla_h \times \mathbf{u})^2 dV$$
$$\equiv \frac{1}{V} \iiint \frac{1}{2} (\nabla_h^2 \psi)^2 dV = \int_0^\infty G(k) dk. \quad (5)$$

∇_h is the horizontal differential operator, ψ is the 2D flow streamfunction ($\mathbf{u} = -\nabla_h \times \psi$), and $G(k)$ is the enstrophy spectrum (assumed isotropic and homogeneous). The enstrophy spectrum has an evolution equation as well,

$$\frac{\partial G(k)}{\partial t} = -\frac{\partial F_G}{\partial k} - \nu k^2 G(k) + S_G(k, t). \quad (6)$$

Kraichnan [1967] notes that there might be an inertial enstrophy cascade analogous to the energy cascade where F_G is constant and denoted by η. Proceeding as above,

$$\eta = \int_0^{k_*} S_G(k, t) dk = \int_{k_*}^\infty \nu k^2 G(k) dk = F_G(k_*).$$
$$[G] = LT^{-2}, \quad [k] = L^{-1}, \quad [\eta] = T^{-3}. \quad (7)$$

Importantly, as $[\varepsilon]/[\eta] = L^2$, it cannot be the case that both ε and η are constant in k and nonzero. Thus, the presence of an energy cascade precludes the presence of an enstrophy cascade and vice versa.

In sum, there are two limits to consider in 2D turbulence. An inertial energy cascade where

$$E \propto \varepsilon^{2/3} k^{-5/3}, \quad G \propto \varepsilon^{2/3} k^{1/3}, \quad \eta = 0, \quad (8)$$

and an inertial enstrophy cascade where

$$E \propto \eta^{2/3} k^{-3}, \quad G \propto \eta^{2/3} k^{-1}. \quad \varepsilon = 0. \quad (9)$$

Kraichnan [1967] also considers the wavenumber triads involved in F_E and F_G, which suggest that the only way a quasi-steady state can exist is if the energy flux direction is inverse in comparison to 3D turbulence (from small scales to large, i.e., $F_E = -|\varepsilon|$) and the enstrophy flux direction is direct (from large scales to small, $F_G = |\eta|$). Both cascades are envisioned to originate in some central range of forcing wavenumber s. In this case, the viscosity is important for dissipating enstrophy—not energy—at small scales and is leading order near the enstrophy dissipation wavenumber

$$k_d = \eta^{1/6} \nu^{-1/2}. \quad (10)$$

For Kraichnan's scenario to work, all of the energy must be dissipated somehow at scales larger than the forcing scales, or there must be no net energy input by forcing (interestingly, with weak dissipation, a flow will often adjust itself to reduce energy and vorticity input [Scott and Straub, 1998; Fox-Kemper, 2004]). Thus, the spectral energy and enstrophy budgets in Kraichnan's scenario become,

$$\eta = \int_{k_{E*}}^{k_{G*}} S_G(k,t)dk = \int_{k_{G*}}^{\infty} \nu k^2 G(k)dk = F_G(k_{G*}), \quad (11)$$

$$\varepsilon = -|\varepsilon| = \int_{k_{E*}}^{k_{G*}} S_E(k,t)dk = F_E(k_{E*})$$

$$= \int_0^{k_{E*}} S_E(k,t) + D_E(k,t)dk. \quad (12)$$

$D_E(k,t)$ is the required large-scale energy dissipation mechanism.

Charney [1971] shows that 2D turbulence and QG turbulence share a great deal in common. As QG scaling applies at the gridscale expected in MOLES—small Rossby number (U/fL) and modest cross-gridcell variations in layer thickness and Coriolis parameter—Charney's result is of great interest to the development here. The governing equations for 2D and 3D QG turbulence are:

$$\frac{\partial q_\alpha}{\partial t} + J(\psi, q_\alpha) = S_{q_\alpha} + D_{q_\alpha}, \quad (13)$$

The 2D relative vorticity is q_{2d}, and q_{QG} is the QG potential vorticity. S_{q_α} is the source of (potential) vorticity and D_{q_α} is the dissipation. The QG streamfunction has the same relationship to the horizontal velocity as the 2D streamfunction ($\mathbf{u} = -\nabla_h \times \psi$), yet it is now a 3D field. In both cases, the streamfunction is related to q_α by a Poisson-like equation,

$$q_{2d} = \nabla_h^2 \psi, \quad (14)$$

$$q_{QG} = \widetilde{\nabla}^2 \psi \equiv \nabla_h^2 \psi + \frac{\partial}{\partial z}\frac{f^2}{N^2}\frac{\partial \psi}{\partial z}. \quad (15)$$

However, in the QG system, the Poisson relation is 3D and is only truly a Poisson equation in a stretched coordinate system. The vertical coordinate is stretched or rescaled by multiplication by the local Prandtl ratio, N/f. The symbol $\widetilde{\nabla}^2$ is used instead of ∇^2 to remember the stretching required. *Hua and Haidvogel* [1986] confirm numerically that in uniform vertical stratification, QG turbulence is isotropic in $\widetilde{\nabla}^2$. *Bracco et al.* [2004] show that many of the Lagrangian statistics of 3D QG turbulence are virtually identical to 2D turbulence for short to intermediate timescales (a few eddy turnovers). At longer times, vertical effects become important, such as vertical vortex alignment [e.g., *Martinsen-Burrell et al.*, 2006].

An important difference between 2D and QG turbulence is the effect of boundaries in the vertical in QG. One result of the boundaries is a split between baroclinic and barotropic modes, *Salmon* [1980] and *Smith and Vallis* [2002]

show that in two-vertical-layer QG turbulence, the combination of baroclinic and barotropic flows complicates the dynamics for scales larger than the baroclinic instability scale, but the enstrophy cascade at smaller scales retains the same form as in Kraichnan's theory. *Hua and Haidvogel* [1986] show that the turbulence in a six-vertical-mode QG system with non-uniform stratification is not isotropic in the Prandtl-stretched coordinate system. However, one expects that boundary effects will be extremely important with such low vertical resolution, and Charney's isomorphism between QG and 2D turbulence relies on boundary effects being unimportant.

Another important boundary effect that differentiates QG from 2D turbulence is the presence of temperature modes on the surface that do not require internal QGPV gradients. The study of systems with no interior PV gradients extends back to *Eady* [1949] and beyond, but study of the dynamics of these surface modes in the ocean has invigorated recently [e.g., *Lapeyre and Klein*, 2006] along with study of near-surface submesoscale dynamics and frontogenesis. Of importance here is the result of *Blumen* [1982] that the inertial cascade of surface modes differs from that of due to interior QGPV anomalies shown by *Charney* [1971] to resemble the 2D turbulent cascade [see also *Pierrehumbert et al.*, 1994; *Held et al.*, 1995]. However, numerical simulations reveal that frontogenesis and mixed-layer eddies may affect the spectral slope near the ocean surface by potential energy conversion at all small scales, which prevents an inertial range of kinetic energy altogether [*Capet et al.*, 2008]. However, these near-surface effects are not well understood presently and have not yet reached the level of a functional parameterization.

Oceanic and atmospheric observations are generally in agreement with the existence of an enstrophy cascade below the deformation radius and away from the surface, but not an inertial inverse energy cascade. *Stammer and Wunsch* [1995] find an $k^{-5/2}$ slope below the deformation radius and a shallower slope at larger and smaller scales. Their acknowledged uncertainty does not preclude $k^{-5/2} \approx k^{-3}$ in the enstrophy cascade region. In the atmosphere, *Nastrom and Gage* [1985] find a clear k^{-3} spectral slope below the deformation radius and a shallower slope near $k^{-5/3}$ at smaller scales. The shallow slope at smaller scales might be an inverse cascade up from smaller scales, or an effect of gravity waves, but may result instead by a surface mode cascade [see *Tulloch and Smith*, 2006, for references]. An inverse energy cascade range does not exist in the atmosphere because the Earth is only slightly larger than the atmospheric deformation radius. There is good evidence of an inverse energy cascade in the ocean, but not an inertial one. The range is limited or nonexistent because the inverse cascade is halted shortly above the deformation radius for reasons that remain unclear [*Scott and Wang*, 2005; *Thompson and Young*, 2007]. The energy flux does decrease rapidly below the deformation radius, which is consistent with an enstrophy cascade, (9), but a small amount of direct energy flux appears to be present, which may be related to interactions between the background internal wave field and the mesoscale QG turbulence (K. Polzin, personal communication). At smaller scale, a shallower slope is observed [*Ferrari and Rudnick*, 2000] and expected due to surface modes and frontogenesis. *LaCasce and Mahadevan* [2006] were able to use a sheer stress transport-constrained numerical model of only surface modes to obtain a favorable model versus in situ data comparison, suggesting that surface modes are relevant in the near-surface ocean at the submesoscale.

Thus, at some distance away from the boundaries, and below the deformation radius, present understanding predicts the dynamics to be close to QG turbulence in an approximate inertial potential enstrophy cascade. These dynamics are assumed to dominate the gridscale dynamics and form the basis of subgrid-scale parameterizations of MOLES.

2.2. Smagorinsky Viscosity

Smagorinsky [1963] proposed a scaling for an "eddy viscosity" for a numerical model whose gridscale lies in the forward energy cascade range of 3D turbulence as proposed by *Kolmogorov* [1941]. Consider integrating the energy spectrum equation (2) over all of the wavenumbers that will be resolved in the model, assuming an approximately steady energy content:

$$\frac{d\langle E_*\rangle}{dt} = -F_E\big|_0^{k_*} - \int_0^{k_*} \nu k^2 E(k)dk + \int_0^{k_*} S_E(k,t)dk. \tag{16}$$

$\langle E_*\rangle$ is the resolved kinetic energy per unit mass of the model, and $k_* = \pi/\Delta x$ is the largest unambiguously resolved wavenumber (the Nyquist wavenumber) of the model gridscale Δx (assumed to be the same in all directions for now). Asterisks will be used to denote quantities that are to be understood as resolved in the model or resolution-dependent. There is negligible friction at and no energy flux through $k=0$ (largest scales), and the resolved energy and forcing are assumed to be relatively steady, so

$$\int_0^{k_*} S_E(k)dk = F_E(k_*) + \int_0^{k_*} \nu k^2 E(k)dk. \tag{17}$$

Resolving the Kolmogorov length scale ($k_d \ll k_*$) guarantees that the viscous sink of energy is resolved and then there would be no inertial transfer through the gridscale ($F_E(k_*) \approx 0$). However, such fine resolution is impractical in ocean modeling as k_d is O(1/cm) in the ocean.

Smagorinsky proposes to use an eddy viscosity, ν_*, that will result in a resolved Kolmogorov scale and satisfaction of the energy budget (17) with an eddy viscosity term replacing the nonlinear energy flux $F_E(k_*)$. To determine the value of ν_*, suppose that k_* lies in the inertial range, so $F_E(k_*) = \varepsilon$, and recall the scaling for the Kolmogorov wavenumber (4). The Kolmogorov scale based on the increased eddy viscosity is set to be proportional to the gridscale, which yields

$$k_* = \Upsilon \varepsilon^{1/4} \nu_*^{-3/4}, \tag{18}$$

$$\varepsilon = F_E(k_*) + \int_0^{k_*} \nu k^2 E(k)dk \equiv \int_0^{k_*} \nu_* k^2 E(k)dk \tag{19}$$

The nondimensional coefficient of proportionality between the gridscale and the Kolmogorov length is Υ, and it is what the user sets.

If the viscous term is evaluated in real space rather than wavenumber space, then

$$\int_0^{k_*} \nu_* k^2 E(k)\,dk = \left\langle \nu_* S_*^{ik} S_{*ik} \right\rangle, \tag{20}$$

where S_{*ik} are the components of the resolved strain rate tensor $2S_{*ik} \equiv (u_{*i,k} + u_{*k,i})$ and angle brackets denote domain average.[1,2] So,

$$\varepsilon = \left\langle \nu_* S^{ik} S_{*ik} \right\rangle. \tag{21}$$

Finally, because the turbulence is assumed to be homogeneous, the domain-averaged friction is replaced with a local value, and the result for ν_* follows.

$$\nu_* = \left(\frac{\Upsilon}{k_*}\right)^2 |D_*|, \tag{22}$$

$$\nu_* = \left(\frac{\Upsilon \Delta x}{\pi}\right)^2 |D_*|, \tag{23}$$

$$|D_*| \equiv \sqrt{S_*^{ik} S_{*ik}}. \tag{24}$$

This assumption of equating domain average to local value is controversial [*Lesieur et al.*, 2005], and it should be kept in mind that the effects of small-scale spatial variation of ν_* are not physical. Furthermore, in a model, the viscous term should be in flux form, $\nabla \cdot \nu_* \nabla$, rather than $\nu_* \nabla^2$, so momentum will not be spuriously created. For simplicity, the viscosity will be assumed to be smooth enough to approximately commute with derivatives for the remainder of this chapter.

The nondimensional coefficient Υ is generally taken to be in the range 0.5–1 for 3D flows [*Smagorinsky*, 1993], although numerical instabilities may result in some discretizations for values that small (*Griffies and Hallberg* [2000] and *Griffies* [2004] recommend 2.2 to 4).

In a review of the use of the Smagorinsky nonlinear viscosity, *Smagorinsky* [1993] carefully works through the stress-strain relations in the case of hydrostatic axially symmetric turbulence (in contrast to 3D isotropic turbulence). In this case, there are two viscosities, a horizontal and a vertical. The horizontal viscosity is similar to the 3D isotropic case above,

$$\nu_{*h} = \left(\frac{\Upsilon_h \Delta x}{\pi}\right)^2 \sqrt{S^{ik}_* S_{*ik}}. \tag{25}$$

except the deformation rate appropriate for hydrostatic flows is

$$|D_{*2d}| = \sqrt{\left(\frac{\partial u_*}{\partial x} - \frac{\partial v_*}{\partial y}\right)^2 + \left(\frac{\partial u_*}{\partial y} + \frac{\partial v_*}{\partial x}\right)^2}. \tag{26}$$

If the horizontal grid is slightly anisotropic, *Griffies* [2004] suggests the geometric mean of Δx and Δy.

Smagorinsky [1993] shows that a corresponding vertical viscosity should be used:

$$\nu_{*v} = \left(\frac{\Upsilon_v \Delta z}{\pi}\right)^2 \sqrt{\left(\frac{\partial u_*}{\partial z}\right)^2 + \left(\frac{\partial v_*}{\partial z}\right)^2}$$

The authors are unaware of any use of this vertical viscosity in a climate-scale simulation. Assuming $\Upsilon_h = \Upsilon_v$ (which would be appropriate if the turbulence were 3D isotropic but the gridscale was not), this viscosity would be comparable to typical background values with a 10 km horizontal grid and an 100 m vertical grid.

Smagorinsky's viscosity is a leap forward in understanding of the interaction of numerical resolution and physics and has proven useful in engineering scale flows since *Deardorff* [1972], yet it is inappropriate for large-scale ocean and atmosphere simulations as it was initially used [*Smagorinsky*, 1963]. *Smagorinsky* [1963] used the earlier work of *Kolmogorov* [1941], not the later work of *Kraichnan* [1967] and *Charney* [1971], who would show that the dynamics rel-

evant at the gridscale of large-scale ocean and atmosphere simulations should be quite different from the scenario envisioned by Kolmogorov. Consider the two possibilities for energy flux in large-scale simulations where QG or 2D turbulence dominates: $\varepsilon = -|\varepsilon|$ in the inverse energy cascade or $\varepsilon = 0$ in the enstrophy cascade. If the Smagorinsky viscosity were to represent the energy flux as in (21), the viscosity would be zero or negative. Furthermore, the Smagorinsky viscosity is proportional to the mean square strain rate ($|D| = \int k^2 E(k)dk$) which is not dominated by the smallest scales in an inertial enstrophy cascade ($|D| = \int k^{-1}dk$) as it is in 3D turbulence ($|D| = \int k^{1/3}dk$). Thus, the Smagorinsky viscosity will not be adequately sensitive to resolution refinement in large-scale atmosphere-ocean flows. Another approach must be used for MOLES.

2.3. Leith Viscosity

Leith [1996] finds an alternative to the Smagorinsky viscosity by focusing on resolving the direct enstrophy cascade in 2D turbulence rather than the direct energy cascade in 3D turbulence. Integrating the enstrophy spectrum equation (6) over resolved wavenumbers yields:

$$\frac{d\langle G_*\rangle}{dt} = -F_G\big|_0^{k_*} - \int_0^{k_*} \nu k^2 G(k)dk + \int_0^{k_*} S_G(k,t)dk, \tag{27}$$

where all notation is the same as above, just with enstrophy swapped for energy. Neglect friction and enstrophy flux through the largest scales, and assuming an approximate steady state yields

$$\int_0^{k_*} S_G(k)\,dk = F_G(k_*) + \int_0^{k_*} \nu k^2 G(k)dk. \tag{28}$$

Again, an eddy viscosity is used to represent the direct inertial cascade with a frictional term whose dissipation lengthscale is resolved.

$$k_* = \Lambda \eta^{1/6} \nu_*^{-1/2}, \tag{29}$$

$$\eta = F_G(k_*) + \int_0^{k_*} \nu k^2 G(k)\,dk \equiv \int_0^{k_*} \nu_* k^2 G(k)\,dk. \tag{30}$$

The nondimensional coefficient Λ, like Υ above, determines the proportionality of the gridscale to dissipation lengthscale.

Estimating the viscous term in real space by the positive definite friction term in the enstrophy equation yields

$$\int_0^{k_*} \nu_* k^2 G(k)\, dk \approx \left\langle \nu_* |\nabla_h q_{*2d}|^2 \right\rangle, \tag{31}$$

This step is a bit more tenuous than in the energy cascade. In the energy cascade, it was argued that the boundary terms vanished with all typical boundary conditions. In the case of vorticity dissipation, the boundary term is proportional to $\nabla^2 q_{2d}^2$, which not only does not vanish at the boundary, it is likely to reach its largest values at the boundary where boundary currents act frictionally [*Fox-Kemper and Pedlosky*, 2004; *Fox-Kemper*, 2004]. Nonetheless, a positive definite viscosity is required, so the positive definite term must suffice.

Assuming that the gridscale lies in the enstrophy cascade inertial range and making the swap of a local value of friction for a global mean, we find

$$\nu_* = \left(\frac{\Lambda}{k_*}\right)^3 |\nabla q_{*2d}| = \left(\frac{\Lambda \Delta x}{\pi}\right)^3 |\nabla q_{*2d}|,$$

$$\nu_* = \left(\frac{\Lambda \Delta x}{\pi}\right)^3 \left|\nabla_h \left(\frac{\partial u_*}{\partial y} - \frac{\partial v_*}{\partial x}\right)\right|. \tag{32}$$

The nondimensional coefficient Λ is, like Υ, an $O(1)$ number.

2.4. Biharmonic Forms of Leith and Smagorinsky

Griffies and Hallberg [2000] emphasize that the Smagorinsky scaling produces too much viscosity in MOLES and suggest that a more scale-selective viscosity is needed for QG turbulence. The Leith viscosity (32) is more scale-selective than the Smagorinsky viscosity (23) because it depends on a higher derivative of the velocity. Another way to get enhanced scale selectivity is to use a biharmonic viscosity as first suggested for oceanography by *Holland* [1978]. A biharmonic viscosity cuts off the energy or enstrophy flux more sharply in wavenumber, but at a cost. Additional artificial boundary conditions must be used with biharmonic viscosities which may strongly affect boundary currents and crucial basin-wide integral balances [*Fox-Kemper and Pedlosky*, 2004; *Fox-Kemper*, 2004]. The appropriate boundary conditions on Laplacian eddy viscosity are unclear at best, but at least, the choices of no-stress or no-slip are plausibly physical.

In practice, biharmonic viscosity and diffusivity allow a less viscous, yet numerically stable, simulation than harmonic viscosity and diffusivity. The arguments here for using a viscous turbulence closure to simply truncate the inertial cascade of energy (in the case of Smagorinsky) or enstrophy (in the case of Leith) at a resolved wavenumber seem not to prefer a method of truncation. However, if one considers Laplacian and biharmonic viscosity to be terms in a Taylor series expansion, then both harmonic and biharmonic terms should occur, and the only question is the choice of coefficients. Using biharmonic viscosity alone implies that one zeros the first nonvanishing term in the Taylor series, which is seems unlikely to occur naturally. However, in the context of Smagorinsky and Leith scalings, this objection is less stringent, as one hopes that the details of the truncation of energy and enstrophy are unimportant in comparison with getting the appropriate flow-dependent scaling based on energy and enstrophy flux.

Griffies and Hallberg [2000] propose that one may scale the biharmonic viscosity with the Laplacian viscosity to avoid a computational mode studied by *Bryan et al.* [1975]. For biharmonic viscosity in a 2D flow, the scaling amounts to

$$\nu_{*4} = \frac{(\Delta x)^2}{8} \nu_* \tag{33}$$

where ν_{*4} is the biharmonic viscosity and ν_* is the Smagorinsky viscosity (23). The same method generalizes the Leith viscosity (32) to a biharmonic form. However, it should be noted that unlike the harmonic forms, the *Griffies and Hallberg* [2000] biharmonic scaling does not directly relate to whether energy-dissipation or enstrophy-dissipation scales are resolved. If similar arguments to those above are used to estimate these scales and scale them to the gridscale, the resulting biharmonic viscosities should be:

$$\nu_{*4Smag} = \left(\frac{\Upsilon_4 \Delta x}{\pi}\right)^5 |\nabla^2 \mathbf{u}_*|$$

$$\nu_{*4Leith} = \left(\frac{\Lambda_4 \Delta x}{\pi}\right)^6 |\nabla^2 q_{*2d}|$$

The scaling differences from Smagorinsky and Leith scaling arise in determining the positive-definite energy- and enstrophy-dissipation operators from $u_i \nabla^4 u^i$ and $q_{2d}\nabla^4 q_{2d}$. Thus, the biharmonic scaling suggested by *Griffies and Hallberg* [2000] implies:

$$\sqrt{S_*^{ik} S_{*ik}} \propto \Delta x |\nabla^2 \mathbf{u}_*| \tag{34}$$

$$|\nabla q_{*2d}| \propto \Delta x |\nabla^2 q_{*2d}| \tag{35}$$

The assumption thus amounts to the curvature in the velocity and vorticity being dominated by the smallest resolved scales. This assumption can be tested by the expected spectral slopes of the energy and enstrophy cascades. In the case of Smagorinsky viscosity in an energy cascade,

$$\frac{|\nabla^2 \mathbf{u}_*|}{\sqrt{S_*^{ik} S_{*ik}}} \propto \frac{\sqrt{\int_0^{k_*} k^4 E(k)dk}}{\sqrt{\int_0^{k_*} k^2 E(k)dk}} \propto \frac{\sqrt{\int_0^{k_*} k^{7/3}dk}}{\sqrt{\int_0^{k_*} k^{1/3}dk}} \propto k_*, \quad (36)$$

and in the case of Leith viscosity in an enstrophy cascade,

$$\frac{|\nabla^2 q_{*2d}|}{|\nabla q_{*2d}|} \propto \frac{\sqrt{\int_0^{k_*} k^4 G(k)dk}}{\sqrt{\int_0^{k_*} k^2 G(k)dk}} \propto \frac{\sqrt{\int_0^{k_*} k^3 dk}}{\sqrt{\int_0^{k_*} k dk}} \propto k_*. \quad (37)$$

Thus, the relevant curvatures are dominated by the gridscale in both cases. Therefore, up to adjustment of Υ and Λ, the relationship between ν_* and ν_{*4} suggested by *Griffies and Hallberg* [2000] is consistent with correct truncation of the enstrophy and energy cascades. Even higher derivative operators will be more and more strongly affected by the gridscale; thus, the generic enstrophy and energy cascading higher order relation is always

$$\nu_{*2n} = \frac{(\Delta x)^{2n}}{N} \nu_*, \quad (38)$$

for any positive integer n. The nondimensional tuning factor N is determined by the number of dimensions and the value of n, but can always be absorbed into Λ_{2n} and Υ_{2n}.

The scaling for higher order (38) seems obvious, but it requires confirmation. In fact, *Griffies and Hallberg* [2000] are in error in proposing the scaling for Smagorinsky in a flow where QG scaling holds and the gridscale lies in a potential enstrophy cascade range ($E(k) \propto k^{-3}$). In that case, the mean squared strain rate ($\int k^2 E(k)dk$) is dominated *by the largest scales*, so the integral in the denominator of (36) does not converge. Thus, the assumption required for the biharmonic Smagorinsky viscosity to scale according to numerical stability, (34), fails.

2.5. Modified Leith Viscosity

The Leith viscosity was implemented in the Massachusetts Institute of Technology general circulation model (MITgcm)

[*Marshall et al.*, 1997], and initial tests on 2D flows, for example the barotropic wind-driven gyre, were successful. However, during the design of the simulations of nonlinear spindown of a submesoscale front [*Boccaletti et al.*, 2007], where large Rossby numbers and strong internal gravity waves resulted from the initial conditions, one of the authors here noted that when using only Leith viscosity a gridscale noise pattern emerged that was difficult to remove by increasing Λ.

Closer analysis revealed that the gridscale noise pattern was checkered in vertical velocity, indicating a divergence/convergence pattern in the horizontal velocity (Figure 1, upper). The Leith viscosity parameterization is derived for purely 2D turbulence where the horizontal flow field is assumed to be divergenceless. However, oceanic flows are only quasi-2D. Divergences in the horizontal velocity are expected to be O(Ro) smaller than the vorticity, a fact that is often used explicitly in scaling arguments to derive "balanced" models where the divergent and vortical horizontal flow are treated separately [e.g., *McWilliams*, 1985]. However, it is possible that a gridscale divergence could arise in the model: through numerical errors, through large gridscale Rossby number ($U/f\Delta x$) and subsequent loss of balance, through forcing, or through internal waves generated by topography or radiated away from large Ro regions. The Leith viscosity only responds to buildup of vorticity at the grid scale, so if this divergent flow happened to have little or no vertical vorticity, it *would be totally undamped*.

A convenient way to fix this problem is to modify the Leith viscosity to add a damping of the divergent velocity. With introspection, one expects something similar to

$$\nu_* = \left(\frac{\Delta x}{\pi}\right)^3 \sqrt{\Lambda^6 |\nabla_h q_{2d}|^2 + \Lambda_d^6 |\nabla_h (\nabla_h \cdot \mathbf{u}_*)|^2}. \quad (39)$$

A physical rationale for this correction is unclear, but the numerical consequences are good. The lower panel of Figure 1 shows that the modified Leith viscosity with $\Lambda_d = \Lambda$ has substantially less checkerboard noise, although the basin mean viscosity is only larger by about 25%. Even doubling Λ with $\Lambda_d = 0$ was less effective in reducing the checkerboard pattern, although this doubling increases the viscosity by a factor of eight.

The divergence in MOLES is typically much smaller than the vorticity, so setting $\Lambda_d = \Lambda$ only slightly increases the viscosity. QG scaling indicates [*Pedlosky*, 1987]

$$\nabla_h \cdot \mathbf{u} \approx -\frac{\beta v}{f_0} - \frac{1}{f_0}\frac{\partial q_{2d}}{\partial t} - \frac{\mathbf{u} \cdot \nabla_h q_{2d}}{f_0},$$

$$\frac{\nabla_h \nabla_h \cdot \mathbf{u}_*}{\nabla_h q_{2d}} = \max\left[O\left(\frac{\beta \Delta x}{f_0}\right), O\left(\frac{\frac{\partial}{\partial t}}{f_0}\right), O\left(\frac{U}{f_0 \Delta x}\right)\right].$$

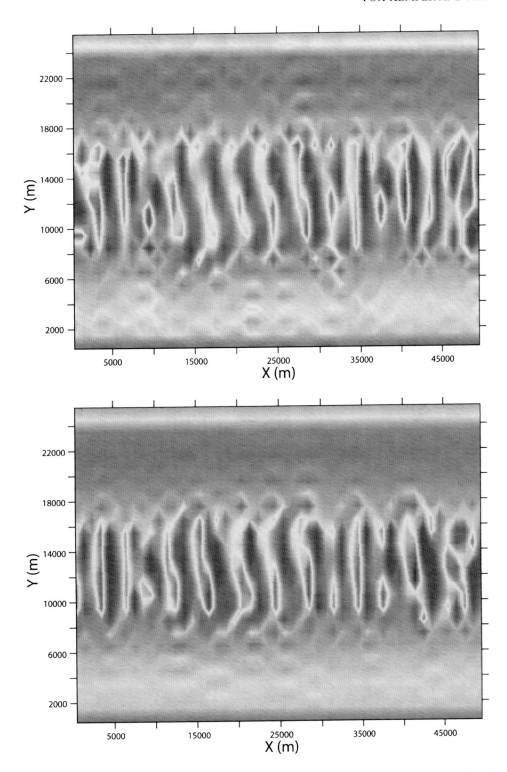

Figure 1. Vertical velocity from a simulation of spindown and instability of a temperature front in a reentrant channel. A simulation with the Leith viscosity applied to the horizontal velocities (upper) and a simulation with the modified Leith viscosity (lower) are shown ($\Lambda_d = \Lambda$). Light colors are near zero; colors represent upward or downward motion.

Therefore, the added divergence-sensing term will have very little effect on the regions where quasi-geostrophic flow dominates. It will have an impact on high-frequency internal waves, but these are typically not well resolved in MOLES in any case. The near-inertial gravity waves will be affected, but only as strongly as the QG flow. Fronts may have large Rossby number, but the expected increase will only be a factor of $\sqrt{2}$ in (39), as the divergence and vorticity contributions should match if gradients in only the cross-front direction dominate.

This scaling seems to indicate that one should expect few physical changes due to the added term, yet when this viscosity acts, it acts where the largest values of vertical velocity are. Because the Courant condition on vertical advection ($\Delta t < \Delta z / w$) is often the numerical constraint that sets the maximum timestep, this viscosity may substantially increase the allowable timestep without severely compromising the simulation. Tests have shown that in some calculations, a timestep three times larger was allowed when $\Lambda = \Lambda_d$ was used instead of $\Lambda_d = 0$.

2.6. High-Resolution Global Ocean Simulations

The modified Leith viscosity scheme has also been tested in a high-resolution global ocean MITgcm configuration described by *Menemenlis et al.* [2005]. This particular configuration employs a cubed-sphere grid projection [*Adcroft et al.*, 2004], which permits relatively even grid-spacing throughout the domain. Each face of the cube comprises 510×510 grid cells for a mean horizontal grid spacing of 18 km. There are 50 vertical levels ranging in thicknesses from 10 m near the ocean surface to 450 m near the ocean bottom. Initial temperature and salinity conditions are from the World Ocean Atlas 2001 [*Conkright et al.*, 2002]. Surface boundary conditions are from the National Center for Environmental Prediction (NCEP) and the National Center for Atmospheric Research (NCAR) atmospheric reanalysis [*Kistler et al.*, 2001] and are converted to heat, freshwater, and wind stress fluxes using the *Large and Pond* [1981, 1982] bulk formulae. Shortwave radiation decays exponentially as per *Paulson and Simpson* [1977]. Vertical mixing follows the method of *Large et al.* [1994] with background vertical diffusivity of 1.5×10^{-5} m^2 s^{-1} and viscosity of 10^{-3} m^2 s^{-1}. A third-order, direct-space-time advection scheme with flux limiter is employed and there is no explicit horizontal diffusivity.

Following a 38-year model spin-up, several additional 1-year (2001) integrations were conducted to test the stability and quality of the modified Leith scheme. Figure 2 displays surface kinetic energy from two such integrations. The first integration uses biharmonic Leith viscosity (*LeithOnly*, top

panel) and the second integration uses biharmonic Leith viscosity modified to sense the divergent flow (*LeithPlus*, bottom panel). Both test integrations use a time step of 600 s to stabilize the *LeithOnly* test case and for more direct comparison with the *LeithPlus* test case. The *LeithOnly* integration has slightly more volume-averaged kinetic energy, 4.36×10^{18} vs 4.23×10^{18} J, but the two simulations are qualitatively very similar.

Figure 3 compares near-surface viscosities from the *LeithOnly* integration (top panel) to viscosities from the *LeithPlus* integration (middle panel). Note that the magnitudes and patterns are similar; the global mean differs by less than 20%, and the time-mean vorticity is nearly identical in the two runs. The divergent flow modification (bottom panel) is modest, being typically an order of magnitude smaller than the total viscosity. The impact on model stability, however, and hence on timestep, is significant. It is found that using biharmonic Leith viscosity, the model can be integrated stably using a maximum time step size of 600 s. By comparison, the modified Leith scheme, which includes damping of divergent motions, can be integrated stably using a maximum time step of 1,200 s.

To illustrate this stability issue, Figure 4 shows daily snapshots of maximum Courant number ($w \Delta t / \Delta z$) in the two simulations. The *LeithOnly* simulation shows spikes dangerously near 1, which can crash the model, while the *LeithPlus* simulation has no such spikes. This indicates that in the *LeithOnly* integration, divergent instabilities occur, which, being unchecked by the Leith viscosity, may render the model unstable. The modified Leith scheme eliminates this problem.

Therefore, in summary, the viscosity for MOLES should be based on an inertial enstrophy cascade that produces the Leith scaling rather than the Smagorinsky scaling resulting from an inertial energy cascade. However, the Leith scaling needs to be adapted so that divergent motions that are present in 3D simulations do not become unstable or overly large to the point that the vertical advection Courant condition is contaminated. Up to this point, the evidence is clear; the remainder of the chapter is more speculative and presents opportunities for improving MOLES in novel ways.

2.7. Nonlinear Diffusivities

Smagorinsky [1963] used equal viscosity and diffusivity, scaled according to (23). He supposed that the stirring by eddies should not distinguish between temperature and velocity at large Reynolds and Péclet number (the Reynolds and Péclet number are the ratio of advection of momentum to friction and advection of temperature to dissipation). This fact, that the eddy Prandtl number (the ratio of eddy viscosity to eddy diffusivity) asymptotes to one at scales much larger

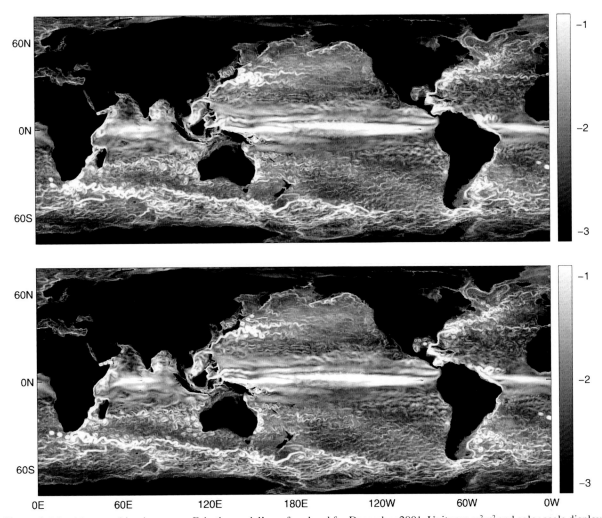

Figure 2. Monthly mean kinetic energy, E, in the model's surface level for December 2001. Units are m^2 s^{-2} and color scale displays $\log_{10}(E)$. Top panel is from the *LeithOnly* integration. Bottom panel is from the *LeithPlus* integration, which is modified to sense the divergent flow.

than the frictional and dissipation scales, has been confirmed for 3D turbulence in laboratory and direct numerical simulations (see chapter 15 of *Sagaut* [2005]). A similar issue arises for diffusivity and viscosity in MOLES, and this section addresses it.

2.7.1. Leith meets Charney. The reasoning behind applying the Leith scaling for viscosity to a 3D flow obeying QG scaling is the similarity of the vorticity equation in the 2D and QG cases (13). This similarity results in similar inverse energy cascades and (potential) enstrophy cascades of the two models. To fully exploit the similarities, however, one should pay attention to the form of the frictional and dissipation terms as well.

In 2D turbulence, the frictional term is easily shown to be

$$D_{q_{2d}} = \nabla_h \times \nu \nabla_h^2 \mathbf{u} = \nu \nabla^2 q_{2d}. \tag{40}$$

This result follows from taking the curl of the 2D Navier-Stokes equation. In QG turbulence, the frictional/dissipation operator is not so easily obtained. But, if one forms the relative vorticity equation and then eliminates the vortex stretching term (section 6.5 of *Pedlosky* [1987]), then one finds

$$D_{q_{QG}} = \nabla_h \times \nu_* \nabla_h^2 \mathbf{u} + \frac{\partial}{\partial z} \frac{f^2}{N^2} \nabla \cdot \mathbf{K}_* \cdot \nabla b. \tag{41}$$

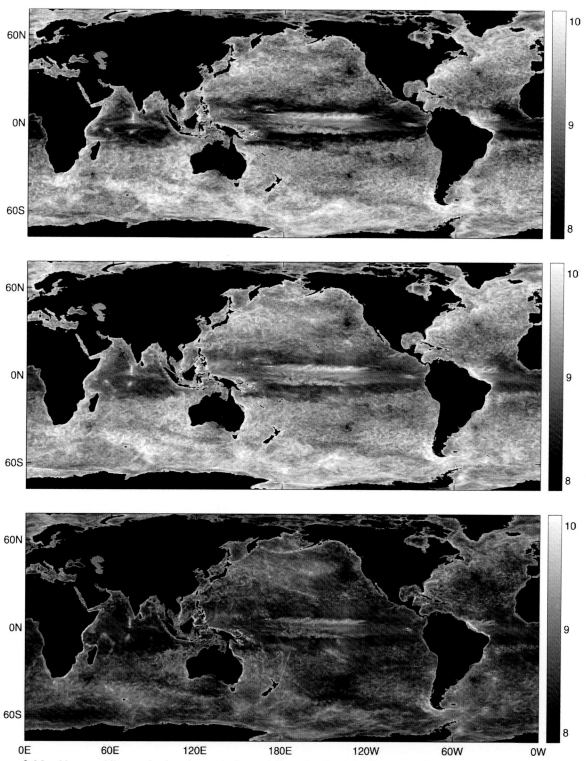

Figure 3. Monthly mean biharmonic viscosity, v_4, in the model's surface level for December 2001. Units are $m^4\ s^{-1}$ and color scale displays $\log_{10}(v_4)$. Top panel is from the *LeithOnly* integration. Middle panel is from the *LeithPlus* integration. Bottom panel shows the divergent modification of the *LeithPlus* integration.

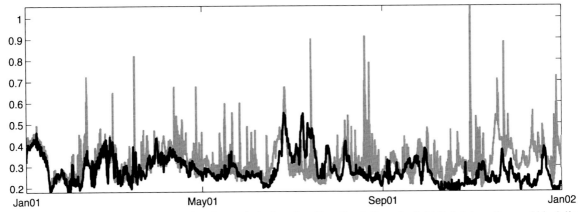

Figure 4. Maximum Courant number, $w\Delta t/\Delta z$, for vertical advection. Gray line is from the *LeithOnly* integration, and black line is from the *LeithPlus* integration.

The buoyancy is $b = g(\rho_0 - \rho)/\rho_0$.[3] Therefore, the horizontal relative vorticity and vertical buoyancy gradients are connected. If the diffusivity is assumed to be horizontally isotropic with a distinct vertical diffusivity, then

$$D_{q_{QG}} = \nu_* \nabla_h^2 q_{2d} + \frac{\partial}{\partial z} \frac{f^2}{N^2} \nabla \cdot \mathbf{K}_* \cdot \nabla b,$$

$$b = \frac{\partial \psi_{QG}}{\partial z},$$

$$D_{q_{QG}} = \nu_* \nabla_h^2 q_{2d} + \frac{\partial}{\partial z} \frac{f^2 \kappa_{*h}}{N^2} \nabla_h^2 \frac{\partial \psi_{QG}}{\partial z} + \frac{\partial}{\partial z} \frac{f^2}{N^2} \kappa_{*v} \frac{\partial^2 b}{\partial z^2},$$

$$D_{q_{QG}} = \nabla_h^2 [\nu_* q_{2d} + \kappa_{*h}(q_{QG} - q_{2d})]$$
$$+ \frac{\partial}{\partial z} \frac{f^2}{N^2} \kappa_{*v} \frac{\partial^2 b}{\partial z^2}. \tag{42}$$

The last step requires the assumption that N^2 and f^2 do not change appreciably in the horizontal in comparison to the eddy perturbations (as in QG scaling). In the spirit of truncating the cascade of potential enstrophy cleanly, that is, not breaking q_{QG} into a part associated with buoyancy diffusion and a part associated with friction, then one would like to assume that the horizontal diffusivity and the horizontal viscosity are equal $\nu_* = \kappa_{*h}$. The vertical diffusivity remains at this stage, without clear guidance as to how to specify it. It will take a little more work to make clear what to do with it.

QG may be derived in either a *z*-level or an isopycnal-layer coordinate framework, and in both cases, the stirring is predominantly horizontal, that is, perpendicular to the vertical coordinate. In QG, the isopycnal slopes must be shallow, and thus, there is little difference between the horizontal velocity and the along-isopycnal velocity. One must be a bit more careful in the primitive equation implementation. *Veronis* [1977] suggests that using an horizontal diffusivity for eddy

parameterizations spuriously mixes adjacent water masses when the isopycnal slope is steep. This mixing effect does not occur in quasi-geostrophic scaling or in primitive equation stirring when buoyancy variance is conserved [*Plumb*, 1979; *McDougall and McIntosh*, 2001a, 2001b]. Reducing the Veronis effect greatly improves the water mass properties in ocean models, and avoiding spurious diabatic mixing in eddy parameterizations and advection schemes is part of that process [*Roberts and Marshall*, 1998; *Griffies et al.*, 2000].

At first, it seems that avoiding spurious diapycnal stirring means that one should interpret the horizontal diffusivity as an along-isopycnal diffusivity and neglect cross-isopycnal diffusivity altogether. In this case, assuming small isopycnal slope [*Redi*, 1982; *Griffies*, 1998],

$$\mathbf{K}_* = \begin{bmatrix} \kappa_{*h} & 0 & \kappa_{*h} \dfrac{-b_{*,x}}{b_{*,z}} (17) \\[2ex] 0 & \kappa_{*h} & \kappa_{*h} \dfrac{-b_{*,y}}{b_{*,z}} (18) \\[2ex] \kappa_{*h} \dfrac{-b_{*,x}}{b_{*,z}} & \kappa_{*h} \dfrac{-b_{*,y}}{b_{*,z}} & \kappa_{*h} \dfrac{b_{*,x}^2 + b_{*,xy}^2}{b_{*,z}^2} \end{bmatrix}, \tag{43}$$

$$D_{q_{QG}} = \nu_* \nabla_h^2 q_{2d}. \tag{44}$$

The dissipation operator is now independent of buoyancy altogether, as $\mathbf{K}^* \cdot \nabla b_* = 0$. That is, if diffusion is oriented along isopycnals, there is no gradient of buoyancy in the along-isopycnal direction to act on, so there is no diffusive flux.

Roberts and Marshall [1998] suggest that the *Gent and McWilliams* [1990] parameterization (hereafter GM) may be used as a subgrid-scale mechanism for the dissipation of enstrophy in eddying models. The GM parameterization is not a diffusion of buoyancy along isopycnals, as the vertical fluxes are reversed in sign to ensure that there is always an extraction of potential energy. However, *Griffies* [1998] shows that the skew form of GM can be combined with the along-isopycnal diffusivity tensor to give a single mixing tensor, $\mathbf{J}_*$, that replaces $\mathbf{K}_*$ in the equations above. Assuming that the Redi horizontal stirring and the GM coefficient are equal, and furthermore that they are equal to ν_*, then

$$
\mathbf{J}_* = \begin{bmatrix} \nu_{*h} & 0 & 0 \\ 0 & \nu_{*h} & 0 \\ \nu_{*h}\dfrac{-2b_{*,x}}{b_{*,z}} & \nu_{*h}\dfrac{-2b_{*,y}}{b_{*,z}} & \nu_{*h}\dfrac{b_{*,x}^2 + b_{*,xy}^2}{b_{*,z}^2} \end{bmatrix}, \quad (45)
$$

$$
D_{q_{QG}} = \nu_* \nabla_h^2 q_{QG} + \frac{\partial}{\partial z}\frac{f^2}{N^2}\frac{\partial}{\partial z}\left(\nu_{*h}\frac{b_{*,x}^2 + b_{*,y}^2}{b_{*,z}}\right). \quad (46)
$$

In this case, the horizontal operator acts on the potential vorticity as desired, but now, there is a new term that can be shown to be a sign-definite sink of potential energy (at least in the case of constant ν_*). In the GM parameterization, this energy sink is understood to represent the conversion of resolved potential energy to kinetic and potential energy of unresolved baroclinic instabilities.

While this energy extraction effect is desirable numerically and desirable in coarse simulations where no eddies are resolved, it may not be desirable in MOLES. The reason is that an energy flux from resolved to unresolved scales is in conflict with the assumption of no scale to scale energy flux that is required for an inertial enstrophy flux. However, it is a flux of potential, not kinetic, energy, so it does not completely invalidate the kinetic energy arguments above, yet it remains worrisome until a more complete theory of inertial cascades in stratified fluids with the joint effects of the spectra of potential and kinetic energy is developed. It may be that this effect realistically represents the forward energy cascade observed below the dominant instability scale found by *Scott and Wang* [2005], yet does not represent enough energy flux to upset the idealization of an inertial enstrophy cascade. On the other hand, *Capet et al.* [2008] find that a similar extraction of potential energy by baroclinic instability allows a powerful forward energy cascade for submesoscale baroclinic instabilities that is able to alter the spectral slope of kinetic energy away from either the *Blumen* [1982] or the Charney/Kraichnan result discussed above.

In the end, it is unclear to what extent one should worry about parameterizing tracer diffusivities in a model with resolved eddies. *Howells* [1960] and *Young* [1987] estimate

that the effective diffusivity for stirring of tracers due to isotropic turbulence at smaller scales than k_o is

$$
\kappa_e(k) \approx \sqrt{2\int_{k_o}^{\infty} k^{-2}E(k)\,dk}. \quad (47)
$$

In MOLES, large eddies will be resolved ($k_o \ll k_*$), and the diffusivity will be dominated by the resolved eddies. This result indicates that one should seek to do as little harm as possible (i.e., spurious mixing) with subgrid-scale diffusive parameterizations. More testing and theory is needed to understand the implications and proper implementation of nonlinear diffusivities in MOLES, but at present, it seems that either using Redi along-isopycnal diffusivity (44) or the GM parameterization (46) is a reasonable handling of tracer diffusivity to complement the Leith scaling of viscosity for MOLES, while using horizontal diffusivity (42) leads to too much spurious mixing of water masses.

3. FILTERING AND DYNAMICAL ADJUSTMENT

The discussion now turns to consideration of dynamical adjustment. It is first necessary to review some of the LES theory on understanding the resolved flow as a filtered form of the total flow.

Often, oceanographers assume that the relationship between the subgrid-scale phenomena and the resolved phenomena is a Reynolds average. A Reynolds average may be considered to be the average over a large ensemble of fluid flows so that the turbulent features are removed. The ensemble average has particular properties: for some variable c, which might be a scalar (e.g., buoyancy) or a vector (e.g., velocity, as used by Reynolds),

$$
c = \bar{c} + c', \quad \bar{\bar{c}} = \bar{c}, \quad \overline{c'} = 0. \quad (48)
$$

Reynolds averaging results in advective fluxes that obey:

$$
\overline{\mathbf{u}c} = \overline{(\bar{\mathbf{u}} + \mathbf{u}')(\bar{c} + c')} = \bar{\mathbf{u}}\,\bar{c} + \mathbf{R}, \quad (49)
$$

$$
\mathbf{R} \equiv \overline{\mathbf{u}'c'}. \quad (50)
$$

In the atmospheric and oceanographic literature, sometimes a zonal mean is used which has similar properties.

However, since *Leonard* [1974], the typical approach used for LES differs in that the gridscale is understood to be a spatiotemporal filter instead of a Reynolds average. The filtering approach had been used for LES by *Lilly* [1967], but Leonard clarified the necessary averaging process. The idea that the gridscale is a filter on high wavenumbers occurs naturally in the context of spectral models where explicit filters are used to avoid aliasing gridscale noise into the advective fluxes [*Orszag*, 1971].

Unlike an ensemble mean or a zonal mean, when most filters are applied, they incompletely smooth the gridscale flow. Thus, repeated applications of the filter do not reproduce the same result: $\bar{\bar{c}} \neq \bar{c}$, $\bar{c}' \neq 0$. Terms that vanish under the Reynolds average do not in the filtered advective flux of c. They are called the Leonard terms, **L**, and the cross terms, **C**:

$$\overline{\mathbf{u}c} = \overline{(\bar{\mathbf{u}} + \mathbf{u}')(\bar{c} + c')} = \bar{\mathbf{u}}\,\bar{c} + \mathbf{L} + \mathbf{C} + \mathbf{R}, \qquad (51)$$

$$\mathbf{L} \equiv \overline{\bar{\mathbf{u}}\,\bar{c}} - \bar{\mathbf{u}}\,\bar{c}, \qquad (52)$$

$$\mathbf{C} \equiv \overline{\bar{\mathbf{u}}\,c'} + \overline{\mathbf{u}'\,\bar{c}}. \qquad (53)$$

The importance of the cross fluxes in a similar context for the ocean is noted by *Berloff* [2005].

The goal of a subgrid-scale model or eddy parameterization is to approximate the difference between the resolved flux and the filtered total flux: $\tau = \overline{\mathbf{u}c} - \bar{\mathbf{u}}\,\bar{c}$.[4] Under Reynolds averaging, the necessary term is just $\tau = \mathbf{R}$, but under the effects of a filter, the necessary term becomes $\tau = \mathbf{L} + \mathbf{C} + \mathbf{R}$.

Different filters are more or less useful for MOLES. Ensemble filters are expensive, but clearly converge to a Reynolds average. Spectral filters are also Reynolds averaging, but they cannot be used in bounded domains. Time averages are easily implemented and suffer no boundary problems, but mesoscale eddies are not the highest frequency motions in MOLES, so the parameterization will need to represent other high-frequency effects as well, for example internal waves. Spatial filters are easily implemented (e.g., boxcar, Gaussian, 1-2-1), but care is needed when boundaries are within the filter stencil and the cross and Leonard term effects must also be parameterized.

3.1. Dynamical Adjustment I: Bardina Similarity

The arguments in section 2 apply most directly to spatial filtering, and the arguments about replacing the inertial transfer of energy and enstrophy through a given wavenumber were not specific about the form of the inertial transfer, so the whole inertial transfer ($\mathbf{L} + \mathbf{C} + \mathbf{R}$) of momentum could be modeled with the Smagorinsky (23) or Leith (32) viscosities. At first glance, then, it would seem that in practice, there is little difference between assuming a Reynolds average or filtering.

However, *Bardina et al.* [1980] note that the filtering operation can be used *explicitly* to allow the subgrid-scale pa-

rameterization to depend on the resolved flow. They propose a subgrid-scale model based on a similarity hypothesis between the resolved and unresolved flow:

$$\overline{\mathbf{u}c} - \bar{\mathbf{u}}\,\bar{c} \approx \overline{\bar{\mathbf{u}}\,\bar{c}} - \bar{\bar{\mathbf{u}}}\,\bar{\bar{c}}. \qquad (54)$$

This model has distinct advantages over the Smagorinsky and Leith models: it can sense heterogeneity in the flow and can approximate backscatter from the unresolved scales to the large. However, *Bardina et al.* [1980] found that using just this model was numerically unstable and so added some Smagorinsky viscosity as well to stabilize it. This combination is called the mixed model.

More accurate results can be obtained if a second filter, called the test filter, is applied in addition to the gridscale filter. In this case, the Bardina model becomes

$$\overline{\mathbf{u}c}^G - \bar{\mathbf{u}}^G \bar{c}^G \approx C_{\mathrm{sim}}\left(\overline{\bar{\mathbf{u}}^G \bar{c}^G}^F - \overline{\bar{\mathbf{u}}^G}^F \, \overline{\bar{c}^G}^F \right). \qquad (55)$$

The G filter is the gridscale filter, and the F filter is the test filter, and C_{sim} is added for flexibility [*Meneveau and Katz*, 2000]. The mixed model has not been used in large-scale ocean modeling but could be implemented easily.

3.2. Dynamical Adjustment II: Germano Identity

Germano [1992] and *Germano et al.* [1991] take the use of the explicit filter a step farther with the so-called Germano identity:

$$\mathcal{L} \equiv \mathbf{T} - \bar{\tau}^F, \qquad (56)$$

$$\mathcal{L} \equiv \overline{\bar{\mathbf{u}}^G \bar{c}^G}^F - \overline{\bar{\mathbf{u}}^G}^F \, \overline{\bar{c}^G}^F \qquad (57)$$

$$\mathbf{T} \equiv \overline{\overline{\mathbf{u}c}}^G{}^F - \overline{\bar{\mathbf{u}}^F \bar{c}^G}^F \qquad (58)$$

$$\bar{\tau}^F \equiv \overline{\overline{\mathbf{u}c}^G - \bar{\mathbf{u}}^G \bar{c}^G}^F. \qquad (59)$$

This identity relates the total subgrid-scale effect at the test filter scale, **T**, to the total sub-gridscale effect at the gridscale, τ, using *only resolved quantities* present in a Leonard tensor L. Therefore, if the Smagorinsky viscosity is to apply at the test filter level and at the gridscale level, then $c = u_j$, and

$$\mathcal{L}_{ij} - \frac{1}{3}\mathcal{L}_{kk}\delta_{ij} = \left(\frac{\Upsilon}{\pi}\right)^2 \mathcal{M}_{ij},$$

$$\mathcal{M}_{ij} = \left(\Delta x_F^2 \overline{\overline{|D|}^G \overline{S_{ij}}^G}^F - \Delta x_G^2 \overline{\overline{|D|}^G \overline{S_{ij}}^G}^F \right),$$

where Δx_F is the test filter scale and Δx_G is the gridscale. This equation is overdetermined for Υ, but *Lilly* [1992] showed that the minimum-error least-squares solution is

$$\Upsilon = \pi \sqrt{\frac{\mathcal{L}_{ij}\mathcal{M}_{ij}}{\mathcal{M}_{ij}\mathcal{M}_{ij}}}. \tag{60}$$

The Germano identity applies equally well to the Leith viscosity, so

$$\mathcal{L}_{ij} - \frac{1}{3}\mathcal{L}_{kk}\delta_{ij} = \left(\frac{\Lambda}{\pi}\right)^3 \mathcal{M}_{ij},$$
$$\mathcal{M}_{ij} = \left(\Delta x_F^3 \overline{|V|}^G \overline{S_{ij}}^G{}^F - \Delta x_G^3 \overline{|V|^G \overline{S_{ij}}^G}^F\right),$$
$$\Lambda = \pi \left[\frac{\mathcal{L}_{ij}\mathcal{M}_{ij}}{\mathcal{M}_{ij}\mathcal{M}_{ij}}\right]^{1/3}. \tag{61}$$

In the 2D turbulence form that Leith originally proposed, one uses

$$V = \nabla_h q_{2d}. \tag{62}$$

The modified Leith form is

$$V = \sqrt{|\nabla_h q_{2d}|^2 + |\nabla_h(\nabla_h \cdot \mathbf{u}_*)|^2}. \tag{63}$$

Dynamical adjustment requires some smoothing of the coefficients Υ and Λ that are determined by the least-squares method. This smoothing is in line with assuming that the viscosities are sufficiently constant to allow them to be moved into and out of averaging operations (a required step above). This smoothness of viscosity requirement is also suggested above in the derivation of Smagorinsky where the global spectral characteristics of the flow were related to the local strain rate. Optimal methods of smoothing will have to be determined for MOLES.

Dynamical adjustment has been shown to substantially improve the Smagorinsky model when the flow is heterogeneous, for example, where there are boundary currents that transition into turbulent ones, because the parameterization can "turn off" when the resolved flow is laminar. In MOLES, this improvement is expected to be particularly dramatic, as a known bias of existing ocean model using Smagorinsky scaling is overly viscous boundary currents [*Jochum et al.*, 2008].

3.3. Dynamical Roberts-Marshall Mixing

Roberts and Marshall [1998] suggest that the GM parameterization and *Redi* [1982] may be used to cut off the cas-cade of tracer variance, just as Smagorinsky and Leith are suggested to cut off the cascade of energy and enstrophy. Above, it was argued that a reasonable choice was to set the eddy Prandtl number to 1, equating the GM/Redi coefficients to the Leith viscosity. The Germano identity may be used instead to produce Prandtl numbers other than 1.

If we assume that the GM coefficient and Redi diffusivity have a Leith-like dependence on the gridscale (i.e., quasi-geostrophic scaling) we find,

$$\kappa = \left(\frac{\Gamma \Delta x_G}{\pi}\right)^3 \overline{|\nabla_h q_{2d}|}^G,$$
$$\mathcal{L}_i = \left(\frac{\Gamma}{\pi}\right)^3 \mathcal{M}_i, \tag{64}$$

$$\mathcal{M}_i = \left(\Delta x_F^3 \overline{|\nabla_h q_{2d}|}^G \overline{\nabla_i c}^G{}^F - \Delta x_G^3 \overline{|\nabla_h q_{2d}|^G \overline{\nabla_i c}^G}^F\right),$$
$$\Gamma = \pi \left[\frac{\mathcal{L}_i \mathcal{M}_i}{\mathcal{M}_i \mathcal{M}_i}\right]^{1/3}. \tag{65}$$

Coefficients for the GM and Redi fluxes could be treated separately by analyzing the fluxes of buoyancy and salinity and temperature.

The dynamical method also allows estimation of phenomena that are not readily treatable with GM/Redi. For example, it is well-known that in the Eulerian average, cross-isopycnal fluxes result where the buoyancy variance varies in time or space even where the Lagrangian average is adiabatic. This effect is exceedingly difficult to anticipate with a parameterization, but in MOLES, diagnosing the cross-isopycnal fluxes of the resolved eddies can be used to guide the parameterized fluxes.[5] Similarly, *Eden et al.* [2007] have shown that fluxes that are in the along-isopycnal plane but perpendicular to the horizontal buoyancy gradient have a large effect in a diagnosis of an ocean model. These fluxes are inherently non-local and would be very difficult to parameterize, but in the presence of some resolved eddies, a dynamical method could be attempted.

One very hopeful result from engineering-scale testing of the dynamical adjustment method should be mentioned. Before implementing dynamical adjustment with active feedback on the flow, *a priori* testing was conducted. That is, the eddy viscosity was diagnosed from high-resolution simulations and compared with the value that would have resulted from the dynamical model scalings. Unexpectedly, using dynamical adjustment with active feedback typically outperforms the level of accuracy expected from the side-by-side comparison of diagnosed and scaled viscosities in *a priori* tests [e.g., *Vreman et al.*, 1995].

4. SUMMARY AND DISCUSSION

Large-basin and global ocean models are beginning to be rich with eddies and are beginning to resolve the potential enstrophy cascade below the instability lengthscale. However, they do not resolve all of the scales necessary to completely remove a dependence on subgrid-scale parameterizations. The tools developed for engineering-scale simulations, where the Smagorinsky viscosity is appropriate, may be adapted to the direct enstrophy cascade of mesoscale flows where the Leith viscosity, or a modified version of it, is appropriate. Dynamical adjustment using the Germano identity seems particularly promising, as it may be readily applied to many extant mesoscale parameterizations. Some of these applications are outlined here, and their validation will surely be an interesting future direction in high-resolution ocean models.

Acknowledgments. Discussions with R. Ferrari, K. S. Smith, D. Marshall, and C. Wunsch are happily recognized as improving this work. W. Smyth gave a preliminary reading of the manuscript that helped enormously. This work was supported by a NOAA Climate and Global Change postdoctoral fellowship and NSF grant OCE-0612143.

Notes

1. The tensor notation is standard, so indices before the comma represent direction, indices after the comma represent partial derivatives, and Einstein summation is implied (i.e., all repeated indices imply that the term is to be summed over all three directions). See *Griffies* [2004] for a more detailed presentation.
2. Incidentally, integration by parts converts the expression based on strain rates to the form encountered by dotting the momentum equation with velocity, $\langle v_i u_i \nabla^2 u^i \rangle$, except for boundary terms that vanish under all typical boundary conditions (periodic, slip, no-slip) or by averaging over a sufficient volume.
3. Salinity and potential temperature are assumed to be diffused similarly and nonlinear equation of state diffusion effects are neglected, both approximations are valid for large Péclet number.
4. In the special case of momentum flux, only the deviatoric stress needs to be modeled because the remaining stresses are solved for along with pressure by enforcing incompressibility.
5. However spurious effects of the z-coordinate advection schemes must be carefully handled [*Griffies et al.*, 2000].

REFERENCES

Adcroft, A., J.-M. Campin, C. Hill, and J. C. Marshall (2004), Implementation of an atmosphere-ocean general circulation model on the expanded spherical cube, *Mon. Weather Rev.*, *132*(12), 2845–2863.

Arbic, B. K., and G. R. Flierl (2004a), Baroclinically unstable geostrophic turbulence in the limits of strong and weak bottom Ekman friction: Application to midocean eddies, *J. Phys. Oceanogr.*, *34*, 2257–2273.

Arbic, B. K., and G. R. Flierl (2004b), Effects of mean flow direction on energy, isotropy, and coherence of baroclinically unstable beta-plane geostrophic turbulence, *J. Phys. Oceanogr.*, *34*, 77–93.

Bardina, J., J. H. Ferziger, and W. C. Reynolds (1980), Improved subgrid scale models for large eddy simulation, *AIAA Pap.*, pp. 80-1357.

Berloff, P. (2005), On dynamically consistent eddy fluxes, *Dyn. Atmos. Oceans, 38*, 123–146.

Blumen, W. (1982), Wave-interactions in quasi-geostrophic uniform potential vorticity flow, *J. Atmos. Sci.*, *39*, 2388–2396.

Boccaletti, G., R. Ferrari, and B. Fox-Kemper (2007), Mixed layer instabilities and restratification, *J. Phys. Oceanogr.*, *35*, 1263–1278.

Bracco, A., J. von Hardenberg, A. Provenzale, J. B. Weiss, and J. C. McWilliams (2004), Dispersion and mixing in quasigeostrophic turbulence, *Phys. Rev. Lett.*, *92*(8), 084,501.

Branscome, L. E. (1983), A parameterization of transient eddy heat flux on a beta-plane, *J. Atmos. Sci.*, *40*, 2508–2521.

Bryan, K., S. Manabe, and R. C. Pacanowski (1975), A global ocean-atmosphere climate model. Part II. The oceanic circulation, *J. Phys. Oceanogr.*, *5*, 30–46.

Burchard, H., O. Petersen, and T. P. Rippeth (1998), Comparing the performance of the Mellor-Yamada and the κ–ε two-equation turbulence models, *J. Geophys. Res.*, *103*, 10,543–10,554.

Capet, X., J. C. McWilliams, M. J. Molemaker, and A. F. Shchepetkin (2008), Mesoscale to submesoscale transition in the California Current System. Part I: Flow structure, eddy flux, and observational tests, *J. Phys. Oceanogr.*, *38*, 29-43, doi:10.1175/2007JPO3671.1.

Charney, J. G. (1971), Geostrophic turbulence, *J. Atmos. Sci.*, *28*, 1087–1095.

Conkright, M. E., R. Locarnini, H. Garcia, T. O'Brien, T. P. Boyer, C. Stephens, and J. Antonov (2002), World Ocean Atlas 2001, *CD-rom documentation*, National Oceanographic Data Center, Silver Spring, MD.

Deardorff, J. W. (1972), Parameterization of the planetary boundary layer for use in general circulation models, *Mon. Weather Rev.*, *100*, 93–106.

Deardorff, J. W. (1980), Stratocumulus-capped mixed layers derived from a three-dimensional model, *Boundary-Layer Meteorol.*, *18*, 495–527.

Eady, E. T. (1949), Long waves and cyclone waves, *Tellus, 1*, 33–52.

Eden, C., R. J. Greatbatch, and J. Willebrand (2007), A diagnosis of thickness fluxes in an eddy-resolving model, *J. Phys. Oceanogr.*, *37*, 727–742.

Ferrari, R., and D. L. Rudnick (2000), Thermohaline variability in the upper ocean, *J. Geophys. Res.*, *105*, 16,857–16,883.

Fox-Kemper, B. (2004), Reevaluating the roles of eddies in multiple barotropic wind-driven gyres, *J. Phys. Oceanogr.*, *35*(7), 1263–1278.

Fox-Kemper, B., and R. Ferrari (2008), Parameterization of mixed layer eddies. II: Prognosis and impact, *J. Phys. Oceanogr.* (in press).

Fox-Kemper, B., and J. Pedlosky (2004), Wind-driven barotropic gyre. I: Circulation control by eddy vorticity fluxes to a region of enhanced removal, *J. Mar. Res., 62*(2), 169–193.

Fox-Kemper, B., R. Ferrari, and R. W. Hallberg (2008), Parameterization of mixed layer eddies. I: Theory and diagnosis, *J. Phys. Oceanogr.* (in press).

Gent, P. R., and J. C. McWilliams (1990), Isopycnal mixing in ocean circulation models, *J. Phys. Oceanogr., 20*, 150–155.

Germano, M. (1992), Turbulence: The filtering approach, *J. Fluid Mech.*, pp. 325–336.

Germano, M., U. Piomelli, P. Moin, and W. H. Cabot (1991), A dynamic subgrid-scale eddy viscosity model, *Phys. Fluids, 3*(7), 1760–1765.

Geurts, B. J., and D. D. Holm (2003), Regularization modeling for large-eddy simulation, *Phys. Fluids, 15*(1), L13–L16.

Griffies, S. M. (1998), The Gent–McWilliams skew flux, *J. Phys. Oceanogr., 28*, 831–841.

Griffies, S. M. (2004), *Fundamentals of Ocean Climate Models*, 522 pp., Princeton University Press, Princeton.

Griffies, S. M., and R. W. Hallberg (2000), Biharmonic friction with a Smagorinsky-like viscosity for use in large-scale eddy-permitting ocean models, *Mon. Weather Rev., 128*(8), 2935–2946.

Griffies, S. M., R. C. Pacanowski, and R. W. Hallberg (2000), Spurious diapycnal mixing associated with advection in a z-coordinate ocean model, *Mon. Weather Rev., 128*, 538–564.

Held, I. M., and V. D. Larichev (1996), A scaling theory for horizontally homogeneous baroclinically unstable flow on a beta plane, *J. Atmos. Sci., 53*, 946–952.

Held, I. M., R. T. Pierrehumbert, S. T. Garner, and K. L. Swanson (1995), Surface quasi-geostrophic dynamics, *J. Fluid Mech., 282*, 1–20.

Holland, W. R. (1978), The role of mesoscale eddies in the general circulation of the ocean—numerical experiments using a wind-driven quasi-geostrophic model, *J. Phys. Oceanogr., 8*, 363–392.

Holloway, G. (1993), The role of the oceans in climate change: A challenge to large eddy simulation, in *Large Eddy Simulation of Complex Engineering and Geophysical Flows*, edited by B. Galperin and S. A. Orszag, pp. 425–440, Cambridge University Press, Cambridge.

Howells, I. D. (1960), An approximate equation for the spectrum of a conserved scalar quantity in a turbulent fluid, *J. Fluid Mech., 9*, 104–106.

Hua, B. L., and D. B. Haidvogel (1986), Numerical simulations of the vertical structure of quasi-geostrophic turbulence, *J. Atmos. Sci., 43*, 2923–2936.

Jochum, M., G. Danabasoglu, M. Holland, Y.-O. Kwon, and W. G. Large (2008), Ocean viscosity and climate, *J. Geophys. Res.*, doi:10.1029/2007JC004515, in press.

Killworth, P. H. (2005), On the parameterisation of eddy effects on mixed layers and tracer transports: a linearised eddy perspective, *J. Phys. Oceanogr., 35*, 1717–1725.

Kistler, R., E. Kalnay, W. Collins, S. Saha, G. White, J. Woollen, M. Chelliah, W. Ebisuzaki, M. Kanamitsu, V. Kousky, H. van den Dool, R. Jenne, and M. Fiorino (2001), The NCEP-NCAR 50-year reanalysis: Monthly means CD-ROM and documentation., *Bull. Am. Meteorol. Soc., 82*, 247–268.

Kolmogorov, A. N. (1941), The local structure of turbulence in incompressible viscous fluid for very large reynolds number, *Dokl. Akad. Nauk. SSSR, 30*, 9–13.

Kraichnan, R. H. (1967), Inertial ranges in two-dimensional turbulence, *Phys. Fluids, 16*, 1417–1423.

LaCasce, J. H., and A. Mahadevan (2006), Estimating subsurface horizontal and vertical velocities from sea-surface temperature, *J. Mar. Res., 64*, 695–721.

Lapeyre, G., and P. Klein (2006), Dynamics of the upper oceanic layers in terms of surface quasigeostrophy theory, *J. Phys. Oceanogr., 36*, 165–176.

Large, W. G., and S. Pond (1981), Open ocean momentum flux measurements in moderate to strong winds, *J. Phys. Oceanogr., 11*, 324–336.

Large, W. G., and S. Pond (1982), Sensible and latent heat flux measurement over the ocean, *J. Phys. Oceanogr., 2*, 464–482.

Large, W. G., J. C. McWilliams, and S. C. Doney (1994), Oceanic vertical mixing: A review and a model with a nonlocal boundary layer parameterization, *Rev. Geophys., 32*, 363–403.

Leith, C. E. (1996), Stochastic models of chaotic systems, *Physica D, 98*, 481–491.

Leonard, A. (1974), Energy cascade in large-eddy simulations of turbulent fluid flows, *Adv. Geophys., 18*, 237–248.

Lesieur, M., O. Métias, and P. Comte (2005), *Large-Eddy Simulations of Turbulence*, 232 pp., Cambridge Univesity Press, Cambridge.

Lilly, D. K. (1967), The representation of small-scale turbulence in numerical simulation experiments, *Proceedings of the IBM Scientific Computing Symposium on Environmental Sciences*, pp. 195–201.

Lilly, D. K. (1992), A proposed modification of the Germano subgrid-scale closure method, *Phys. Fluids A, 4*, 633–635.

Marshall, J., A. Adcroft, C. Hill, L. Perelman, and C. Heisey (1997), A finite-volume, incompressible Navier-Stokes model for studies of the ocean on parallel computers, *J. Geophys. Res., 102*(C3), 5753–5766.

Martinsen-Burrell, N., K. Julien, M. R. Petersen, and J. B. Weiss (2006), Merger and alignment in a reduced model for three-dimensional quasigeostrophic ellipsoidal vortices, *Phys. Fluids, 18*, 057,101–1–14.

McDougall, T. J., and P. C. McIntosh (2001a), The temporal-residual mean velocity. Part I: Derivation and the scalar conservation equations, *J. Phys. Oceanogr., 31*, 1222–1246.

McDougall, T. J., and P. C. McIntosh (2001b), The temporal-residual mean velocity. Part II: Isopycnal interpretation and the tracer and momentum equations, *J. Phys. Oceanogr., 31*, 1222–1246.

McWilliams, J. C. (1985), A uniformly valid model spanning the regimes of geostrophic and isotropic, stratified turbulence: Balanced turbulence, *J. Atmos. Sci., 42*, 1773–1774.

McWilliams, J. C., P. P. Sullivan, and C.-H. Moeng (1997), Langmuir turbulence in the ocean, *J. Fluid Mech., 334*, 1–30.

Mellor, G., and T. Yamada (1982), Development of turbulence closure model for geophysical fluid problems., *Rev. Geophys. Space Phys., 20*(4), 851–875.

Menemenlis, D., C. Hill, A. Adcroft, J. Campin, B. Cheng, B. Ciotti, I. Fukumori, P. Heimbach, C. Henze, A. Koehl, T. Lee, D. Stammer, J. Taft, and J. Zhang (2005), NASA supercomputer improves prospects for ocean climate research, *Am. Geophys. Union*, *86*(9), 89, 95–96.

Meneveau, C., and J. Katz (2000), Scale-invariance and turbulence models for large-eddy simulation, *Ann. Rev. Fluid Mech.*, *32*, 1–32.

Moeng, C. H. (1984), A large-eddy-simulation model for the study of planetary boundary-layer turbulence, *J. Atmos. Sci.*, *41*, 2052–2062.

Nastrom, G. D., and K. S. Gage (1985), A climatology of atmospheric wavenumber spectra of wind and temperature observed by commercial aircraft, *J. Atmos. Sci.*, *42*, 950–960.

Orszag, S. A. (1971), On the elimination of aliasing in finite-difference schemes by filtering high-wavenumber components, *J. Atmos. Sci.*, *28*, 1074.

Paulson, C. A., and J. J. Simpson (1977), Irradiance measurements in the upper ocean, *J. Phys. Oceanogr.*, *7*, 952–956.

Pedlosky, J. (1987), *Geophysical Fluid Dynamics*, second ed., 710 pp., Springer, Berlin.

Pierrehumbert, R. T., I. M. Held, and K. L. Swanson (1997), Spectra of local and nonlocal two-dimensional turbulence, *Chaos Solitons Fractal*, *4*(6), 1111–1116.

Plumb, R. A. (1979), Eddy fluxes of conserved quantities by small-amplitude waves, *J. Atmos. Sci.*, *36*, 1699–1704.

Redi, M. H. (1982), Oceanic isopycnal mixing by coordinate rotation, *J. Phys. Oceanogr.*, *12*, 1154–1158.

Roberts, M. J., and D. P. Marshall (1998), Do we require adiabatic dissipation schemes in eddy-resolving models?, *J. Phys. Oceanogr.*, *28*, 2050–2063.

Sagaut, P. (2005), *Large Eddy Simulation for Incompressible Flows*, 3rd ed., 556 pp., Springer, Heidelberg.

Salmon, R. (1980), Baroclinic instability and geostrophic turbulence, *Geophys. Astrophys. Fluid Dyn*, *10*, 25–52.

Scott, R. B., and D. N. Straub (1998), Small viscosity behavior of a homogeneous, quasigeostrophic, ocean circulation model, *J. Mar. Res.*, *56*, 1225–1258.

Scott, R. B., and F. Wang (2005), Direct evidence of an oceanic inverse kinetic energy cascade from satellite altimetry, *J. Phys. Oceanogr.*, *35*, 1650–1666.

Skyllingstad, E. D., W. D. Smyth, J. N. Moum, and H. Wijesekera (1999), Turbulent dissipation during a Westerly wind burst: A comparison of large-eddy simulation results and microstructure measurements, *J. Phys. Oceanogr.*, *29*, 5–29.

Smagorinsky, J. (1963), General circulation experiments with the primitive equations: I. The basic experiment., *Mon. Weather Rev.*, *91*, 99–164.

Smagorinsky, J. (1993), Some historical remarks on the use of non-linear viscosities, in *Large Eddy Simulation of Complex Engineering and Geophysical Flows*, edited by B. Galperin and S. A. Orszag, pp. 3–36, Cambridge University Press, Cambridge.

Smith, K. S., and G. K. Vallis (2002), The scales and equilibration of midocean eddies: forced-dissipative flow, *J. Phys. Oceanogr.*, *32*(6), 1699–1720.

Steiner, N., G. Holloway, R. Gerdes, Häkkinen, D. Holland, M. Karcher, F. Kauker, W. Maslowski, A. Proshutinsky, M. Steele, and J. Zhang (2004), Comparing modeled streamfunction, heat and freshwater content in the Arctic Ocean, *Ocean Model.*, *6*, 265–284.

Stone, P. H. (1972), A simplified radiative-dynamical model for the static stability of rotating atmospheres, *J. Atmos. Sci.*, *29*, 405–418.

Thompson, A. F., and W. R. Young (2006), Scaling baroclinic eddy fluxes: Vortices and energy balance, *J. Phys. Oceanogr.*, *36*, 720–738.

Thompson, A. F., and W. R. Young (2007), Two-layer baroclinic eddy heat fluxes: zonal flows and energy balance, *J. Atmos. Sci.*, *64*(9), 3214.

Tulloch, R., and K. S. Smith (2006), A theory for the atmospheric energy spectrum: Depth-limited temperature anomalies at the tropopause, *Proc. Natl. Acad. Sci.*, *103*(40), 14,690–14,694.

Veronis, G. (1977), Use of tracers in circulation studies, in *The Sea, vol. 6: Marine Modeling*, edited by E. D. Goldberg, pp. 169–188, Wiley, New York.

Visbeck, M., J. C. Marshall, T. Haine, and M. Spall (1997), Specification of eddy transfer coefficients in coarse resolution ocean circulation models, *J. Phys. Oceanogr.*, *27*, 381–402.

Vreman, A. W., B. J. Geurts, and J. G. M. Kuerten (1995), A priori tests of large eddy simulation of the compressible plane mixing layer, *J. Eng. Math.*, *29*, 299–327.

Wang, D., J. C. McWilliams, and W. G. Large (1998), Large-eddy simulation of the diurnal cycle of deep equatorial turbulence, *J. Phys. Oceanogr.*, *28*, 129–148.

Wunsch, C., and D., Stammer, (1995), The global frequency-wavenumber spectrum of oceanic variability estimated from TOPEX/POSEIDON altimetric measurements, *J. Geophys. Res.*, *100*(C12), 24,895–24,910.

Young, W. R. (1987), Baroclinic theories of the wind driven circulation, in *General Circulation of the Oceans*, edited by D. I. Arbarbanel and W. R. Young, pp. 134–201, Springer, Berlin.

B. Fox-Kemper, Cooperative Institute for Research in the Environmental Sciences and Department of Atmospheric and Oceanic Sciences, University of Colorado, Boulder, CO 80309, USA. (baylor@alum.mit.edu)

D. Menemenlis, Jet Propulsion Laboratory, California Institute of Technology, Pasadena, California, USA.

Lateral Mixing in the Eddying Regime and a New Broad-Ranging Formulation

Matthew W. Hecht, Mark R. Petersen, and Beth A. Wingate

Computational Physics Group, CCS Division, Los Alamos National Laboratory, Los Alamos, New Mexico, USA

Elizabeth Hunke and Mathew Maltrud

Fluid Dynamics Group, Theoretical Division, Los Alamos National Laboratory, Los Alamos, New Mexico, USA

We survey a number of issues associated with lateral dissipation in eddy-resolving ocean models and present two effective techniques. The first is a specification of lateral viscosity that is closely related to that of *Chassignet and Garraffo* [2001] involving the combined application of biharmonic and Laplacian forms of viscosity. The specification can in principle be applied across a broad range of model resolution, although our testing was performed only at eddy-resolving scale where a relatively simple form suffices. The second is the implementation of the Lagrangian averaged Navier–Stokes alpha (LANS-α) subgridscale turbulence scheme in a primitive equation ocean model, with our presentation here being largely a summary of the recent work of *Hecht et al.*, [2008] and *Petersen et al.* [2008]. As an inherently non-dissipative turbulence parameterization, one can understand the higher levels of eddy variability with LANS-α as coming about through an increase in the effective Rossby radius of deformation.

1. INTRODUCTION

More of the mixing processes that occur in the ocean are explicitly included in what we would refer to as eddy-resolving models, yet the parameterization of mixing remains of foremost concern. Oceanic tracer mixing may be decomposed into diabatic mixing, which tends to be very weak except under conditions of unstable stratification and a much more vigorous adiabatic mixing. Therefore, one of the great challenges of ocean modeling is the minimization of any projec-

tion of strong adiabatic mixing of tracers onto the diapycnal direction.

Whereas lateral mixing of tracers in non-eddying ocean models is nearly always formulated in an isopycnal plane, to mimic the way in which the unresolved eddies mix heat, salt, and other tracers, it remains a matter of judgment whether to do so at the strongly eddying resolutions that we address. Along with this question, we take up the issue of more effective formulation of lateral viscosity for eddy-resolving ocean modeling. In the case of lateral viscosity, we do not distinguish between application of this strong mixing in the horizontal or isopycnal planes. We restrict our focus to the consideration of lateral mixing, both in the momentum and the tracer transport equations, as nearly all of what is known about vertical mixing has been worked out in lower resolution, non-eddy-resolving models. Vertical or diabatic

Ocean Modeling in an Eddying Regime
Geophysical Monograph Series 177
This paper is not subject to U.S. copyright. Published in 2008 by the American Geophysical Union.
10.1029/177GM020

mixing only enters our discussion as an unintended side effect of the specification of lateral tracer mixing.

Biharmonic forms of lateral dissipation (both viscosity and diffusivity) are most often used in simulations intended to resolve some significant level of mesoscale variability, with the simulation of the North Atlantic at 37-km scale of *Semtner and Mintz* [1977] an early example. The reason for this transition from the more physically justifiable Laplacian dissipation used at lower resolutions to the higher order biharmonic form is a pragmatic one: The level of Laplacian dissipation required to provide adequate noise control may also strongly suppress the eddies and diffuse the jets. The reason for the limitation of hyperviscosity (or hyperdiffusivity) to the fourth-order biharmonic form in finite-difference or finite-volume ocean general circulation models is another pragmatic concern, that of maintaining a compact stencil size.

A more sophisticated usage of biharmonic lateral dissipation is the Smagorinski form presented by *Griffies and Hallberg* [2000], which allows for low values of viscosity in the interior while maintaining the higher values needed in the more problematic western boundary current regions, where these higher values may be needed either for numerical stability, to more fully represent the viscous balance effected by eddies if this effect is not adequately resolved.

Exceptions to the common usage of biharmonic dissipation may be found: For instance, the Ocean Circulation and Climate Advanced Modeling Project model uses Laplacian diffusion even at high resolution [*Lee et al.*, 2002], although this is used with a third-order upwind form of transport for which the truncation error can be shown to be equivalent to a velocity-dependent fourth-order diffusion [*Holland et al.*, 1998]. The anisotropic formulation of Laplacian viscosity introduced by *Smith and McWilliams* [2003] allows for acceptable levels of eddy activity and well-defined jets with judicious selection of parameters, including zero or nearly so, for the cross-stream component, as demonstrated by *Smith and Gent* [2004] (a paper that we bring in later for its introduction of an anisotropic form of adiabatic tracer mixing).

It is known that biharmonic dissipation can produce upgradient fluxes that are shown by *Delhez and Deleersnijder* [2007] to resemble the Gibbs phenomenon that occurs in spectral modeling. Nevertheless, biharmonic forms of dissipation continue to be used most frequently in eddy-resolving ocean modeling.

A number of simulations have used a value of biharmonic viscosity similar to that of *Smith et al.* [2000], a one tenth degree, Mercator-grid, regional calculation in which the Gulf Stream/North Atlantic Current system compared better with observations, indicating that the variability allowed within the model, while not fully resolved in a formal sense, was

sufficient to effect a considerable change in the character of the mean circulation (see *Hurlburt and Hogan.* [2000] and *Bryan et al.* [2007] for discussion of convergence or lack thereof). Smith et al. scaled the coefficient of biharmonic friction with grid cell area as

$$\nu_4 = \nu_4(\text{area}_0)(\text{area}/\text{area}_0)^{3/2} \qquad (1)$$

to maintain a constant grid Reynolds number across the grid and hence uniform capacity to suppress noise generated by the advective operator; this point is important to later discussion (here, $\nu_4(\text{area}_0)$ refers to a reference value of biharmonic viscosity in a grid cell of area area_0). The lateral biharmonic tracer diffusivity was also scaled as $\text{area}^{3/2}$ by *Smith et al.* [2000].

Simulations at this one tenth degree scale remain very sensitive, however, to the level of dissipation, as illustrated within the North Atlantic regional domain by *Bryan et al.* [2007] (hereafter BHS07). In that study, the range in parameter space that produced both realistic separation of the Gulf Stream and downstream penetration of the North Atlantic Current (NAC) into the region of the Northwest Corner was seen to be very limited. These same features are also sensitive to grid discretization and boundary conditions, topics touched on by *Hecht and Smith* [2008].

In the Miami Isopycnal Coordinate Ocean Model, *Chassignet and Garraffo* [2001] (hereafter CG01) found excessive variability in the Gulf Stream upstream from Cape Hatteras, with the Gulf Stream tending to undergo early separation when using the same viscous prescription as *Smith et al.* [2000]; use of higher values of viscosity, on the other hand, was seen to be overly dissipative. The authors settled on a combined isotropic approach, with biharmonic viscocity for noise control and a Laplacian form providing the viscous contribution to the balance of larger scale features. This approach has been carried over to the hybrid coordinate version of the model [*Bleck*, 2002]. Their use of biharmonic and Laplacian viscosities together has led to our reconsideration of lateral viscosity within a simple prescription applicable across a broad range of model resolution.

We proceed to introduce this new prescription for lateral dissipation. Short review discussions of adiabatic tracer mixing (section 3) and of the implied diffusion associated with tracer advection (section 4) follow. A promising and very different approach to subgridscale turbulence parameterization, one that is inherently non-dissipative, is covered in section 5. This approach is based on the first implementation of the Lagrangian averaged Navier–Stokes alpha model (LANS-α) [*Foias et al.*, 2001] in a primitive equation ocean model and raises the possibility of bringing out more realistic levels of oceanic mesoscale variability at what would otherwise be only eddy-admitting resolutions, thereby providing a very significant cost savings over that of conventional eddy-resolving modeling.

2. A NEW PRESCRIPTION FOR LATERAL VISCOSITY

Our development was motivated by the success of the anisotropic form of Laplacian viscosity and by the experience of CG01 in combining isotropic forms of biharmonic and Laplacian viscosity. Sole use of biharmonic dissipation has proven more adequate in our simulations in the z-coordinate Parallel Ocean Program model (POP) [*Smith and Gent*, 2002]; even so, we were led to reconsider the physical and numerical roles of viscosity and how one might articulate an overarching prescription for viscosity that would be dominated by biharmonic dissipation at high resolution and by Laplacian dissipation in the low-resolution limit.

In CG01, the relative strength of biharmonic and Laplacian viscosities is presented in terms of a damping time analysis. Damping times for monochromatic waves, for Laplacian, and biharmonic dissipation are

$$\tau_2 = v_2^{-1} \left(\frac{2}{\Delta x} \sin \left(\frac{k\Delta x}{2} \right) \right)^{-2}, \qquad (2)$$

$$\tau_4 = v_4^{-1} \left(\frac{2}{\Delta x} \sin \left(\frac{k\Delta x}{2} \right) \right)^{-4} \qquad (3)$$

[see *Griffies*, 2004]. Here, v_2 and v_4 are the Laplacian and biharmonic viscous coefficients, respectively; k is the wave number of the monochromatic wave that happens to be aligned, in this case, with the grid axis associated with spacing Δx. Using a small wavenumber approximation, as would hold for $\frac{k\Delta x}{2} \ll \frac{\pi}{2}$ and setting the two times to be equal, we solve for the wavenumber of the crossover point. In terms of a crossover lengthscale $\lambda_c = 2\pi/k_c$, we have

$$\lambda_c = 2\pi \sqrt{\frac{v_4}{v_2}}. \qquad (4)$$

For higher wavenumbers, v_4 provides the more rapid damping; for lower wavenumbers, v_2 is dominant. A choice of crossover length scale of around 80 km at middle latitudes was made in CG01.

Whereas CG01 scaled both v_4 and v_2 for constant grid Reynolds number, resulting in the area$^{3/2}$ scaling of equation (1) and a slower area$^{1/2}$ scaling for v_2, it suffices for one of the two terms to provide noise control, and so we reconsider the scaling of the Laplacian term. We note that the level of Lapla-

cian viscosity in CG01 was sufficient to span the width of the viscous Munk layer of the western boundary [*Munk*, 1950],

$$\delta_M = \left(\frac{\pi}{\sqrt{3}} \right) \left(\frac{v_2}{\beta} \right)^{1/3}, \qquad (5)$$

with three grid lengths at the latitude of Cape Hatteras. Indeed, one would expect this to be the case, or very nearly so, as this is a second requirement of the model viscosity, more physically based, in addition to the purely numerical requirement of noise control.

We propose a more general prescription for the use of Laplacian and biharmonic viscosity together, consisting of:

1. Laplacian viscosity scaled as Δx^3 to span the width of the Munk layer with a fixed number of grid lengths (of order two or three) at mid-latitudes, independent of overall model resolution, and
2. biharmonic viscosity scaled as area$^{3/2}$, for noise control, as in the work of *Smith et al.* [2000] and CG01.

If the grid is one that maintains uniform aspect ratio, with $\Delta x = \Delta y$, then the scaling of the Laplacian coefficient of prescription (1) is identical to that of the biharmonic coefficient of prescription (2), and the crossover point [where damping times for Laplacian and biharmonic terms are equal, defined in equation (4)] becomes independent of grid resolution, not just within some limited mid-latitude range on a particular grid but from the eddy-resolving regime treated in this volume to the coarse-resolution (two, three, or four degree) models used for long timeline paleoclimate study. In contrast, the approach of CG01 would prescribe a crossover length that grows as grid resolution in coarsened, and so is inconsistent with the widespread practice of using Laplacian viscosity at low resolution and biharmonic forms at high resolution. We adopt their combined application of biharmonic and Laplacian viscosity, but with a modified and broader ranging spatial scaling.

We present results based on use of a crossover lengthscale [equation (4)] of around 87 km with 2½ grid lengths across the viscous Munk layer [equation (5)] at a latitude of 35°N. The coefficients used, given in Table 1, are slightly lower than those of CG01, even with our grid resolution 20% more coarse than theirs, reflecting the level of difference in explicit viscosity required to produce satisfactory results with two very different models (we first tried values more similar to those of CG01). If our crossover lengthscale (or that of CG01) is interpreted in terms of grid length, one sees that it falls toward the coarse end of what would be referred to as "eddy-permitting." Our prescription results in an active, primary role for the biharmonic viscosity operator in eddy-resolving applications and in a secondary role in non-eddy-resolving application.

Table 1. Base coefficients of horizontal dissipation.

Case	v_4 (m⁴/s)	v_2 (m²/s)	κ_4 (m⁴/s)
14a	27.0×10^9	–	9.0×10^9
14b	13.5×10^9	–	4.5×10^9
14c	6.75×10^9	–	2.25×10^9
14x	6.75×10^9	35.5	2.25×10^9

Cases 14a, 14b, and 14c are from BHS07; case 14x is our test case with the addition of Laplacian viscosity. Spatial scaling of coefficients is as per equation (1).

The values of biharmonic viscosity and diffusivity used in our test are equal to the lowest considered in BHS07; three cases from that paper, along with our test case, are listed in Table 1 where our test case is referred to as 14x. With the inclusion of Laplacian viscosity, the global mean kinetic energy of case 14x is seen to drop to around the same level as that of case 14b of BHS07 where the fourth-order dissipative coefficients were twice as large (see Figure 1).

The sensitivity of ocean simulations at around 1/10° is seen not only in levels of kinetic energy, but also in qualitative features, some of which would appear to be important to climate system response. The recent work of *Hallberg and Gnanadesikan* [2006] demonstrates the very different response of the Southern Ocean to changes in wind stress forcing in a simulation with a vigorous field of eddies, as compared to simulations in which the eddies are parameterized. In the North Atlantic, much attention has been paid to Gulf Stream separation, but

downstream features may be more important, as discussed in this same volume by *Hecht and Smith* [2008]. *Weese and Bryan* [2006] examined North Atlantic Current bias and atmospheric circulation, attributing much of the cause of an anomalous stationary wave pattern and excessive Icelandic Low in their climate model to the sea surface temperature bias.

Not only are some of the largest sea surface temperature biases in climate simulations to be found in this southern region of the Labrador Sea, where the North Atlantic Current penetrates into the region known as the Northwest Corner (*Rossby* [1996]), but there is also the issue raised by *Hecht et al.* [2006], and *Hecht and Smith* [2008]: The dense waters that convect in winter, forming a vulnerable link in the thermohaline circulation are, in ocean climate models, subject to biased conditioning as a consequence of this incorrect path.

The observed path of the North Atlantic Current is evident in satellite-based observations of altimetry, as in Plate 1 (panel at upper right). This branch of North Atlantic circulation is also produced in all the 0.1° model simulations shown except for the case with the highest values of dissipation (14a, upper left). Lower resolution simulations generally do not produce penetration of the North Atlantic Current into the Northwest Corner (again, see BHS07 and references therein). Our test case does well in this respect (lower right-hand panel of Plate 1).

One of the points made in BHS07 is that deep penetration of eddy kinetic energy is correlated with reattachment of the Stream as it encounters the topography of the Southeast Newfoundland Rise (at around 48°W). The physical tendency for greater topographic control with stronger levels of eddy kinetic energy near that topography, or the converse of

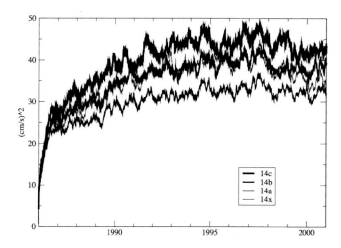

Figure 1. Mean kinetic energy, with the most viscous biharmonic case (14a) having the lowest kinetic energy, the least viscous case (14c) having the highest kinetic energy. Our test case (14x) has the same biharmonic dissipation as case 14c, but with the addition of a Laplacian viscosity; its level of mean kinetic energy is similar to that of the intermediate case 14b. The four cases are described in Table 1.

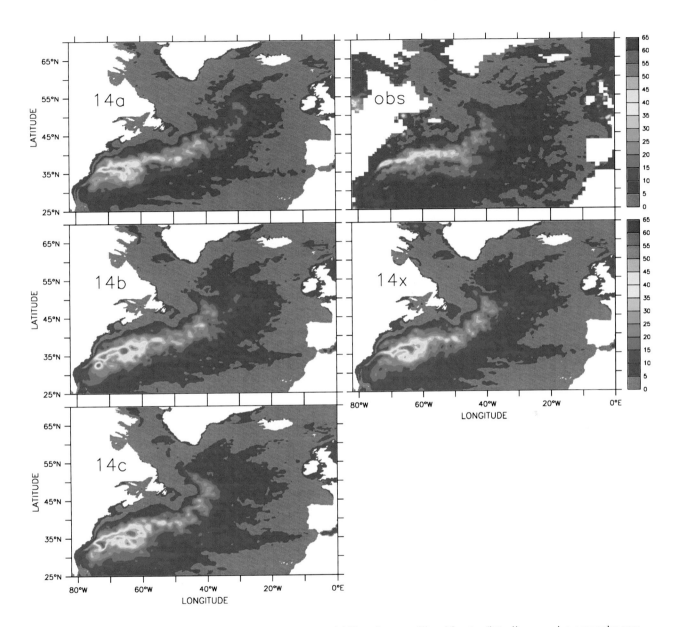

Plate 1. Time-averaged North Atlantic sea surface height variability, from satellite altimetry (http://www.aviso.oceanobs.com, upper right), from the three biharmonic cases of BHS07 (left), and from our test case (lower right), all for the 3-year period 1998–2000. The anomalously high variability seen near the Gulf Stream's separation point in the less viscous 14b and 14c biharmonic cases is reduced in the more viscous case 14a, and in the 14x test case; the North Atlantic Current turns northward around the Grand Banks in the latter case.

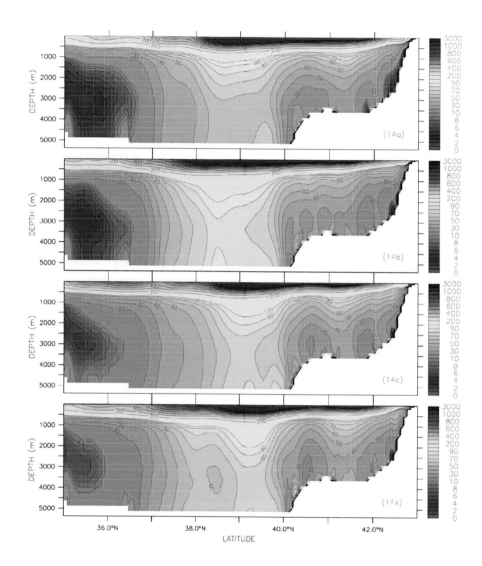

Plate 2. Eddy kinetic energies at 50°W. Our test case 14x, despite having a less energetic (and more realistic) separation of the Gulf Stream from the North American coast at Cape Hatteras (Plate 1), appears to have sufficient vertical penetration of eddy kinetic energy at this downstream point to successfully reattach to the continental topography at the Grand Banks. In contrast, the most viscous biharmonic case (14a) has less vertical penetration of kinetic energy and fails to make the northward turn around the Grand Banks (as was seen, again, in the earlier figure).

detachment made possible through lower levels of deep eddy kinetic energy and weaker topographic control, was cited by Özgökmen et al. [1997] to explain the leading-order physics upon which Gulf Stream separation depended in their somewhat idealized modeling study. This factor of greater topographic control was found, in BHS07, to be more relevant to the reattachment of the Stream at the Grand Banks than its detachment from the continental slope at Cape Hatteras. There is no obvious evidence of excessive suppression of deep eddy kinetic energy with inclusion of Laplacian viscosity in the 14x test case seen in Plate 2 at 50°W; the only one of the four cases with significantly weaker penetration of deep eddy kinetic energy is the more dissipative case 14a, the one case in which the NAC failed to make the downstream turn northward around the topography of the Grand Banks (Plate 2).

In the two less dissipative of the three cases from BHS07, there is excessive variability at the Cape Hatteras separation point, as evident in the sea surface height variability of Plate 2. The intermediate case 14b was identified as providing the best compromise between Gulf Stream separation and strong penetration of the NAC into the Northwest Corner region in BHS07. Our presciption for combined use of biharmonic and Laplacian viscosities produces an improved solution: The excess variability near Hatteras is much reduced in case 14x, even as the Gulf Stream and North Atlantic Current remain well represented. This finding of a reduction in variability at the separation point of the Gulf Stream is consistent with that of CG01.

3. ADIABATIC TRACER MIXING

The assumption that "resolution solves all problems" is sometimes mistakenly applied to ocean modeling. Any such problem-free regime remains remote. At model resolutions of around 0.1°, horizontal tracer mixing is often used, although Roberts and Marshall [1998] argue that spurious cross-frontal mixing should be expected to remain substantial. They consider this cross-frontal mixing (the "Veronis effect") [Veronis, 1975] in terms of the vorticity gradient associated with the front, explaining that high levels of spurious diabatic mixing persist, even at high resolution, due to horizontal parameterization of the turbulent cascade. Model simulations at resolutions as high as 1/8° are offered in support of the authors' case.

The z-coordinate model simulations that employ adiabatic dissipation in the work of Roberts and Marshall [1998] require only the thickness diffusion, or eddy-induced transport term associated with the so-called GM90 scheme of Gent and McWilliams [1990], because of their specification of constant salinity (if the equation of state becomes dependent only upon temperature then mixing along isopycnal surfaces reduces to zero). With only the thickness diffusion term active, it is straightfor-

ward to replace the usual Laplacian form of thickness diffusion with a more scale-selective biharmonic term, preserving the eddy-resolving aspect of the 1/8° simulation. Indeed, this is an advantage held by isopycnal or hybrid z-coordinate/isopycnal models where a complete implementation of GM90 requires that only the thickness diffusion term and a biharmonic form is readily and routinely used.

A full implementation of a biharmonic form of GM90 becomes problematic in a z-coordinate model, as discussed by Smith and Gent [2004], who present an anisotropic Laplacian form of the scheme as an alternative to the numerically difficult biharmonic form. With this approach also motivated by the success of anisotropic forms of viscosity, they show that their solution compares favorably in terms of energetic levels, despite its Laplacian formulation. They report modest changes in the wind-driven circulation and more substantial improvements in poleward heat transport and meridional overturning circulation, as compared with solutions based on a biharmonic form of horizontal tracer mixing. They make the interesting observation that an anisotropic tracer mixing formulation with zero cross-stream component, as was the case in their work, will not degrade an idealized front with uniform along-front values of temperature and salinity.

The work of Smith and Gent [2004] is not only highly relevant to the subject of this book but deserves further investigation. It would be useful to consider anisotropic forms of viscosity and adiabatic mixing separately at 0.1° resolution, and the benefit of avoiding the Veronis effect could be evaluated more thoroughly in relation to the cost of the schemes. They also point out the need to evaluate the inherent degree of anisotropy in a strongly eddying ocean; the recent work of Eden et al. [2007] identifies some degree of anisotropy in eddy-driven mixing associated with Rossby wave propagation, but the extent of the effect remains to be determined.

The cost of an adiabatic tracer mixing scheme of anisotropic form is substantial, and so simple and efficient horizontal tracer mixing remains a viable choice at eddy-resolving resolution despite the demonstrated benefit of using the more sophisticated mixing scheme. At lower, eddy-admitting resolutions, however, the cost/benefit analysis tilts in favor of adiabatic mixing.

Hunke et al. [2008] found that the anisotropic form of adiabatic tracer mixing allowed for a relatively energetic circulation in the Arctic region of their global simulation where cells on their grid tend to be significantly smaller than in other regions. They also evaluated the use of the more conventional isotropic adiabatic tracer mixing, considering horizontal biharmonic tracer mixing, scaled as in equation (1), as well. With the use of isotropic adiabatic tracer mixing, they found the incorporation of spatial scaling of the mixing coefficient with grid-cell area, based on the same criteria of constant grid-Reynolds number that underlies equation (1), to be essential to bringing out an energetic model circulation in the Arctic. This scaling with grid-cell area resulted in a more modest but still significant level of improvement with the anisotropic form of the scheme.

4. ADVECTION, TRUNCATION ERROR, AND MIXING

Upwind-biased advection schemes will tend to add to the overall level of mixing in a numerical ocean circulation model through a leading-order truncation error that is dissipative. Centered advection schemes, while non-dissipative, produce dispersive errors that in turn generate artificially high levels of dissipation by causing the diffusive operator, if in the tracer transport equations, to act strongly to dissipate the "ringing" associated with the dispersive error [*Hecht et al.*, 1995]. The same is true in the momentum equations where the viscous operator will tend to act to suppress dispersive errors generated by a centered discretization of momentum transport.

Griffies et al. [2000] present a thorough testing and analysis of the truncation errors associated with the most widely used tracer transport schemes of the time, evaluating the diffusion associated with truncation error in applications as high as 1/9° in grid resolution. They show that there are competing influences as the grid spacing is refined. The magnitude of the error tends to fall off as Δx^2. On the other hand, the lower values of explicit diffusion that one would use on a more refined grid allow for a more vigorous flow field, generating more implicit mixing from the transport scheme.

Schemes that have been adopted for use in ocean models more recently, such as that based on the work of *Hundsdorfer and Trompert* [1994; see *Adcroft et al.*, 2005; (K. Lindsay, A class of flux limited advection schemes applied to ocean models, submitted to *Ocean Modelling*, 2007, hereinafter referred to as Lindsay, submitted manuscript, 2007] produce significantly lower levels of implicit dissipation in these same tests (S. Griffies, personal communication). There are, however, some unsettling, if yet unresolved, questions associated with the use of such a scheme at eddy-resolving scale. In a revised configuration of the global model of *Maltrud and McClean* [2005], now using a tripolar grid [*Smith and Gent*, 2002] and partial cell representation of bottom topography [*Adcroft et al.*, 1997], a scheme from this newer class (Lindsay, submitted manuscript, 2007) degraded the equatorial jet structure in the Pacific.

A recent line of discussion among ocean model developers concerns the question of whether adiabatic tracer mixing schemes have any physical basis at higher resolutions or whether they are simply accomplishing the suppression of noise generated by inadequate advection schemes (S. Griffies, personal communication). In the near future, we will certainly see careful consideration of more advanced transport schemes in strongly eddying ocean models in pursuit of this question.

We express a note of caution, however, that the answer to this question is not yet entirely certain. The satisfactory results found by *Roberts and Marshall* [1998] and *Smith and Gent* [2004] most likely imply that dispersive errors associated with centered advection tend, in the vicinity of strong fronts, to be aligned in an along-frontal orientation. If it were not so, then we would see a more significant Veronis effect in eddy-resolving models. Indeed, it should not be entirely unexpected that the spurious mixing which results from dispersive advective error should be oriented in this fashion, as the advecting velocity will also along tend to orient parallel to the front where geostrophic balance is dominant.

5. NON-DISSIPATIVE PARAMETERIZATION OF TURBULENCE FOR ENHANCED VARIABILITY: LANS-α

In this section, we give an overview of a new approach to turbulence parameterization for ocean modeling. The turbulence model itself is known as the Lagrangian averaged Navier–Stokes α model (LANS-α) [*Holm et al.*, 1998; *Holm*, 1999, 2002]. This turbulence parameterization, closely related to the generalized Lagrangian mean theory of *Andrews and McIntyre* [1978], is derived from an averaging along mean Lagrangian trajectories (following fluid parcels) in Hamilton's Principle, yielding modified dynamical equations. The modified transport equations, which now include turbulent effects of scales smaller than α, maintain the Kelvin Circulation Theorem, meaning that the circulation around a loop embedded in the fluid neither spins up nor spins down as a result of inclusion of LANS-α. On a more pragmatic level, it means that the model is non-dissipative. Consequently, the conservation of certain quantities including total energy and potential enstrophy is maintained, making for a model with characteristics quite different from those of more familiar closures, as explained below. The model may be used in conjunction with more conventional closures, such as eddy viscosity, to dampen grid scale noise.

The filter width, α, can be thought of as representing the smallest active scale in the solution below which the dynamics at smaller scales is modeled as passive. These small scales are "dragged" or "swept" by the fluid motion of the large scales, instead of being diffused as occurs in many other methods.

The LANS-α model has been studied extensively in the context of theory [*Holm et al.*, 1998; *Holm*, 1999, 2002; *Foias et al.*, 2001], direct numerical simulations [*Chen et al.*, 1998, 1999a, 1999b], fluid instability [*Holm and Wingate*, 2005], quasi-geostrophy [*Holm and Nadiga*, 2003], large eddy simulations [*Domaradzki and Holm*, 2001; *Geurts and Holm*, 2003], and the shallow water equations [*Wingate*, 2004; for a review, see *Holm et al.* [2005].

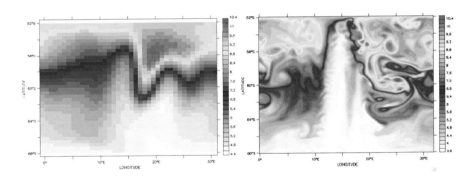

Plate 3. Sea surface temperature of the periodic channel problem used to evaluate POP-α, at low (0.8°, left) and high (0.1°, right) resolutions (but without inclusion of LANS-α here). Mesoscale eddies form if the resolution is sufficiently high.

One feature of this model that is particularly relevant to ocean modeling was discussed by *Holm and Wingate* [2005] who showed that for a two-layer baroclinic instability problem, the alpha model had the effect of moving the Rossby deformation radius to lower wave number such that instability can occur on a coarser mesh. Essentially, one can bring energetic mesoscale variability into an ocean model at what would otherwise be only eddy-admitting resolution.

The first implementation of LANS-α in a primitive equation ocean model has been described by *Hecht et al.* [2008] with further investigation of efficiency by Petersen, *Petersen et al* [2008]. In brief, the primitive equations are modified as

$$\frac{d\mathbf{v}}{dt} + \sum_j v_j \nabla u_j + \mathbf{f} \times \mathbf{u} = -\frac{1}{\rho_0}\nabla\pi + \mathcal{F}(\mathbf{v}), \quad (6)$$

$$\frac{d\varphi}{dt} = \mathcal{D}(\varphi), \quad (7)$$

$$\frac{d}{dt} = \frac{\partial}{\partial t} + \mathbf{u}\cdot\nabla + u_3\partial_z \quad (8)$$

$$\sum_j v_j\frac{\partial u_j}{\partial z} + \frac{\partial\pi}{\partial z} = -\rho g, \quad (9)$$

$$\nabla\cdot\mathbf{u} + \partial_z u_3 = 0, \quad (10)$$

$$\mathbf{u} = \left(1 - \alpha^2\nabla^2\right)^{-1}\mathbf{v}, \quad (11)$$

$$\pi = p - \frac{1}{2}|\mathbf{u}|^2 - \frac{\alpha^2}{2}|\nabla\mathbf{u}|^2, \quad (12)$$

where π is a modified pressure, φ is a tracer, and $\mathcal{F}$ and $\mathcal{D}$ are diffusion operators. One should note that there are now two velocities in the ocean model, a smooth velocity $\mathbf{u}$, only

containing scales greater than α, that performs the transport, and a rough velocity $\mathbf{v}$ that is transported by the flow. These two velocities are related through the Helmholtz operation of equation (11).

In the case when the density is a linear function of salinity and temperature, rotation is constant, and in the absence of dissipation, these equations have the following conservation laws

$$\frac{d}{dt}\int_V \left(|\mathbf{u}|^2 + \alpha^2|\nabla\mathbf{u}|^2 + \rho^2\right) = 0 \quad (13)$$

$$\frac{d}{dt}\int_V q = 0 \quad (14)$$

where

$$q = \nabla_3\rho\cdot\left((\nabla_3\times\mathbf{v}_H) + f\right). \quad (15)$$

The LANS-α turbulence parameterization was implemented in the POP primitive equation ocean model (referred to collectively as POP-α) with the addition of the smooth advecting velocity $\mathbf{u}$. This velocity may be obtained using the Helmholtz inversion of equation (11) or by a convolution filter that simply smoothes the $\mathbf{v}$ velocity by averaging over neighboring grid cells. The filter is more computationally efficient and has been shown to produce results equivalent to those from the Helmholtz inversion [*Petersen et al.*, 2008]. The proper implementation of LANS-α in the barotropic solver of the POP model presented a particularly difficult challenge. The barotropic solver includes an iterative solution for the surface elevation, and the formal derivation of LANS-α in the POP algorithm requires smoothing steps within each iteration. That proved to be too expensive, so a reduced algorithm was designed that avoids the smoothing within the iterative step, thereby improving efficiency greatly while producing results nearly identical to those found with the full LANS-α model [*Hecht et al.*, 2008].

Figure 2. Comparison of turbulence statistics produced by POP, POP-α with smoothing by the Helmholtz inversion, and POP-α with smoothing by convolution filters. Identifiers such as 0.8 refer to an $0.8°\times 0.8°$ grid. The α parameter controls the strength of smoothing by the Helmholtz inversion, while the stencil width (F3–F9) controls smoothing strength for the filters. In general, the LANS-α model improves all turbulence statistics with only minor addition of computational cost in the more efficient implementations. These figures reprinted from *Petersen et al.* [2008].
(a) Kinetic energy, Helmholtz inversion
(b) Kinetic energy, convolution filters
(c) Eddy kinetic energy, Helmholtz inversion
(d) Eddy kinetic energy, filters
(e) 6°C isotherm depth, filters
(f) Computation time

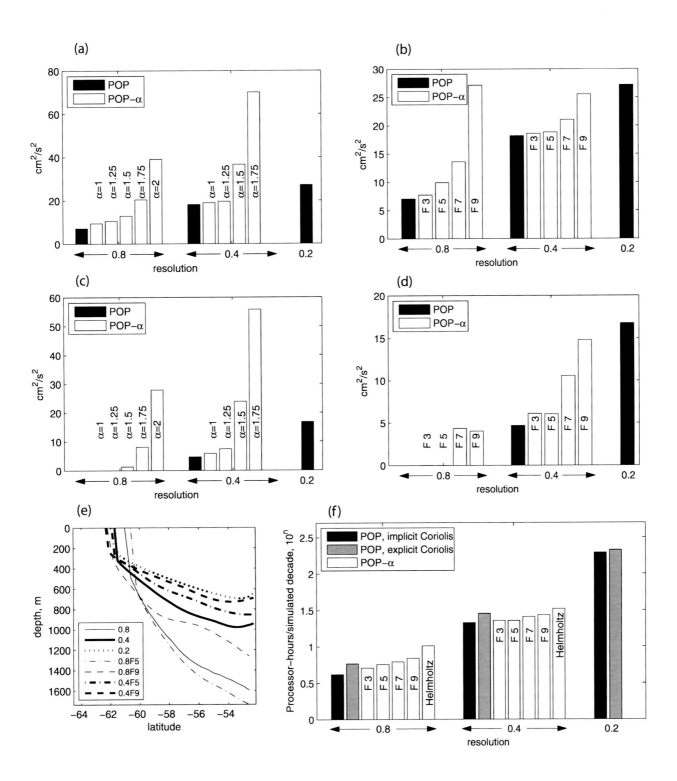

The POP-α algorithms and smoothing methods were tested using an idealized configuration that induces baroclinic instability (Plate 3). The domain is a zonally periodic channel with a meridional deep-sea ridge, westerly wind forcing, and surface thermal forcing that is warm in the north and cool in the south. These conditions are similar to those in the Southern Ocean and cause the isopycnals to tilt downward from south to north. In the real ocean, mesoscale eddies transfer heat and flatten the isopycnals, thereby converting the potential energy of tilted isopycnals to the kinetic energy of the eddies themselves. In ocean models, this only occurs if the resolution is sufficient for eddies to appear in the simulations (Figures 2e and 3). Thus, the slope of the isopycnals is an additional diagnostic to measure eddy activity in this test problem.

LANS-α improves the representation of turbulence in all statistics measured in the channel configuration. In general, the turbulence model becomes more effective as the smoothing operator on **u** is strengthened. This can be accomplished by increasing the α parameter in the Helmholtz operator of equation (11) or by using a larger filter stencil. Figure 2 shows that both methods flatten the isopycnals and increase kinetic energy and eddy kinetic energy. The ability to increase eddy activity is a unique feature of the LANS-α model; other turbulence closure models, such as eddy viscosity and hyperviscosity, reduce kinetic and eddy kinetic energy. *Holm and Wingate* [2005] showed that LANS-α preserves the value of forcing required for the onset of baroclinic instability, while eddy viscosity models increase the required forcing so that eddies are less likely to appear.

It is possible to bring out an unrealistic level of variability with the use of an overly large filter width α, as demonstrated by *Holm and Nadiga* [2003]. Tuning is required for optimal results, as with other parameterizations. We have indeed found that POP-α can produce turbulence statistics similar to a doubling of resolution or greater and only requires an additional 27% computational time (in the most efficient implementation) as compared to a factor of eight to ten in computational time to double the horizontal resolution [*Petersen et al.*, 2008].

6. SUMMARY DISCUSSION

We have presented in Section 2 a simple method of combining Laplacian and biharmonic forms of viscosity and have found this prescription to be effective in reducing the excessive variability near the Gulf Stream's separation point of Cape Hatteras, consistent with the experience of CG01. Our prescription is designed for use over a broad range of resolutions, with biharmonic viscosity providing noise control and Laplacian viscosity providing viscous balance over the western boundary current regions. This method

has been demonstrated in the context of viscous dissipation in an eddy-resolving regional simulation of the North Atlantic Ocean. The basic scheme can be extended readily to more sophisticated anisotropic schemes where our prescription for the scaling of the coefficient of Laplacian viscosity with grid resolution would provide the cross-stream or zonal component in the western boundary regions. Use in non-eddy-resolving models will most likely require such an extension.

Our method of scaling biharmonic and Laplacian viscous coefficients can be readily extended to the scaling and joint application of biharmonic and Laplacian diffusive coefficients, with the caveat that biharmonic forms of GM90 are difficult to implement in z-coordinate models.

We have touched in brief on the issue of adiabatic tracer mixing, as it remains of concern even at eddy-resolving scale. Issues of mixing cannot be entirely isolated from the choice of transport scheme, and this topic also has been touched on briefly in our effort to survey a number of issues associated with the choice of lateral mixing parameterizations for eddy-resolving ocean modeling.

The LANS-α model offers an entirely different approach to subgrid-scale turbulence parameterization and should be seen as a more sophisticated option for use in conjunction with conventional viscous and diffusive parameterizations to bring out more realistic levels of variability at what would otherwise be eddy-admitting resolutions. It is important here to understand that effective use of the POP-α model does not parameterize the effect of eddies but more readily allows for the inclusion of eddies. Eddy flux parameterizations based on the work of *Gent and McWilliams* [1990] have been a tremendous boon to ocean and climate modeling, and yet, they do not capture all of the important effects of eddies, as discussed, for instance, in *Eden et al.* [2007] where eddy tracer fluxes are identified as being upgradient over much of the Gulf Stream region.

An open question in eddy-resolving ocean modeling, partially addressed in BHS07, is the question of how much of the benefit of grid refinement is associated with increased levels of mesoscale eddy variability and how much may be attributable simply to better resolution of the topography. Unquestionably, the former is of considerable importance, and to the extent this is true, the LANS-α model offers a possible means to obtain more realistic feedback of the eddy variability on the mean at around half the resolution as required for a conventional eddy-resolving ocean model. The order-of-magnitude savings in computational cost associated with this step back in resolution makes this method attractive for use in climate science.

Acknowledgments. We wish to acknowledge Robert Hallberg for a conversation that eventually led to our reconsideration of the joint use of biharmonic and Laplacian viscosity, Stephen Griffies for insightful comments as a reader, and two anonymous reviewers for their essential contributions. This work was supported by the Department of Energy's Office of Science. Los Alamos National Laboratory is operated by Los Alamos National Security, LLC for the Department of Energy.

REFERENCES

Adcroft, A., C. Hill, and J. Marshall (1997), Representation of topography by shaved cells in a height coordinate ocean model, *Mon. Weather Rev., 125*, 2293–2315.

Adcroft, A., J. Campin, P. Heimbach, C. Hill, and J. Marshall (2005), MIT-gcm manual, *Tech. Rep.*, MIT, http://mitgcm.org/pelican/online_documents/manual.html.

Andrews, D. B., and M. E. McIntyre (1978), An exact theory of nonlinear waves on a Lagrangian-mean flow, *J. Fluid Mech., 89*, 609–646.

Bleck, R. (2002), An oceanic general circulation model framed in hybrid isopycnic-Cartesian coordinates, *Ocean Model., 37*, 55–88.

Bryan, F. O., M. W. Hecht, and R. D. Smith (2007), Resolution convergence and sensitivity studies with North Atlantic circulation models. Part I: The western boundary current system, *Ocean Model., 16*(3–4), 141–159, doi:10.1016/j.ocemod.2006.08.005.

Chassignet, E. P., and Z. D. Garraffo (2001), Viscosity parameterization and Gulf Stream separation, in *Proceedings of the 'Aha Huliko'a Hawaiian Winter Workshop*, edited by P. Muller and D. Henderson, pp. 367–374, University of Hawaii.

Chen, S., C. Foias, D. D. Holm, E. Olson, E. S. Titi, and S. Wynne (1998), Camassa–Holm equations as a closure model for turbulent channel and pipe flow, *Phys. Rev. Lett., 81*, 5338–5341, doi:10.1103/PhysRevLett.81.5338.

Chen, S., C. Foias, D. D. Holm, E. Olson, E. S. Titi, and S. Wynne (1999a), The Camassa–Holm equations and turbulence, *Physica D, 133*, 49–65.

Chen, S., D. D. Holm, L. G. Margolin, and R. Zhang (1999b), Direct numerical simulations of the Navier–Stokes alpha model, *Physica D, 133*, 66–83.

Delhez, E. J. M., and E. Deleersnijder (2007), Overshootings and spurious oscillations caused by biharmonic mixing, *Ocean Model., 17*, 183–198, doi:10.1016/j.ocemod.2007.01.002.

Domaradzki, J. A., and D. D. Holm (2001), Navier–Stokes-alpha model: LES equations with nonlinear dispersion, *Special LES volume of ERCOFTAC Bulletin, 48*.

Eden, C., R. J. Greatbatch, and J. Willebrand (2007), A diagnosis of thickness fluxes in an eddy-resolving model, *J. Phys. Oceanogr., 37*(3), 727–742, doi:10.1175/JPO2987.1.

Foias, C., D. D. Holm, and E. S. Titi (2001), The Navier–Stokes-alpha model of fluid turbulence, *Physica D, 152–153*, 505–519.

Gent, P. R., and J. C. McWilliams (1990), Isopycnal mixing in ocean circulation models, *J. Phys. Oceanogr., 20*, 150–155.

Geurts, B. J., and D. D. Holm (2003), Regularization modeling for large-eddy simulation, *Phys. Fluids, 15*(L13–L16).

Griffies, S., R. Pacanowski, and R. Hallberg (2000), Spurious diapycnal mixing associated with advection in a *z*-coordinate ocean model, *Mon. Weather Rev., 128*, 538–564.

Griffies, S. M. (2004), *Fundamentals of Ocean Climate Models*, Princeton University Press, New Jersey.

Griffies, S. M., and R. W. Hallberg (2000), Biharmonic friction with a Smagorinsky-like viscosity for use in large-scale eddy-permitting ocean models, *Mon. Weather Rev., 128*, 2935–2946.

Hallberg, R., and A. Gnanadesikan (2006), The role of eddies in determining the structure and response of the wind-driven Southern Hemisphere overturning: Results from the Modeling Eddies in the Southern Ocean (MESO) Project, *J. Phys. Oceanogr., 36*, 2232–2252.

Hecht, M. W., and R. D. Smith (2008), Towards a physical understanding of the North Atlantic: A review of model studies in an eddying regime, this volume.

Hecht, M. W., D. D. Holm, M. R. Petersen, and B. A. Wingate (2008), Implementation of the LANS-α turbulence model in a primitive equation ocean model, *J. Comp. Physics*, doi:10.1016/j.jcp.2008.02.018.

Hecht, M. W., W. R. Holland, and P. J. Rasch (1995), Upwind-weighted advection schemes for ocean tracer transport: An evaluation in a passive tracer context, *J. Geophys. Res., 100*, 20,763–20,778.

Hecht, M. W., F. O. Bryan, M. Maltrud, and R. D. Smith (2006), The Northwest Corner of the Atlantic and rapid climate change, Los Alamos National Laboratory report, LA-UR-06-7346; available at http://www.noc.soton.ac.uk/rapid/rapid2006/ic06pres.php.

Holland, W. R., F. O. Bryan, and J. C. Chow (1998), Application of a third-order upwind scheme in the NCAR Ocean Model, *J. Clim., 11*, 1487–1493.

Holm, D., C. Jeffery, S. K. Kurien, D. Livescu, M. Taylor, and B. Wingate (2005), The LANS-α model for computing turbulence: Origins, results, and open problems, *LA Science, 29*, 152.

Holm, D. D. (1999), Fluctuation effects on 3d Lagrangian mean and Eulerian mean fluid motion, *Physica D, 133*, 215–269.

Holm, D. D. (2002), Averaged Lagrangians and the mean effects of fluctuations in idea fluid dynamics, *Physica D, 2875*, 1–34.

Holm, D. D., and B. T. Nadiga (2003), Modeling mesoscale turbulence in the barotropic double-gyre circulation, *J. Phys. Oceanogr., 33*, 2355–2365.

Holm, D. D., and B. A. Wingate (2005), Baroclinic instabilities of the two-layer quasigeostrophic alpha model, *J. Phys. Oceanogr. 35*, 1287–1296.

Holm, D. D., J. E. Marsden, and T. S. Ratiu (1998), Euler–Poicaré models of ideal fluids with nonlinear dispersion, *Phys. Rev. Lett, 80*, 4173–4177.

Hundsdorfer, W., and R. A. Trompert (1994), Method of lines and direct discretization: a comparison for linear advection, *Appl. Numer. Math., 13*(6), 469–490.

Hunke, E., M. Maltrud, and M. Hecht (2008), On the grid dependence of lateral mixing parameterizations for global ocean simulations, *Ocean Model., 20*, 115–133, doi:10.1016/j.ocemod.2007.06.010.

Hurlburt, H. E., and P. J. Hogan (2000), Impact of 1/8° to 1/64° resolution on Gulf Stream model—data comparisons in basin-scale subtropical Atlantic Ocean models, *Dyn. Atmos. Ocean, 32*(3–4), 283–329.

Lee, M.-M., A. C. Coward, and A. J. G. Nurser (2002), Spurious diapycnal mixing of the deep waters in an eddy-permitting global ocean model, *J. Phys. Oceanogr., 32*(5), 1522–35.

Maltrud, M. E., and J. L. McClean (2005), An eddy resolving 1/10° ocean simulation, *Ocean Model., 8*(1–2), 31–54.

Munk, W. H. (1950), On the wind-driven ocean circulation, *J. Meteorol., 7*(2), 79–93.

Özgökmen, T. M., E. P. Chassignet, and A. M. Paiva (1997), Impact of wind forcing, bottom topography, and inertia on midlatitude jet separation in a quasigeostrophic model, *J. Phys. Oceanogr., 27*, 2460–2476.

Petersen, M. R., M. W. Hecht, and B. A. Wright (2008), Efficient form of the LANS-α turbulence model in a primitive ocean model, *J. Comp. Physics*, doi:10.1016/j.jcp.2008.02.017.

Roberts, M., and D. Marshall (1997), Do we require adiabatic dissipation schemes in eddy-resolving ocean models?, *J. Phys. Oceanogr., 28*(10), 2050–2063.

Rossby, T. (1996), The North Atlantic Current and surrounding waters: At the crossroads, *Rev. Geophys., 34*(4), 463–481.

Semtner, A. J., and Y. Mintz (1977), Numerical simulation of the Gulf Stream and mid-ocean eddies, *J. Phys. Oceanogr., 7*(2), 208–230.

Smith, R. D., and P. Gent (2002), Reference manual of the Parallel Ocean Program (POP), *Los Alamos National Laboratory report LA-UR-02-2484*, Los Alamos National Laboratory, Los Alamos, NM.

Smith, R. D., and P. R. Gent (2004), Anisotropic GM parameterization for ocean models, *Ocean Model., 34*, 2541–2564, doi:10.1175/JPO2613.1.

Smith, R. D., and J. C. McWilliams (2003), Anisotropic horizontal viscosity for ocean models, *J. Phys. Oceanogr., 5*, 129–156.

Smith, R. D., M. E. Maltrud, F. O. Bryan, and M. W. Hecht (2000), Numerical simulation of the North Atlantic Ocean at 1/10°, *J. Phys. Oceanogr., 30*, 1532–1561.

Veronis, G. (1975), *Numerical Models of Ocean Circulation*, chap. The role of models in tracer studies, pp. 133–146, Natl. Acad. Sci..

Weese, S. R., and F. O. Bryan (2006), Climate impacts of systematic errors in the simulation of the path of the North Atlantic Current, *Geophys. Res. Lett., 33*, L19708, doi:10.1029/2006GL027669.

Wingate, B. A. (2004), The maximum allowable time step for the shallow water α model and its relation to time-implicit differencing, *Mon. Weather Rev., 132*, 2719–2731.

M. W. Hecht, Computational Physics Group, CCS Division, Mail Stop B296, Los Alamos National Laboratory, Los Alamos, NM 87545, USA. (mhecht@lanl.gov)

E. C. Hunke, Fluid Dynamics Group, Theoretical Division, Mail Stop B216, Los Alamos National Laboratory, Los Alamos, NM, 87545 USA. (eclare@lanl.gov)

M. E. Maltrud, Fluid Dynamics Group, Theoretical Division, Mail Stop B216, Los Alamos National Laboratory, Los Alamos, NM 87545, USA. (maltrud@lanl.gov)

M. R. Petersen, Computational Physics Group, CCS Division, Mail Stop B296, Los Alamos National Laboratory, Los Alamos, NM 87545, USA. (mpetersen@lanl.gov)

B. A. Wingate, Computational Physics Group, CCS Division, Mail Stop D413, Los Alamos National Laboratory, Los Alamos, NM 87545, USA. (wingate@lanl.gov)

Eddy-Resolving Global Ocean Prediction

Harley E. Hurlburt[1], Eric P. Chassignet[2], James A. Cummings[1], A. Birol Kara[1],
E. Joseph Metzger[1], Jay F. Shriver[1], Ole Martin Smedstad[3], Alan J. Wallcraft[1],
and Charlie N. Barron[1]

[1]Naval Research Laboratory, Oceanography Division, Stennis Space Center, Mississippi, USA

[2]Center for Ocean–Atmospheric Prediction Studies, Florida State University, Florida, USA

[3]Planning Systems, Inc., Stennis Space Center, Mississippi, USA

Global prediction of the ocean weather (e.g., surface mixed layer, meandering currents and fronts, eddies, and coastally trapped waves) has been feasible only since the turn of the century when sufficient computing power and real-time data became available. Satellite altimetry is the key observing system for mapping ocean eddies and current meanders, but sea surface temperature, temperature and salinity profiles, and atmospheric forcing are also essential. The multinational Global Ocean Data Assimilation Experiment (GODAE) has fostered the development of eddy-permitting and eddy-resolving basin-scale to global prediction systems in several countries. In this work, the focus is on eddy-resolving global ocean prediction, the value of ocean model skill in dynamical interpolation of data and in forecasting, plus the increased skill gained from using eddy-resolving versus eddy-permitting models. In ocean prediction, one must consider the classes of ocean response to atmospheric forcing for different phenomena and regions. In ocean nowcasting, this is important in considering the relative impact of the atmospheric forcing and different ocean data types. In forecasting, these classes impact whether or not the timescale for oceanic predictive skill is limited by the timescale for atmospheric predictive skill or whether a coupled ocean-atmosphere model would be advantageous. Results from the existing eddy-resolving global ocean prediction systems demonstrate forecast skill up to 1 month, globally and over most subregions. Outside surface boundary layers and shallow water regions, the forecast skill typically is only modestly impacted by reverting toward climatological forcing after the end of the atmospheric forecast versus using analysis-quality forcing for the duration.

1. INTRODUCTION

Adequate real-time data input, computing power, numerical ocean models, data assimilation capabilities, atmospheric forcing, and bottom topography are key elements required for successful eddy-resolving global prediction of the "ocean weather", such as the surface mixed layer, equatorial and

Ocean Modeling in an Eddying Regime
Geophysical Monograph Series 177

coastally trapped waves, upwelling of cold water, meandering of currents and fronts, eddies, Rossby waves, and the associated temperature, salinity, currents, and sea surface height (SSH). Only since the turn of the century have all the key elements finally reached a status that makes global ocean weather prediction feasible. The focus of this chapter is on the roles that eddy-resolving ocean models perform in accurate nowcasting (estimating the present state) and forecasting of the ocean weather on timescales up to a month. However, much of the content is also relevant to climate prediction.

In the remainder of the introduction, the relevance of this chapter to climate is briefly discussed in section 1.1. Section 1.2 outlines the history of the first 10 years of basin-scale to global ocean weather prediction, the product of a multinational effort largely fostered by the Global Ocean Data Assimilation Experiment (GODAE). In section 1.3, ocean prediction is discussed in relation to four classes of ocean response to atmospheric forcing, how they impact ocean prediction, and how they affect the involvement of ocean models. Section 2 presents results illustrating the roles that ocean models play in nowcasting and forecasting with emphases on results from eddy-resolving global ocean models and prediction systems and the relation to classes of responses to atmospheric forcing. These include dynamical interpolation of assimilated data, the value of atmospheric forcing as a data type, sensitivity to ocean model resolution (e.g., eddy-resolving versus eddy-permitting and impact on the space scales that can be mapped using nadir beam satellite altimetry), use of mean SSH from an eddy-resolving global ocean model in creating a mean SSH field for addition to the anomalies from satellite altimetry, and the use of simulated data from eddy-resolving ocean models in testing, evaluating, and improving data assimilation skill. The latter is also an example of observing system simulation and assessment.

1.1. Relevance to Climate

To date, multidecadal climate prediction models and seasonal to interannual climate forecasts (focused primarily on El Niño/La Niña conditions) have used coarse-resolution ocean models that are unable to depict realistic ocean fronts. As climate predictions become increasingly region-specific, eddy-resolving ocean models will be required for success in many regions, for example, for accurate positioning, strength, and eastward penetration of major ocean currents and fronts. As specific examples, eddy-resolving ocean models are required for realistic prediction of the Kuroshio/Oyashio current system in the western North Pacific and the Gulf Stream current system in the western North Atlantic. In the Pacific, such models must simulate a strongly inertial Kuroshio

with sufficient eastward penetration and advection of heat as well as cold southward currents along the east coast of Japan rather than the warm northward flow found in coarse-resolution models. Eddy-resolving models are also needed to obtain the sharp ocean fronts that span the entire North Pacific [Hurlburt et al., 1996; Hurlburt and Metzger, 1998]. In the North Atlantic, eddy-resolving models are required to accurately simulate the Gulf Stream pathway between Cape Hatteras and the Grand Banks and to obtain the associated large nonlinear recirculation gyres in this region [Hurlburt and Hogan, 2000; Bryan et al., 2007]. These large nonlinear gyres change the large-scale shape of the subtropical gyre into a C-shape. East of the Grand Banks, the Gulf Stream turns northward as the North Atlantic Current, and an eddy-resolving model is essential for the model to advect sufficient heat northward and then eastward across the Atlantic at around 50°N [Smith et al., 2000]. These are examples of regions where errors caused by coarse resolution in ocean global circulation models (OGCMs) would lead to large errors in climate anomaly or climate change predictions for North America, Europe, the Arctic, and Japan.

1.2. Ocean Weather Prediction, a Brief History of the First 10 Years

GODAE [Smith, 2000, 2006] has been a major driving force behind a multinational effort working toward the development of eddy-resolving global ocean prediction capabilities in different countries. Australia, Britain, France, Japan, and the United States are all sponsoring efforts that are presently in various stages toward reaching this goal (see the book by Chassignet and Verron [2006] for an overview).

The Forecasting Ocean Assimilation Model (FOAM) system, developed by the Met Office UK, became the first operational global ocean prediction system in 1997, but at 1° resolution it was non-eddy-resolving [Bell et al., 2000]. Ten years later, the availability of sufficient computer resources is still a significant limiting factor in ocean weather prediction. Hence, a variety of approaches and trade-offs have been used to work toward the eddy-resolving global goal. All of the remaining global- and basin-scale prediction systems considered here are either eddy-permitting or eddy-resolving, and all include some form of altimeter data assimilation. SSH from satellite altimetry is the key available data type for mapping the ocean weather in deep water [Hurlburt, 1984].

Basin-scale prediction systems have been a common step toward the global goal, either nested in a coarser resolution global system (the Met Office UK) or standalone (the remainder). Such systems started at 1/3° resolution in January 2001 (British FOAM Atlantic/Arctic and Indian Ocean,

French MERCATOR North and tropical Atlantic). These were soon followed by higher resolution basin-scale systems (1/9° British FOAM North Atlantic/Mediterranean Sea since 2002, operational in 2004 [*Bell et al.*, 2006], the French MERCATOR 1/15° North Atlantic plus 1/16° Mediterranean Sea operational in January 2003 [*Bahurel*, 2006], and a 1/12° Atlantic system run in demonstration mode by the HYbrid Coordinate Ocean Model (HYCOM) Consortium in the United States since July 2002 [*Chassignet et al.*, 2007]).

Some of the basin-scale systems have used variable resolution focused on a region of interest. The Norwegian DIADEM (later TOPAZ) Atlantic and Arctic system [*Bertino and Evensen*, 2002] has been running since 2001 using a curvilinear grid with 18- to 35-km resolution (increased to 11–16 km in July 2007), with the highest resolution focused on the northeast Atlantic and Nordic Seas. Since 2001, the Japan Meteorological Agency has been running a daily operational North Pacific assimilation-prediction system COMPASS-K (12°N to 55°N) with variable latitude and longitude resolution as high as 1/4° in the western North Pacific around Japan [*Kamachi et al.*, 2004a, 2004b], and in 2007, the Japanese Meteorological Research Institute began running a real-time ocean data assimilation and prediction system MOVE/MRI.COM-NP covering the Pacific Ocean (15°S–65°N) with variable latitude and longitude resolution as high as 1/10° near Japan [*Usui et al.*, 2006a, 2006b, 2008; *Tsujino et al.*, 2006]. The US National Oceanic and Atmospheric Administration/National Centers for Environmental Prediction (NOAA/NCEP) has been running a real-time HYCOM equatorial and North Atlantic prediction system with assimilation of satellite altimeter data added on June 28, 2007. The horizontal resolution is about 4 to 18 km, with the highest resolution focused in the Gulf of Mexico and along the east coast of North America.

Crosnier and Le Provost [2007] performed an intercomparison study of five basin-scale prediction systems run in five different countries covering the North Atlantic and/or the Mediterranean Sea [the 1/9° British FOAM North Atlantic/Mediterranean Sea system, the French MERCATOR 1/15° North Atlantic/1/16° Mediterranean Sea system, the US 1/12° Atlantic HYCOM system, the HYCOM-based Norwegian TOPAZ Atlantic/Arctic system, and the Italian MFS 1/8° (later 1/16°) Mediterranean Sea system]. Of these, three use fixed-depth z-level models and two use HYCOM. The latter has a generalized vertical coordinate that allows Lagrangian isopycnal layers, pressure levels (~z levels), terrain-following, and other vertical coordinates [*Bleck*, 2002; *Chassignet et al.*, 2003; *Halliwell*, 2004; *Chassignet et al.*, 2007].

Another approach is that of the Shallow-Water Analysis and Forecast System (SWAFS) North World developed at the Naval Oceanographic Office (NAVOCEANO). SWAFS used the Princeton Ocean Model with terrain-following coordinates [*Blumberg and Mellor*, 1987]. This system covered the world ocean over the latitude range 20°S to 80°N with ~1/5° (~23 km) mid-latitude resolution. Effectively, it was a convenient way to run multiple basin-scale models within one model domain, as the basins are disconnected at the southern boundary and, although the Indonesian Archipelago is included, the Indonesian throughflow is largely blocked because of the southern boundary. The SWAFS North World system was run at NAVOCEANO from 1998 to 2006, but never became operational. A SWAFS system of similar design for the Mediterranean Sea [*Horton et al.*, 1997] did become an operational prediction system.

Global ocean prediction systems that make various compromises to cope with limitations in available computing power have also been developed. A 1/4° global French MERCATOR system with 46 levels in the vertical has been running operationally since October 2005 [*Bahurel*, 2006]. The Australian BLUElink global ocean prediction system has 1/10° resolution on a 90° latitude and longitude sector surrounding Australia, variable resolution as coarse as 2° elsewhere and 47 levels in the vertical [*Schiller and Smith*, 2006]. The BLUElink system became operational in July 2007. All of the preceding systems have 20 to 47 coordinate surfaces in the vertical direction.

The US Navy has used a still different approach in its present operational global ocean prediction systems, a linked two-model approach [*Rhodes et al.*, 2002]. One model has high horizontal resolution but low vertical resolution, while the other has high vertical resolution but lower horizontal resolution. The earliest version of this system used only the model with low vertical resolution, the Naval Research Laboratory (NRL) Layered Ocean Model (NLOM). It had six Lagrangian layers in the vertical and 1/4° resolution globally, excluding the Arctic and most shallow water [*Metzger et al.*, 1998a, 1998b]. It was the first global ocean weather prediction model to assimilate real-time satellite altimeter data (from TOPEX/Poseidon and ERS-2) and ran daily in real time from 1997 to 2000 at the Fleet Numerical Meteorology and Oceanography Center (FNMOC) but never became operational. In 2000, NAVOCEANO became the US Navy center for numerical prediction of the ocean weather. A mixed layer (seventh layer) [*Wallcraft et al.*, 2003] was added to the NLOM-based component of a system for NAVOCEANO, which began running daily in real time with 1/16° resolution on October 18, 2000. It ran operationally from September 27, 2001 to March 11, 2006, thereby becoming the first operational eddy-resolving global ocean prediction system [*Smedstad et al.*, 2003]. A 1/32° version of this system began running in near real time on November 1,

2003, real time since March 1, 2005, and replaced the 1/16° system as the operational system on March 6, 2006 [*Shriver et al.*, 2007].

The second component of the NAVOCEANO system uses the Navy Coastal Ocean Model (NCOM) and has 1/8° (~15 km) mid-latitude resolution, 1/6° (~19 km) tropical resolution, 40 levels in the vertical, and is fully global [*Barron et al.*, 2006; *Kara et al.*, 2006]. It has been running daily in real time since October 2001 and became operational on February 19, 2006. The NLOM component is used to assimilate SSH data along altimeter tracks with a model forecast as the first guess for each analysis cycle and it is used to make 30-day ocean weather forecasts. It also assimilates model-independent analyses of sea surface temperature (SST). The NCOM component assimilates steric SSH anomalies from NLOM in the form of synthetic temperature and salinity (T & S) profiles [*Rhodes et al.*, 2002]. These profiles are obtained from surface dynamic height and SST anomalies by means of regression equations for subsurface temperature and T & S relations for salinity, both using statistics derived from the historical hydrographic data base [*Carnes et al.*, 1990; *Fox et al.*, 2002]. This two-model global system requires less computer power than a single global model with both high horizontal and high vertical resolution, which is the goal of most GODAE participants.

The US Navy began running the first eddy-resolving global system with high vertical resolution in near real time on December 22, 2006, and it has been running in real time since February 16, 2007. This system uses a fully global configuration of HYCOM with 1/12° equatorial (~7 km mid-latitude) resolution and 32 layers in the vertical. Each day, it performs a 5-day hindcast (assimilation of data before the real-time data window) at 1-day increments to pick up delayed data, and it makes a five-day forecast. In addition, tests of 30-day forecasts have been performed.

Although limitations on ocean model domain size and resolution have been essential in dealing with limitations in available computer power, computing requirements have also been economized through the choice of data assimilation techniques and assimilated data sets. In addition, while most of the preceding systems run daily in real time, computing requirements for other systems have been reduced by running in near real time and updating once a week, rather than daily.

1.3. Ocean Prediction in Relation to Classes of Ocean Response to Atmospheric Forcing

Table 1 is a summary of some useful classes of ocean response to atmospheric forcing that cover many phenomena of potential interest in ocean prediction. In developing an ocean

prediction system, it is essential to consider these classes and their implications for (1) data requirements, (2) the timescales of oceanic predictive skill, and (3) prediction system design.

In the case of class 1 phenomena, characterized by strong and rapid development (less than a week), ocean model simulation skill plays a critical role in converting atmospheric forcing into oceanic information. The timescale for oceanic forecast skill is largely determined by the timescale for atmospheric predictive skill, and the influence of anomalies in the initial state rapidly decreases. The timescale for oceanic predictive skill may exceed that for the atmosphere to some extent because of lagged responses, the influence of geometric or topographic constraints, or the initiation of class 4 free waves or ocean eddies. Such phenomena are commonly found in shallow water where rapid barotropic responses are particularly important, within the surface mixed layer or Ekman layer, in equatorial and coastal wave guides, and in the vicinity of hurricanes or other strong wind events (e.g., wind-driven eddies in the Gulf of Tehuantepec). In section 2.7, some examples are given in which class 1 anomalies lead to class 4 anomalies or where class 1 and class 4 phenomena interact to generate new class 4 features.

Nowcasting and forecasting of class 1 phenomena place high demands on the accuracy of the ocean model and the atmospheric forcing. In shallow water and near coastal boundaries, there are additional demands for accurate topography and coastlines, tides, and fine-resolution atmospheric forcing. Near coastlines, it is important that atmospheric values over land are not used as forcing at ocean model grid points, a common problem in fine-resolution ocean models forced by a coarser resolution atmospheric product [*Kara et al.*, 2007]. When ocean data are assimilated, the model time and the observation time must closely match because of the rapid evolution; but even so, the ocean data may rapidly lose impact after assimilation.

Class 2 phenomena (slower and indirectly forced) tend to be a largely nondeterministic response to atmospheric forcing due to flow instabilities, for example, the generation of mesoscale eddies and the meandering of ocean currents and fronts. Thus, ocean data assimilation is essential to nowcast and forecast these features, and the nowcast and forecast results are relatively insensitive to the atmospheric forcing, allowing forecasts up to a month or more. SSH anomalies measured by satellite altimetry are the key data that all of the GODAE participants use to map these mesoscale features, as well as many other features, including El Niño and La Niña events. SSH is a particularly useful data type observable from space because (1) it is geostrophically related to surface currents, (2) it is an integral measure of subsurface variations in T & S (e.g., the depth of the thermocline) plus a bottom

Table 1. Classes of Ocean Response to Atmospheric Forcing.[a]

Class	Examples	Implications
1. Strong, rapid (<1 week), and direct	Surface waves, storm surges, and rapid variations within the upper mixed layer, beneath hurricanes, in coastal and equatorial upwelling, and often the onset of equatorial and coastal trapped waves	Forecasts are short range; limited by atmospheric predictive skill. Less sensitive to errors in the initial state; more sensitive to errors in forcing
2. Slower (weeks to months) and indirect	Mesoscale eddies, meandering currents, frontal locations, features related to flow instabilities on the mesoscale	Forecast skill up to a month or more; more sensitive to errors in the initial state; less sensitive to errors in forcing; ocean data assimilation is essential; variability statistics may be predicted via simulation
3. Slow (weeks to years) direct integrated response	El Niño; much of the tropical ocean circulation, gyres, persistent features associated with geometric and topographic constraints	Long range forecasts possible; sensitive only to errors in forcing on long time scales; nowcasting and forecasting feasible using ocean models with sparse ocean data
4. Free propagation of existing features (weeks to years)	Equatorial and coastally trapped waves, Rossby waves and stable isolated eddies generated under Classes 1, 2, and 3	Reduced sensitivity to onset mechanisms and extended forecast skill, weeks to years

[a]This table summarizes the classes of ocean response to atmospheric forcing most relevant to ocean climate and/or ocean weather prediction, the first three in order of increasing time scale. The fourth class is initiated by anomalies generated under the preceding three classes. Class 2 is a close analog of atmospheric cyclones, anticyclones, and jet streams, but with larger timescales and much smaller space scales, the latter largely due to the relative atmospheric and oceanic first internal radii of deformation. A fifth class includes tides and tsunamis that are (mostly) not in response to the atmosphere. Other classes are not considered here, for example, most inertia-gravity waves and fine-scale flow instabilities that may result from atmospheric forcing. Adapted from *Hurlburt* [1984].

pressure anomaly, and (3) the oceanic vertical structure is generally low mode. Thus, it has been possible to develop a number of techniques to project the SSH anomalies downward, in some cases including the bottom pressure anomaly (the nonsteric contribution to SSH). These techniques range from model dynamics [*Hurlburt*, 1986], to model statistics [*Hurlburt et al.*, 1990], to a potential vorticity constraint [*Cooper and Haines*, 1996], to data assimilation covariances, to synthetic T & S profiles estimated from a combination of SSH and SST related by regression to the historical hydrographic data base [*DeMey and Robinson*, 1987; *Carnes et al.*, 1990; *Fox et al.*, 2002].

The regions of the global ocean dominated by class 2 variability can be estimated by calculating the fraction of variability of different ocean variables [e.g., SSH, SST, eddy kinetic energy (EKE), and bottom pressure] that is a nondeterministic response to atmospheric forcing in eddy-resolving model simulations [*Metzger et al.*, 1994; *Hurlburt and Metzger*, 1998; *Metzger and Hurlburt*, 2001; *Melsom et al.*, 2003; *Hogan and Hurlburt*, 2005]. Although areas that are highly nondeterministic are not necessarily regions of high SSH variability or EKE, it is relevant to evaluate the model against its ability to simulate regions of high variability. In addition, ocean forecast skill beyond the range of atmospheric predictive skill can be verified. The sensitivity of ocean forecast skill to at-

mospheric forcing can be tested versus persistence (a forecast of no change) and by comparing the skill of forecasts performed using analysis quality forcing versus forecasts reverting toward climatological forcing beyond the end of the forecast forcing [*Smedstad et al.*, 2003; *Shriver et al.*, 2007, section 2 of this chapter] (relevant to all four classes of oceanic response).

Class 3 phenomena (slow, direct integrated responses to atmospheric forcing on timescales of weeks to years) are perhaps more relevant to seasonal to interannual prediction (e.g., El Niño and La Niña events) using coupled ocean-atmosphere models. However, class 3 phenomena are also relevant to uncoupled models and shorter range predictions. They are relevant to both types of prediction when considering the role of uncoupled models in determining the mean SSH added to SSH anomalies from satellite altimetry, a topic discussed in section 2.6. The required mean SSH contains a wide range of space scales, some of which are adequately resolved only in simulations by a high-resolution model. This places a heavy burden on the accuracy of the ocean model and the atmospheric forcing.

In addition, class 3 phenomena are important in nowcasting and forecasting some ocean features that are inadequately observed, for example, for an eddy-resolving model, the seasonal cycle of mixed layer depth (MLD) and anomalies in MLD lasting longer than the class 1 timescale of skillful atmospheric weather prediction, examples where a class 3 response to atmospheric forcing will have a much larger impact than assimilation of ocean data. Presently, ~3000 Argo floats [*Roemmich et al.*, 2004] profile T & S down to typical depths of 2000 m with high vertical resolution. These floats constitute an accurate observing system for measuring subsurface T & S, including MLD, but one that, while suitable for climate, is too sparse horizontally and temporally to constrain an eddy-resolving ocean model and too sparse to constrain MLDs in any model on class 1 timescales. As discussed by *Hurlburt et al.* [2001], 3000 Argo floats, arrayed like altimeter tracks, could provide horizontal coverage at a single level (e.g., MLD or surface dynamic height) which is only ~4% of that from one Jason-1 altimeter over its 10-day orbital repeat cycle. This comparison is based on placing Argo floats along over-water altimeter tracks at 50-km intervals and profiling at 10-day intervals.

Class 4 phenomena consist of free propagation of ocean features that exist in the initial state of the forecast, that is, features generated earlier as class 1, 2, or 3 phenomena. In addition to examples mentioned in the discussion of class 1 phenomena, Kelvin waves generated by El Niños have propagated cyclonically around the North Pacific as far as Alaska and the Kamchatka Peninsula of Russia in both observations and models [*Metzger et al.*, 1998a; *Melsom et al.*, 2003]. *Jacobs et al.* [1994] even demonstrated a decade-long forecast of a trans-Pacific nondispersive internal Rossby wave, generated from the 1982–1983 El Niño, which eventually displaced the Kuroshio Extension east of Japan. The forecast used climatological atmospheric forcing (a corresponding simulation with interannual forcing was also run), and the forecast was verified using satellite altimetry and other data. In addition to wind-generated eddies (like Tehuantepec eddies), stable isolated eddies that are products of flow instability, but which escape any region of strong flow instability, may belong to class 4. Such eddies can last for over a year [*Lai and Richardson*, 1977], although they are unlikely to be predictable for nearly that long.

Nowcasting is an essential step in ocean monitoring and prediction of all the classes of oceanic response in Table 1. To perform a nowcast, a numerical model forecast is used as the first guess for assimilation of new data. This use of the forecast allows the model (1) to help fill in the temporal and spatial gaps between observations (including temporal gaps between the repeat cycles of satellite altimetry) and make effective use of delayed data by exploiting its predictive skill, (2) to help convert the better observed surface fields into subsurface structure [*Hurlburt*, 1986], (3) to convert the better observed atmospheric forcing functions into useful oceanic responses, and (4) to apply bottom

topography, coastline geometry, and the ocean surface as physical and dynamical constraints with associated boundary layers. Partially in conjunction with a barotropic model [*Carrere and Lyard*, 2003], the model component of the ocean prediction system can also help separate the steric and nonsteric contributions to SSH, a significant capability in the assimilation of satellite altimeter data [*Rhodes et al.*, 2002]. Examples of the preceding model contributions to nowcast and forecast skill (without a reference) are illustrated in section 2.

There is an important difference between data assimilation in oceanography and meteorology. In oceanography, there is a greater burden on the models to extract information from the data, including use of the models to extract oceanic responses to atmospheric forcing as well as converting surface information into subsurface information, which is sparsely observed. In addition, an eddy-resolving global ocean model should play a major role in determining the mean SSH used in assimilation of satellite altimeter data (discussed in section 2), and it may be used as a source of covariances and other statistics for data assimilation. Therefore, models play a crucial role in making effective use of the data sources and in reducing data requirements in ocean nowcasting in addition to their essential role in ocean forecasting.

Other applications of eddy-resolving ocean models in ocean prediction include understanding model and ocean dynamics, assessing and improving the realism of the ocean model, studying ocean model resolution requirements, performing oceanic predictability studies, investigating the classes of response to atmospheric forcing in different dynamical regimes, and providing realistic simulated data (a) for data assimilation studies, (b) for studies of data base requirements and observing system simulation experiments, and (c) for spinning up models to the initial time when data assimilation begins. Eddy-resolving models can also be used to investigate model sensitivity to the choice of atmospheric forcing product, although a linear barotropic numerical model with realistic deep water boundaries (essentially *Sverdrup* [1947] interior flow with *Munk* [1950] viscous western boundary layers consistent with the *Godfrey* [1989] island rule) can be very useful for this purpose with much lower computational requirements [*Townsend et al.*, 2000; *Hogan and Hurlburt*, 2005]. A number of the issues raised in this subsection are illustrated and discussed in section 2.

2. EDDY-RESOLVING GLOBAL OCEAN PREDICTION SYSTEMS

Here, we focus on the only two existing eddy-resolving global ocean prediction systems to date, the 1/16° (later

1/32°) NLOM-based system with low vertical resolution and the more recent 1/12° global HYCOM-based system, which has both high horizontal and high vertical resolution. HYCOM is fully global, unlike NLOM, which excludes the Arctic and most shallow water. The results presented are used in the illustration and discussion of some key issues in ocean weather prediction.

2.1. The HYCOM and NLOM Prediction Systems

Both HYCOM and NLOM have a free surface and use a Lagrangian vertical coordinate to model the ocean interior. However, simplifications are made in NLOM that make it much more computationally efficient, but which limit its range of application and versatility, while the generalized vertical coordinate and the allowance of zero thickness layers make HYCOM extremely versatile and suitable for a wide range of ocean modeling applications.

General details of NLOM design and progressive development are discussed by *Hurlburt and Thompson* [1980], *Wallcraft* [1991], *Wallcraft and Moore* [1997], *Moore and Wallcraft* [1998], and *Wallcraft et al.* [2003]. The last of these discusses the development of NLOM as a thermodynamic model with a *Kraus and Turner* [1967]-type bulk mixed layer and SST. This version of NLOM was used in the 1/16° and 1/32° global prediction systems listed in section 1.2. More specifically, the resolution for each variable is 1/16° (1/32°) in latitude and 45/512° (45/1024°) in longitude or ~7 km (3–4 km) at mid-latitudes in the 1/16° (1/32°) system. The NLOM domain is nearly global, extending from 72°S to 65°N, with the model boundary generally following the 200-m isobath with a few exceptions, like the shallow straits connecting the Sea of Japan (East Sea) to the Pacific Ocean. NLOM uses vertically compressed but otherwise realistic bottom topography confined to the lowest layer [Hurlburt et al., Pathways of upper ocean currents and fronts: Steering by the topographically-constrained abyssal circulation and the role of flow instabilities, applicability of a two-layer theory to low versus high vertical resolution models, submitted to *Dynamics of the Atmospheres and Oceans*, 2008 hereinafter referred to as Hurlburt et al., submitted manuscript, 2008]. Sill depths of straits that are shallower than the bottom layer are maintained by constraining the flow to small values below the sill depth [*Metzger and Hurlburt*, 1996]. The reduced amplitude of the topography generally does not have a significant impact on the abyssal current steering of upper ocean current pathways (an essential capability of eddy-resolving ocean models which simulate eddy-driven abyssal currents) because even low-amplitude topographic features can constrain the pathways of abyssal flow [*Hurlburt and Metzger*, 1998; Hurlburt et al., submitted manuscript, 2008]. Deep wa-

ter formation outside the model domain is included via prescribed observationally based transports through four straits in the northern boundary of the North Atlantic, northward in the upper layers and southward in the bottom layer [*Shriver and Hurlburt*, 1997].

The history of NLOM use in ocean prediction systems is discussed in section 1.2. Each day, the NLOM prediction system performs a hindcast at daily increments to pick up delayed altimeter data (a 5-day hindcast for the 1/32° system, a 3-day hindcast for the former 1/16° system), and it makes a 4-day forecast (plus a 30-day forecast once a week) [*Smedstad et al.*, 2003; *Shriver et al.*, 2007]. NLOM assimilates along-track satellite altimeter data using the model as a first guess for the SSH analysis, and it assimilates SST in the form of daily operational model-independent SST analyses from NAVOCEANO [*Barron and Kara*, 2006]. The SSH assimilation process consists of an optimum interpolation (OI) deviation analysis from the model first guess and an empirical orthogonal function regression technique based on model statistics to project the SSH updates downward, including to the abyssal layer [*Hurlburt et al.*, 1990]. These updates are geostrophically balanced outside the equatorial region and inserted incrementally to further reduce inertia-gravity wave generation. Anisotropic, spatially varying mesoscale covariance functions determined from TOPEX/Poseidon and ERS-2 altimeter data [*Jacobs et al.*, 2001] are used in the OI analysis. The SST assimilation consists of relaxing the analysis into the model.

HYCOM has been developed as a collaborative multi-institutional effort starting from the Miami Isopycnic Coordinate Ocean Model (MICOM) using the theoretical foundation set forth by *Bleck and Boudra* [1981], *Bleck and Benjamin* [1993], and *Bleck* [2002]. HYCOM uses a generalized vertical coordinate that is normally isopycnal in the open stratified ocean, with a dynamically smooth transition to pressure coordinates (~z-levels) in the unstratified mixed layer and to σ (terrain-following) coordinates in shallow water, but it is not limited to these types. The transition between coordinate types is made dynamically in space and time using the layered continuity equation and a hybrid coordinate generator. This generalized coordinate approach retains particular advantages associated with the different vertical coordinate types: (1) the retention of water mass characteristics for centuries (a characteristic of the isopycnal coordinate interior), (2) high vertical resolution in the surface mixed layer and unstratified or weakly stratified regions of the ocean (a characteristic of z-level coordinates), and (3) high vertical resolution in coastal regions (a characteristic of terrain-following coordinates) [*Chassignet et al.*, 2003]. The generalized coordinate also facilitates accurate transition between deep and shallow water and allows the use of a sophisticated embedded mixed-layer model in an ocean model with an isopycnal interior, for example, a K-profile parameterization [*Large et al.*, 1997] in the ocean prediction system. For the 1/12° global HYCOM, the model resolution south of 47°N is exactly 0.08° by 0.08 cos(latitude)° for each variable (~7 km at mid-latitudes) and is described by the approximate equatorial resolution. A bipolar grid is used north of 47°N, resulting in a tripole global grid [*Murray*, 1996], which gives ~3.5-km resolution in the Arctic. The model includes 32 hybrid coordinate surfaces in the vertical direction and is run with thermobaricity [*Sun et al.*, 1999; *Chassignet et al.*, 2003; *Hallberg*, 2005] using a reference depth of 2000 db for potential density, a model configuration known as σ_2^*.

The 1/12° global HYCOM prediction system has been running daily in near real time since December 22, 2006 and in real time since February 16, 2007. Even more than for the NLOM system, because HYCOM includes shallow water, doubling the resolution is desirable when the computing power becomes available. The impact of model resolution is a recurring theme in the rest of section 2. The data are assimilated into HYCOM using the Navy Coupled Ocean Data Assimilation (NCODA) system of *Cummings* [2005] with a modification described in section 2.8. Within NCODA, a HYCOM forecast is used as the first guess for a multivariate optimum interpolation (MVOI) [*Daley*, 1991] analysis of SSH track data from satellite altimetry, SST, and T & S profiles. The analysis is then inserted incrementally into HYCOM. In the analysis, the method of *Cooper and Haines* [1996] is used for downward projection of SSH below the mixed layer, although downward projection of SSH and SST using synthetic T & S profiles [*Fox et al.*, 2002] is also an option in NCODA. The updates to the mass field are geostrophically balanced outside an equatorial band.

2.2. 1/16° Eddy-Resolving Versus 1/4° Eddy-Permitting Models in Nowcasting and Forecasting the Kuroshio Extension

Plate 1 illustrates the relative dynamical interpolation (nowcast) and forecast skill of 1/4° eddy-permitting versus 1/16° eddy-resolving six-layer versions of NLOM in the Kuroshio Extension region east of Japan, as verified by NAVOCEANO operational 1/8° Modular Ocean Data Assimilation System (MODAS) model-independent SST analyses [*Barron and Kara*, 2006] with the color bar designed to highlight the Kuroshio pathway. The 1/16° Pacific prediction system discussed here is the forerunner of the 1/16° global NLOM prediction system which later became operational at NAVOCEANO, except that it lacked a mixed layer and SST [*Hurlburt et al.*, 2000; *Metzger et al.*, 2000]. The 1/4° global system is one of two versions run in real time at FNMOC

(section 1.2). Results for January 1 and January 15, 1999 are shown for each product. Two-week forecasts from January 1 for each model can be compared with the three analyses for January 15. The model nowcasts were updated daily with assimilation of altimeter track data from TOPEX/Poseidon and ERS-2 using a 3-day window (red tracks overlaid on two of the plots) and the methodology outlined in section 2.1 for NLOM. Because the geoid of the Earth is not adequately known, only the altimetric deviations from their own mean are useful in the assimilation. In the model assimilations, a slightly modified 1993–1997 mean from the corresponding model was added to these deviations. Generally, the two means agree closely, but the 1/16° model gives a sharper depiction of mean currents and a sharper depiction of the Kuroshio in the nowcast and forecast results.

Several features are clearly evident in the SST and the model results. Here, we focus on the sharp meander between 155°E and 160°E on January 1, 1999, which evolved into a distinctive eddy detachment in progress on January 15. This sequence is captured with striking agreement between the SST analyses, the 1/16° model nowcasts, and the 1/16° model forecast. The 1/4° model represents the sharp meander on January 1 quite well, but the 1/4° nowcast for the detaching eddy on January 15 is poor and the January 15 1/4° model forecast for this feature is a very unrealistic meander contraction.

These results are a demonstration of ocean model eddy-resolving nowcast and forecast skill using satellite altimeter data. In particular, the results demonstrate (1) that satellite altimetry can be an effective observing system for mapping and forecasting mesoscale ocean features, (2) that an ocean model with high enough horizontal resolution can be a skillful dynamical interpolator of satellite altimeter data in depicting mesoscale oceanic variability, and (3) that an eddy-resolving ocean model can provide skillful forecasts of mesoscale variability when a model with assimilation of altimeter data is used to define the initial state, in this example for the duration of the 31-day forecast [*Hurlburt et al.*, 2000]. It is worth noting that this was the very first forecast performed with the 1/16° Pacific model used here, not the best of many. In contrast, the 1/4° global model had insufficient horizontal resolution, lower skill as a dynamical interpolator, and poor forecast skill, with a strong tendency to damp the variability. The broadening of the Kuroshio in the 1/4° model means the current speeds are lower, and dynamically, the current is not as inertial, a clear indication that one should anticipate a substantial impact on model forecast skill and skill as a dynamical interpolator.

The 1/16° Pacific model used for assimilation of real altimeter data in Plate 1 was also used in testing assimilation of simulated altimeter track data derived from the model along

ERS-2, GFO, and both sets of TOPEX/Poseidon repeat tracks. Simulated altimeter data from model year 1994 were assimilated into the model starting from an initial state in model year 1997, a time when the model Kuroshio pathway was quite different. Compared to no assimilation or a model mean, large error reduction was obtained in all six layers of the model, including the abyssal layer, after about a month of simulated altimeter data assimilation. This was true even when simulated data from only one of the altimeters (any one of them) were assimilated. The error was significantly reduced by adding a second altimeter, with the tandem tracks of Jason-1 and TOPEX/Poseidon clearly performing the best. Only modest error reduction was obtained by adding a third altimeter [*Hurlburt et al.*, 2000; *Smedstad et al.*, 2003].

2.3. Impact of Resolution that is Eddy-Resolving Versus Eddy-Permitting on Model Dynamics and Simulation Skill

Because of its relative computational efficiency and simplicity, NLOM has been used to perform a number of studies investigating the impact of increasing the horizontal resolution as well as the impact of other aspects of model design on (a) model dynamics, (b) simulation skill, (c) convergence of the mean circulation and variability in relation to mesoscale features, and (d) agreement with observations in different regions of the world ocean. Most of these studies were performed without ocean data assimilation.

For the Kuroshio near Japan (west of ~155°E), *Hurlburt et al.* [1996, 1997] found that 1/8° (~15 km) mid-latitude resolution was sufficient to simulate a realistic Kuroshio pathway, including the mean meanders observed east of Japan. This simulation included vertically compressed but otherwise realistic bottom topography. Corresponding models with 1/4° resolution or with 1/8° resolution and a flat bottom gave unrealistic simulations of the Kuroshio in this region. The mean Kuroshio pathway was realistic in the 1/8° simulation with topography because it simulated upper ocean-topographic coupling via mixed baroclinic-barotropic instabilities [*Hurlburt et al.*, 1996]. This coupling occurs when baroclinic instability transfers energy to the abyssal layer where it is constrained to follow the geostrophic contours of the bottom topography. The eddy-driven mean abyssal currents, in turn, advect upper ocean current pathways, especially where these currents intersect at large angles. In the advection process, there is often a natural tendency toward barotropy because the current pathway advection is reduced when this occurs. In the example considered here, a modest seamount complex just east of Japan played a critical role in steering a mean meander in the Kuroshio pathway immediately east of Japan, thus affecting the separation of the Kuroshio from the east coast of Japan.

Most of the references in this subsection discuss upper ocean-topographic coupling via flow instabilities and a theory for steering of upper ocean current pathways by abyssal currents, which is discussed most comprehensively by Hurlburt et al. (submitted manuscript, 2008). East of 155°E, *Hurlburt et al.* [1996] and *Hurlburt and Metzger* [1998] found that 1/8° resolution is insufficient and 1/16° is sufficient to simulate the eastward penetration of the Kuroshio as an inertial jet, the bifurcation of the Kuroshio at the Shatsky Rise, and sharp fronts that span the northern Pacific (the subarctic front and a front associated with the Kuroshio Extension). *Tilburg et al.* [2001] also found that 1/8° resolution is insufficient and 1/16° is sufficient to simulate the observed mean meanders of the Tasman Front caused by eddy-driven mean abyssal currents along the slopes of ridges and trenches beneath this front. Increasing the resolution to 1/32° gave only modest improvement.

Hurlburt and Hogan [2000] investigated the impact of 1/8° (~14 km), 1/16° (~7 km), 1/32° (~3.5 km), and 1/64° (~1.8 km) resolution on a wide range of Gulf Stream model data comparisons. They found that 1/16° resolution was the minimum for realistic results, and substantial improvement was obtained by an increase to 1/32° resolution, but only modest improvement occurred with a further increase to 1/64° resolution. These improvements include realistic Gulf Stream separation from the coast at Cape Hatteras and a realistic pathway to the Grand Banks (but a more inertial current at 1/32° and 1/64° resolution), realistic nonlinear recirculation gyres that contribute to the C-shape of the subtropical gyre, realistic upper ocean and Deep Western Boundary Current (DWBC) transports, realistic patterns and amplitude of SSH variability surrounding the Gulf Stream, realistic warm core ring diameters, population and rings generated per year on the north side of the Gulf Stream, and realistic patterns and amplitude of abyssal EKE. Hurlburt and Hogan (H. E. Hurlburt and P. J. Hogan, The Gulf Stream pathway and the impacts of the eddy-driven abyssal circulation and the Deep Western Boundary Current, submitted to *Dynamics of the Atmosphere and Oceans*, 2008) found that steering by the topographically constrained, eddy-driven mean abyssal circulation alone is sufficient to obtain realistic Gulf Stream separation from the coast at Cape Hatteras when using 1/32° and higher resolution, but assistance by the DWBC is required to obtain Gulf Stream separation in the model with 1/16° resolution.

The Sea of Japan (East Sea; JES) is characterized by weaker currents and a smaller internal Rossby radius of deformation than the preceding regions [*Emery et al.*, 1984; *Chelton et al.*, 1998; *Oh et al.*, 2000]. In the JES, upper ocean-topographic coupling via flow instabilities has a large impact on the mean upper ocean circulation, including the pathway of the subpolar front and, depending on the choice of atmospheric forcing product, on the separation of the East Korea Warm Current from the Korean coast. In the resolution sequence 1/8°, 1/16°, 1/32°, 1/64°, this coupling does not occur at resolutions coarser than 1/32°, but the mean circulation and variability statistics are quite similar when the resolution is increased from 1/32° to 1/64° [*Hogan and Hurlburt*, 2000, 2005]. Recently, anticyclonic intrathermocline eddies (ITEs; characterized by a submerged lens of nearly homogeneous water with a bowl-shaped bottom and a domed top) were discovered in the JES [*Gordon et al.*, 2002; *Talley et al.*, 2004]. Such eddies are far from ubiquitous in the world ocean and were not found in the four-layer NLOM simulations. However, they were found in 15-layer, 1/25° JES HYCOM simulations (~3.5-km resolution, nearly the same as the 1/32° NLOM; see section 2.1), and these simulations were used to help explain why ITEs form in the JES [*Hogan and Hurlburt*, 2006]. Like the eddies in the 1/32° NLOM simulations, the eddies and ITEs in the JES tend to be persistent and topographically constrained to a limited range of movement. Further, mean SSH, surface currents, and abyssal currents from 1/25° JES HYCOM and 1/32° JES NLOM showed close quantitative agreement when they used the same wind stress forcing [Hurlburt et al., submitted manuscript, 2008].

In a data-assimilative example, *Chassignet et al.* [2005] found that the dynamical interpolation skill of the operational 1/16° NLOM prediction system was very poor in the northern Gulf of Mexico. Because of insufficient constraint of the flow below the sill depth, that system had spurious southward leakage of a small fraction of the DWBC though the Florida Strait, which then flowed cyclonically around the continental slope in the Gulf of Mexico. Advection by this spurious abyssal current is consistent with the excessively rapid westward propagation of the upper ocean features in the northern Gulf of Mexico and the poor dynamical interpolation skill in the assimilation of altimeter data in that area. The spurious DWBC leakage was eliminated in the 1/32° NLOM prediction system, and the dynamical interpolation skill in the northern Gulf of Mexico was greatly improved.

The preceding results in this subsection and section 2.2 point out the striking differences between the performance of eddy-permitting and eddy-resolving ocean models with or without ocean data assimilation. A common theme is the importance of mesoscale flow instabilities in allowing eddy-driven abyssal currents, constrained by bottom topography, to advect the pathways of mid-latitude upper ocean currents, and thus influence their mean pathway. This upper ocean-topographic coupling requires that mesoscale variability and related flow instabilities be very well resolved to obtain sufficient coupling, for example, to achieve accurate representation of the vortex stretching and compression associated

with baroclinic instability in the presence of realistic topography. Thus, this eddy-driven topographic effect is missed at coarser resolution, for example, in eddy-permitting models, and can lead to unexplained errors in simulations of mid-latitude mean upper ocean current pathways and false conclusions about the influence of topography. Although based on a two-layer theory, upper ocean-topographic coupling has been demonstrated in multilayer models using NLOM and in simulations with high vertical resolution using HYCOM [Hurlburt et al., submitted manuscript, 2008]. This theory applies in mid-latitude regions with low vertical mode structure in areas where the topography does not intrude significantly into the stratified ocean.

Resolving the internal Rossby radius of deformation has been used as a criterion for distinguishing between eddy-permitting and eddy-resolving models because of its relation to the predominant space scale for baroclinic instability. But how well does it need to be resolved, and what other criteria need to be satisfied? The preceding results provide some very useful input. To be eddy-resolving, a model needs to (1) realistically simulate the eastward penetration of inertial jets and associated recirculation gyres where they exist, (2) demonstrate near convergence for the mean strength and pathways of both upper ocean and abyssal currents plus their mesoscale variability statistics and, when attempting to simulate the real ocean, it must (3) provide realistic comparisons to observations of mean upper ocean and abyssal currents and their variability statistics, including the basic characteristics of eddies and their behavior (adapted from Hurlburt et al., submitted manuscript, 2008).

2.4. Can Satellite Altimetry Be Used to Map Small Eddies? What Is the Lower Bound on Eddy Size and its Dependence on Model Resolution?

Plate 2a shows a SeaWiFS ocean color image in the northwestern Arabian Sea and Gulf of Oman. Such a cloud-free image during a very large outbreak of high chlorophyll is an unusual event in this region. The image shows numerous eddies ranging in diameter from 1/4° to > 2°, which makes it suitable for use in a quantitative evaluation showing the ability of satellite altimeter data to map mesoscale variability when assimilated by four global ocean models, models with differing horizontal and vertical resolution, model design, and data assimilation techniques, that is, 1/16° NLOM (8-km resolution at 20°N; Plate 2b), 1/32° NLOM (4 km; Plate 2c), 1/8° NCOM (18 km; Plate 2d), and 1/12° HYCOM (8 km; Plate 2e). 1/32° NLOMn (Plate 2f) shows results from a control run designed to demonstrate the impact of the data assimilation. This run is a repeat of the run shown in Plate 2c, but with no assimilation of ocean data. Based on the results presented in

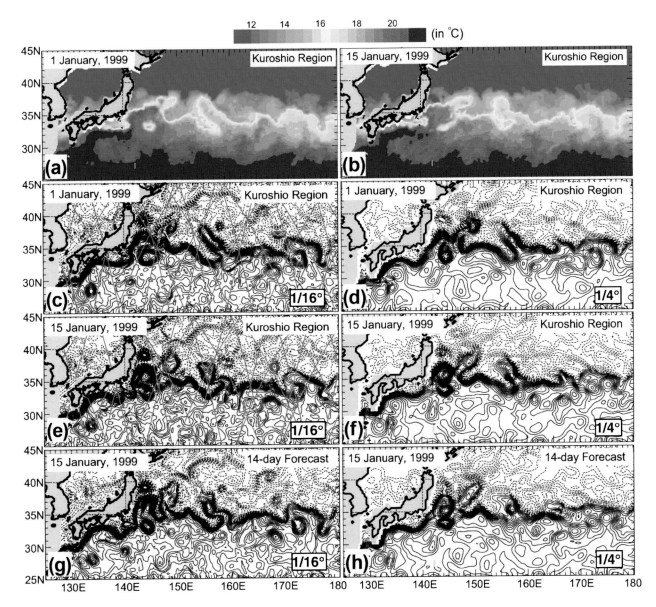

Plate 1. Kuroshio region (a, b) NAVOCEANO operational 1/8° MODAS SST analyses for (a) January 1, 1999 and (b) January 15, 1999 with the color bar (0.5°C interval) designed to highlight the Kuroshio pathway, (c-f) SSH analyses (5-cm interval) with assimilation of altimeter track data from TOPEX/Poseidon and ERS-2 using the model forecast as the first guess from (c, e) 1/16° Pacific NLOM and (d, f) 1/4° global NLOM for (c, d) January 1, 1999 and (e, f) January 15, 1999. The altimeter tracks assimilated on the date given (using a 3-day data window) are overlaid in red on (c, e). (g, h) The 14-day forecast SSH from (g) 1/16° Pacific NLOM and (h) 1/4° global NLOM. Adapted from *Hurlburt et al.* [2000].

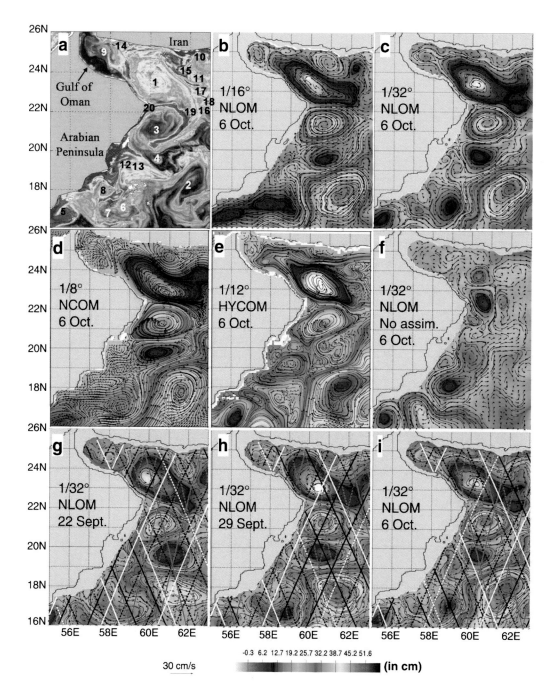

Plate 2. A comparison of eddies and currents seen in (a) chlorophyll concentration from SeaWiFS (October 2–6, 2002, latest cloud-free pixel composite with most data from October 6) with (b-e) data-assimilative model nowcast SSH and currents on October 6, 2002 from (b) 1/16° global NLOM, (c) 1/32° global NLOM, (d) 1/8° global NCOM, and (e) 1/12° global HYCOM. (f) same as (c) but without ocean data assimilation. In (a), eddies with clearly-defined eddy centers are numbered in order of decreasing size for use in Table 2. The color of the number varies only for visual clarity. 1/32° NLOM SSH and currents for (g) September 22, 2002; (h) September 29, 2002; (i) October 6, 2002 [same as (c)] with the most recent week of observed altimeter tracks overlaid as solid lines and the remaining tracks as dashed lines, red for ERS-2, black for GFO, and white for Jason-1. At the time, Jason-1 data were not available in real time at NAVOCEANO and thus not assimilated by the operational 1/16° NLOM system which provided the results used in (b). Note that a few of the smallest eddies are not visible due to limitations of the small plots or because they were slightly east of 63°E in a model result. NLOM results are adapted from *Shriver et al.* [2007].

section 2.3, with 18-km resolution at 20°N and ~15 km at mid-latitudes, the 1/8° NCOM system is eddy-permitting, and the rest of the systems are eddy-resolving. All of the model runs include atmospheric forcing; all of the data-assimilative runs include assimilation of SST, and 1/12° HYCOM also assimilates T & S profiles, but satellite altimetry is the key observing system for mapping the eddies. Results of this comparison for 1/16° and 1/32° NLOM can be found in *Shriver et al.* [2007]. All of the systems assimilated altimeter data from the ERS-2, GFO, and Jason-1 missions, except that the operational 1/16° NLOM system omitted Jason-1 because the data were not yet available in real time at NAVOCEANO. *Shriver et al.* [2007] found that the Jason-1 data added only minimal value in this particular case, which is consistent with the results of *Smedstad et al.* [2003], who used simulated data and found that adding a third altimeter provided only modest error reduction.

A quantitative analysis of model eddy center location error in comparison to the ocean color image (Plate 2a) is shown in Table 2. All eddies with a clearly defined center were used except for a cyclonic eddy located inside anticyclonic eddy 3. The eddy center locations in the models and the ocean color were determined as independently as possible and compared only to the extent needed to match them up (location, rotation direction, and size). The accuracy in determining the locations of the eddy centers is typically 10–15 km for both the ocean color and the models and thus is 15–20 km in determining the eddy center location error. The eddies in Table 2 are listed in the order of decreasing diameter, as seen in the ocean color, to assist in determining the minimum eddy size that could be mapped by assimilation of the altimeter data in each model. The first four eddies are ~2° or more in diameter, eddies 5–8 are ~1°, 12–16 are ~1/2°, and 17–20 are ~1/4°. With a few exceptions, model eddies 11 through 20 are typically ~50% or more larger than seen in the ocean color, but the rest are generally consistent with the ocean color imagery. Using model-independent analyses of TOPEX/Poseidon and ERS-1 and 2 altimeter data, *Ducet et al.* [2000] estimated ~75 km as the minimum eddy diameter that could be mapped using satellite altimeter data. Thus, 75 km was used to divide the observed eddies into large and small eddies, making eddies 11 through 20 the small eddies.

Table 2 summarizes the results of comparing eddy center locations in the models with those in the ocean color. In this comparison, the ability to map eddies as small as 1/4° in diameter is not limited by the satellite altimetry, but by the model resolution. The 1/32° NLOM system depicted 90% of the eddies, 80% of the small eddies, and all four of the 1/4° eddies (three of the 1/4° eddies with an accurate location) versus the 1/8° NCOM system, which depicted 55% of

the eddies, 80% of the larger eddies (1–10) and 30% of the small eddies. Note that the 1/8° NCOM system assimilated the steric SSH anomalies from the 1/32° NLOM system in the form of synthetic T & S profiles [*Rhodes et al.*, 2002 and section 1.2]. Generally, about seven grid intervals was the minimum eddy diameter the models could realistically depict, that is, 128 km for 1/8° NCOM, 56 km for 1/12° HYCOM and 1/16° NLOM, and 28 km for 1/32° NLOM. In some cases, eddies were depicted by increasing the diameter of small eddies to at least this minimum size, for example, eddy 11 (but not eddy 18) in 1/8° NCOM and eddy 18 in 1/16° NLOM. Consistent with the seven grid interval minimum, 1/16° NLOM was able to map eddies 1/2° and larger, but not the 1/4° eddies. All of the eddies seen in the ocean color were mapped by at least one of the data-assimilative systems.

In comparison to 1/32° NLOMn (with no ocean data assimilation), the impact of the altimeter data is strikingly evident (Plate 2 and Table 2). As 1/32° NLOMn depicts numerous small eddies (Plate 2f) and five of these match up with 50% of the small eddies seen in the ocean color, one might question the impact of the altimeter data assimilation in mapping these eddies. However, Table 2 clearly shows the impact of the assimilation on the depiction of small eddies, for example, by comparing 1/32° NLOM and 1/32° NLOMn, we find that the depiction of small eddies is 80% versus 50%, the more accurate eddy center location is 67% versus 33%, the median eddy center location error is 22.5 versus 42 km, and the minimum eddy center location error is 11 versus 26 km. Furthermore, 62.5% (87.5%) of the small eddies in 1/32° NLOM have eddy center location errors less than the minimum of 26 km (median of 42 km) in 1/32° NLOMn.

Clearly, the minimum eddy diameter mapped using satellite altimetry is much less than 75 km when the altimeter data are assimilated by an eddy-resolving ocean model (all except the eddy-permitting 1/8° NCOM in Plate 2 and Table 2). The 1/32° NLOM system (Plates 2g–2i) is used to illustrate how a model can assist in the depiction of eddies and currents by using its dynamical interpolation skill. The altimeter tracks crossing the region are overlaid on each panel (the most recent week as solid lines and the rest as dashed lines), although only a 3-day data window is used in the daily assimilation cycle. In total, these panels show the most recent three weeks of altimeter data, September 16 through October 6, 2002.

A striking example of model dynamical interpolation skill is provided by a current that enters the southern boundary of the region near 60°E and ultimately wraps around eddy 12. It is marked by a ribbon of blue (lower chlorophyll) in the ocean color and flows northward along ~60°E, then westward north of 18°N and then northward along ~58.5°E until it wraps around eddy 12, which is 1/2° in diameter. This current is seen in all of the data-assimilative models. Eddy 12 and adjacent eddy 13 are depicted in all the data-assimilative models except for eddy 13 in 1/8° NCOM. The non-assimilative 1/32° NLOMn depicts neither the current nor the eddies. During the week leading up to October 6, 2002, the entire system is unobserved by the altimetry except for the southernmost portion of the current. The system is quite well observed during September 16–22, when it is significantly different. From September 16 to October 6, eddies 12 and 13 are never more than peripherally observed, and eddy 13 does not even exist

Table 2. Eddy Center Location Errors in Ocean Prediction Models Compared to Ocean Color From SeaWiFS in the Northwestern Arabian Sea and Gulf of Oman.

Ocean Color Eddy ID#	A or C	Ocean Color Eddy Center Location		1/16° NLOM	1/32° NLOM	1/8° NCOM	1/12° HYCOM	1/32° NLOMn No Assim
		°N	°E	Eddy Center Position Error, km				
1	C	23.55	60.2	35	18	80	48	NP
2	A	18.3	62.0	103	28	66	48	NP
3	A	21.15	60.25	44	58	48	37	NP
4	C	19.65	60.4	43	12	15	59	45
5	C	16.8	55.65	35	42	75	72	31
6	C	17.0	58.9	42	17	38	68	NP
7	C	16.7	57.9	53	79	NP	97	NP
8	A	18.0	57.6	NP	40	36	68	NP
9	C	25.1	57.6	NP	39	91	76	NP
10	A	24.7	62.5	30	35	NP	NP	NP
11	C	23.7	62.3	[a]	22	37	52	42
12	C	19.3	58.8	30	30	50	65	NP
13	A	19.225	59.35	35	11	NP	26	NP
14	A	25.3	58.55	36	33	NP	28	30
15	C	24.1	61.75	55	NP	NP	NP	47
16	C	22.25	62.7	14	[a]	NP	48	[a]
17	C	23.1	62.55	NP	13	NP	44	NP
18	A	22.5	62.85	18	51	47	NP	44
19	C	22.05	62.05	NP	23	NP	NP	26
20	C	22.2	59.95	NP	12	NP	22	NP
		% of Eddies Present in the Model						
	All eddies			70	90	55	80	35
	Large eddies, 1–10			80	100	80	90	20
	Small eddies, 11–20			60	80	30	70	50
		Median Eddy Center Position Error, km						
	All eddies			35.5	29	48	50	42
	Large eddies			42.5	37	57	68	38
	Small eddies			32.5	22.5	47	44	42
		% of eddies With Most Accurate Position						
	All eddies			22.5	52.5	5	10	10
	Large eddies			20	50	10	10	10
	Small eddies			25	55	0	10	10

on September 22 and 29. A GFO track does appear to play an important role in narrowing the westward bulge of eddy 2 from September 22 to 29, which assists in the detachment of eddy 13. However, model dynamics are required to move eddy 13 northward and partially merge it with an existing anticyclonic eddy north of eddy 12. This process gave an 11-km eddy center location error for eddy 13 in the 1/32° NLOM system.

In a second example, eddy 19 was well observed by an ERS-2 track during the week leading up to September 22. The result was a strong small eddy in 1/32° NLOM and a weak one in 1/16° NLOM. The 1/32° NLOM system maintained this nearly stationary eddy through October 6, but it quickly dissipated in the 1/16° NLOM system. During the week leading to October 6, eddy 16 was well observed by GFO and ERS-2 altimeter tracks, and it was represented in 1/16° NLOM with a 14-km eddy center location error. Eddy 16 was not represented in 1/32° NLOM because eddy 19 was too large and the observations of eddy 16 were treated as observations of the eastern edge of eddy 19, thereby making it a fusion of eddies 16 and 19.

Cyclonic eddy 20 is the smallest of the 1/4° eddies (24 km in diameter in the 1/32° NLOM system). It lies adjacent to the coast and a strong current that separates from the coast. In the ocean color image, this current wraps around large anticyclonic eddy 3. Eddy 20 was not observed by altimetry, but forms in 1/32° NLOM and 1/12° HYCOM due to the current separation from the boundary and proximity to the boundary. This small eddy does not form in the coarser resolution 1/16° NLOM and 1/8° NCOM, although they depict eddy 3 and the separating boundary current with comparable realism. The separating boundary current is shown with less

accuracy in 1/12° HYCOM, but HYCOM does simulate a cyclonic eddy similar to eddy 20 adjacent to the coast.

The preceding results demonstrate that an eddy-resolving ocean model can be skillful in assimilating data along multiple altimeter tracks from multiple satellites and in dynamically interpolating it through time to build coherent mesoscale features that are consistent with observations, including small eddies. They also demonstrate that increasing the horizontal resolution of the model increases that skill, especially in mapping small eddies 25–75 km in diameter. In addition, the results show that a model with dynamical interpolation skill can use the data assimilation to produce an environment more conducive to the formation of poorly observed eddies than the same model without the assimilation, a capability that is especially useful in representing small eddies, as demonstrated here.

2.5. Eddy-Resolving Ocean Model Forecast Skill and the Impacts of Atmospheric Forcing, Model Choice, and Model Resolution

Several metrics are used routinely to assess model forecast skill based on comparison to a verifying analysis. To demonstrate useful skill, a forecast must be superior to persistence (a forecast of no change) and to either climatology or some minimum value of the metric. The metrics are root mean square (rms) error, anomaly correlation (AC), skill score, and for the Kuroshio and the Gulf Stream, axis error. See *Smedstad et al.* [2003] for discussion and examples of results from each of these. Normalized rms error, normalized by the standard deviation of the anomalies, is another useful forecast metric. Here, only the SSH is evaluated and only the AC is used, a metric commonly used in verifying weather

---

Notes to Table 2:

[a]A single "fused" model eddy represents two eddies in the ocean color and the model eddy center lies between the two observed eddies. The model eddy center position error is listed only under the closer ocean color eddy.

A, anticyclonic eddies; C, cyclonic eddies; NP, eddy not present.

The ocean color eddy ID numbers are plotted in Plate 2a. Eddies are listed in order of decreasing size as depicted by the ocean color. Eddy position measurement error is 10–15 km in both the ocean color and the models.

1/32° NLOM: 1/32° global NLOM 7-layer prediction system with assimilation of altimeter track data from ERS-2, GFO and Jason-1 altimeters (presently operational at NAVOCEANO).

1/16° NLOM: Then operational 1/16° global NLOM seven-layer prediction system with assimilation of real-time altimeter track data from the ERS-2 and GFO altimeters (Jason-1 was not in the operational data stream at that time).

1/8° NCOM: 1/8° global NCOM 40-level prediction system with assimilation of 1/32° NLOM SSH via synthetic T & S profiles (presently operational at NAVOCEANO).

1/12° HYCOM: .08° global HYCOM prediction system with 32 hybrid layers and assimilation of altimeter track data from ERS-2, GFO and Jason-1 altimeters plus T & S profiles (presently pre-operational and running in real time at NAVOCEANO).

1/32° NLOMn: 1/32° global NLOM with no assimilation of ocean data, only atmospheric forcing.

Some model versions of eddy 18 lie east of 63°E and thus are not depicted in Plate 2.

forecasts. It is calculated with the same methodology used by *Smedstad et al.* [2003] and *Shriver et al.* [2007]:

$$AC(f,a) = \frac{\sum (f-\bar{f})(a-\bar{a})}{\left(\sum (f-\bar{f})^2 \sum (a-\bar{a})^2\right)^{1/2}}$$

where f is the forecast value and a is the analysis value at a given time and model grid point. Here, the mean of the forecast $\bar{f}$ and the analysis $\bar{a}$ are identical, that is, the modified model mean SSH used in assimilating SSH anomalies observed by satellite altimetry.

Some of the real-time forecast models perform 30-day forecasts once a week. For these forecasts, the real-time atmospheric forcing reverts toward climatology after the end of the forecast atmospheric forcing (after 5 days in these examples). To evaluate the impact of real-time versus analysis-quality atmospheric forcing on ocean model forecast skill, two sets of "forecasts" were performed and evaluated using 1/12° global HYCOM and 1/32° global NLOM, but only real-time forcing in the 1/16° global NLOM forecasts. In addition, the SSH forecasts were verified against tide gauge data, an independent, unassimilated data set.

The results of forecast verification versus analyses and tide gauge data are presented in Plate 3. The global forecast verification covers only regions with assimilation of altimeter data, from 50°S to 65°N for NLOM and 60°S to 47°N in these early results for HYCOM. The Gulf Stream subregion was chosen to illustrate forecast skill in a highly nonlinear region with strong flow instabilities, current meanders, and generation of mesoscale eddies. These are class 2 (Table 1) phenomena which are insensitive to errors in the atmospheric forcing on a 30-day timescale. The equatorial Pacific has significant elements from all four classes, including class 1 phenomena such as the onset of wind-driven equatorial waves and variations in equatorial upwelling. Thus, 30-day forecasts in this region are more sensitive to errors in the atmospheric forcing. The region in the NW Arabian Sea and Gulf of Oman was chosen to show forecast skill, a prerequisite for dynamical interpolation skill, over nearly the same region shown in Plate 2. The forecast results are quite typical of the world ocean with modest sensitivity to the errors in atmospheric forcing beyond the first 2 weeks.

A dramatically different result is seen in the forecast results for the Persian Gulf from the 1/12° global HYCOM system. Here, persistence demonstrates forecast skill (AC >0.6) for only 2 days and, similarly, the forecast with real-time forcing lasts only 2 days beyond the end of the 2 days of forecast forcing (represented by 5 days of analysis-quality forcing in these tests). In contrast, forecast skill remains high for the 30-day duration of the forecasts when analysis quality forcing is used. The Persian Gulf is a shallow-water basin where the SSH exhibits a largely deterministic class 1 response to atmospheric forcing. As a result, the forecast skill is characterized by rapid loss of initial state impact, and it is strongly determined by the timescale of accurate atmospheric forcing (i.e., it is limited by the timescale of atmospheric predictive skill in a real-time prediction system). Similar results were obtained for the shallow Yellow and Bohai Seas region north of 30°N. *Shriver et al.* [2007] show forecast results from more subregions, using the 1/32° global NLOM system, including results for two deep-water semi-enclosed seas. For these, the forecast skill is more characteristic of other deep-water regions.

Overall, the 30-day forecast results show low sensitivity to the difference between the climatological and analysis-quality atmospheric forcing fields in deep-water regions. Outside the surface mixed layer and shallow water, the evolution of the forecasts is more sensitive to the initial state than to the atmospheric forcing anomalies over most of the global ocean. Much of this is due to the fact that mesoscale eddies are not confined to highly energetic regions of the world ocean like the Gulf Stream and Kuroshio, but are nearly ubiquitous in the world ocean, as demonstrated by model-independent analyses of satellite altimeter data [*Le*

Plate 3. (Opposite) Verification of 30-day ocean forecasts, (a-i) median SSH anomaly correlation versus forecast length in comparison with the verifying analysis for (a, e), the global domain (45°S–45°N) and four subregions (b-d, f-i) defined in Table 3. (a-d) are for 1/32° NLOM (solid lines) and 1/16° NLOM (dashed lines) and (e-i) are for 1/12° HYCOM. The red curves verify forecasts using operational atmospheric forcing which reverts toward climatology after five days. The green curves verify "forecasts" with analysis-quality forcing for the duration, and the blue curves verify forecasts of persistence (i.e., no change from the initial state). The plots for NLOM give the median statistics over twenty-two 30-day forecasts initialized during the period from June 1, 2001 through May 31, 2002, and for HYCOM twenty 30-day forecasts initialized during the period from January 2004 through December 2005, both periods when data from three nadir-beam altimeters in corresponding orbits were assimilated. The same forecast results were used to obtain (j-l) median correlation between forecast and observed SSH fluctuations from 1/12° HYCOM with operational forcing during the forecast (red lines), 1/12° HYCOM with analysis quality forcing for the duration (green lines), 1/16° (blue lines), and 1/32° (black lines) NLOM (both with operational atmospheric forcing) at (j) 23 (49) open ocean island tide gauge stations for HYCOM (NLOM), (k) 91 (29) coastal tide gauges for HYCOM (NLOM), and (l) all 114 (78) tide gauges for HYCOM (NLOM). A 13-day moving average was applied to filter time scales not resolved by the altimeter data. Tide gauge SSH data are not assimilated by the ocean prediction systems. NLOM results are adapted from *Shriver et al.* [2007].

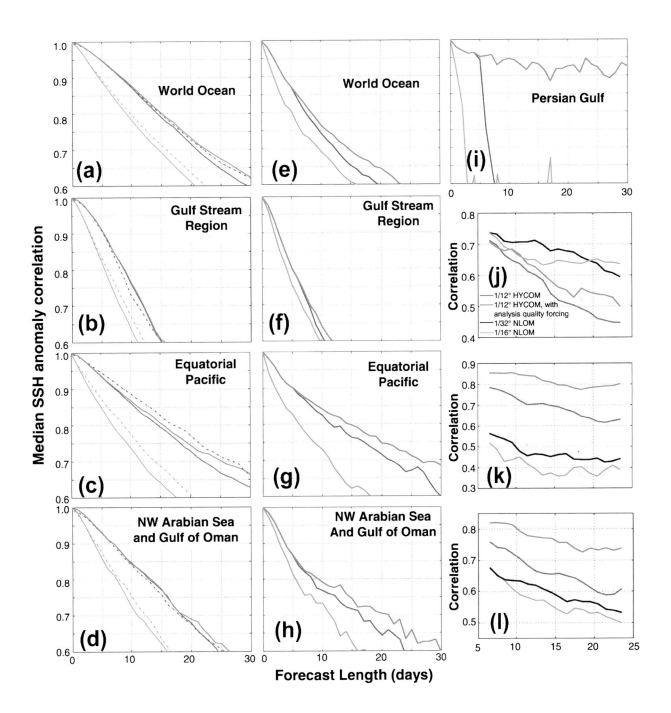

Table 3. Subregions Depicted in Plate 3

Subregion	Latitude Range	Longitude Range
World Ocean	45°S–45°N	All
Gulf Stream region	35°N–45°N	76°W–40°W
Equatorial Pacific	20°S–20°N	109°E–77°W
NW Arabian Sea and Gulf of Oman	15°N–26°N	51°E–65°E

Traon et al., 1998] and as illustrated in Plate 2a for a range of space scales in a region without high SSH variability.

Comparing AC versus time from the 30-day forecasts performed using 1/16° NLOM and 1/32° NLOM, one finds the perhaps surprising result that globally and for many subregions, the AC is higher in the 1/16° model [*Shriver et al.*, 2007]. The higher resolution model has more smaller scale features with larger amplitude that can get out of phase on shorter timescales. Consistent with this difference between the two models, persistence also has less skill in the 1/32° NLOM system than in the 1/16° system. Thus, the best way to judge performance is not by the AC for the forecast, but by the spread between the AC of the forecast and persistence. For a given day, that is always greater for the 1/32° NLOM system.

The preceding results also highlight the need to assess forecast skill using comparisons to independent, unassimilated data sets. Thus, Plates 3j–3l show evaluation of SSH from 1/12° global HYCOM and the two global NLOM systems by comparison to unassimilated tide gauge data. For HYCOM that also includes a comparison of "forecasts" using operational and analysis-quality forcing. A 13-day moving average was applied to filter timescales not resolved by the assimilated altimeter data. Again, these comparisons indicate that the 1/32° NLOM system has greater forecast skill than the 1/16°. HYCOM forecasts give greater skill at coastal stations (Plate 3k) than island stations (Plate 3j), while the reverse is true for the NLOM systems. HYCOM forecasts demonstrate greater sensitivity to analysis-quality versus operational-quality atmospheric forcing at coastal stations than at island stations, as expected from the differ-

ing classes of response to atmospheric forcing in coastal versus deep ocean regions (Table 2). HYCOM outperformed NLOM at coastal stations because it includes shallow water and because the HYCOM numerical grid generally extends closer to the tide-gauge locations. In contrast, NLOM outperformed HYCOM at island stations on forecast timescales longer than 2 weeks where model grid proximity to the tide gauges is similar for both models. In addition, HYCOM forecast accuracy decreased more rapidly than NLOM's at island stations. As the HYCOM and NLOM forecasts were performed for different years and there are large differences in the tide gauge stations used, only large HYCOM-NLOM differences were considered in the preceding comparisons between the systems.

2.6. Mean SSH for Assimilation of Altimeter Data: Roles of Model and Observation Based Means

The capability to simulate mean SSH accurately is important for an ocean model because (1) an accurate mean SSH field is required for addition to the deviations provided by satellite altimetry, and a sufficiently accurate model can be used in providing it, and (2) if the mean SSH is not well simulated by the model, the model forecasts will tend to drift with a bias toward the erroneous model mean, reducing the AC scores of the forecasts (discussed in section 2.5). In a data-assimilative ocean model, high accuracy in SSH fields is essential to properly represent ocean currents and fronts and to avoid contributing to errors in subsurface T & S. Thus, accurate mean SSH is a critical application of a class 3 ocean response to atmospheric forcing in ocean data assimilation. Model and observation-based means each have their advantages and disadvantages for this purpose. Fortunately, the two types of mean are converging. The standard deviation of the difference between the observation-based *Maximenko and Niiler* [2005] SSH mean (Plate 4a) and a 1/12° global HYCOM mean (Plate 4b) is only 9.3 cm, and the bias between their global areal-averaged means is 7.1 cm. The detailed agreement of the frontal segments between 30°S and 60°S is particularly striking.

Plate 4. (Opposite) Mean SSH (a) from *Maximenko and Niiler* [2005] based on satellite altimeter, drifting buoy and wind data spanning 1992–2002 and an improved geoid, GRACE (Gravity Recovery and Climate Experiment) Gravity Model 01 (GGM01) and (b) simulated by 1/12° global HYCOM using ECMWF climatological wind and thermal forcing. The standard deviation of the difference between them is 9.3 cm, and the bias between their global areal-averaged mean is 7.1 cm. (c) MODAS mean steric height anomaly relative to 1,000 m based solely on historical hydrography. Zoom views of (d) the Maximenko and Niiler mean SSH and (e) the 1/12° global HYCOM mean SSH in the western equatorial Pacific and adjacent marginal seas. (f, g) Mean currents in the northeast Pacific with mean current speed in color from 1/8° global NCOM (f) without ocean data assimilation and (g) with assimilation of SSH anomalies analyzed by NLOM from satellite altimetry using the MODAS mean SSH in (c). Without the assimilation, the Alaska Stream, marked by the ribbon of red in (f), is robustly simulated by 1/8° global NCOM, but it is greatly weakened in (g) when the data assimilation is added because the mean SSH associated with the Alaska Stream is poorly represented in the MODAS mean. Panels (f) and (g) are adapted from *Barron et al.* [2007].

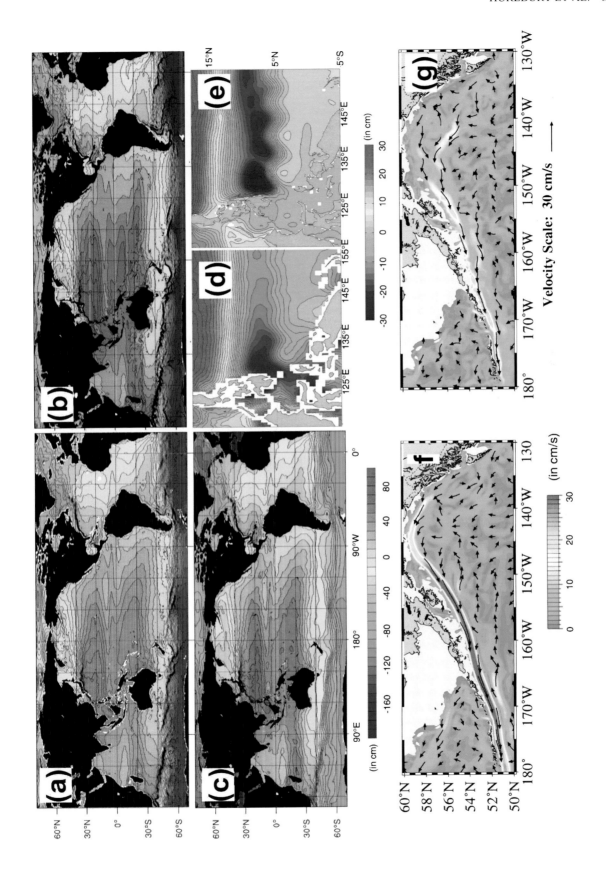

Velocity Scale: 30 cm/s ⟶

The HYCOM simulation was initialized from a hydrographic climatology, and the mean over years 9–13 is shown in Plate 4b. Although ocean data assimilation could have been included, this simulation used only climatological atmospheric forcing, except for weak relaxation to sea surface salinity (in addition to surface salinity forcing from evaporation, precipitation, and rivers). The monthly atmospheric climatology is based on the European Centre for Medium-Range Weather Forecasts (ECMWF) reanalysis (ERA-15) [*Gibson et al.*, 1997] with the 10-m winds converted to wind stress using the bulk formulation of *Kara et al.* [2005]. Submonthly fluctuations were added to the wind stress to energize the mixed layer.

The MODAS mean steric height anomaly relative to 1,000 m (Plate 4c) is based solely on historical hydrography. Such means have insufficient data to accurately map sharp features in the mean, such as boundary currents and ocean fronts. This deficiency causes serious problems in ocean data assimilation as illustrated in Plates 4f and 4g showing mean currents in the NE Pacific from 1/8° global NCOM. Plate 4f shows that without ocean data assimilation, NCOM simulates a robust Alaska Stream (red ribbon of high current speed), a current with realistic transport in comparison to 9 years of observations presented by *Onishi and Ohtani* [1999] and *Onishi* [2001]. However, when the MODAS mean is used as the reference for assimilation of altimetric anomalies, the Alaska Stream in NCOM is greatly weakened. This problem was rectified by using a modified model mean.

Maximenko and Niiler [2005] derived a 1/2° absolute SSH mean (i.e., not relative to a reference depth) based on data from drifting buoys, satellite altimetry, winds, and the GRACE (Gravity Recovery and Climate Experiment) Gravity Model 01 (GGM01) geoid [see also *Niiler et al.*, 2003]. It tends to have more sharply defined features than the MODAS mean surface dynamic height relative to 1000 m from hydrography. However, mean fronts and boundary currents tend to be the most sharply defined in the model mean. In addition, some boundary currents are poorly represented in the Maximenko and Niiler mean, for example, the Mindanao Current along the east coast of the southern Philippines (Plate 4d compared to the model mean shown in Plate 4e) and the unrealistic features around the island of Borneo.

Thus, it is desirable to use a mean SSH field from an accurate eddy-resolving ocean model that was initialized from climatology. Mean SSH features have a wide range of amplitudes and space scales, and an eddy-resolving model should have an advantage in representing mean SSH associated with smaller gyres and weaker currents and features associated with complex geometry, such as the Indonesian and Philippine archipelagos (Plates 4d, 4e). However, models are prone to errors in the strength and pathways of interior inertial jets, larger scale SSH biases in gyres caused by errors in subsurface thermal structure, including the mixed layer depth and temperature. Such errors are caused by the atmospheric forcing as well as by the model, and the atmospheric forcing is a significant source of error in the model mean SSH, for example, the cleavage in the large-scale South Pacific gyre structure, not seen in either observation-based mean, is caused by a longstanding problem with winds from ECMWF. These winds drive a South Equatorial Countercurrent that is too strong and extends too far east [*Metzger et al.*, 1992], a problem that continues in the 1/12° HYCOM simulation with ECMWF ERA-15 forcing. In the North Atlantic, it drives a western boundary current that is too weak.

Because of the preceding problems, it is necessary to make corrections to the model-based means using ocean data, including observation-based means, mean frontal analyses based on satellite infrared imagery, SSH variability, the limited number of nearly simultaneous air-dropped expendable bathythermograph (XBT) under-flights of altimeter tracks (very useful for the Gulf Stream; *Chassignet and Marshall* [2008] this volume), and the limited number of often-repeated high-resolution lines of XBT or acoustic Doppler current profiler data. This was done to a limited extent in the HYCOM mean used in the data-assimilative 1/12° global HYCOM (but not the mean shown in Plate 4). In addition, the best available atmospheric forcing is chosen, and corrections are made to the wind field, for example, unrealistic wind curl features around Hawaii were corrected to eliminate unrealistic zonal currents west of Hawaii, and corrections have been made to avoid having wind values over land from coarser resolution atmospheric fields force sea points in a finer resolution ocean model [*Kara et al.*, 2007]. Scatterometer data could also be used in making corrections to the wind speeds from atmospheric prediction centers.

2.7. Atmospheric Forcing, a Significant Data Type in Ocean Prediction

In the preceding subsection, atmospheric forcing played an essential class 3 role in helping eddy-resolving 1/12° global HYCOM simulate an accurate mean SSH field for use in altimeter data assimilation. In section 2.5, it was shown that forecasts of SSH in the Persian Gulf and at coastal tide gauge locations are sensitive to the quality of atmospheric forcing due to class 1 responses in shallow water and in coastal and equatorial wave guides. In a model without ocean data assimilation, *Zamudio et al.* [2006] showed that verifiable eddies can be generated by a combination of Tehuantepec winds (class 1) and coastal trapped waves originating in the equatorial Pacific (class 4), including subsequent class 4 eddy propagation ~1,000 km offshore. In another non-assimilative

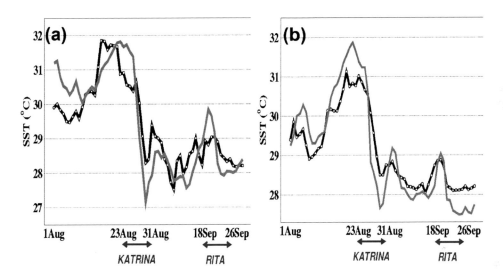

Plate 5. Daily observed buoy SST (black lines) versus 1/12° global HYCOM (red lines) using three-hourly FNMOC NOGAPS wind and thermal forcing. The time series span from August 1 to September 30, 2005 and cover the passage of Hurricanes Katrina and Rita at (a) NDBC buoy 42040 south of Mobile Bay and (b) buoy 42036 southeast of Pensacola, FL.

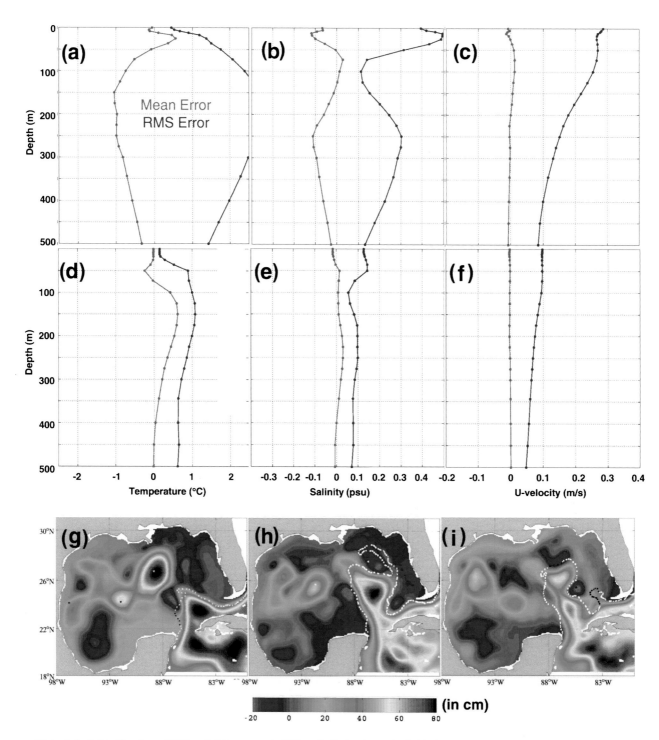

Plate 6. (a-f) Verification of SST and altimeter track SSH assimilation in 1/12° Gulf of Mexico HYCOM using NCODA and data simulated by the model during a different time period as sampled at observed SST and altimeter track locations during that time. Mean error (red) and rms error (blue) versus depth to 500 m for (a, d) temperature, (b, e) salinity, and (c, f) zonal component of velocity at the beginning of the assimilation period on August 29, 1999, and 50 days later on October 18, 1999. (g-i) Results of assimilating real SST, temperature profile, and altimeter track data into 1/12° Gulf of Mexico HYCOM and verification against operational NAVOCEANO frontal analyses of satellite infrared imagery overlaid on HYCOM SSH fields for (g) October 18, 1999, (h) March 6, 2000, and (i) May 4, 2000. Black dotted segments of the frontal analyses are based on infrared imagery more than 4 days old.

model simulation, *Zamudio et al.* [2002] demonstrated that Hurricane Juliette drove coastal downwelling (class 1), which then propagated along the coast and cyclonically around the Gulf of California as a coastal trapped wave (class 4). Both of these are examples where the generation was a class 1 response to atmospheric forcing, which then became a class 4 feature that propagated away from the region of origin.

Simulation of SST and MLD are examples where a combination of class 1 and class 3 responses are significant [*Kara et al.*, 2003; *Kara and Hurlburt*, 2006; Kara et al., An eddy-resolving ocean model for the Pacific Ocean: Part 2: Daily and monthly SST variability over the Pacific Ocean from 1990 to 2004, submitted to *Journal of Geophysical Research*, 2007; hereinafter referred to as Kara et al., submitted manuscript, 2007], class 1 for the sub-monthly fluctuations, and class 3 for the seasonal cycle and its interannual variations [*Kara et al.*, 2004]. With no ocean data assimilation, HYCOM and NLOM simulate daily time series of SST with a median rms error of ~0.8°C in comparison to year-long daily time series of moored buoy data [*Kara and Hurlburt*, 2006; Kara et al., submitted manuscript, 2007]. However, with assimilation of SST from advanced very high-resolution radiometer (AVHRR) data in the form of model-independent analyses, the median nowcast SST error in 1/16° and 1/32° global NLOM decreased to 0.35°C [*Smedstad et al.*, 2003; *Shriver et al.*, 2007] when verified against unassimilated daily buoy time series, data which HYCOM assimilates. While the assimilation of SST and SSH data affects the MLD in a data-assimilative model, their assimilation is insufficient to accurately constrain MLD, and the quantity of subsurface data is insufficient to perform that role. As a result, there is a heavy burden on the model to use a combination of atmospheric forcing and ocean data assimilation in nowcasting MLD.

In Plate 5, daily SST time series from 1/12° global HYCOM (without ocean data assimilation) are compared with Gulf of Mexico buoy data from two locations during August-September 2005. Both show realistic model response to the passage of Hurricanes Katrina and Rita. Global HYCOM simulates the observed drop in SST after the passage of both storms, indicating realistic upwelling and mixing of subsurface water, as well as sufficiently accurate atmospheric wind and heat flux forcing to achieve this result. This is an important capability for an eddy-resolving ocean model that could make it very useful in coupled air-ocean prediction of hurricane intensity.

2.8. Use of Simulated Data From an Eddy-Resolving Ocean Model in Testing Ocean Data Assimilation

Another significant use of eddy-resolving ocean models is to provide realistic simulated data for a variety of applica-tions, as discussed in section 1.3. Providing simulated data for testing ocean data assimilation techniques is a very useful example in eddy-resolving global ocean prediction. In addition, the data-assimilative results can be used in observing system simulation and assessment. Here, simulated data from 1/12° Gulf of Mexico HYCOM run with six-hourly interannual forcing was used in testing the ability of the NCODA system [*Cummings*, 2005] to reconstruct the 3D fields of temperature, salinity, and velocity when only SSH along altimeter tracks and SST were assimilated. Complete model fields from the simulation were available to verify the results of the assimilation.

The 3D NCODA MVOI ocean analysis variables include temperature, salinity, geopotential, and velocity, which are analyzed simultaneously using a HYCOM forecast from the previous assimilation cycle as the first guess. In support of HYCOM, a new analysis variable was added to NCODA that corrects the model isopycnal layer pressures based on differences between the depths of density surfaces predicted by the model and those from observations. The NCODA horizontal correlations are multivariate in geopotential and velocity, thereby permitting adjustments to the mass field to correlate with adjustments to the velocity field. The velocity adjustments are in geostrophic balance with the geopotential increments, and the geopotential increments are in hydrostatic agreement with the T & S increments. Synthetic T & S profiles obtained using the approach of *Cooper and Haines* [1996] were used for downward projection of SSH below the mixed layer. HYCOM is updated by adding a fraction of the analysis increments over a chosen number of time steps.

In the test, the HYCOM simulation was sampled along the observed altimeter tracks of TOPEX/Poseidon and ERS-2 and at observed locations of multi-channel SST data during the period of the assimilation, August 29, 1999–October 18, 1999. The initial condition for the simulated data assimilation experiment was the simulated state a year later, that is, August 29, 2000, when the Gulf of Mexico Loop Current and eddies were in an extremely different configuration, for example, deep penetration of the Loop Current on August 29, 1999, compared to almost no penetration on August 29, 2000. Plates 6a-6c show the domain-averaged rms error of temperature, salinity, and the zonal component of velocity versus depth at the beginning of the assimilation period (August 29, 1999) and Plates 6d-6f at the end (October 18, 1999). By the end of the assimilation period, the error has been greatly reduced.

The test was then repeated with assimilation of the real altimeter and SST data, and with the addition of temperature profile data. The test was initialized from the atmospherically forced simulation on August 29, 1999 and run for 1 year. SSH analyses from the HYCOM/NCODA assimilation were compared with the operational NAVOCEANO frontal analyses based on AVHRR imagery (dashed lines overlaid on the SSH fields; Plates 6g-6i). These comparisons indicate the success of the assimilation in mapping mesoscale variability in the Gulf of Mexico. See *Chassignet et al.* [2005] for a comparison of five different data-assimilative systems in mapping the Gulf of Mexico Loop Current and eddies in comparison with ocean color imagery, including an early version of 1/12° Atlantic HYCOM with a different data assimilation approach, the 1/16° and 1/32° global NLOM systems, 1/8° global NCOM, and a 1/24° Intra-Americas Sea NCOM. In other testing, HYCOM has been used to improve the performance of error covariances in NCODA, and ocean models play a crucial role in determining error covariances in some data assimilation methods, such as ensemble Kalman filtering [*Evensen and van Leeuwen*, 1996; *Brusdal et al.*, 2003] and ensemble OI [*Oke et al.*, 2007].

3. SUMMARY AND CONCLUSIONS

Eddy-resolving global ocean prediction only became feasible near the turn of the century when sufficient data and computing power became available to nowcast and forecast the ocean weather, for example, the surface mixed layer, ocean eddies, the meandering ocean currents and fronts, and coastally trapped waves. Computing requirements for predicting the global ocean weather are much greater than for atmospheric weather prediction because the characteristic diameter of ocean eddies (~100 km) is 20 to 30 times smaller than for atmospheric highs and lows. Real-time satellite altimetry is the key observing system that allows global prediction of the ocean weather associated with meandering currents and eddies (deep ocean mesoscale variability), but sea surface temperature, temperature and salinity profiles, and atmospheric forcing are also vital real-time data sets for ocean prediction.

Since 1998, the multinational GODAE has fostered the development of global and basin-scale prediction systems of the ocean weather in several countries [*Smith*, 2000, 2006]. Because of limitations in computing power, most of these are eddy-resolving basin-scale or eddy-permitting global prediction systems. As of 2007, only two eddy-resolving global ocean prediction systems have been developed.

In this chapter, we have focused on the two existing eddy-resolving global ocean prediction systems, especially the critical roles that ocean models play in these systems, and the need for models that are eddy-resolving, not just eddy-permitting. Resolving the first internal Rossby radius of deformation has been used as a criterion to distinguish between eddy-permitting and eddy-resolving models because of the relation of the Rossby radius to the predominant space scale for baroclinic instability. This criterion is useful but does not

provide a sufficient definition of "eddy-resolving," and additional criteria are needed. To be eddy-resolving, a model without ocean data assimilation needs to (1) realistically simulate the eastward penetration of inertial jets and associated recirculation gyres where they exist, (2) demonstrate near-convergence for the mean strength and pathways of both upper ocean and abyssal currents plus their mesoscale variability statistics, and (3) when used to simulate the real ocean, it must provide realistic comparisons to observations of the mean upper ocean and abyssal circulation and their variability statistics, including the basic physical characteristics and behavior of eddies and meandering currents. Additional common characteristics of eddy-resolving models, not seen in eddy-permitting models, are a basin-wide explosion of eddies and eddy-driven abyssal currents constrained by the bottom topography which, in turn, advect the pathways of mid-latitude upper ocean currents, and thus influence their mean pathways. This upper ocean-topographic coupling requires that mesoscale variability and related flow instabilities be very well resolved to obtain sufficient coupling.

In an ocean prediction system, the dynamical interpolation skill and forecast skill of an eddy-resolving ocean model are much greater than in an eddy-permitting model, as demonstrated in this chapter. In most regions, an ocean model with 1/12° equatorial and 1/16° (or ~7 km) mid-latitude resolution for each variable has the eddy-resolving characteristics described above, while a model with 1/4° or 1/8° mid-latitude resolution does not. A doubling of the mid-latitude resolution to 1/32° further improves the dynamical interpolation and forecast skill of the model, even at low latitudes. At ~20°N, a 1/32° model with assimilation of altimeter data from three satellites reliably mapped eddies as small as 1/4° in diameter when compared to an unusually clear ocean color image.

Model dynamical interpolation skill affects the quality of the nowcast in several ways. In performing a nowcast, a model forecast is often used as the first guess for assimilation of new data. Through this use of the forecasts, (1) the model helps make effective use of delayed data and helps fill in the space-time gaps between observations (including temporal gaps between the repeat cycles of satellite altimetry) by using model predictive skill; (2) the model helps convert the better observed surface fields into subsurface structure; (3) it helps convert the better observed atmospheric forcing functions into useful oceanic responses, and (4) it helps apply bottom topography, coastline geometry, and the ocean surface as physical and dynamical constraints with associated boundary layers. As additional examples, an eddy-resolving model should also play a major role (a) in determining the mean SSH used in altimeter data assimilation, (b) in separating eddy-driven steric and non-steric contributions to SSH, and (c) in testing and improving data assimilation techniques and error covariances by providing realistic simulated data.

In ocean prediction, it is essential to consider the classes of oceanic response to atmospheric forcing for the phenomena and regions of interest (Table 1). In ocean nowcasting, this is important in considering the relative impact of the atmospheric forcing and different oceanic data types. In ocean forecasting, these classes impact whether or not the timescale for oceanic predictive skill is limited by the timescale for atmospheric predictive skill, and also whether a coupled ocean-atmosphere model would be advantageous. Results from the existing eddy-resolving global ocean prediction systems demonstrate forecast skill for about 1 month, globally and over most subregions. Outside surface boundary layers and shallow water regions the forecast skill typically is only modestly impacted by reverting toward climatological forcing after the end of the atmospheric forecast versus using analysis-quality forcing during the entire forecast. The quality of the atmospheric forcing during the forecasts generally has greater than average impact in equatorial regions.

Surface boundary layers and shallow water regions can respond rapidly to the atmospheric forcing, and in these locations, the timescale for oceanic predictive skill can be limited by the timescale for atmospheric predictive skill. In some cases, the need for coupled ocean-atmosphere prediction is strongly indicated, for example, (1) seasonal to interannual prediction of phenomena like El Niño, (2) hurricane prediction where SST can respond strongly to the hurricane and the hurricane to the SST change, and (3) long-range climate prediction where long oceanic adjustment timescales can be important. As seasonal to interannual and long-range climate predictions become increasingly region-specific, eddy-resolving ocean models will be needed to get realistic current pathways, heat and mass transports, and variability statistics, all of which are essential for that purpose.

User interest in real-time ocean products has been high, as evidenced by more than 36 million hits to the NRL Oceanography Division web pages during 2005. Numerous and diverse research, commercial, military, and recreational applications have been reported [*Hurlburt et al.*, 2002; *Smedstad et al.*, 2003; *Johannessen et al.*, 2006]. Ice prediction is included in one of the existing eddy-resolving global ocean prediction systems, but more extensive capabilities and a greater range of applications are possible. Increased capabilities include adding tides, more advanced data assimilation techniques, ensemble nowcasting and forecasting (to obtain better estimates of uncertainty in the ability of the data and the atmospheric forcing to constrain the ocean model nowcasts and forecasts, and to help identify the potential for extreme events), coupled ocean-atmosphere-ice-surface wave prediction with resolution sufficient for hurricanes, global biogeochemical optical tracer and contaminant prediction, and ultimately earth system prediction. Many of these capabilities would greatly increase computational requirements. Based on the results presented here and in some of the references, 1/25° equatorial or 3-4 km mid-latitude resolution is appropriate for a pause in further resolution increases for global ocean prediction systems to use increases in computing power to add such capabilities, and tides should be included in a 1/25° system. Finer resolution, needed in coastal regions, can be obtained by model nesting.

Acknowledgments. This work was sponsored by the National Ocean Partnership Program (NOPP) through two projects, HYCOM Consortium for Data-Assimilative Ocean Modeling and US GODAE: Global-Ocean Prediction with the HYbrid Coordinate Ocean Model (HYCOM) and by the Office of Naval Research through the 6.1 project, Global Remote Littoral Forcing via Deep Water Pathways. The Department of Defense High Performance Computing Modernization Program provided grants of computer time at Major Shared Resource Centers operated by the Naval Oceanographic Office, Stennis Space Center, MS, USA and the Army Engineering Research and Development Center (ERDC) in Vicksburg, MS, USA. NAVOCEANO runs the operational NLOM and NCOM global ocean prediction systems and provides operational computer time for the real-time pre-operational global HYCOM system. We thank acknowledged reviewer Mike Bell of the Met Office UK for his numerous comments on the submitted manuscript. His comments and those from an anonymous reviewer led to significant improvements in this chapter. In addition, we thank our numerous colleagues who provided input for the brief history of ocean prediction. To the best of our knowledge and ability, it is complete, accurate, and up-to-date as of November 2, 2007. We also thank Charlene Parker for her assistance. This is contribution NRL/BC/7304-07-7170.

REFERENCES

Bahurel, P. (2006), MERCATOR OCEAN global to regional ocean monitoring and forecasting, in *Ocean Weather Forecasting: An Integrated View of Oceanography*, edited by E. P. Chassignet and J. Verron, pp. 381–396, Springer, Netherlands.

Barron, C. N., and A. B. Kara (2006), Satellite-based daily SSTs over the global ocean, *Geophys. Res. Lett.*, *33*(15), L15603, doi:10.129/2006GL026356.

Barron, C. N., A. B. Kara, P. J. Martin, R. C. Rhodes, and L. F. Smedstad (2006), Formulation, implementation and examination of vertical coordinate choices in the global Navy Coastal Ocean Model (NCOM), *Ocean Model.*, *11*(3–4), 347–375.

Barron, C. N., L. F. Smedstad, J. M. Dastugue, and O. M. Smedstad (2007), Evaluation of ocean models using observed and simulated drifter trajectories: Impact of sea surface height on synthetic profiles for data assimilation, *J. Geophys. Res.*, *112*, C07019, doi:10.1029/2006JC003982.

Bell, M. J., R. M. Forbes, and A. Hines (2000), Assessment of the FOAM global data assimilation system for real-time operational ocean forecasting, *J. Mar. Syst.*, *25*, 1–22.

Bell, M. J., R. Barciela, A. Hines, M. Martin, A. Sellar, and D. Storkey (2006), The Forecasting Ocean Assimilation Model (FOAM) system, in *Ocean Weather Forecasting: An Integrated View of Oceanography*, edited by E. P. Chassignet and J. Verron, 397–412, Springer, Netherlands.

Bertino, L., and G. Evensen (2002), The DIADEM/TOPAZ monitoring and prediction system for the North Atlantic, Proc. of the 3rd International Conference on EuroGOOS, *Elsevier Oceanogr. Ser.*, *69*, Building the European Capacity in Operational Oceanography, Athens, Greece, 3–6 Dec 2002.

Bleck, R. (2002), An oceanic general circulation model framed in hybrid isopycnic-Cartesian coordinates, *Ocean Model.*, *4*, 55–88.

Bleck, R., and S. G. Benjamin (1993), Regional weather prediction with a model combining terrain-following and isentropic coordinates, 1. Model description, *Mon. Weather Rev.*, *121*(6), 1770–1785.

Bleck, R., and D. B. Boudra (1981), Initial testing of a numerical ocean circulation model using a hybrid (quasi-isopycnic) vertical coordinate, *J. Phys. Oceanogr.*, *11*(6), 755–770.

Blumberg, A., and G. Mellor, A description of a three-dimensional ocean circulation model (1987), in *Three-Dimensional Coastal Ocean Circulation Models, Coastal Estuarine Sci.*, Vol. 4, edited by N. S. Heaps, pp. 1–16, AGU, Washington, DC.

Brusdal, K., J. M. Brankart, G. Halberstadt, G. Evensen, P. Brasseur, P. J. van Leeuwen, E. Dombrosky, and J. Verron (2003), A demonstration of ensemble-based assimilation methods with a layered OGCM from the perspective of operational ocean forecasting systems, *J. Mar. Syst.*, *40–41*, 253–289.

Bryan, F. O., M. W. Hecht, and R.D. Smith (2007), Resolution convergence and sensitivity studies with North Atlantic circulation models. Part 1: The western boundary current system, *Ocean Model.*, *16*(3–4), 141–159.

Carnes, M. R., J. L. Mitchell, and P. W. de Witt (1990), Synthetic temperature profiles derived from Geosat altimetry: Comparison with air-dropped expendable bathythermograph profiles, *J. Geophys. Res.*, *95*(C10), 17,979–17,992.

Carrère, L., and F. Lyard (2003), Modeling the barotropic response of the global ocean to atmospheric wind and pressure forcing—Comparisons with observations, *Geophys. Res. Lett.*, *30*(6), 1275, doi:10.1029/2002GL016473.

Chassignet, E. P. and D. P. Marshall (2007), Gulf Stream separation in numerical ocean models, this volume.

Chassignet, E. P., and J. Verron (Eds.) (2006), *Ocean Weather Forecasting: An Integrated View of Oceanography*, 577 pp., Springer, Netherlands.

Chassignet, E. P., L. T. Smith, G. R. Halliwell, and R. Bleck (2003), North Atlantic simulations with the HYbrid Coordinate Ocean Model (HYCOM): Impact of the vertical coordinate choice, reference pressure, and thermobaricity, *J. Phys. Oceanogr.*, *33*(12), 2504–2526.

Chassignet, E. P., H. E. Hurlburt, O. M. Smedstad, C. N. Barron, D. S. Ko, R. C. Rhodes, J. F. Shriver, A. J. Wallcraft, and R. A. Ar-

none (2005), Assessment of ocean prediction systems in the Gulf of Mexico using ocean color, in *Circulation in the Gulf of Mexico: Observations and models*, AGU Monograph Series, vol. 161, edited by W. Sturges and A. Lugo-Fernandez, pp. 87–100, AGU, Washington, DC.

Chassignet, E. P., H. E. Hurlburt, O. M. Smedstad, G. R. Halliwell, P. J. Hogan, A. J. Wallcraft, R. Baraille, and R. Bleck (2007), The HYCOM (HYbrid Coordinate Ocean Model) data assimilative system, *J. Mar. Syst.*, *65*, 60–83.

Chelton, D. B., R. A. deSzoeke, and M. G. Schlax (1998), Geographical variability of the first baroclinic Rossby radius of deformation, *J. Phys. Oceanogr.*, *28*(3), 433–460.

Cooper, M., and K. Haines, (1996), Altimetric assimilation with water property conservation, *J. Geophys. Res.*, *101*(C1), 1059–1077.

Crosnier, L., and C. Le Provost (2007), Inter-comparing five forecast operational systems in the North Atlantic and Mediterranean basins: The MERSEA-Strand 1 methodology, *J. Mar. Syst.*, *65*, 354–375.

Cummings, J. A. (2005), Operational multivariate ocean data assimilation, *Q. J. R. Meteorol. Soc.*, *131*(613), 3583–3604.

Daley, R. (1991), *Atmospheric Data Analysis*, 457 pp., Cambridge University Press, Cambridge.

DeMey, P., and A. R. Robinson (1987), Assimilation of altimeter eddy fields in a limited-area quasi-geostrophic model, *J. Phys. Oceanogr.*, *17*(12), 2280–2293.

Ducet, N., P. Y. Le Traon, and G. Reverdin (2000), Global high-resolution mapping of ocean circulation from TOPEX/Poseidon and ERS-1 and -2, *J. Geophys. Res.*, *105*(C8), 19,477–19,498.

Emery, W. J., W. G. Lee, and L. Magaard (1984), Geographic and seasonal distributions of Brunt-Vaisala frequency and Rossby radii in the North Pacific and North Atlantic. *J. Phys. Oceanogr.*, *14*, 294–317.

Evensen, G., and P. J. van Leeuwen (1996), Assimilation of Geosat altimeter data for the Agulhas Current using the ensemble Kalman filter with a quasigeostrophic model, *Mon. Weather Rev.*, *124*, 85–96.

Fox, D. N., W. J. Teague, C. N. Barron, M. R. Carnes, and C. M. Lee (2002), The Modular Ocean Data Assimilation System (MODAS), *J. Atmos. Oceanic Technol.*, *19*(2), 240–252.

Gibson, J. K., P. Källberg, S. Uppala, A. Hernandez, A. Nomura, and E. Serrano (1997), *ECMWF Re-Analysis Project Report Series: 1. ERA Description*, 71 pp., ECMWF, Reading, Berkshire, UK.

Godfrey, J. S. (1989), A Sverdrup model of the depth-integrated flow for the world ocean allowing for island circulations, *Geophys. Astrophys. Fluid Dyn.*, *45*(1–2), 89–112.

Gordon, A. L., C. F. Giulivi, C. M. Lee, H. H. Furey, A. Bower, and L. Talley (2002), Japan/East Sea intrathermocline eddies, *J. Phys. Oceanogr.*, *32*(6), 1960–1974.

Hallberg, R. (2005), A thermobaric instability of Lagrangian vertical coordinate ocean models, *Ocean Model.*, *8*(3), 279–300.

Halliwell, G. R. (2004), Evaluation of vertical coordinate and vertical mixing algorithms in the HYbrid Coordinate Ocean Model (HYCOM), *Ocean. Model.*, *7*(3–4), 285–322.

Hogan, P. J., and H. E. Hurlburt (2000), Impact of upper ocean-topographic coupling and isopycnal outcropping in Japan/East Sea models with 1/8° to 1/64° resolution, *J. Phys. Oceanogr.*, *30*, 2535–2561.

Hogan, P. J., and H. E. Hurlburt (2005), Sensitivity of simulated circulation to surface wind forcing in the Japan/East Sea, *Deep-Sea Res.*, *52*, 1464–1489.

Hogan, P. J., and H. E. Hurlburt (2006), Why do intrathermocline eddies form in the Japan/East Sea? A modeling perspective, *Oceanography*, *19*(3), 134–143.

Horton, C., M. Clifford, J. Schmitz, and L. H. Kantha (1997), A real-time oceanographic nowcast/forecast system for the Mediterranean Sea, *J. Geophys. Res.*, *102*(C11), 25,123–25,156.

Hurlburt, H. E. (1984), The potential for ocean prediction and the role of altimeter data, *Mar. Geod.*, *8*(1–4), 17–66.

Hurlburt, H. E. (1986), Dynamic transfer of simulated altimeter data into subsurface information by a numerical ocean model, *J. Geophys. Res.*, *91*(C2), 2372–2400.

Hurlburt, H. E., and P. J. Hogan (2000), Impact of 1/8° to 1/64° resolution on Gulf Stream model-data comparisons in basin-scale subtropical Atlantic Ocean models, *Dyn. Atmos. Oceans*, *32*(3–4), 283–329.

Hurlburt, H. E., and E. J. Metzger (1998), Bifurcation of the Kuroshio at the Shatsky Rise, *J. Geophys. Res.*, *103*(C4), 7549–7566.

Hurlburt, H. E., and J. D. Thompson (1980), A numerical study of Loop Current intrusions and eddy shedding, *J. Phys. Oceanogr.*, *10*, 1611–1651.

Hurlburt, H. E., D. N. Fox, and E. J. Metzger (1990), Statistical inference of weakly correlated subthermocline fields from satellite altimeter data, *J. Geophys. Res.*, *95*(C7), 11,375–11,409.

Hurlburt, H. E., A. J. Wallcraft, W. J. Schmitz, Jr., P. J. Hogan, and E. J. Metzger (1996), Dynamics of the Kuroshio/Oyashio current system using eddy-resolving models of the North Pacific Ocean, *J. Geophys. Res.*, *101*(C1), 941–976.

Hurlburt, H. E., E. J. Metzger, and P. J. Hogan (1997), The impact of upper ocean - topographic coupling on the Kuroshio pathway south and east of Japan, *Int. WOCE Newsl.*, *25*, pp. 19–25.

Hurlburt, H. E., R. C. Rhodes, C. N. Barron, E. J. Metzger, O. M. Smedstad, and J.-F. Cayula (2000), A feasibility demonstration of ocean model eddy-resolving nowcast/forecast skill using satellite altimeter data, Naval Research Laboratory tech. report NRL/MR/7320--00-8235, 23 pp., Naval Research Laboratory, Stennis Space Center, MS.

Hurlburt, H. E., R. C. Rhodes, O. M. Smedstad, A. J. Wallcraft, E. J. Metzger, J. F. Shriver, and A. B. Kara (2001), A real-time eddy-resolving 1/16° global ocean prediction system, in *Report of the High-Resolution Ocean Topography Science Working Group meeting*, edited by D. B. Chelton, Oregon State University College of Oceanic and Atmospheric Sciences, Reference 2001-4, pp. 52–60.

Hurlburt, H. E., M. J. Bell, G. Evensen, C. N. Barron, A. Hines, O. M. Smedstad, and D. Storkey (2002), Operational global ocean prediction systems. *Proceedings of the "En route to GO-DAE" International Symposium*, pp. 97–105, 13–15 June 2002, Biarritz, France.

Jacobs, G. A., H. E. Hurlburt, J. C. Kindle, E. J. Metzger, J. L. Mitchell, W. J. Teague, and A. J. Wallcraft (1994), Decadal-scale trans-Pacific propagation and warming effects of an El Niño anomaly, *Nature*, *370*, 360–363.

Jacobs, G. A., C. N. Barron, and R. C. Rhodes (2001), Mesoscale characteristics, *J. Geophys. Res.*, *106*(C9), 19,581–19,595.

Johannessen, J. A., P. –Y. Le Traon, I. Robinson, K. Nittis, M. J. Bell, N. Pinardi, and P. Bahurel (2006), Marine environment and security for the European area—Toward operational oceanography, *Bull. Am. Meteorol. Soc.*, *87*(8), 1081–1090.

Kamachi, M., T. Kuragano, S. Sugimoto, K. Yoshita, T. Sakurai, T. Nakano, N. Usui, and F. Uboldi (2004a), Short-range prediction experiments with operational data assimilation system for the Kuroshio south of Japan, *J. Oceanogr.*, *60*, 269–282.

Kamachi, M., et al. (2004b), Operational data assimilation system for the Kuroshio south of Japan: Reanalysis and validation, *J. Oceanogr.*, *60*, 303–312.

Kara, A. B., and H. E. Hurlburt (2006), Daily inter-annual simulations of SST and MLD using atmospherically-forced OGCMs: Model evaluation in comparison to buoy time series, *J. Mar. Syst.*, *62*, 95–119.

Kara, A. B., A. J. Wallcraft, and H. E. Hurlburt (2003), Climatological SST and MLD simulations from NLOM with an embedded mixed layer, *J. Atmos. Ocean. Technol.*, *20*, 1616–1632.

Kara, A. B., H. E. Hurlburt, P. A. Rochford, and J. J. O'Brien (2004), The impact of water turbidity on interannual sea surface temperature simulation in a layered global ocean model, *J. Phys. Oceanogr.*, *34*, 345–359.

Kara, A. B., H. E. Hurlburt, and A. J. Wallcraft (2005), Stability-dependent exchange coefficients for air-sea fluxes, *J. Atmos. Ocean. Technol.*, *22*, 1080–1094.

Kara, A. B., C. N. Barron, P. J. Martin, R. C. Rhodes, and L. F. Smedstad (2006), Validation of interannual simulations from the 1/8° Global Navy Coastal Ocean Model, *Ocean Model.*, *11*(3–4), 376–398.

Kara, A. B., A. J. Wallcraft, and H. E. Hurlburt (2007), A correction for land contamination of atmospheric variables near land-sea boundaries, *J. Phys. Oceanogr.*, *37*, 803–818.

Kraus, E. B., and J. S. Turner (1967), A one-dimensional model of seasonal thermocline, 2. general theory and its consequences, *Tellus*, *19*(1), 98–106.

Lai, D. Y., and P. L. Richardson (1977), Distribution and movement of Gulf Stream rings, *J. Phys. Oceanogr.*, *7*(5), 670–683.

Large, W. G., G. Danabasoglu, S. C. Doney, and J. C. McWilliams (1997), Sensitivity to surface forcing and boundary layer mixing in a global ocean model: Annual-mean climatology, *J. Phys. Oceanogr.*, *27*(11), 2418–2447.

Le Traon, P. Y., F. Nadal, and N. Ducet (1998), An improved mapping of multisatellite altimeter data, *J. Atmos. Ocean. Technol.*, *15*(2), 522–534.

Maximenko, N. A., and P. P. Niiler (2005), Hybrid decade-mean global sea level with mesoscale resolution, in *Recent Advances in Marine Science and Technology, 2004*, edited by N. Saxena, pp. 55–59, Honolulu: PACON International.

Melsom, A., E. J. Metzger, and H. E. Hurlburt (2003), Impact of remote oceanic forcing on Gulf of Alaska sea levels and mesoscale circulation, *J. Geophys. Res.*, *108*(C11), 3346, doi:10.1029/2002JC001742.

Metzger, E. J., and H. E. Hurlburt (1996), Coupled dynamics of the South China Sea, the Sulu Sea, and the Pacific Ocean, *J. Geophys. Res.*, *101*, 12,331–12,352.

Metzger, E. J., and H. E. Hurlburt (2001), The nondeterministic nature of Kuroshio penetration and eddy-shedding in the South China Sea, *J. Phys. Oceanogr.*, *31*, 1712–1732.

Metzger, E. J., H. E. Hurlburt, J. C. Kindle, Z. Sirkes, and J. M. Pringle (1992), Hindcasting of wind-driven anomalies using a reduced-gravity global ocean model, *Mar. Technol. Soc. J.*, *26*(2), 23–32.

Metzger, E. J., H. E. Hurlburt, G. A. Jacobs, and J. C. Kindle (1994), Hindcasting wind-driven anomalies using reduced-gravity global models with 1/2° and 1/4° resolution, Naval Research Laboratory tech. report NRL/FR/7323-93-9444, 22 pp., Naval Research Laboratory, Stennis Space Center, MS.

Metzger, E. J., H. E. Hurlburt, J. C. Kindle, R. C. Rhodes, G. A. Jacobs, J. F. Shriver, and O. M. Smedstad (1998a), The 1997 El Niño in the NRL Layered Ocean Model, *NRL Review*, Naval Research Laboratory, Washington, DC, pp. 63–71.

Metzger, E. J., R. C. Rhodes, D. S. Ko, and H. E. Hurlburt (1998b), Validation Test Report for Oceans 1.0, Naval Research Laboratory tech. report NRL/FR/7323-97-9673, 31 pp., Naval Research Laboratory, Stennis Space Center, MS.

Metzger, E. J., O. M. Smedstad, H. E. Hurlburt, A. J. Wallcraft, R. C. Rhodes, J. F. Shriver, C. N. Barron, J.-F. Cayula, and A. B. Kara (2000), A real-time 1/16° Pacific Ocean nowcast/forecast system, Proceedings 2000 Marine Technology Society, Gulf Coast Chapter, 6–7 Sept. 2000, Planning Systems Incorporated, Stennis Space Center, MS, USA, pp. 97–102.

Moore, D. R., and A. J. Wallcraft (1998), Formulation of the NRL Layered Ocean Model in spherical coordinates, Naval Research Laboratory tech. report NRL/CR/7323--96-0005, 24 pp., Naval Research Laboratory, Stennis Space Center, MS.

Munk, W. H. (1950), On the wind-driven ocean circulation, *J. Meteorol.*, *7*(2), 79–93.

Murray, R. J. (1996), Explicit generation of orthogonal grids for ocean models, *J. Comput. Phys.*, *126*, 251–273.

Niiler, P. P., N. A. Maximenko, G. G. Panteleev, T. Yamagata, and D. B. Olson (2003), Near-surface dynamical structure of the Kuroshio Extension, *J. Geophys. Res.*, *108*(C6), 3193, doi:10.1029/2002JC001461.

Oh, I. S., V. Zhurbas, and W. Park (2000), Estimating horizontal diffusivity in the East Sea (Sea of Japan) and the northwest Pacific from satellite-tracked drifter data, *J. Geophys. Res.*, *105*(C3), 6483–6492.

Oke, P. R., P. Sakov, and S. P. Corney (2007), Impacts of localisation in the EnKF and EnOI: Experiments with a small model, *Ocean Dyn*, *57*, 32–45, doi:10.1007/s10236-006-0088-8.

Onishi, H. (2001), Spatial and temporal variability in a vertical section across the Alaska Stream and Subarctic Current, *J. Oceanogr.*, *57*, 79–91.

Onishi, H., and K. Ohtani (1999), On seasonal and year to year variation in flow of the Alaska Stream in the central North Pacific, *J. Oceanogr.*, *55*, 597–608.

Rhodes, R. C., H. E. Hurlburt, A. J. Wallcraft, C. N. Barron, P. J. Martin, O. M. Smedstad, S. L. Cross, E. J. Metzger, J. F. Shriver, A. B. Kara, and D. S. Ko (2002), Navy real-time global modeling systems, *Oceanography*, *15*(1), 29–43.

Roemmich, D., S. Riser, R. Davis, and Y. Desaubies (2004), Autonomous profiling floats: Workhorse for broad-scale ocean observations, *Mar. Technol. Soc. J.*, *38*(2), 21–29.

Schiller, A., and N. Smith (2006), BLUElink: Large-to-coastal scale operational oceanography in the southern hemisphere, in *Ocean Weather Forecasting: An Integrated View of Oceanography*, edited by E. P. Chassignet and J. Verron, pp. 427–440, Springer, Netherlands.

Shriver, J. F., and H. E. Hurlburt (1997), The contribution of the global thermohaline circulation to the Pacific to Indian Ocean throughflow via Indonesia, *J. Geophys. Res.*, *102*, 5491–5511.

Shriver, J. F., H. E. Hurlburt, O. M. Smedstad, A. J. Wallcraft, and R. C. Rhodes (2007), 1/32° real-time global ocean prediction and value-added over 1/16° resolution, *J. Mar. Syst.*, *65*, 3–26.

Smedstad, O. M., H. E. Hurlburt, E. J. Metzger, R. C. Rhodes, J. F. Shriver, A. J. Wallcraft, and A. B. Kara (2003), An operational eddy-resolving 1/16° global ocean nowcast/forecast system, *J. Mar. Syst.*, *40–41*, 341–361.

Smith, N. R. (2000), The Global Ocean Data Assimilation Experiment, *Adv. Space Res.*, *25*, 1089–1098.

Smith, N. R. (2006), Perspectives from the Global Ocean Data Assimilation Experiment, in *Ocean Weather Forecasting: An Integrated View of Oceanography*, edited by E. P. Chassignet and J. Verron, pp. 1–18, Springer, Netherlands.

Smith, R. D., M. E. Maltrud, F. O. Bryan, and M. W. Hecht (2000), Numerical simulation of the North Atlantic Ocean at 1/10°, *J. Phys. Oceanogr.*, *30*(7), 1532–1561.

Sun, S., R. Bleck, C. Rooth, J. Dukowicz, E. Chassignet, and P. Killworth, (1999), Inclusion of thermobaricity in isopycnic-coordinate ocean models, *J. Phys. Oceanogr.*, *29*(10), 2719–2729.

Sverdrup, H. U. (1947), Wind-driven currents in a baroclinic ocean—with application to the equatorial currents of the eastern Pacific, *Proc. Natl. Acad. Sci. U. S. A.*, *33*(11), 318–326.

Talley, L. D., P. Tishchenko, V. Luchin, A. Nedashkovskiy, S. Sagalaev, D. J. Kang, M. Warner, and D. H. Min (2004), Atlas of Japan (East) Sea hydrographic properties in summer, 1999, *Prog. Oceanogr.*, *61*(2–4), 277–348.

Tilburg, C. E., H. E. Hurlburt, J. J. O'Brien, and J. F. Shriver (2001), The dynamics of the East Australian Current system: The Tasman Front, the East Auckland Current and the East Cape Current, *J. Phys. Oceanogr.*, *31*, 2917–2943.

Townsend, T. L., H. E. Hurlburt, and P. J. Hogan (2000), Modeled Sverdrup flow in the North Atlantic from 11 different wind stress climatologies, *Dyn. Atmos. Oceans*, *32*, 373–417.

Tsujino, H., N. Usui, and H. Nakano (2006), Dynamics of Kuroshio path variations in a high-resolution general circulation model, *J. Geophys. Res.*, *111*, C11001, doi:10.1029/2005JC003118.

Usui, N., Y. Fujii, S. Ishizaki, H. Tsujino, T. Yasuda, and M. Kamachi (2006a), Introduction of the Meteorological Research Institute Multi-Variate Ocean Variational Estimation System (MOVE-System), *Adv. Space Res.*, *37*, 806–822.

Usui, N., H. Tsujino, Y. Fujii, and M. Kamachi (2006b), Short-range prediction experiments of the Kuroshio path variabilities south of Japan, *Ocean Dyn.*, *56*, 607–623, DOI 10.1007/s10236-006-0084-z.

Usui, N., H. Tsujino, Y. Fujii, and M. Kamachi (2008), On the Kuroshio large meander in 2004: Generation of a trigger meander, *J. Geophys. Res.*, in press.

Wallcraft, A. J. (1991), The Navy Layered Ocean Model users guide, NOARL Report 35, 21 pp., Naval Research Laboratory, Stennis Space Center, MS.

Wallcraft, A. J., and D. R. Moore (1997), The NRL Layered Ocean Model, *Parallel Comput.*, 23, 2227–2242.

Wallcraft, A. J., A. B. Kara, H. E. Hurlburt, and P. A. Rochford (2003), The NRL Layered Ocean Model (NLOM) with an embedded mixed layer sub-model: Formulation and tuning, *J. Atmos. Ocean. Technol.*, 20, 1601–1615.

Zamudio, L., H. E. Hurlburt, E. J. Metzger, and O. M. Smedstad (2002), On the evolution of coastally trapped waves generated by Hurricane Juliette along the Mexican west coast, *Geophys. Res. Lett.*, 29(23), 2141, doi:10.1029/2002GL014769.

Zamudio, L., H. E. Hurlburt, E. J. Metzger, S. L. Morey, J. J. O'Brien, C. Tilburg, and J. Zavala-Hidalgo (2006), Interannual variability of Tehuantepec eddies, *J. Geophys. Res.*, 111, C05001, doi:10.1029/2005JC003182.

Harley E. Hurlburt, James A. Cummings, A. Birol Kara, E. Joseph Metzger, Jay F. Shriver, Alan J. Wallcraft, and Charlie N. Barron, Naval Research Laboratory, Oceanography Division, Stennis Space Center, MS 39529-5004, USA.

Eric P. Chassignet, Center for Ocean-Atmospheric Prediction Studies, Florida State University, 200 R. M. Johnson Bldg., Tallahassee, FL 32306-2840, USA.

Ole Martin Smedstad, Planning Systems, Inc., Stennis Space Center, MS 39529-5004, USA.

Unstructured Adaptive Meshes for Ocean Modeling

Matthew D. Piggott, Christopher C. Pain, and Gerard J. Gorman

Applied Modelling and Computation Group, Department of Earth Science and Engineering, Imperial College London, UK

David P. Marshall

Atmospheric, Oceanic and Planetary Physics, University of Oxford, Clarendon Laboratory, Oxford, UK

Peter D. Killworth

Ocean Modelling and Forecasting, National Oceanography Centre, Southampton; Empress Dock, Southampton, UK

This article presents an overview of unstructured and adaptive mesh methods for numerical ocean modeling. These methods are able to simultaneously resolve coupled dynamics at both the small and large scale, while smoothly varying resolution and conforming to complex geometries. The finite element method is the favored discretization method to be used with unstructured meshes here; other options include the finite volume and spectral element methods, but are not covered in the interest of brevity. Many of the techniques employed derive from the computational fluid dynamics literature, but there exist significant challenges when applying them to the oceanographic regime. Some of the topics discussed include: unstructured mesh generation and data structures; background to the finite element method including weak forms of the underlying equations and assembly procedures; error measures and anisotropy; and the stability of finite element discretizations, especially for advection dominated flows and in the context of balance in strongly stratified rapidly rotating regimes. Numerical examples conducted with a three-dimensional nonhydrostatic adaptive unstructured mesh ocean model are also presented.

1. INTRODUCTION

In this section, unstructured meshes, mesh adaptivity, and the finite element method (FEM) are discussed with reference to the oceanographic and computational fluid dynamics (CFD) literature. In section 2, data structures used with unstructured meshes are introduced and techniques for generating meshes in complex domains discussed. The finite element method is briefly reviewed in section 3. Error measures, adaptive algorithms, and load balancing for parallel

Ocean Modeling in an Eddying Regime
Geophysical Monograph Series 177

computations are covered in section 4. The stability of discretization methods for incompressible advection-dominated large-scale geophysical flows is discussed in section 5. Some example applications of the model under development by the authors [the Imperial College Ocean Model (ICOM); *Ford et al.*, 2004a; *Pain et al.*, 2005] are presented in section 6. Finally, a discussion on the future outlook for the methods presented here is given in section 7.

1.1. Unstructured Meshes

Unstructured meshes offer great geometric flexibility and hence are often used when modeling problems in complex domains. Their use is especially prevalent in many fields of engineering with the need to represent the interior and exterior of pipework [*Moulinec et al.*, 2005; *Benhamadouche and Laurence*, 2003], the exterior of aircraft wings and vehicles [*Peraire and Morgan*, 1997; *Weatherill et al.*, 2002], for example, being crucial for accurate and efficient calculations, see also *Zienkiewicz et al.* [2005].

Three-dimensional (3D) unstructured meshes derive flexibility from their ability to tessellate a region of interest in an arbitrary manner with variable sized and shaped nonoverlapping nonempty convex polyhedra. An example of a mesh representing the North Atlantic is given in Figure 2 (details of its construction are given in section 2.2). The polyhedra making up a mesh could comprise a combination of tetrahedra (to fit complex geometry), pentahedra (typically used as transitional elements), and hexahedra, forming the so-called hex-dominant, or mixed-element, unstructured meshes. While hex-dominant meshes are attractive for many reasons and the ultimate goal of many, they are, in general, difficult to generate and work with within an adaptive algorithm. Tetrahedral meshes are more common and are the only form considered from herein. In the context of finite element discretization methods, these polyhedra are termed the elements of the mesh; for finite or control volume disctretizations, they are often also termed finite or control volumes. For a thorough overview of meshes, see *Thompson et al.* [1999].

The flexibility afforded by unstructured meshes is in contrast to structured meshes that are typically used with finite difference discretization methods. Structured meshes are so called because they possess a uniform topological structure, for example, one knows implicitly that mesh nodes with indices $(i - 1, j), (i + 1, j - 1)$ etc. are connected to node (i, j). Structured meshes tend to comprise a largely uniform resolution and have a limited ability to orient or align with coastlines and bathymetry in all but the simplest situations, let alone aligning with dynamic solution structures.

The complexity of the domains necessary for high-quality ocean modeling and their important influence on circulation and dynamics implies that unstructured meshes might be the natural choice as the discrete framework for a numerical model. Surprisingly, however, and in contrast to many other applications areas, unstructured meshes have been relatively underused in favor of models based on structured meshes. Overall, the potential advantages of unstructured mesh usage in ocean models are significant. In addition to simply representing the domain more accurately, efficiently, and with enhanced regional preference (i.e., the analogue of nesting), unstructured meshes also enable the use of increased resolution in the direction normal to the coastline than that used tangentially. This is exactly what would be required to resolve boundary layers (e.g., western boundary currents or surface/bottom Ekman layers) if one did not want, or could not afford, to use increased resolution globally and/or isotropically. This philosophy may be generalized to, for example, place more emphasis on accurately representing shallow regions, while the deep ocean's bathymetry may be heavily subsampled or smoothed (as in Figure 2), for example, perhaps when modeling all or certain shelf seas more accurately within a global context. Another possibility which may be attractive, as it removes problems associated with open boundaries, would be to use a global model as default, and if the focus of the study was actually only in a limited region, of the Atlantic say, use extremely coarse resolution in other oceans and seas with an appropriately smooth variation of resolution as the region of interest is approached. While this would obviously introduce errors to the system, it would be interesting to examine these in comparison to the errors introduced through the use of open boundaries. The use of an unstructured mesh on the sphere also removes pole and coordinate singularity problems.

Much research has gone into developing methods to handle topography in structured mesh ocean models. It is now well recognized that the simplest approach of a discontinuous, or "staircase," representation of topography generally leads to substantial errors when used to model flow over continuous topography. A popular alternative with *z*-coordinate models is to employ shaved or partial cells at the bottom of the water column. These methods have been used successfully by *Adcroft et al.* [1997] and have much in common with the Cartesian cut cell technique [*Ingram et al.*, 2003] from CFD. Sigma coordinates have been used, for example, by *Haidvogel et al.* [1991], but these can lead to substantial errors in the calculation of the horizontal pressure gradient, especially where the topography is steep [*Mellor et al.*, 1998]. Note that similar pressure gradient errors can occur with unstructured meshes, see section 5.3. The sigma coordinate approach suffers from the fact that the number of vertical levels is held constant over the entire domain. This could mean that resolution is either too coarse in the deep waters or too fine in the shallow waters or perhaps both, leading to either inaccuracies or inefficiencies. A recent development in structured mesh models is to make use of a hybrid vertical coordinate [*Griffies et al.*, 2000; *Chassignet et al.*, 2003], which follows sigma coordinates in the shallow regions, z-coordinates in the mixed layer and convective unstratified regions, and isopycnal coordinates in stratified regions. This approach is in part motivated by the fact that each of the classical vertical coordinate choices can perform best for different solution structures which may occur simultaneously at different locations within a domain.

Far less algorithmic work has gone into improving the representation of coastal boundaries in structured mesh models, although one could imagine the cull cell methodology being extended to cover this case. Generally, the coastline is represented to the resolution of the mesh cells in a staircase manner with "land" cells masked in the calculation. This can lead both to memory inefficiencies and to the imposition of spurious drag in an analogous manner to staircase bathymetry [*Adcroft and Marshall*, 1998].

As an intermediate step between structured and completely unstructured meshes, nested techniques have been pursued by the ocean modeling community for many of the same reasons that advocates of unstructured meshes use. That is to make use of preferential resolution in regions of particular importance. Ignoring problems with one- or two-way coupling and the reflection of wave structures at the interface between resolutions; this approach assumes that one knows a priori both the fixed location that extra resolution should be placed and the total required computational degrees of freedom for the whole simulation. For some examples, see the work of *Laugier et al.* [1996] and *Ly and Luong* [1999].

Resolving developing flow features (e.g., fronts, eddies, overflows, convection events) whose positions are not necessarily known a priori and generally evolve with time requires the use of adaptive mesh resolution to place degrees of freedom in regions where the most efficient use of finite computational resources may be made. While mesh adaptivity is possible within a structured mesh framework (see below), it is more straightforward to construct powerful adaptivity algorithms based on unstructured meshes, especially when one desires a non-hierarchical mesh, with anisotropic smoothly varying resolution for example.

1.2. Mesh Adaptivity

Ocean dynamics occurs at a huge range of spatial and temporal scales which are often strongly coupled. This coupling means that failure to accurately model a particular process at small scale could have disastrous implications for the entire simulation. Some of these processes are currently, and will

still need to be, parameterized in models, but others such as western boundary currents, regions of intense eddy activity, convection events, and overflows may be more appropriately resolved using a model that has the ability to dynamically vary its resolution. The use of an adapting variable resolution mesh represents a practical means to faithfully represent the physical processes at work, both locally in isolation as well as in relation to their coupling with larger scales.

Ocean models based on structured meshes use a horizontal resolution that is essentially uniform. While significant improvements in the quality of simulations result from increasing resolution in the horizontal (and using additional layers in the vertical), especially at relatively coarse starting resolutions, this becomes prohibitive at higher resolution, generally requiring a new computational platform to become available. In addition, once a resolution has been reached when structures such as boundary layers and eddies are being resolved in the flow, the dynamics become highly variable, nonuniform, and anisotropic. It is then that the strength of adaptive methods should become apparent. This is because adaptive algorithms are able to alter the total number of computational degrees of freedom (proportional to the number of nodes in the mesh) used at different times in the simulation depending on solution complexity and also to optimally locate them to best represent variable and transient solution structures.

It should be noted that numerical methods based on unstructured meshes often possess significant computational overheads, generally over and above structured mesh models. Models using adaptive mesh algorithms also need to compute high-quality error measures to guide the adaptivity process, perform local modifications to the mesh so that the error is controlled, and interpolate solution fields between the old and new meshes. It is not surprising then that per degree of freedom adaptive mesh methods, especially those based on unstructured meshes, are more expensive than fixed-structured mesh-based methods, perhaps by more than 1 order of magnitude. However, the motivation for pursuing this approach is that for certain problems, the adaptive approach is more cost-effective. That is, a given simulation can be performed within appropriate error bounds with sufficiently fewer degrees of freedom to outweigh the additional computational overheads. It remains an open question as to which class of problems the adaptive approach will be the method of choice. Certainly, most likely those problems where a coupled range of anisotropic scales in evolving locations are important. One could argue that this covers the vast majority of applications in oceanography and many other disciplines. This perhaps goes someway to explain the popularity of unstructured meshes and adaptivity in other areas and the growing interest in oceanography.

Another interesting application of adaptive methods is to problems where the required degrees of freedom vary in both time and space. In a given simulation, the optimal number of degrees of freedom may vary significantly, typically increasing during spin-up and later, as full complexity in the flow is achieved, varying by a relatively small amount. One can also envisage situations where this number may vary greatly, for example, with seasonal variations in biological activity and strength of stratification and convection. There will then be inherent inefficiencies in any fixed mesh model for such scenarios.

An interesting open question is to what extent the horizontal and vertical directions should be dealt with differently in a numerical model. To date, virtually all ocean models exploit the differences between the two directions within their algorithms. This is in contrast to standard CFD which generally treats all three spatial directions comparably. There are advantages to their distinction in geophysical fluid dynamics, but there are also advantages to using unstructured meshes and adaptivity in all three directions. Layer-like meshes, unstructured in the horizontal and with uniform structure in the vertical, remain a subset and may be appropriate for certain problems. An intelligent error measure guiding adaptivity should revert to a distinction between directions in the mesh automatically where this is appropriate, for example resulting in a form of optimal hybrid coordinate system in the vertical. An advantage to treating all three directions equally is the ability to appropriately couple the dynamics and optimize resolution between very shallow and very deep regions. Also, to use different mesh structures at the same horizontal location but different depths. For example, to optimally represent both the northward moving Gulf Stream close to the upper surface and the southward moving Deep Western Boundary Current at depths, each of which may require different resolution and different mesh alignment.

For some applications of mesh adaptivity to ocean modeling from an unstructured mesh perspective with the ICOM model, see *Piggott et al.* [2005, 2008], *Pain et al.* [2005], and *Power et al.* [2006]. Further details of ICOM are given in the next subsection, and applications are presented in section 6. A similar approach is used by *Bernard et al.* [2007], where a mesh optimization algorithm in 2D is applied to shallow water flow.

In the context of structured mesh modeling, *Blayo and Debreu* [1999] and *Debreu et al.* [2005] present an adaptive mesh refinement (AMR)-based method [*Berger and Oliger*, 1984] within a finite difference ocean model. In the work of *Popinet and Rickard* [2007], a finite volume ocean model is presented which uses quadtree refinement in the horizontal. Quadtree refinement (a form of AMR) is a hierarchical method where the quadrilateral structured mesh cells can be subdivided into four, each of which may then be individu-

ally subdivided and so on. In the work of *Popinet and Rickard* [2007], the refinement is projected down through layers with a cut-cell approach used to more faithfully represent bathymetry. AMR-based methods have much in common with mesh nesting approaches. They are, however, far more powerful, because there can be many levels of refinement and they can evolve dynamically based on transient solution structures. However, AMR-based methods do share the problem of discrete jumps in resolution, the need to interpolate at boundaries where hanging nodes are present, and the inability to anisotropically orient resolution based on geometry or solution structures.

Adaptive mesh methods are also being investigated for atmospheric modeling. For an excellent recent overview of the area, see *Behrens* [2006]. Early efforts included the studies of *Dietachmayer and Droegemeier* [1992], *Skamarock et al.* [1989], and *Skamarock and Klemp* [1993]; these works probably represent the first application of dynamic adaptivity to the atmosphere or oceans. More recently, in the study of *Bacon et al.* [2000], a finite-volume-based model on an unstructured mesh in the horizontal is described and is now used for hurricane forecasting. In the work of *Jablonowski et al.* [2006], an AMR-based method (in the horizontal) is described on the sphere and tracer advection experiments performed.

One could imagine that it will not be long before adaptive algorithms and unstructured meshes are investigated in the context of coupled atmosphere-ocean dynamics and climate.

1.3. Finite Elements and Ocean Models

The use of finite element methods in oceanography has been the subject of research for the last three decades. Notable early work included the studies of *Fix* [1975], *Haidvogel et al.* [1980], *Davies* [1980], and *Dumas et al.* [1982]. Finite elements have been used successfully for some time in tidal and coastal problems, for example, by *Lynch and Gray* [1979], *Le Provost and Vincent* [1986], *Le Provost et al.* [1994], *Lynch et al.* [1996], and *Lyard et al.* [2006]. These have primarily focused on shallow water models. Larger scale nontidal applications initially focused on diagnostic models [e.g., *Myers and Weaver*, 1995; *Nechaev et al.*, 2003]. Research into the application of finite elements to larger scales and 3D nonhydrostatic processes is now thriving. A brief summary of some of the models based on unstructured meshes is now given.

The QUODDY finite element model (e.g., *Lynch et al.* [1996]; see also their excellent review of the literature) is a 3D hydrostatic model used for coastal/shelf applications. It uses an unstructured mesh of linear triangles in the horizontal with sigma coordinates in the vertical.

The finite element ocean model (FEOM) [*Danilov et al.*, 2004] is a 3D hydrostatic model using linear tetrahedral elements unstructured in the horizontal with a combination of *z* and sigma layers in the vertical. The method uses Galerkin least-squares discretization methods for stability and an elementwise constant vertical velocity. An evaluation of the model in the North Atlantic is given by *Danilov et al.* [2005], who used a barotropic/baroclinic mode split.

A version of the SLIM finite element model is presented by *White et al.* [2007]. The model is 3D hydrostatic, unstructured in the horizontal with a *z*-level prismatic mesh in the vertical using the equivalent of shaved cells to represent bathymetry. The model presented is for barotropic applications and uses the $P_1^{NC}P_1$ representation in the horizontal (see section 5.2).

SUNTANS [*Fringer et al.*, 2006] is a 3D finite volume model which uses an analogue of the C grid, is unstructured in the horizontal with prismatic *z* levels in the vertical and an immersed boundary method to represent bathymetry. It is of particular note, as it is one of the few models that includes nonhydrostatic dynamics and has been designed originally to simulate internal waves.

FVCOM [*Chen et al.*, 2003] is a 3D hydrostatic finite volume model designed for use in coastal and estuarine applications. It is unstructured in the horizontal with sigma coordinates in the vertical and uses a mode splitting for barotropic and baroclinic components. Horizontal components of velocity are stored at element centroids, while all other solution components are stored at element vertices.

SEOM [*Iskandarani et al.*, 2003] is a 3D hydrostatic spectral element model which uses high-order Lagrangian interpolants whose collocation points are the Gauss-Lobatto roots of the Legendre polynomials on unstructured hexahedral elements with sigma coordinates in the vertical.

Finally, ICOM is a 3D nonhydrostatic ocean model designed for both large- and small-scale applications and their coupling. It has various novel aspects; in particular, it uses unstructured meshes in all three dimensions and is the only model to make use of parallel load-balanced anisotropic adaptive meshes in 3D (i.e., it is not restricted by a layered philosophy in the vertical). It also removes the balance-induced stability issue (section 5.3), allowing the use of a wider range of element types. Work is currently ongoing on advanced implicit error measures, adjoint data assimilation, mesh movement, new element types, subgrid-scale regularization methods, and fast solvers.

2. DATA STRUCTURES AND MESH GENERATION

2.1. Data Structures

A schematic of a 2D unstructured mesh of triangles is shown in Figure 1. To fully describe this mesh, in addition to node locations, a "connectivity matrix" is required. From this, all other information such as which elements are adjacent to one another can easily be derived. The connectivity

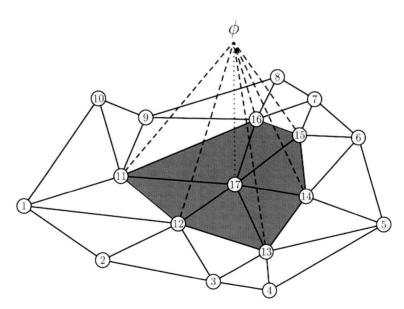

Figure 1. Schematic of an unstructured mesh with global node numbers given at the nodes. A patch of elements is shaded gray upon which the piecewise-linear continuous basis function (ϕ) corresponding to node 17 varies between zero and one. It is zero outside this region.

matrix defines the global node numbers of each element. In this case, this matrix takes the form

$$
\begin{pmatrix}
1 & 2 & 1 & 1 & 12 \dots \\
2 & 3 & 12 & 11 & 17 \dots \\
12 & 12 & 11 & 10 & 11 \dots
\end{pmatrix}
$$

which may be interpreted as a mapping from local coordinate numbers (the rows of the matrix) and elements (the columns) to the global coordinate number. Therefore, the first column has entries 1, 2, 12, which correspond to the global numbers of the nodes making up the bottom left element of the mesh given in Figure 1.

2.2. Unstructured Mesh Generation

As discussed in section 1.1, a high-quality discrete representation of the computational domain is vital. In practice, there is a trade-off between how close the discretized domain is to reality and how appropriate it is for numerical modeling with finite computational resources. Ideally, this representation should not be altered during the course of a simulation, that is, the original surface should not be coarsened during an adaptive simulation to avoid changing the integrity of the domain and associated conservation issues. In an adaptive simulation, the mesh may be refined without changing the bathymetry representation where/when necessary to follow the flow. Therefore, an initial computational mesh should be optimized to represent the geometry to a specified error appropriate for the simulation in hand, with a minimum number of nodes.

Unstructured mesh generators used in oceanography are often based upon 2D Delaunay algorithms, for example as used by codes such as TriGrid [Henry and Walters, 1993] and Triangle [Shewchuk, 2003]. This approach has been used in oceanographic applications on the sphere by Legrand et al. [2000]. Many of these methods provide techniques for inserting extra nodes in shallow regions [Hagen and Parrish, 2002] or where there is a steep gradient or curvature [Gorman et al., 2006].

A typical bathymetry conforming unstructured mesh is presented in Figure 2 for the North Atlantic. The mesh generation approach taken here employs similar anisotropic mesh adaptivity methods as described in section 4. To conform to the bathymetry, the field whose accuracy is being optimized is the depth of the ocean. Thus, the quality of elements is measured with respect to a metric constructed from the curvature of the bathymetry. This, combined with appropriate coastline recovery, results in a 3D anisotropic mesh that focuses resolution where it is required to optimally represent

the bathymetry of the domain. For further details and examples, see Gorman et al. [2006, 2007].

3. BRIEF BACKGROUND ON FINITE ELEMENTS

3.1. Weak Forms of the Equations

Consider the following advection-diffusion equation,

$$
\frac{\partial u}{\partial t} + \boldsymbol{a} \cdot \nabla u = \nabla \cdot (\kappa \nabla u) + f \quad \text{in } \Omega, \tag{1}
$$

with $\boldsymbol{a}$ a transport velocity, κ a tensor of diffusivities, f a source term, and with the boundary condition

$$
\boldsymbol{n} \cdot (\kappa \nabla u) = 0 \quad \text{on } \partial \Omega. \tag{2}
$$

The weak form of (1) is obtained by multiplying by an arbitrary function ϕ from some "test" space and integrating over the domain to yield

$$
\int_\Omega \frac{\partial u}{\partial t} \phi \, d\Omega + \int_\Omega (\boldsymbol{a} \cdot \nabla u) \phi \, d\Omega
$$
$$
= \int_\Omega (-\kappa \nabla u \cdot \nabla \phi + f \phi) \, d\Omega. \tag{3}
$$

Here, integration by parts has been invoked on the second-order terms and the boundary condition (2) used to remove the corresponding surface integral. Note that a boundary condition that allows one to simply drop surface integrals resulting from integration by parts is termed a "natural" boundary condition for that weak form of the equation. The ability to include the boundary condition naturally in the formulation is one reason that integration by parts was invoked. More importantly, however, is the fact that the smoothness required of u to satisfy (3) has been weakened; hence, so-called weak solutions of (3) can be sought that are not necessarily classical solutions [i.e., solutions of the original equation (1)]. This will be seen later to have important consequences when constructing basis functions to represent the numerical solution in the finite element method. For further details, see Gresho and Sani [1998], Elman et al. [2005], and Zienkiewicz et al. [2005], and references therein.

3.2. The Finite Element Method

The finite element method is now obtained by assuming that the solution takes the discrete (finite dimensional) form $u = \sum_{j=1}^{N} u_j \psi_j$ where ψ_j is a so-called "trial" function taken from an N-dimensional subspace that generally lies within

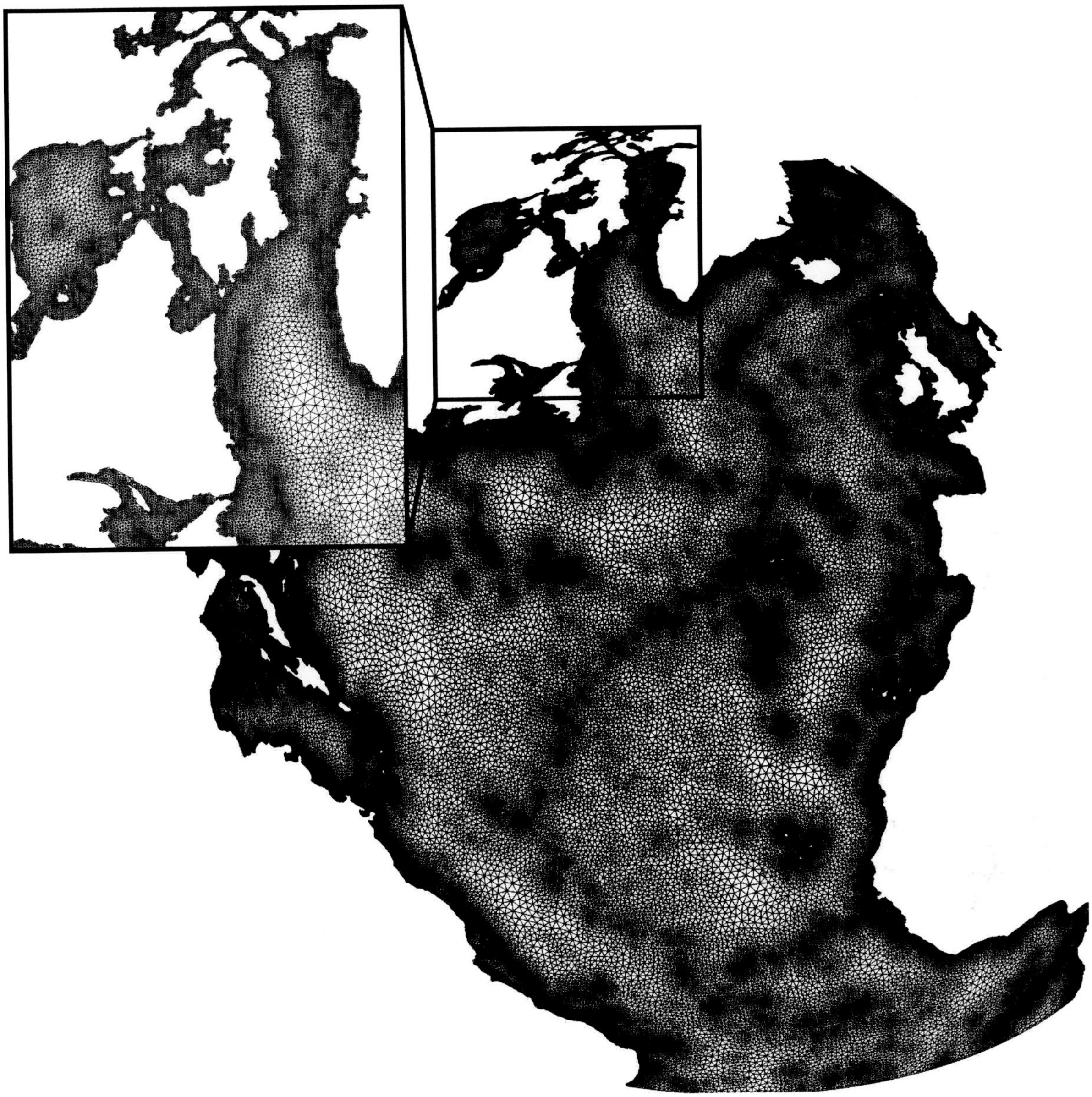

Figure 2. Unstructured mesh of the North Atlantic created using the methods described in section 2.2. Surface mesh shown is of a one element deep 3D mesh comprising 211,560 nodes and 612,882 elements. The mesh was optimized from initial GEBCO data subsampled to 7′ in −110 ≤ longitude ≤ 20, −10 ≤ latitude ≤ 80. A relative (to the local depth) interpolation error has been used to focus resolution in the shallows. The mesh has been constructed and plotted on the WGS84 reference ellipsoid. A magnification of the Labrador Sea and Canadian archipelago region is shown to demonstrate the ability of unstructured meshes to represent complex domains, allowing flow through narrow channels, for example, with relatively few degrees of freedom.

the infinite dimensional space in which the exact solution of the weak form of the equation may be located. This form for a candidate solution is then substituted into (3). This is termed the weighted residual form of the problem. In the finite element method, both ϕ and ψ have compact support over a limited number of elements. A finite-dimensional subset of the test functions is taken, for example, in the *Galerkin* finite element method, this subspace is the set of discrete trial functions introduced above (i.e., $\psi = \phi$). The resulting matrix system for the unknowns u_j, $j = 1, \ldots, N$, takes the form

$$\sum_{j=1}^{N} \underbrace{\int_{\Omega} \phi_i \phi_j \, d\Omega}_{M_{ij}} \frac{du_j}{dt} + \sum_{j=1}^{N} \underbrace{\int_{\Omega} \phi_i \boldsymbol{a} \cdot \nabla \phi_j \, d\Omega}_{T_{ij}} u_j$$

$$+ \sum_{j=1}^{N} \underbrace{\int_{\Omega} \kappa \nabla \phi_i \cdot \nabla \phi_j \, d\Omega}_{D_{ij}} u_j = \underbrace{\int_{\Omega} f \phi_i \, d\Omega}_{F_i},$$

for $i = 1, \ldots, N$, which may be written in matrix form as

$$\boldsymbol{M} \frac{du}{dt} + \boldsymbol{T}u + \boldsymbol{D}u = \boldsymbol{F},$$

where $\boldsymbol{u} = (u_1, \ldots, u_N)^T$.

Discretization in time may now be performed; the simplest option is to use finite differences. For example, use of the theta method [*Gresho and Sani*, 1998] results in the fully discrete system

$$(\boldsymbol{M} + \Delta t \theta (\boldsymbol{T} + \boldsymbol{D})) u^{n+1}$$
$$= (\boldsymbol{M} - \Delta t (1 - \theta)(\boldsymbol{T} + \boldsymbol{D})) u^n + \boldsymbol{F}^{n+\theta},$$

where the superscripts refer to the time level, with the values at level n assumed known, and the source term is assumed to be calculable at any time level. The standard choices $\theta = 0, 1/2, 1$ correspond, respectively, to forward Euler, Crank-Nicolson (trapezoidal), and backward Euler discretizations. The method is second-order accurate in time for the choice $\theta = 1/2$ and otherwise first-order. If one desired higher order accuracy in time, then standard ordinary differential equation methods such as linear multistep and Runge-Kutta methods may be employed. Fractional step [*Turek*, 1999] and Taylor-Galerkin [*Donea and Huerta*, 2003] methods are also popular in the finite element literature. Note that for problems where $\boldsymbol{a}$ or κ are time-dependent, or for nonlinear problems such as the Navier-Stokes momentum equations, the left hand side matrix needs to be recalculated at every time step for $\theta \neq 0$. For the nonlinear case, a set of nonlinear discrete equations must be solved. For this reason, a semi-implicit approach is sometimes taken where some of the terms are treated explicitly. However, for variable resolution meshes such as those one would expect with adaptive mesh methods, the implicit approach is often preferred [*Löhner*, 2001].

3.3. Basis Functions

In Figure 1, a patch of elements is identified around node 17. The piecewise-linear continuous basis function ϕ is defined so that it has support over the patch only, takes the value 1 at the central node of the patch, and is zero at all other nodes of the mesh. The discrete solution can now be seen to be defined over the entire domain and taking the form of a continuous piecewise-linear function taking the value u_j at node j. Many additional options for basis functions are possible, including higher order and discontinuous functions. Mixed methods that use different basis functions for different solution variables, for example, velocity and pressure, are also important. See section 5.2 for further details.

3.4. Assembly Procedure and Quadrature

Assembly is the procedure where the matrices and vectors (i.e., the integrals) above are calculated. For example, the entries of matrix $\boldsymbol{T}$ may be calculated as follows

$$\int_{\Omega} \phi_i \boldsymbol{a} \cdot \nabla \phi_j \, d\Omega = \sum_{e=1}^{E} \int_{\Omega_e} \phi_i \boldsymbol{a} \cdot \nabla \phi_j \, d\Omega,$$

where e refers to an element, and E is the total number of elements in the mesh. Note that the integral over an individual element above only needs to be calculated if both ϕ_i and $\nabla \phi_j$ are nonzero. The assembly procedure loops over the elements and only adds contributions into the $(i, j)^{th}$ entry of $\boldsymbol{T}$ if the integral is known to have a non-zero integrand.

To calculate the integrals over an element, quadrature (numerical integration) is required. The derivatives of the basis functions also need to be calculated. These operations are generally performed by considering transformations from the element in question to a reference element (local coordinates). An isoparametric transformation between the current and reference element may be written (assuming summation over repeated indices) $x = x_i \phi_i(\xi)$, $y = y_i \phi_i(\xi)$, $z = z_i \phi_i(\xi)$, where for tetrahedral linear elements, for example, (x_i, y_i, z_i), $i = 1, \ldots, 4$, are the coordinates of the vertices of the element, $\boldsymbol{\xi} \equiv (\xi, \eta, \zeta)^T$ are coordinates on the reference element, and ϕ_i is the basis function that takes the value 1 at node i. Note that to invert this relationship to find ξ for a given element and point $\boldsymbol{x} \equiv (x, y, z)^T$, the summation to unity property, $\sum_{i=1}^{4} \phi_i \equiv 1$, can be used. This operation is often used to determine which element an arbitrary spatial point lies within, for the purposes of interpolation for example. Superparametric transformations may also be used to represent curved boundaries with linear basis functions; this may be useful for representing the free surface on a globe. Derivatives of functions on the current element, in particular derivatives of basis functions required to evaluate the integral above, can be calculated via the chain rule using

$$\nabla_{\xi} \phi = J \nabla_x \phi, \quad \text{where} \quad J = \frac{\partial(x, y, z)}{\partial(\xi, \eta, \zeta)}.$$

and hence $\nabla_x \phi = J^{-1} \nabla_\xi \phi$. The volume element for the purposes of integration over the reference element transforms as $dx = \det(J)d\xi$, and numerical quadrature may then be performed by transforming to the reference element

$$\int_e f(x)\,dx = \int_r f(x(\xi))\det(J)\,d\xi \approx \sum_{g=1}^{ng} f(x(\xi_g))\det(J)w_g,$$

where subscripts e and r refer to integration over the physical and reference element, respectively, and where in the final step numerical quadrature has been performed. Here, ng is the number of quadrature points, w_g are the quadrature weights, and ξ_g are the locations of the quadrature points in the reference element.

4. MESH ADAPTIVITY/OPTIMIZATION

4.1. Error Measures

The starting point for any adaptive mesh method is a means to guide where the mesh should be refined and coarsened. This requires some measure which indicates regions in the domain where solution complexity warrants the use of additional computational resources and those regions where it does not. Such measures may take the form of physics-based indicators which are constructed so as to be large in the presence of solution structures whose theory and experience indicate that numerical discretization methods may exhibit large errors or instabilities. Examples include error indicators based on solution gradients or vorticity. More advanced error measures can be constructed to give local estimates of discretization error. Mesh optimization algorithms seek to control this error, refining in regions where the measure estimates errors which are higher than those set by the user and coarsening in regions where it is lower. The inclusion of anisotropic information is also desirable to optimize the quality of the mesh and to minimize the degrees of freedom required to achieve a given error.

4.1.1. A priori (or explicit) measures.
Interpolation error estimates are an excellent starting point for an error measure. This is because they are cheap to compute, yield directional information, and give a good guide to (in some cases bounding) the discretization error (for background, see D'Azevedo [1991]; Apel [1999]; Frey and Alauzet [2005]). For example, over a mesh of tetrahedra in 3D, assume a sufficiently smooth function $u \equiv u(x)$ is approximated by its piecewise-linear Lagrange interpolant $\prod_h u$ on element e. The interpolation error on element e satisfies

$$\varepsilon_e \equiv \left\| u - \prod_h u \right\|_{\infty,e} \le \tilde\gamma \max_{x\in e}\ \max_{v\,|\,x+v\in e}\left\{ v^T |H(x)| v \right\} \quad (4)$$

where $\tilde\gamma$ is an $O(1)$ constant independent of the mesh, $\|\cdot\|_{\infty,e}$ is the max-norm on element e, and $H \equiv \nabla^T \nabla u$ is the Hessian. Note that $v^T H(x)v$ is the curvature of the field u at point x and in the direction of v. Importantly, the directional information available here may be used to construct optimal *anisotropic* meshes. The Hessian may be diagonalized as $H = Q\Lambda Q^T$, where Λ is the diagonal matrix consisting of the eigenvalues of H (the principal curvatures of the solution field), and Q is an ortho-normal matrix of eigenvectors. Q may be interpreted geometrically as a rotation and Λ a rescaling. For use as an error indicator to guide adaptivity, it is appropriate to take the absolute values of the eigenvalues, that is, $|H| = Q|\Lambda|Q^T$. The components of the second order Hessian matrix can be obtained at the mesh nodes using a process of reconstruction (e.g., repeated L2 projection) or using procedures such as superconvergent patch recovery [*Zienkiewicz and Zhu*, 1992]. For a comparison of available techniques, see *Vallet et al.* [2007].

Construction of an adapted mesh which equidistributes the error throughout the domain can be seen from (4) to be equivalent to the construction of a uniform mesh of equilateral tetrahedra, that is, all element edge lengths are unity, in a metric space where the metric is given by $M = |H|/\varepsilon$ (assuming $\tilde\gamma = 1$), where ε is a user-defined, possibly spatially varying desired error. For an overview of this approach, see *George* [1998]. One can simplify (4) for use as an error measure by replacing the max operator over vectors in the element by the maximum over element edges [*Frey and Alauzet*, 2005]. One can then see that when measured with respect to this, element edge lengths of unity imply a desired error of ε over element e. Here, the transformation $v = Q\sqrt{|\Lambda|}^{-1}\,\tilde v$ may be used to transform a uniform isotropic element in a computational space into an adapted anisotropic element in the physical domain, achieving the desired interpolation error everywhere (Figure 3). This can be seen from

$$1 = v^T M v = \left(\sqrt{|\Lambda|}Q^T v\right)^T\left(\sqrt{|\Lambda|}Q^T v\right) = \tilde v^T \tilde v.$$

By noting that the eigenvalues (λ_i, $i = 1,2,3$) of the Hessian are proportional to the inverse of the desired edge length squared in physical space, maximum (h_{max}) and minimum (h_{min}) bounds on the edge lengths may be built into the metric definition. This is achieved by limiting the minimum and maximum eigenvalues, respectively, via

$$\tilde\lambda_j = \min\left\{ \frac{1}{h_{min}^2}, \max\left\{ \frac{|\lambda_j|}{\varepsilon}, \frac{1}{h_{max}^2} \right\} \right\}, \quad j = 1,2,3,$$

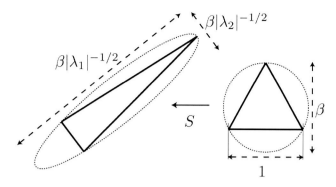

Figure 3. Transformation ($S = Q\sqrt{|\Lambda|}^{-1}$ in section 4.1.1) of an element of unit size in metric space to an anisotropic element in physical space. $\beta = 2/\sqrt{3}$ in 2D and $\beta = \sqrt{6}/2$ in 3D.

At this stage, the user-defined interpolation error ε and the absolute values of the eigenvalues are also incorporated. For high aspect ratio domains, a straightforward generalization allows different max/min values to be used in different directions.

Typically, multiple solution fields will be adapted to, for example, the three components of velocity, temperature, and salinity, in which case, metric fields are constructed for each with their own interpolation errors ($\varepsilon_u, \varepsilon_v, \varepsilon_w, \varepsilon_T, \varepsilon_S$), and these are then superimposed so that the minimum desired edge-lengths for each individual metric are respected.

4.1.2. A posteriori (or implicit) measures. A posteriori error estimation is an active area of research; for an overview, see Ainsworth and Oden [2000] and Bangerth and Rannacher [2003]. Error measures based on a posteriori information represent a step change in power over the methods presented in the previous section, but at additional numerical costs and issues with the inclusion of anisotropic information. They are likely to become more important with the developing use of adaptivity and crucial to the long-term success of adaptive algorithms in oceanography. For a first example of their application to oceanographic problems, see Power et al. [2006].

4.2. h-Adaptive Mesh Optimization Methods

Mesh optimization algorithms represents a generalization of classical h-adaptivity or AMR. In the previous section, a measure of the interpolation error was derived in terms of the Hessian matrix and a metric tensor field reflecting desired edge lengths constructed. Given a mesh and an error metric tensor field, a local optimization algorithm may be used to improve the overall quality of the mesh by visiting each element in turn and attempting to optimize its size and

shape. Equivalently, the mesh may be considered as being constructed in a metric space with lengths measured in terms of the weighted norm $\|v\|_M = (v^T M v)^{1/2}$, with a mesh of unit sized equilateral tetrahedra the goal. For simplicity, when calculating the length of an edge via the metric, M is assumed constant, with value taken to be the average of its values at the two endpoints of the edge. Quantities such as element volume and in-sphere radius in metric space are also used to help gauge mesh quality, and here, an element centered metric is used, which is defined as the average of the metric at the four element nodes.

An objective functional is constructed to gauge mesh quality and whose minimization is the goal of the optimization algorithm. Here, it is given by the max-norm over all elements of the local element quality measure:

$$F_e = \frac{1}{2} \sum_{l \in \mathcal{L}_e} (r_l - 1)^2 + \left(\frac{\alpha}{\rho_e} - 1 \right)^2, \tag{5}$$

defined for element e, where $\mathcal{L}_e$ is the set of edges of element e, r_l is the length of edge l, ρ_e is the radius of the element's inscribed sphere, and $\alpha = \sqrt{6}/12$ (or $1/2\sqrt{3}$ in 2D) is the radius of the inscribed sphere for a unit equilateral tetrahedron (see Figure 4).

Mesh optimization is now performed on the mesh via a sequence of local topological and geometrical operations [*Pain et al.*, 2001; *Freitag and Ollivier-Gooch*, 1997] which seek to efficiently minimize the objective functional. The algorithm proceeds in a Gauss-Seidel iterative manner, each

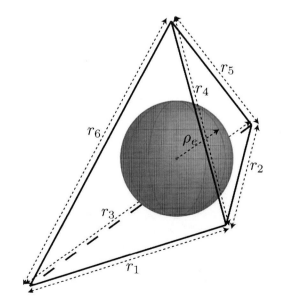

Figure 4. 3D example of an element with edge lengths $r_l, l = 1, \ldots, 6$ and the inscribed sphere of radius ρ_e.

of the elements is visited in turn and the local mesh structure of the element and its neighbors altered via the operations below.

The operations employed in the examples presented here include: *Edge split*—a node is inserted at the center of an edge and surrounding elements created (Figure 5). *Edge collapse*—all elements belonging to the edge are deleted and the two nodes of the edge collapsed to its center (Figure 5). In cases where one of the nodes is used to define the domain geometry, the nodes are collapsed to that point; if both nodes define some geometrical structure, then this operation is not permitted. *Edge-face swap*—if two tetrahedra share a common face and provided their combined interior is convex, the face is deleted and a new edge introduced between the two nodes not shared, thus producing three tetrahedra with different alignment (Figure 6), the inverse operation where an edge is replaced by a face is also allowed. *Edge-edge swap*—S elements are assumed to lie around an edge which may be replaced by a different edge resulting in $2S - 4$ elements with different alignment (Figure 7). *Freitag and Ollivier-Gooch* [1997] found that for increasing S, the number of transformations that improve the mesh decline dramatically; for this reason and due to computational overhead, this operation is limited to $S \leq 4$. *Node movement*—the local topology of the mesh is preserved, but mesh quality is improved by visiting each node and moving it to the centroid of all surrounding nodes, that is, ensuring that when measured in metric space, the lengths of all edges attached to this node are approximately equal (Figure 6).

More extensive node movement, for example a Lagrangian component to track isopycnal layers or eddies, or arbitrary movement to help minimize (5) continually so that topological operations are required less frequently would be attractive. For further discussions, see *Piggott et al.* [2005].

4.3. Parallel Adaptivity and Load Balancing

The ultimate aim of this work includes eddy-resolving ocean modeling and so large computational requirements are inevitable. Therefore, parallel computers and algorithms are required to provide the simulation power required. The principal challenges associated with developing a parallel mesh adaptivity method include consistency of shared node and element information between subdomains, minimization of communication overhead costs, and dynamic load balancing. Mesh data consistency refers to the requirement that all processors that have information regarding particular mesh entities (e.g., nodes, elements) have that information correctly updated as part of the mesh adaptivity process and that the complete mesh remains conformal after the application of mesh optimization (i.e., no hanging nodes are allowed). After mesh optimization has been performed, load imbalances are to be expected, as nodes and elements are, in general, added and deleted according to solution requirements, and thus, the distribution of work may vary across subdomains. Therefore, there is a need for dynamic load balancing where the mesh is repartitioned across the processors; here, the graph partitioner ParMetis [*Karypis et al.*, 2002] is used to achieve this.

Generally, an adaptation involves a change in the local mesh configuration. If any mesh entity is to be adapted, then all domains that contain information specific to that entity must be updated. The very nature of the mesh optimization method used precludes the independent prediction on different partitions of the local connectivity of the final mesh. This poses a problem for the modification of mesh entities at the interface between subdomains.

The parallel strategy presented here addresses the issues of load imbalances and the difficulty in adapting elements at the interface between subdomains simultaneously in a conceptually straightforward, yet efficient, manner. First, the serial mesh optimization method is applied to each subdomain

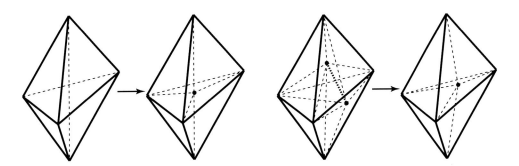

Figure 5. Example of the edge split (left) and edge collapse (right) operations in 3D. The edge is split at its midpoint, and the newly created node is then connected to the vertices of the elements containing the edge, thus creating additional elements. For edge collapse, the elements surrounding the edge are deleted and a node placed at the midpoint of the removed edge.

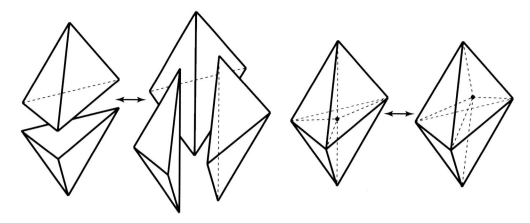

Figure 6. An example of the face-to-edge and edge-to-face swapping operation (left) and node movement (right) in 3D.

independently while disallowing any modification to be made to the subdomain interfaces. Second, load imbalances will occur because of local refinement and coarsening and so a graph repartitioning is required to rebalance the number of mesh nodes per domain. Assuming that all elements in the mesh are close to optimal after optimization, and thus have volumes close to the volume of the optimal tetrahedron in metric space, $\gamma = 1/\sqrt{72}$, then the number of elements, E_{new}, after optimization can be estimated using:

$$E_{\text{new}} = \sum_{e=1}^{E_{\text{old}}} \tilde{V}_e/\gamma, \quad \tilde{V}_e = (\det(M_e))^{1/2} V_e,$$

where E_{old} is the number of elements in the original mesh, $\tilde{V}_e$ is the volume of an element in metric space, and V_e is the volume of the element in Cartesian space. This relative measure is used to define node weights on the mesh before graph partitioning, thus approximate load balancing can be obtained before the mesh optimization operation has completed. In addition to this, weights are applied to the graph edges such that edges associated with poor elements have high weights and are therefore less likely to be split by a partition. This choice of edge weight, when used with diffusive repartitioning methods, has the effect of perturbing partitions away from elements that require optimization while still balancing the load.

Applying high edge weights to regions which were previously locked (because of a graph partition) can lead to unnecessary migration between processors, as not all regions which were previously locked are in need of optimization. Hence, edge weights are defined as a function of the functional (5). As high values of F_l identify regions of poor mesh quality, high edge weight values are applied to discourage the graph partition cutting these edges. When the graph partitioning method calculates the *edge cut* (one of the metrics to be minimized by the graph partition) of a graph partition, it sums the edge weights rather than counting the number of edges cut by the partition. The strategy employed here uses *diffusive* repartitioning by default to minimize data migration and also applies *scratch remapping* periodically to remove any islands that can occur.

Mesh adaptivity is then reapplied to the mesh to address any suboptimal elements. The whole procedure is then iterated if necessary (Figure 8). Because a repartitioning is required to address the load imbalance caused by mesh adaptivity, the computational cost of this graph partition perturbation and data migration overlaps with the requirements of parallel mesh optimization. Experiments indicate that the cost of data migration is low when compared to the cost of the solver stage, so the penalty associated with forcing a data migration step will not generally be significant. Further details may be found in the work of Gorman et al. [Combined parallel tetrahedral mesh optimisation and dynamic load-balancing, manuscript in review, *Comp. Methods Mech. Eng.*, 2007].

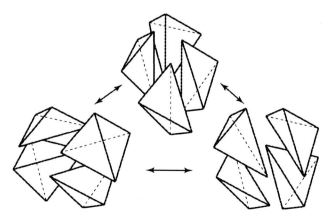

Figure 7. An example of the topological operation of edge to edge swapping in 3D.

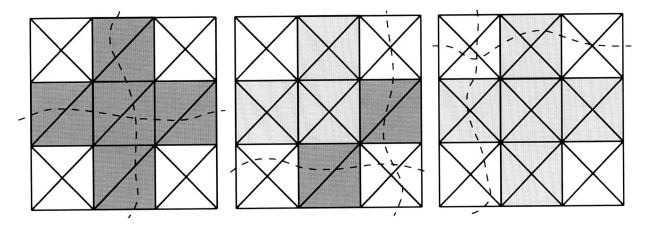

Figure 8. In this 2D repartitioning example, two successive data remapping steps are required. The suboptimal restricted region, darker gray, first forms a polyline of elements and then a set of disjointed elements. In 3D, an additional iteration would be required.

5. DISCRETIZATION STABILITY

5.1. Advective Stability and Subgrid-Scale Modeling

All areas of CFD, including ocean modeling, require accurate and robust algorithms for simulating advection (or transport) processes, for example, *Donea and Huerta* [2003] and *Hanert et al.* [2004]. Complications arise due to the conflicting demands of accuracy and stability/boundedness. For all mesh types and discretization approaches, first-order schemes are highly numerically diffusive. Conversely, high-order schemes, although formally accurate, tend to be dispersive, giving rise to numerical oscillations in regions where there are large solution gradients.

5.1.1. Petrov-Galerkin methods. The Bubnov-Galerkin (using the same trial functions as test functions, as in section 3) discretization approach for transport is dispersive, often leading to growing instabilities which globally pollute the system. An alternative class of method is the Petrov-Galerkin (PG) discretization which uses a different test function to the trial function. For example, the popular SUPG (streamline upwind Petrov-Galerkin, sometimes streamline-diffusion) [Hughes and Brooks, 1979] method places more emphasis on the upstream direction by using $\phi + \tau\mathbf{a} \cdot \nabla\phi$ as the test function in which $\mathbf{a}$ is the transport velocity for the field being solved for, and τ is a so-called intrinsic timescale function. These methods are still prone to oscillate in the direction orthogonal to the flow, as their dissipation is streamline-based. The common cure is to introduce additional dissipation in the cross-stream direction [Hughes and Mallet, 1986]. These methods are non-oscillatory, but typically at the expense of excessive dissipation. The Galerkin least-squares method [Hughes et al., 1989] is another popular stabilization method

in the PG family. It is the approach employed in the FEOM model by Danilov et al. [2004], for example.

5.1.2. Control volume finite elements methods. Control volume (CV)-based methods [Gresho and Sani, 1998] integrate transport terms over surfaces of CVs to gauge the quantity of incoming and outgoing information. The accuracy of this approach is limited by the accuracy of the information evaluated at CV boundaries. High-order schemes are typically realized by placing polynomials through local CV values so that accurate values on the CV boundaries can be established. These high-order polynomials can put more emphasis on interpolating upstream CV cell values such as in the QUICK [Leonard, 1979] scheme.

CV-based methods can easily switch to a non-oscillatory method when a solution extrema is detected, and thus yield non-oscillatory and boundedness properties. The Sweby and normalized variable diagram approaches [*Sweby*, 1984; *Leonard*, 1988] are particularly successful at this and typically switch between a number of low- and high-order schemes defined by a curve on the diagrams. This curve can be changed to optimize the compressive properties of the scheme, for example, used to yield sharp interface preserving properties for use in interface tracking of isopycnals. A form of CV-FEM is employed for tracer transport in the applications presented here.

The discontinuous finite element method is a popular related approach which has many advantages for advection-dominated flows and has recently become popular in the oceanography literature [*Levin et al.*, 2006]. Advantages include high-order accuracy, solution discontinuity capturing properties, efficiency, and scalability [*Cockburn et al.*, 2000]. Jumps in the solution at element boundaries can also be used as an error measures [*Bernard et al.*, 2007].

5.1.3. Subgrid-scale modeling. Subgrid-scale (SGS) modeling becomes a highly nonlinear, complex, and fascinating topic when used in conjunction with adaptive mesh methods that continually vary which length scales are subgrid-scale. One interesting possibility is the ability to include both discretization and SGS modeling errors in the overall error measure guiding adaptivity (section 4.1.2). The model should then be able to decide whether to explicitly resolve a process (if feasible) using high resolution or allow it to be modeled as a SGS process so as to minimize the overall error in the simulation [*Hoffman and Johnson*, 2006]. For example, suppose that the accurate simulation of heat transport in the North Atlantic is the goal, then boundary current separation at Cape Hatteras and the generation of geostrophic eddies transporting heat might be highly resolved by the model. Boundary current separation and eddy activity in other parts of the domain could then be represented using lower resolution, with SGS models accounting for their impact on the larger resolved scale.

The Smagorinksy large eddy simulation (LES) method has been extended to work with the anisotropic inhomogeneous meshes produced here in adaptive simulations. Classical Smagorinsky LES adds an additional eddy viscosity tensor that takes the form of a length scale (l_s) squared multiplied by the strain rate tensor. The length scale is typically taken to be a constant (e.g., $C_s \approx 0.1$) multiplied by the filter width—generally a multiple of the local mesh size. Here, this is variable throughout the domain and locally variable with direction due to mesh anisotropies. As the mesh is constructed to approximately satisfy a unit edge length criteria in metric space, the standard Smagorinsky model is replaced by one where the filter width is anisotropic and inhomogeneous and given by twice the length scale implied by the metric tensor field, that is in the notation of section 4.1.1, $l_s = 2C_s Q \sqrt{|\Lambda|}^{-1}$. A further development used to improve the scale selectivity of this approach is to couple it with a fourth-order rather than second-order dissipative term.

5.2. Element Choice and Stability

In an analogous manner to structured A to E grids which place their velocity and pressure unknowns at different locations within the computational cell [*Arakawa and Lamb*, 1977], there exist many possible finite element pairs. These have varying stability properties in the incompressible flow limit [often termed LBB stability; *Brezzi and Fortin*, 1991] and capabilities in representing various oceanographic structures such as inertia-gravity and Rossby waves. For background related to standard structured mesh methods, see *Haidgovel and Beckmann* [1999] and *Griffies et al.* [2000] and references therein. Some of the most popular low-order choices are now briefly discussed.

The notation of *Gresho and Sani* [1998] is used where P and Q refer to polynomials over tetrahedral and quadrilateral elements, respectively, with subscripts representing the order of the polynomial. The hexahedral (Q_1Q_1) and tetrahedral (P_1P_1) elements employ a shared mesh for velocity and pressure and have similarities with the A grid. Stabilization of this element typically takes the form of a second-order dissipation of pressure term introduced into the right hand side of the discrete continuity equation [*Norburn and Silvester*, 1998]. A fourth-order version of the filter, motivated in part by *Sotiropoulos and Abdallah* [1991], is employed in the simulations presented in following sections. Introducing bubble functions to the approximation of velocity (which enriches the discrete space and is often termed the MINI element) can also cure the stability issues [e.g., *Fortin and Fortin*, 1985]. The MINI element has been employed in oceanography in the FINEL3D model [*Labeur and Pietrzak*, 2005].

The hexahedral (Q_1Q_0) element with trilinear velocity and piecewise-constant pressure is equivalent to the B grid and thus has the property of representing well internal waves when the Rossby deformation radius is large in comparison to the local mesh resolution. It is popular in practice [*Gresho and Sani*, 1998] but does support singular pressure modes which seem largely not to affect the accuracy of velocity, especially with the limited use of Dirichlet velocity boundary conditions. It was used by *Ford et al.* [2004a, 2004b] to simulate stratified flow past a tall isolated seamount with ICOM.

There has been much analysis of more exotic finite element pairs, for velocity and free surface elevation in shallow water systems in particular. For example, the triangular Raviart-Thomas element (RT_0) is the analogue of the C grid and consists of a piecewise-linear basis function centered on element edges and used to represent the normal component of velocity, along with a piecewise-constant pressure. It was shown by *Raviart and Thomas* [1977] to be free from spurious free surface pressure modes, see also *Hanert et al.* [2002]. The triangular $P_1^{NC}P_1$ element is the analogue of the CD grid and consists of a piecewise-linear basis function centered on element edges for velocity and a standard piecewise-linear basis function centered on vertices for pressure. It was shown by *Le Roux et al.* [1998], along with many other low-order pairs (e.g., the E grid analogue $P_1^{NC}P_0$), to have some problems representing geostrophic balance accurately. In the work of *Hanert et al.* [2005], the $P_1^{NC}P_1$ element, however, was used to accurately simulate the propagation of eddies in nonlinear shallow-water models. *Le Roux* [2005] analyzed the element further and showed that it also properly modeled the dispersion of inertia-gravity waves. Many of these elements need to be considered further in the context of 3D models. Higher order element pairs, including discontinuous representations of some solution variables, need to be considered and analyzed further. For additional material, see also *Fortin and Fortin* [1985], *Gunzburger* [1989], and *Brezzi and Fortin* [1991] and references therein.

5.3. Large-Scale Balance

When using an unstructured mesh for large-scale geophysical problems, a problem occurs which is related to the relative magnitude of terms in the momentum equations. In particular, the Coriolis and buoyancy accelerations are largely balanced by a pressure gradient of similar magnitude, while the other terms in the momentum equations can be of much smaller magnitude. Mathematically, this balance between dominant terms can be written

$$2\Omega \times u + \rho g \approx -\nabla p_{gh} \qquad (6)$$

where Ω is the rotation vector, ρ is the perturbation pressure (assume for now the Boussinesq approximation has been made), g is the gravitational vector, and p_{gh} is the component of pressure which approximately balances the Coriolis and buoyancy accelerations. With any standard finite element discretization, an immediate problem is evident. One typically represents velocity and density with piecewise polynomials of the same order [*Gunzburger*, 1989] and pressure to the same or lower order (section 5.2). As the gradient of this pressure is taken, and thus the order reduced, this balance is poorly represented using classical finite elements. While this is also the case with all engineering fluids, it is the strength of rotation and stratification that causes a problem for geophysical problems. Any small relative error in the model's ability to satisfy (6) can result in large spurious accelerations. Experience has shown that this problem is exacerbated on unstructured meshes which are not aligned with the rotational and gravitational directions.

In the following sections, two related techniques are described that seek to reduce this spurious behavior. The first deals with buoyancy only; the second is more general and can deal with both buoyancy and Coriolis accelerations.

5.3.1. A hydrostatic balance inspired solver. In this approach, which deals with buoyancy only, pressure is split into two parts: a "hydrostatic pressure" p_h and a "nonhydrostatic pressure" p_{nh}, so that the total pressure p is $p = p_{nh} + p_h$. The hydrostatic pressure p_h is required to balance the buoyancy acceleration in the momentum equations:

$$\nabla p_h = -\rho g. \qquad (7)$$

Following the decomposition of pressure, the nonhydrostatic momentum equations may then be written

$$\frac{Du}{Dt} + 2\Omega \times u = -\nabla p_{nh} - \nabla p_h + \nabla \cdot \tau, \qquad (8)$$

where τ is a stress tensor used to represent viscous effects [*Ford et al.*, 2004a], and (7) has been used to replace the buoyancy acceleration by a pressure gradient. Assuming piecewise-linear density and velocity, a method for calculating p_h is constructed which results in a piecewise-linear representation of ∇p_h. This component of pressure is then free to balance the buoyancy acceleration, and hence minimize spurious behavior, while the nonhydrostatic component of pressure still plays the role of enforcing incompressibility and can be chosen to ensure a stable velocity-pressure element as in section 5.2. Given ρg, this is achieved here by either solving for the "horizontal" gradient of p_h directly by differentiating (7) and using low-order elements (and so an implicit higher order representation of p_h is effectively employed) or by calculating p_h with higher order elements so that its gradient is of sufficient order to satisfy (7) accurately.

Assuming for now that the gravitational vector takes the form $g = gk$ for a constant acceleration due to gravity g, and the unit vertical vector k, then (7) yields

$$\frac{\partial}{\partial z}\frac{\partial p_h}{\partial x} = -\frac{\partial(\rho g)}{\partial x}, \quad \frac{\partial}{\partial z}\frac{\partial p_h}{\partial y} = -\frac{\partial(\rho g)}{\partial y}. \qquad (9)$$

To achieve high spatial accuracy on unstructured meshes, the discontinuous Galerkin method [*Johnson*, 1987; *Cockburn et al.*, 2000] is used to obtain ∇p_h directly using discontinuous P_1 basis functions (as in Figure 9) or p_h using discontinuous P_2 basis functions. The method is often used to solve equations of the form $\nabla \cdot (au) = s$, where a is the transport velocity, u is the quantity being transported, and s is a source. Here, the original expression (7) takes this form if $a = -k$, $u = p_h$, and $s = \rho g$. Alternatively, (9) takes this form if $a = -k$, $u = (\partial p_h/\partial x, \partial p_h/\partial y)^T$ and $s = (\partial(\rho g)/\partial x, \partial(\rho g)/\partial y)^T$. To discretize, the equations are first multiplied by a discontinuous basis function $\tilde{\phi}$. The discontinuous basis function

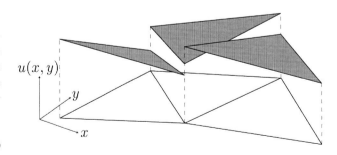

Figure 9. A discontinuous piecewise-linear representation of a function in 2D.

corresponding to the piecewise-linear basis function introduced in Section 3 is defined over a single element not a patch of elements as in the continuous case (see Figure 9). This means that for a mesh of N nodes and E elements, continuous piecewise-linear basis functions would result in N degrees of freedom, but the discontinuous piecewise-linear basis function have $4E$ degrees of freedom in 3D. For an unstructured mesh, E may be far larger than N (typically $E \approx 5N$); hence, the approximation power for the same mesh is higher. The resulting expression it then integrated over the element e in which $\tilde{\phi}$ is nonzero to ultimately yield, for any $\tilde{\phi}$,

$$
\int_e [\nabla \cdot (a u) - s] \tilde{\phi} \, dV
$$
$$
+ \int_{\{\Gamma_e : a \cdot n < 0\}} a \cdot n (u_{in} - u) \tilde{\phi} \, dS = 0, \qquad (10)
$$

where $\{\Gamma_e : a \cdot n < 0\}$ represents the restriction of the boundary of element e to that part on which $a \cdot n < 0$.

Here, the direction of integration is down from the ocean surface in the direction of gravity. In the interior of the fluid, information is transferred between neighboring elements according to (10). At the ocean's upper surface, homogeneous Dirichlet boundary conditions ($u_{in} = 0$) are used for both p_h or ∇p_h which is satisfactory for a rigid lid where air pressure is ignored. Otherwise, more complex boundary conditions must be employed.

System (10) is solved element by element in a Gauss-Seidel iterative manner. Due to the "flow of information" from the ocean surface to ocean floor, the exact solution within an element may be obtained once the solution in the elements "upstream" of it have been found. An optimal numbering of the order in which to visit elements may be found so as to reduce the number of iterations required to achieve convergence. If a complete renumbering can be found, the solver converges in a single iteration. In some situations, such an optimal numbering cannot be found, in which case, the ordering is broken up into a disjoint series which results in a method that converges in small number of iterations. Here, the renumbering is constructed via an element–element data structure and a breadth first search algorithm starting at the upper surface.

Once the gradient of p_h has been obtained, it is substituted in as a new source term to the momentum equations, and p_{nh} then takes the role of the pressure which ensures the velocity is divergence-free and is solved for in a standard manner, see, for example the discussion of a projection method in the work of *Piggott et al.* [2008]. Note that the "full" pressure $p_h + p_{nh}$ is never formed in the discretization, although it could be as a postprocessing step following projection of

one of the terms into the other's discrete space. Hence, there are no pressure modes caused by the use of the higher order representation of p_h.

5.3.2. A solver for both Coriolis and buoyancy. An alternative method can be derived by noting that the Coriolis and buoyancy accelerations are largely matched by a pressure gradient for states close to geostrophic and hydrostatic balance. The method proceeds via a Helmholtz decomposition of the form

$$
2\Omega \times \mathbf{u} + \rho \mathbf{g} = \nabla \times \mathbf{A} - \nabla p_{hg}, \qquad (11)
$$

where p_{hg} is the "hydrostatic-geostrophic" component of the full pressure. In states close to balance, $\nabla \times \mathbf{A}$ is a small residual term, the difference of two large terms, which, for standard finite element representations of velocity, density, and pressure, could exhibit large spurious errors for the reasons outlined in section 5.3. As in section 5.3.1, an improved method results from calculating $\mathbf{A}$, or rather $\nabla \times \mathbf{A}$, more accurately and then substituting into the momentum equations instead of the problematic Coriolis and buoyancy accelerations. There are two obvious ways to proceed, either by taking the curl or divergence of (11). The method used in Table 1 uses the approach which takes the divergence of (11) to yield the elliptic equation

$$
\nabla^2 p_{hg} = -\nabla \cdot (2\Omega \times \mathbf{u} + \rho \mathbf{g}),
$$

with natural Neumann boundary conditions obtained by taking the dot product of (11) with the unit normal vector. Following the same philosophy as the hydrostatic pressure solver presented above, the equation is solved for p_{hg} using piecewise-polynomials of higher order than those used to represent velocity and density. In the simulations presented presently, continuous piecewise-quadratic basis functions are used. The new source term $\nabla \times \mathbf{A}$ is then recovered from (11) by taking the difference of two known terms.

5.3.3. Testing the representation of balance. A standard test for errors related to pressure gradients in a numerical model is to initialize with a stable density profile which depends only on the vertical direction z and to impose no inflow or outflow at boundaries. The model should remain in rest, and any deviation from this is due to discretization errors.

Here, a problem similar to that used by *Ford et al.* [2004a] is considered. The geometry is a channel 560 km long, 514 km wide, and 4.5 km deep. The topography is given by an isolated Gaussian seamount of maximum height 90% that of the channel and a full-width at half-maximum value of

Table 1. Spurious kinetic energy in the seamount balance test case.

Method	Resolution (number of nodes)						
	A: 8,238	B: 15,087	C: 26,860	D: 49,155	E: 95,598	F: 25,400	G: 52,048
I	3.56×10^{-1}	4.45×10^{-1}	3.76×10^{-1}	3.38×10^{-1}	1.89×10^{-1}	4.14×10^{-2}	6.39×10^{-2}
II	6.38×10^{-5}	4.28×10^{-5}	3.11×10^{-5}	1.69×10^{-5}	8.35×10^{-6}	1.73×10^{-6}	2.42×10^{-6}
III	6.68×10^{-5}	4.64×10^{-5}	3.52×10^{-5}	1.96×10^{-5}	1.03×10^{-5}	1.89×10^{-6}	4.17×10^{-6}
IV	1.41×10^{-4}	1.09×10^{-4}	9.82×10^{-5}	7.00×10^{-5}	1.83×10^{-5}	3.48×10^{-6}	5.25×10^{-6}

Experiments A to E comprise progressively finer mesh edge lengths. Experiments F and G have preferentially more refinement in the vertical. Method I is the standard continuous piecewise-linear finite element discretization. Method II is that presented in section 5.3.1, with discontinuous piecewise-linear basis functions used to compute ∇p_h directly. Method III is that presented in section 5.3.1, but with discontinuous piecewise-quadratic basis functions used to compute p_h. Method IV uses the elliptic equation approach of section 5.3.2.

41.63 km. The initial mesh follows sigma coordinates in the vertical so the bathymetry is approximated accurately and smoothly. The initial mesh has eight layers in the vertical, 26,803 nodes, and 135,625 elements in total, with increased horizontal resolution over the seamount. The flow is initialized with a state of rest, closed free slip boundaries with an initial quadratic density variation with zero slope at the deepest part of the domain and a Burger number, representing the ratio of the buoyancy and Coriolis terms, of $Bu = 1.0$. The model is run for 11 time steps of size 432 s in total, and the mesh is adapted following the first time step so that ten time steps are performed on a mesh which is irregular in 3D. The criteria used for this adaptation of the mesh are based entirely on a maximum allowed element size; to minimize the adaptivity functional (5), the optimization algorithm seeks to achieve this maximum element size as well as it can throughout the domain. Here, the maximum element size has been chosen to take the value h_x in the horizontal and a different value h_z in the vertical. Five experiments of increasing resolution (labeled A to E) have been performed where $h_x = 0.32\alpha^i$, $h_z = 1.28 \times 10^{-2}\alpha^i$, $i = 5, \dots, 1$, where $\alpha = 1.25$. Table 1 presents the kinetic energy at the end of the simulation in non-dimensional units. Four methods are considered: the first (I) uses a standard $P_1 P_1$ discretization with no stabilization and no special treatment of the Coriolis and buoyancy terms (the effect of Coriolis is minimal in this problem where the velocities present are relatively small; it will be considered in additional detail in future work); the second (II) method solves for the horizontal gradients of the hydrostatic pressure directly using discontinuous linear elements; the third (III) solves for hydrostatic pressure using quadratic discontinuous elements; and the fourth (IV) method solves the elliptic problem described in section 5.3.2.

The results show the standard finite element discretization performing very poorly. A least-squares linear regression analysis of the kinetic energies shows that the three "bal-ance"-motivated methods produce errors which are quadratic in mesh spacing, with method II marginally the best followed closely by methods III and then IV. The improved performance of method II over III, although III uses higher order elements, could be explained by noting that at the discrete level, it is generally more accurate to differentiate and then integrate rather than vice versa. The improvement over IV may be due to the additional approximation power of the discontinuous approach used in methods II and III. But note that this method must be used in situations where Coriolis is important. Two additional experiments were performed (F and G) where $h_z = 10^{-2}$ and $h_x = 0.75, 0.5$, respectively. These show the interesting result that at high vertical resolution, additional horizontal resolution seems to yield worse kinetic energy levels, also that these values are lower than in experiments A to E, which used coarser resolution in the vertical. This indicates that resolution in the vertical is crucial and that there should be a link between horizontal and vertical resolutions. These findings are consistent with the results presented by *Ford et al.* [2004a] where a sigma coordinate structured hexahedral mesh was tested.

6. APPLICATIONS

6.1. Comparison With Laboratory Experiments

In this section, sloping convection in a differentially heated rotating annulus is simulated. This is an ideal test problem with which to validate a new ocean model. It encompasses non-hydrostatic dynamics and the effects of rotation, buoyancy, external forcing, and non-trivial geometry. It also does not require open boundary conditions or domain periodicity. Comparisons can be made qualitatively against published experiments [e.g., *Hide and Mason*, 1975] or quantitatively against laboratory data. To match *Hide and*

Mason [1975] (in particular their Figure 4), the domain is annular with height $d = 0.135$ m and cylindrical horizontal boundaries at radii $a = 0.04$ m and $b = 0.0864$ m with lateral walls parallel to the vertical. The rotation (of magnitude Ω_z) and gravity (of magnitude $g = 9.8\,\mathrm{ms}^{-2}$) vectors are also parallel to the vertical direction. The lateral walls are maintained at temperatures T_a and T_b where here $T_b - T_a = 9\mathrm{K}$. The mean density of the fluid is $\rho_0 = 1{,}037$ kg m^{-3}, with the thermal expansion coefficient set to $\alpha = 2.5 \times 10^{-4}$ K^{-1}. The kinematic viscosity is set to $\mu = 1.56 \times 10^{-6}$ m^2 s^{-1}, with thermal diffusivity a factor 7 lower. Two simulations have been performed at rotation rates $\Omega_z = 1.21$ rad s^{-1} and $\Omega_z = 3.91$ rad s^{-1} (Figure 10). The faster rotation rate results in a mode five solution as expected from experiments. The slower rotation rate results in a mode three pattern initially as expected, but soon settles down into a mode two state. Further experimentation is ongoing considering the effect of numerical parameters. The number of nodes used by the simulations is presented in Figure 11. The number of nodes follows the same pattern as the kinetic energy in the system in each case.

6.2. Lock Exchange

In this example, the simulations conducted by *Özgökmen et al.* [2007] using a high-order accurate spectral method have been followed. *Özgökmen et al.* [2007] compared direct numerical simulations with Richardson-number-dependent LES methods. The nondimensionalization of *Özgökmen et al.* [2007] has been followed and a Reynolds number of 1,000 used here. The domain is 3D with a small length scale in the y direction: $10 \times 0.1 \times 2$ units in size. The time step is 0.01, the mesh adapts every 0.25 time units with min/max

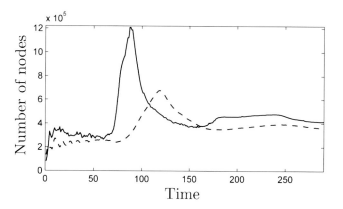

Figure 11. Total number of nodes in the simulation against model time for the rotating annulus with $\Omega_z = 3.91$ rad s^{-1} (solid line) which reaches a maximum of 1.21 M nodes after 87 s, and $\Omega_z = 1.21$ rad s^{-1} (dashed line) which reaches a maximum of 676 K nodes at 120 s.

element bounds of 0.003 and 0.5, respectively, and the result is shown after 7.4 time units in Figure 12. The number of nodes used by the simulation is presented in Figure 13.

6.3. Overflows and Gravity Currents

In this simulation, a 3D domain following *Özgökmen* [2003] is set up. The model is initialized with a region of cooler water in the embayment which is forced via a sponge region to remain 5° lower than the background homogenous temperature of the fluid. The domain is on an *f*-plane with $f = 10^{-4}$s^{-1}, the time step is 8 min, and the mesh is adapted every 3.5 h. The viscosity in the horizontal is 60 m^2 s^{-1} and is

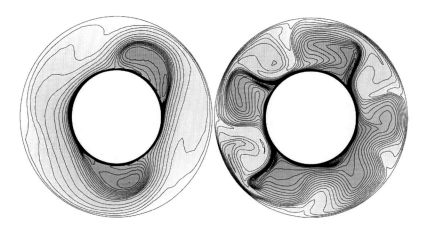

Figure 10. Rotating annulus simulation at a rotation rate of $\Omega_z = 1.21$ rad s^{-1} (left) and $\Omega_z = 3.91$ rad s^{-1} (right) at time 240 s. Contour plot of density shown on a plane 0.5 cm from the top of the domain.

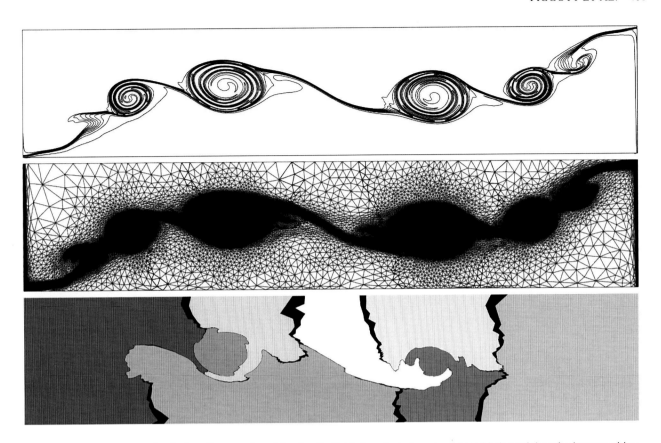

Figure 12. Lock exchange simulation. Contour plot of the density field, the adapted mesh, and load-balanced domain decomposition.

2 orders of magnitude smaller in the vertical. The minimum and maximum bounds on element sizes in the horizontal are 400 m and 100 km, respectively, and 4 m and 1 km in the vertical. A snapshot from the simulation is shown in Figure 14.

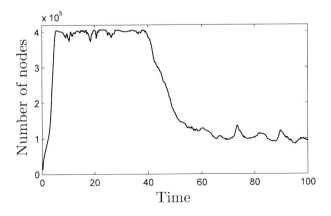

Figure 13. Total number of nodes (degrees of freedom) in the simulation against model time for the lock exchange problem.

6.4. Eddying Western Boundary Currents

In this problem, a flat-bottomed domain of square horizontal extent 1,000-km and 500-m depth on a beta plane with $\beta = 1.8 \times 10^{-11}$ m^{-1} s^{-1} is forced with a wind stress of strength $\tau_0 = 0.1$ N m^{-2}, which varies as a sine of latitude and sets up an anticyclonic gyre. Due to the variable strength of rotation, an intense western boundary is formed. Figure 15 presents results after approximately 1 year of simulated time. In Figure 16, a similar simulation is presented, but here, the geometry has been replaced with a coastline from the Mediterranean Sea.

6.5. M2 Tides in the North Atlantic

Figure 17 shows the tidal range in a simulation of the astronomically forced M_2 tide on the mesh produced in Section 2.2. A constant quadratic drag of magnitude 0.0025 has been applied. The time step is 10 min, and results are shown after seven days, following a 2-day spin-up period. The simulations shows good overall agreement with *Lyard et al.* [2006], including amplification in the Bay of Fundy, for example, whose structured mesh models would have trouble representing.

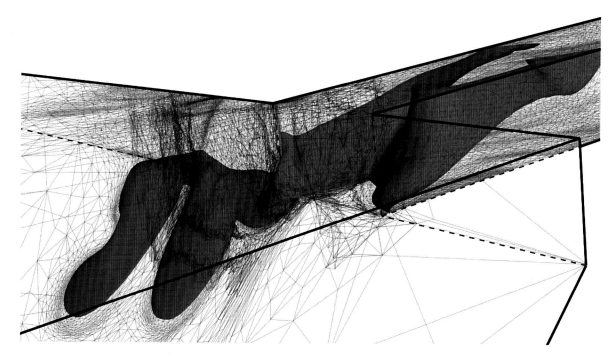

Figure 14. Density-driven overflow in the presence of rotation. The mesh on the domain surface and an isosurface of density are shown.

7. FUTURE OUTLOOK

In previous sections, new numerical techniques for simulating the oceans at a range of scales have been discussed. Here, these techniques revolve around unstructured and adaptive mesh methods. The next several years is expected to be a period of rapid progression for these novel methods, as they have now reached a maturity where they can be applied in earnest to both idealized and realistic geophysical problems. As new models exploiting unstructured and adaptive meshes are developed and applied in more complex applications, it will become clearer to the community what their applicability is likely to be, where their strengths and weaknesses lie, and whether investment in moving to the use of these methods is warranted. In addition, if shown to be successful for a range of application areas, some of these ideas may begin to be incorporated into today's state-of-the-art operational models where possible. Also, within the community developing these new methods, there may well be somewhat of a convergence on approaches which have been demonstrated to perform well, just as has been the case with structured mesh models. It is hoped that the development process may be hastened by adopting an open source development strategy.

When considering the future for these new numerical methods, it must be remembered that they are still in their infancy when compared to the number of person-years of effort that

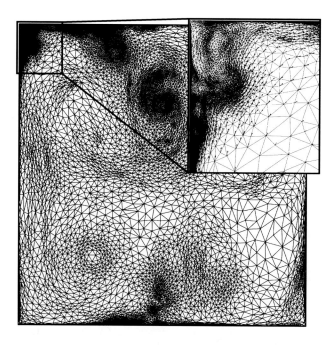

Figure 15. Western boundary current and subsequent generation of nonlinear eddies. A blowup of the northwest corner is also shown with velocity vectors overlaid.

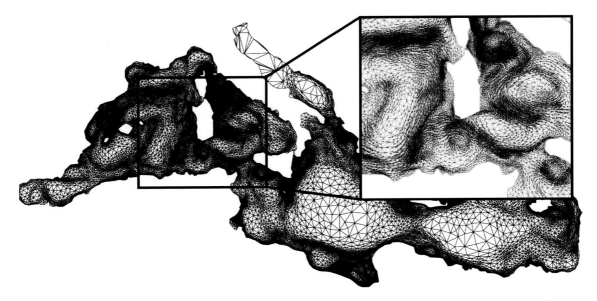

Figure 16. Wind-driven circulation in the Mediterranean. The adapted mesh is shown, along with velocity vectors in the magnified region.

has gone into the development and validation of structured grid models. The methods discussed here have come on a great deal in the past decade, from very few examples of their use (other than in tidal and a few coastal applications) to now many groups around the world pursuing them. The

Tidal Range (m)

| 0.00 | 0.500 | 1.00 | 1.50 | 2.00 | 2.50 | 3.00 | 3.50 | 4.00 |

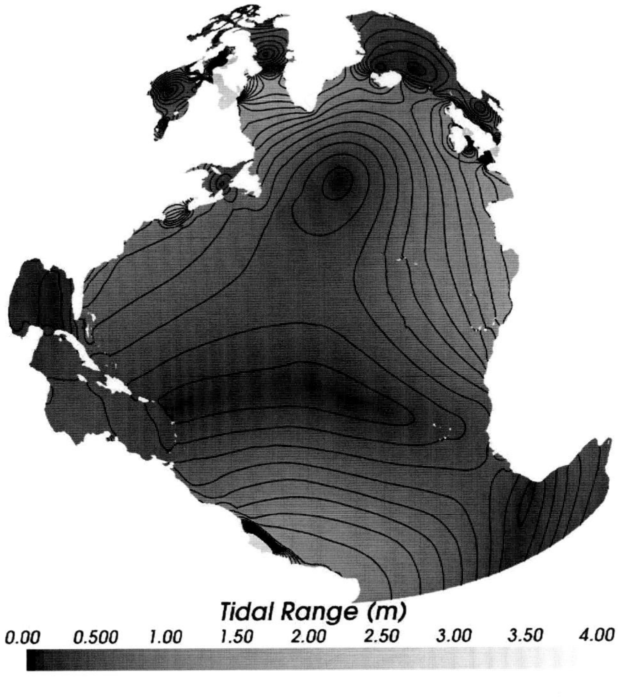

Figure 17. M2 tidal simulation in the North Atlantic with the mesh given in Figure 2.

growth in research and capabilities of models is expected to increase further over the coming years and also include atmospheric modeling and coupled climate modeling.

An aim might be for at least one model based upon unstructured, and possibly adaptive, technology, to be sufficiently developed, validated, and robust that it will be considered for inclusion in the next Intergovernmental Panel on Climate Changes (IPCC) assessment. A genetic diversity in algorithms, that is the use of algorithms based on fundamentally distinct design philosophies, is of real importance when these methods are used in coupled climate models for IPCC assessments for example. This is simply because several models based on similar design philosophies and discretization methods, etc., giving similar output only tells you so much, having a radically different model giving a consistent result gives much more confidence in that result.

Full consideration, development, and validation of new methods is always a worthy endeavor. It is particularly appropriate for methods based upon unstructured adaptive meshes to be pursued for oceanographic application, as these methods have a number of powerful advantages over structured grid models. Many believe that these advantages will more than outweigh possible disadvantages. In particular, models based upon unstructured meshes and a dynamically adaptive mesh capability have advantages which stem from having full flexibility in the computational mesh so that the mesh becomes an aid to a model's simulation power rather than as a constraint, which is usually how it is seen. The flexibility means that smooth variations in resolution can be easily achieved without the need to revert to nesting. Meshes can be optimized to best represent the level of detail in coastlines and bathymetry that is required for the simulation at hand. It is feasible to take a multiscale modeling approach and couple the global and coastal scales along with admitting important process scale phenomena in a priori unknown locations, given large enough computational resources of course. Note also that in allowing full flexibility, the model may revert back to a structured mesh in a subset or the entire domain if appropriate, for example, if the dynamics dictate this or for facilitating intercomparisons of discretization methods only. Full flexibility of the mesh in the vertical as well as horizontal means that an analogue of the hybrid coordinate approach is possible with flexibility to locally alter the vertical resolution.

An important part of the discussion here has been in respecting anisotropy present in the dynamics in the adapted mesh so that every possible efficiency saving is built into the overall approach. An interesting extension to the use of a single optimized mesh for "multiphysics" applications is the use of different adapted meshes for different solution components, for example, solving the forward fluids equations on one mesh, while biology, chemistry, sediments, ice, and the adjoint equations are solved on different meshes adapted to their own particular solution characteristics and interpolating between them when needed. One immediate issue with this is how one balances the computational load of the overall model effectively in parallel [*Plimpton et al.*, 2004]. New models obviously need to be shown to scale well up to large numbers of processors on supercomputing platforms if they are going to one day be used in an operational mode.

The principal issue that must be addressed by the developers of unstructured and adaptive mesh methods before their models find widespread use in all but idealized applications is computational cost. As discussed in previous sections, numerical methods based upon unstructured meshes can take up to an order of magnitude hit in terms of computational cost per degree of freedom. This is somewhat of an estimate at this stage, and comparisons with today's state-of-the-art ocean models will be required in the coming years to obtain more accurate values for this overhead in relevant applications. Of course, the saving grace of models based on unstructured adaptive meshes is that they are able to produce superior results using fewer degrees of freedom. At the time of writing, it is an open question for which application areas this advantage of unstructured mesh models will outweigh their additional costs, that is, given a choice between structured and unstructured mesh model which would a user be advised to use for their given problem. The developers of new models are of course confident in their methods, but quite rightly the user community and developers of classical models need convincing. For large-scale climate-related simulations, levels of diapycnal mixing and conservation

of key quantities are two other important gauges of model quality which need to be quantified and compared to levels present in today's leading models.

An interesting issue that must be addressed involves the interaction of adaptivity (i.e., continually changing resolution) with subgrid-scale models or parameterizations. This will require the use of flexible scale-dependent SGS models (such as inhomogeneous anisotropic LES) and error estimators which can decide whether certain features in the dynamics need to be resolved or can be modeled. In addition, with fully flexible meshes and discretization methods where a combination of mesh optimization (h-adaptivity), mesh movement (r-adaptivity), and local changes to the discretization order (p-adaptivity) is possible, error estimators must be sufficiently sophisticated so that they are able to guide the model as to what the optimal mesh and discretization is in a particular region. This is also the case if mixed meshes made up of tetrahedra, pentahedra, and hexahedra are used together, again, the error estimator will need a means to guide which element type to use. Thus, error estimators that really can identify the most important processes for the problem at hand will need to be constructed.

It is worth noting that much of the work presented here has benefited from the cross-fertilization of ideas and techniques pioneered in other disciplines. It is hoped and expected that this will continue and that the exchange will be two-way. In pursuing a model development approach that takes advantage of techniques from computational fluid dynamics research, a wide range of algorithms and tools may be taken advantage of. Future examples of developments which may be exploited for oceanographic application include model reduction (e.g., proper orthogonal decomposition) and the inclusion of uncertainty in the model calculation (e.g., through the use of stochastic discretization methods).

8. CONCLUDING REMARKS

Adaptive unstructured meshes have been presented in the context of ocean modeling. While these methods have a number of potential advantages for this application area, much work on their development and validation is still needed. Principal developments required include: the acceleration of the matrix solvers, for example, through the use of improved preconditioners; the use of more advanced error measures and mesh movement methods; new element and discretization choices including discontinuous methods; and subgrid-scale models which can work naturally with variable resolution meshes. The use of data assimilation and adjoints with adaptive mesh models [e.g., *Fang et al.*, 2006] also needs further development if these methods are going to become operational. Appropriate open data standards, and associated

APIs, for unstructured meshes need to be agreed upon by the community. The extraction of diagnostic quantities on unstructured meshes is not a fundamental problem [*Cotter and Gorman*, 2008], but the array of diagnostic tools that exist for structured grid methods must be developed. With regards visualization, this application area can make use of the various open source packages developed for the CFD community. As an increase in model complexity typically leads to a more difficult user environment, much effort needs to go into good quality pre/post-processing tools and training material. Finally, a large amount of code validation and benchmarking is still required for a range of analytical and laboratory-based test problems, extending up to comparisons with structured mesh models on a range of relevant applications.

Due to space limitations, detailed discussions of many exciting possible future developments have been omitted here, for example, the use of p-refinement (locally varying the order of the numerical approximation) to complement mesh optimization and mesh movement; the ability to use mixed meshes comprising hexahedral, tetrahedral, and pentahedral (pyramids, wedges) shaped elements; mesh movement with the flexibility to track solution structures, for example, eddies or density levels; and data assimilation methods where one may seek to increase resolution where increased data is available, and also to use an inconsistent adjoint to open up the possibility of using a different adapted mesh for the adjoint calculation.

Acknowledgments. The authors would like to acknowledge the funding of NERC under grants NE/C52101X/ and NE/C51829X/1 and the support of colleagues past and present at Imperial College London, the University of Oxford, the National Oceanography Centre Southampton, the University of Reading, and the Proudman Oceanographic Laboratory. Animations of presented results are available at http://amcg.ese.ic.ac.uk/~piggott/public/agu/

REFERENCES

Adcroft, A., and D. Marshall (1998), How slippery are piecewise-constant coastlines in numerical ocean models?, *Tellus*, *50A*, 95–108.

Adcroft, A., C. Hill, and J. Marshall (1997), Representation of topography by shaved cells in a height coordinate ocean model, *Mon. Weather Rev.*, *125*, 2293–2315.

Ainsworth, M., and J. T. Oden (2000), *A Posteriori Error Estimation in Finite Element Analysis*, Wiley, New York.

Apel, T. (1999), *Anisotropic Finite Elements: Local Estimates and Applications*, Teubner, Stuttgart.

Arakawa, A., and V. R. Lamb (1977), Computational design of the basic dynamical processes of the UCLA general circulation model, *Methods Comput. Phys.*, *17*, 174–267.

Bacon, D. P., et al. (2000), A dynamically adapting weather and dispersion model: The operational multiscale environment model with grid adaptivity (OMEGA), *Mon. Weather Rev.*, *128*, 2044–2076.

Bangerth, W., and R. Rannacher (2003), *Adaptive Finite Element Methods for Differential Equations*, Birkhäuser, Lectures in Mathematics, ETH Zürich.

Behrens, J. (2006), *Adaptive Atmospheric Modeling*, Springer, Berlin.

Benhamadouche, S., and D. Laurence (2003), LES, coarse LES, and transient RANS comparison on the flow across a tube bundle, *Int. J. Heat Fluid Flow*, *24*, 470–479.

Berger, M., and J. Oliger (1984), Adaptive mesh refinement for hyperbolic partial differential equations, *J. Comp. Phys.*, *53*, 484–512.

Bernard, P. E., N. Chevaugeon, V. Legat, E. Deleersnijder, and J. F. Remacle (2007), High-order h-adaptive discontinuous Galerkin methods for ocean modeling, *Ocean Dyn.*, *9*, 03497.

Blayo, E., and L. Debreu (1999), Adaptive mesh refinement for finite-difference ocean models: first experiments, *J. Phys. Oceanogr.*, *29*, 1239–1250.

Brezzi, F., and M. Fortin (1991), *Mixed and Hybrid Finite Element Methods*, Springer, Berlin.

Chassignet, E. P., L. T. Smith, G. R. Halliwell, and R. Bleck (2003), North Atlantic simulations with the hybrid coordinate ocean model (HYCOM): Impact of the vertical coordinate choice, reference pressure, and thermobaricity, *J. Phys. Oceanogr.*, *33*, 2504–2526.

Chen, C., H. Liu, and R. C. Beardsley (2003), An unstructured grid, finite-volume, three-dimensional, primitive equations ocean model: Application to coastal ocean and estuaries, *J. Atmos. Oceanic Techno.*, *20*, 159–186.

Cockburn, B., G. E. Karniadakis, and C. W. Shu (Eds.) (2000), *Discontinuous Galerkin Methods: Theory, Computation and Applications*, Springer, Berlin.

Cotter, C. J., and G. J. Gorman (2008), Diagnostic tools for 3D unstructured oceanographic data, *Ocean Model.*, *20*, 170–182.

Danilov, S., G. Kivman, and J. Schröter (2004), A finite element ocean model: Principles and evaluation, *Ocean Model.*, *6*, 125–150.

Danilov, S., G. Kivman, and J. Schröter (2005), Evaluation of an eddy-permitting finite-element ocean model in the North Atlantic, *Ocean Model.*, *10*, 35–50.

Davies, A. M. (1980), Application of the Galerkin method to the formulation of a three-dimensional nonlinear hydrodynamic numerical sea model, *Appl. Math. Model.*, *4*, 245–256.

D'Azevedo, E. F. (1991), Optimal triangular mesh generation by coordinate transformation, *SIAM J. Sci. Stat. Comput.*, *12*, 755–786.

Debreu, L., E. Blayo, and B. Barnier (2005), A general adaptive multi-resolution approach to ocean modelling: Experiments in a primitive equation model of the north Atlantic, in *Adaptive Mesh Refinement—Theory and Applications. Proceedings of the Chicago Workshop on Adaptive Mesh Refinement Methods, Sept. 3–5, 2003*, pp. 303–313, Springer, Berlin.

Dietachmayer, G. S., and K. K. Droegemeier (1992), Application of continuous dynamic grid adaptation techniques to meteorological modeling. Part I: Basic formulation and accuracy, *Mon. Weather Rev.*, *120*, 1675–1706.

Donea, J., and A. Huerta (2003), *Finite Element Methods for Flow Problems*, Wiley, New York.

Dumas, E., C. Le Provost, and A. Poncet (1982), Feasibility of finite element methods for oceanic general circulation modelling, *Proceedings of the Fourth International Conference on Finite Elements in Water Resources*, pp. 543–555, Hanover, Germany.

Elman, H., D. Silvester, and A. Wathen (2005), *Finite Elements and Fast Iterative Solvers*, Prentice-Hall, NJ.

Fang, F., M. Piggott, C. Pain, G. Gorman, and A. Goddard (2006), An adaptive mesh adjoint data assimilation method, *Ocean Model.*, *15*, 39–55.

Fix, G. J. (1975), Finite element models for ocean circulation problems, *SIAM J. Appl. Math.*, *29*, 371–387.

Ford, R., C. C. Pain, M. D. Piggott, A. J. H. Goddard, C. R. E. de Oliveira, and A. P. Umbleby (2004a), A non-hydrostatic finite element model for three-dimensional stratified oceanic flows. Part I: Model formulation, *Mon. Weather Rev.*, *132*(12), 2816–2831.

Ford, R., C. C. Pain, M. D. Piggott, A. J. H. Goddard, C. R. E. de Oliveira, and A. P. Umbleby (2004b), A non-hydrostatic finite element model for three-dimensional stratified oceanic flows. Part II: Model validation, *Mon. Weather Rev.*, *132*(12), 2832–2844.

Fortin, M., and A. Fortin (1985), Newer and newer elements for incompressible flow, in *Finite Elements in Fluids—Volume 6*, edited by R. H. Gallagher, G. F. Carey, J. T. Oden, and O. C. Zienkiewicz, pp. 171–187, Wiley, New York.

Freitag, L. A., and C. Ollivier-Gooch (1997), Tetrahedral mesh improvement using swapping and smoothing, *Int. J. Numer. Methods Eng.*, *40*, 3979–4002.

Frey, P. J., and F. Alauzet (2005), Anisotropic mesh adaptation for CFD computations, *Comput. Methods Appl. Mech. Eng.*, *194*, 5068–5082.

Fringer, O. B., M. Gerritsen, and R. L. Street (2006), An unstructured-grid, finite-volume, nonhydrostatic, parallel coastal-ocean simulator, *Ocean Model.*, *14*, 139–278.

George, P. L. (1998), *Delaunay Triangulation and Meshing: Application to Finite Elements*, Editions HERMES, Paris.

Gorman, G. J., M. D. Piggott, C. C. Pain, C. R. E. de Oliveira, A. P. Umpleby, and A. J. H. Goddard (2006), Optimal bathymetric representation through constrained unstructured mesh adaptivity, *Ocean Model.*, *12*, 436–452.

Gorman, G. J., M. D. Piggott, and C. C. Pain (2007), Shoreline approximation for unstructured mesh generation, *Comput. and Geosci.*, *33*, 666–677.

Gresho, P. M., and R. L. Sani (1998), *Incompressible Flow and the Finite Element Method*, Wiley, New York.

Griffies, S. M., C. Böning, F. O. Bryan, E. P. Chassignet, R. Gerdes, H. Hasumi, A. Hirst, A. M. Treguier, and D. Webb (2000), Developments in ocean climate modelling, *Ocean Model.*, *2*, 123–192.

Gunzburger, M. (1989), *Finite Element Methods for Viscous Incompressible Flows: A Guide to Theory, Practice, and Algorithms*, Academic, New York.

Hagen, S. C., and D. M. Parrish (2002), Unstructured mesh generation for the western North Atlantic tidal model domain, *Eng. Comp.*, *20*, 136–146.

Haidvogel, D. B., and A. Beckmann (1999), *Numerical Ocean Circulation Modeling*, Imperial College Press, London.

Haidvogel, D. B., A. R. Robinson, and E. E. Schulman (1980), The accuracy, efficiency, and stability of three numerical models with application to open ocean problems, *J. Comput. Phys.*, *34*, 1–53.

Haidvogel, D. B., J. L. Wilkin, and R. Young (1991), A semi-spectral primitive equation ocean circulation model using vertical sigma coordinates and orthogonal curvilinear horizontal coordinates, *J. Phys. Oceanogr.*, *94*, 151–185.

Hanert, E., V. Legat, and E. Deleersnijder (2002), A comparison of three finite elements to solve the linear shallow water equations, *Ocean Model.*, *5*, 17–35.

Hanert, E., D. Y. Le Roux, V. Legat, and E. Deleersnijder (2004), Advection schemes for unstructured grid ocean modelling, *Ocean Model.*, *7*, 39–58.

Hanert, E., D. Y. Le Roux, V. Legat, and E. Deleersnijder (2005), An efficient eulerian finite-element method for the shallow water equations, *Ocean Model.*, *10*, 115–136.

Henry, R., and R. Walters (1993), Geometrically based, automatic generator for irregular triangular networks, *Commun. Numer. Methods. Eng.*, *9*, 555–566.

Hide, R., and P. J. Mason (1975), Sloping convection in a rotating fluid, *Adv. Phys.*, *24*, 47–100.

Hoffman, J., and C. Johnson (2006), A new approach to computational turbulence modeling, *Comput. Methods Appl. Mech. Eng.*, *195*, 2865–2880.

Hughes, T. J. R., and A. N. Brooks (1979), A multi-dimensional upwind scheme with no crosswind diffusion, in *Finite Element Methods for Convection Dominated Flows, AMD*, vol. 3, edited by T. J. R. Hughes, pp. 19–35, ASME, New York.

Hughes, T. J. R., and M. Mallet (1986), A new finite element formulation for computational fluid dynamics: IV. A discontinuity-capturing operator for multidimensional advective-diffusion systems, *Comput. Methods Appl. Mech. Eng.*, *58*, 329–336.

Hughes, T. J. R., L. P. Franca, and G. M. Hulbert (1989), A new finite element formulation for computational fluid dynamics: VIII. The Galerkin/least-squares method for advective-diffusive equations, *Comput. Methods Appl. Mech. Eng.*, *73*, 173–189.

Ingram, D. M., D. M. Causon, and C. G. Mingham (2003), Developments in Cartesian cut cell methods, *Math. Comput. Simul.*, *61*, 561–572.

Iskandarani, M., D. B. Haidvogel, and J. C. Levin (2003), A three-dimensional spectral element method for the solution of the hydrostatic primitive equations, *J. Comput. Phys.*, *186*, 397–425.

Jablonowski, C., M. Herzog, J. E. Penner, R. C. Oehmke, Q. F. Stout, B. van Leer, and K. G. Powell (2006), Block-structured adaptive grids on the sphere: Advection experiments, *Mon. Weather Rev.*, *134*, 3691–3713.

Johnson, C. (1987), *Numerical Solution of Partial Differential Equations by the Finite Element Method*, Cambridge University Press, Cambridge.

Karypis, G., K. Schloegel, and V. Kumar (2002), Parmetis: Parallel graph partitioning and sparse matrix ordering library, http://www-users.cs.umn.edu/~karypis/metis/.

Labeur, R. J., and J. D. Pietrzak (2005), A fully three dimensional unstructured grid non-hydrostatic finite element coastal model, *Ocean Model.*, *10*, 51–67.

Laugier, M., P. Agnot, and L. Mortier (1996), Nested grid methods for an ocean model: a comparative study, *Int. J. Numer. Methods Fluids*, *23*, 1163–1195.

Legrand, S., V. Legat, and E. Deleersnijder (2000), Delaunay mesh generation for an unstructured-grid ocean general circulation model, *Ocean Model.*, *2*, 17–28.

Leonard, B. P. (1979), A stable and accurate convection modeling procedure based on quadratic upstream interpolation, *Comput. Methods Appl. Mech. Eng.*, *19*, 59–98.

Leonard, B. P. (1988), Simple high accuracy resolution program for convective modeling of discontinuities, *Int. J. Numer. Methods Fluids*, *8*, 1291–1318.

Le Provost, C., and P. Vincent (1986), Some tests of precision for a finite element model of ocean tides, *J. Comput. Phys.*, *110*, 273–291.

Le Provost, C., M. L. Genco, F. Lyard, P. Vincent, and P. Canceil (1994), Spectroscopy of the world ocean tides from a finite element hydrodynamic model, *J. Geophys. Res.*, *99*(C12), 24,777–24,797.

Le Roux, D. Y. (2005), Dispersion relation analysis of the $P_1^{NC} - P_1$ finite-element pair in shallow-water models, *SIAM J. Sci. Comput.*, *27*, 394–414.

Le Roux, D. Y., A. Staniforth, and C. A. Lin (1998), Finite elements for shallow-water equation ocean models, *Mon. Weather Rev.*, *126*, 1931–1951.

Levin, J. C., M. Iskandarani, and D. B. Haidvogel (2006), To continue or discontinue: Comparisons of continuous and discontinuous Galerkin formulations in a spectral element ocean model, *Ocean Model.*, *15*, 56–70.

Löhner, R. (2001), *Applied CFD Techniques*, Wiley, New York.

Ly, L. N., and P. Luong (1999), Numerical multi-block grids in ocean circulation modeling, *Appl. Math. Model.*, *23*, 865–879.

Lyard, F., F. Lefevre, T. Letellier, and O. Francis (2006), Modelling the global ocean tides: modern insights from FES2004, *Ocean Dyn.*, *56*, 394–415.

Lynch, D. R., and W. R. Gray (1979), A wave equation model for finite element tidal computations, *Comput. Fluids*, *7*, 207–228.

Lynch, D. R., J. T. C. Ip, C. E. Naimie, and F. E. Werner (1996), Comprehensive coastal circulation model with application to the Gulf of Maine, *Cont. Shelf Res.*, *16*, 875–906.

Mellor, G. L., L.-Y. Oey, and T. Ezer (1998), Sigma coordinate pressure gradient errors and the seamount problem, *J. Atmos. Ocean Technol.*, *15*, 1122–1131.

Moulinec, C., S. Benhamadouche, D. Laurence, and M. Peric (2005), LES in a U-bend pipe meshed by polyhedral cells, in *Engineering Turbulence Modelling and Experiments 6*, edited by W. Rodi, pp. 48–60, Elsevier, Amsterdam.

Myers, P. G., and A. J. Weaver (1995), A diagnostic barotropic finite-element ocean circulation model, *J. Atmos. Ocean. Technol.*, *12*, 511–526.

Nechaev, D., J. Schröter, and M. Yaremchuk (2003), A diagnostic stabilized finite-element ocean circulation model, *Ocean Model.*, *5*, 37–63.

Norburn, S., and D. Silvester (1998), Stabilised vs. stable mixed methods for incompressible flow, *Comput. Methods Appl. Mech. Eng.*, *166*, 131–141.

Özgökmen, T. M. (2003), DOME, www.rsmas.miami.edu/personal/tamay/DOME/dome.html.

Özgökmen, T. M., T. Iliescu, P. F. Fischer, A. Srinivasan, and J. Duan (2007), Large eddy simulation of stratified mixing in two-dimensional dam-break problem in a rectangular enclosed domain, *Ocean Model.*, *16*, 106–140.

Pain, C. C., A. P. Umpleby, C. R. E. de Oliveira, and A. J. H. Goddard (2001), Tetrahedral mesh optimisation and adaptivity for steady-state and transient finite element calculations, *Comput. Methods Appl. Mech. Eng.*, *190*, 3771–3796.

Pain, C. C., M. D. Piggott, A. J. H. Goddard, F. Fang, G. J. Gorman, D. P. Marshall, M. D. Eaton, P. W. Power, and C. R. E. de Oliveira (2005), Three-dimensional unstructured mesh ocean modelling, *Ocean Model.*, *10*(1–2), 5–33.

Peraire, J., and K. Morgan (1997), Unstructured mesh generation including directional refinement for aerodynamic flow simulation, *Finite Elem. Anal. Des.*, *25*, 343–356.

Piggott, M. D., C. C. Pain, G. J. Gorman, P. W. Power, and A. J. H. Goddard (2005), h, r, and hr adaptivity with applications in numerical ocean modelling, *Ocean Model.*, *10*(1–2), 95–113.

Piggott, M. D., G. J. Gorman, C. C. Pain, P. A. Allison, A. S. Candy, B. T. Martin, and M. R. Wells (2008), A new computational framework for multi-scale ocean modelling based on adapting unstructured meshes, *Int. J. Numer. Methods Fluids*, *56*, 1003–1015, doi:10.1002/fld.1663.

Plimpton, S., B. Hendrickson, and J. Stewart (2004), A parallel rendezvous algorithm for interpolation between multiple grids, *J. Parallel Distrib. Comput.*, *64*, 266–276.

Popinet, S., and G. Rickard (2007), A tree-based solver for adaptive ocean modelling, *Ocean Model.*, *16*, 224–249.

Power, P. W., M. D. Piggott, F. Fang, G. J. Gorman, C. C. Pain, D. P. Marshall, A. J. H. Goddard, and I. M. Navon (2006), Adjoint goal-based error norms for adaptive mesh ocean modelling, *Ocean Model.*, *15*, 3–38.

Raviart, P., and J. Thomas (1977), A mixed finite element method for second-order elliptic problems, in *Mathematical Aspects of the Finite Element Method, Lecture Notes in Mathematics vol. 606*, pp. 292–315, Springer, Berlin.

Shewchuk, J. (2003), Triangle, http://www-2.cs.cmu.edu/~quake/triangle.html.

Skamarock, W., J. Oliger, and R. L. Street (1989), Adaptive grid refinement for numerical weather prediction, *J. Comput. Phys.*, *80*, 27–60.

Skamarock, W. C., and J. B. Klemp (1993), Adaptive grid refinement for two-dimensional and three-dimensional nonhydrostatic atmospheric flow, *Mon. Weather Rev.*, *121*, 788–804.

Sotiropoulos, F., and S. Abdallah (1991), The discrete continuity equation in primitive variable solutions of incompressible flow, *J. Comput. Phys.*, *95*, 212–227.

Sweby, P. K. (1984), High resolution schemes using flux limiters for hyperbolic conservation laws, *SIAM J. Numer. Anal.*, *21*, 995–1011.

Thompson, J. F., N. P. Weatherill, and B. K. Soni (1999), *Handbook of Grid Generation*, CRC, Florida.

Turek, S. (1999), *Efficient Solvers for Incompressible Flow Problems: An Algorithmic and Computational Approach*, Springer, Berlin.

Vallet, M.-G., C.-M. Manole, J. Dompierre, S. Dufour, and F. Guibault (2007), Numerical comparison of some Hessian recovery techniques, *Int. J. Numer. Methods Eng*, *72*, 987–1007.

Weatherill, N. P., O. Hassan, K. Morgan, J. W. Jones, B. G. Larwood, and K. Sorenson (2002), Aerospace simulations on parallel computers using unstructured grids, *Int. J. Numer. Methods Fluids*, *40*, 171–187.

Zienkiewicz, O. C., and J. Z. Zhu (1992), The superconvergant patch recovery and a posteriori error estimates. part 1: The recovery technique, *Int. J. Numer. Methods Eng.*, *33*, 1331–1364.

Zienkiewicz, O. C., R. L. Taylor, J. Z. Zhu, and P. Nithiarasu (2005), *The Finite Element Method: Its Basis and Fundamentals; for Solid and Structural Mechanics; for Fluid Mechanics*, Sixth ed., Elsevier, Butterworth-Heinemann.

M. D. Piggott, Applied Modelling and Computation Group, Department of Earth Science and Engineering, Imperial College London, SW7 2AZ, UK. (m.d.piggott@imperial.ac.uk)

Index